Material Forming
ESAFORM 2023

Part 2

The 26th International ESAFORM Conference on Material Forming – ESAFORM 2023 – held in Kraków, Poland, April 19-21, 2023

Editors
Lukasz Madej
Mateusz Sitko
Konrad Perzynski

Peer review statement

All papers published in this volume of "Materials Research Proceedings" have been peer reviewed. The process of peer review was initiated and overseen by the above proceedings editors. All reviews were conducted by expert referees in accordance to Materials Research Forum LLC high standards.

Published under License by **Materials Research Forum LLC**
Millersville, PA 17551, USA

Published as part of the proceedings series
Materials Research Proceedings
Volume 28 (2023) Part 1

ISSN 2474-3941 (Print)
ISSN 2474-395X (Online)

ISBN 978-1-64490-246-2 (Print Part 1)
ISBN 978-1-64490-248-6 (Print Part 2)
ISBN 978-1-64490-249-3 (Print Part 3)
ISBN 978-1-64490-247-9 (eBook)

This book contains information obtained from authentic and highly regarded sources. Reasonable efforts have been made to publish reliable data and information, but the author and publisher cannot assume responsibility for the validity of all materials or the consequences of their use. The authors and publishers have attempted to trace the copyright holders of all material reproduced in this publication and apologize to copyright holders if permission to publish in this form has not been obtained. If any copyright material has not been acknowledged please write and let us know so we may rectify in any future reprint.

Distributed worldwide by
Materials Research Forum LLC
105 Springdale Lane
Millersville, PA 17551
USA
https://www.mrforum.com

Manufactured in the United State of America
10 9 8 7 6 5 4 3 2 1

Table of Contents

Friction and wear in metal forming

Incremental and sheet metal forming

Innovative joining by forming technologies

Lionel fourment ms on optimization and inverse analysis in forming

Machining and cutting

Material behaviour modelling

The following chapters are including in Part 1 sold separately

Additive manufacturing

Composites forming processes

Extrusion and drawing

Forging and rolling

The following chapters are including in Part 3 sold separately

New and advanced numerical strategies for material forming

Non-conventional processes

Polymer processing and thermomechanical properties

Sustainability on material forming

Property-controlled forming

Preface

Proceedings of the 26th International ESAFORM Conference on Material Forming (ESAFORM 2023)

The 26th International ESAFORM Conference 2023 was held in the historic city of Kraków, Poland, between 19-21 April 2023 at AGH University of Science and Technology.

ESAFORM is an association with the mission to stimulate applied and fundamental research in the broad field of material forming. Its annual conference, the International ESAFORM Conference on Material Forming, is used to achieve one of the main goals of ESAFORM: to spread scientific and technological information related to material forming within academia and industry. The most important idea of this event is to bring together researchers in the field to encourage discussion, collaboration and interchange of ideas.

ESAFORM 2023 was hosted by AGH University of Science and Technology. AGH is a modern state university that develops partner collaborations with colleges and universities in Poland, Europe, and all over the world. AGH is one of Poland's best technical universities, with approx. 18,000 students at 16 Faculties (various engineering faculties, science, sociology, business and management). AGH University widely collaborates with international scientific and industrial partners.

The ESAFORM 2023 was attended by more than 300 participants from 27 countries. During the event, approx. 240 presentations were delivered across sixteen organized mini-symposia focused on a broad spectrum of forming-related topics.

The proceedings present papers from MS01: Additive Manufacturing, MS02: Composites Forming Processes, MS03: Extrusion and Drawing, MS04: Forging and Rolling, MS05: Formability of Metallic Materials, MS06: Friction and Wear in Metal Forming, MS07: Incremental and Sheet Metal Forming, MS08: Innovative Joining by Forming Technologies, MS09: Lionel Fourment MS on Optimization and Inverse Analysis in Forming, MS10: Machining and Cutting, MS11: Material Behaviour Modelling, MS12: New and Advanced Numerical Strategies for Material Forming, MS13: Non-Conventional Processes, MS14: Polymer Processing and Thermomechanical Properties, MS15: Sustainability on Material Forming, MS16: Property-Controlled Forming.

The ESAFORM 2023 was organized under the patronage of the Dean of the Faculty of Industrial Computer Science and Modelling of the AGH University, the Mayor of the City of Kraków and the Minister of Science and Education in Poland. The conference was also co-financed from the state budget under the programme of the Minister of Education and Science called Excellent Science and by many supporting companies.

Committees

Esaform 2023 Chair

Lukasz Madej
AGH University of Science
and Technology, Poland

Organizing committee

Konrad Perzynski
AGH University of Science
and Technology, Poland

Mateusz Sitko
AGH University of Science
and Technology, Poland

Anna Smyk
Secretariat of the Conference
AGH University of Science
and Technology, Poland

Scientific committee

Prof. Alexander Brosius, Technische Universität Dresden, Germany
Prof. Anne-Marie Habraken, University of Liège, Belgium
Dr. Anne Mertens, University of Liège, Belgium
Prof. Antonello Astarita, University of Naples Federico II, Italy
Prof. António Gil Andrade-Campos, University of Aveiro, Portugal
Prof. Barbara Reggiani, University of Modena and Reggio Emilia, Italy
Prof. Beatriz Silva, University of Lisbon, Portugal
Dr. Benjamin Klusemann, Leuphana Universität Lüneburg, Germany
Prof. Bernd-Arno Behrens, University of Hannover, Germany
Prof. Carpoforo Vallellano, Universidad de Sevilla, Spain
Prof. Celal Soyarslan, University of Twente, Netherlands
Dr. Cédric Courbon, École nationale d'ingénieurs de Saint-Étienne, France
Dr. Chris Valentin Nielsen, Technical University of Denmark, Denmark
Dr. Christian Grandfils, University of Liège, Belgium
Dr. Daniel Cooper, University of Michigan, USA
Dr. Denise Bellisario, University of Rome Tor Vergata, Italy
Prof. Dermot Brabazon, Dublin City University, Ireland
Prof. Domenico Umbrello, University della Calabria, Italy
Prof. Dorel Banabic, Universitatea Tehnică din Cluj-Napoca, Romania
Dr. Gabriela Vincze, University of Aveiro, Portugal
Prof. Gianluca Buffa, University of Palermo, Italy
Prof. Giuseppe Ingarao, Università degli Studi di Palermo, Italy
Prof. Giusy Ambrogio, Università della Calabria, Italy
Dr. Guenael Germain, Arts et Métiers ParisTech, France
Dr. Hans-Peter Schulze, Leukhardt Schaltanlagen Systemtechnik GmbH, Magdeburg,Germany
Prof. Hinnerk Hagenah, Friedrich-Alexander-Universität Erlangen-Nürnberg, Germany
Dr. Holger Aretz, Hydro Aluminium Rolled Products GmbH Bonn Germany
Prof. Halilovič Miroslav, University of Ljubljana, Slovenia
Dr. Javad Hazrati, University of Twente, Netherlands
Prof. Katia Mocellin, MINES ParisTech, France
Prof. Laurentiu Slatineau, Gheorghe Asachi Technical University of Iasi, Romania

Esaform Board of Directors

Formability of metallic materials

Material Forming - ESAFORM 2023   Materials Research Forum LLC
Materials Research Proceedings 28 (2023) 695-704   https://doi.org/10.21741/9781644902479-75

Total strain on the outer surface of steel sheets in air bending

POKKA Aki-Petteri[1,a] *, KESTI Vili[2,b] and KAIJALAINEN Antti[1,c]

[1]University of Oulu, Materials and Mechanical Engineering, P.O. Box 4200, 90014 Oulu, Finland

[2]SSAB Europe, Rautaruukintie 115, 92101 Raahe, Finland

[a]aki-petteri.pokka@oulu.fi, [b]vili.kesti@ssab.com, [c]antti.kaijalainen@oulu.fi

Keywords: Air Bending, Steel, Digital Image Correlation, Strain Distribution, Total Strain, Bend Shape

Abstract. Air bending is a commonly used method for sheet-metal forming. However, several challenges exist around the bending behavior of materials with poor global formability, that are difficult to study using conventional bending test methods, and thus may not be fully understood. In this study, nine thermomechanically rolled steel grades with various strengths and ductility properties are bent using three different punch radii. The strain distributions on the outer surface are measured using Digital Image Correlation (DIC). The relationships between the strain distribution, peak strain, and total strain (area under the strain distribution curve) are determined. The total strain is observed to be independent from the peak strain and the shape of the strain distribution. The total strain is found to depend on the bend angle and sheet thickness. An analytical formula for approximating the total strain is derived. Potential for further approximations of the total strain, strain distribution and bend shape are discussed.

Introduction

Air bending is a commonly used forming method for sheet metals due to its flexibility, cost-effectiveness, and speed. A large range of bend angles can be achieved without tool changes just by controlling the punch displacement. The radius of the bent sheet is usually controlled by changing the punch radius. However, the shape of the bend is also dependent on the material properties. The distribution of strain and curvature on the outside bend surface can vary significantly depending on the work-hardening properties of the material. When bending materials with poor global formability (i.e., low work-hardening and uniform elongation), the inside radius of the bent sheet may decrease far below the punch radius, leading to high local strains and bend shapes that may be difficult to predict [1]. Various terms have been used referring to this phenomenon in previous studies: multi-breakage, gap formation, punch-sheet separation, punch detachment, punch-sheet-liftoff, loss of contact with the punch/sheet etc. [2-8]. The phenomenon has been known for a long time. However, research on the root of the phenomenon, i.e., the causes and effects of the strain distribution development in air bending, has been relatively scarce. Strain distributions in air bending have been studied by several authors [1,9-14], but not for a large selection of materials and punch radii, that would allow thorough analysis of the effects of certain tool parameters and material properties. Research in this area is therefore necessary.

In this paper, nine steel grades are bent in a 3-point bending setup using three different punch radii. The development of the strain distribution on the outer curvature is measured from each test using a Lavision Digital Image Correlation (DIC) system. The aim of the paper is to study the relationships between the strain distributions, peak strains, and total strains (area under the strain distribution curve).

Material Forming - ESAFORM 2023
Materials Research Forum LLC
Materials Research Proceedings 28 (2023) 695-704
https://doi.org/10.21741/9781644902479-75

Experimental Procedure

Nine thermomechanically rolled steel grades are tested in this study. Table 1 provides the sheet thicknesses, tensile properties of the investigated materials, along with the typical minimum bend radii for corresponding grades, provided by a steel manufacturer. The tensile data was measured from ISO 6892 compliant tests, using dog-bone specimens with straight sections of $6 \times 8 \times 45$ mm^3, and strain rates of 0.0025 1/s (to yield point) and 0.008 1/s (after yield point). Both longitudinal and transverse directions were tested, relative to the material's rolling direction. In terms of the direction of the major strains relative to the rolling direction, the longitudinal (0°) tensile test corresponds to the transverse (TD) bend test and the transverse (90°) tensile test corresponds to the longitudinal (RD) bend test.

Table 1. Sheet thicknesses (t) and mechanical properties of the tested materials. The yield strength (R_e), ultimate tensile strength (R_m), uniform elongation (A_g) and total elongation with 40 mm gage length (A_{40}) were measured from the performed tensile tests. The minimum bend radii to 90° bend angle (R_{min}) were provided by a steel manufacturer for corresponding grades.

Material	t [mm]	R_e [MPa]	R_m [MPa]	A_g [%]	A_{40} [%]	R_{min} (to 90°)
St355 (0°/90°)	5.94	438/480	496/502	16.4/15.4	32.7/31.2	0.3 t
St500 (0°/90°)	5.92	585/628	653/669	11.7/10.4	26.9/25.4	0.8 t
St700 (0°/90°)	6.04	785/818	864/907	6.0/4.5	18.2/15.2	1 t
St900_1 (0°/90°)	5.99	974/1024	1042/1142	3.2/2.2	13.4/8.2	3 t
St900_2 (0°/90°)	5.93	1034/1067	1130/1151	3.9/3.1	13.9/11.4	3.5 t
St1100_1 (0°/90°)	6.01	1132/1127	1161/1172	4.7/4.7	15.4/13.8	3.5 t
St1100_2 (0°/90°)	5.93	1098/1111	1253/1271	2.9/2.4	12.1/10.8	3 t / 4 t
St1500 (0°/90°)	5.68	1537/1500	1761/1780	3.4/3.1	10.6/9.4	6 t
St1700 (0°/90°)	6.09	1740/1680	1958/1975	2.7/2.7	9.6/9.0	6 t

The bending tests are conducted in both longitudinal (RD) and transverse (TD) orientations. In this paper, the longitudinal bend orientation (RD) refers to the bend axis being parallel to the rolling direction, and the transverse orientation (TD) refers to the bend axis being perpendicular to the rolling direction. The bend specimens were cut from the 6 mm thick sheets into rectangular strips with a width of 80 mm. The specimen width is small enough to prevent exceeding the force limit of the used Zwick 100 kN universal tensile test machine, while still ensuring plane strain conditions at the center of the bend.

The specimens were bent in room temperature using purpose-built bending tools, shown in Fig. 1a. Using the tensile test machine for the bend tests allows accurate measurements of the bending force and punch displacement. The punch displacement is then used for calculating the bend angle according to ISO 7438 [15]. The measured vertical force is used to adjust the punch displacement and bend angle calculations for the vertical elasticity of the setup (51.9 kN/mm).

The shape of the die (lower bend tool) is illustrated in Fig. 1b. The openings allow an unobstructed line-of-sight between the DIC cameras and the measured specimen surface. The die width, i.e., the distance between the centers of the two shoulders, is also adjustable with this tool. The shoulders rotate freely in their sockets. To minimize the effect of friction even further, a PTFE lubricant spray is applied to the shoulders before each test.

Fig. 2 shows a schematic representation of the bending geometry used in this paper. Three bending tool setups are used in this study, in order to achieve a large variety of strain distributions, and to observe their effects on the behavior of the materials. The parameters for each setup are presented in Table 2. In each setup, the specimens were bent until they either fractured or reached the bend angle of α_{end}. Two repeat tests were conducted for each combination of tool setup, material, and bend orientation. Therefore, a total number of 108 tests are included in this study.

Material Forming - ESAFORM 2023   Materials Research Forum LLC
Materials Research Proceedings 28 (2023) 695-704   https://doi.org/10.21741/9781644902479-75

A DIC system, Strainmaster by Lavision, was used in this study for measuring the deformations on the outer bend surface. The system was equipped with two monochrome CCD cameras with a resolution of 2456 × 2058 pixels. The captured images were processed in the DaVis 8.4 software, which uses an iterative least squares matching (LSM) algorithm for displacement and strain calculation. The DIC recording and processing parameters are presented in Table 3. After calculating the strain maps from each image of each test, the values of the major strain were extracted from three sections positioned at the center of the bend, as illustrated in Fig. 1c. The average of the peak values of the three sections were also calculated at each point in time, as well as the average total strain (Riemann sums) of the three sections.

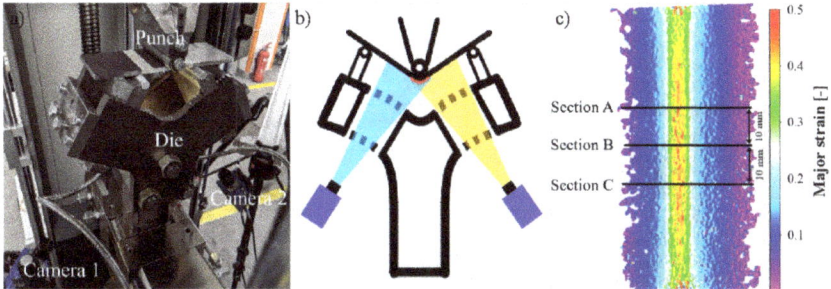

Fig. 1. a) The bending test setup, b) positioning of the DIC cameras, and c) sections A – C on a strain map of the St1100_1 (RD) at 90° bend angle, using a punch radius $R_p = 2\ t$.

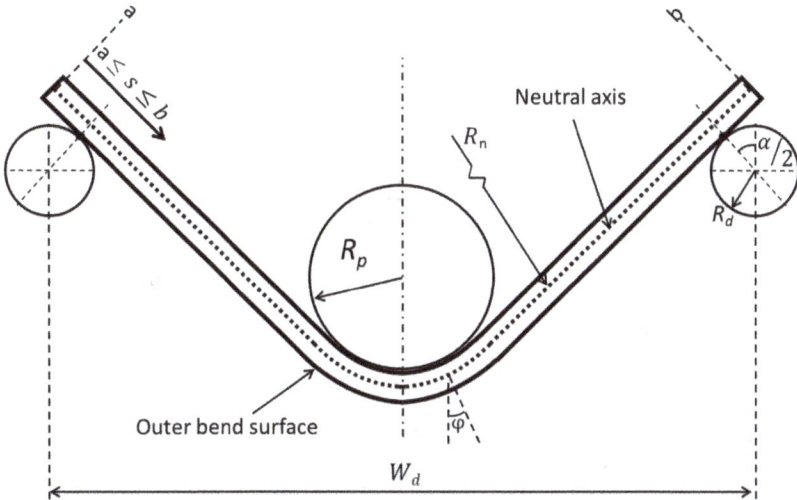

Fig. 2. The bending tool geometry and parameters involved in the analytical solution for the total strain $A_{\varepsilon b}$.

Table 2. Bending tool setup parameters: Punch radius (R_p), die width (W_d), radius of the die shoulders (R_d), vertical speed of the punch (V_p) and the final bend angle before unloading (α_{end}).

Tool setup	R_p	W_d [mm]	R_d [mm]	V_p [mm/s]	α_{end} [°]
1	4 mm ≈ 0.66 t	90	6	1	120
2	12 mm ≈ 2 t	90	6	1	100
3	24 mm ≈ 4 t	110	6	1	110

Table 3. Recording and processing parameters for the DIC system.

DIC system	Lavision Strainmaster (Stereo DIC)
Sensor and digitization	2456 × 2058, 12-bit
Lens, imaging distance	35 mm C-mount, 0.37 – 0.57 m
Imaging rate	2 Hz
Subset size	15x15 pixels
Step size	5 pixels
Strain window, smoothing method	5x5 data points, 2nd order polynomial fit
Virtual strain gage size	35 pixels
Image scale	21 – 30 pixels/mm
St.dev of principle strain	450 - 1700 microstrain
Interpolation, shape function, algorithms	6th order spline interpolation; affine shape function; LSM (iterative least squares matching) algorithm based on optical flow estimation

Before each test, a black-and-white speckle pattern was spray painted on the specimen surfaces to improve the DIC measurement reliability and to minimize glare. To minimize the effect of paint peeling on the measurements, the surfaces were cleaned with ethanol before painting and the tests were conducted as soon as possible after painting to prevent the paint from becoming excessively dry and brittle.

Results and Discussion

Fig. 3 shows the strain distributions of St700 in the RD direction measured at 30°, 60° and 90° bend angles and with three different punch radii. Fig. 3b illustrates the area under the strain distribution curve at 30°, i.e., the total strain $A_{\varepsilon b}$, as well as the peak strains ε_{b_max} at 30° and 90° angles. The distributions are similar for all three punches at 30° but as the bend angle increases, the effect of the punch radius on the strain distribution is clear. With the punch radius $R_p = 0.66$ t, the deformation increases mostly at the center, creating a narrow strain distribution with large peak strain at the center. In contrast, when using larger punches, as in Figs. 3b and 3c, the deformation is spread more evenly, creating a wider strain distribution with lower peak strains at the center. In fact, when using the largest punch ($R_p = 4$ t), the peak strain does not increase at all between 60° and 90° bend angles, as the deformation increases solely from the sides.

Fig. 4a presents the developments of the peak strain ε_{b_max} (average peak values of sections A, B and C) for St355 (TD) and St700 (RD) using three different punches. The peak strains develop almost identically in the first 15°, but after around 15 – 20° bend angle (0.05 – 0.10 peak strain) the effects of the punch radii and the material properties start to show. With the smallest punch radius ($R_p = 0.66$ t), the peak strain increases almost linearly between 30 – 90° bend angles for both materials, with only a slight stagnation in peak strain towards the end of the test. For $R_p = 2$ t, the stagnation of peak strain seems to start at around 30 – 40° bend angle, and for $R_p = 4$ t, around 15 – 20° bend angle.

The differences in material properties can also be seen in Fig. 4a. St355 produces lower peak strains than St700 with all three punches. This is due to the greater strain-hardening and ductility

Material Forming - ESAFORM 2023   Materials Research Forum LLC
Materials Research Proceedings 28 (2023) 695-704   https://doi.org/10.21741/9781644902479-75

of St355 compared to St700, indicated by their values of A_g and A_{40} in Table 1. In general, materials with greater work-hardening have greater resistance to strain localization, which leads to wider strain distributions and lower peak strains in bending.

Fig. 4b presents the developments of the total strain $A_{\varepsilon b}$ for St355 and St700. The total strain $A_{\varepsilon b}$ was calculated as the average of the Riemann sums of the strains in sections A, B and C shown in Fig. 1c. For clarification, the total strain $A_{\varepsilon b}$ represents the area under the strain distribution curve, as illustrated in Fig. 3b, and could also be described as the difference in length between the outer surface and the neutral axis. As can be seen in Figs. 4a and 4b, the total strain seems to grow at a similar, constant rate in all six tests, despite the differences in the peak strain development. The linear growth of the total strain continues until around 80°, after which some stagnation can be seen for all curves. It is assumed, that the apparent stagnation is mostly caused by loss of data at the later stages of the test, due to parts of the measured surface going beyond the field-of-view of the DIC cameras due to increased curvature. This can be seen in Figs. 3b and 3c, where the distributions at 90° angle are "cut off" at the edges, meaning that the increase of strain at the edges could not be fully measured. As the strain distribution is wider when using a larger punch, more data is lost at the edges with larger punches compared to smaller punches.

The peak strains and total strains from all tests at 50° and 80° angles are presented in Fig. 5. Despite the large number of tests and the variety in the peak strain values, no correlation between the total strain and peak strain or punch radius can be seen. At any given bend angle, the total bend strain value seems to be roughly constant, regardless of the peak strain or punch radius.

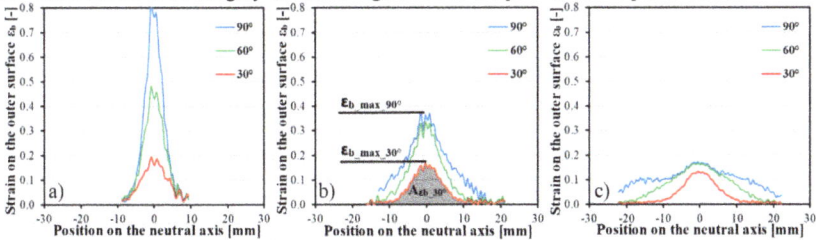

Fig. 3. Strain distributions at 30°, 60° and 90° bend angle, extracted from section B of St700 in the longitudinal direction (RD), using a punch radius of a) $R_p = 0.66\,t$, b) $R_p = 2\,t$ and c) $R_p = 4\,t$. The position on the neutral axis corresponds to a value of the arc coordinate s.

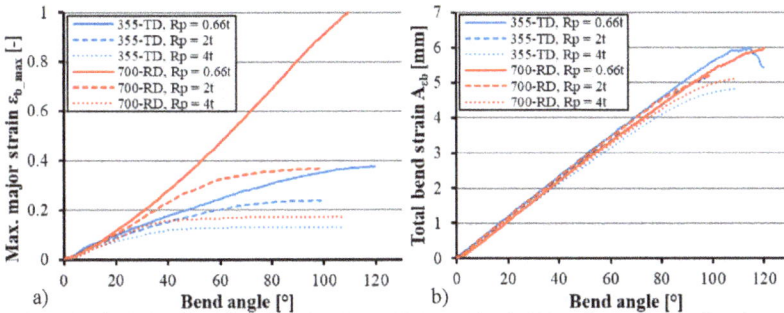

Fig. 4. Development of a) the peak strain and b) the total strain of St355 (TD) and St700 (RD), with punch radii of $R_p = 0.66\,t$, $R_p = 2\,t$, and $R_p = 4\,t$.

Material Forming - ESAFORM 2023

Materials Research Proceedings 28 (2023) 695-704

Materials Research Forum LLC

https://doi.org/10.21741/9781644902479-75

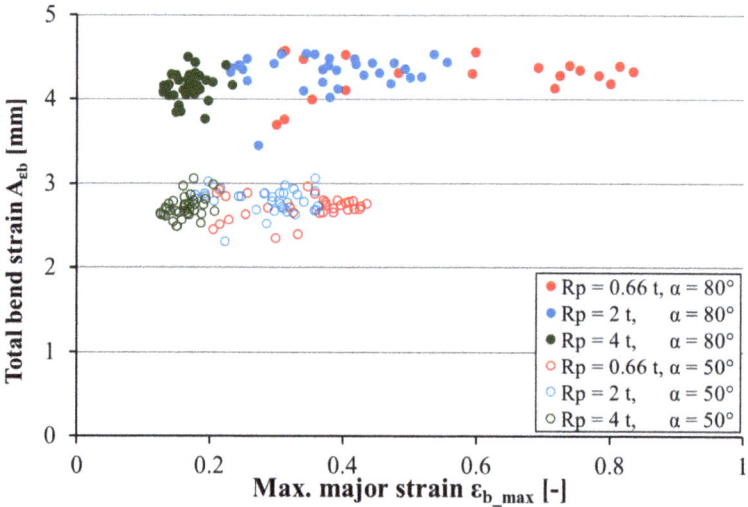

Fig. 5. Relationship between the total bend strain $A_{\varepsilon b}$ and the peak strain ε_{b_max}.

The total strain $A_{\varepsilon b}$ seems to be independent of the material properties, punch radius, the shape of the strain distribution or bend geometry. Assuming that the apparent stagnation after 80° is indeed caused by the limitations in the DIC field-of-view, the total strain seems to be a linear function of the bend angle. It should be possible to find an analytical solution for this relationship.

The parameters involved in this section are presented in Fig. 2. The bending angle α can be expressed as a sum of the rotation angles ϕ of the incremental arc lengths ds of the neutral axis, between the points a and b at the end of each flange, as follows:

$$\alpha = \int_a^b \frac{d\varphi(s)}{ds}\, ds \tag{1}$$

where s is an arc coordinate indicating the position on the neutral axis, and a and b are the end points of the neutral axis. The radius of the neutral axis R_n can also be expressed as

$$\frac{d\varphi}{ds} = \frac{1}{R_n}. \tag{2}$$

If the neutral axis is assumed to be positioned at the mid-thickness of the sheet, the strain on the outer surface ε_b can be given as

$$\varepsilon_b = \frac{t}{2R_n} \tag{3}$$

Combining Eq. 1, 2, and 3, the total strain on the outer bend surface (i.e., the strain distribution area) can be calculated as

$$A_{\varepsilon_b} = \int_a^b \varepsilon_b\,(s)\, ds = \int_a^b \frac{t}{2R_n}(s)\, ds = \frac{t}{2}\int_a^b \frac{1}{R_n}(s)\, ds = \frac{t}{2}\int_a^b \frac{d}{ds}\varphi\,(s)ds = \frac{\alpha t}{2} \tag{4}$$

Material Forming - ESAFORM 2023   Materials Research Forum LLC
Materials Research Proceedings 28 (2023) 695-704   https://doi.org/10.21741/9781644902479-75

The total strain $A_{\varepsilon b}$ seems to be a function of the bend angle and sheet thickness. If the total strain is constant for a given bend angle and sheet thickness, and all strain distributions in air bending are assumed to follow a known shape, e.g., a triangular function, the peak strain could be used to approximate the width of the strain distribution and the shape of the bent sheet. The approximate shape of the sheet could then be used for calculating the bend allowance and bend deduction, as well as approximating the effects of multi-breakage (loss of punch-sheet contact). Therefore, the peak strain ε_{b_max} at a given bend angle could be considered a decent stand-alone measure for the strain distribution. Furthermore, the effects of different material and tool parameters on the strain distribution development can then be investigated through simple linear regression, using the measured peak strain at a specific bend angle as the dependent variable.

In Fig. 6, the average total strains of all tests for each punch radius are plotted against the bend angle. The analytical total strain, calculated using Eq. 4, is included as a reference. Again, the total strain increases linearly for all punch radii until around 80°, after which a stagnation can be seen, presumably due to the limits of the DIC field-of-view (FOV). Although the analytical approximations are mostly within the standard deviation of the measured average values, the approximation seems to slightly underestimate the total strains. This could be due to the assumption of a fixed neutral axis in the analytical solution. If the neutral axis shifts towards the inside surface of the bend, the total strain on the outer surface will increase. For more precise analytical solutions and approximations in future works, the effects of the neutral axis shift could be considered.

Other factors that affect these measurements include the potential errors in the bend angle calculation, variation in sheet thickness and the loss of data due to fractures. If the ISO 7438 bend angle calculation overestimates the real bend angle, as was found by Cheong et al. [10], this would cause a downward bias to the measured $A_{\varepsilon b} - \alpha$ curves. However, the calculation of bend angle is hugely dependent on the tool geometry, so the previous findings may not necessarily apply for the geometry used in this study. Nevertheless, the calculation error of the bend angle introduces an additional unknown variable, and direct measurement of the bend angle would be preferable in future works.

As the total strain seems to be directly proportional to the sheet thickness, variations in sheet thickness will increase the scatter in the measured total strain. For more precise measurements and predictions, the sheet thickness variation should also be taken into account.

Material Forming - ESAFORM 2023
Materials Research Proceedings 28 (2023) 695-704

Materials Research Forum LLC
https://doi.org/10.21741/9781644902479-75

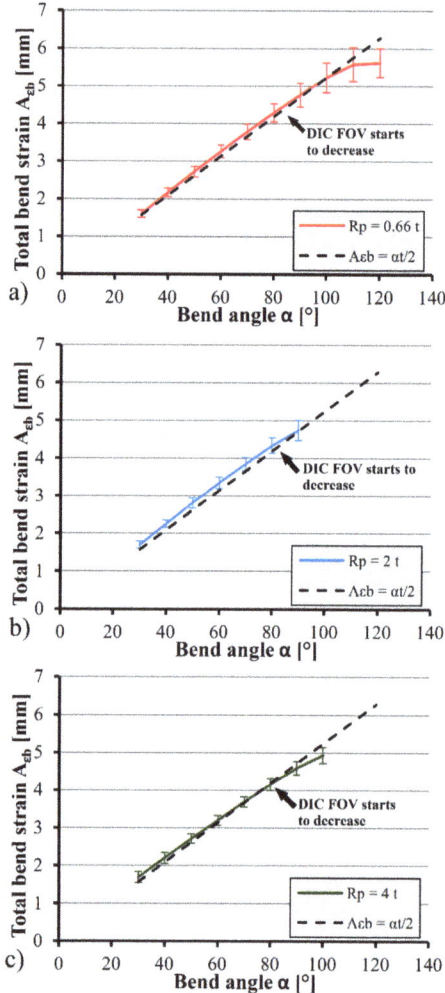

Fig. 6. Average total strain $A_{\varepsilon b}$, measured from bends with punch radius of a) $R_p = 0.66\ t$, b) $R_p = 2\ t$ and c) $R_p = 4\ t$. Analytical total strain is included as a reference, calculated as $A_{\varepsilon b} = \alpha t/2$, where $t = 6$ mm and α is the bend angle in radians.

Out of the 36 specimens tested with the punch radius $R_p = 0.66\ t$, 16 fractured before 80° bend angle. For $R_p = 2\ t$, four specimens fractured before 80°. As the tests were stopped at fracture, no data was gathered from these specimens after their fracture. Consequently, the sample size is decreased for the smaller punches and larger angles, meaning that the remaining specimens have more weight on the results. This increases the effect of random variation on the results, and could potentially introduce a bias, as only the specimens with the greatest ductility remain.

Part of the scatter and deviations in the total bend strain $A_{\varepsilon b}$, seen in Figs. 5 and 6 could be reduced if the DIC strain measurement setup and procedures were optimised for capturing the entire strain distribution. The setup and procedures used in this study were developed prioritizing the measurement of the peak strain, leading to increased noise and loss of data around the edges of the strain distributions. This could also be refined in future works.

Summary

The aim of this study was to characterize nine thermomechanically rolled steel grades with various strengths and ductility properties by means of their bendability properties using three different punch radii. The strain distributions on the outer curvature are measured throughout the test with a DIC system. The target was to understand and determine the relationships between the strain distribution, peak strain and total strain in air bending. The main observations and conclusions of the work can be summarized as follows:

- The strain distribution and peak strain had no measurable effect on the total strain on the outside surface. The total strain was found to grow at a similar rate throughout the test regardless of the punch radius or material properties.
- The total strain on the outside surface of the bend seems to be a function of the bend angle and sheet thickness. A decent approximation for the total strain was achieved with an analytical formula.
- The peak strain could be considered a decent stand-alone value for describing or approximating the strain distribution for a certain bend angle and sheet thickness. Possible use cases and benefits of such an approximation are discussed. These could include estimating the bend shape, multi-breakage, bend allowance and bend deduction, as well as allowing the use of simple linear regression for investigating the effects of certain material and tool parameters on the strain distribution development.

Acknowledgments

Financial assistance of the Business Finland, project FOSSA– Fossil-Free Steel Applications, is acknowledged.

References

[1] A.-P. Pokka, A.-M. Arola, A. Kaijalainen, V. Kesti, J. Larkiola, Strain distribution during air bending of ultra-high strength steels, ESAFORM 2021 (2021). https://doi.org/10.25518/esaform21.2509

[2] S.D. Benson, Press Brake Technology: A Guide to precision sheet metal bending, Society of Manufacturing Engineers, Dearborn, Michigan, 1997.

[3] V. Vorkov, R. Aerens, D. Vandepitte, J.R. Duflou, The Multi-Breakage Phenomenon in Air Bending Process, Key Eng. Mater. 611–612 (2014) 1047-1053. https://doi.org/10.4028/www.scientific.net/KEM.611-612.1047

[4] A. Väisänen, K. Mäntyjärvi, J.A. Karjalainen, Bendability of Ultra-High-Strength Steel, Key Eng. Mater. 410–411 (2009) 611-620. https://doi.org/10.4028/www.scientific.net/KEM.410-411.611

[5] J. Heikkala, A. Väisänen, Usability Testing of Ultra High-Strength Steels, Proc. ASME 2012 11th Biennial Conference on Engineering Systems Design and Analysis 4 (2012) 163-173. https://doi.org/10.1115/ESDA2012-82770

[6] V. Kesti, A. Kaijalainen, A. Väisänen, A. Järvenpää, A. Määttä, A.-M. Arola, K. Mäntyjärvi, R. Ruoppa, Bendability and Microstructure of direct quenched Optim® 960QC, Mater. Sci. Forum 783-786 (2014) 818-824. https://doi.org/10.4028/www.scientific.net/MSF.783-786.818

[7] A.-M. Arola, V. Kesti, R. Ruoppa, The effect of punch radius on the deformation of ultra-high strength steel in bending, Key Eng. Mater. 639 (2015) 139-146.

Material Forming - ESAFORM 2023   Materials Research Forum LLC
Materials Research Proceedings 28 (2023) 695-704   https://doi.org/10.21741/9781644902479-75

[8] L. Wagner, H. Schauer, H. Pauli, J. Hinterdorfer, Improved bendability characterization of UHSS sheets, IOP Conf. Ser. Mater. Sci. Eng. 651 (2019) 012019. https://doi.org/10.1088/1757-899X/651/1/012019

[9] M. Kaupper, M. Merklein, Bendability of advanced high strength steels - A new evaluation procedure, CIRP Ann. - Manuf. Technol. 62 (2013) 247-250. https://doi.org/10.1016/j.cirp.2013.03.049

[10] K. Cheong, K. Omer, C. Butcher, R. George, J. Dykeman, Evaluation of the VDA 238-100 Tight Radius Bending Test using Digital Image Correlation Strain Measurement, J. Phys. Conf. Ser. 896 (2017) 12075. https://doi.org/10.1088/1742-6596/896/1/012075

[11] A.-M. Arola, A. Kaijalainen, V. Kesti, A.-P. Pokka, J. Larkiola, Digital image correlation and optical strain measuring in bendability assessment of ultra-high strength structural steels, Procedia Manuf. 29 (2019) 398-405. https://doi.org/10.1016/j.promfg.2019.02.154

[12] S. Gothivarekar, S. Coppieters, A. Van de Velde, D. Debruyne, Advanced FE model validation of cold-forming process using DIC: Air bending of high strength steel, Int. J. Mater. Form. 13 (2020) 409-421. https://doi.org/10.1007/s12289-020-01536-1

[13] R. Ruoppa, R. Toppila, V. Kesti, A. Arola, Bendability tests for ultra-high-strength steels with optical strain analysis and prediction of bending force, Proc. METNET Seminar 2014 Moscow (2014) 68-78.

[14] C. Suppan, T. Hebesberger, A. Pichler, J. Rehrl, O. Kolednik, On the microstructure control of the bendability of advanced high strength steels, Mater. Sci. Eng. A 735 (2018) 89-98. https://doi.org/10.1016/j.msea.2018.07.080

[15] SFS-EN ISO 7438:2016: Metallic materials – Bend test.

Material Forming - ESAFORM 2023   Materials Research Forum LLC
Materials Research Proceedings 28 (2023) 705-710   https://doi.org/10.21741/9781644902479-76

Prediction of forming limit diagram of automotive sheet metals using a new necking criterion

PHAM Quoc Tuan [1,a] *, ISLAM Md Shafiqul [1,b], SIGVANT Mats[1,2,c] and CARO Perez Lluis [3,d]

[1]Department of Mechanical Engineering, Blekinge Institute of Technology, Karlskrona 37141, Sweden

[2]Volvo Cars, Department 81110 Strategy & Concept, Olofström, Sweden

[3]RISE Research Institutes of Sweden, Component Manufacturing Unit, Olofström, Sweden

[a]quoc.tuan.pham@bth.se, [b]shafiqul.islam@bth.se, [c]mats.sigvant@bth.se, [d]lluis.perez.caro@ri.se

Keywords: Forming Limit Diagram, DP800, AA6016, MMFC2, MK Method

Abstract. A theoretical model for predicting the forming limit diagram of sheet metal, named MMFC2, was recently proposed by the authors based on the modified maximum force criterion (MMFC). This study examines the application of MMFC2 for two automotive sheets, DP800 and AA6016, which are widely used in making car body parts. Uniaxial tensile and bulge tests are conducted to calibrate constitutive equations for modeling the tested materials. The developed material models are employed into different frameworks such as MMFC, MMFC2, and Marciniak-Kuczynski (MK) models to forecast the forming limit curve (FLC) of the tested materials. Their predictions are validated by comparing with an experimental one obtained from a series of Nakajima tests. It is found that the derived results of MMFC2 are comparable to that of MK model and agreed reasonably with experimental data. Less computational time is the major advantage of MMFC2 against the MK model.

Introduction

The formability of sheet metal is commonly evaluated by using a forming limit diagram (FLD). In this concept, a graphical representation of the forming limit, the forming limit curve (FLC), is used to separate the safe forming region in the strain space. Different testing setup, such as the Nakajima and Marciniak tests, can be used to determine the FLC experimentally [1]. In order to reduce the cost of experimental methods, huge efforts have been made to predict the FLC theoretically.

Publications of Swift [2] and Hill [3] are considered the pioneering works for this task. Storen and Rice [4] presented an alternative theoretical approach for FLC prediction. In this approach, the bifurcation analysis of deformation velocity was coupled with the deformation theory to determine the initiation of localized necking. Based on observations of geometrical imperfection of the tested coupon, Marciniak and Kuczynski [5] introduced a numerical procedure for predicting the FLC. Later, Hora et al. [6] formulated a criterion (MMFC) for neck determination based on the strain path change at the onset of diffuse neck.

Previous studies published in the literature demonstrate that these theoretical models may provide good predictions for FLC of some particular materials. Reviewing studies pointed out that there is not exist a universal model that can be applied for any metallic sheets. Recently, the MMFC is attractive because of its validation, which can be done with the use of digital cameras and correlation techniques (DIC). However, the application of the MMFC is limited due to some numerical issues, as discussed in references [7,8]. Recently, the authors discussed the reliability of theoretical assumptions implied in the MMFC, which leads to an improved version named MMFC2 [9].

Material Forming - ESAFORM 2023   Materials Research Forum LLC
Materials Research Proceedings 28 (2023) 705-710   https://doi.org/10.21741/9781644902479-76

This study presents comparisons between the applications of the original MMFC and the new version (MMFC2) for predicting the FLC of two automotive sheet metals, DP800 and AA6016. In addition, the predictions of MK model are also put into the comparison. Experimental FLC determined from the Nakajima tests are provided to discuss the reliability of these theoretical predictions.

The MMFC2 Model

Based on experimental observations of the strain path change beyond the diffuse neck, Hora et al. [6] introduced the condition for neck initiation as follows.

$$\frac{\partial \sigma_1}{\partial \varepsilon_1} + \frac{\partial \sigma_1}{\partial \beta}\frac{\partial \beta}{\partial \varepsilon_1} \geq \sigma_1 \tag{1}$$

In this equation, σ_1 and ε_1 denotes the principal major stress and the principal major strain, respectively; $\beta = \Delta \varepsilon_2 / \Delta \varepsilon_1$ is the ratio between the minor and major strain increments. This equation results in an evolution of β, which is expressed as follows.

$$\Delta \beta = \frac{\sigma_1 - \frac{\partial \sigma_1}{\partial \varepsilon_1}}{\frac{\partial \sigma_1}{\partial \beta}} \Delta \varepsilon_1 \tag{2}$$

Hence, the strain localization is determined as soon as β approaches zero, indicating the plane-strain forming mode.

The condition is formulated based on the condition of maximum loading force observed during a uniaxial tensile test. In the MMFC, the strain path change is enforced to keep the maximum force unchanged. However, numerous experiments pointed out that the force drop may be up to 10% after the maximum for many automotive sheet metals [10]. Therefore, a scaling factor, ξ is proposed to slowdown the changing rate of β. Consequently, the evolution of β is updated as follows.

$$\Delta \beta = \xi \frac{\sigma_1 - \frac{\partial \sigma_1}{\partial \varepsilon_1}}{\frac{\partial \sigma_1}{\partial \beta}} \Delta \varepsilon_1 \tag{3}$$

In order to determine the value of ξ, a finite element simulation for uniaxial tensile test should be conducted to compare the simulated β evolution with the curve predicted by the MMFC for the uniaxial tension mode. In this study, a recommended value ($\xi = 0.5$) is adopted in the subsequent calculation without any further calibration.

For ductile materials, necking may occur before the fracture. Therefore, theoretical assumption of plane-strain forming mode at necking required in MMFC seems to overreach. An alternative condition for detecting the strain localization is presented as follows.

$$\left|\frac{d(\Delta \beta)}{d\varepsilon}\right| \tag{4}$$

In other words, the localized neck is supposed to occur when the acceleration of strain path change reaches its minimum.

Calculation of the Forming Limit Curves

This study investigates two automotive sheet metals, i.e., DP800 and AA6016. The thickness of both materials is 1.2 mm. A series of Nakajima tests are conducted following the ISO 12004-2 standard to determine the experimental FLC for the examined materials. During these tests, the ARAMIS digital image correlation (DIC) system is used to monitor the strain field distribution on

Material Forming - ESAFORM 2023
Materials Research Proceedings 28 (2023) 705-710

Materials Research Forum LLC
https://doi.org/10.21741/9781644902479-76

the surface of the deformed specimens. The experimental FLC is calculated using the FLC built-in function in ARAMIS.

A proper material model is needed to calculate the theoretical FLC. For this purpose, constitutive models used to describe material's behavior under complex stress states were previously calibrated by Barlo et al. [11] and not be repeated here. The Swift hardening law [1] is adopted for DP800 whereas the Voce model [2] is applied for AA6016. Moreover, the Yld2000-2d yield function proposed by Barlat et al. [12] is applied to capture the yield surface of these tested materials. The formulations of these functions are expressed below.

$$\text{Swift}: H(\underline{\varepsilon}) = C(\varepsilon_0 + \underline{\varepsilon})^n \tag{5}$$

$$\text{Voce}: H(\underline{\varepsilon}) = S - a.exp(-b\underline{\varepsilon}) \tag{6}$$

$$\text{Yld2000-2d}: \underline{\sigma} = \left\{ \frac{1}{2}\left(\left|X_1' - X_2'\right|^m + \left|2X_1'' + X_2''\right|^m + \left|X_1'' + 2X_2''\right|^m \right) \right\}^{1/m} \tag{7}$$

where C, ε_0, n, S, a, and b are hardening parameters; $X_{1,2}^{',''}$ are the principal values of two linearly transformed tensors which contain eight anisotropic parameters ($\alpha_1 - \alpha_8$) and the exponent parameter m. The calibrated parameters of each hardening law and yield function for the investigated materials are reported in Table 1 and Table 2, respectively.

Table 1. Hardening parameters of the tested materials.

Hardening law	Swift			Hardening law	Voce		
	C [MPa]	ε_0	n		S [MPa]	a [MPa]	b
DP800	1322.89	0.0012	0.165	AA6016	336.84	215.9	8.179

Table 2. Parameters of the Yld2000-2d function calibrated for the tested materials.

Yield function	α_1	α_2	α_3	α_4	α_5	α_6	α_7	α_8	m
DP800	0.9037	1.0270	1.0546	1.0055	1.0165	0.9367	0.9917	1.0306	6
AA6016	0.9633	0.9996	0.9438	1.0244	1.0134	0.9906	0.9683	1.1448	8

Fig. 1 depicts the calibrated hardening laws for both tested materials in comparison with the experimental data obtained from uniaxial tensile tests. Fig. 2 compares the predicted yield locus of these materials. These material models are employed in the framework of MMFC2 to predict their FLC. Moreover, the predictions of the original MMFC and MK models for the tested materials are also calculated for comparison purposes. A detailed description of these models can be found in Pham et al. [9].

Material Forming - ESAFORM 2023
Materials Research Proceedings 28 (2023) 705-710

Materials Research Forum LLC
https://doi.org/10.21741/9781644902479-76

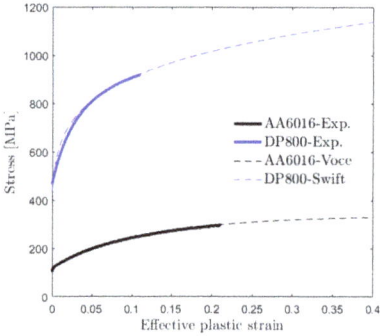

Fig. 1. Calibrated hardening law of the tested materials DP800 and AA6016.

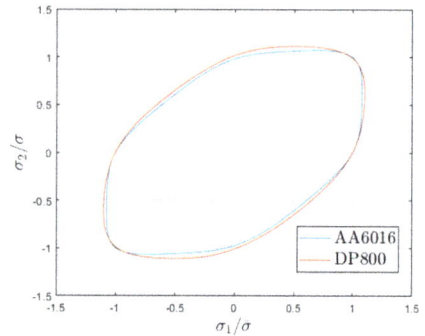

Fig. 2. Yield locus predicted by Yld2000-2d function for the tested materials.

Discussion

The predicted FLCs of DP800 and AA6016 sheets, based on different models, are compared with experimental data in Fig. 3 and Fig. 4, respectively. To simplify the comparison of the predicted and experimental FLCs, the experimental data were offset so the lowest forming limit point was moved to the plane strain region.

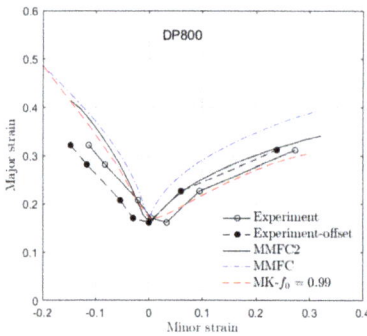

Fig. 3. Experimental and predicted FLCs of DP800.

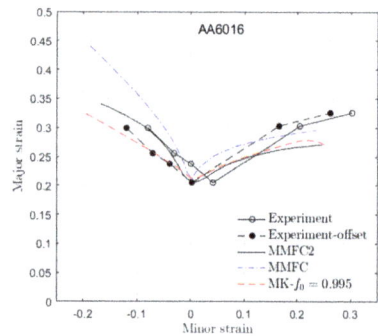

Fig. 4. Experimental and predicted FLCs of AA6016.

It can be seen in Fig. 3 that the MK model provides an excellent prediction of the experimental data of DP800, especially the region comprised between plane strain and biaxial tension. The MMFC model overestimates the experimental data, except for the plane strain point. Between uniaxial tension and plane strain regions, the MMFC2 agrees with MK model prediction, which is slightly higher than the experimental data. However, the results of MMFC2, in the right-side of the curve, lie in between these two other predictions.

In the case of AA6016 shown in Fig. 4, all theoretical models underestimate the experimental data in the biaxial tension regimes, probably due to the use of the Voce hardening law, which

Material Forming - ESAFORM 2023   Materials Research Forum LLC
Materials Research Proceedings 28 (2023) 705-710   https://doi.org/10.21741/9781644902479-76

shows a saturation stress in large strain ranges. In addition, the selected value of the exponent parameter, m influences the derived results, as discussed in [8]. Again, the MK model provides the best prediction for the left-side of the experimental curve. In contrast, the result of MMFC is significantly higher than the experiment. On the one hand, the MMFC2 presents an intermediate prediction for the left-side curve, whereas on the other hand, the result of this model for the right-side curve is close to that of the MK model.

The comparisons reveal that the prediction of the MK model is the most reliable among the considered theoretical models. The good performance of the MK model has been demonstrated in the literature for different materials subjected to different loading scenarios [13]. This study affirms the conclusion. The MMFC seems to overestimate the measured data of the tested materials. The MMFC2 improves the accuracy of the MMFC. Compared to the MK model, the MMFC2 derives comparable predictions for the FLC of the tested materials with an around five times faster computational time. The reason for the computational benefit of MMFC2 is due to its formulation, which does not require solving any system of non-linear equations as the MK model does. Details on the computational time of each theoretical model implemented in this study are reported in Table 3. According to this table, the computational efficiency of the MMFC2 seems negligible since the MK computational time is very good. However, the advantage gains more attractive when either a more complex material, such as, a distortional hardening model is adopted or a huge number of simulations is inquired, for example, in a data-driven application.

Table 3. Comparison between the computational time of theoretical models.

Theoretical model	MMFC	MMFC2	MK
Computational time (s)	20	25	120

Summary

This work aims to verify the potential of the newly proposed model MMFC2 for theoretically predicting the FLC of sheet metal. Two automotive sheets, DP800 and AA6016, were investigated by comparing the theoretical predictions of three different models, i.e., MMFC, MMFC2, and MK, with the experimental FLC obtained from Nakajima tests. The following conclusions can be made after this study.
•   All theoretical models predict similar forming limits at the plane strain region, which are close to the experimental measurement.
•   The MMFC model presents remarkably higher forming limits than the measured FLC.
•   For both investigated materials, the predictions of the MMFC2 are comparable to those of MK model within a significant reduction of computational time.

Acknowledgement

The authors gratefully acknowledge the financial support from KK-stiftelsen (grant number 20200125) and VINNOVA in the Sustainable Production subprogram within Vehicle Strategic Research and Innovation (FFI) program (grant number 2020-02986).

References

[1] D. Banabic, Sheet Metal Forming Processes: Constitutive Modelling and Numerical Simulation, Springer Berlin, Heidelberg, 2010. https://doi.org/10.1007/978-3-540-88113-1
[2] H.W. Swift, Plastic instability under plane stress. J. Mech. Phys. Solid. 1 (1952) 1-18. https://doi.org/10.1016/0022-5096(52)90002-1
[3] R. Hill, On discontinuous plastic states, with special reference to localized necking in thin sheets, J. Mech. Phys. Solid. 1 (1952) 19-30. https://doi.org/10.1016/0022-5096(52)90003-3
[4] S. Storen, J. Rice, Localized necking in thin sheets, J. Mech. Phys. Solid. 23 (1975) 421-441. https://doi.org/10.1016/0022-5096(75)90004-6

Material Forming - ESAFORM 2023
Materials Research Proceedings 28 (2023) 705-710

Materials Research Forum LLC
https://doi.org/10.21741/9781644902479-76

[5] Z. Marciniak, K. Kuczynski, Limit strains in the processes of stretch-forming sheet metal. Int. J. Mech. Sci. 9 (1967) 609–620. https://doi.org/10.1016/0020-7403(67)90066-5

[6] P. Hora, L. Tong, B. Berisha, Modified maximum force criterion, a model for the theoretical prediction of forming limit curves, Int. J. Mat. Form. 6 (2013) 267-279. https://doi.org/10.1007/s12289-011-1084-1

[7] H. Aretz, Numerical restrictions of the modified maximum force criterion for prediction of forming limits in sheet metal forming, Model. Simu. Mater. Sci. Eng. 12 (2004) 677. https://doi.org/10.1088/0965-0393/12/4/009

[8] Q.T. Pham, B.H. Lee, K.C. Park, Y.S. Kim, Influence of the post-necking prediction of hardening law on the theoretical forming limit curve of aluminium sheets, Int. J. Mech. Sci. 140 (2018) 521-536. https://doi.org/10.1016/j.ijmecsci.2018.02.040

[9] Q.T. Pham, M.S. Islam, M. Sigvant, L.P. Caro, M.-G. Lee and Y.-S. Kim: submitted to International Journal of Solids and Structures (2022).

[10] N. Manopulo, P. Hora, P. Peters, M. Gorji, F. Barlat, An extended modified maximum force criterion for the prediction of localized necking under non-proportional loading, Int. J. Plast. 75 (2015) 189-203. http://doi.org/10.1016%2Fj.ijplas.2015.02.003

[11] A. Barlo, M. Sigvant, B. Endelt, On the Failure Prediction of Dual-Phase Steel and Aluminium Alloys Exposed to Combined Tension and Bending, IOP Conf. Ser.: Mater. Sci. Eng. 651 (2019) 012030. https://doi.org/10.1088/1757-899X/651/1/012030

[12] F. Barlat, J. Brem, J.W. Yoon, K. Chung, R. Dick, D. Lege, F. Pourboghrat, S.-H. Choi, E. Chu, Plane stress yield function for aluminum alloy sheets—part 1: theory, Int. J. Plast. 19 (2003) 1297-1319. https://doi.org/10.1016/S0749-6419(02)00019-0

[13] M.C. Butuc, G. Vincze, The performance of Marciniak – Kuczynski approach on prediction of plastic instability of metals subjected to complex loadings, Numisheet 2018, IOP Publishing, Journal of Physics: Conference Series 1063 (2018) 12061. http://doi.org/10.1088/1742-6596/1063/1/012061

Material Forming - ESAFORM 2023
Materials Research Proceedings 28 (2023) 711-716

Materials Research Forum LLC
https://doi.org/10.21741/9781644902479-77

Anisotropic deformation behavior during cup drawing at room temperature of a ZX10 magnesium alloy sheet

HAMA Takayuki[1,a] *, HIGUCHI Koichi[1,b] and NAKATA Yuto[1,c]

[1]Graduate School of Energy Science, Kyoto University, Yoshida-honmachi,
Sakyo-ku, Kyoto 606-8501, Japan

[a]hama@energy.kyoto-u.ac.jp, [b]koichi.higuchi.72a@gmail.com,
[c]nakata.yuto.64r@st.kyoto-u.ac.jp

Keywords: Magnesium Alloy, Room-Temperature Cup Drawing, Twinning, Lankford Value, Thickness Strain

Abstract. Magnesium (Mg) alloy sheets have low density and high specific strength; thus, they are expected to facilitate weight reduction of structural components. However, because of the strong crystal anisotropy of the hexagonal structure and the strong basal texture observed in typical rolled Mg alloy sheets, their press formability at room temperature is low. To improve the room-temperature press formability, ZX series Mg alloy sheets that weakened the basal texture have recently been developed. The plastic deformation behavior of a rolled Mg-1.5mass%Zn-0.1mass%Ca (ZX10Mg) alloy sheet was studied in a previous study [7], and it was reported that the plastic deformation behavior showed strong in-plane anisotropy and differed notably from that of AZ series rolled Mg alloy sheets. In the present study, cylindrical cup drawing of a ZX10Mg alloy sheet was performed at room temperature and the drawability was examined in terms of cup height distributions, strain evolution, and texture evolution. The cup height differed significantly between the rolling and transverse directions. The thickness at the cup edge was the largest in the rolling direction and the smallest in the transverse direction. The magnitude relationship of the thickness correlated with the Lankford values under compression. The mechanism that yielded the difference in texture evolution was also discussed.

Introduction

Magnesium (Mg) alloy sheets have low density and high specific strength; thus, they are expected to facilitate weight reduction of structural components [1]. However, because of the strong crystal anisotropy of the hexagonal structure and the strong basal texture observed in typical rolled Mg alloy sheets, their press formability at room temperature is low [2,3]. To improve the press formability, ZX series Mg alloy sheets that weakened the basal texture have recently been developed. Chino et al. [4-6] conducted the room-temperature Erichsen cupping test of ZX series Mg alloy sheets and reported that their stretchability is superior to that of AZ series rolled Mg alloy sheets. Hama et al. [7] reported that the work-hardening behavior of a rolled Mg-1.5mass%Zn-0.1mass%Ca (ZX10Mg) alloy sheet shows strong in-plane anisotropy, which differs notably from those of typical AZ series rolled Mg alloy sheets. Nakata et al. [8] conducted room-temperature V-bending tests of a ZX10 Mg alloy sheet and studied the bendability in terms of activities of slip and twinning systems. It is expected that ZX series Mg alloy sheets exhibit not only better stretchability and bendability but also superior drawability, which is also one of the important properties for industrial applications. It is also presumed from the previous study that anisotropic deformation is exhibited during drawing because of the strong in-plane anisotropy of the work-hardening behavior. However, room-temperature drawability of ZX series Mg alloy sheets have not been studied yet.

In the present study, cylindrical cup drawing of a ZX10Mg alloy sheet was performed at room temperature and the drawability was discussed in terms of cup height and strain distributions.

Material Forming - ESAFORM 2023   Materials Research Forum LLC
Materials Research Proceedings 28 (2023) 711-716   https://doi.org/10.21741/9781644902479-77

Texture evolution was also measured to discuss the correlation between the macroscopic and mesoscopic deformation behaviors.

Experimental Methods

A rolled ZX10 Mg alloy sheet with 1.0 mm thickness produced by Sumitomo Electric Industries, Ltd. [7] was used. Circular cup drawing tests were conducted at room temperature. The experimental procedures were the same as those reported in a literature [9]. A photograph of the experimental setup for cup drawing test is shown in Fig. 1. The diameters of punch and die cavity were 27.8 mm and 30 mm, respectively. The punch and die shoulder radii were 5.0 mm and 7.0 mm, respectively. The diameter of the circular specimen was 40 mm, which corresponds to the drawing ratio of 1.33. The distance between the die and the blank holder was set to 1.1 mm and kept unchanged during forming. A solid lubricant (Moly Paste, Sumico Lubricant Co.) was used for lubrication between the sheet and the dies. The punch was penetrated to a stroke of 16 mm with a punch speed of 5 mm/min.

Fig. 1. Experimental setup for the cylindrical cup drawing.

Thickness and cup height of the products were measured using a micrometer and a two-dimensional laser displacement sensor (LJ-V7080, Keyence Co.), respectively. Electron backscattered diffraction (EBSD) measurements were conducted to measure the texture at some points. Fig. 2 shows the pole figure of the initial sample [7]. In the (0001) pole figure, strong peaks appeared in the vicinity of the normal direction (ND), but at the same time weak peaks appeared near the transverse direction (TD), showing that the basal texture is less pronounced than that of typical AZ series rolled Mg alloy sheets [10, 11].

Results and Discussion

Figs. 3 (a) and 3 (b) show the photograph of a drawn cup and the distributions of cup height in the circumferential direction at different punch strokes, respectively. In Fig. 3 (b), the vertical axis denotes the relative height in which the height at the rolling direction (RD) was set to zero in order to exclude the effect of the curvature at the cup bottom. Moreover, considering the symmetry of the deformation, only the results in the first quadrant is shown.

The cup height was larger in the TD than in the RD already at the stroke of 8 mm. The increasing trend is pronounced from the angle of 15° to 60°. The cup height increased with increasing the punch stroke. This trend was more pronounced in the vicinity of the angle of 60°, which eventually

Material Forming - ESAFORM 2023
Materials Research Proceedings 28 (2023) 711-716

Materials Research Forum LLC
https://doi.org/10.21741/9781644902479-77

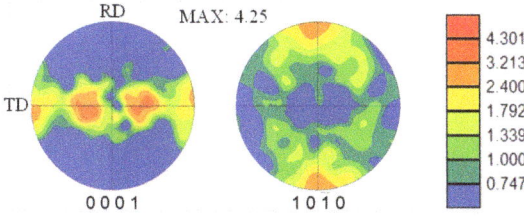

Fig. 2. Initial pole figures of the sample.

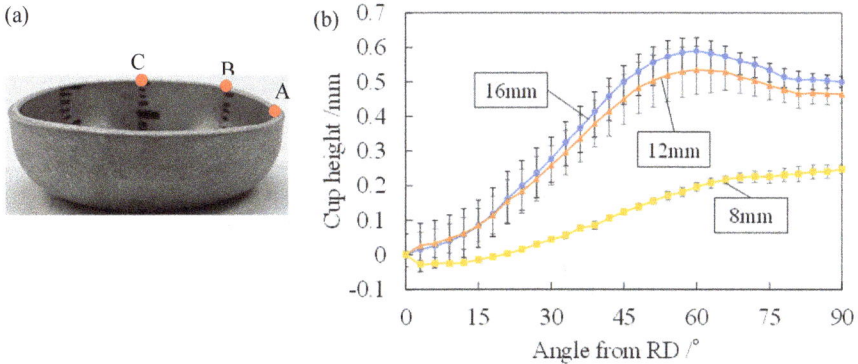

Fig. 3. Geometry of the drawn cup. (a) Photograph of a drawn cup, and (b) distribution of cup height.

resulted in earing formation in the vicinity of the angle of 60°. The tendency of the cup height distribution at the punch stroke of 16 mm remained almost unchanged from that of the punch stroke of 12 mm. It is noted that the cup heights differed largely between the RD and TD also in the final product.

Fig. 4 shows the evolution of thickness strain at the points near the sheet edge in the RD, diagonal direction (DD), and TD, i.e., points A, B, and C in Fig. 3 (a), respectively. The thickness strains increased largely from the punch stroke of 4 mm to 13 mm. The thickness strain was the largest in the RD and that of the DD was larger than that of the TD until the stroke of 13 mm. Thereafter, the thickness strains decreased because of ironing between the punch and the die. Eventually, the thickness strains at the RD and DD became almost the same.

Fig. 5 shows the pole figures measured at the points A and B of the drawn cup. Compared to the initial pole figure shown in Fig. 2, strong peaks near the ND did not appear at both points. In contrast, strong peaks appeared near the TD and RD at the points A and B, respectively. These results show that the texture evolution depended largely on the region.

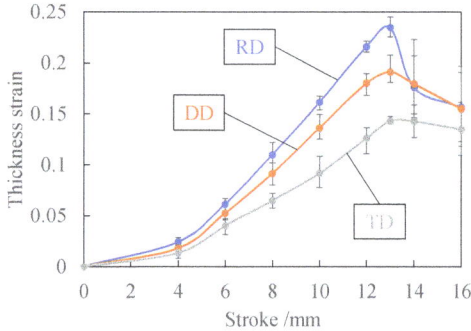

Fig. 4. Evolution of thickness strain at the sheet edge.

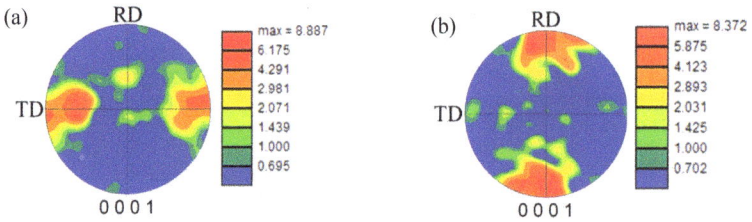

Fig. 5. (0001) pole figures measured at the points (a) A and (b) B.

Results and Discussion

Thickness distributions. The mechanism that the thickness strain was the largest in the RD and the smallest in the TD is discussed. Large thickening occurred at the sheet edge because the sheet edge was subjected to circumferential compression before the sheet edge was drawn into the die cavity. Specifically, during drawing, the regions near the sheet edge in the RD, DD, and TD are subjected to compression in the TD, DD, and RD, respectively. This indicates that thickness increase would be the largest in TD compression and the smallest in RD compression. It may be useful to consider the correlation with Lankford value to examine the difference in thickness increase depending on the direction. However, Lankford values of the Mg alloy sheet would be different between tension and compression because of the polar character of twinning [12]. Therefore, the Lankford values were measured under both tension and compression. In the compression test, comb-shaped dies were attached to the specimen to avoid occurrence of out-of-plane buckling. In both tension and compression tests, longitudinal and width strains were measured using a digital image correlation (DIC) method. In the DIC method, the software GOM Correlate Professional V8 (GOM) was used. Plastic strains were evaluated by subtracting elastic strains estimated using Young's modulus and Poisson ratio from the measured total strains.

Figs. 6 shows the evolution of Lankford values as a function of longitudinal strain. Under tension, the Lankford value was the largest in the RD at small strains, and that in the DD was larger than that of the TD. However, the Lankford values in the DD and TD tended to increase as the strain increased, while the Lankford value in the RD remained almost unchanged; thus, that of the DD became the largest at large strains. Under compression, the Lankford value in the RD was the largest, and that of the DD was larger than that of the TD. This trend remained unchanged in the

Material Forming - ESAFORM 2023   Materials Research Forum LLC
Materials Research Proceedings 28 (2023) 711-716   https://doi.org/10.21741/9781644902479-77

strain range tested in this work. Apparently, the magnitude relationship of the Lankford value differed between tension and compression. Moreover, the magnitude relationship of the Lankford value under compression at large strains is consistent with that of thickening on the drawn cup before the sheet was subjected to ironing. These results show that, for the ZX10 Mg alloy sheet, the thickness strain distribution should be evaluated not by the Lankford value under tension but by that of compression. Moreover, these results also suggest that the difference in the Lankford value between the RD and TD also yielded the notable difference in the cup height between the RD and the TD.

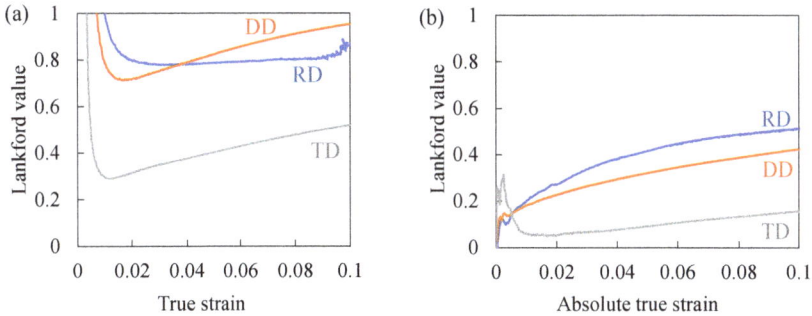

Fig. 6. Evolution of Lankford value under (a) tension and (b) compression.

Texture evolution. Next, the difference in texture evolution depending on the region is discussed. As discussed in a literature [7], under TD compression, twinning is active at the grains whose basal planes oriented near the ND, whereas it is not at the grains whose basal planes oriented near the TD. As a result, the peaks near the ND observed in the initial pole figure tend to disappear and strong peaks appear near the TD. This texture evolution is consistent with that observed at the point A which experienced TD compression

Similarly, under RD compression, twinning is easy to active at both grains whose basal planes oriented near the ND and the TD; thus, the peaks near the TD and ND observed in the initial pole figure disappeared and the peaks near the RD appeared [7]. This is also consistent with the result at the point B.

These results suggest that the texture evolution observed at the sheet edge was primarily governed by circumferential compression and, moreover, one of the reasons that the texture evolution differed depending on the region is the difference in the twinning activity.

Summary

Room-temperature cylindrical cup drawing of a ZX10Mg alloy sheet was conducted and the drawability was discussed. The results obtained in this study can be summarized as follows.

(1) The cup height started increasing largely after the punch stroke exceeded 8 mm. The maximum height appeared at 60° from the rolling direction. The cup height was larger in the transverse direction than in the rolling direction.

(2) The thickness at the cup edge was the largest in the rolling direction and the smallest in the transverse direction before the sheet was subjected to ironing. The magnitude relationship of the thickness correlated with the Lankford values under compression.

(3) The difference in twinning activity was one of the reasons that yielded the difference in the texture evolution depending on the direction.

Acknowledgements

The authors would like to thank Sumitomo Electric Industries, Ltd. for providing the ZX10 Mg alloy sheet used in this study. The authors gratefully acknowledge Mr. Sohei Uchida of TRI Osaka for his kind assistance in performing the EBSD measurements. This work was supported by Japan Society for the Promotion of Science JSPS, KAKENHI, grant number JP20H02480 and the AMADA Foundation (AF-2019004-A3).

References

[1] A.I. Taub, A.A. Luo, Advanced lightweight materials and manufacturing processes for automotive applications, MRS Bull. 40 (2015) 1045-1053. https://doi.org/10.1557/mrs.2015.268

[2] F.-K. Chen, T.-B. Huang, C.-K. Chang, Deep drawing of square cups with magnesium alloy AZ31 sheets, Int. J. Mach. Tools Manuf. 43 (2003) 1553-1559. https://doi.org/10.1016/S0890-6955(03)00198-6

[3] E. Doege, K. Dröder, Sheet metal forming of magnesium wrought alloys — formability and process technology, J. Mater. Process. Technol. 115 (2001) 14-19. https://doi.org/10.1016/S0924-0136(01)00760-9

[4] Y. Chino, X. Huang, K. Suzuki, M. Mabuchi, Enhancement of stretch formability at room temperature by addition of Ca in Mg-Zn alloy, Mater. Trans. 51 (2010) 818-821. http://doi.org/10.2320/matertrans.M2009385

[5] Y. Chino, T. Ueda, Y. Otomatsu, K. Sassa, X. Huang, K. Suzuki, M. Mabuchi, Effects of Ca on tensile properties and stretch formability at room temperature in Mg-Zn and Mg-Al alloys, Mater. Trans. 52 (2011) 1477-1482. http://doi.org/10.2320/matertrans.M2011048

[6] Y. Chino, K. Sassa, X. Huang, K. Suzuki, M. Mabuchi, Effects of Zinc concentration on the stretch formability at room temperature of the rolled Mg-Zn-Ca alloys, J. Japan Inst. Metals 75 (2011) 35-41. (in Japanese)

[7] T. Hama, K. Higuchi, H. Yoshida, Y. Jono, Work-hardening behavior of a ZX10 Magnesium alloy sheet under monotonic and reverse loadings, Key Eng. Mater. 926 (2022) 926-932. https://doi.org/10.4028/p-7bgcsj

[8] T. Nakata, T. Hama, K. Sugaya, S. Kamado, Understanding room-temperature deformation behavior in a dilute Mg-1.52Zn-0.09Ca (mass%) alloy sheet with weak basal texture, Mater. Sci. Eng. A 852 (2022) 143638. https://doi.org/10.1016/j.msea.2022.143638

[9] T. Hama, K. Hirano, R. Matsuura, Cylindrical cup drawing of a commercially pure titanium sheet: experiment and crystal plasticity finite-element simulation, Int. J. Mater. Form. 15 (2022) 8. https://doi.org/10.1007/s12289-022-01655-x

[10] T. Hama, Y. Tanaka, M. Uratani, H. Takuda, Deformation behavior upon two-step loading in a magnesium alloy sheet, Int. J. Plast. 82 (2016) 283-304.

[11] T. Hama, T. Suzuki, S. Hatakeyama, H. Fujimoto, H. Takuda, Role of twinning on the stress and strain behaviors during reverse loading in rolled magnesium alloy sheets, Mater. Sci. Eng. A 725 (2018) 8-18. http://doi.org/10.1016/j.msea.2018.03.124

[12] D Steglich, X. Tian, J. Bohlen, T. Kuwabara, Mechanical testing of thin sheet Magnesium alloys in biaxial tension and uniaxial compression, Exp. Mech. 54 (2014) 1247-1258. https://doi.org/10.1007/s11340-014-9892-0

Material Forming - ESAFORM 2023
Materials Research Proceedings 28 (2023) 717-726

Materials Research Forum LLC
https://doi.org/10.21741/9781644902479-78

Modelling failure of joining zones during forming of hybrid parts

WESTER Hendrik[1,a] *, STOCKBURGER Eugen[1,b], PEDDINGHAUS Simon[1,c],
UHE Johanna[1,d] and BEHRENS Bernd-Arno[1,e]

[1]Institute of Forming Technology and Machines (IFUM), Leibniz Universität Hannover,
An der Universität 2, 30823 Garbsen, Germany

[a]wester@ifum.uni-hannover.de, [b]stockburger@ifum.uni-hannover.de,
[c]peddinghaus@ifum.uni-hannover.de, [d]uhe@ifum.uni-hannover.de,
[e]behrens@ifum.uni-hannover.de

Keywords: Pre-Joined Hybrid Semi-Finished Parts, Numerical Modelling, Cohesive Zones, Local Tensile Tests

Abstract. Combining diverse materials enables the use of the positive properties of the individual material in one component. Hybrid material combinations therefore offer great potential for meeting the increasing demand on highly loaded components. The use of hybrid pre-joined semi-finished products simplifies joining processes through the use of simple geometries. However, the use of pre-joined hybrid semi-finished products also results in new challenges for the following process chain. For example, the materials steel and aluminium may form brittle intermetallic phases in the joining zone, which can be damaged in the following forming process under the effect of thermo-mechanical loads and thus lead to a weak point in the final part. Due to their small thickness as well as their position in the component, the analysis of the joining zone is only possible by complex destructive testing methods. FE simulation therefore offers an efficient way to analyse the development of damage in the process design and to reduce damage by process modifications. Therefore, within this study a damage model based on cohesive zone elements is implemented in the FE software MSC Marc 2018 and calibrated using experimental local tensile tests performed under process relevant conditions.

Introduction

Given the constant shortage of energy as well as raw material resources and the associated costs, the conservation of resources in all areas is of great importance to society as a whole. The manufacturing industry in particular offers a wide range of potentials that extend across development, production and subsequent part usage. The resulting steadily increasing demands on components in terms of strength, functional integration and weight can no longer be met by monolithic materials. In this context, hybrid components show great potential, due to the possibility to locally use suitable materials at the right place. In recent years, composite forging has developed considerably for the production of hybrid components such as gear elements [1], whereby the semi-finished products are joined during the forging process. For example, hybrid bulk metal components such as connecting rods, wheel hubs and control arms are manufactured by Leiber Group GmbH & Co. KG, where the material composite is produced in particular by frictional and interlocking connection [2].

Groche et al. studied the influence of process variables on the welding of steel and aluminium by cold extrusion [3]. The billets' initial microstructural states, height ratio as well as the treatment of welding surfaces were investigated. They found that the precipitation hardened state of aluminium alloys are more suited for processing than the soft annealed state if the intention is to obtain a sound bonding behaviour with steel. An innovative manufacturing route for the production of hybrid components named Tailored Forming was developed using hybrid pre-joined semi-finished [4]. The material combination of steel and aluminium has great potential for industrial

Material Forming - ESAFORM 2023
Materials Research Proceedings 28 (2023) 717-726

Materials Research Forum LLC
https://doi.org/10.21741/9781644902479-78

application in hybrid components. On the one hand, the high lightweight potential of aluminium alloys can be exploited. On the other hand, components that are subject to special requirements in terms of strength and load-bearing capacity can be made of steel. However, use of hybrid components also result in new challenges with regard to component design and the development as well as implementation of manufacturing processes. When using dissimilar metallic material combinations, the bonding zone in particular represents a critical point during forming and in the final component. Therefore, it requires special consideration. Previous investigations show that during impact extrusion critical tensile stresses occur within the joining zone which can lead to local rupture [5]. However, the joining zone within the process as well as in the final component can only be analysed at great expense and in destructive tests. Therefore, this work focuses on the experimental analysis of bonding strength of hybrid pre-joined semi-finished parts consisting of steel and aluminium taking into account process relevant temperatures. Based on the results a newly developed material model [6] for describing the damage development within the joining zone is calibrated and implemented in the commercial FE software MSC Marc.

Materials and Methods
Tailored Forming process chain.
The processing of pre-joined hybrid semi-finished products requires adapted process chains. In particular, the diverging material properties and the resulting requirements on the forming process must be taken into account for material combinations of different types, such as steel and aluminium. The Tailored Forming process chain for manufacturing a hybrid demonstrator shaft in the material combination 1.3505 (100Cr6) and 3.2315 (EN AW-6082) is shown schematically in Fig. 1 and described in detail in the following. The joining of the steel and aluminium is carried out by friction welding. Afterwards, the part is heated inductively. Previous studies have shown that temperature control is of particular importance for the forming of hybrid steel-aluminium semi-finished products. Therefore, an inhomogeneous temperature field is set via induction heating [7]. The aim is to heat the steel as much as possible and keep the aluminium as cold as possible, to match the flow stresses of both materials. After heating, forming is carried out by an impact extrusion process on a screw press (Lasco SPR 500). The process takes 0.12 s, whereby the diameter is reduced by 31.5 %. Finally, the part is finished by turning and a high-performance component is created.

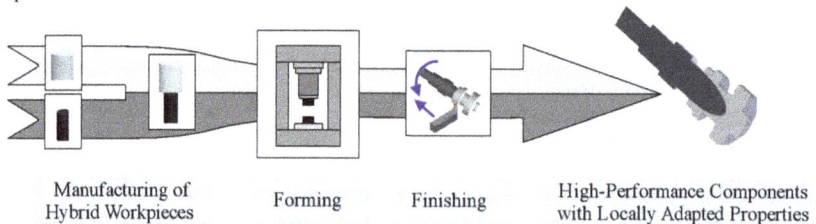

| Manufacturing of Hybrid Workpieces | Forming | Finishing | High-Performance Components with Locally Adapted Properties |

Fig. 1. Tailored Forming process chain for the production of hybrid shafts according to [8].

Thermo-mechanical material modelling.
In order to describe the material behaviour of the used aluminium EN AW-6082, flow curves are taken from previous work [5]. For the 100Cr6, flow curves from Simufact Forming 16.0 database are used to model the steel. A comparison of the used flow curves for different temperatures and a strain rate of 1 s^{-1} is given in Fig. 2. The comparison of the yield stresses of the materials clearly shows that inhomogeneous heating is required to adjust the yield stress levels. Comparable flow stress levels can be defined between 600°C and 1000°C in the case of 100Cr6 and between 20°C and 300°C in the case of the EN AW-6082. The temperature and strain rate

Material Forming - ESAFORM 2023   Materials Research Forum LLC
Materials Research Proceedings 28 (2023) 717-726   https://doi.org/10.21741/9781644902479-78

dependent flow curves were implemented in the FE software Simufact Forming 16 and MSC MARC Mentat 2018 for a simulation of the inductive heating and the impact extrusion.

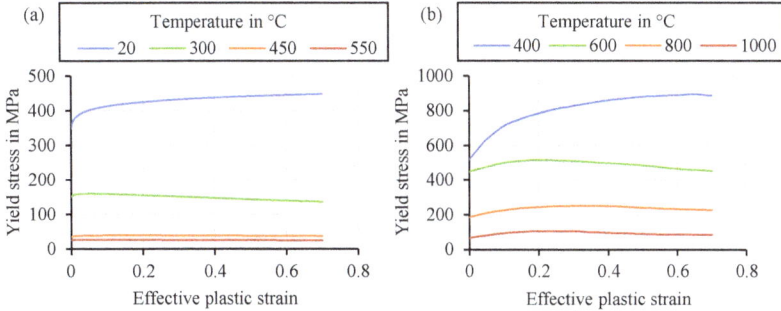

Fig. 2. Flow curves for various temperatures and a strain rate of 1 s⁻¹ for EN AW-6082 (a) and 100Cr6 (b).

Analysis of local bonding strength.

The resulting local bonding strength after friction welding is investigated by local tensile tests (as shown in Fig. 3) using miniaturised sample geometries on the forming dilatometer DIL805A/D+T. The use of miniaturised samples with a thickness of 1 mm allows local analysis of bond strength as a function of the semi-finished product diameter. In order to physically map the critical thermo-mechanical stresses of the hybrid semi-finished products, which lead to failure of the joining zone, the tensile tests are carried out at various testing temperatures. The temperatures for tensile testing are determined based on numerically calculated temperatures in the joining zone during impact extrusion. The tensile tests are performed at a quasi-static strain rate of 0.001 s⁻¹.

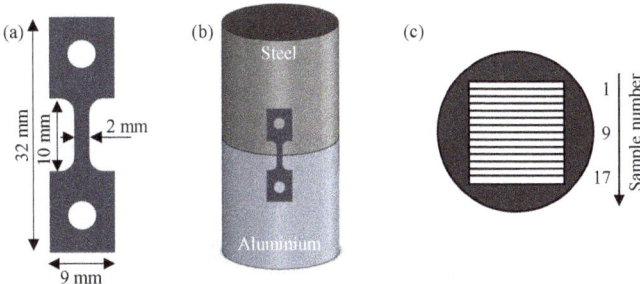

Fig. 3. Procedure for characterising the local bonding strength of joining zones: sample geometry (a) and sectional view (b) as well as top view (c) of the sample extraction area.

Numerical model of induction heating and impact extrusion.

Numerical simulation allows a realistic determination of temperature distribution during the impact extrusion process. To determine process relevant testing temperature for the tensile tests a numerical model representing induction heating as well as the forming process was first developed using the commercial FE software Simufact Forming 16. The model for induction heating consists of the hybrid billet placed within the induction coil. During heating, only the steel is heated

Material Forming - ESAFORM 2023   Materials Research Forum LLC
Materials Research Proceedings 28 (2023) 717-726   https://doi.org/10.21741/9781644902479-78

inductively, the aluminium is heated by thermal conduction. Between steel and aluminium an adhesive contact was defined and an intermetallic layer of 0.07 μm was assumed [9] with a thermal conductivity of 10 W/mK according to Nacke [10]. Boundary conditions like current, time and frequency were taken from experiments and set as input. Thermal material data was defined based on Simufact Forming 16 database. The relative permeability was determined by Behrens et al. using the method described in [7]. The calculated inhomogeneous temperature field after induction heating is transferred to the semi-finished product of the impact extrusion model. The elastic-plastic material behaviour of steel and aluminium parts is modelled as shown before. The impact extrusion tool system consists of a die and a punch, which are modelled as rigid heat conducting bodies with an initial temperature of 250 °C. Friction is modelled by the combined friction model consisting of the Tresca formulation with $m = 0.3$ and the Coulomb formulation with $\mu = 0.1$. The press kinematics were derived from experimental data of the used screw press and considered within the numerical model. However, the modelling of the joining zone requires cohesive zone elements. Since these are not available in Simufact Forming, another model of the extrusion process was created in the FE software MSC Marc 2018.

Damage modelling.

The brittle material behaviour of the joining zone is of particular importance in the forming of hybrid semi-finished products. However, the very low thickness of the intermetallic phase compared to the dimensions of the semi-finished products is challenging. Conventionally, the damage in thin layers is modelled by flat cohesive zone elements [11]. Their constitutive behaviour is described via traction separation laws, which take into account expansion under normal or shear loading. However, cohesive zone elements are flat and therefore cannot be used to describe the constitutive behaviour in bulk metal forming. In this context, Töller et al. developed a new concept for damage modelling of thin joining zones [6]. As shown in Fig. 4, the model combines the damage to cohesive zone elements described by traction separation behaviour with damage induced by severe membrane deformation such as stretching, which can occur during forming of hybrid semi-finished parts within the joining zone. Membrane deformation is considered using the Internal Thickness Extrapolation (InTEx) concept. This enables the use of cohesive zone elements for bulk forming simulation by consideration of a thickness within the element formulation. Tractions within the cohesive elements are calculated from resulting separations δ and stiffness k with a linear relationship. As proposed by Töller, a large but finite artificial stiffness k is introduced. The stiffness can be interpreted as a penalty parameter to model the brittle behaviour of thin intermetallic phases. Due to the use of effective tractions the tractions are reduced by local prevailing damage D. Thus, a local pre-damage as well as the damage by membrane deformation during forming process can be taken into account. A detailed description of the developed model can be found in [6]. To model the damage in joining zones of semi-finished hybrid parts during forming processes within the Tailored Forming process chain, the modelling concept was implemented in the cohesive element formulation of the commercial FE software MSC Marc 2018 using the user subroutine UCOHESIVE. This software was selected because, contrary to Simufact Forming, it supports the use of cohesive zone elements. The presented work deals with the development of a methodology determining the model parameters taking into account the thermo-mechanical process boundary conditions. The calibration of the model was done inversely by numerical simulation of the local tensile tests. The elastic-plastic material behaviour of steel and aluminium was described as shown before. The displacements are applied via the nodes on the end sample faces according to experimental tensile test. The isothermal test temperatures are imposed on all nodes of the model via a thermal boundary condition. The joining zone in the centre of the sample is modelled via cohesive zone elements. The failure time determined from the experimental force-displacement curves and the applied force are used as the target variables. Since the focus is on determining the value $t_{t,max}$ and only samples with a joining zone normal to the direction of loading are analysed, the value for $t_{s,max}$ is specified analogously to the procedure in [6] via the ratio $t_{t,max}/t_{s,max} = 1.22$. Damage due to membrane deformation is described via the Lemaitre damage model [12]. The parameters $s = 1$ and $S = 0.34$ for the description of the Lemaitre damage model are chosen accordingly to [6].

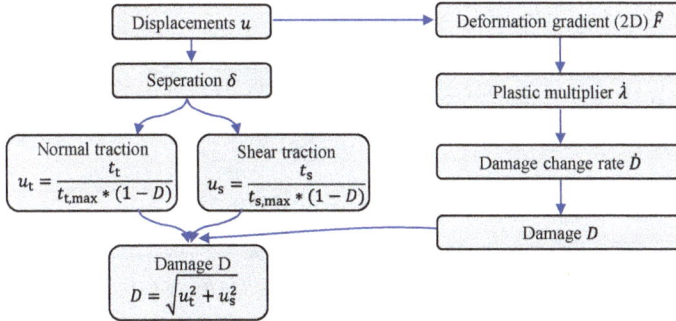

Fig. 4. Schematic representation of damage model for thin intermetallic phase.

Results

Numerical analysis of induction heating and impact extrusion.

As shown in Fig. 2, the flow curves of the materials differ significantly demonstrating the need for inhomogeneous heating to adjust the material flow properties for impact extrusion. According to the flow curves, the steel semi-finished product should be heated to 900°C and the aluminium semi-finished product to 300°C for homogenous material flow properties. The adjustability of the temperature field was investigated by experimental and numerical analyses of the inductive heating. The numerical simulation is used for the process analysis regarding the temperature development in order to define proper testing temperatures, as shown in Fig. 5 (a). Analogously to the experimental process, the semi-finished steel product is inductively heated for 25 s. A maximum temperature of 900°C is reached. During the transfer period of 5 s until the start of the impact extrusion, the maximum temperature drops to approx. 800°C due to heat transfer processes. The aluminium is heated only by heat conduction from the steel and reaches a maximum temperature of approx. 400°C after 30 s before forming. The maximum temperature of the joining zone is 420°C before the forming process starts. Due to the short forming process period of only 0.12 s, no significant temperature change is observed during forming, as shown in Fig. 5 (b). Fig. 5 (c) and (d) display the temperature distribution of the inhomogeneous heated semi-finished product after the transfer period from the induction heating and before the impact extrusion, both at the process time of 30 s. Based on this analysis, the bond strength of the joining zone is investigated in local tensile tests at the process relevant temperatures of 400°C and 450°C. The results of the tensile tests are used to determine the bond strength of the joint zone.

Material Forming - ESAFORM 2023
Materials Research Proceedings 28 (2023) 717-726

Materials Research Forum LLC
https://doi.org/10.21741/9781644902479-78

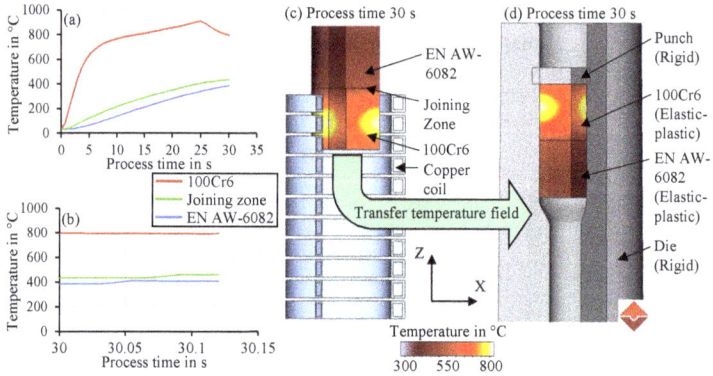

Fig. 5. Temperature development during induction heating (a) and impact extrusion (b) as well as coupled numerical process chain consisting of induction heating process (c) and impact extrusion (d).

Local bond strength and damage model calibration.

Fig. 6 shows the determined bond strengths as a function of temperature and the sample extraction area from the semi-finished product for a load normal to the joining zone. The sample extraction area can be found in Fig. 3 (c). Therefore, the bond strengths of samples with low and high numbers correspond to the edge area and the sample no. 9 to no. 11 to the bond strength in the centre of the friction-welded semi-finished product. For sample no. 10 no results could be determined at room temperature, because the samples failed directly.

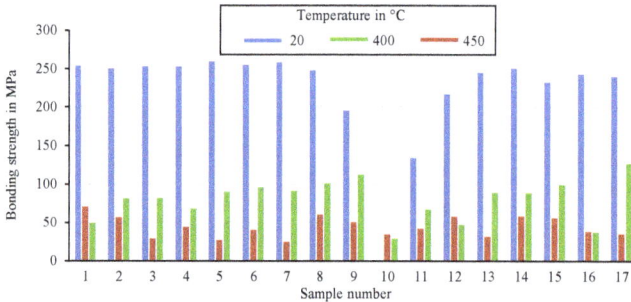

Fig. 6. Bond strength for normal loading depending on temperature and sample extraction area.

In particular, the tests at room temperature show a strong dependence on the removal position. Thus, the samples from the edge region exhibit a significantly higher bond strength than samples from the middle region. The tests also show a clear dependence on temperature. With increasing temperature, the bond strength decreases significantly. An optical analysis of the fracture surfaces showed a failure in the joining zone for all samples. Thus, at the process relevant temperatures compared to room temperature, even low tensile loads are critical with regard to a possible local failure of the joining zone and must be particularly taken into account in the process design.

Material Forming - ESAFORM 2023   Materials Research Forum LLC
Materials Research Proceedings 28 (2023) 717-726   https://doi.org/10.21741/9781644902479-78

As the stress-strain curve in Fig 7 (a) shows exemplarily for the sample 4 at 20°C, the samples from the edge region exhibit significant plastic deformation. An optical analysis of the samples shows a clear necking in the aluminium region and a failure of the monolithic material aluminium. Thus, the bond strength is greater than the tensile strength of the aluminium. Samples from the middle region like sample 9 in Fig 7 (b) show lower bond strengths and no pronounced plastic deformation. An optical analysis of the fracture zone shows failure in the joining zone without significant plastic deformation of the monolithic materials. The bond strength is thus below that of the weaker monolithic material aluminium and can be determined directly. The strong dependence of the bond strength is due to the process characteristics of friction welding. While high relative velocities prevail in the edge zone, which cause the oxide layer on the aluminium to break up and flow out, the relative velocity decreases toward the centre of the semi-finished product [13].

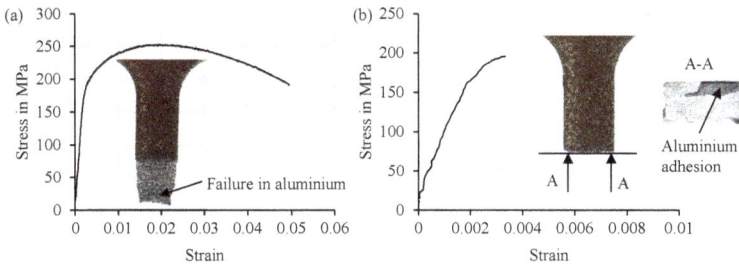

Fig. 7. Stress-strain curves and optical images after failure for sample 4 (a) and sample 9 (b) at 20°C.

For the determination of $t_{t,max}$ to calibrate the damage model, the performed tensile tests were mapped in the FE software MSC Marc 2018 with the corresponding experimental boundary conditions. A representation of the numerical model of the tensile test is given in Fig. 8. The temperature dependent values of $t_{t,max}$ were determined numerically iteratively by comparison with the experimental force-displacement curves and the time of failure. For the calibration, the sample no. 8 is used. This specimen exhibited the highest bond strength at failure in the joining zone without significant plastic deformation of the aluminium part. Thus, the calculated stress can be directly related to the bonding strength.

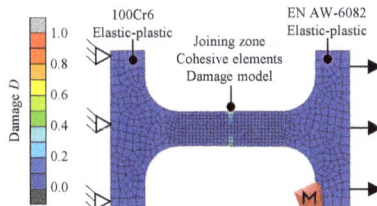

Fig. 8. Numerical model of tension test with joining zone modelled by means of cohesive elements.

A comparison of the force-displacement curves for the temperatures considered using the corresponding value for the damage parameter $t_{t,max}$ can be found in Fig. 9 (a) to (c). The

Material Forming - ESAFORM 2023
Materials Research Proceedings 28 (2023) 717-726

Materials Research Forum LLC
https://doi.org/10.21741/9781644902479-78

numerically inverse determined values can be found in Fig. 9 (d). The force-displacement curves agree well, so that the model and also the methodology can be considered validated. The force level at fracture is well predicted. However, there are small deviations regarding the length change at fracture between experiment and simulation. This is probably due to a slight plastic deformation of the specimens in the aluminium part, which is underestimated in the simulation. For a later application of the numerical representation of the impact extrusion process considering the damage of the joining zone, the local distribution of the maximal bond strength according to the findings of the local tensile tests has to be taken into account by defining local initial values for the pre-damage D and maximal bond strength $t_{t,max}$. For samples extracted at larger diameters of the semi-finished product, which show plastic deformation and failure of the aluminium, the achievable bond strength $t_{t,max}$ is increased by 25 % for later application. The decrease of the bonding strength with decreasing diameter towards the centre of the semi-finished product can be taken into account in the application by the damage parameter D as pre-damage, which is scaled according to the results in Fig. 6 taking into account the local position.

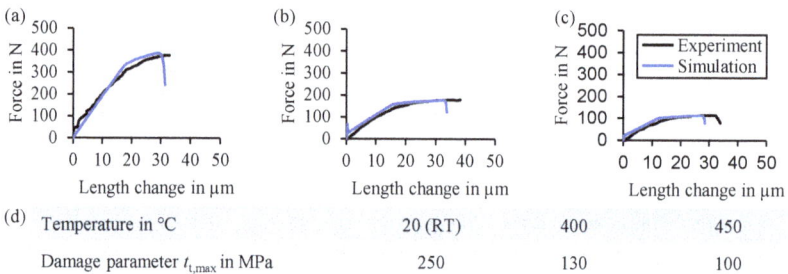

Fig. 9. Force-displacement curves from experiment and simulation at a temperature of 20°C (a), 400°C (b) and 450°C (c) for sample no. 8 as well as the determined temperature dependent damage parameter $t_{t,max}$ (d).

For a numerical analysis of the impact extrusion taking the joining zone into account, the process was also modelled in MSC Marc 2018. The boundary conditions were selected analogously to the described procedure for the tensile test simulations. As in the tensile tests, the joining zone itself is represented by cohesive zone elements. Since the damage model requires a three-dimensional mapping, a quarter model is used. The inhomogeneous temperature distribution is taken from the inductive heating in analogy to the procedure in Simufact Forming and imposed as initial temperature on the nodes in the Marc model. The extrapolation of temperature results from the nodes within the 2D space of the induction heating simulation (Simufact Forming) to the 3D space of the impact extrusion process model (MARC) is done via the radial relation $r = \sqrt{x^2 + x^2}$. The von Mises stress distribution after forming is shown in Fig. 10 (a). Due to heating, very low stresses are present, especially in the aluminium. In the steel part, higher stresses are present at the transition area to the aluminium. This is due to the physically induced inhomogeneous temperature distribution also in the steel component. As can be seen in Fig. 5, lower temperatures occur in the transition area of the steel due to heat conduction effects into the aluminium. This cannot be completely prevented even by adapted induction heating. Fig. 10 (b) shows the distribution of damage before and after forming. As can be seen, initial damage to the joining zone is already present before forming for small diameters in order to map the influence of the friction welding process on the local bond strength. After forming, damage can be seen in the

Material Forming - ESAFORM 2023
Materials Research Proceedings 28 (2023) 717-726

Materials Research Forum LLC
https://doi.org/10.21741/9781644902479-78

central area of the joining zone. This damage has already been observed by Behrens et al. for impact extrusion of the steel/aluminium combination 20MnCr5/EN AW-6082 pre-joined by laser welding, as shown in Fig. 10 (c) [14]. For this material combination, it was experimentally proven that an applied counterpressure prevents tearing of the joining zone by reducing the tensile stresses present. In subsequent investigations, other parameters of the damage model must therefore be further identified using the methodology described here and finally validated on the basis of experimental impact extrusion tests.

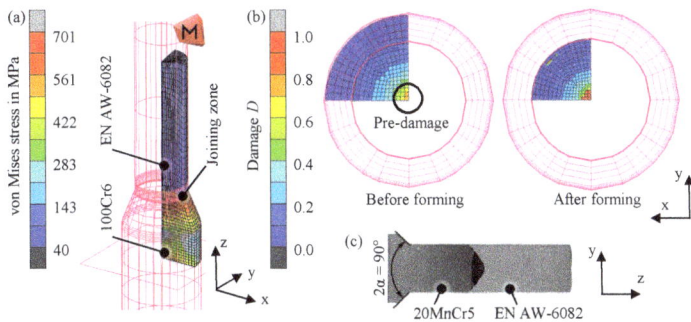

Fig. 10. von Mises stress after impact extrusion (a), damage within the joining zone before and after forming (b) as well as metallographic cross-section view of the component from [14] (c).

Summary

The scope of this work was the development of a methodology for the numerical inverse calibration of a damage model to describe the failure of a joining zone in hybrid pre-joined semi-finished products made of 100Cr6 and EN AW-6082. For this purpose, local tensile tests were performed at room temperature with miniature tensile samples taken from hybrid semi-finished products. To account for the effect of process relevant forming temperatures, the tests were also carried out at elevated temperatures. The test temperatures were defined based on numerical simulations of the inhomogeneous induction heating and the impact extrusion process. The results showed a strong influence of the sample extraction area and temperature on the achievable bond strength. The damage parameter $t_{t,max}$ was determined inversely by a numerical simulation of the performed tensile tests. The numerically calculated force-displacement curves showed good agreement with the experimental results.

An initial numerical simulation of the impact extrusion process shows that damage occurs in the joining zone. In further investigations, the other parameters of the model are calibrated by supplementary tensile tests with varying the angle between the loading direction and the joining zone of EN AW-6082 as well as 100Cr6 to improve the model quality. Finally, the model will be validated by means of experimental extrusion tests regarding the failure behaviour of the joining zone. The validated model will then be used to evaluate and design concepts such as a counterpressure concept in order to ensure damage-free production of hybrid components.

Acknowledgements

The results presented in this paper were obtained within the Collaborative Research Centre 1153 "Process chain to produce hybrid high-performance components by Tailored Forming" in the subproject C01. The authors would like to thank the German Research Foundation (Deutsche

Forschungsgemeinschaft, DFG, 252662854) for the financial and organisational support of this project. The authors would like to thank the subproject B03 for friction welding of hybrid semi-finished parts and the subproject C04 for support regarding the damage model.

References

[1] P. Wu, B. Wang, J. Lin, B. Zuo, Z. Li, J. Zhou, Investigation on metal flow and forming load of bi-metal gear hot forging process, Int. J. Adv. Manuf. Technol. 88 (2017) 2835-2847. https://doi.org/10.1007/s00170-016-8973-x

[2] R. Leiber, Hybridschmieden bringt den Leichtbau voran, Aluminium Praxis (2011) 7-8.

[3] P. Groche, S. Wohletz, A. Erbe, A. Altin, Effect of the primary heat treatment on the bond formation in cold welding of aluminum and steel by cold forging, J. Mat. Process. Technol. 214 (2014) 2040-2048. https://doi.org/10.1016/j.jmatprotec.2013.12.021

[4] B.-A. Behrens, J. Uhe, Introduction to tailored forming, Prod. Eng. Res. Devel. 15 (2021) 133-136. https://doi.org/10.1007/s11740-021-01022-w

[5] B.-A. Behrens, D. Duran, D., J. Uhe, T. Matthias, Numerical investigations on the influence of the weld surface and die geometry on the resulting tensile stresses in the joining zone during an extrusion process, 24th Int Conf Mat Form, Liège, Belgique (2021). https://doi.org/10.25518/esaform21.919

[6] F. Töller, S. Löhnert, P. Wriggers, Membrane mode enhanced cohesive zone element, Eng. Comp. 39 (2022) 722-743. https://doi.org/10.1108/EC-08-2020-0489

[7] B.-A. Behrens, H. Wester, S. Schäfer, C. Büdenbender, Modelling of an induction heating process and resulting material distribution of a hybrid semi-finished product after impact extrusion, 24th Int Conf Mat Form, Liège, Belgique (2021). https://doi.org/10.25518/esaform21.574

[8] R. Goldstein, B.-A. Behrens, D. Duran, Role of Thermal Processing in Tailored Forming Technology for Manufacturing Multi-Material Components, Heat Treat Conference (2017).

[9] S. Herbst, H. Aengeneyndt, M.J. Maier, F. Nürnberger, Microstructure and Mechanical Properties of Friction Welded Steel-Aluminum Hybrid Components after T6 Heat Treatment, Mater. Sci. Eng: A 696 (2017) 33-41. https://doi.org/10.1016/j.msea.2017.04.052

[10] B. Nacke, Ein Verfahren zur numerischen Simulation induktiver Erwärmungsprozesse und dessen technische Anwendung. Dissertation, Hannover (1987).

[11] G. Lélias, E. Paroissien, F. Lachaud, J. Morlier, Experimental characterization of cohesive zone models for thin adhesive layers loaded in mode I, mode II, and mixed-mode I/II by the use of a direct method, Int. J. Solid. Struct. 158 (2019) 90-115. https://doi.org/10.1016/j.ijsolstr.2018.09.005

[12] J. Besson, Continuum Models of Ductile Fracture: A Review, Int. J. Dam. Mech. 19 (2010) 3-52. https://doi.org/10.1177/1056789509103482

[13] S. D. Meshram, T. Mohandas, G. M. Reddy, Friction welding of dissimilar pure metals, J Mat Proc Tech 184 (2007) 330-337. https://doi.org/10.1016/j.jmatprotec.2006.11.123

[14] B.-A. Behrens, J. Uhe, F. Süer, D. Duran, T. Matthias, I. Ross, Fabrication of steel-aluminium parts by impact extrusion, Mater. Today: Proceedings 59 (2022) 220-226. https://doi.org/10.1016/j.matpr.2021.11.093

Material Forming - ESAFORM 2023
Materials Research Proceedings 28 (2023) 727-736

Materials Research Forum LLC
https://doi.org/10.21741/9781644902479-79

Influence of process parameters and process set-up on damage evolution during stretch drawing of u-shaped profiles

MÜLLER Martina[1,a] *, WEISER Ingo Felix[1,b], HERRIG Tim[1,c]
and BERGS Thomas[1,2,d]

[1]Laboratory for Machine Tools and Production Engineering WZL of RWTH Aachen University, Campus-Boulevard 30, 52074 Aachen, Germany

[2]Fraunhofer Institute for Production Technology IPT, Steinbachstr. 17, 52074 Aachen, Germany

[a]martina.mueller@wzl.rwth-aachen.de, [b]f.weiser@wzl.rwth-aachen.de, [c]t.herrig@wzl.rwth-aachen.de, [d]t.bergs@wzl.rwth-aachen.de

Keywords: Damage, Stretch Drawing, Dual Phase Steel

Abstract. The damage state in the form of voids and lattice defects of a sheet metal component has a substantial impact on the performance of a component in service regarding fatigue or crash behaviour. Therefore, managing the damage evolution during forming, especially the accumulation and distribution of damage, by targeted changes of the process parameters and set-up enables to improve component performance by influencing the stress-strain state [1]. The evolution of the stress-strain state during the forming process and along the process route represents the most significant factor influencing the resulting damage state. This paper focuses on the influence of the damage state of sheet metal components in order to improve the performance of a component regarding fatigue and crash behavior. Considering a variation of the process parameter (drawing die radius) and change in process set-up (singlestep, multistep, reverse stretch drawing) the damage accumulation and distribution within the component is analyzed using a calibrated LEMAITRE damage model. For the consideration of this paper, an u-shaped geometry of dual phase steel DP800, which is often found as an element in vehicle body construction, is used.

Introduction

The automotive sector is one of the main contributors to CO_2 emissions [1]. Due to the climate goals of the European Union to become a net zero greenhouse gas emitting economy by 2050 [2], it is necessary to rethink the approaches in this sector and to find a way to significantly reduce CO_2 emissions. One approach is lightweight design [3]. Lightweight design enables to reduce CO_2 emissions in every life cycle by means of resource efficiency.

Stretch drawing, often combined with deep drawing, is one of the most important processes for the production of three-dimensional sheet metal components and is used in particular in the manufacturing of vehicle bodies in the automotive sector [4]. Since 40% of the total mass is attributed to the vehicle body, stretch drawing is a lever for saving CO_2 by implementing lightweight design measures [5]. One possibility to introduce lightweight design measures in stretch drawing is to reduce the thickness of the sheet metal in order to save weight. In the automotive industry in particular, however, the components used have to meet stringent requirements in order to ensure the safety of passengers in the best possible way [6].

In the field of lightweight design, high-strength materials or topology optimization are currently used in order to save material and thus weight [7]. What is currently only insufficiently exploited are the possibilities offered by the control of ductile damage along the process chain and by ductile damage meaning *the formation, growth and coalescence of voids in the microstructure. These voids are formed in structural materials by decohesion at interfaces such as phase or grain*

Material Forming - ESAFORM 2023   Materials Research Forum LLC
Materials Research Proceedings 28 (2023) 727-736   https://doi.org/10.21741/9781644902479-79

boundaries and inclusions, or by the formation of new surfaces within phases or inclusions. This damage causes degradation of the performance of the corresponding component [8]. Damage control then enables either the damage accumulation or the damage distribution to be adapted by adjusting the process parameters or process set-up in such a way that components achieve a higher performance in their subsequent application or can be designed lighter [9].

On the one hand, damage development is dependent on the strain state. However, ductile damage only occurs in combination with plastic strain. On the other hand, damage is dependent on the stress state. Li has thereby shown in the context of forming that it is not sufficient to use equivalent stress alone to model damage [10]. The three-dimensional stress state in the form of the VON MISES stress σ_{vm}, the lode parameter θ and the triaxiality η must be used (Eq. 1- 3). Whereby I_1 is the first invariant of the CAUCHY stress tensor σ and J_2 and J_3 are the second and third invariants of the deviatoric stress tensor s.

$$\sigma_{vM} = \sqrt{3J_2} \tag{1}$$

$$\eta = \frac{I_1}{\sigma_{vM}} \tag{2}$$

$$L = -\left(\frac{J_3}{\sigma_{vM}}\right)^3 \tag{3}$$

Lode parameter θ and triaxiality η are proven to have an influence on the damage and in particular also on the shape of the resulting voids. While the triaxiality η has a particular effect on the size of the voids, the lode parameter θ influences the geometry of the voids. Both parameters have in common, that they cannot be measured during stretch drawing [11]. Therefore numerical or analytical modelling is necessary.

Materials and Methods

Approach. In this paper the objective is to investigate the damage development in the form of damage accumulation and distribution. For this purpose, classical stretch drawing (SD, Fig. 1a)) with different drawing die radii r_{DR} as well as multistep stretch drawing (MSSD, Fig. 1b)) and reverse stretch drawing (RSD, Fig. 1c)) are considered. The aim is to generate the same geometry on different process routes in order to identify the influence of different load paths on damage development, so that in the future the damage development can be specifically adjusted in the form of accumulation and distribution in order to achieve a higher component performance. Due to the dependence of the damage on the stress and strain states, the load paths are first analyzed in terms of the time- and location-dependent states, which requires numerical modelling performed in this work. Afterwards the load paths are correlated with damage accumulation and distribution to disclose cause effect relations.

Stretch drawing, reverse stretch drawing and multi-step stretch drawing process. In the present work, the focus is on a u-shaped geometry from DP800 dual phase steel with a sheet thickness of $s = 1.5$ mm. The punch velocity v_p was set to $v_p = 15$ mm/s and the blank holder was position controlled due to different necessary blank holder forces to prevent the sheet metal from bypassing the drawing groove. The different processes are schematically depicted in Fig. 1.

Material Forming - ESAFORM 2023
Materials Research Proceedings 28 (2023) 727-736

Materials Research Forum LLC
https://doi.org/10.21741/9781644902479-79

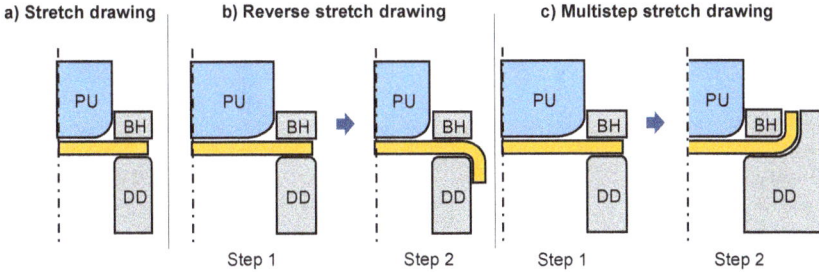

Fig 1. SEQ Figure * ARABIC 1. Schematic illustration of a) stretch drawing, b) reverse stretch drawing and c) multistep stretch drawing.

Process model. The SD, MSDD and RDD processes were modelled using Abaqus/explicit. To save computation time, the geometry was reduced to one quarter using symmetry. A general contact with a uniform COULOMB friction coefficient of $\mu = 0.05$ was applied to model the contact of the part, as identified in prior strip drawing tests. The parameters of the mesh as well as the types of the bodies can be seen in table 1. A LEMAITRE damage model was utilized in this paper which is based on the work of SPRAVE [12]. The model parameters of the calibrated material model are consistent with prior work [8]. The constitutive model was implemented as a VUMAT subroutine and was successfully used and validated in earlier work [13, 14].

The LEMAITRE damage model is based on the definition of the damage variable D. The parameter reflects the influence of voids and microcracks on the plastic behavior of the material [15]. The damage parameter D is defined as the ratio of the damage-free area increment ∂S and the area increment affected by damage ∂S_D (Eq. 4). As the damage grows the area increment affected by damage ∂S_D becomes greater. Thus the forces F respectively the stresses σ applied to the material are not distributed over the entire material cross section. Therefore the concept of effective stress $\tilde{\sigma}$ is introduced (Eq. 5) and used to estimate the response of the material to applied stress under consideration of damage evolution [16].

$$D = \frac{\partial S_D}{\partial S} \tag{4}$$

$$\tilde{\sigma} = \frac{\partial F}{\partial S - \partial S_D} = \frac{\sigma}{1-D} \tag{5}$$

Table 1. Parameter of the mesh and types of bodies for process model.

Name	Mesh	Type of body
Sheet	C3D8R	Solid deformable body
Punch, blank holder, die	R3D4	Discrete rigid

Reference process, reference position and area under investigation. As reference process the SD with a drawing die radius of $r_{DR} = 3$ mm is considered. The triaxiality η, lode parameter θ and damage parameter D analysis is performed at a point in the outer bend of the component (reference point (RP)), which is in principle the most sensitive area for performance in later applications. For the comparison of the die radii r_{DR}, a punch path s of $s = 25$ mm is used as reference position. This punch path s corresponds to the punch path shortly before the first component fails. For the comparison of the different process set-ups, the reference position of $s = 7.5$ mm is determined accordingly. Due to material limitations, the process set-up only considers the area in which the components have the same geometry or have the same cup height as a functional measure of the geometry respectively.

Results

Influence of drawing die radius r_{DR} on load path and damage evolution. The damage distribution of the damage parameter D in dependence of the drawing die radius r_{DR} in the cross section of the components is shown in Fig. 2a), the course of triaxiality η and lode parameter θ in dependence of equivalent plastic strain φ_{pl} in Fig. 2b) and the course of damage parameter D in dependence of equivalent plastic strain φ_{pl} in Fig. 2c). Considering the course of the lode parameter θ as a function of the equivalent plastic strain φ_{pl} in Fig. 2b), the curves for the different die radii r_{DR} differ, in particular for the mean value ($\theta_{med.r_{DR=3}}$ =0.121, $\theta_{med.r_{DR=6}}$ = 0.118, $\theta_{med.\,r_{DR=9}}$ = 0.116). All curves initially start at a value of almost $\theta = 1$ and then drop and continue to progress in an oscillating manner. The course of $r_{DR} = 3$ mm, however, shows a steep peak before ending. The course of the triaxiality η, however, differs. The course of die radius $r_{DR} = 3$ mm of the triaxiality η has a higher value on average ($\eta_{med.r_{DR=3}}$ = 0.6, $\eta_{med.r_{DR=6}}$ = 0.519, $\eta_{med.r_{DR=9}}$ = 0.518). Due to the non-steady shaped courses, a clear and explicit description of the courses is not possible respectively. However, what stands out concisely is the decreasing peak of $\eta_{r_{DR=3}}$ at the end. The different components also reach different equivalent plastic strains in the reference point at the reference position ($\varphi_{pl.r_{DR=3}}$ = 0.258, $\varphi_{pl.r_{DR=6}}$ = 0.272, $\varphi_{pl.r_{DR=9}}$ = 0.277). The courses of the damage parameter D as a function of the equivalent plastic strain φ_{pl} in Fig. 2c) show a quadratic course with an offset with ascending drawing die radius r_{DR} for all three die radii r_{DR}. The difference in the die radius of $r_{DR} = 3$ mm is particularly pronounced. The damage parameter D is consistently higher for the same equivalent plastic strain φ_{pl} .

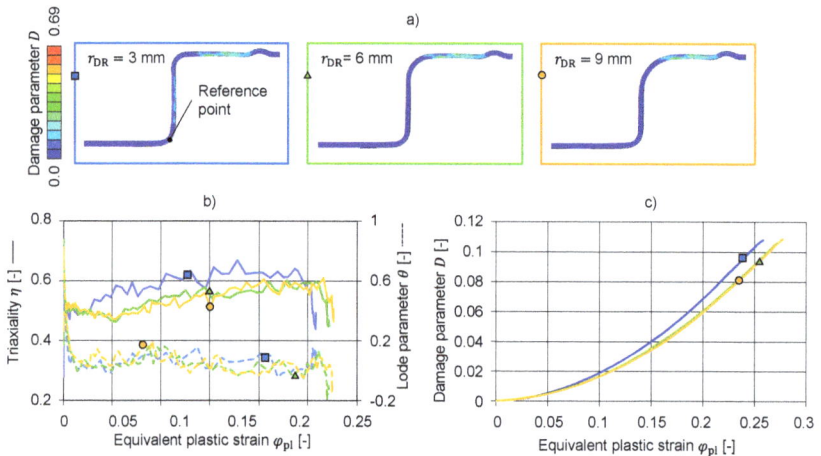

*Fig. 2. a) Damage parameter **D** distribution in cross section of components; b) course of triaxiality η and lode parameter Θ in dependence of equivalent plastic strain φ_pl at RP; c) course of damage parameter **D** in dependence of equivalent plastic strain φ_pl at RP considering different drawing die radii **r_ddr**.*

Fig. 3a) shows the course of VON MISES stress σ_{vm} in dependence of punch path s, Fig. 3b) shows the course of damage parameter D in dependence of punch path s, Fig. 3c) the damage distribution D in the cross section of the component with a drawing die radius $r_{DR} = 3$ mm and Fig. 3d) the course of damage parameter D in dependence of the distance x to the symmetry plane. The VON MISES stresses σ_{vm} of the individual curves all start at $\sigma_{vm} = 0$ MPa. A reverse square increase of all curves follows. With increasing die radius r_{DR}, however, a smaller gradient appears. After achieving the maximum and a slight decrease, all curves then decline to reach a quasi-stationary value to oscillate around. The drop occurs with increasing die radius r_{DR} at a later punch path s ($\sigma_{med.vM\ r_{DR} = 3\ mm} = 806.51$ MPa, $\sigma_{med.vM\ r_{DR} = 3\ mm} = 826.98$ MPa, $\sigma_{med.vM\ r_{DR} = 3\ mm} = 832.58$ MPa). The progression of the damage variable D as a function of the punch path s starts at $s = 0$ mm for all three curves until the first contact between blank holder and sheet and punch and sheet has been established. From a punch path of $s = 2.2$ mm, the curve of the die radius $r_{DR} = 3$ mm begins to increase, followed by the curve of the die radius $r_{DR} = 6$ mm and finally the curve of the die radius $r_{DR} = 9$ mm. Irrespective of the offset of the curves, the course of the three curves initially shows a quadratic increase, which changes to a steep linear increase and finally approaches a final value in a quadratic manner. At the reference position, the damage parameter D reaches a value of $D_{r_{DR} = 3\ mm} = 0.108$, $D_{r_{DR} = 6\ mm} = 0.106$ and $D_{r_{DR} = 9\ mm} = 0.108$. Considering the damage parameter D in dependence of the distance x to the symmetry plan, the three courses of the different die radii r_{DR} in the area of the wall have a nearly identical course. The same can be seen in the transition area and after the wall. In the area of the wall, in contrast, the three curves show a different course. There is a local maximum of the damage variable D with the smallest die radius $r_{DR} = 3$ mm. Equivalent to this is the maximum value of the damage parameter D in this area with decreasing die radius r_{DR}.

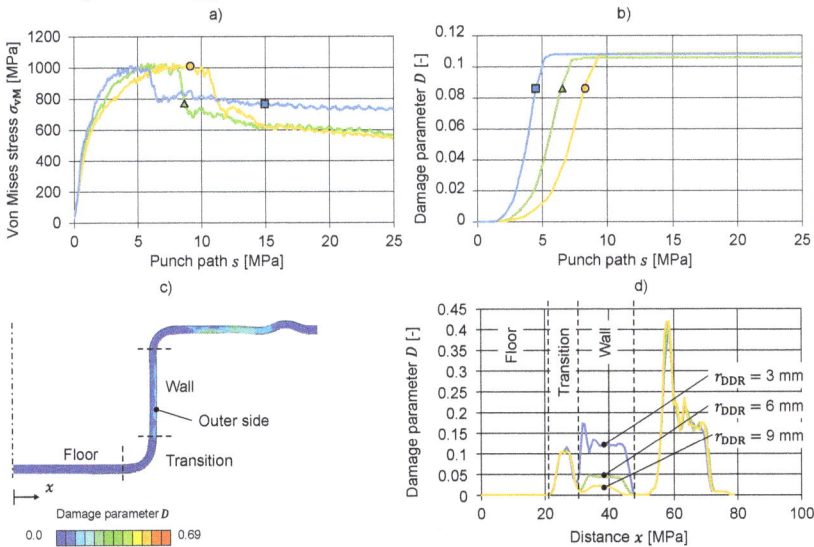

Fig. 3. a) Course of VON MISES stress σ_{vm} in dependence of punch path s at RP; b) course of damage parameter D in dependence of punch path s at RP; c) damage distribution D in cross section of component $r_{DDR} = 3$ mm; d) course of damage parameter D in dependence of distance x to symmetry plane considering different drawing die radii r_{ddr}.

Material Forming - ESAFORM 2023
Materials Research Proceedings 28 (2023) 727-736

Materials Research Forum LLC
https://doi.org/10.21741/9781644902479-79

Influence of process set-up on load path and damage evolution. The distribution of the damage parameter D in the cross section of the components from SD, MSSD and RSD are depicted in Fig. 4a). Considering the different heights of the cups shown in the figure, only the area up to the transition of the wall into the die radius r_{DR} was selected for the analysis of the results. This is a consequence of the fact that the maximum drawing depth was limited due to material constraints and the objective is to compare geometrically almost identical components. The courses of η_{SD}, θ_{SD} and D_{SD} are the same than in Fig. 2 since no damage respectively change of stress and strain state occurs in the reference position after a punch path of $s = 6$ mm. The damage distribution, however, differs.

The course of triaxiality η and lode parameter θ in dependence of equivalent plastic strain φ_{pl} are shown in Fig. 4b). The outer side of the outer bend, which is under consideration, experiences different equivalent plastic strains φ_{pl} to obtain the same geometry ($\varphi_{pl\,SD\,max.} = 0.26$, $\varphi_{pl\,MSSD\,max.} = 0.52$, $\varphi_{pl\,RSD\,max.} = 0.56$). The courses of the triaxiality η of MSSD and RSD show a similar behavior with mean values of $\eta_{med.MSDD} = 0.523$ and $\eta_{med.RDD} = 0.53$. Both courses start with a steep rise. After a quadratic drop, an almost linear curve follows, which drops at 0.34 and then rises again. Considering the lode parameter θ of MSDD, the course begins with a value of almost $\theta \approx 1$. The course then drops with a steep gradient into the negative range followed by an oscillating curve. This results in an average value of $\theta_{med.MSSD} = 0.21$. The course of the lode parameter θ of RSD starts similar to the course of MSDD. However, when the course drops with a steep gradient the curve remains in the positive range and starts to oscillate around a value of $\theta_{med.RSD} = 0.19$. The course of damage parameter D in dependence of equivalent plastic strain φ_{pl} considering different process set-ups is shown in Fig. 4c). The curves of MSSD and RSD depict a similar course. SD, however, has a slightly higher level of damage parameter D. The maximum value of damage parameter D obtained are $D_{MSSD} = 0.52$ and $D_{RSD} = 0.41$.

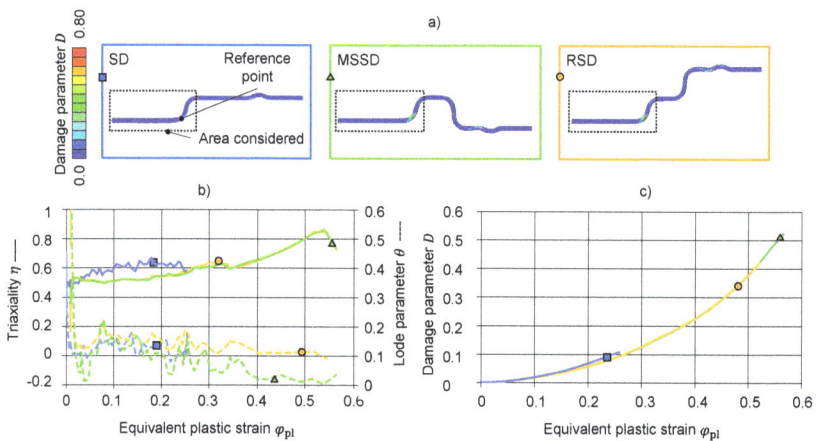

Fig. 4. a) Damage parameter **D** distribution in cross section of components; b) course of triaxiality η and lode parameter Θ in dependence of equivalent plastic strain φ_{pl} at RP; c) course of damage parameter **D** in dependence of equivalent plastic strain φ_{pl} at RP considering different process set-ups.

The course of VON MISES stress σ_{vm} in dependence of punch path s can be seen in Fig. 5a). The initial values of the RSD and MSSD processes differ. Since a U-profile is produced in the first step and is rotated 180° or is not rotated depending on the process route, the courses start differently. Despite the different start, both courses show a reverse square increase which is very similar. The mean values of VON MISES stress σ_{vm} obtained are $\sigma_{med.vM\ MSSD} = 882.52$ MPa and $\sigma_{med.vM\ RSD} = 852.58$ MPa. Fig. 5b) shows the course of damage parameter D in dependence of punch path s. The courses of MSSD and RSD depict a very similar quadratic course which start to elevate from a punch path s of $s = 3$ mm. The course of SD, on the other hand, shows a s-shaped course, which depicts a stronger elevation at $s = 3$ mm followed by a steady course of $D_{SD} = 0.108$. The Course of damage parameter D in dependence of distance x to symmetry plane considering different process set-ups can be seen in Fig. 5c). All three courses begin with a damage parameter $D = 0$. With the start of the transition from floor to wall, the damage parameter D increases for the first time in the courses. For MSSD and RSD, the curves show an initial quadratic rise, which changes to a linear rise with a very high gradient, and then fall off inversely. The course of SD shows a course with also a quadratic rise, but with a lower gradient as well as rise and descent shifted further to higher x values. Considering the wall of the three components, there is a small rise in damage parameter D for MSSD and RSD.

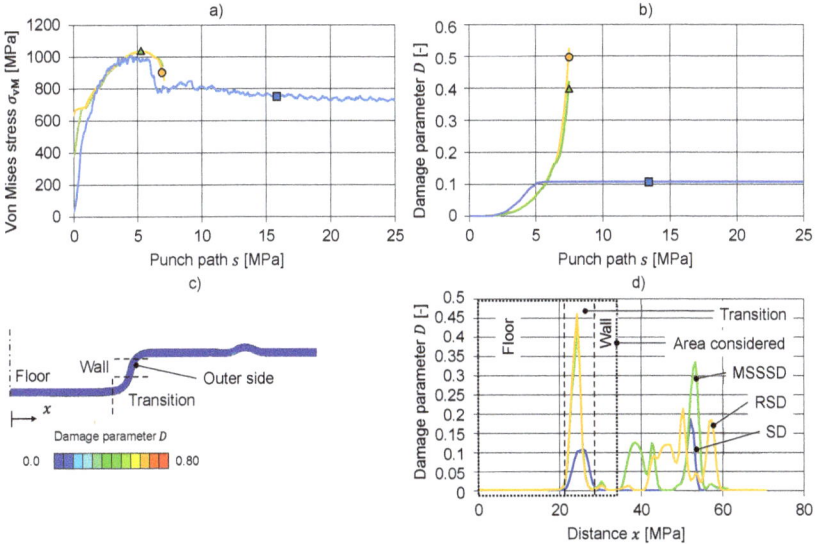

Fig. 5. a) Course of VON MISES stress σ_{vm} in dependence of punch path s at RP; b) course of damage parameter D in dependence of punch path s at RP; c) damage distribution D in cross section of component $r_{DDR} = 3$ mm; d) course of damage parameter D in dependence of distance x to symmetry plane mm considering different process set-ups.

Discussion

Change of die radii r_{DR}. The comparison of the load paths and the damage development at the reference point using different die radii r_{DR} shows a distinction, especially for $r_{DR} = 3$ mm. As early as from an equivalent plastic strain φ_{pl} of $\varphi_{pl} = 0.025$, a higher value of the triaxiality η

Material Forming - ESAFORM 2023   Materials Research Forum LLC
Materials Research Proceedings 28 (2023) 727-736   https://doi.org/10.21741/9781644902479-79

becomes apparent, which can be consulted moderately for the damage development. Analogously, the damage parameter D differs from the other two curves at the same point in strain. Due to the different equivalent strains φ_{pl} needed, however, all three achieve an almost identical damage parameter value D in the end. When observing the course of the damage parameter D as a function of the punch path s, a later onset of damage development with increasing die radius r_{DR} becomes apparent. It is concluded that this is due to the time dependent course of the bending radius of the sheet. With a smaller die radius r_{DR}, the final bending radius is reached sooner than with a higher die radius r_{DR}. Accordingly, the damage development begins at a later stage, since damage is always associated only with plastic deformation.

Considering the different equivalent plastic strains φ_{pl} at the reference point depending on the die radius r_{DR}, it is noticeable that although the bending radius is significantly smaller at $r_{DR} = 3$ mm, the equivalent plastic strain φ_{pl} is lower than with a greater die radius r_{DR}. It is assumed that the larger contact area of the sheet and the die with increasing die radius r_{DR} results in a larger frictional surface. This larger frictional surface leads to a higher flow resistance of the sheet, which results in higher punch forces. The higher punch forces required also result in higher stresses above the yield point at the reference point, which are reflected in a higher equivalent plastic strain φ_{pl}. With regard to the damage distribution in the component, it is assumed that the smaller die radius r_{DR} and the correspondingly smaller bending radius result in correspondingly higher damage parameter D in the wall of the component. This is brought into the context of greater predominant stresses in this area due to the smaller bending radius.

Change of process set-up. The course of triaxiality η of RSD and MSSD show a very similar course. The two courses differ only in the equivalent plastic strain φ_{pl} achieved. Similar to the observation of the different die radii r_{DR}, the necessary punch forces and the associated higher stresses at the reference point also seem to lead to more deformation being introduced into the area under consideration. A similar result can be seen in the course of the damage parameter D over the equivalent strain respective of the punch path s. The two curves do not differ in their characteristics, but in the final values obtained due to different equivalent strains φ_{pl}. However, severe thinning takes place in the reference position respectively the bend of the component.

Summary

A total of six different process routes for the production of a geometrically similar component were investigated. On the one hand, different drawing ring radii r_{DR} ($r_{DR} = \{3, 6, 9$ mm$\}$) and, on the other hand, different process set-ups (stretch drawing, multistep stretch drawing, reverse stretch drawing) were analyzed. The individual routes were compared with each other on the basis of their load paths and the associated damage development.

When comparing the die radii r_{DR}, a larger die radius r_{DR} at the selected reference point was found not to result in a lower damage parameter D. However, the damage development starts at a later stage with respect to the punch path s or the equivalent plastic strain φ_{pl}. Furthermore, the die radius r_{DR} was shown to have an influence on the damage distribution in the component. Particularly in the wall, a larger die radius r_{DR} resulted in less damage, which can be used for later applications to increase the performance of the component in respect of the use case. Considering the different process setups, the RSD and MSSD show very similar load paths despite different process routes, but differ in the equivalent plastic strain φ_{pl} and thus the values of the damage parameter D are different. The deformation at the reference point, however, is so severe that there is significant sheet thinning and a high damage parameter D. Due to this, the RDS and MSDD are not recommended as a suitable process route.

To complement the numerical analysis of the load paths and damage development, experimental investigations will be carried out. For this purpose, the process routes of the different die radii r_{DR} will be selected on the basis of the present results. The numerically determined load paths will then be correlated with the results regarding the damage development. Nevertheless, the present results have already enabled a pre-selection of possible process routes to be considered and the corresponding load paths to be determined.

Acknowledgements

This research was funded by Deutsche Forschungsgemeinschaft (DFG, German Research Foundation; Project number 278868966 – TRR 188; *Damage Controlled Forming Processes*). The authors would also like to thank A. Erman Tekkaya, Till Clausmeyer, Alexander Schowtjak and Jan Gerlach from the Institute of Forming Technology and Lightweight Components of TU Dortmund, sub-project S01 of TRR 188, for the usage of their implementation of the LEMAITRE damage model. Simulations were performed with computing resources granted by RWTH Aachen University under project rwth0907.

References

[1] Information on https://www.europarl.europa.eu

[2] Information on https://climate.ec.europa.eu

[3] H.J. Kim, C. McMillan, G.A. Keoleian, S.J. Skerlos, Greenhouse Gas Emissions Payback for Lightweighted Vehicles Using Aluminum and High-Strength Steel, J. Ind. Ecol. 14 (2010) 929-946. https://doi.org/10.1111/j.1530-9290.2010.00283.x

[4] F. Klocke, Manufacturing Processes: 4. Forming, Springer Berlin Heidelberg, Berlin, Heidelberg, 2013.

[5] N.P. Lutsey, Review of technical literature and trends related to automobile mass-reduction technology, University of California, Davis, 2010.

[6] Information on https://ec.europa.eu/commission/presscorner/detail/en/IP_22_4312

[7] J. Tschorn, D. Fuchs, T. Vietor, Potential impact of additive manufacturing and topology optimization inspired lightweight design on vehicle track performance, Int. J. Interact. Des. Manuf. (IJIDeM) 15 (2021) 499-508. https://doi.org/10.1007/s12008-021-00777-x

[8] M. Müller, I.F. Weiser, T. Herrig, T. Bergs, Numerical prediction of the influence of process parameters and process set-up on damage evolution during deep drawing of rectangular cups, International Conference on Accuracy in Forming Technology (ICAFT), 2022

[9] G. Hirt, E. Tekkaya, T. Clausmeyer, J. Lohmar, Potential and status of damage controlled forming processes, Prod. Eng. 14 (2020) 1-4. https://doi.org/10.1007/s11740-019-00948-6

[10] Y. Li, M. Luo, J. Gerlach, T. Wierzbicki, Prediction of shear-induced fracture in sheet metal forming, J. Mater. Process. Technol. 210 (2010) 1858-1869. http://doi.org/10.1016/j.jmatprotec.2010.06.021

[11] C.C. Roth, D.Mohr, Ductile fracture experience with locally proportional loading histories, Int. J. Plast. (2016) 328-354. https://doi.org/10.1016/J.IJPLAS.2015.08.004

[12] L. Sprave, A. Schowtjak, R. Meya, T. Clausmeyer, A.E. Tekkaya, A. Menzel, On mesh depend-encies in finite-element-based damage prediction: application to sheet metal bending, Prod. Eng. 14 (2020) 123-134. https://doi.org/10.1007/s11740-019-00937-9

[13] M. Nick, M. Müller, H. Voigts, I.F. Weiser, T. Herrig, T. Bergs, Effect of Friction Modelling on Damage Prediction in Deep Drawing Simulations of Rotationally Symmetric Cups, Defect and Diffusion Forum 414 (2022) 103-109. https://doi.org/10.4028/p-29o20d

[14] M. Nick, A. Feuerhack, T. Bergs, T. Clausmeyer, Numerical Investigation of Damage in Single-step, Two-step, and Reverse Deep Drawing of Rotationally Symmetric Cups from DP800 Dual Phase Steel, Procedia Manuf. 47 (2020) 636-642. https://doi.org/10.1016/j.promfg.2020.04.195

Material Forming - ESAFORM 2023
Materials Research Proceedings 28 (2023) 727-736

Materials Research Forum LLC
https://doi.org/10.21741/9781644902479-79

[15] J. Lian, Y. Feng, S. Münstermann, A modified Lamaitre Damage Model Phenomenologically Accounting for the Lode Angle 31 Effect on Ductile Fracture, Procedia Mater. Sci. 3 (2014) 1841-1847. https://doi.org/10.1016/j.mspro.2014.06.297

[16] T.S. Cao, Models for ductile damage and fracture prediction in cold bulk metal forming processes: a review, Int. J. Mater. Form. 10 (2017) 139-171. https://doi.org/10.1007/s12289-015-1262-7

Material Forming - ESAFORM 2023
Materials Research Proceedings 28 (2023) 737-746

Materials Research Forum LLC
https://doi.org/10.21741/9781644902479-80

Improved failure characterisation of high-strength steel using a butterfly test rig with rotation control

STOCKBURGER Eugen[1,a,*], WESTER Hendrik[1,b],
JEGATHEESWARAN Vithusaan[1,c], DYKIERT Matthäus[1,d]
and BEHRENS Bernd-Arno[1,e]

[1]Institute of Forming Technology and Machines (IFUM), Leibniz Universität Hannover,
An der Universität 2, 30823 Garbsen, Germany

[a]stockburger@ifum.uni-hannover.de, [b]wester@ifum.uni-hannover.de,
[c]jegatheeswaran@ifum.uni-hannover.de, [d]dykiert@ifum.uni-hannover.de,
[e]behrens@ifum.uni-hannover.de

Keywords: CP800, Butterfly Specimen, Experimental-Numerical Procedure

Abstract. A forming limit diagram is the standard method to describe the forming capacity of sheet materials. It predicts failure due to necking by limiting major and minor strains. For failure due to fracture, the fracture forming limit diagram is used, but fracture caused by plastic deformation at a shear-dominated stress state cannot be predicted with a conventional fracture forming limit diagram. Therefore, stress-based failure models are used as an alternative. These models are describing the fracture of sheet materials based on the failure strain and the stress state. Material-specific parameters must be determined, but a standardised procedure for the calibration of stress-based failure models is currently not established. Most test procedures show non-constant stress paths and varying stress states in the crack initiation area, which leads to uncertainties and inaccuracies for modelling. Therefore, a new test methodology was invented at the IFUM: a prior presented butterfly test rig was extended to enable an online rotation to adapt the loading angle while testing. First, butterfly tests with CP800 were performed for three fixed loading conditions. The tests were modelled numerically with boundary conditions corresponding to the tests. Based on the numerical results, the stress state as well as failure strain were identified and the stress state deviations were calculated. Afterwards, the necessary angular displacements to compensate the stress state deviations for the adaptive test rig were iteratively determined with numerical simulations using an automatised Python script. Finally, the butterfly tests were performed experimentally with the determined adaptive loading angles to identify the specimen failure and compared to the simulations for validation.

Introduction

The forming limit diagram (FLD) defines the critical major and minor strains of a sheet metal, which lead to material failure if exceeded. The FLD is conventionally used to describe forming limits due to necking. Failure due to fracture can be described with the fracture forming limit diagram (FFLD). However, the FLD and the FFLD give an inadequate prediction of the failure for high-strength steel (HSS) sheets, since it is valid for linear strain paths and for the strain states between uniaxial tension and biaxial tension [1]. Modern forming processes and HSS sheets often show non-linear strain paths. Furthermore, failure due to fracture initiation caused by plastic deformation at a shear-dominated stress state cannot be predicted by the FLD or the FFLD [2]. As an alternative to the FFLD, failure models based on the stress state are used. Those failure models describe the failure of the sheet metal under consideration of the stress state by means of the stress-state-dependent equivalent plastic strain at failure ε_f, which allows statements in the case of non-linear strain paths and shear-dominated stress states. To describe the three-dimensional stress state,

Material Forming - ESAFORM 2023   Materials Research Forum LLC
Materials Research Proceedings 28 (2023) 737-746   https://doi.org/10.21741/9781644902479-80

the stress triaxiality η and the normalised Lode angle θ are used. A common stress-based model is the Modified Mohr-Coulomb (MMC) failure model [3], which can be written as

$$\varepsilon_{f,MMC} = \left[c_1 \cdot \left(\eta + \frac{1}{3} \cdot sin\left(\frac{\theta\pi}{6}\right) \right) + c_2 \cdot \left(\sqrt{\frac{1+c_1^2}{3}} \cdot cos\left(\frac{\theta\pi}{6}\right) \right) \right]^{-\frac{1}{n}} \tag{1}$$

whereby c_1, c_2 and n are material-specific parameters. For the calibration of the failure models, failure characterisation experiments are performed, but a standardised procedure for testing and calibrating is currently not available. Used failure characterisation experiments mostly result in non-constant stress paths in the area of the specimen failure [4]. In Fig. 1 (A) common failure characterisation specimen such as the Miyauchi, holed tensile as well as waisted tensile specimen are shown and in Fig. 1 (B) the corresponding non-constant stress paths are visible from those specimens. Only the hydraulic bulge test leads to a constant stress path. In order to minimise the resulting uncertainties and inaccuracies of the failure characterisation for stress-based failure models, a new experimental-numerical testing methodology for an improved failure characterisation with constant stress paths is engineered and presented within the scope of this work.

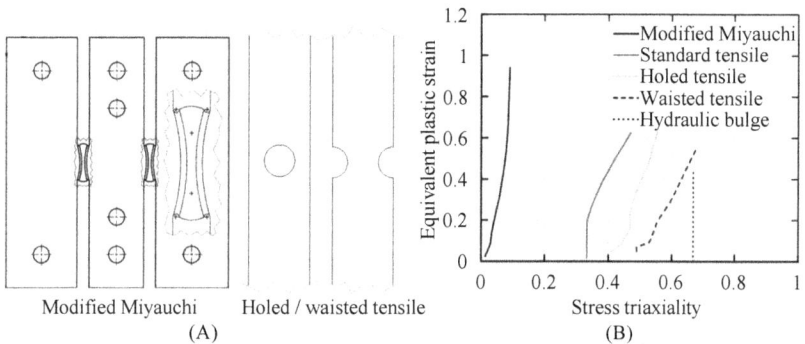

Fig. 1. Failure characterisation specimen and resulting non-constant stress paths based on the work from Behrens et al. [4].

New Experimental-Numerical Failure Characterisation Methodology
The module for the new adaptive test methodology is based on a prior developed test module shown by Stockburger et al. in [5]. In this test module, according to the test principle of Arcan et al. [6], a butterfly specimen can be tested under different orientations or loading angles α to the normal loading direction of the tensile testing machine, whereby different stress states can be achieved in the test area of the specimen. Using the butterfly specimen has the advantage compared to conventional failure characterisation specimen that the specimen failure is not initialised at the specimen edge, but in the centre of the specimen. Nevertheless, as for other specimen the stress paths of the butterfly specimens are not entirely constant. To estimate the characteristic stress state and the equivalent plastic strain at failure which is needed to calibrate models like the MMC failure model, the butterfly specimen must be numerically modelled considering boundary conditions from the experiments.

Material Forming - ESAFORM 2023
Materials Research Proceedings 28 (2023) 737-746

Materials Research Forum LLC
https://doi.org/10.21741/9781644902479-80

In order to solve the problem of non-constant stress paths in the experiment, a new adaptive test module was developed. Fig. 2 shows an illustration of the module for the new adaptive test methodology. The new test module consists of two separate specimen holders, which each have an outer ring gear and a recess for the butterfly specimen. Two mounting plates are used to attach the butterfly specimen to the specimen holder. The outer ring gears from the upper and lower specimen holders are each in contact with a worm shaft, while the upper and lower worm shafts are connected to two electric motors by two bevel gears. A rotational movement of the specimen holders is generated when the electric motors are activated and thus the current loading angle α can be changed by an adaptive loading angle α_{adap}. The rotation is regulated by a control device with a sensor, that determines the angular position. The entire test module is attached to a tensile testing machine by means of two connecting plates. Overall, the tensile testing machine is applying the load, which is transferred from the two separate specimen holders to the butterfly specimen. With the presented testing module, loading angles α are continuously adjustable during testing in order to obtain constant stress states.

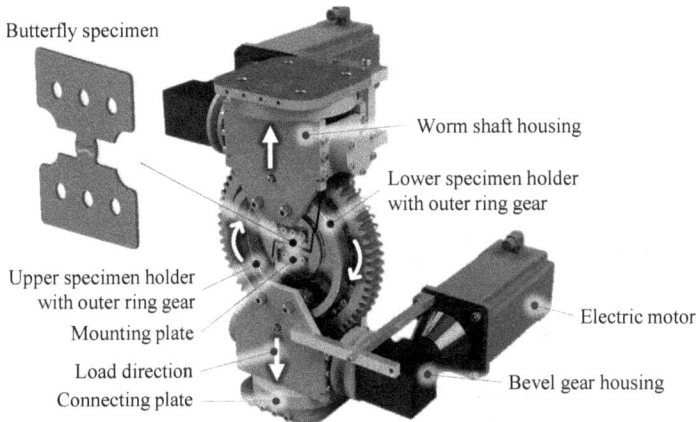

Fig. 2. Module for the new adaptive test method with butterfly specimen.

The design of the adaption loading angle α_{adap} is carried out iteratively with the aid of a numerical model of the test module. For this purpose, tests with a fixed loading condition α are first carried out experimentally and then mapped numerically. The experimental displacements up to specimen failure serve as boundary conditions for the simulation model, which is shown in Fig. 3 (A). Based on the results, the required adaption loading angle α_{adap} is now calculated numerically. For this purpose, the adaption loading angle α_{adap} is varied automatically within a sensitivity study using a Python script as shown in Fig 3 (B). First, a variation range of the adaption loading angle α_{adap} and the step size of it is specified. Consequently, a simulation is carried out iteratively for each adaption loading angle α_{adap} in the variation range. In order to determine the time of the angle adaption at which the simulations start, the stress triaxiality is taken as an indicator. The maximal deviation of the stress triaxiality was subjectively defined as 0.03. As soon as the equivalent plastic strain-stress triaxiality path showed a change higher than the maximal deviation, the iterative simulations for the adaption loading angle α_{adap} begin. Depending on the outcome, the procedure is repeated for further adaption. After the numerical design, experimental tests are performed using the numerically-iterative determined angle adaption. Those results are

Material Forming - ESAFORM 2023   Materials Research Forum LLC
Materials Research Proceedings 28 (2023) 737-746   https://doi.org/10.21741/9781644902479-80

used to identify the material failure under constant stress states and to verify the numerical procedure. Finally, the MMC failure model can be calibrated as well as compared for the fixed and the new adaptive test methodology.

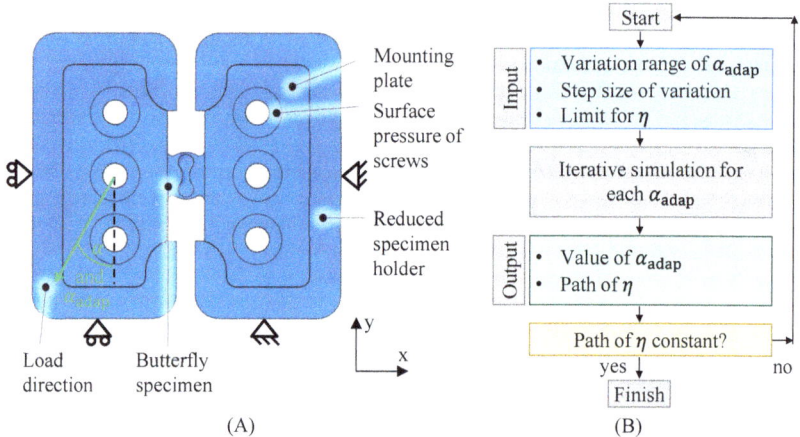

Fig. 3. Simulation model of the butterfly tests (A) and sequence plan of the Python script (B).

Application of the New Failure Characterisation Methodology

The new failure characterisation methodology was applied to the complex-phase HSS CR570Y780T-CP (CP800) from voestalpine Stahl GmbH. The outer contour of the specimen was cut by waterjet and the thickness reduction in the examination area of the specimen was carried out by milling. The new butterfly test module was installed in the tensile testing machine S100/ZD from DYNA-MESS Prüfsysteme GmbH. First, butterfly tests for the fixed configuration of the loading angles -3°, 28° and 74.5° were performed. For measuring the specimen displacement and estimation of fracture occurrence in the examination area, the optical measurement system Aramis from Carl Zeiss GOM Metrology GmbH was used. For each loading angle the tests were repeated three times. The testing temperature corresponded to room temperature and the testing speed was set to 0.05 mm/s.

Afterwards, the experiments with fixed loading conditions were numerically modelled to estimate the equivalent plastic strain at failure and the stress triaxiality as well as the normalised Lode angle paths. The FE models were generated in Abaqus/Standard for each loading angle with the corresponding displacements until failure in x- and y-direction from the experiments. The boundary conditions were extracted from the optical measurements using digital image correlation and directly applied to the specimen holder. The elastic behaviour of the material was described with a modulus of elasticity of 210 GPa and a Poisson's ratio of 0.3. The data required for modelling the plastic flow behaviour, such as flow curve and anisotropy parameters, are summarised in Table 1 and were previously published by Behrens et al. in [7]. The hardening behaviour was modelled by the Hockett-Sherby approach [8] and the anisotropy by the Hill48 yield function [9]. The Hockett-Sherby approach describes the flow stress k_f of the material as a function of the equivalent plastic strain ε and the four material parameters a, b, c as well as d. The specimen holder was geometrically reduced and modelled as rigid body with 210 GPa as elastic modulus and 0.3 as Poisson's ratio. A discretisation of the specimen was done with linear reduced-integrated hexahedral elements with hourglass control, referred to as C3D8R in Abaqus. The examination area of the specimens was meshed with an element edge length of 0.1 mm and the

Material Forming - ESAFORM 2023   Materials Research Forum LLC
Materials Research Proceedings 28 (2023) 737-746   https://doi.org/10.21741/9781644902479-80

thickness of the specimens by five elements. The surface pressures of the screws acting on the mounting plates were modelled with a surface pressure of 204 MPa, which results from the tightening torque of the screws.

Table 1. Parameters for the extrapolation approach and anisotropy function from [7].

Equation			a in MPa	b in MPa	c	d
Hockett-Sherby	$k_f = a - (a - b) \cdot e^{c \cdot \varepsilon^d}$	(2)	4.96 E-05	502.5	0.823	0.170
	F	G	H	L	M	N
Hill'48	0.495	0.531	0.469	1.5	1.5	1.634

Further, the numerical models were evaluated with regard to the equivalent plastic strain, the stress triaxiality path and the normalised Lode angle path. The models were analysed on the surface of the investigation area. The equivalent plastic strain was plotted as a function of the triaxiality and of the normalised Lode angle. Based on these curves, the characteristic stress state was calculated by the area-weighted centroid. The equivalent plastic strain at failure and the characteristic stress state from the fixed configuration are later compared to the values of the adaptive tests.

Next, the design of the adaption loading angle α_{adap} was carried out. The procedure is shown in Fig. 4 exemplarily for the adaptions of the loading angle 28°. First, the course of the equivalent plastic strain-stress triaxiality curve for the fixed loading condition α was considered. Based on the illustrated curve, a value for the stress triaxiality was manually selected, at which the angular adaption begins. When selecting the limit value, care was taken to intervene early enough to correct the curve with regard to a more constant path. In Fig. 4 (A), the limit value is marked orange at a stress triaxiality of 0.135. Now the Python script was used to design the angle adaption. During pre-processing, only the displacement boundary conditions were automatised redefined by a subroutine to change the loading. With the changed input variables, the simulation was proceeded. In the example given, the adaption loading angle α_{adap} was varied between 5° and 6°. The corresponding influence of the angle adaption on the stress state is displayed in Fig. 4 (B). 6° was selected as the adaption loading angle α_{adap} and the limit for the stress triaxiality was set to 0.139. When choosing the values, care was taken again to ensure that the path remains as constant as possible.

Material Forming - ESAFORM 2023

Materials Research Proceedings 28 (2023) 737-746

Materials Research Forum LLC

https://doi.org/10.21741/9781644902479-80

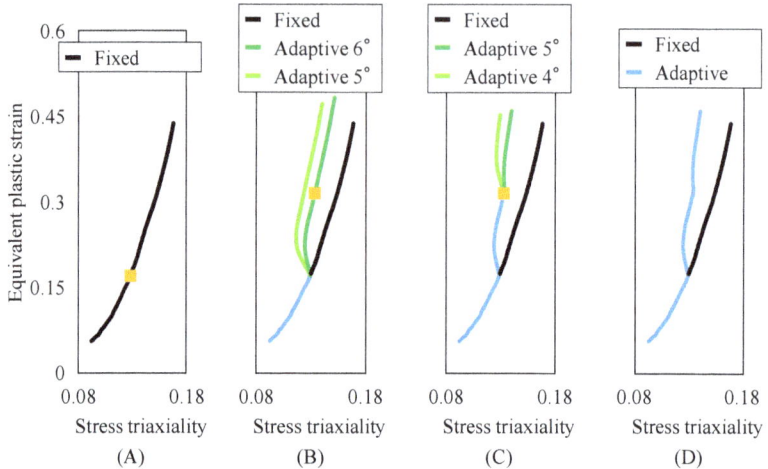

Fig. 4. Iterative design of the angular adaptions for the loading angle 28°.

In order to design the second angle adaption, a next adaption loading angle α_{adap} were chosen manually. Now the subroutine proceeded further in the same way as for the first angle adaption. Next, the adaption loading angle α_{adap} 4° and 5° were compared. The equivalent plastic strain-stress triaxiality paths after the second angle adaption are illustrated in Fig. 4 (C). With regard to a constant curve, the adaption loading angle α_{adap} of 5° was selected. Fig. 4 (D) compares the final equivalent plastic strain-stress triaxiality path of the designed adaptive test for a loading angle α of 28° with the path for a fixed loading condition α.

Additionally, butterfly experiments with the designed angular adaptions were carried out with the new butterfly test module for the three loading angles α -3°, 28° and 74.5°. Here, the limit values for the stress triaxialities served as boundary conditions with displacements in x- and y-direction. A comparison of the experimental and numerical force-displacement curves for the adaptive test methodology is shown in Fig. 5 (A). The comparison shows that the properties as well as boundary conditions used to describe the flow behaviour and the structure of the FE model represented the experimental tests sufficiently accurate and were therefore considered validated in this range. Based on the comparison, the models were considered suitable for the evaluation of the stress state. For increasing the loading angle, the force rises and the displacement until failure reduces.

Materials Research Forum LLC
https://doi.org/10.21741/9781644902479-80

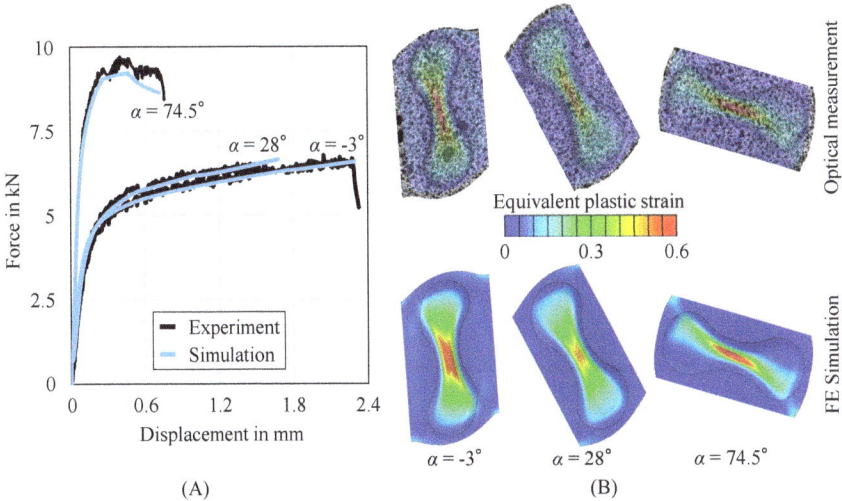

Fig. 5. Comparison of the experimental as well as numerical force-displacement curves (A) and equivalent plastic strain distribution of the butterfly specimen (B) for the adaptive test methodology.

Fig. 5 (B) illustrates the equivalent plastic strain distribution directly before failure of the butterfly tests with the adaptive test methodology. The deformation is accumulated in the centre of all three specimen for both, the experiments and the simulations. Comparing the optical measurement from the experiments and simulation results, the equivalent plastic strain distribution is very similar. In general, the equivalent plastic strain reduces from a loading angle of -3° to 28° and rises again for 74.5°. In the centre of the specimen, the equivalent plastic strain and the stress state were further evaluated.

Fig. 6 (A) illustrates the comparison of the numerically determined equivalent plastic strain-stress triaxiality paths at the location of fracture initiation. It shows that the angular adaptions in the adaptive tests compared to the tests with fixed loading condition α result in significantly more constant paths with regard to the equivalent plastic strain-stress triaxiality path. Further, it is evident in Fig. 6 (A) that higher equivalent plastic strains are achieved in the adaptive tests compared to the fixed tests. This is due to higher displacements until failure for the adaptive tests. This is also shown in the comparison of the characteristic stress states in Fig. 6 (B). Due to the more constant equivalent plastic strain-stress triaxiality path in the adaptive tests, smaller characteristic values are achieved than in the fixed tests. This is due to the fact that the area under the curves decreases because of the more constant stress path. As a result, the centre of gravity of the area shifts to smaller values and the characteristic values decrease. In the tests with a fixed loading condition α, the stress state change during the tests is large and non-constant stress paths are present.

Materials Research Forum LLC
https://doi.org/10.21741/9781644902479-80

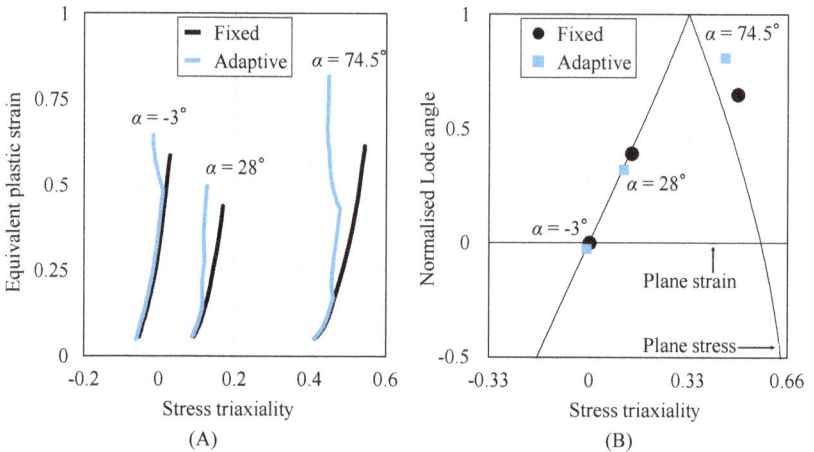

Fig. 6. Equivalent plastic strain-stress triaxiality paths (A) and characteristic stress states (B) for the fixed and adaptive test methodology.

A comparison of the calibrated MMC failure models for the test methodology with fixed loading condition α and for the new adaptive test methodology is presented in Fig. 7. The failure models are shown for plane stress in Fig. 7 (A) and plane strain state in Fig. 7 (B). The failure curves in the plane stress and plane strain state are of similar course for both methods. The major difference is the higher level of the failure curve achieved with the adaptive test methodology due to higher equivalent plastic strains of each test. A minor difference is that the failure curve from the fixed test methodology has a lower slope than the failure curve of the adaptive test methodology. Anyway, the higher accuracy of the failure model for the adaptive test methodology compared to the fixed test methodology needs to be proven by the simulation of forming processes, which is planned for future investigations.

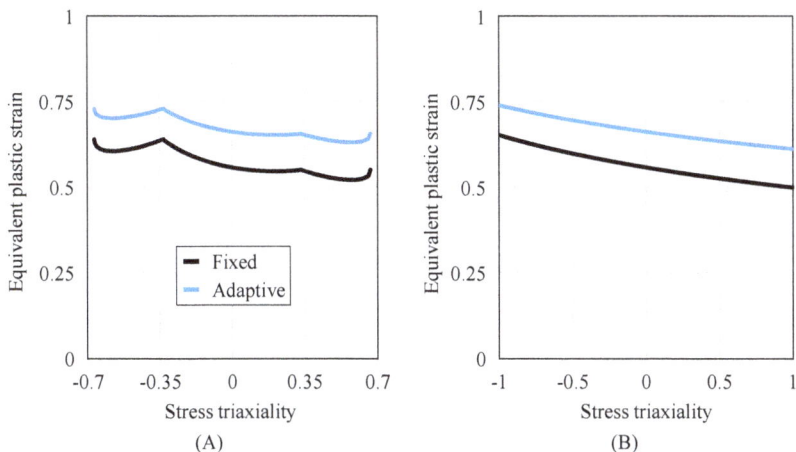

Fig. 7. Plane stress (A) and plane strain state (B) of the fitted MMC-model for the fixed and adaptive test methodology.

Summary

In this work a new experimental-numerical failure characterisation methodology is presented. Conventional failure characterisation tests show a non-constant stress path and therefore inaccuracies for modelling. Butterfly tests for the fixed loading conditions -3°, 28° and 74.5° were performed and numerically modelled. Based on the results for the stress path of the simulations, angle adaptions while testing were iteratively-numerical determined to correct the stress path into a constant path. Afterwards, butterfly experiments were performed using the estimated adaptive angles and compared to the simulations. Therefore, it was possible to achieve a more constant stress path and to minimise the change in the stress state at the location of fracture. Finally, MMC failure model were calibrated for both methodologies and compared to each other, which showed the influence of the new failure characterisation methodology.

In future investigations, more loading angles for the butterfly tests will be experimentally performed and numerically modelled to enhance the data for the failure model parametrisations. Further, it is planned to optimise the Python script to estimate the angle adaptions not subjectively anymore but automatically based on general criteria. Then, demonstrators in the style of a B-pillar as shown in [10] will be experimentally produced and the tests will be reproduced numerically using the MMC failure models from the fixed as well as adaptive test methodology. Based on the results, the failure prediction of the new adaptive test methodology will be evaluated regarding its potential for a higher accurate failure modelling.

Acknowledgements

The authors would like to thank the German Research Foundation (Deutsche Forschungsgemeinschaft, DFG) for the financial and organisational support of this project with the number 405334714. Furthermore, the authors would like to thank voestalpine Stahl GmbH for providing the material under investigation.

References

[1] Standard ISO 6892-1:2020-06, Metallic materials - Tensile testing – Part 1: Method of test at room temperature, Beuth Verlag GmbH, Berlin, Germany, 2020.

[2] Y. Li, M. Luo, J. Gerlach, T. Wierzbicki, Prediction of shear-induced fracture in sheet metal forming, J. Mat. Proc. Tech. 210 (2010) 1858-1869. https://doi.org/10.1016/j.jmatprotec.2010.06.021

[3] Y. Bai, T. Wierzbicki, A new model of metal plasticity and fracture with pressure and Lode dependence, Int. J. Plast. 24-6 (2008) 1071-1096. https://doi.org/10.1016/j.ijplas.2007.09.004

[4] B.-A. Behrens, A. Bouguecha, M. Vucetic, I. Peshekhodov, Characterisation of the quasi-static flow and fracture behaviour of dual-phase steel sheets in a wide range of plane stress states, Arch. Civil Mech. Eng. 12 (2012) 397-406. https://doi.org/10.1016/j.acme.2012.06.017

[5] E. Stockburger, H. Vogt, H. Wester, S. Hübner, B.-A. Behrens, Evaluating Material Failure of AHSS Using Acoustic Emission Analysis, Mat. Res. Proc. 25 (2023) 379-386. https://doi.org/10.21741/9781644902417-47

[6] M. Arcan, Z. Hashin, A. Voloshin, A method to produce uniform plane-stress states with applications to fiber-reinforced materials, Exp. Mech. 18 (1978) 141-146. https://doi.org/10.1007/BF02324146

[7] B.-A. Behrens, K. Brunotte, H. Wester, M. Dykiert, Fracture Characterisation by Butterfly-Tests and Damage Modelling of Advanced High Strength Steels, Key Eng. Mat. 883 (2021) 294-302. https://doi.org/10.4028/www.scientific.net/kem.883.294

[8] J.E. Hockett, O.D. Sherby, Large strain deformation of polycrystalline metals at low homologous temperatures, J. Mech. Phys. Solid. 23 (1975) 87-98. https://doi.org/10.1016/0022-5096(75)90018-6

[9] R.A Hill, Theory of the yielding and plastic flow of anisotropic metals, Proc. R. Soc. A 193 (1948) 281-297. https://doi.org/10.1098/rspa.1948.0045

[10] B.-A. Behrens, S. Jüttner, K. Brunotte, F. Özkaya, M. Wohner, E. Stockburger, Extension of the Conventional Press Hardening Process by Local Material Influence to Improve Joining Ability, Proc. Manuf. 47 (2020) 1345-1352. https://doi.org/10.1016/j.promfg.2020.04.258

Material Forming - ESAFORM 2023
Materials Research Proceedings 28 (2023) 747-754

Materials Research Forum LLC
https://doi.org/10.21741/9781644902479-81

Investigation of the influence of loading direction after pre-forming on the formability and mechanical properties of DP600

NORZ Roman[1,a]*, VALENCIA Fabuer R.[2,b], VITZTHUM Simon[2,c],
UNGUREANU Bogdan[2,d], GERKE Steffen[2,e], BRÜNIG Michael[2,f]
and VOLK Wolfram[1,g]

[1]Chair of Metal Forming and Casting, Technical University of Munich, Walther-Meissner-Str. 4, 85748 Garching, Germany

[2]Universität der Bundeswehr, Institut für Mechanik und Statik, Werner-Heisenberg-Weg 39, Neubiberg, 85579 Germany

[a]roman.norz@utg.de, [b]fabuer.ramon@unibw.de, [c]simon.vitzthum@utg.de, [d]simon.vitzthum@utg.de, [e]steffen.gerke@unibw.de, [f]michael.bruenig@unibw.de, [g]wolfram.volk@utg.de

Keywords: Non-Proportional Load Paths, Cross-Loading, Onset of Yielding, Formability

Abstract. The material behaviour after non-proportional strain paths is subject of numerous current investigations. These investigations show that in addition to the stress state, a change in the direction of loading also has a significant influence on the material behaviour. In this work, the influence of non-proportional strain paths on the mechanical properties such as tensile strength, uniform strain, elongation at fracture and yield strength as well as the influence on formability is determined. For this purpose, a dual-phase steel CR330Y590-DP, with a thickness of 0.8 mm, is investigated in more detail. The material is pre-strained to different strain values and then further examined with Nakajima and tensile tests. The influence of the loading direction is determined by five different post-strain directions (0°, 22,5°, 45°, 67.5° and 90°) to the initial pre-strain direction. In addition to the method according to the standard, the yield strength is determined by a temperature-dependent determination method, which is based on the thermoelastic effect. This method has already been qualified for simple uniaxial tensile tests and a relation to the microstructural behaviour was proven. Thus, it provides valuable conclusions on the microstructural behavior for the tests performed within this study with a change in the loading direction.

Introduction

Non-proportional load paths are being investigated for several years. Many researchers focus on the effect on the mechanical properties by using tensile or shear tests but also on the formability using Nakajima experiments. One of the most looked at materials are dual-phase steels with different strengths. By using shear tests, Hérault et al. [1] analysed the influence of uniaxial pre-strain on the elasto-plastic behavior. To describe the material behavior using the Homogeneous Anisotropic Hardening model (HAH-model), experiments with a pre-forming level of at least $\varepsilon = 2\%$ are required due to the significant material behaviour evolution below that level. Larsson et al. [2] investigated the influence of pre-forming height and a change in loading direction on Docol 600DP and Docol 1200DP using tensile and shear tests with pre-formed material. The pre-forming height, as well as a change in loading direction not only influence the onset of yielding but also the formability in the tensile test experiments. A change in loading direction by 45° or 90° after a pre-forming of $\varepsilon = 10\%$ lead to a reduced formability. Liao et al. [3] used SEM, EBSD and TEM to investigate the microstructural effects of a proportional, pseudo cross and cross-loading in DP500, DP600 and DP800. For the experiments with a change in loading direction, yielding

Material Forming - ESAFORM 2023
Materials Research Forum LLC
Materials Research Proceedings 28 (2023) 747-754
https://doi.org/10.21741/9781644902479-81

occurred at lower stress compared to the monotonic experiments. This effect is enhanced at higher pre-forming levels. The evolution of texture was found to be weak for all three DP-steels and cannot explain the effects on the mechanical properties. Nevertheless, the presence of a soft ferritic and a hard martensitic phase leads to a very inhomogeneous plastic deformation of the material. At the onset of yielding the soft and weakly strain hardened ferrite grains control the mechanical behaviour. For a DP800 the effect on the onset of yielding takes place already at very low strains of $\varepsilon = 1\%$ and increases as the pre-forming height reaches higher levels [4]. This effect was also captured by other researchers [5]. Yu and Shen [6] looked at various pre-forming states for a DP590. They investigated the influence of a uniaxial, plane-strain and equi-biaxial pre-forming state on the tensile test results. The pre-forming state also influences the mechanical properties. The Young's – modulus is reduced with increasing pre-forming height, where the equi-biaxial pre-forming state lead to the strongest decrease.

The aim of this study is to determine the influence of three different plane-strain pre-forming levels and three different post-loading directions on the mechanical properties using tensile tests. The influence of the pre-forming on the formability will be examined by standard Nakajima experiments. The experiments in this paper are conducted with a CR330Y590-DP (DP600) steel with an initial thickness of 0.8 mm.

Experimental Procedure

The pre-forming of the material took place on a hydraulic Dieffenbacher press in combination with a modified Marciniak-tool developed by Weinschenk and Volk [7]. To ensure a homogenous plane-strain pre-forming, the specimen geometry was adapted from prior investigations [8,9] and was optimized for the used DP600 steel. The change in loading direction is obtained by a rotation of the tensile test and Nakajima specimens. For the tensile tests, five tensile test samples can be manufactured from one single pre-formed specimen, while for the Nakajima experiments only one specimen can be extracted. All pre-forming specimens are cut out under 0° to the initial rolling direction, see Fig. 1.

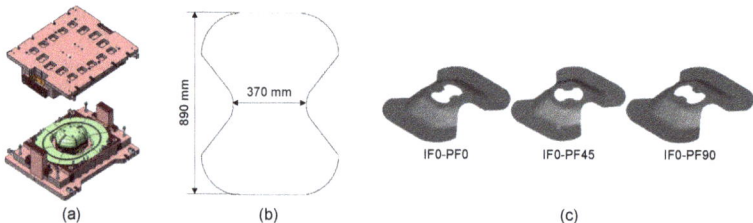

Fig. 1. (a) The used modified Marciniak tool, (b) Specimen geometry used for the plane-strain pre-forming. (s) Specimen extraction and nomenclature of the cut-out specimens.

For the tensile tests specimens according to DIN 50125 – geometry H with a width of 12.5 mm are used. The strain measurement is done by a laser extensometer with a gauge length of 50 mm and a constant strain rate of 0.001 1/s. To determine the onset of yielding, the temperature based method according to Vitzthum et al. [10] was employed. This method makes use of the thermo-elastic effect where the temperature decreases during the elastic deformation and increases during the plastic deformation. A closer description of the method can be found in [10].

The Nakajima experiments are carried out on a BUP1000 in combination with an Aramis 4M DIC system. The punch speed was set to 1 mm/s and a measuring frequency of the DIC-System of 10 Hz is applied. To minimize the friction between the punch and the specimen a PVC-pad with lubricant is used. To assess the onset of necking the Time Dependent Evaluation Method (TDEM)

Material Forming - ESAFORM 2023   Materials Research Forum LLC
Materials Research Proceedings 28 (2023) 747-754   https://doi.org/10.21741/9781644902479-81

proposed by Volk and Hora [11] is applied. Four different Nakajima specimens are tested, ranging from 30 mm for the uniaxial strain state to 235 mm for the biaxial strain state.

Experimental Results

The results of the experiments with the initial material show a low dependency of the loading direction on the stress-strain curves (Fig. 2) as well as the forming limit curves (Fig. 7). All investigated directions differ only marginally. This changes for the pre-formed experiments. The influence of a change in loading direction can clearly be seen. At a pre-forming level of $\varepsilon_{v.Mises} = 0.066$ and $\varepsilon_{v.Mises} = 0.097$ specimens with a change in loading direction of 45° show the lowest fracture strain, while the elastic-plastic transition is different for all specimens with a change in loading direction. This effect increases for higher pre-forming levels. At a pre-forming level of $\varepsilon_{v.Mises} = 0.121$ the fracture elongation of the specimen with a change in loading direction is almost the same.

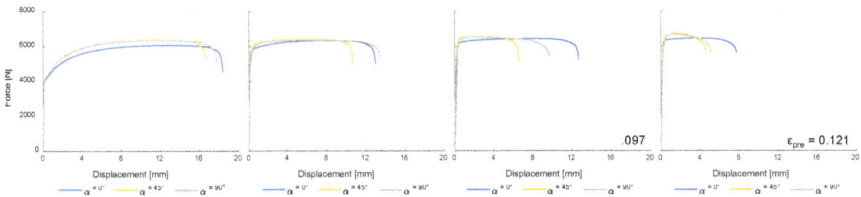

Fig. 2. Force-Displacement curves of the initial material and the three investigated pre-forming heights.

As many researchers have found a significant influence of the pre-forming and a change in loading direction on the elastic-plastic transition, the onset of yielding is further examined. To determine the onset of yielding, three different methods are employed. Next to the classic $R_{p0.2\%}$, depicted as $YS_{0.2\%}$ two temperature based methods are also applied. The temperature is measured throughout the experiment using a PT1000 thermometer, see Fig. 3 (a). The onset of yielding was then determined at the temperature minimum (YS_{Tmin}) and at zero plastic strain (YS_0). YS_0 is calculated by fitting two regression lines into the first derivative of the temperature signal, see Fig. 3 (b). One line is fitted into the elastic region where the temperature is decreasing while the other line is fitted into the area around the temperature minimum where the derivative is zero. The onset of yielding is then determined at the intersection of the two lines and via an angle bisector. A closer description of this method can be found in Vitzthum et al. [10].

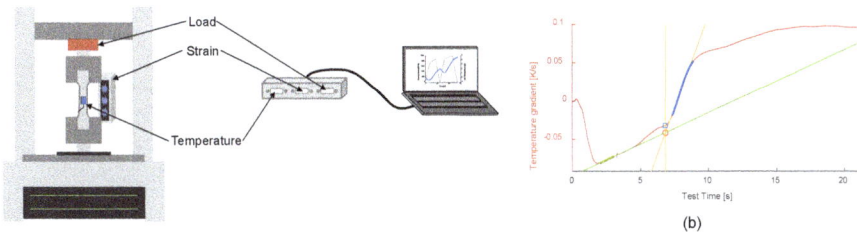

(b)

Fig. 3. (a) Experimental setup using the thermometer, (b) Determination of YS0 using the temperature derivative.

Material Forming - ESAFORM 2023
Materials Research Proceedings 28 (2023) 747-754

Materials Research Forum LLC
https://doi.org/10.21741/9781644902479-81

The onset of yielding is strongly dependent on a change in loading direction and the pre-forming level. The higher the pre-forming level, and when there is no change in loading direction, the closer the onset of yielding assessed by the three different methods is moving together. As soon as there is a change in loading direction, the temperature signal changes significantly. While specimens with no change in loading direction show a sharp temperature signal, almost like a V-shape, the specimens with a change in loading direction have a round temperature signal, almost like an U-shape, shown in Fig. 4.

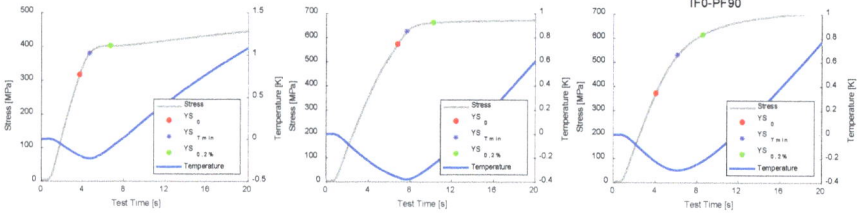

Fig. 4. (a) Onset of yielding for an initial specimen in 0° to the rolling direction, (b) Pre-formed specimen with no change in loading direction (IF0-PF0), (c) Pre-formed specimen with a change in loading direction by 90° (IF0-PF90).

Looking at all results for the different angles and pre-forming heights, it can be seen, that the onset of yielding for the $YS_{0.2\%}$ method is steadily increasing with increasing pre-forming level almost untouched by a change in loading direction, see Fig. 5 (a). The temperature-based results however, show a strong dependency on a change in loading direction as soon as a pre-forming took place, Fig. 5 (b) and (c). These results indicate that when a change in loading direction occurs after a pre-forming, some grains start to yield a lot earlier than others do. This leads to the smooth and round temperature signal. While the grains of specimens with no change in loading direction yield almost at the same time, leading to this sharp V-like shape. This effect is more substantial, the higher the pre-forming level is.

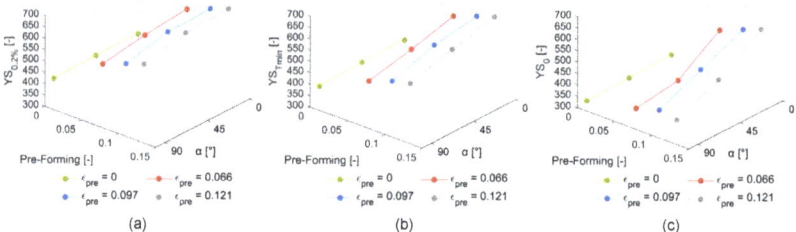

Fig. 5. Results for the $YS_{0.2\%}$ method (a), (b) for the temperature minimum and (c) for the YS_0-method.

In the Nakajima experiments, for high pre-forming levels and a change in loading direction, shear failure is observed. When using the Time Dependent Evaluation Method (TDEM-method) a significant overestimation of the formability for such specimens is found, shown in Fig. 6 (b). The

Material Forming - ESAFORM 2023   Materials Research Forum LLC
Materials Research Proceedings 28 (2023) 747-754   https://doi.org/10.21741/9781644902479-81

evaluation of specimens with such a shear failure is performed by a modified TDEM-method, see Fig. 6 (a), which was already used in [8]. In Fig. 6, the two methods are shown for a 30 mm wide specimen with a change in loading direction of 45° (IF0-PF45) and a pre-forming level of ε_{pre} = 0.193.

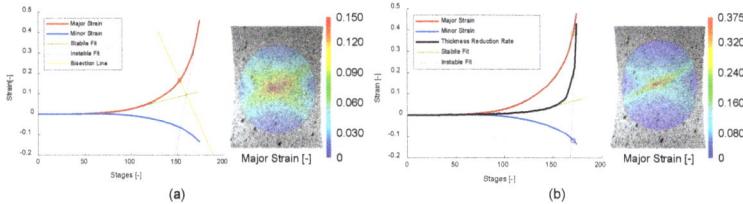

Fig. 6. (a) Method used in Volk et al. [8] for Shear-Failure, (b) Time Dependent Evaluation Method [11].

The higher the pre-forming level, the more a change in loading direction influences the formability, see Fig. 7. For the pre-forming level with ε_{pre} = 0.151, the IF0-PF45 specimens show the lowest formability, this is different to the results for the HC340LAD [8]. The most severe influence was observed on the plane-strain post-forming and a change in loading direction by 90° (IF0-PF90) after a pre-forming of ε_{pre} = 0.193. Here, almost instant necking occurs.

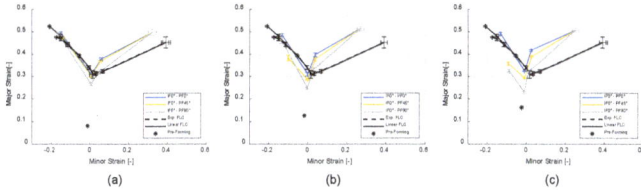

Fig. 7. (a) Forming Limit Curves for ε_{pre} = 0.097, (b) Forming Limit Curves for ε_{pre} = 0.151 (c) Forming Limit Curves for ε_{pre} = 0.193.

In comparison to the results for the micro-alloyed steel HC340LAD and the aluminium alloy AA6016-T4, the effect of a change in loading direction is somewhere in between. The HC340LAD showed a stronger dependency on the formability, see [8], while the AA6016-T4 showed almost no influence of a change in loading direction on the formability [12].

One reason for such a dependency on a change in loading direction might be the presence of voids. Asik et al. [13] found, that void growth and nucleation for a DP600 steel already takes place before the uniform elongation is reached. These voids primarily occur at inclusions and at the ferrite-martensite interface but also martensite cracking leads to the creation of voids. Due to the banded structure of the martensite inside the ferritic matrix, the damaged area is dependent on the loading direction. Gerstein et al. [14] investigated the voids after deformation of a DP600 using SEM – microscopy. Voids caused by plastic deformation have shown an elongated shape in the direction of loading. By using in-situ X-Ray microtomography Maire et al. [15] also found, that the biggest cavities in tensile test showed an elongated shape in the direction of loading. If the stress state has a higher triaxiality than that in the tensile test, the cavities are likely to become more isotropic. As the pre-forming state in this study is plane-strain, the voids might show that elongated appearance in the direction of the first loading step. This could explain the anisotropic damage behaviour in the Nakajima experiments for specimens with a change in loading direction. In Fig. 8 the different fracture surfaces are shown. While the specimens with no change in loading direction (Fig. 8 a and d) show a ductile fracture with many small voids, the 30 mm wide specimens with a change in loading direction show almost a brittle fracture surface (Fig. 8 b and c). The number of voids is significantly reduced, compared to (Fig. 8 a). This can explain the reduced formability of the material as the ductility is reduced after a change in loading direction. For the plane-strain specimens with a width of 100 mm a different behaviour is observed. While the specimens with no change in loading direction (Fig. 8 d) show again a ductile fracture surface with many small voids, the specimens with a change in loading direction have considerably big voids caused by hard martensitic inclusions which can be seen in the big void in (Fig. 8 f). In between the vast voids, a shear behaviour is found where the big voids started to coalescence. The increased void growth and the coalescence of the voids could lead to the significant reduction in formability.

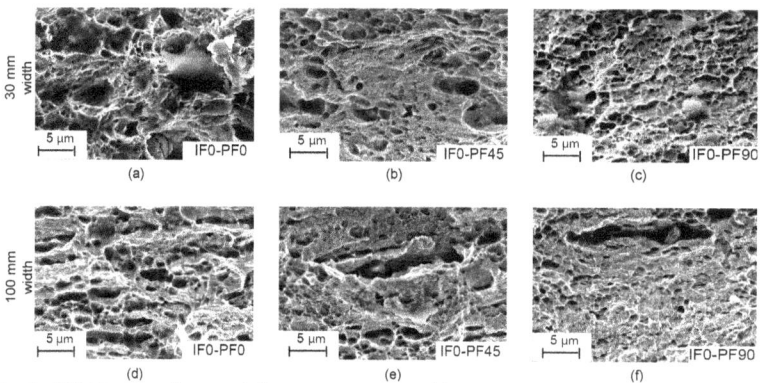

Fig. 8. SEM-images for two different specimen widths (30 mm and 100 mm) and the three investigated post-loading directions IF0-PF0 (a,d), IF0-PF45 (b,e) and IF0-PF90 (c,f) for the pre-forming height of $\varepsilon_{pre} = 0.193$.

Summary

In this paper, the influence of a plane-strain preforming at different levels as well as a change in loading direction has been investigated. By using tensile and Nakajima tests, the impact on the mechanical properties like yield strength and fracture elongation as well as the Forming Limit Curve has been quantified. It is found, that a change in loading direction strongly affects the onset of yielding in tensile test by using temperature-based methods. This effect is not captured by using the classical $YS_{0.2\%}$ method. This might be caused by the fact that some grain orientations start to yield earlier than others and therefore lead to a continuous increase in temperature and a U-shaped temperature signal. For specimens with no change in loading direction no such effect is observed. In fact, the two temperature-based methods, temperature minimum YS_{Tmin} and yield strength at zero strain YS_0, are moving closer together. This indicates that the grains with different orientations almost yield at the same time, leading to a sharp elastic-plastic transition.

For the Nakajima experiments, a strong effect on the formability after a certain level of pre-forming is noticed. Especially on the left-hand side of the Forming Limit Curve, this influence was determined. As some of the specimens, mostly those with a significant loss of formability did not show a ductile necking failure but instead a shear failure, the Time Dependent Evaluation Method (TDEM-method) was not applicable. To determine the limit strains for those specimens, the same procedure as for the TDEM-method, but instead of the thickness reduction rate $\dot{\varphi}_3$ the Major Strain

Material Forming - ESAFORM 2023
Materials Research Proceedings 28 (2023) 747-754

Materials Research Forum LLC
https://doi.org/10.21741/9781644902479-81

is used to fit the two regression lines. This method showed better results in comparison to the TDEM-method, which overestimates the limit strains for shear failure specimens.

As voids caused by the pre-forming process might be the cause for this anisotropic behaviour, SEM-microscopy will be carried out with the pre-formed material in order to detect voids inside the material. In addition, the disposal of the voids is of interest as some researchers found that voids are aligned in certain directions.

Acknowledgements

The authors would like to thank the German Research Foundation (DFG) for their financial support under the grant numbers 455960756. Furthermore, the support of Wolfgang Saur (Institut für Werkstoffe des Bauwesens, Universität der Bundeswehr München) in performing the scanning electron micrographs is gratefully acknowledged.

References

[1] D. Hérault, S. Thuillier, S.-Y. Lee, P.-Y. Manach, F. Barlat, Calibration of a strain path change model for a dual phase steel, Int. J. Mech. Sci. 194 (2021) 106217. https://doi.org/10.1016/j.ijmecsci.2020.106217.

[2] R. Larsson, O. Björklund, L. Nilsson, K. Simonsson, A study of high strength steels undergoing non-linear strain paths—Experiments and modelling, J. Mater. Process. Technol. 211 (2011) 122-132. https://doi.org/10.1016/j.jmatprotec.2010.09.004

[3] J. Liao, J.A. Sousa, A.B. Lopes, X. Xue, F. Barlat, A.B. Pereira, Mechanical, microstructural behaviour and modelling of dual phase steels under complex deformation paths, Int. J. Plast. 93 (2017) 269-290. https://doi.org/10.1016/j.ijplas.2016.03.010

[4] V. Tarigopula, O.S. Hopperstad, M. Langseth, A.H. Clausen, Elastic-plastic behaviour of dual-phase, high-strength steel under strain-path changes, European Journal of Mechanics - A/Solids 27 (2008) 764-782. https://doi.org/10.1016/j.euromechsol.2008.01.002

[5] F. Barlat, G. Vincze, J.J. Grácio, M.-G. Lee, E.F. Rauch, C.N. Tomé, Enhancements of homogenous anisotropic hardening model and application to mild and dual-phase steels, Int. J. Plast. 58 (2014) 201-218. https://doi.org/10.1016/j.ijplas.2013.11.002

[6] H.Y. Yu, J.Y. Shen, Evolution of mechanical properties for a dual-phase steel subjected to different loading paths, Mater. Des. 63 (2014) 412-418. https://doi.org/10.1016/j.matdes.2014.06.003

[7] A. Weinschenk, W. Volk, FEA-based development of a new tool for systematic experimental validation of nonlinear strain paths and design of test specimens, Penang, Malaysia, Author(s), 2017, pp. 20009.

[8] W. Volk, R. Norz, M. Eder, H. Hoffmann, Influence of non-proportional load paths and change in loading direction on the failure mode of sheet metals, CIRP Annals 69 (2020) 273-276. https://doi.org/10.1016/j.cirp.2020.03.009

[9] R. Norz, S. Vitzthum, W. Volk, Failure behaviour of various pre-formed steel sheets with respect to the mechanical grain boundary properties, Int. J. Mater. Form. 15 (2022) 1215. https://doi.org/10.1007/s12289-022-01700-9

[10] S. Vitzthum, J. Rebelo Kornmeier, M. Hofmann, M. Gruber, E. Maawad, A.C. Batista, C. Hartmann, W. Volk, In-situ analysis of the thermoelastic effect and its relation to the onset of yielding of low carbon steel, Mater. Des. 219 (2022) 110753. https://doi.org/10.1016/j.matdes.2022.110753

[11] W. Volk, P. Hora, New algorithm for a robust user-independent evaluation of beginning instability for the experimental FLC determination, Int. J. Mater. Form. 4 (2011) 339-346. https://doi.org/10.1007/s12289-010-1012-9

Material Forming - ESAFORM 2023
Materials Research Proceedings 28 (2023) 747-754

Materials Research Forum LLC
https://doi.org/10.21741/9781644902479-81

[12] R. Norz, F.R. Valencia, S. Gerke, M. Brünig, W. Volk, Experiments on forming behaviour of the aluminium alloy AA6016, IOP Conf. Ser.: Mater. Sci. Eng. 1238 (2022) 12023. https://doi.org/10.1088/1757-899X/1238/1/012023

[13] E.E. Aşık, E.S. Perdahcıoğlu, A.H. van den Boogaard, Microscopic investigation of damage mechanisms and anisotropic evolution of damage in DP600, Mater. Sci. Eng. A 739 (2019) 348-356. https://doi.org/10.1016/j.msea.2018.10.018

[14] G. Gerstein, H.-B. Besserer, F. Nürnberger, L.A. Barrales-Mora, L.S. Shvindlerman, Y. Estrin, H.J. Maier, Formation and growth of voids in dual-phase steel at microscale and nanoscale levels, J. Mater. Sci. 52 (2017) 4234–4243. https://doi.org/10.1007/s10853-016-0678-x

[15] E. Maire, O. Bouaziz, M. Di Michiel, C. Verdu, Initiation and growth of damage in a dual-phase steel observed by X-ray microtomography, Acta Mater. 56 (2008) 4954-4964. https://doi.org/10.1016/j.actamat.2008.06.015

Material Forming - ESAFORM 2023
Materials Research Proceedings 28 (2023) 755-760

Materials Research Forum LLC
https://doi.org/10.21741/9781644902479-82

Heat assisted V-bending characteristics of high strength S700MC steel

GÖRTAN Mehmet Okan[1,a] *, CAGIRANKAYA Fatih[2,b] and OZSOY Oguz[2,c]

[1]Hacettepe University, Mechanical Engineering Department, Ankara, Turkey

[2]TEKNOROT Steering & Suspension Parts, Düzce, Turkey

[a]okangortan@hacettepe.edu.tr, [b]fatih.cagirankaya@teknorot.com, [c]oguz.ozsoy@teknorot.com

Keywords: V-Bending, Springback, Mild Temperature Forming, High Strength Steel

Abstract. Increasingly stringent regulations force automotive manufacturers to reduce their product weights without compromising passenger safety. To overcome this problem, high strength steels are used that enable a higher payload capacity whilst reducing the plate thickness and overall weight. However, high strength steels are generally associated with lower formability and significantly greater springback compared to mild steels. In the current study, bendability characteristics of high strength S700MC steel in mild temperatures are investigated using V-bending tests. A temperature range between 25°C and 650°C is analysed. Springback characteristics as well as hardness distribution along the thickness direction in the bent area is investigated for several bent samples.

Introduction

Due to the stricter demands in terms of greenhouse gas emission regulations, automotive manufacturers are forced to use lightweight solutions in their design. Two major approaches have been developed in sheet metal forming industry to reduce the thickness and hence weight of parts without compromising the safety and toughness [1]. The first one is to replace conventional materials with their higher strength counterparts and so to be able to carry the same loads using less material. However, due to the limitations in ductility of advanced high strength steels (AHSS) and ultra-high strength steels (UHSS), this strategy can be only used in comparatively simple geometries such as bumper reinforcements, door beams or seat tracks, which can be cold-formed using multiple stage bending or roll-forming processes. The second strategy is the application of a press hardening process where austenitized steels are hot-formed in a press and quenched in the same die system. That way, martensitic microstructure is obtained, resulting in strengths that exceed 1500 MPa in the produced parts without major formability problems [2]. Nevertheless, press hardening process is significantly more energy demanding than conventional stamping or deep drawing processes. In best-case scenario, more than 69% of the energy related to forming is consumed during the heating and related operations in press hardening [3].

There are multiple studies about formability and especially about bendability of steels. Recent ones focus on advanced high strength steels. Majority of the articles investigate critical bending angle and fracture formation depending on the used steels microstructure and mechanical properties. Four different failure types are defined for bending of steels, namely waviness with a flattening on the outside, ductile cuts, fine cracks and instable cracks [4]. Furthermore, it is reported that fracture formation starts during bending before reaching the maximum force in the process [5]. Moreover, regardless of microstructure, bendability of steels improves as strain hardening increases and resistance against crack formation getting higher [6-8]. However, all these studies are conducted at room temperature.

To increase the energy efficiency of the forming operations and to be able to use high strength steels in the manufacturing of complex parts, it is intended to develop mild temperature forming

Material Forming - ESAFORM 2023
Materials Research Proceedings 28 (2023) 755-760

Materials Research Forum LLC
https://doi.org/10.21741/9781644902479-82

strategies for conventional high strength steels. The idea behind the mild temperature forming is to shape steels at a temperature level below austenitizing region. That way a decrease in strength and increase in formability is expected. Hence, springback can be reduced. Moreover, since the temperatures are kept below austenitizing region, minimum change is foreseen in microstructure and mechanical properties after forming process is finished and parts are cooled to room temperature. In the current study, bending characteristics of high strength steel S700MC is investigated at mild temperatures ranging from 25°C to 650°C. Springback after bending is evaluated. Moreover, mechanical properties are characterized using hardness measurements.

Material and Methodology

A hot-rolled structural high-strength steel sheet S700MC made for cold forming with a thickness of 4 mm has been used in the study. Mechanical properties and chemical composition of the material are given in Table 1 and Table 2, respectively.

Table 1. Mechanical properties of S700MC steel.

Yield Strength	Tensile Strength	Elongation
792.2 MPa	941.2 MPa	13.2%

Table 2. Chemical composition of S700MC steel (in mass %).

C	Si	Mn	Ni	Mo	Fe
0.78	0.083	1.94	0.15	0.168	balance

Bending samples are prepared using shear-cutting with a length and width of 150 mm and 50 mm, respectively. Width of the samples coincides with the rolling direction of the steel sheets. V-bending test are conducted on those samples according to the ISO 7438:2020 standard. A bending die with an angle of 90° is prepared for this purpose. Since mild steels are usually bent with a r/t ratio of 1, the same ratio is used in the experiments. Die system is shown in Fig. 1.

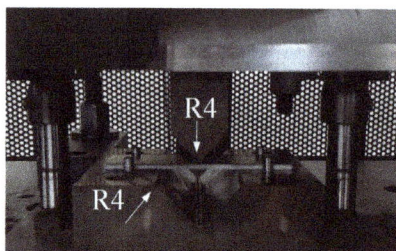

Fig. 1. V-bending die system.

Bending tests are conducted at room temperature and additionally at elevated temperatures ranging from 150°C to 650°C with increments of 100°C. Induction heating is used to heat the samples in bending region. Temperature of the samples are measured with a laser sensor during heating. A hydraulic press with a capacity of 400 tons is used for forming tests. Motion control mode is used and ram speed is set constant to 40 mm /sec. Dwell time at lower dead center is set to 2 seconds. All samples are cleaned with acetone prior to bending. No lubrication is used.

Material Forming - ESAFORM 2023   Materials Research Forum LLC
Materials Research Proceedings 28 (2023) 755-760   https://doi.org/10.21741/9781644902479-82

All bending tests are repeated at least three times. Springback on all samples are measured using mechanical angle gauge with a precision of 0.1°. Crack formation on outer side of the bent samples are investigated with optical microscopy and dye-penetrant testing.

All bent samples are cut in the middle section. The cross section is ground with sand papers between P400 and P2500 grit. Afterwards, surfaces are polished with 6 μm and 1 μm diamond solution. Prepared surfaces are etched with 4% Nital solution for approximately 10 seconds. Microstructure of the bent section is investigated using optical microscopy (Nikon Eclipse LV150).

Strain hardening behavior of the samples in the bending region along the thickness direction is characterized using Vickers hardness measurements according to ISO 6507-1 with a load of 500 g and dwell time of 15 seconds (Future-Tech FM-700e).

Results and Discussion

Outside section of samples bent at different temperatures are shown in Fig. 2. A clear crack formation is observed on samples formed at temperatures below 550°C. There are multiple reasons for the crack formation. At room temperature, ductility of the material is not sufficient to accomplish the bending. However, when the temperature is increased, material enters the blue brittleness region. Therefore, it becomes highly brittle and blue cracks are observed. Apparently, the investigated material exits the blue brittleness region at 550°C and crack free samples are generated at temperatures above that level.

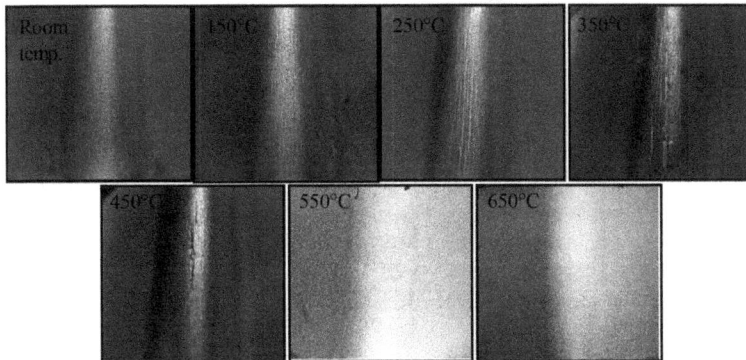

Fig. 2. Outside section of the bent samples at different temperatures.

Side view of the bent samples are shown in Fig. 3. Thanks to the low r/t ratio, springback value of sample bent at room temperature is comparatively low at 90.6°. At temperatures higher than 150°C, no springback is observed in the bent samples. In the die-design process, every bending angle above 90° is a challenge. Even though springback at room temperature of the investigated case is low, it necessitates a certain over-bending. Since there is no springback at elevated temperatures, it is seen as major advantage of mild temperature forming of S700MC steel.

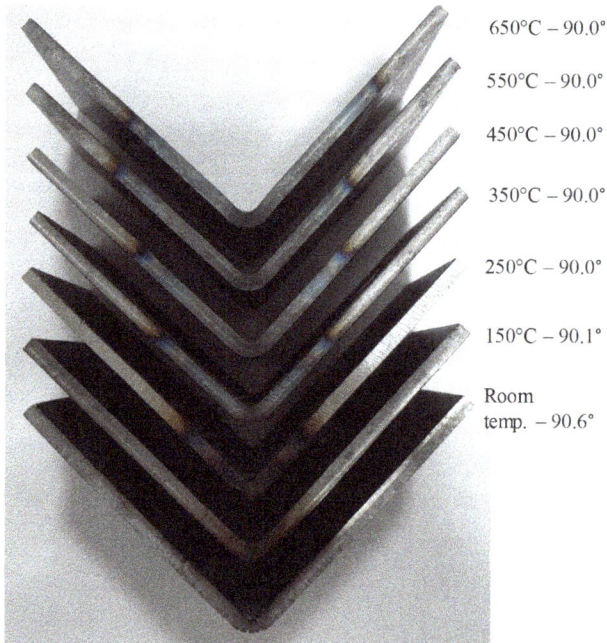

650°C – 90.0°

550°C – 90.0°

450°C – 90.0°

350°C – 90.0°

250°C – 90.0°

150°C – 90.1°

Room
temp. – 90.6°

Fig. 3. Side view of bent samples at different temperatures.

Optical microscopy results are shown in Fig. 4. Initial microstructure of S700MC steel is predominantly in ferritic-pearlitic form. This structure remains the same after bending at all temperatures. Since all investigated temperatures were below A_1 level, austenitizing of the microstructure was not expected. Moreover, there isn't any significant change in the grain size. This is attributed to the fast heating and cooling of the material prior and after bending. During heating of steels below A_1 temperature, recovery and recrystallization is expected which yields in dislocation density drop and grain growth. However, both of those processes are time dependent. Thanks to rapid heating and cooling, recrystallization wasn't observed in the microstructure.

Optical microscopy investigations also reveal that there are buckling marks on the inside of samples bent at room temperature. Those marks may act as stress concentration points in repeated loading conditions and reduce fatigue life significantly. However, no sign of buckling is observed at samples bent at 550°C and 650°C. It is seen as an important advantage of forming at mild temperature for S700MC steel.

Fig. 4. Microstructure of bent samples.

Material Forming - ESAFORM 2023
Materials Research Proceedings 28 (2023) 755-760

Materials Research Forum LLC
https://doi.org/10.21741/9781644902479-82

An important aspect of mild temperature bending of high strength steels is the mechanical properties after forming. Steels formed at room temperature are expected to exhibit a certain amount of strain hardening. As a result, static strength of the parts increases. Strength change can be characterized using hardness measurements [9]. In the current study, Vickers hardness of bent samples along thickness direction in bent are is measured. Results are shown in Fig. 5.

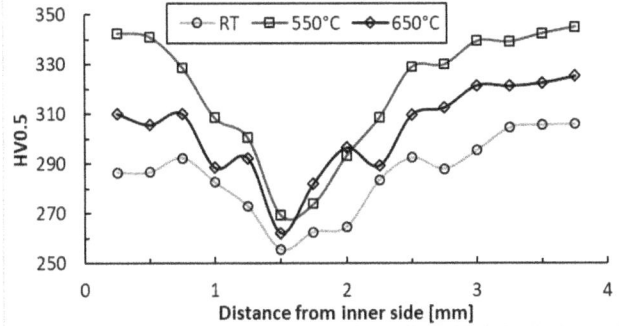

Fig. 5. Hardness distribution along thickness direction.

It is clear from Fig. 5 that the neutral axis of all investigated samples are shifted about 0.5 mm inwards. As we move away from the neutral axis both inwards and outwards, strength of the investigated steel increases due to strain hardening. Since strain value on the outside is higher due to neutral axis shift, hardness level on the outside is approximately 7% higher compared to inside section. When the bending temperature is increased to 550°C, a more significant strain hardening is observed in the bent area. Usually, if temperature of steels is increased, first recovery and later recrystallization is expected. As a result, decline in dislocation density and grain growth occurs. Therefore, strength of steels decrease. However, in the investigated case, a more significant strength increase is observed in samples bent at 550°C compared to the ones bent at room temperature. Underlying reason for that increase should be investigated more in detail using microstructural analysis.

When hardness distribution of the sample bent at 650°C is compared to the one bent at 550°C, a drop in hardness is observed which is less than 10%. All these investigations suggest a strain hardening in the investigated steel even at elevated temperatures.

Summary
In the current study, bending characteristics of hot-rolled structural high-strength steel sheet S700MC made for cold forming is investigated at mild temperatures. Studied temperatures are kept below austenitizing temperature to decrease the energy requirement of the bending operation while increasing the formability characteristics. Following conclusions are drawn:
- Increase in temperature enables better formability and reduced crack formation risk for S700MC steel.
- Crack formation on the outside of the samples formed at temperatures up to 450°C is observed which is due to the blue brittleness of the material in that region
- Reduced springback is observed as the bending temperature increases.
- A significant hardness increase is observed in the samples bent at 550°C and 650°C compared to the ones bent at room temperature.

Material Forming - ESAFORM 2023   Materials Research Forum LLC
Materials Research Proceedings 28 (2023) 755-760   https://doi.org/10.21741/9781644902479-82

- In the outer and inner region of all bent samples, an increase in hardness is observed which indicates a strain hardening despite the rise in temperatures.
- No grain growth is observed in the samples bent at elevated temperatures.

Acknowledgements

The conducted work was under a project funded under the SMART EUREKA CLUSTER on Advanced Manufacturing Program. Moreover, the authors thank the Turkish Research Council for financially supporting the project 9210025.

References

[1] S. Keeler, M. Kimchi, P.J. Mooney (Eds.), Advanced High-Strength Steels Application Guidelines Version 6.0, 2017.

[2] H. Karbasian, A.E. Tekkaya, A review on hot stamping, J. Mater. Process. Technol. 210 (2010) 2103-2118. https://doi.org/10.1016/j.jmatprotec.2010.07.019

[3] E. Meza-Garcia, A. Rautenstrauch, M. Braunig, V. Krausel, D. Landgrebe, Energetic evaluation of press hardening processes, Proc. Manuf. 33 (2019) 367-374. https://doi.org/10.1016/j.promfg.2019.04.045

[4] M. Kaupper, M. Merklein, Bendability of advanced high strength steels—A new evaluation procedure, CIRP Ann. - Manuf. Technol. 62 (2013) 247-250. http://doi.org/10.1016/j.cirp.2013.03.049

[5] A. Mishra, S. Thuillier, Investigation of the rupture in tension and bending of DP980 steel sheet, Int. J. Mech. Sci. 84 (2014) 171-181. http://doi.org/10.1016/j.ijmecsci.2014.04.023

[6] C. Suppan, T. Hebesberger, A. Pichler, J. Rehrl, A. Kolednik, On the microstructure control of the bendability of advanced high strength steels, Mater. Sci. Eng. A 735 (2018) 89-98. https://doi.org/10.1016/j.msea.2018.07.080

[7] D. Rèchea, T. Sturel, O. Bouaziz, A. Col, A.F. Gourgues-Lorenzon, Damage development in low alloy TRIP-aided steels during air-bending, Mater. Sci. Eng. A 528 (2011) 5241-5250. https://doi.org/10.1016/j.msea.2011.03.042

[8] K. Benedyk, A. Pichler, T. Kurz, O. Kolednik, Die Versagensmechanismen pressgehärteter und hochfester Stähle im Dreipunkt-Biegeversuch, BHM 159 (2014) 122-129. https://doi.org/10.1007/s00501-014-0237-1

[9] M.D. Taylor, K.S. Choi, X. Sun, D.K. Matlock, C.E. Packard, L. Xu, F. Barlat, Correlations between nanoindentation hardness and macroscopic mechanical properties in DP980 steels, Mater. Sci. Eng. A 597 (2014) 431-439. http://doi.org/10.1016/j.msea.2013.12.084

Material Forming - ESAFORM 2023   Materials Research Forum LLC
Materials Research Proceedings 28 (2023) 761-770   https://doi.org/10.21741/9781644902479-83

A stress-based proposal for wrinkling criterion of clamped surfaces

BORBÉLY Richárd[1,a], KÖLÜS Martin L.[1,b] and BÉRES Gábor J.[1,c] *

[1]John von Neumann University, GAMF Faculty of Engineering and Computer Science, Izsáki road 10. 6000, Kecskemét, Hungary

[a]borbely.richard@gmaf.uni-neumann.hu, [b]kolus.martin@gamf.uni-neumann.hu, [c]beres.gabor@gamf.uni-neumann.hu

Keywords: Wrinkling Limit, Clamped Surfaces, Stress-Based Diagram

Abstract. This study presents a possible stress-based limit theory regarding the wrinkling occurrence of clamped surfaces. There are several theorems that describe the buckling and/or the wrinkling phenomenon in different forms and conditions, but a general description, which would represent the limit state of wrinkling in sheet metal forming still does not exist according to the authors knowledge. This could be particularly important for the finite element simulations that are mostly used for process minoring purposes. However, some software work with body elements is suitable for the representation of wrinkles, users do not receive information about how close a process is to the wrinkling limit, or how it is affected by the input parameters. This is even less estimable if shell, or membrane elements are used in a finite element code. In this work, a purely analytical calculation for the wrinkling limit stress of clamped surfaces is carried out, i.e., when blank holder tool acts on the sheet. To take into consideration the effect of the normal pressure, Wang and Cao's proposal was used. After expressing the critical stress by its major and minor principal components using anisotropic yield criteria, a novel illustration method of the wrinkling limit has become available and is published in this article.

Introduction

During the manufacturing of complex sheet parts in the press shop, necking or facture and the geometrical defects (like wrinkling and springback) are the most frequent failure modes [1]. Necking or fracture and wrinkling take place during the forming process, however springback can be considered as a post-forming defect. Although the occurrence of necking and fracture are deeply studied and different limit theorems are already introduced with the aim of process monitoring, wrinkling is less researched. Or at least, an accepted wrinkling limit theory for general cases (e.g., when blank holder also works) is not yet available, with the use of which one can monitor how close a process is to the risk of wrinkling. In this study a proposal can be seen for the stress-based wrinkling limit determination of clamped surfaces, using the basis of Wang and Cao's theory [2,3].

Most of the studies that cover the limits of formability, where the failure of a component is considered under different stress conditions and then the material behavior is summarized into one diagram, are directed to the classical forming limit diagram (FLD). This tool is accepted for necking evaluation in general by the sheet forming society. The FLD indicates the limit of global formability from shearing up to biaxial failure [4,5], i.e., the failure risk in negative (compression) stress states is less discussed in detail. Nevertheless, remarkable improvements have been made in the past, both on the practical determination [6] and the theoretical description [7] of the FLD. To this article, the development of the stress-based forming limit diagram proposed by Stoughton and Zhu [8] has a special importance, which calculation method partially forms the basis of the stress components' calculations applied in this manuscript.

In the respect of compression generated failure modes, the wrinkling is particularly problematic on sheet workpieces that require aesthetic appearance. Besides, wrinkles formed in the first drawing step can cause unpleasant complications in a subsequent press forming operation, too.

Some of the parameters affecting this phenomenon are the stress state, the initial work piece geometry and the blank holder force. The latter one was analytically described by Ju and Johnson in an energy-based model [9] and was later used for discussing the behavior of axisymmetric drawn parts by Agrawal et al. [10]. The biggest drawback of this model is that it only works for cylindrical geometry.

Wang and Cao's article [2] lighted up a different, but also energy-based theorem, in which they proposed the calculation of the deformation energy of a completely flat sheet and a buckled sheet to judge the critical wrinkling conditions. The main advantage of this model is that it does not depend directly on the geometry, thus it is useful to any component geometry. They defined both the critical wrinkling stress as well as the optimal blank holder pressure that need to eliminate wrinkling, using the equations of anisotropic plasticity, expanded for different stress states in [3].

It should be also noted that, although the definition of rupture is perhaps exact even now, the onset of wrinkling is still based on individual subjective judgement. It is therefore necessary to define a preliminary boundary condition that is considered as the criterion of wrinkling. For example, in Wang and Cao's work, the transition from semi-sinusoidal to completely sinusoidal characteristic of surface wrinkles was nominated to the analytical condition. However, Hutchinson and Neale [11] developed an exact, purely theoretical equation (not detailed here), which states that wrinkling occurs when the sum of the bending stress and the stress resultants from buckling and stretching is zero, it is still not yet implemented in practice.

In this study, a possible way for wrinkling risk calculation is presented to define a wrinkling limit diagram that can serve as an input boundary for the design of forming processes. We base our claims on previous experimental investigations and numerical simulations, which are only partially addressed here.

Description of the Wrinkling Behavior

The analytical calculation of wrinkling limit criterion was carried out based on the proposals of Wang and Cao [2,3]. In this methodology, the critical, equivalent wrinkling stress can be obtained by the strain energy difference of a perfectly flat sheet (E_0) and a buckled sheet (E_b), during a deformation process. If negative stress acts on the edge of the component, i.e., for example on the flange of a drawn workpiece, wrinkling can develop if the strain energy of buckling is the larger one. However, when blank holder is applied during a deep drawing process, the optimum of the external work of the blank holder (W) is exactly the same as the difference of the two mentioned energy terms:

$$W = E_0 - E_b \tag{1}$$

Assuming that the external work of the blank holder can be mathematically described in the knowledge of the normal force and the buckling deflection, as well as the normal force itself is a non-linear function of the buckled height (δ), the normal pressure can be expressed in the following form, according to [2]:

$$p = \frac{3(E_0 - E_b)}{4\delta Lw} \tag{2}$$

This is the case of a simplified, rectangular flat blank, which has L length, w width and s thickness (see Fig. 1). Applying the deduction of the energy terms based on [2,3], in which Swift hardening law [12] and Hill48 anisotropic plastic potential [13] was used, each energy members can be obtained according to Eq. 3 and Eq. 4.

$$E_0 = \frac{1}{w}\iint \bar{\sigma}\, d\bar{\varepsilon}\, dV = \frac{KLs}{n+1}(\varepsilon_0 + c_1\varepsilon_{1_0})^{n+1} \tag{3}$$

$$E_b = \frac{2Ks}{n+1}\left[\frac{c_2 s}{2} + (\varepsilon_0 + c_3)\left(\frac{1}{m^2\delta} + \frac{s}{2}\right)\right]^{n+1} \cdot \left[\frac{1}{m^2\delta} + \frac{s}{2}\right]^{-n} \cdot tan^{-1}(m\delta) \tag{4}$$

In these functions, K, ε_0 and n refer to the constants of the Swift hardening law, while c_1, c_2 and c_3 are material parameters considering the plastic anisotropy and the stress state. The frequency of the wave mode is given by m.

With combining Eq. 1-4, the optimal blank holder pressure can be obtained by a purely analytical formula, which contains practically understandable material parameters as well as the u_t edge displacement and the L length of the blank:

$$p = \frac{3Ks}{4(n+1)\delta}\left\{\left[\varepsilon_0 + c_1 ln\left(1 - \frac{2u_t}{L}\right)\right]^{n+1} - \frac{2}{L}\left[\frac{c_2 s}{2} + (\varepsilon_0 + c_3) \cdot \left(\frac{1}{m^2\delta} + \frac{s}{2}\right)\right]^{n+1} \cdot \left[\frac{1}{m^2\delta} + \frac{s}{2}\right]^{-n} \cdot tan^{-1}(m\delta)\right\} \tag{5}$$

It means that the critical edge displacement (or the critical, normalized edge displacement u_{tcr}/L) for a given component length has to be known to obtain the critical normal pressure, which needs to eliminate wrinkling. Wang and Cao defined this critical value at the transition point, where half sinusoid wave form changes to a complete sinusoid wave form. With performing the above analysis (Eq. 1-5) by continuously changing L lengths (L_1, L_2, ..., L_i), intersecting curves will be output, at which the transition points will determine the edge displacement dependent function of the critical normal pressure:

$$p_1\left(L_1, \frac{u_{tcr}}{L}\right) - p_2\left(L_2, \frac{u_{tcr}}{L}\right) = 0 \tag{6}$$

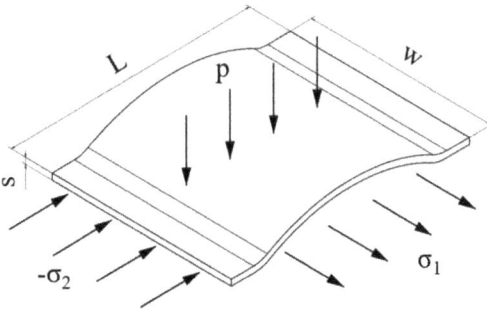

Fig. 1. A simplified, rectangular buckled sheet, affected by tension and compression stresses (the ut edge displacement is interpreted along the L direction).

Assuming that the equivalent plastic strain is direct function of the edge displacement and the stress state, it can be expressed from the associated flow rule that belongs to the Hill48 yield criterion, i.e.,

$$\varepsilon^{pl} = c_1 \cdot ln\left(1 - \frac{2u_{tcr}}{L}\right) \tag{7}$$

In Eq. 7, c_1 parameter follows from the anisotropic criterion of yielding and embodies the effect of the stress state (α), too.

Material Forming - ESAFORM 2023   Materials Research Forum LLC
Materials Research Proceedings 28 (2023) 761-770   https://doi.org/10.21741/9781644902479-83

$$c_1 = \sqrt{\frac{1+r}{1+2r} \cdot \left(1 + \frac{r(1+\alpha)^2 + (\alpha + \alpha r - r)^2}{(1+r-\alpha r)^2}\right)} \qquad (8)$$

The stress ratio is calculated as the ratio of the minor and the major principal stresses:

$$\alpha = -\frac{\sigma_2}{\sigma_1} \qquad (9)$$

Now, the critical wrinkling stress can be considered as the equivalent stress, and it can be given in the function of the normal pressure. This fact also means that the critical pressure needs to eliminate wrinkling became known for any equivalent stress values that may be experienced in a deformable workpiece.

$$\sigma_{cr} = K \sqrt{\frac{1+r}{1+\alpha^2 + r(1-\alpha)^2}} \left[\varepsilon_0 + c_1 \cdot ln\left(1 - \frac{2u_{tcr}}{L}\right)\right]^n \qquad (10)$$

Proposal for the Wrinkling Limit Stress Diagram
For a given material, the hardening parameters (K, ε_0, n) and the average anisotropy coefficient (r) are constants in Eq. (10). Therefore, the slope of the σ_{cr}-p_{cr} curve only depends on the stress state. Fig. 2 shows an example for the calculated σ_{cr}-p_{cr} values of DC04 steel sheet, in which diagrams the equivalent stress is the same but the α stress ratio changes from -1/0.01 up to -1/1.0. The applied material parameters (K, ε_0, n and r) are listed slightly below in Table 1.

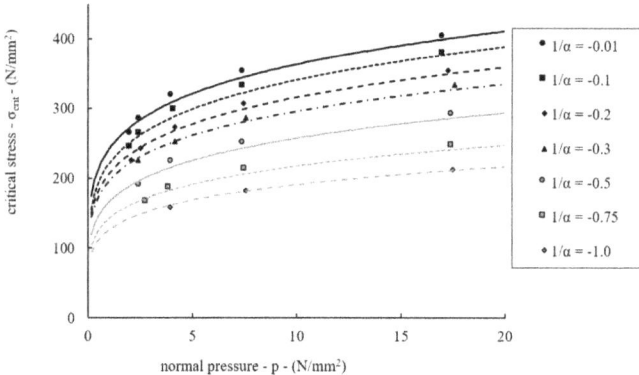

Fig. 2. The critical stress – critical pressure functions for different stress ratios.

Now it can be observed that α = -1/0.01 belongs to the nearly pure compression (the denominator cannot be zero) that exists on the outer sheet edge. With α = -1/1.00, the tension stress component is the same than the compression stress. At the surrounding regions of the inner edge of a drawn part, α = -1/0.20...+1/2.00 are typical values according to finite element simulations.

Considering the above phenomenon, the critical stress can therefore be determined for any blank holder pressure and for any stress ratio. Since the stress ratio is known for all coordinate points of a real workpiece in a finite element simulation, the risk of wrinkling can now be expressed as the function of the location coordinate. This does not only mean that the process monitoring can be achieved after implementing the curves of Fig. 2 into a finite element code (which is very time

Material Forming - ESAFORM 2023   Materials Research Forum LLC
Materials Research Proceedings 28 (2023) 761-770   https://doi.org/10.21741/9781644902479-83

consuming and complicated), but a clear and well-understandable limit diagram can also be created similar to the FLD. To do this, first, the stress ratio dependent relationship between the critical stress and the normal pressure should be determined analytically. Based on our experiences, it can be solved by simple power laws (Eq. 11), since the normal pressure is directly dependent on the amount of the plastic strain (see Eq. 3 – 5).

$$\sigma_{crit}(\alpha) = \gamma \cdot p^\zeta \tag{11}$$

In Eq. 11, the stress state dependent critical wrinkling stress is only expressed by two parameters. This model (continuous lines in Fig. 2) has relatively good agreement with the calculated points (dots in Fig. 2) in the practical range of the blank holder pressure. The regression coefficients of the power law curves are about $R^2 = 0.984...0.994$ for the different stress states.

Knowing the critical stress and the stress ratio, the associated, principal stress components can be calculated using an arbitrary yield criterion. Here we assume plain stress state, since the binder pressure is close to zero in practice ($\sim 2 ... 8 \frac{N}{mm^2}$), thus it is much smaller than the tension (σ_1) and much higher than the compression ($-\sigma_2$) stresses (only exception is the outer edge where $\sigma_1 \approx 0$). Because Wang and Cao used the Hill48 quadratic yield function during the derivations of the energy terms, it is obvious to use it now, too. Although, there is not much difference in using other yield function, for example Yld89, as we shall see later.

Hill derived his suggestion for the yielding of anisotropic sheet metals on the basis of the von Mises yield criterion [1]. If the anisotropy axes coincide with the principal axes, the proposed model in plane stress state can be expressed in the following form:

$$2f(\boldsymbol{\sigma}) = H(\sigma_2 - \sigma_1)^2 + G\sigma_1^2 + F\sigma_2^2 = 1 \tag{12}$$

The relationship between the yield constants and the plastic strain ratios can be obtained by using the associated flow rule,

$$\frac{H}{G} = r_0; \frac{H}{F} = r_{90} \tag{13}$$

The yielding point at uniaxial loading ($\bar{\sigma} = \sigma_1 \ and \ \sigma_2 = 0$) can be taken into account as

$$2f(\boldsymbol{\sigma}) = H\left(1 + \frac{1}{r_0}\right)\bar{\sigma}^2 \tag{14}$$

and hence the yield criterion can be defined through the plastic strain ratios:

$$\bar{\sigma}^2(r_0 + 1) = r_0(\sigma_2 - \sigma_1)^2 + \sigma_1^2 + \frac{r_0}{r_{90}}\sigma_2^2 \tag{15}$$

After some rearrangement and introducing α, we get

$$\bar{\sigma} = \sigma_1 \sqrt{\left\{1 + \alpha^2 \left[\frac{r_0(1+r_{90})}{r_{90}(1+r_0)}\right] - 2\alpha \frac{r_0}{1+r_0}\right\}} \tag{16}$$

Finally, the minor principal stress can be obtained by Eq. 9.

In the case of the Yld89 yield criterion, the determination of the principal stress components is originated from the equation of

$$f(\boldsymbol{\sigma}) = a|K_1 + K_2|^M + a|K_1 - K_2|^M + c|2K_2|^M = 2\bar{\sigma}^M \tag{17}$$

in which

$$K_1 = \frac{\sigma_1 + h\sigma_2}{2} \tag{18}$$

$$K_2 = \sqrt{\left(\frac{\sigma_1 - h\sigma_2}{2}\right)^2 + p^2\sigma_{12}^2} \tag{19}$$

If the coordinate system coincides with the principal directions, the shear stress and hence $p^2\sigma_{12}^2$ can be neglected, and the terms in brackets are simplified as follows:

$$K_1 + K_2 = \sigma_1$$

$$K_1 - K_2 = h\sigma_2 \tag{20}$$

Incorporating Eq. 20 into Eq. 17, the Yld89 yield theory will reduce to

$$f(\boldsymbol{\sigma}) = a|\sigma_1|^M + a|h\sigma_2|^M + c|\sigma_1 - h\sigma_2|^M = 2\bar{\sigma}^M \tag{21}$$

and the relationship between the major principal stress and the equivalent stress can be expressed using the stress ratio (α) again:

$$\sigma_1 \left\{ \frac{1}{2}[a + a|h\alpha|^M + c|1 - h\alpha|^M] \right\}^{\frac{1}{M}} = \bar{\sigma} \tag{22}$$

The constants a, c and h can be determined from the strain ratios [14], while the minor principal stress component can be calculated from Eq. 10 again.

As a result of the derivations detailed above, the wrinkling criterion can be edited in the stress space analytically, which, like the FLD, provides a transparent method for process monitoring. As an example, the results obtained by the principal stresses' calculations for both the Hill48 and the Yld89 criteria can be seen in Fig. 3.

To this figure, the mechanical parameters of the applied DC04 sheet were characterized by tensile tests, and the necessary values are summarized in Table 1. (The data acquisition is not detailed here.)

Table 1. applied material parameters.

	K	ε_0	*n*	r_0	r_{90}
DC04	578	0.017	0.220	1.823	2.380

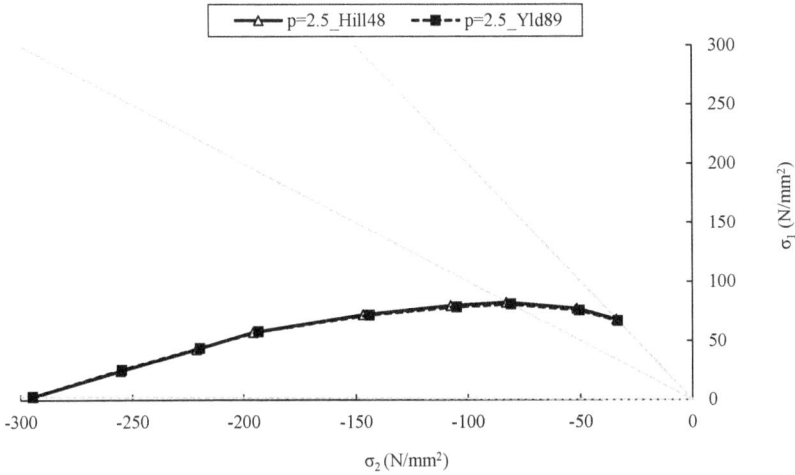

Fig. 3. Stress-based wrinkling limit curves calculated based on the Hill48 and the Yld89 yield criteria. The normal pressure value is 2.5 N/mm² for both cases.

It is worth mentioning that the calculated curves in Fig. 3 belong to $p = 2.5 \frac{N}{mm^2}$ normal pressure. Since it was introduced that the wrinkling limit stress is the function of the blank holder pressure, the wrinkling limit diagram should be interpreted as a curve family, instead of one curve. With the shifting of the limit curve, both the effect of the stress state and the blank holder pressure can be visualized, as well as the limit values can be compared with numerical results. Fig. 4 shows the calculated limit curves of the DC04 sheet for three different blank holder pressures. In the graph, numerically obtained stress points from the outer edge of a drawn cup are also seen, just after the time step when wrinkles appear in the simulation at lower blank holder forces.

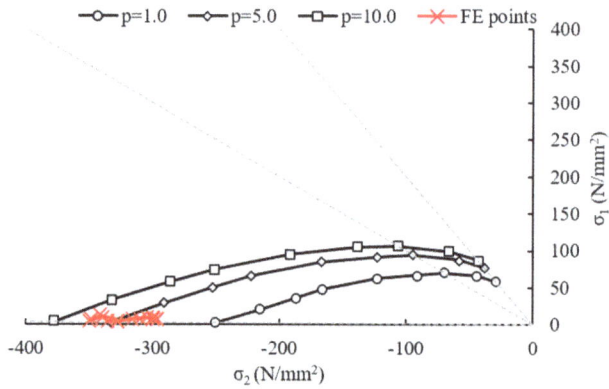

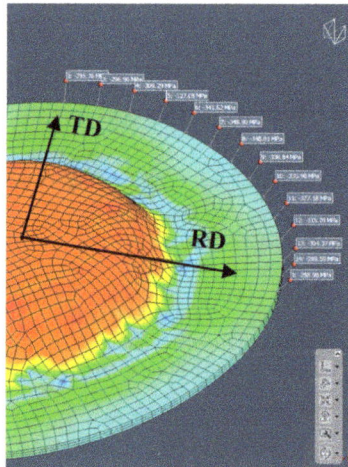

Fig. 4. Stress-based wrinkling limit curves for three different blank holder pressures and numerical points from a drawn part's flange. The pressure values are given in N/mm² in the graph, the applied blank holder pressure was 5 N/mm² in the simulation (no visible wrinkles).

The simulation was carried out in Simufact Forming 2021.1 software. The input material parameters are listed in Table 1, next to which Coulomb friction coefficient was used with the value of 0.12 on the die side and 0.20 on the punch side. The initial element size was 1.4 mm, and the sheet was discretized to three layers in the thickness direction that resulted in a total number of 6696 sheet mesh elements for a diameter Ø66 mm initial blank with 1 mm sheet thickness. The simulation results' validation was done by cup drawing tests supported by digital imagine correlation system, but a detailed insight to the measurements is not possible due to the limited content of the paper. The geometrical data of the tests are following. Punch diameter and radius: Ø33.0 mm and 5 mm; die diameter and radius: Ø35.4 mm and 5 mm (1.2 mm clearance). The applied blank holder pressure for the simulation results of Fig. 4 was 5 N/mm².

Material Forming - ESAFORM 2023   Materials Research Forum LLC
Materials Research Proceedings 28 (2023) 761-770   https://doi.org/10.21741/9781644902479-83

Consequently, it can be observed that the numerically obtained points form the outer perimeter of the flange are far above the limit curve of 1 N/mm² blank holder pressure, thus wrinkling is expected next to this value. At the same time, the stress points more or less fall below the limit curve of 5 N/mm², and wrinkles exist neither in the simulation, nor in the practice with this setting. Naturally, this analysis can be performed for any location point of a drawn part, although above α = 1/1.0 stress ratio, we do not think it make sense in terms of wrinkling.

It is also visible that the critical stress seems to be reached earlier at the inner region of the flange, since the decrease of the stress ratio causes the decrease of the compression stress, too. In this way authors recognize that this may lead to controversy and are currently working on the possible development of the theory.

Summary

With the calculation and then the allocation of the critical equivalent stress that causes wrinkling at a given blank holder pressure as indicated in this article, an easy-to-understand diagram was edited in the major and minor stresses' coordinate system. This diagram is logically consistent with the forming limit curve (either strain-, or stress-based), since it is also stress state dependent, but takes into consideration the blank holder pressure, too. Despite the simplifications during the calculations, the results that show the wrinkling risk at any specimen locations are comparable to the practical blank holder values. In addition, applying this criterion in a finite element code, the process monitoring from the perspective of wrinkling could become possible.

References

[1] D. Banabic, H.J. Bunge, K. Pöhlandt, A.E. Tekkaya, Formability of Metallic Materials, Springer-Verlag Berlin Heidelberg, 2000.

[2] X. Wang, J. Cao, An Analytical Prediction of Flange Wrinkling in Sheet Metal Forming, J. Manuf. Process. 2 (2000) 100-107. https://doi.org/10.1016/S1526-6125(00)70017-X

[3] J. Cao, X. Wang, An analytical model for plate wrinkling under tri-axial loading and its application, Int. J. Mech. Sci. 42 (2000) 617-633. https://doi.org/10.1016/S0020-7403(98)00138-6

[4] S. Keeler, W.A. Backofen, Trans. ASM 56 (1963) 25-48.

[5] G.M. Goodwin: Society of Automotive Engineers No. 680093 (1968) 380-387.

[6] M. Merklein, A. Kuppert, M. Geiger, Time dependent determination of forming limit diagrams, CIRP Annals – Manuf. Technol. 59 (2010) 295–298. https://doi.org/10.1016/j.cirp.2010.03.001

[7] D. Banabic, F. Barlat, O, Cazacu, T. Kuwabara, Advances in anisotropy and formability, Int. J. Mater. Form. 3 (2010) 165-189. https://doi.org/10.1007/s12289-010-0992-9

[8] T.B. Stoughton, X. Zhu, Review of theoretical models oft he strain-based FLD and their relevance to the stress-based FLD, Int. J. Plast. 20 (2004) 1463-1486. https://doi.org/10.1016/j.ijplas.2003.11.004

[9] T.X. Yu, W. Johnson, The buckling of annular plates in relation to the deep-drawing process, Int. J. Mech. Sci. 24, (1982) 175-188. https://doi.org/10.1016/0020-7403(82)90036-4

[10] A. Agrawal, N.V. Reddy, P.M. Dixit, Determination of optimum process parameters for wrinkle free products in deep drawing process, J. Mater. Process. Technol. 191 (2007) 51–54. http://doi.org/10.1016/j.jmatprotec.2007.03.050

[11] J.W. Hutchinson, K.W. Neale, Proceedings of Int. Symp. on Plastic Instability, Paris, France (1985) 1841–1914.

[12] H.W. Swift, Plastic instability under plane stress, J. Mech. Phys. Solid. 1 (1952) 1-18. https://doi.org/10.1016/0022-5096(52)90002-1

Material Forming - ESAFORM 2023
Materials Research Proceedings 28 (2023) 761-770

Materials Research Forum LLC
https://doi.org/10.21741/9781644902479-83

[13] R. Hill, A theory of the yielding and plastic flow of anisotropic metals, The hydrodynamics of non-Newtonian fluids I, Proceedings of the Royal Society of London. Series A, Mathematical and Physical Sciences 193 (1948) 281-297.

[14] F. Barlat, J. Lian, Plastic behavior and stretchability of sheet metals. Part I: A yield function for orthotropic sheets under plane stress conditions, Int. J. Plast. 5 (1989) 51-66. https://doi.org/10.1016/0749-6419(89)90019-3

Material Forming - ESAFORM 2023
Materials Research Proceedings 28 (2023) 771-778

Materials Research Forum LLC
https://doi.org/10.21741/9781644902479-84

Evaluation of hot and warm forming of age-hardenable aluminium alloys into manufacturing of automotive safety critical parts

MYROLD Benedikte[1,a] *, JENSRUD Ola[1,b] and HOLMESTAD Jon[1,c]

[1]SINTEF Manufacturing, Box 163, 2831 Raufoss, Norway

[a]benedikte.myrold@sintef.no, [b]ola.jensrud@sintef.no, [c]Jon.Holmestad@sintef.no

Keywords: Hot and Warm Forming, Formability, Hardening of Aluminium Alloys

Abstract. Lightweight solutions will become increasingly necessary for the next generation of passenger cars. Applications of age-hardenable high strength aluminium material can be superior to traditional low strength alloys, especially if the processing of the parts itself is cost effective. Sheet forming of aluminium at elevated temperatures has been addressed in the current work. Warm forming directly after solution heat treatment can eliminate several process steps and reduce cycle time in the production of light weight load bearing car components. However, this is not straight forward as 6xxx alloys need minimum critical cooling rate to achieve the full potential of precipitation hardening. The technical forming methods must take this into the thermomechanical setup during the process steps of handling, lubrication, deformation, and final ageing. Elevated forming temperatures, more likely warm than hot show high formability of the alloys. Further, a short artificial ageing time demonstrate required strength and fatigue capacity in real components. Reduction of the deformation temperature to below 300°C is still favourable when lubrication for serial production is the issue. The modified thermomechanical method is appropriate for achieving accurate geometrical tolerances, uniform properties, and high productivity for aluminium automotive parts. The paper describes a feasibility study and method development for forming of sheet material of hardenable aluminium alloys at intermediate temperatures.

Introduction

Forming at elevated temperatures (>500°C) has been demonstrated earlier by reference [1] and [2] and showed advantages compared to traditional cold forming with respect to spring back, formability, and microstructural control. Forming of aluminium blanks is commonly carried out cold in soft annealed temper (O-temper), which provides very good and tight tolerances. Annealing for O-temper is time consuming and thus requires high energy consumption. Additionally, O-temper parts needs new heat treatment to reach T6 properties with challenges to keep tolerances and microstructure within expectations. Jensrud et al. developed a new profile shaping technique for manufacturing of automotive components in as quenched condition (W-temper). This gives a good formability and a reduction in processing time due to in-line heat treatment integrated with the mechanical operation [3]. This method is a measure of efficiency.

Material Forming - ESAFORM 2023   Materials Research Forum LLC
Materials Research Proceedings 28 (2023) 771-778   https://doi.org/10.21741/9781644902479-84

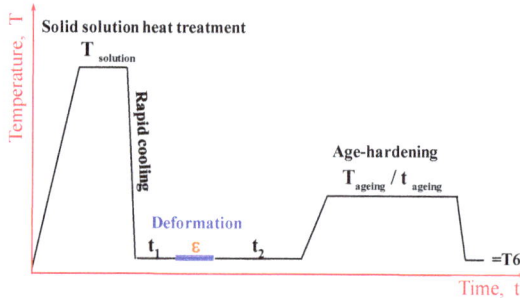

Fig. 1. The process route of forming in W-temper [5].

Forming prior to age hardening may also advantageously be carried out at elevated temperatures. Hot forming of aluminium has been widely explored and gives a further improved formability and productivity. A process with integrated forming and hardening has been proposed where the blank is simultaneously formed and quenched from solution heat treatment before artificial ageing [1,4]. This proved to be an effective process in terms of formability, efficiency, spring-back and properties. However, challenges with lubrication are encountered that results in sticking and thinning.

This research has been conducted in the framework of the above-mentioned work, except that hot forming is replaced by warm forming. Hot forming is defined as deformation at elevated temperatures below the recrystallization temperature [5]. This new proposal includes the integrated forming and quenching step with the alteration that the forming and final quenching happens as the material has reached an intermediate temperature of 200-300°C. Warm forming will provide a more controllable friction regime compared to hot forming and still obtains a low flow restriction and sufficient ductility. The deformation hardening mechanism combined with precipitation hardening is not well understood. The key mechanisms in age hardening of alloys deformed in W-temper are the interactions between dislocations, vacancies and formation of Guinier Preston (GP) zones, i.e. the first stage of precipitation hardening [6,7]. The contribution of each mechanism in age hardening after warm forming should be considered. As self-diffusion of dislocations increases with increasing temperatures, it would be reasonable to believe that the dislocations are annihilated during warm forming and therefore does not provide any strength contribution. However, whether the exposure time to high temperatures after forming are long enough for the dislocations to annihilate is not known.

In this paper, the quench sensitivity of relevant alloys has been carefully considered to obtain the required mechanical properties of load bearing automotive components. The limitations and latitude of the quenching step has been identified.

The sheet material evaluated is within AA6010 with a thickness of 5.3 mm. In this work it has been developed a process by a press quench tooling set-up integrated with heat treatment capability. The formability of the sheets is investigated by forming experiments performed in flat tool with subsequent testing of mechanical properties [8]. In this thermo-mechanical process, the solution heat treatment sequence of the material is a pre-process of the blank before the forming step with direct ageing from the press operation as the last step.

Material Forming - ESAFORM 2023
Materials Research Proceedings 28 (2023) 771-778

Materials Research Forum LLC
https://doi.org/10.21741/9781644902479-84

Method Development

The experimental setup requires proper control of time and temperature. During this method development, the cooling rate was measured with a thermo-logger that collected data with 0.1s intervals before the data was used for time control of the cooling. To select parameters for the forming trials, several experiments were designed and conducted at a laboratory scale. The experiments were intended to map cooling rate, quench sensitivity and resulting properties of the selected material. The cooling rates are so fast that it is difficult to manually stop the cooling or form at the correct temperatures during cooling. A manageable cooling rate had to be found that would not compromise the hardening of the material. Small sheet samples were cooled with different media; water, steel, compressed air, still air and insulated material. It was found that cooling the sheet between two steel plates gives a suitable cooling rate for the lab simulation of the process. The measured cooling rates for the different cooling media are listed in Table 1.

Table 1. Measured cooling rates for samples cooled from SHT in different cooling media.

Cooling medium	Cooling rate (°C/s)
Water	310
Closed steel	37
Open steel	18
Compressed air	10
Insulation	1.7

The selected cooling method still gives a high cooling rate so that manual handling of the sample during the experiments is difficult. To enable forming at elevated temperatures the cooling rate had to be slightly reduced to maintain temperature control when approaching the forming temperatures of 200-300°C, this can be achieved by cooling the samples in heated steel plates. A cooling tool was built as two steel plates with hinges and a handle that is heated in the oven to 100°C. After solution heat treatment, the samples are placed between the heated steel plates and cooled to the selected intermediate temperatures before being transported to the press for deformation and subsequent ageing. This process was validated by comparing the mechanical properties with a traditional process route shown in Table 2. The mechanical properties are not severely affected with the new proposed process route. Illustrations of the hot forming process route and the warm forming process route are shown in Fig. 2 and Fig. 3, respectively.

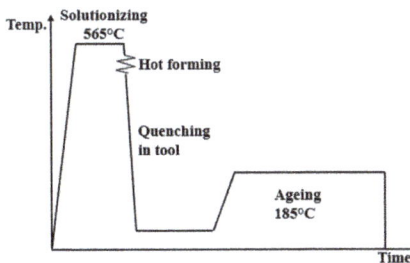

Fig. 2. Process route for hot forming.

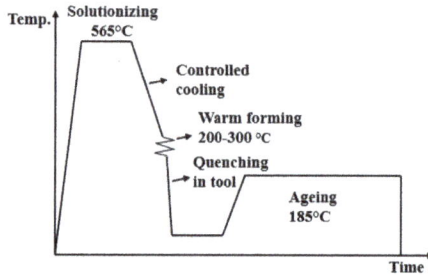

Fig. 3. Process route for warm forming.

*Table 2. Comparison of mechanical properties after hot forming, warm forming and T6
condition with no deformation as reference.*

	Yield strength (MPa)	Tensile strength (MPa)	Ag (%)	A (%)	Hardness HV10
Warm forming route	360	382	7.1	11.5	132
Hot forming route	354	373	7.9	13.9	130
Reference	364	382	7.5	12.6	135

At this point most process parameters have been selected and verified. The parameters for solution heat treatment, natural ageing and artificial ageing were selected based on the current industrial processes. Furthermore, the optimal forming temperature had to be found by investigation of all aspects of the material response to the process of forming at intermediate temperatures. The following experiments were conducted to find the influence on strength, ductility, anisotropy, and formability. After solution heat treatment, samples were cooled in a suitable manner and formed at 25°C (water quench), 100°C (between dies in the press), 200°C (cooling tool), 300°C (cooling tool) and 400°C (dies in the press). The samples were formed in a flat pressing tool by compression of 15% with a forming speed of 59 mm/s.

After solution heat treatment the different conditions were analysed by tensile testing, shear testing and tear testing. The presented results are the averages from 3 parallel samples with standard deviations represented by error bars.

Results and Discussion
The tensile test results from samples being formed at different temperatures are shown in Fig. 4 and Fig. 5. The results show stable tensile strength and significant variations in yield strength. The elongation does not seem to have been severely affected by the different cases. This indicates that there is no significant reduction in tensile strength due to an altered cooling sequence. The contributions of hardening mechanisms are complex, and it would require TEM investigations to identify the strengthening precipitates for each deformation temperature. As the forming temperatures are varying, the cooling sequences are also different. The precipitation mechanisms might differ for the different conditions. It is reasonable to assume that the distributions of precipitation and dislocations vary with temperature, which may explain the reduction in yield strength for the case of deformation at 200°C. However, for this alloy and the tested conditions the varying contribution of mechanisms do not seem to have diminished the final properties of the material.

Material Forming - ESAFORM 2023
Materials Research Proceedings 28 (2023) 771-778

Materials Research Forum LLC
https://doi.org/10.21741/9781644902479-84

Another possible explanation to the small variations in the final properties of the material after deformation at elevated temperatures might be the high cooling rate during forming. The temperatures of the material at the end of deformation are not known. The cooling rate between cold dies and with applied force is high, and there is a possibility the samples have reached room temperature before the deformation is finished. In that case, the same hardening mechanisms would apply for all cases. However, with a die speed of 59mm/s and a deformation of 15%, the deformation is finished after 0.01s. This is not enough time for the samples to reach room temperature before the deformation is complete. Although, if the material temperature would be low at the end of deformation, the same would apply in a forming operation of larger scale as the sheet thickness would be the equal and have the same temperature gradient as in these experiments.

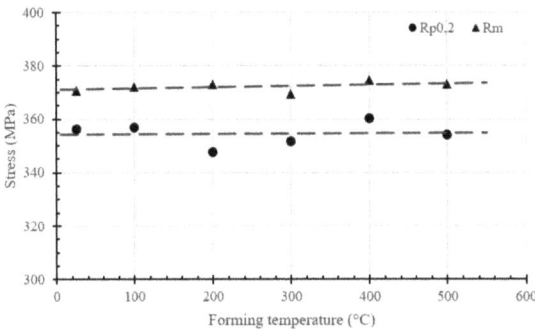

Fig. 4. Results from tensile tests of different forming conditions. Yield strength ($R_{p0.2}$) and tensile strength (R_m).

The elongation shown in Fig. 5 is not significantly affected by deformation temperature compared to cold forming. Ideally, the ductility should be improved at elevated forming temperatures, but it seems that hot forming is required for a slight increase in elongation. However, it is beneficial that severe reduction in elongation is not observed.

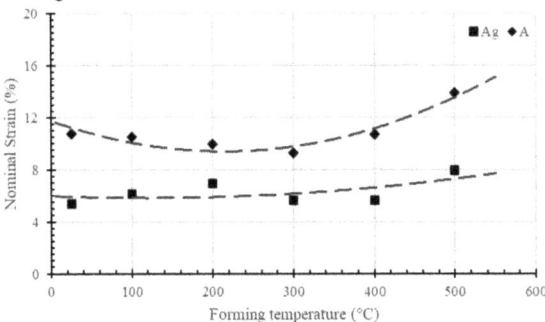

Fig. 5. Results from tensile tests of different forming conditions. Fracture elongation (A) and elongation at maximum force (Ag).

Shear testing were performed according to ASTM B831-05. A notched, rectangular specimen is subjected to a single shear force to failure in a fixture using a tensile test machine. The shear

Material Forming - ESAFORM 2023   Materials Research Forum LLC
Materials Research Proceedings 28 (2023) 771-778   https://doi.org/10.21741/9781644902479-84

strength is the maximum force required to fracture the sample. Three parallels were tested for each condition. The results from shear testing are shown in Fig. 6, i.e. the average shear strength with standard deviation for the different quenching and forming conditions. The shear test shows the fracture shear strength. Insignificant differences are observed for the shear strength of the different conditions. There is a large variation for the undeformed condition which was water quenched.

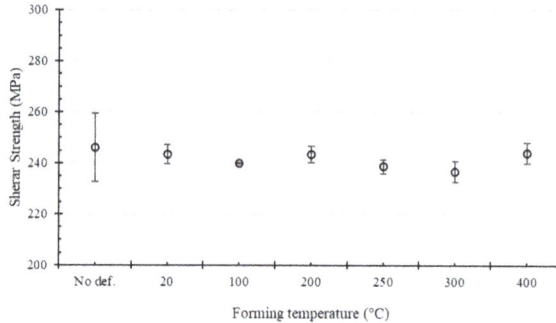

Fig. 6. Shear test results from samples deformed at different temperatures. Average shear strength with standard deviation.

Table 3. Results from shear tests.

Shear tests		
Forming temperature	Cooling medium	Shear Strength (MPa)
No def.	Water	246
20°C	Water	243
100°C	Cold dies	240
200°C	Cooling tool	243
250°C	Cooling tool	239
300°C	Cooling tool	237
400°C	Cold dies	244

Tear tests were performed according to ASTM B871-1 and involves a single edge notched specimen that is statically loaded through pin loading holes. The test is an indicator of toughness and provides a comparative measure of resistance to unstable fracture. The tear test shows the ability of a material to absorb plastic deformation and can reveal significant reduction in toughness. The tear strength shown in Fig. 7 is reduced for intermediate forming temperatures. The samples being water quenched and quenched in a cold tool have a higher tear strength than the samples being pre-cooled in the cooling tool. This indicates that the cooling rate affects the toughness of the material as the cooling rate is slower for the samples cooled in the cooling tool.

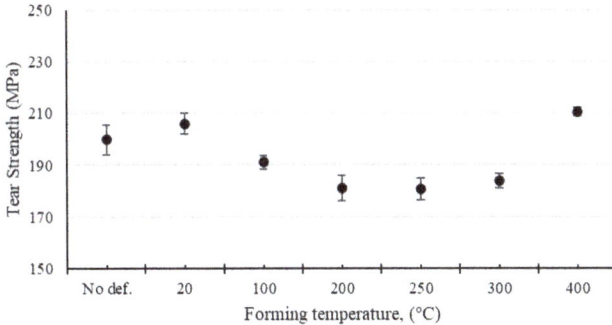

Fig. 7. Tear test results from samples formed at different temperatures. Average tear strength with standard deviation.

Table 4. Tear test results, average tear strength.

Tear tests		
Forming temperature	Cooling medium	Tear Strength (MPa)
No def.	Water	200
20°C	Water	206
100°C	Cold dies	191
200°C	Cooling tool	181
250°C	Cooling tool	181
300°C	Cooling tool	184
400°C	Cold dies	210

When comparing the tear test results with the tensile test results, the reduction in tear strength applies to the same conditions that show increased work hardening. Hence, the total performance of the material at all forming temperatures argue for a capable method and material. There are no test results that suggest that the proposed method inflict detrimental effects on the alloy properties. However, further testing of the method will be conducted, including hot tensile testing, microstructure investigations and additional alloys.

Summary

A method for warm forming of an AA6010 sheet material at intermediate temperatures have been developed. The proposed process route results in acceptable material properties in terms of strength, ductility, and toughness. The high strength material holds a potential of above 350 MPa in tensile strength, which is well above the common requirement for age hardenable aluminium alloys. In addition to satisfying material properties, the method enables sheet forming with good formability and extended possibilities in terms of lubrication.

Acknowledgement

The research has been carried out as a part of the Norwegian Research Council funded BIA-IPN project SUFICCS (Superior Fatigue Stressed Chassis Components). The project involves the following partners: Raufoss Technology AS, Hydro Aluminium AS, AP&T, SINTEF Manufacturing AS and SINTEF Industry. The experiments were carried out at SINTEF Manufacturing AS.

References

[1] R.P. Garrett, J. Lin, T.A. Dean, Solution Heat treatment and Cold Die Quenching in Forming AA 6xxx Sheet Components: Feasibility Study, Adv. Mater. Res. 6-8 (2005) 673-680. http://doi.org/10.4028/www.scientific.net/AMR.6-8.673

[2] O. Jensrud, K.E. Snilsberg, A. Kolbu, Conceptual testing of Press Form Hardening of High-Strength Aluminium Alloys, Aluminium Extr. Finish. (2016) 16-20.

[3] O. Jensrud, T. Høiland, A new advanced profile shaping technique for design and manufacturing of automotive components, The ExtruForm®, ET08 Orlando 13-16. May 2008, Proceedings of the ET08, Vol. 2 2008.

[4] B. Myrold, O. Jensrud, K.E. Snilsberg, The Influence of Quench Interruption and Direct Artificial Ageing on the Hardening Response in AA 6082 during Hot Deformation and In-Die Quenching, Metals 10 (2020) 935. https://doi.org/10.3390/met10070935

[5] ASM International, Metals Handbook, Vol. 14: Forming and Forging, 1988.

[6] O. Jensrud, K. Pedersen, Warm forming as an energy efficient process in manufacturing of high strength aluminium parts, Metall. Sci. Technol. 31 (2013).

[7] A. Latkowski, J. Grytiecki, M. Bronicki, Arch. Metal. 32 (1989) 313-325.

[8] K.E. Snilsberg, T. Welo, K.E. Moen, B. Holmedal, O. Jensrud, C. Koroschetz, A New Tribological System Test for Integrated Hot Forming and Die Quenching of Aluminium Alloy Sheets, ESAFORM Dublin, 2017.

Material Forming - ESAFORM 2023
Materials Research Proceedings 28 (2023) 779-786

Materials Research Forum LLC
https://doi.org/10.21741/9781644902479-85

Nonlinear stress path experiment using mild steel sheet for validation of material model

TAKADA Yusuke[1,a] and KUWABARA Toshihiko[1,b] *

[1]Tokyo Univ. Agriculture and Technology, 2-24-16 Nakacho, Koganei-shi, Tokyo 184-8588, Japan

[a]takada.no12@gmail.com, [b]kuwabara@cc.tuat.ac.jp

Keywords: Mechanical Test, Biaxial Stress, Mild Steel Sheet, Yield Function, Linear Stress Path, Nonlinear Stress Path, Contour of Plastic Work, Biaxial Tube Expansion Test

Abstract. A linear stress path (LSP) experiment was performed using uniaxial and biaxial tensile tests with a cold-rolled mild steel sheet (SPCD; nominal thickness: 0.8 mm) as the test material. In the LSP experiment, nine LSPs were applied to the specimens to measure the contours of plastic work and the directions of the plastic strain rates, β, for a plastic strain range of $0.002 \le \varepsilon_0^\mathrm{p} \le 0.234$. Then, the Yld2000-2d yield function (Barlat et al., 2003) was used to identify a material model that accurately reproduces the experimental data observed in the LSP experiment. Furthermore, a nonlinear stress path (NLSP) experiment was performed. The NLSP consists of two linear stress paths with $\sigma_x : \sigma_y$ = 4:1 and 1:1, and a curved stress path connecting the LSPs. The measured work hardening behavior and β values were compared with those calculated using the Yld2000-2d yield function identified from the LSP experiment. It was found that the deformation behavior of the test sample predicted by the material model determined from the LSP experiment clearly shows some deviation from that observed for the NLSP experiment.

Introduction

One of the most influential factors that affect the accuracy of sheet metal forming simulations is a material model. In order to guarantee the accuracy of the forming simulation, it is necessary to guarantee the accuracy of the material model used for the analysis. Guaranteeing the accuracy of the material model means that the deformation behavior of the material in real forming operations should be reproduced by the material model.

To experimentally determine a material model, many linear stress paths (LSPs) are applied to the material using biaxial tensile tests with cruciform test pieces [1,2] and/or the biaxial tube expansion test (BTET) with tubular specimens, to which an axial force and an internal pressure are simultaneously applied [3]. However, the stress paths in real sheet metal forming processes are generally nonlinear. Therefore, verifying that the material model determined using an LSP experiment can reproduce the deformation behavior of the material under nonlinear stress paths (NLSPs) is crucial for establishing an accurate material model.

Many studies have investigated the deformation behavior of materials subjected to NLSPs. However, most of them measured the subsequent yield surface of the materials by first applying a specific plastic strain and then, after unloading, subsequent loading paths in loading directions different from the pretrain direction to the test sample [4-8]. Of note, such NLSP experiments have been utilized to verify the validity of distortional plasticity models [9-13]. There have been several studies that have used NLSPs without unloading to verify the accuracy of material models for test samples [14-17]. However, the plastic strain applied to the test samples in these studies are less than several percent.

Takada and Kuwabara [18] investigate the deformation behavior of a cold-rolled mild steel sheet for deep drawing (SPCD) subjected to LSPs and NLSPs. The deformation behavior of the test sample is precisely measured until specimen fracture. The objective of this study is to clarify

Material Forming - ESAFORM 2023
Materials Research Proceedings 28 (2023) 779-786

Materials Research Forum LLC
https://doi.org/10.21741/9781644902479-85

whether the deformation behavior of a test sample under a NLSP can be reproduced using the material model determined from an LSP experiment. In this study the deviation of the NLSP from the LSP is much larger than that used in [18].

Experimental Method

The test material used in the present study was a 0.8-mm-thick mild steel sheet, SPCD, with deep drawing quality. The work hardening characteristics and r-values at $0°$, $45°$, and $90°$ (transverse direction, TD) with respect to the rolling direction (RD) are shown in Table 1. It is the same material as used in [18].

Table 1. Mechanical properties of test material (SPCD).

Tensile direction /°	$\sigma_{0.2}$ /MPa	ε^{p}_{TS} *	c^{**} /MPa	n^{**}	α^{**}	r-value***
0	146	0.241	560	0.269	0.0030	1.91
45	158	0.227	586	0.267	0.0038	1.63
90	155	0.236	563	0.271	0.0038	2.25

* Logarithmic plastic strain giving the maximum tensile load.
** Approximated using $\sigma = c(\alpha + \varepsilon^{p})^{n}$ for $\varepsilon^{p} = 0.002 \sim \varepsilon^{p}_{TS}$
***Measured at uniaxial nominal strain of $\varepsilon_{N}=0.1$

Two types of biaxial tensile test were performed to measure the plastic deformation behavior of the test material from initial yield to fracture. Fig. 1 (a) shows a schematic diagram of the cruciform specimen used for the biaxial tensile tests with the as-received sheet sample. The geometry of the specimen is that recommended in ISO 16842 [19]. The specimen arms were parallel to the RD and TD of the material. Each arm of the specimen had seven slits (length: 30 mm, width: 0.2 mm) at 3.75-mm intervals to remove the geometric constraint on the deformation of the 30×30 mm2 square gauge area. The slits were fabricated using laser cutting.

Fig. 1 (b) shows a schematic diagram of the tubular specimen used for the BTET. The specimens were fabricated by bending a sheet sample into a cylindrical shape and CO2 laser welding the sheet edges together to fabricate a tubular specimen with an inner diameter of 47.1 mm, a length of 230 mm, and a gauge length (distance between the grips of the testing machine) of 150 mm. Two types of tubular specimen were fabricated. For type I specimens, the RD was in the axial direction; for type II specimens, the RD was in the circumferential direction. Type I specimens were used for the tests with $\sigma_x \leq \sigma_y$ and type II specimens were used for the tests with $\sigma_x \geq \sigma_y$; the maximum principal stress direction was always taken to be the circumferential direction. Because the measurement results for the BTET include the effect of prestrain due to bending during the preparation of the circular tube test pieces, they were corrected using the data obtained from the cruciform test piece. For details, refer to [3].

Material Forming - ESAFORM 2023   Materials Research Forum LLC
Materials Research Proceedings 28 (2023) 779-786   https://doi.org/10.21741/9781644902479-85

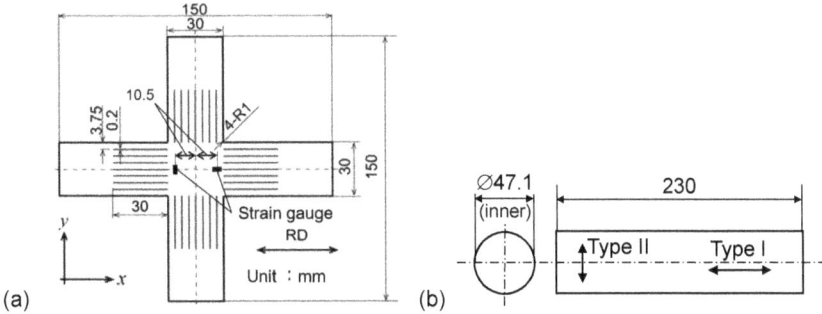

Fig. 1. Specimens used for biaxial tensile tests [19]. (a) Cruciform specimen and (b) tubular specimen. In (b), ↔ indicates the RD of the original sheet sample. The RD is taken in the axial direction for type I specimens and in the circumferential direction for type II specimens.

For details of the servo-controlled tension-internal pressure testing machine used in the experiment, see [3]. For details of the measurement method of the axial and circumferential strain components, ε_ϕ and ε_θ, the axial radius of curvature, R_ϕ, and the calculation method of the axial and circumferential stress components, σ_ϕ and σ_θ, see [20].

Linear Stress Path Experiment. For the LSP experiment, both cruciform and tubular specimens were subjected to seven LSPs, namely $\sigma_x : \sigma_y$ = 4:1, 2:1, 4:3, 1:1, 1:2, 3:4, and 1:4, where σ_x and σ_y are the true stress in the RD and TD, respectively. For $\sigma_x : \sigma_y$ =1:0 and 0:1, the uniaxial tensile test specimens standardized in JIS13B were used. The equivalent plastic strain rate was controlled to be approximately constant at 5×10^{-4} s^{-1}.

Nonlinear Stress Path Experiment. Fig. 2 shows the NLSP applied to the test sample in the NLSP experiment. The NLSP was determined by assuming that the material deforms following the IH model based on the Yld2000-2d yield function [21] as determined in Fig. 3. The first stress path was the LSP with $\sigma_x : \sigma_y$ = 4:1. When ε_0^p reached 0.07, the stress path was changed to the second loading. ε_0^p gradually increased along the second loading. Then, when ε_0^p reached 0.09, the stress path followed the LSP with $\sigma_x : \sigma_y$ =1:1. The equivalent plastic strain rate was controlled to be approximately constant at 5×10^{-4} s^{-1}. Two specimens, Exp.1 and Exp.2, were used for the NLSP experiment.

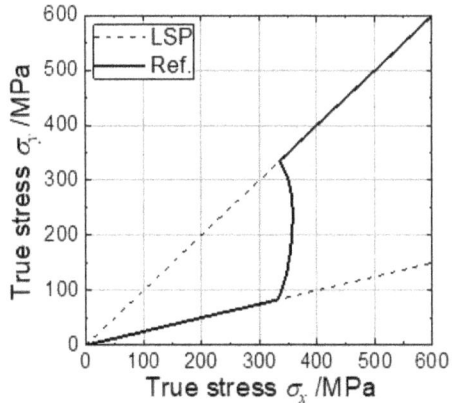

Fig. 2. Nonlinear stress path.

In the NLSP experiment, the evolution of $\varepsilon_x^p(t)$ and $\varepsilon_y^p(t)$ was approximated using a polynomial function for every stress path section between abrupt stress path change points, and the instantaneous value of β was calculated as $\tan^{-1}\{d\varepsilon_y^p(t)/d\varepsilon_x^p(t)\}$.

Experimental Results

The contours of plastic work in the stress space [22,23] were measured to identify appropriate material models for the test samples subjected to uniaxial and biaxial tension. With the true stress-logarithmic plastic strain curve measured for the RD used as reference data for work hardening, the plastic work per unit volume, W_0, and uniaxial true stress, σ_0, associated with particular values of ε_0^p (referred to as the reference plastic strain hereafter) were determined. Next, from the biaxial and uniaxial stress-strain curves, the stress points that give the same plastic work as W_0 were plotted in the principal stress space to determine the contour of plastic work associated with ε_0^p.

Fig. 3 shows the stress points that form work contours. Each stress point represents an average of two specimen data points; the difference between the two points was less than 1.4% of the flow stress for all data points. We measured up to $\varepsilon_0^p = 0.234$ for all LSPs. To quantitatively evaluate the shape change of the work contours with increasing ε_0^p, the nondimensional work contours were determined by dividing the value of the stress points that formed each work contour by the σ_0 value belonging to the work contour. It was confirmed that the stress points along the stress paths of $\sigma_x : \sigma_y = 4{:}1$, 2:1, 4:3, 1:1, and 3:4 do not fall on a single point with increasing ε_0^p; therefore,

the test material exhibited differential hardening (DH). Fig. 3 also includes the yield loci calculated using the selected yield functions. The solid line is that based on the Yld2000-2d yield function, the exponent, M, and parameters, $\alpha_1 - \alpha_6$, of which were determined to approximate the shape of the work contour and the directions of the plastic strain rate, $\mathbf{D}^p$, for $\varepsilon_0^p = 0.10$. The evolution of the work contours with ε_0^p is well reproduced by the DH model based on the Yld2000-2d yield function, as indicated by the red and black dotted lines for $\varepsilon_0^p = 0.002$ and 0.10, respectively.

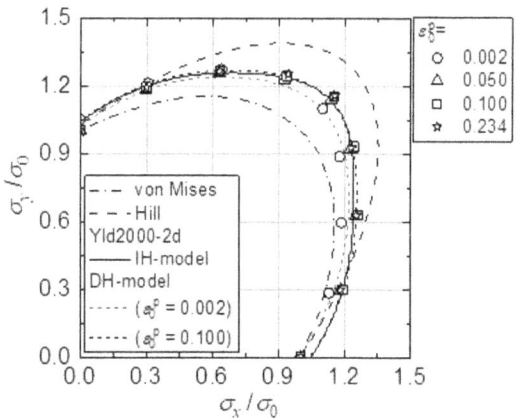

Fig. 3. Measured contours of plastic work normalized by σ_0 belonging to same group of work contours.

Fig. 4 compares the measured directions of $\mathbf{D}^p$, β, with those predicted using the selected yield functions. Again, the measured data are well reproduced by the Yld2000-2d yield function for all loading directions. Moreover, it is noteworthy that the experimental tendency for β to decrease for $\sigma_x{:}\sigma_y = 4{:}3$ and increase for $\sigma_x{:}\sigma_y = 3{:}4$ with increasing ε_0^p was well reproduced by the DH model.

In summary, the evolution of the contours of plastic work and the directions of the plastic strain rates with increasing ε_0^p measured in the LSP experiment are accurately reproduced by the DH model based on the Yld2000-2d yield function.

Material Forming - ESAFORM 2023
Materials Research Proceedings 28 (2023) 779-786

Materials Research Forum LLC
https://doi.org/10.21741/9781644902479-85

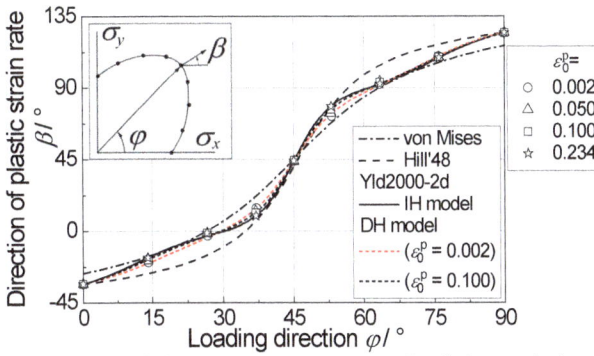

Fig. 4. Measured directions of plastic strain rates compared with those calculated using selected yield functions.

Fig. 5 shows the $\sigma_x - \varepsilon_x^p$ and $\sigma_y - \varepsilon_y^p$ curves observed for the NLSP shown in Fig. 2. Two specimens were used, and the reproducibility of the two tests was good. The small protrusion in the black line was caused by the change in σ_x along the second (curved) stress path.

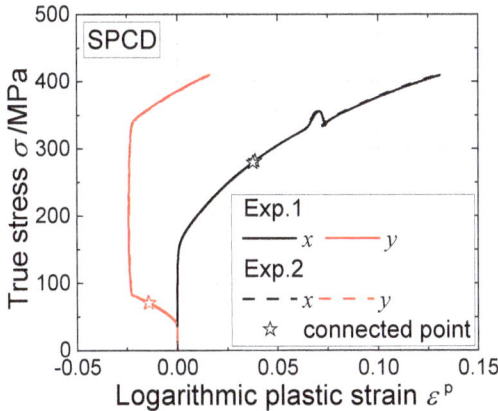

Fig. 5. $\sigma_x - \varepsilon_x^p$ and $\sigma_y - \varepsilon_y^p$ curves observed for the NLSP shown in Fig. 2. Those for the lower strain range than the star marks were measured using a cruciform specimen, and those for the higher strain range than the star marks were measured using the BTET.

Fig. 6 shows the variation of loading direction, φ, with increasing ε_0^p for the stress path shown in Fig. 2. The largest deviation of the experimental values from the prediction by the Yld2000-2d yield function with DH was $\Delta\varphi = 5.8°$ at $\varepsilon_0^p = 0.08$.

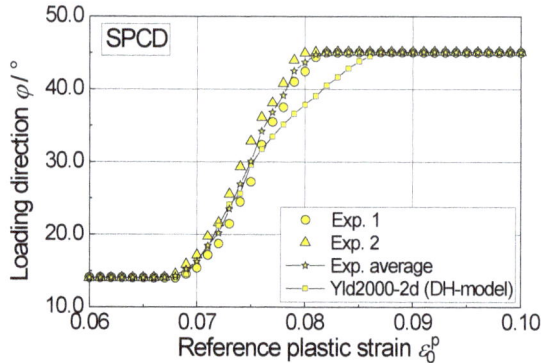

Fig. 6. Variation of loading direction, φ, with increasing ε_0^p for the stress path shown in Fig. 2.

Fig. 7 shows the change in the direction of plastic strain rate, β, with increasing ε_0^p for the stress path shown in Fig. 2. The largest deviation of the experimental values from the prediction by the Yld2000-2d yield function with DH was $\Delta\beta = 20.4°$ at $\varepsilon_0^p = 0.08$. It is noted that β reached 35° at $\varepsilon_0^p = 0.08$, and remained almost constant at $\beta = 35°$ for $\varepsilon_0^p \geq 0.08$; it was 9° lower than the prediction by the Yld2000-2d yield function. Considering that the stress state almost reached equibiaxial tension at $\varepsilon_0^p = 0.08$ in the experiment, this difference in β between the NLSP and LSP was possibly caused by the change in texture during the NLSP for $\varepsilon_0^p \leq 0.08$. Therefore, we can conclude that the deformation behavior predicted by the material model determined from the LSP experiment clearly shows some deviation from that observed for the NLSP experiment.

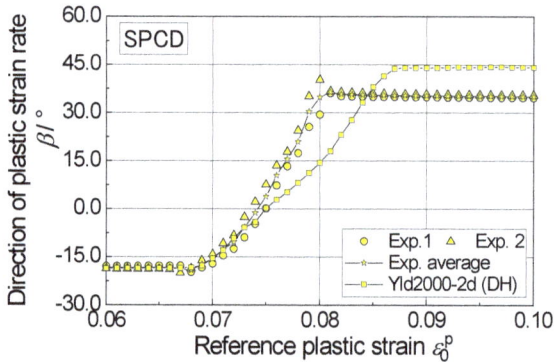

Fig. 7. Change in direction of plastic strain rate, β, with increasing ε_0^p for the stress path shown in Fig. 2.

Summary

The deformation behavior of a mild steel sheet was observed both for the LSPs and NLSP. A proper material model was determined from the LSP experiment using the Yld2000-2d yield

Material Forming - ESAFORM 2023   Materials Research Forum LLC
Materials Research Proceedings 28 (2023) 779-786   https://doi.org/10.21741/9781644902479-85

function. Moreover, a NLSP experiment was performed for the test sample, and the deformation behavior of the test sample during the NLSP experiment was precisely measured, and compared with that predicted using the Yld2000-2d yield function. It was found that the deformation behavior of the test sample predicted by the material model determined from the LSP experiment clearly shows some deviation from that observed for the NLSP experiment.

References

[1] T. Kuwabara, S. Ikeda, K. Kuroda, Measurement and analysis of differential work-hardening in cold-rolled steel sheet under biaxial tension, J. Mater. Process. Technol. 80-81 (1998) 517-523. https://doi.org/10.1016/S0924-0136(98)00155-1

[2] T. Kuwabara, A. Van Bael, E. Iizuka, Measurement and analysis of yield locus and work-hardening characteristics of steel sheets with different R-values, Acta Mater. 50 (2002) 3717-3729. https://doi.org/10.1016/S1359-6454(02)00184-2

[3] T. Kuwabara, F. Sugawara, Multiaxial tube expansion test method for measurement of sheet metal deformation behavior under biaxial tension for a large strain range. Int. J. Plast. 45 (2013) 103-118. http://doi.org/10.1016/j.ijplas.2012.12.003

[4] S.S. Hecker, Yield surfaces in prestrained aluminum and copper, Metal. Trans. 2 (1971) 2077-2086. https://doi.org/10.1007/BF02917534

[5] K. Ikegami, A historical perspective of experimental study on subsequent yield surfaces for metals (part 1), J. Soc. Mat. Sci., Jpn. 24 (1975) 491-504 (in Japanese) https://doi.org/10.2472/jsms.24.491

[6] K. Ikegami, A historical perspective of experimental study on subsequent yield surfaces for metals (part 2), J. Soc. Mat. Sci., Jpn., 24 (1975) 709-719 (in Japanese). https://doi.org/10.2472/jsms.24.709

[7] Y. Ohashi, K. Kawashima, T. Yokochi, Anisotropy due to plastic deformation of initially isotropic mild steel and its analytical formulation, J. Mech. Phys. Solid. 23 (1975) 277-294. https://doi.org/10.1016/0022-5096(75)90029-0

[8] E. Shiratori, K. Ikegami, K. Kaneko, The influence of the Bauschinger effect on the subsequent yield condition, Bull. JSME 16 (1973) 1482-1493. https://doi.org/10.1299/jsme1958.16.1482

[9] F. Barlat, S.Y. Yoon, S.Y. Lee, M.S. Wi, J.H. Kim, Distortional plasticity framework with application to advanced high strength steel, Int. J. Solids Struct. 202 (2020) 947-962. https://doi.org/10.1016/j.ijsolstr.2020.05.014

[10] J.J. Ha, M.-G. Lee, F. Barlat, Strain hardening response and modeling of EDDQ and DP780 steel sheet under non-linear strain path, Mech. Mater. 64 (2013) 11-26. http://dx.doi.org/10.1016/j.mechmat.2013.04.004

[11] H. Kim, F. Barlat, Y. Lee, S.B. Zaman, C.S. Lee, Y. Jeong, A crystal plasticity model for describing the anisotropic hardening behavior of steel sheets during strain-path changes, Int. J. Plast. 111 (2018) 85-106. https://doi.org/10.1016/j.ijplas.2018.07.010

[12] J. Qin, B. Holmedal, O.S. Hopperstad, A combined isotropic, kinematic and distortional hardening model for aluminum and steels under complex strain-path changes, Int. J. Plast. 101 (2018) 156-169. https://doi.org/10.1016/j.ijplas.2017.10.013

[13] M.S. Wi, S.Y. Lee, J.H. Kim, J.M. Kim, F. Barlat, Experimental and theoretical plasticity analyses of steel materials deformed under a nonlinear strain path, Int. J. Mech. Sci. 182 (2020) 105770. https://doi.org/10.1016/j.ijmecsci.2020.105770

[14] S.S. Hecker, Experimental Investigation of Corners in the Yield Surface, Acta Mech. 13 (1972) 69-86. https://doi.org/10.1007/BF01179659

[15] T. Kuwabara, M. Kuroda, V. Tvergaard, K. Nomura, Use of abrupt strain path change for determining subsequent yield surface: experimental study with metal sheets, Acta Mater. 48 (2000) 2071-2079. https://doi.org/10.1016/S1359-6454(00)00048-3

Material Forming - ESAFORM 2023   Materials Research Forum LLC
Materials Research Proceedings 28 (2023) 779-786   https://doi.org/10.21741/9781644902479-85

[16] K. Yoshida, T. Tsuchimoto, Plastic flow of thin-walled tubes under nonlinear tension-torsion loading paths and an improved pseudo-corner model, Int. J. Plast. 104 (2018) 214-229. https://doi.org/10.1016/j.ijplas.2018.02.013

[17] T. Hama, S. Yagi, K. Tatsukawa, Y. Maeda, Y. Maeda, H. Takuda, Evolution of plastic deformation behavior upon strain-path changes in an A6022-T4 Al alloy sheet, Int. J. Plast. 137 (2021) 102913. https://doi.org/10.1016/j.ijplas.2020.102913

[18] Y. Takada, T. Kuwabara, Nonlinear biaxial tensile stress path experiment without intermediate elastic unloading for validation of material model, Int. J. Solid. Struct. 257 (2022) 111777. https://doi.org/10.1016/j.ijsolstr.2022.111777

[19] ISO 16842: 2014 Metallic materials −Sheet and strip −Biaxial tensile testing method using a cruciform test piece.

[20] T. Hakoyama, T. Kuwabara, Effect of biaxial work hardening modeling for sheet metals on the accuracy of forming limit analyses using the Marciniak-Kuczynski approach, in: H. Altenbach, T. Matsuda, D. Okumura (Eds.), From Creep Damage Mechanics to Homogenization Methods, Springer, 2015, pp. 67-95. https://doi.org/10.1007/978-3-319-19440-0_4

[21] F. Barlat, J.C. Brem, J.W. Yoon, K. Chung, R.E. Dick, D.J. Lege, F. Pourboghrat, S.H. Choi, E. Chu, Plane stress yield function for aluminum alloy sheets-part 1: theory, Int. J. Plast. 19 (2003) 1297-1319. https://doi.org/10.1016/S0749-6419(02)00019-0

[22] R. Hill, J.W. Hutchinson, Differential hardening in sheet metal under biaxial loading: a theoretical framework, J. Appl. Mech. 59 (1992) S1-S9. https://doi.org/10.1115/1.2899489

[23] R. Hill, S.S. Hecker, M.G. Stout, An investigation of plastic flow and differential work hardening in orthotropic brass tubes under fluid pressure and axial load, Int. J. Solid. Struct. 31 (1994) 2999-3021. https://doi.org/10.1016/0020-7683(94)90065-5

Material Forming - ESAFORM 2023
Materials Research Proceedings 28 (2023) 787-798

Materials Research Forum LLC
https://doi.org/10.21741/9781644902479-86

Damage prediction in roll forming of the high strength aluminum alloy AA7075

SUCKOW Timon[1,a] * and GROCHE Peter[1,b]

[1]Institute for Production Engineering and Forming Machines, TU Darmstadt, Otto-Berndt-Str. 2, 64287 Darmstadt, Germany

[a]suckow@ptu.tu-darmstadt.de, [b]groche@ptu.tu-darmstadt.de

Keywords: Roll Forming, Aluminum, 7075, Material Failure, Damage, FEM

Abstract. The high-strength AA7075 alloy offers great potential for lightweight construction thanks to its high specific strength. However, high strength and low ductility are challenging for forming the material in the peak-aged T6-condition in terms of material failure and springback. Therefore, this alloy is usually formed in temperature-supported process routes, which poses major challenges for process design. For cold forming of the alloy in the T6-condition, a reliable prediction of material failure is required in terms of process design. Within this study, this is achieved by applying the modified Mohr-Coulomb (MMC) criterion and an increment based damage evolution rule to the FE-model. To validate the failure prediction and verify the general applicability to different profile geometries, two U-profiles and a V-profile are roll formed. Failure occurs during forming for all profile geometries and the experimental results show a good agreement with the failure prediction. The quality of the damage prediction strongly depends on the calibration for the MMC criterion and the setup of the FE-model, depending on the mesh size and the element type used.

Introduction

Recent developments show that roll forming of high-strength aluminum alloys has become a focus of research and development [1-3]. Roll forming is an economical process for the production of long profiles, particularly in mass production [4]. For commercial vehicles, the use of high-strength aluminum is possible for the side impact beam, A- and B-pillars and other impact-absorbing components. The versatile shaping potential in roll forming compared to die or swivel bending processes allow additional design freedom for the optimal design of the profiles [5]. Additionally, smaller bending radii can be achieved during roll forming, which contributes to an increase in design freedom [6,7].

However, high strength and low ductility of lightweight materials, such as the AA7075 alloy, are challenging for process design. This is expressed in large safety factors and compromises in terms of the target geometry of components, especially with regard to achievable bending radii and bending angles. Process design is of particular relevance in roll forming due to the three-dimensional forming, as various production-related defects can occur. In roll forming of high-strength materials, the occurrence of cracks in the outer fibre of the bending zone and a high springback are critical [8]. The occurrence of cracks is particularly critical for small ratios of the bending radius r to the sheet thickness s and for large bending angles. A plane strain state is present in the outer fibre of the bending zone where fracture occurs [9].

Damage Prediction. Accurate prediction of damage is necessary to improve process design. The modelling of damage in forming technology has been developed over the last decades and is used in the design of forming processes for the optimal utilisation of process windows and for the prediction of material failure [10]. Failure is the final stage of damage and appears in forming processes through intolerable and function-restricting cracks [11]. In general, the damage of metals can be described using uncoupled fracture criteria or coupled damage models [11]. In coupled

Material Forming - ESAFORM 2023   Materials Research Forum LLC
Materials Research Proceedings 28 (2023) 787-798   https://doi.org/10.21741/9781644902479-86

damage models, the elastic and/or plastic material behaviour is linked to the evolution of damage. Due to the higher practicability, uncoupled fracture criteria are widespread in the field of metal forming [11]. Metals usually fail as the result of nucleation, growth and coalescence of microscopic voids within the metal structure during forming processes [12]. All three mechanisms are dependent on the stresses and strains during forming. Developments in modelling damage have shown that damage is dependent on the stress triaxiality η [12,13]. The stress triaxiality η is defined as the ratio of the hydrostatic stress σ_m to the equivalent stress $\bar{\sigma}$:

$$\eta = \frac{\sigma_m}{\bar{\sigma}} \tag{1}$$

where hydrostatic stress,

$$\sigma_m = \frac{\sigma_1 + \sigma_2 + \sigma_3}{3} \tag{2}$$

equivalent stress,

$$\bar{\sigma} = \sqrt{\frac{(\sigma_1 - \sigma_2)^2 + (\sigma_2 - \sigma_3)^2 + (\sigma_3 - \sigma_1)^2}{2}} \tag{3}$$

and σ_1, σ_2 and σ_3 are the principal stresses. Another significant parameter for the extent of damage in three-dimensional stress states is the Lode angle parameter $\bar{\theta}$. This is relevant especially for small stress triaxialities and is employed for example in the modified Mohr-Coulomb (MMC) criterion by Bai and Wierzbicki [13] and the Lou-Huh criterion [12]. All stress directions (loading conditions) can be characterized by the parameters η and $\bar{\theta}$,.

MMC Criterion. The equivalent plastic strain at fracture $\bar{\varepsilon}_f$ depends on the stress triaxiality and the lode angle parameter in the MMC criterion [14]:

$$\bar{\varepsilon}_f = \left\{ \frac{A}{c_2} [1 - c_\eta (\eta - \eta_0)] \times \left[c_\theta^s + \frac{\sqrt{3}}{2 - \sqrt{3}} (c_\theta^{ax} - c_\theta^s) \left(\sec\left(\frac{\bar{\theta}\pi}{6}\right) - 1 \right) \right] \left[\sqrt{\frac{1 + c_1^2}{3}} \cos\left(\frac{\bar{\theta}\pi}{6}\right) + c_1 \left(\eta + \right. \right. \right.$$

$$\left. \left. \left. \frac{1}{3}\sin\left(\frac{\bar{\theta}\pi}{6}\right) \right) \right] \right\}^{-\frac{1}{n}} \tag{4}$$

Eight parameters (A, n, c_η, η_0, c_θ^s, c_θ^{ax}, c_1 and c_2) need to be found for the calibration of the MMC criterion, but only two have to be determined from fracture tests. An explanation of the parameters can be found in [14]. The effects of c_η and c_1 are similar in terms of stress triaxiality, which reduces Eq. 4 to:

$$\bar{\varepsilon}_f = \left\{ \frac{A}{c_2} \left[c_\theta^s + \frac{\sqrt{3}}{2 - \sqrt{3}} (c_\theta^{ax} - c_\theta^s) \left(\sec\left(\frac{\bar{\theta}\pi}{6}\right) - 1 \right) \right] \left[\sqrt{\frac{1 + c_1^2}{3}} \cos\left(\frac{\bar{\theta}\pi}{6}\right) + c_1 \left(\eta + \frac{1}{3}\sin\left(\frac{\bar{\theta}\pi}{6}\right) \right) \right] \right\}^{-\frac{1}{n}} \tag{5}$$

The parameters A and n are determined from the power hardening curve: $\bar{\sigma} = A\varepsilon^n$ [14]. The parameters c_1 and c_2 are determined from material tests up to material fracture, where the fracture strain $\bar{\varepsilon}_f$, stress triaxiality η and lode angle parameter $\bar{\theta}$ are determined numerically and/or experimentally. For accurate damage prediction, the tests should cover load paths that are close to the investigated forming process [11]. The parameters c_θ^{ax} and c_θ^s could be determined from careful plasticity tests and determination of the yield surface [14]. Potential simplifications are the assumption of $c_\theta^{ax}=1$ and $c_\theta^s = 1$ or the determination of the values by the material tests up to

Material Forming - ESAFORM 2023   Materials Research Forum LLC
Materials Research Proceedings 28 (2023) 787-798   https://doi.org/10.21741/9781644902479-86

fracture [14]. The final function provides a three-dimensional fracture locus in the space of stress triaxiality and lode angle parameter. A relationship between the equivalent plastic strain $\bar{\varepsilon}_p$ during the forming process and the equivalent plastic strain at fracture $\bar{\varepsilon}_f = f(\eta, \bar{\theta})$ from the fracture locus provides the damage evolution rule, where material failure occurs when $D \geq 1$ [13]:

$$D(\bar{\varepsilon}_p) = \int_0^{\bar{\varepsilon}_p} \frac{d\bar{\varepsilon}_p}{f(\eta,\bar{\theta})} \tag{6}$$

Damage Prediction in Roll Forming. Wang et al. [15] provide an example for the application of damage models in roll forming with the application of the Oyane model. Shear stresses in the forming zone are responsible for material failure [15]. Furthermore, Wang et al. [15] hypothesize that roll forming seems to be a sheet forming process but is intrinsically a bulk forming process in terms of stress and strain state. Deole et al. [8] use the fracture criterion of Lou and Huh [16] for failure prediction. For calibration, additional bending tests are carried out to represent the plane strain state over the entire loading path [8]. In contrast to [15], the damage and final material failure are related to the plane strain tension state in the outer edge fibre in the bending radius [8]. In the work of Talebi et al. [9], fracture criteria (Rice-Tracey, Crockoft-Latham and Ayada) are used with two different approaches to calibrate the models. The basic approach is the calibration of the model with a uniaxial tensile test. In the extended calibration, with additional material tests, the failure prediction error is reduced from 88 % to 8 % [9]. Lee et al. [2] provide another study on fracture and damage modelling on the roll forming of a bumper beam made of the high-strength aluminum alloy AA7075-T6 using the GISSMO damage model. The summary of the previous approaches shows contradictory statements on the dominant failure-critical stress states and emphasizes the relevance of the calibration and application of fracture criteria and damage models in roll forming. All approaches have in common, that only one profile flower is investigated. Within this study, three different profile flowers are investigated. Due to the application-oriented research, an industrially widespread FE-solver (Marc Mentat) for roll forming is used for the numerical simulation, as well as for calibration of the MMC criterion. All tests for calibration of the MMC criterion are conducted on a universal testing machine. The aim of the study is to apply an easy-to-use fracture criterion in combination with a damage evolution rule on the one hand. On the other hand, a low experimental and numerical effort with a simultaneously accurate damage and failure prediction is strived for.

Results
Calibration of the MMC Criterion. To calibrate the MMC criterion for the peak-aged AA7075-T6 alloy, material tests up to fracture are performed on a Zwick Roell 100 universal testing machine for measuring the equivalent plastic strain at fracture $\bar{\varepsilon}_f$. Plastic strain is measured by optical strain measurement with the GOM Aramis system, after applying a stochastic pattern on the surface of the specimens. Three specimens were tested for each geometry. Before testing, the specimens were milled out of the sheet metal at a 90 ° angle to the rolling direction. Additionally, FE-simulations of the material tests are carried out for each geometry in Marc Mentat 2012 for determining the average values of the stress triaxiality η_{avg} and the lode angle parameter $\bar{\theta}_{avg}$ during plastic deformation. Within the material tests, different loading conditions in terms of stress triaxiality and lode angle parameter are applied. Fig. 1a shows the strain paths for different specimens and

the parameter set, obtained by the material tests and Fig. 1b shows the specimens for calibration of the MMC criterion.

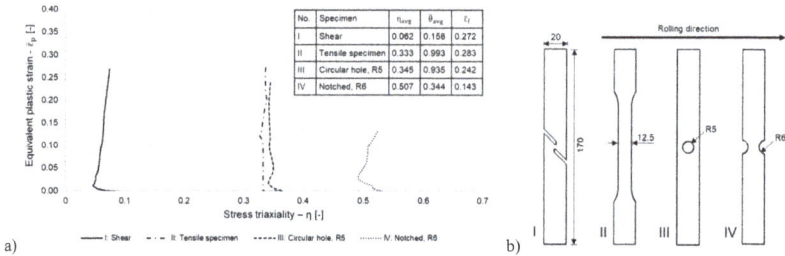

Fig. 1. a) Plastic strain ε_f and stress triaxiality η during the calibration experiments. b) Specimens for calibration of the MMC criterion: I: Shear; II: Tensile specimen; III: Circular hole, R5; IV: Notched, R6.

The set of parameters is the basis for calculating the constants of the MMC criterion. Since the lode angle parameter $\bar{\theta}$ is greater than zero for all specimens, c_θ^{ax} is set to 1, as recommended by Bai and Wierzbicki [14]. The remaining parameters c_1, c_2 and c_θ^s are determined by the nonlinear least-squares solver "lsqcurvefit" via Matlab:

$$\min_{(c_1,c_2,c_\theta^s)} (\Delta\bar{\varepsilon}_f)^2 = \min_{(c_1,c_2,c_\theta^s)} \left[\frac{1}{N} \sum_{i=1}^N \left(\bar{\varepsilon}_{f,i} - \hat{\bar{\varepsilon}}_{f,i}(c_1, c_2, c_\theta^s) \right)^2 \right] \tag{7}$$

The values for $\bar{\varepsilon}_{f,i}$ are determined from the calibration tests, while $\hat{\bar{\varepsilon}}_{f,i}(c_1, c_2, c_\theta^s)$ are the function values of the optimized MMC criterion (Eq. 5). The parameters n = 0.0862 (hardening exponent) and A = 784.96 are obtained from the flow curve from uniaxial tensile tests. In addition, the tensile strength R_m = 590 MPa and yield strength $R_{p0,2}$ = 528 MPa are determined in these tests. Fig. 2a shows the calibrated fracture locus in the $(\eta, \bar{\theta}, \bar{\varepsilon})$-space and Fig. 2b in the $(\eta, \bar{\varepsilon})$-space.

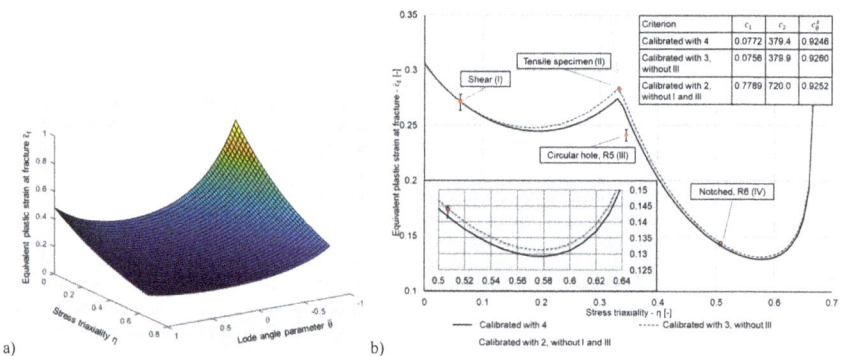

Fig. 2. Representation of the MMC fracture criterion. a) In the $(\eta, \bar{\theta}, \bar{\varepsilon})$-space, calibrated with 4 specimens. b) In the $(\eta, \bar{\varepsilon})$-space for plane stress condition. Calibrated with 2, 3 and 4 specimens.

To investigate the influence of various specimen geometries on the shape of the fracture locus, the fracture locus is illustrated in $(\eta, \bar{\varepsilon})$-space, which is representative for plane stress condition. In plane stress condition, η and $\bar{\theta}$ are dependent on each other, as shown in Eq. 8:

$$\bar{\theta} = 1 - \frac{2}{\pi} \cdot \cos^{-1}\left(-\frac{27}{2}\eta\left(\eta^2 - \frac{1}{3}\right)\right) \tag{8}$$

The shape of the fracture locus is influenced by removing material tests for calibration of the fracture locus (Fig. 2b). Calibrating the fracture criterion with two, three, four or five specimens does not have a high impact on the shape of the fracture locus for stress triaxialities above $\eta = 0.\bar{3}$. During roll forming, bending stresses are expected to be dominant during forming. This would correspond to a stress triaxiality of $\eta = 0.577$. The max. relative error of the fracture locus in plane stress condition for the stress triaxiality of $\eta = 0.577$ is 1.1 % ($\bar{\varepsilon}_f = 0.0014$). For the stress triaxiality of $\eta = 0.\bar{3}$, the relative error is 3.3 % ($\bar{\varepsilon}_f = 0.009$). Less material tests for the calibration of the MMC criterion result in reduced experimental effort and thus in a more efficient calibration of the criterion. The effect of a reduced number of material tests on the damage prediction during roll forming is discussed within this study.

Experimental Setup. Within this study, three different profiles are investigated in order to apply different loading conditions in the bending zone, as shown in Fig. 3a. The profile flowers are presented in each case up to the stage where failure occurs. The difference between the U-profiles is the forming strategy (constant radius vs. constant length) and the flange length. The experiments were performed on two different roll forming machines. The U-profiles were formed on a VOEST P 450/4 roll forming machine and the V-profile on another conventional roll forming machine of the Deakin University, Australia. Feeding speed of the sheet was 1 m/min, the initial sheet length 1500 mm and no lubrication was used. Within the roll forming experiments, the cracks occurred in the bending radii for all profile flowers. Fig. 3b shows the experimental setup for roll forming of a U-Profile and the occurrence of cracks after the third pass.

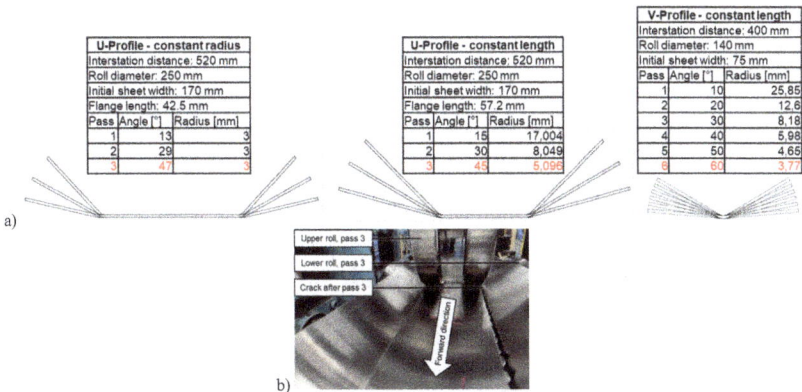

Fig. 3. Profile flowers, investigated within this study. Fracture occurred in the last pass for each profile within the roll forming experiments (marked red).b) Experimental setup and occurrence of cracks (U-Profile - constant radius).

Material Forming - ESAFORM 2023
Materials Research Proceedings 28 (2023) 787-798

Materials Research Forum LLC
https://doi.org/10.21741/9781644902479-86

FE Model. The aim of the FE model is to support the design of the roll forming process with regard to the final geometry and manufacturing-related defects such as material failure. By applying the MMC criterion in combination with a damage evolution rule in the FE model, a prediction of the damage and thus a determination of the process limits is possible. Accurate prediction of damage requires accurate prediction of local stresses and strains in the FE model. The setup of the FE model including material parameters and boundary conditions is shown in Fig. 4. Mesh size and element type are varied with regard to the conflict of objectives between a good prediction of the stresses and strains and the computation time. The variation of the mesh size is similar for all profiles. The mesh size in the bending zone is small, compared to the remaining part of the sheet. The flow curve is obtained from uniaxial tensile tests with a strain rate of 0.002 1/s.

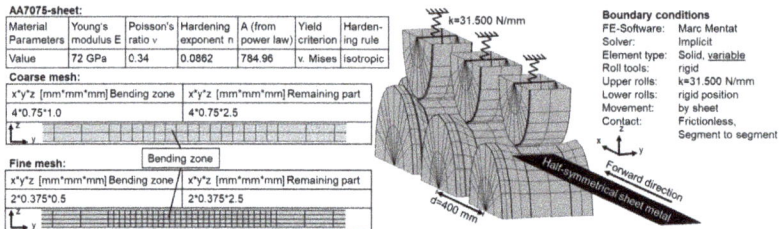

Fig. 4. Setup of the FE model based on the example of the U-profile (constant radius).

Damage Prediction. For damage prediction, the MMC-criterion is applied to the critical node in the FE-model, which is on the outer fibre of the bending zone. Fig. 5a shows the evaluation node for investigating the roll forming process and damage prediction within this study. For all profile flowers, the node is located at the point with the highest plastic strain in the failure-critical forming stage. The node is located in the steady-state area of the sheet at a length of 600 mm (half sheet length). Plastic deformation in the bending zone occurs primarily, when the sheet passes through the forming rolls (Fig. 5b). The stress exceeds the yield stress, which leads to a permanent plastic deformation.

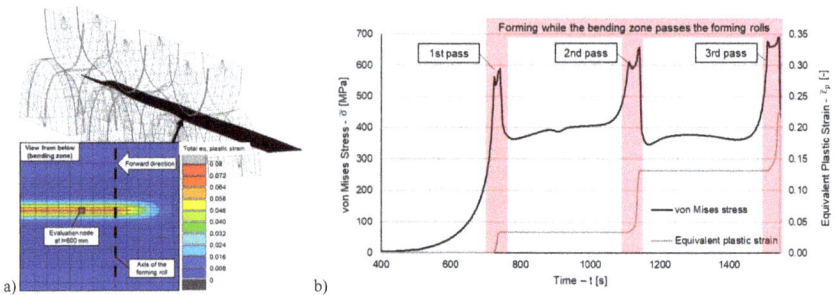

Fig. 5. a) Location of the evaluation node in the bending zone in the example of the U-profile (constant radius). b) Development of equivalent plastic strain and von Mises stress during the roll forming process.

Material Forming - ESAFORM 2023   Materials Research Forum LLC
Materials Research Proceedings 28 (2023) 787-798   https://doi.org/10.21741/9781644902479-86

For nonlinear loading paths, a nonlinear incremental rule must be applied for accurate prediction of damage, as proposed by Bai and Wierzbicki [13]. Within this study, the loading path is considered in the damage evolution rule by summing up the damage value for each increment in the FE-simulation, while considering the loading condition in terms of stress triaxiality and lode angle parameter. For an accurate prediction of damage, the time step for each increment should be small enough to represent the current stress- and strain state and thus the loading path during forming. In this study, one increment is evaluated every 2 mm in the forward direction of the sheet. Increasing the distance to 10 mm per increment leads to a relative error in damage prediction of up to 15 %, due to the inaccurate representation of the loading path. The incremental damage D_{inc} for each node and time increment in the FE-model is calculated by the following equation:

$$D_{inc,\ t} = \frac{\bar{\varepsilon}_{p,t} - \bar{\varepsilon}_{p,t-1}}{\frac{1}{2}\left(\bar{\varepsilon}_f(\eta,\bar{\theta})_t + \bar{\varepsilon}_f(\eta,\bar{\theta})_{t-1}\right)} \tag{9}$$

where $\bar{\varepsilon}_{p,t}$ is the equivalent plastic strain at the evaluated node in the FE-model and $\bar{\varepsilon}_f(\eta, \bar{\theta})_t$ is the equivalent strain at fracture, calculated by the loading condition $(\eta, \bar{\theta})$ of the evaluation node. The total Damage D is defined as the sum of the incremental damage D_{inc}, where n is the total number of increments up to the evaluated time t in the FE-model:

$$D = \sum_{t=1}^{n} D_{inc,t} \tag{10}$$

In the following section, the damage prediction is applied for different profile flowers, mesh and element types. Table 1 shows the naming scheme for the FE-models. Fig. 6 depicts the loading condition for the different profile flowers in order to analyze the differences between the different profile types. The stress states correspond to the plane-stress state and the calibration tests cover a large section of the load path, but the difference between the profile flowers is small. Furthermore, for all profile flowers, the non-linear loading paths follow a similar trend, with increasing stress triaxiality during bending in roll forming in each pass. At the beginning of the plastic deformation in each stage, the loading condition is between shear and uniaxial tension and develops towards stress states between plane strain tension and biaxial tension. An additional test for matching the biaxial tension would be helpful to cover all loading conditions, but is not possible with a universal testing machine.

Table 1. Naming scheme for the FE-models.

Profile	Mesh	Element type
U-cR: U - constant radius	C: coarse	FI: Fully integrated 8-node-Element (Type 7)
U-cL: U - constant length	F: fine	RI: 8-node Element with reduced integration / hourglass control (Type 117)
V: V-Profile		

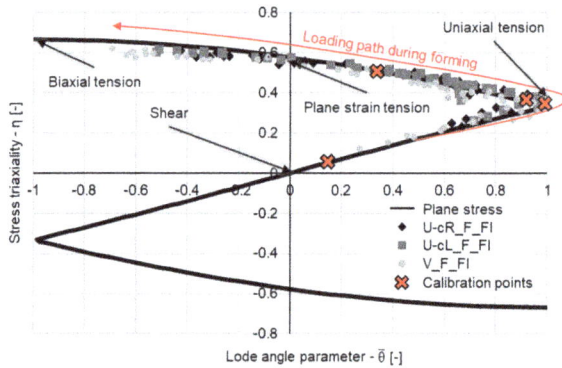

Fig. 6. Representation of the loading condition during plastic forming in the outer fibre of the bending zone for different profile flowers compared to the plane stress condition and the calibration tests.

After a short introduction to the typical stress states during roll forming, various parameters in the FE models are analyzed. The aim is an accurate and efficient prediction of the damage. Efficient means with low experimental and numerical effort. If it is not explicitly described, the prediction of the damage is based on the MMC criterion, based on four material tests. Table 2 and Fig. 7 show the results for different profile flowers, varying the element type and the mesh size in order to overcome the conflict of objectives. It is shown, that the damage is underestimated when the mesh is too coarse, which is fatal for a process design. The same tendency is shown for the reduced integrated elements. This is mainly due to the underestimated plastic strain and stress triaxiality. For the V-profile, the trend is reversed, although the differences are significantly smaller. Furthermore, this leads to a conservative process design and is therefore not too critical. In summary, the fully integrated elements with fine meshing provide the best results in terms of damage prediction. However, the computation time is high, particularly for the fine meshed simulations: 55.200 elements (153 h) instead of 10.800 elements (16 h) for the coarse mesh. The reduced integrated elements decrease the computation time by approx. 5 %. Finer meshing is not practicable due to the already high computing time.

Table 2. Comparison of ductile fracture (experimental) and predicted damage (numerical) for different mesh and element types. Damage underestimated numerically: marked in red. *Fracture estimated in pass 5 already.

Profile flower	U-Profile (constant radius)				U-Profile (constant length)				V-Profile			
Mesh	C		F		C		F		C		F	
El. Type	FI	RI	FI	RI	FI	RI	FI	RI	FI	RI	FI	RI
Ductile fracture (exp.)	3rd pass				3rd pass				6th pass			
Damage at pass of experimental fracture	1.08	0.96	1.52	1.42	0.91	0.75	1.01	0.99	1.16	1.19	1.03*	1.09

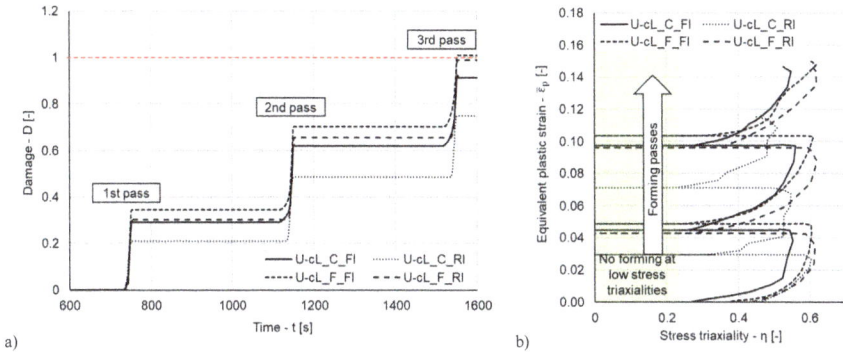

Fig. 7. Numerical prediction of damage (a) and loading paths (b) in the $(\eta, \bar{\varepsilon})$-space during roll forming of the U-Profile (constant length), for different mesh sizes and element types.

In Fig. 8, the different profile types are compared to each other in terms of damage prediction with the best-suited mesh and element type: fully integrated elements with a fine mesh. Damage prediction for the U-Profiles is accurate, while material failure for the V-Profile is predicted too early. Damage is 1.03 in pass 5 already, although there is no material failure experimentally until the 6th pass. This corresponds to an error in the damage prediction of at least 3 %.

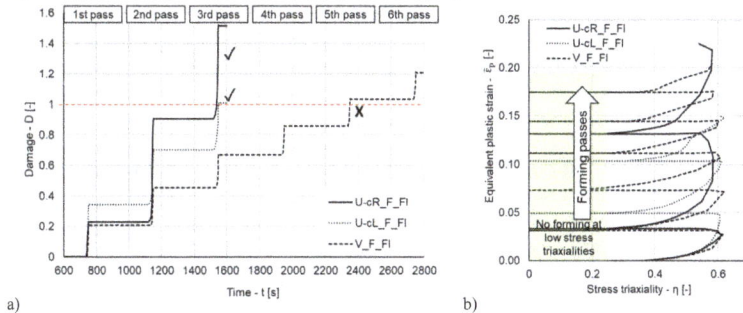

Fig. 8. Numerical prediction of damage (a) and loading paths (b) in the $(\eta, \bar{\varepsilon})$-space during roll forming for all profiles.

In the following section, the MMC criteria, calibrated with a varying number of material tests (Fig. 2b), are compared in terms of failure prediction. Table 3 and Fig. 9 show the results in relation to the number of material tests for calibration of the MMC criterion. Excluding material tests to calibrate the MMC criterion leads to a less accurate prediction of damage. In this case, the damage is underestimated. The reason is the insufficiently represented loading condition in the calibrated MMC criterion, which is replaced by model-based assumptions. The exclusion of a single sample (III: circular hole) leads to a small but decisive deviation in the damage prediction for the U-profile (constant length). The additional exclusion of the shear specimen (I), however, leads to a high deviation of the damage prediction of up to 16 %. This highlights the fact that not only the plane strain condition should be captured to calibrate the damage criterion, but further loading conditions, even $\eta < 0.33$, occur during roll forming.

Table 3. Comparison of ductile fracture (experimental) and predicted damage (numerical) for different calibrations. Damage underestimated numerically: marked in red. *Fracture estimated in pass 5 already.

Profile flower	U-Profile (constant radius)			U-Profile (constant length)			V-Profile		
Calibrated with X material tests	4	3	2	4	3	2	4	3	2
Damage at pass of experimental fracture	1.52	1.49	1.31	1.01	0.99	0.96	1.03*	1.01*	0.95

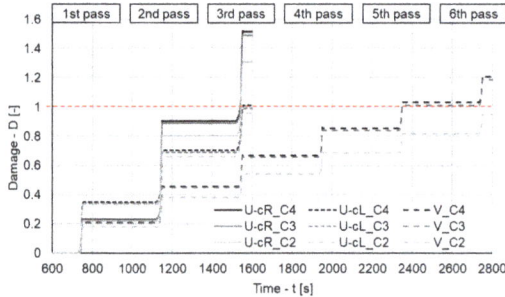

Fig. 9. Numerical prediction of damage for different calibrations of the MMC criterion.

Summary

The investigations have shown that roll forming of the high-strength aluminum alloy AA7075 in the peak-aged T6-condition is challenging, as material failure occurs in an early stage of the process. The application of the MMC criterion in combination with the used incremental damage evolution rule is suitable for predicting material failure during roll forming of the AA7075 alloy. This is even possible for a simple setup of the FE-model, with a von Mises yield criterion, an isotropic hardening rule and neglecting the friction. The roll forming process covers a wide range of loading conditions in terms of stress triaxiality η and Lode angle parameter $\bar{\theta}$, independent of the profile flower. The loading paths are not linear, as they include loading conditions from shear/uniaxial tension up to plane strain tension and biaxial tension, but the used incremental damage evolution rule was shown to be suitable for damage and failure prediction.

The calibration of the MMC criterion is based on material tests with optical strain measurement for determining fracture strain $\bar{\varepsilon}_f$ and FE-simulations for determining the stress triaxiality and the Lode angle parameter. The FE solver Marc Mentat proved to be suitable for performing the material tests and the roll forming simulation. For calibration of the MMC criterion it is recommended to use at least a tensile specimen, a notched specimen and a shear specimen. Otherwise, the loading conditions occurring during roll forming are only insufficiently represented in the MMC criterion. Adding a specimen with a circular hole leads to further improvement of damage prediction. Another result of the study is the recommendation of the element type and meshing that is best suited for predicting the damage. A prediction of the global material behavior, e.g. the forming force or springback is given with a coarse mesh [1]. However, for accurate prediction of damage, the prediction of local stresses and strains must also be accurate, but too fine meshing results in excessive computation time. Therefore, a meshing with four elements in thickness direction with fully integrated elements is recommended.

The inaccurate damage prediction for the V-profile may be due to the high non-linearity of the loading path compared to the other profile flowers. Given that the tests were carried out with one material, further tests should be carried out with different materials to confirm the suitability of the MMC criterion in combination with the incremental damage evolution rule used.

Acknowledgement

The authors gratefully acknowledge financial support from the Hessen State Ministry for Higher Education, Research and the Arts- Initiative for the Development of Scientific and Economic Excellence (LOEWE) for the Project ALLEGRO (Subproject A1). In addition, the authors would like to thank the DAAD for funding the international exchange programme "Laser heat assisted roll forming of new high strength aluminium alloys" and the Institute for Frontier Materials at Deakin University for the support in conducting the roll forming tests (V-profile) within this project. Further thanks go to my colleague Dirk Molitor for programming the Matlab code.

References

[1] T. Suckow, P. Groche, Evaluation of Cold Roll Forming Strategies for the Production of a High-Strength Aluminum Hat Profile, Key Eng. Mater. 926 (2022) 690-699. https://doi.org/10.4028/p-y5090o

[2] S. Lee, J. Lee, J. Song, J. Park, S. Choi, W. Noh, G. Kim, Fracture simulation of cold roll forming process for aluminum 7075-T6 automotive bumper beam using GISSMO damage model, Procedia Manuf. 15 (2018) 751-758. https://doi.org/10.1016/j.promfg.2018.07.314

[3] T. Suckow, J. Schroeder, P. Groche, Roll forming of a high strength AA7075 aluminum tube, Product. Eng. 15 (2021) 573-586. https://doi.org/10.1007/s11740-021-01046-2

[4] K. Sweeney, U. Grunewald, The application of roll forming for automotive structural parts, J. Mater. Process. Technol. 132 (2003) 9-15. https://doi.org/10.1016/S0924-0136(02)00193-0

[5] O. Röcker, Untersuchungen zur Anwendung hoch- und höchstfester Stähle für walzprofilierte Fahrzeugstrukturkomponenten, PhD Thesis, Technische Universität Berlin, 2008.

[6] K. Mäntyjärvi, M. Merklein, J. Karjaleinen, UHS Steel Formability in Flexible Roll Forming, Key Eng. Mater. 410 (2009) 661-668. http://doi.org/10.4028/www.scientific.net/KEM.410-411.661

[7] O.M. Badr, B. Rolfe, P.D. Hodgson, M. Weiss, Forming of high strength titanium sheet at room temperature, Mater. Des. 66 (2015) 618-626. http://doi.org/10.1016/j.matdes.2014.03.008

[8] A.D. Deole, M.R. Barnett, M. Weiss, The numerical prediction of ductile fracture of martensitic steel in roll forming, Int. J. Solid. Struct. 144-145 (2018) 20-31. https://doi.org/10.1016/j.ijsolstr.2018.04.011

[9] H. Talebi-Ghadikolaee, H.M. Naeini, M.J. Mirnia, M.A. Mirnia, M.A. Mirzai, S. Alexandrov, M.S. Zeinali, Modeling of ductile damage evolution in roll forming of U-channel sections, J. Mater. Process. Technol. 283 (2020) 116690. http://doi.org/10.1016/j.jmatprotec.2020.116690

[10] T.S. Cao, Models for ductile damage and fracture prediction in cold bulk metal forming processes: a review, Int. J. Mater. Form. 10 (2017) 139-171. https://doi.org/10.1007/s12289-015-1262-7

[11] A.E. Tekkaya, O.-O. Bouchard, S. Bruschi, C.C. Tasan, Damage in metal forming, CIRP Annals. 69 (2020) 600-623. https://doi.org/10.1016/j.cirp.2020.05.005

[12] Y. Lou, S. Lim, K. Pack, New ductile fracture criterion for prediction of fracture forming limit diagrams of sheet metals, Int. J. Solid. Struct. 49 (2012) 3605-3615. https://doi.org/10.1016/j.ijsolstr.2012.02.016

[13] Y. Bai, T. Wierzbicki, A new model of metal plasticity and fracture with pressure and Lode dependence, Int. J. Plast. 24 (2008) 1071-1096. https://doi.org/10.1016/j.ijplas.2007.09.004

[14] Y. Bai, T. Wierzbicki, Application of extended Mohr-Coulomb criterion to ductile fracture, Int. J. Fract. 161 (2010) 1-20. https://doi.org/10.1007/s10704-009-9422-8

Material Forming - ESAFORM 2023
Materials Research Proceedings 28 (2023) 787-798

Materials Research Forum LLC
https://doi.org/10.21741/9781644902479-86

[15] H. Wang, Y. Yan, F. Jia, F. Han, Investigations of fracture on DP980 steel sheet in roll forming process, J. Manuf. Process. 22 (2016) 177-184. https://doi.org/10.1016/j.jmapro.2016.03.008

[16] Y. Lou, H. Huh, Prediction of ductile fracture for advanced high strength steel with a new criterion: Experiments and simulation, J. Mater. Process. Technol. 213 (2013) 1284-1302. https://doi.org/10.1016/j.jmatprotec.2013.03.001

Material Forming - ESAFORM 2023
Materials Research Proceedings 28 (2023) 799-806

Materials Research Forum LLC
https://doi.org/10.21741/9781644902479-87

Multi-scale modeling of the effect of crystallographic texture

REVIL-BAUDARD Benoit[1,a] * and CAZACU Oana[1,b]

[1]Department of Mechanical & Aerospace Engineering, University of Florida, REEF, Shalimar, USA

[a]revil@ufl.edu; [b]cazacu@reef.ufl.edu

Keywords: Crystallographic Texture, Polycrystalline Model, Macroscopic Plasticity, Metal Forming

Abstract. Among processes involving plastic deformation, sheet metal forming requires a most accurate description of plastic anisotropy. One of the main sources of mechanical anisotropy is crystallographic texture, which induces directionality in the macroscopic plastic properties of the polycrystalline metallic alloy sheets (e.g. anisotropy in yield stresses, Lankford coefficients). Recently, we develop a single-crystal yield criterion that satisfies the intrinsic symmetries of the constituent crystals and the condition of insensitivity to hydrostatic pressure [1]. Moreover, this single-crystal criterion is defined for any 3-D stress state. It was shown that the use of this single-crystal criterion for the description of the plastic behavior of the constituent crystals in conjunction with appropriate homogenization procedures leads to an improved prediction of the plastic anisotropy in macroscopic properties under uniaxial loading for polycrystalline aluminum alloys. In this paper, using this polycrystalline model, we simulate the deformation response of sheets of various crystallographic textures. Examples demonstrate the predictive capabilities of the model to describe the influence of the crystallographic texture on the macroscopic behavior and on the final shape of parts obtained using deep-drawing.

Introduction

Elastic/plastic constitutive models based on macroscopic orthotropic yield criteria are usually used to describe the mechanical behavior of metallic materials and leads to accurate predictions when applied to sheet forming operations (e.g. see [2]). Another approach is to use multi-scale models which explicitly account for the plastic response of the constituent at the grain scale as well as a statistical description of the texture of the material. Usually the grain-level behavior is modeled using a viscoplastic approach based on a power-type law or a rate-independent model based on the Schmid law or a regularized form of Schmid law (e.g. see [3–6]).

Recently, Cazacu, Revil, and Chandola [1] developed an analytical yield criterion for single-crystals. For any 3-D stress state, this yield function is continuous and differentiable and satisfies the symmetries requirements associated with the cubic lattice. Consequently, this yield criterion accounts for the specificities of the plastic flow of the crystal. For general loadings, four anisotropy coefficients are involved in this yield criterion. It was shown that the use of this single-crystal criterion for the description of the plastic behavior of the constituent crystals in conjunction with appropriate homogenization procedures leads to an improved prediction of the plastic anisotropy in macroscopic properties under uniaxial loading for polycrystalline aluminum alloys (see [7,8]).

While one ingredient of a multi-scale model is the constitutive model at the crystal scale level, another ingredient is the texture of the material. For aluminum alloys, the texture plays an important role and induce specific effects on forming properties, resulting in the formation of specific earing profile during cup deep drawing [9]. Depending on the rolling reduction and rolling temperature, the textures of rolled aluminum vary between the typical cube recrystallization texture and the typical rolled Aluminum texture. For a same aluminum alloy, changing the texture

Material Forming - ESAFORM 2023   Materials Research Forum LLC
Materials Research Proceedings 28 (2023) 799-806   https://doi.org/10.21741/9781644902479-87

induce a change in the earing profile [10]. In this paper, using our polycrystalline model, we simulate the deformation response of sheets of various crystallographic textures.

Polycrystalline model based on Cazacu et al. [1] single-crystal law

In our model, the polycrystalline material is represented by a finite set of grains characterized by orientation and volume fraction to reproduce the material texture. Elastic deformations are modeled using Hooke's law for the type of symmetry shown by cubic crystals. The crystal plastic behavior is modeled using the Cazacu et al. single-crystal criterion [1], normality rule, and isotropic hardening described by a Swift-type law. The effective stress of the single-crystal is expressed in terms of cubic stress-invariants and relative to the Cartesian coordinate system $Ox_1x_2x_3$ associated with the crystal axes is given by:

$$\bar{\sigma}_{grain} = \frac{3}{\left(27 - 4cn_1^2\right)^{1/6}} \left\{ \begin{array}{l} \left[\frac{1}{2}m_1\left(\sigma_{11}'^2 + \sigma_{22}'^2 + \sigma_{33}'^2\right) + m_2\left(\sigma_{12}'^2 + \sigma_{13}'^2 + \sigma_{23}'^2\right)\right]^3 \\ -c\left[n_1\sigma_{11}'\sigma_{22}'\sigma_{33}' - n_3\left(\sigma_{33}'\sigma_{12}'^2 + \sigma_{11}'\sigma_{23}'^2 + \sigma_{22}'\sigma_{13}'^2\right) + 2n_4\sigma_{12}'\sigma_{13}'\sigma_{23}'\right]^2 \end{array} \right\}^{1/6} \quad (1)$$

where σ' denotes the Cauchy stress deviator, m_1, m_2, n_1, n_3, n_4 are anisotropy coefficients and c is a parameter that describes the relative importance of the second-order and third-order cubic stress-invariants on yielding. The plastic strain-rate of each crystal $\mathbf{d}_{grain}^p$ is uniquely defined for any stress state and can be easily calculated as:

$$\mathbf{d}^P = \dot{\lambda}\frac{\partial\bar{\sigma}_{grain}}{\partial\boldsymbol{\sigma}} \quad (2)$$

where $\dot{\lambda}$ is the plastic multiplier, and $\bar{\sigma}$ is given by Eq.1.

The multi-scale model using the single-crystal law (1) was implemented in a finite-element (FE) framework. In the FE calculations, the polycrystal behavior is obtained by considering 250 grains per element. It is considered that the total strain-rate of each grain belonging to a given element is equal to the overall strain-rate $\mathbf{D}$. At the time increment (n), the stress in each grain is computed by solving the governing equations, namely:

$$\mathbf{D}_{grain}^{(n)} = \left(\mathbf{R}^{(n)}\right)^T \mathbf{D}^{(n)} \mathbf{R}^{(n)}$$

$$\mathbf{D}_{grain}^{(n)} = \mathbf{D}_{grain}^{e(n)} + \mathbf{D}_{grain}^{p(n)}$$

$$\boldsymbol{\sigma}_{grain}^{(n)} = \boldsymbol{\sigma}_{grain}^{(n-1)} + \left(\mathbf{C}^{el} : \mathbf{D}_{grain}^{e(n)}\right)dt \quad (3)$$

$$\bar{\sigma}_{grain}^{(n)} - Y\left(\bar{\varepsilon}_{grain}^{p\,(n)}\right) \leq 0$$

$$\mathbf{D}_{grain}^{p(n)} = \dot{\lambda}_{grain}\frac{\partial\bar{\sigma}_{grain}^{(n)}}{\partial\boldsymbol{\sigma}}$$

where $\mathbf{D}_{grain}^{(n)}$, $\mathbf{D}_{grain}^{p(n)}$ and $\mathbf{D}_{grain}^{e(n)}$ are respectively the crystal's total strain-rate, the plastic and elastic strain-rate with $\mathbf{C}^{el}$ being the fourth-order elasticity tensor , $\boldsymbol{\sigma}_{grain}^{(n-1)}$ and $\boldsymbol{\sigma}_{grain}^{(n)}$ are the stress tensors at the beginning and end of the increment, respectively, while $\bar{\varepsilon}_{grain}^{p\,(n)}$ is the equivalent plastic strain in the given grain, $Y\left(\bar{\varepsilon}_{grain}^{p\,(n)}\right)$ is the hardening law, and $\dot{\lambda}_{grain}$ the plastic multiplier. The stress of the polycrystal at the end of the increment is given by:

$$\boldsymbol{\sigma}^{(n)} = \left(\sum_i w_i \mathbf{R}_i^{(n)}\left(\boldsymbol{\sigma}_{grain}^{(n)}\right)_i \left(\mathbf{R}_i^{(n)}\right)^T\right) / \left(\sum_i w_i\right) \quad (4)$$

Material Forming - ESAFORM 2023   Materials Research Forum LLC
Materials Research Proceedings 28 (2023) 799-806   https://doi.org/10.21741/9781644902479-87

where $\left(\sigma_{grain}^{(n)}\right)_i$ is the stress tensor of grain i, and $\mathbf{R}_i^{(n)}$ is the transformation matrix for passage from the crystal axes of grain i to the loading frame axes, while w_i is the weight of the grain i. To describe the macroscopic response of FCC polycrystals, the model given by Eq. 1-4 was implemented in the commercial FE solver Abaqus Standard (implicit solver, see Abaqus (2014)). A polycrystalline aggregate composed of N crystals was associated with each FE integration point. The set of governing equations are solved for each of the constituent crystals using a fully-implicit backward Euler method.

Influence of the Initial Texture of the Material on The Earing Profile for Aluminum Alloy
Hirsch [10] has shown that for an Al-Mg-Mn alloys, it is possible by changing the rolling reduction and rolling temperature to change the texture of the material from a typical cube recrystallization texture to a ß-fiber rolling texture. This change in texture components leads to a change in the earing profile during cup deep-drawing. For a typical recrystallized aluminum alloy, a four ear profile with a minimum height at 45° to the rolling direction (RD) is observed, while for a ß-fiber rolling texture, a four ear profile with maximum height at 45° to RD is obtained (see [9]).

The polycrystalline model described in the previous section was used to assess it capability to accurately describe the influence of the initial texture of the material on the earing profile. For this purpose, the coefficients involved in the yield criterion [1] are kept fixed (i.e. $m_1 = 0.36$, $m_2 = 0.18$, $n_1 = 0.21$, $n_1 = 0.11$ $n_4 = 0.08$ and $c = 1.227$) and only the initial texture of the material is changed to reflect different rolling conditions. Two generic textures were generated using the software MTex [11] and the experimental observations reported in [10]. The main components and their weights are summarized in Table 1.

Table 1. Texture components and their weights.

Cube recrystallized Texture				Rolling Texture				
Texture component	Euler Angles			weight	Texture component	Euler Angles			weight
	ϕ_1	ψ	ϕ_2			ϕ_1	ψ	ϕ_2	
Cube	0	0	0	55%	C	90	35	45	20%
R	63	31	60	25%	S	63	31	60	20%
Goss	70	45	0	10%	B	35	45	0	60%
P	45	15	10	5%					
Q	0	45	0	5%					

For the cube recrystallized texture material, the pole figures as well as the anisotropy in uniaxial yield stresses and r-values obtained with the polycrystalline model with the given set of parameters are plotted in Fig. 1. It is to be noted that the polycrsytalline model predicts a minimum r-value of 0.6 along the direction 45° from RD and two maxima which are located along RD and TD. Concerning the yield stresses, the maximum is predicted at 45° from RD at minima are along RD and TD.

Material Forming - ESAFORM 2023
Materials Research Proceedings 28 (2023) 799-806

Materials Research Forum LLC
https://doi.org/10.21741/9781644902479-87

(a)

(b)

(c)

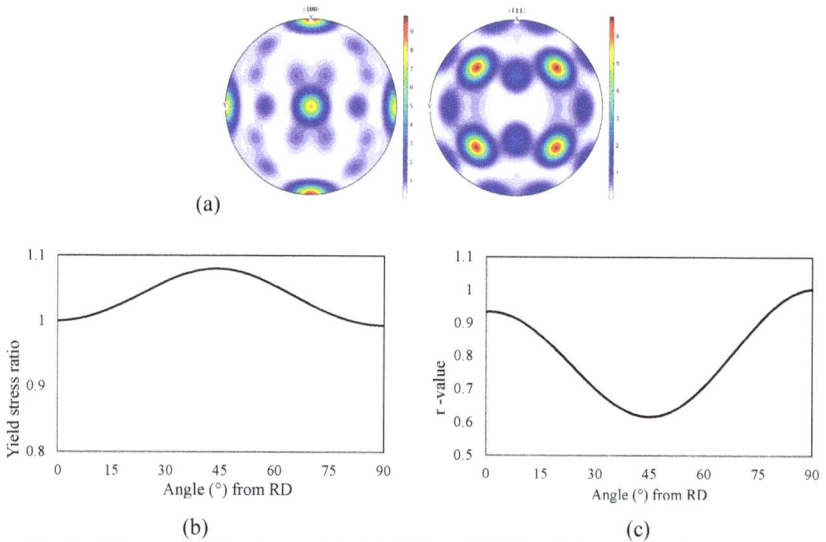

Fig. 1. Cube recrystallized material: (a) (100) and (111) pole figure; predictions of the anisotropy using the polycrystalline model based on the single-crystal law [1]: b) Uniaxial tensile flow stresses; (c) Lankford coefficients (r-values).

For the rolling texture material, the pole figures as well as the anisotropy in uniaxial yield stresses and r-values obtained with the polycrystalline model are shown in Fig. 2. Note that for a rolling texture, the polycrsytalline model predicts a maximum r-value along the direction ~ 45° from RD (r = 1.51), while the two minima are located along RD and TD.

Comparisons between the anisotropy predictions obtained with the polycrystalline model based on the single crystal yield criterion [1] with the same single crystal anisotropy coefficients show the influence of the initial texture of the material on its mechanical response. While for typical cube texture, it is predicted that the minimum r-value is obtained for a tensile test at an angle of 45 ° from RD, for a typical rolling texture, a maximum r-value is obtained for the same orientation. It is also worth noting that the amplitude of the r-value variation predicted also depends on the initial texture.

Material Forming - ESAFORM 2023
Materials Research Proceedings 28 (2023) 799-806

Materials Research Forum LLC
https://doi.org/10.21741/9781644902479-87

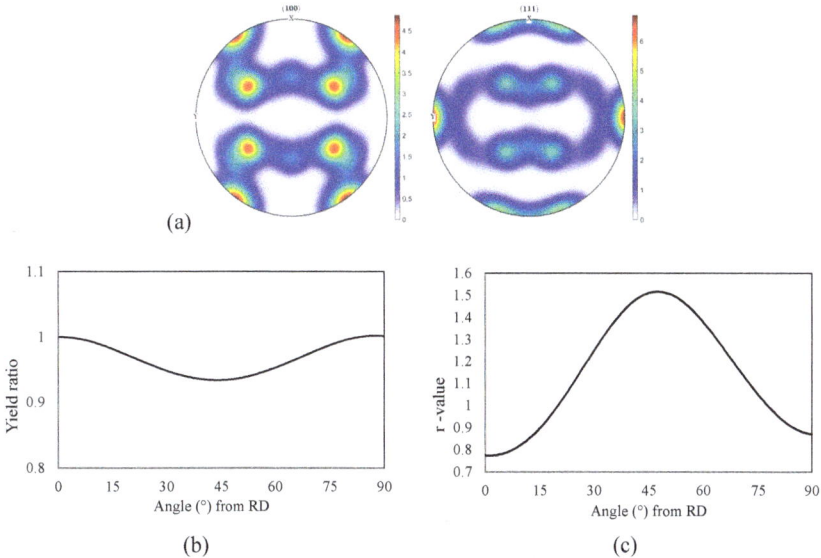

Fig. 2. Rolling texture material : (a) (100) and (111) pole figure; predictions of the anisotropy using the polycrystalline model based on the single-crystal law [1]: (b) Uniaxial tensile flow stresses; (c) Lankford coefficients (r-values).

The polycrystalline model based on the Cazacu et al. [1] single crystal criterion (see Eq.(1)) was applied to study the influence of the initial texture of the material on the forming of a cylindrical cup. A blank of thickness 1 mm and radius of 50 mm was drawn by a punch of radius 30 mm into a die of opening radius of 31.2 mm. The blank-holder force was of 40 kN. In all the FE simulations presented hereafter, a polycrystalline aggregate composed of 250 orientations representative of the overall texture of the material is associated with each FE integration point. In the FE simulations of the circular cup, a total of 10900 reduced integration elements (Abaqus C3D8R) was used to mesh a quarter of the blank, resulting in the consideration of 2 725 000 grains in the FE simulation. In terms of computational time, one simulation of the cup drawing process performed using 6 cores takes about 3h40 on a desktop computer (Intel Core i7-4770 / 16GB RAM).

For the cube recrystallized texture material, the predicted isocontours of the equivalent plastic strain of the fully drawn cup using the polycrystalline model is shown in Fig.3 along with the predicted earing profile. For this initial texture, it is predicted a 4 ears profile with the maximum height being obtained for RD and the minimum height being obtained at 45° from RD. For the rolling texture material, using the same anisotropy coefficients (i.e. $m_1 = 0.36$, $m_2 = 0.18$, $n_1 = 0.21$, $n_1 = 0.11$ $n_4 = 0.08$ and $c = 1.227$), the polycrystalline model based on the single crystal criterion [1] also predicts a four ears profile, but with the maximum height obtained for the 45° from RD direction and the minimum height obtained at RD and TD.

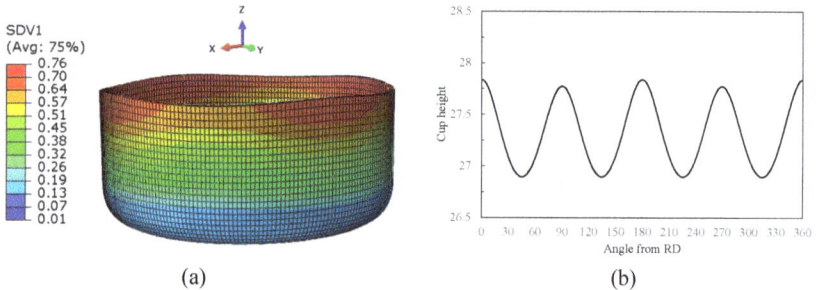

(a)   (b)

Fig. 3. FE results obtained with the polycrystalline model based on the Cazacu et al. [1] single crystal criterion (with $m_1 = 0.36$, $m_2 = 0.18$, $n_1 = 0.21$, $n_1 = 0.11$ $n_4 = 0.08$ and $c = 1.227$) for an aluminium alloy with a cube recrystallized texture : (a) Isocontours of the equivalent plastic strain; (b) earing profile.

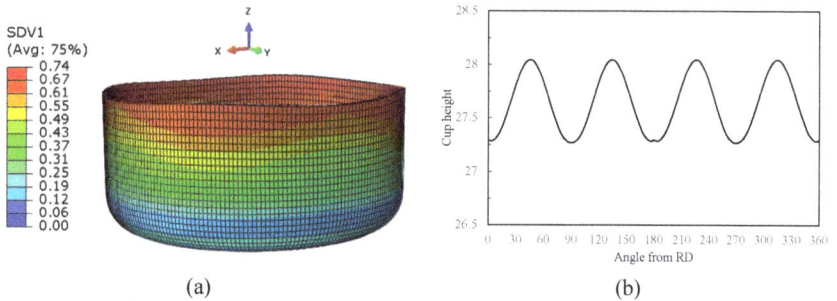

(a)   (b)

Fig. 4. FE results obtained with the polycrystalline model based on the Cazacu et al. [1] single crystal criterion (with $m_1 = 0.36$, $m_2 = 0.18$, $n_1 = 0.21$, $n_1 = 0.11$ $n_4 = 0.08$ and $c = 1.227$) for an aluminium alloy with a rolling texture: (a) Isocontours of the equivalent plastic strain; (b) earing profile.

To further compare the predictions of the polycrystalline model for the two different materials, in Fig 5 are superposed the predictions of the earing profile and the punch load. It is worth recalling that in the simulations, only the initial texture was different, the material parameters and deep drawing process parameters (friction, type of elements, blankholder forces) being the same. It is to be noted that the polycrystalline model is able to capture the influence of the initial texture on the mechanical response of the material and furthermore on the earing profile obtained during cup deep-drawing. Furthermore, the trends seen experimentally in a Al-Mg-Mn alloy by [9] are also accurately predicted by the polycrystalline model, that is for a typical recrystallized aluminum alloy, a four ear profile with a minimum height at 45° to RD is predicted, while for a rolling texture, a four ear profile with the maximum height at 45° to RD is predicted, which is similar to experimental observations. Furthermore, the polycrystalline model predicts an slightly higher punch force for the material with a rolling texture than for the material with a cube recrystallized texture.

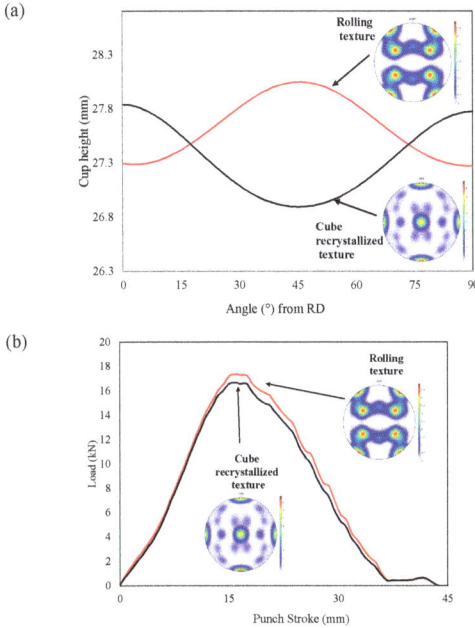

Fig. 5. Comparison of the predictions obtained with the polycrystalline model based on the Cazacu et al. [1] single crystal criterion (with $m_1 = 0.36$, $m_2 = 0.18$, $n_1 = 0.21$, $n_1 = 0.11$ $n_4 = 0.08$ and $c = 1.227$) for aluminium alloys with a cube recrystallized texture and a rolling texture, respectively: (a) earing profile; (b) forming force vs. punch stroke.

Summary

In this paper, we further illustrated the capabilities of the polycrystalline model [7] to predict the mechanical response of aluminum alloys in forming operations. Key in the formulation of this polycrystalline model is the use for the description of the plastic behavior at the crystal scale, the recent single-crystal yield criterion [1]. This cubic single-crystal yield criterion is defined for any stress state and involves the correct number of anisotropy coefficients required to satisfy the intrinsic symmetries of the cubic lattice and the condition of yielding insensitivity to hydrostatic pressure.

Using this polycrystalline model, simulation of the cup drawing process have been simulated for two aluminum alloy with different initial textures in order to investigate the influence of the initial texture on the shape of the formed part. The first material considered is characterized by a cube recrystallized texture. For this material, the polycrystalline model predicts a minimum r-value at 45° orientation and the maximum for TD. This in turn results in a predicted four ears profile with a minimum height at 45° to RD. Using the same set of material parameters, but changing the initial texture to a typical rolling texture, the polycrystalline model predict a completely different anisotropy for the material. For a rolling texture, it is predicted a minimum r-value for RD and TD while the maximum r-value is predicted for the 45° direction. Similarly, the polycrystalline model is also able to account for the influence of the initial texture on the final shape of the cup. For a rolling texture, a four ear profile with the maximum height at 45° to RD is predicted. These predictions are in agreement with experimental observations on Al alloys, i.e. the transition from

Material Forming - ESAFORM 2023   Materials Research Forum LLC
Materials Research Proceedings 28 (2023) 799-806   https://doi.org/10.21741/9781644902479-87

a cube recrystallized texture to a rolling texture induce a transition between an ear profile with the minimum height at 45° to an earing profile with the maximum height at 45°. Furthermore, the polycrystalline model predict a slightly higher punch force for the material with a rolling texture than a material with a cube recrystallized texture.

References

[1] O. Cazacu, B. Revil-Baudard, N. Chandola, A yield criterion for cubic single crystals, Int. J. Solid. Struct. 151 (2018) 9-19. https://doi.org/10.1016/j.ijsolstr.2017.04.006

[2] O. Cazacu, B. Revil-Baudard, Plasticity of Metallic Materials: Modeling and Applications to Forming, Elsevier, 2020.

[3] O. Cazacu, B. Revil-Baudard, N. Chandola, Plasticity-Damage Couplings: From Single Crystal to Polycrystalline Materials, Springer, 2019.

[4] N. Chandola, O. Cazacu, B. Revil-Baudard, Prediction of plastic anisotropy of textured polycrystalline sheets using a new single-crystal model, Comptes Rendus Mécanique 346 (2018) 756-769.

[5] E. Schmid, W. Boas, Plasticity of Crystals (Translation by F.A. Hughes & Co. Limited), Chapman & Hall Ltd, London, 1951.

[6] C.F. Elam, Distortion of metal crystals, Clarendon Press, 1935.

[7] B. Revil-Baudard, O. Cazacu, N. Chandola, Finite Element Analysis of AA 6016-T4 Sheet Metal Forming Operations Using a New Polycrystalline Model, Key Eng. Mater. 926 (2022) 1067–1074. http://doi.org/10.4028/p-3b1ez2

[8] A.M. Habraken, T.A. Aksen, J.L. Alves, R.L. Amaral, E. Betaieb, N. Chandola, L. Corallo, D.J. Cruz, L. Duchêne, Bernd Engel, E. Esener, M. Firat, P.Frohn-Sörensen, J. Galán-López, H. Ghiabakloo, L.A.I. Kestens, J. Lian, R. Lingam, W. Liu, J. Ma, L.F. Menezes, T. Nguyen-Minh, S.S. Miranda, D.M. Neto, A.F.G. Pereira, P.A. Prates, J. Reuter, B. Revil-Baudard, C. Rojas-Ulloa, B. Sener, F. Shen, A. Van Bael, P. Verleysen, F. Barlat, O. Cazacu, T. Kuwabara, A. Lopes, M.C. Oliveira, A.D. Santos, G. Vincze, Analysis of ESAFORM 2021 cup drawing benchmark of an Al alloy, critical factors for accuracy and efficiency of FE simulations, Int. J. Mater. Form. 15 (2022) 1–96. https://doi.org/10.1007/s12289-022-01672-w

[9] J. Hirsch, Texture and anisotropy in industrial applications of aluminium alloys, Arch. Metall. Mater. 50 (2005) 21-34.

[10] J. Hirsch, Texture evolution and earing in aluminium can sheet, Mater. Sci. Forum 495 (2005) 1565-1572. http://doi.org/10.4028/www.scientific.net/MSF.495-497.1565

[11] F. Bachmann, R. Hielscher, H. Schaeben, Texture analysis with MTEX–free and open source software toolbox, Solid state phen. 160 (2010) 63-68. https://doi.org/10.4028/www.scientific.net/SSP.160.63

[12] O. Cazacu, New yield criteria for isotropic and textured metallic materials, Int. J. Solid. Struct. 139–140 (2018) 200-210. https://doi.org/10.1016/j.ijsolstr.2018.01.036

Material Forming - ESAFORM 2023
Materials Research Proceedings 28 (2023) 807-816

Materials Research Forum LLC
https://doi.org/10.21741/9781644902479-88

Experimental identification of uncoupled ductile damage models and application in flow forming of IN718

VURAL Hande[1,a], ERDOGAN Can[1,b], KARAKAŞ Aptullah[2,c],
FENERCIOGLU Tevfik Ozan[2,d] and YALÇINKAYA Tuncay[1,e] *

[1]Department of Aerospace Engineering, Middle East Technical University, 06800 Ankara,
Turkey

[2]Repkon Machine and Tool Industry and Trade Inc., 34980 Şile, Istanbul, Turkey

[a]vural.hande@metu.edu.tr, [b]cane@metu.edu.tr, [c]aptullah.karakas@repkon.com.tr,
[d]ozan.fenercioglu@repkon.com.tr, [e]yalcinka@metu.edu.tr

Keywords: Ductile Fracture, Fracture Locus, Flow Forming Process

Abstract. The aim of this study is to calibrate the parameters of the Johnson-Cook (JC) and modified Mohr-Coulomb (MMC) ductile failure models for Inconel 718 and predict the formability limit in the flow forming process using the aforementioned uncoupled damage models. Uniaxial tensile tests are performed on four different specimen geometries to cover a variety of stress states. A hybrid methodology combining finite element simulations and experimental findings is used to calibrate the JC and MMC damage models. The models are implemented in the finite element solver Abaqus using a user-defined subroutine. Results show that the calibrated models agree well with the experimental data in all tensile tests. In shear dominant loads, the MMC model is found to be more capable of showing accurate crack propagation. In flow forming simulations, a significant difference is observed between the JC and MMC models in the prediction of damage. Lode parameter-dependent damage models, such as the MMC, are found to be more suitable for the prediction forming limits in the flow forming process.

Introduction

Flow forming, or spinning, is an incremental metal forming process used to produce axisymmetric parts and it has been widely adopted in the automotive and defense industries (see e.g. [1]). The thickness of a tubular preform material is incrementally reduced by utilizing one or more rotating rollers. It has several benefits such as low scrap material, low cost due to simple tool processing, smooth surface quality, and high geometrical accuracy. The preform thickness can be reduced by up to 80 % throughout the process. Such high deformations under complex forces potentially lead to surface defects or complete rupture. As in many forming applications, it is crucial to identify the regions that are more susceptible to crack initiation or defects and to estimate the formability limits in flow forming.

In computational structural analysis, ductile fracture models have been widely adapted to describe the ductile failure of metallic materials. According to their relationship with constitutive equations, damage models employed in finite element (FE) simulations can be separated into coupled and uncoupled models. Coupled models in which the damage parameters and constitutive equations are coupled, and the stress is influenced by the damage evolution (e.g. [2-4]). Uncoupled models exclude the effect of damage evolution from constitutive relations. They are often referred to through a failure criterion, which consists of plastic strain, strain rate, stress state, and temperature (e. g. [5,6]). Uncoupled models are more common in engineering applications because of their simplicity in implementation and parameter calibration, despite coupled models providing more realistic failure predictions. It has been discussed that a ductile failure criterion depends only on stress triaxiality (T) could be insufficient for shear dominant ductile failures (see [7]). To

Material Forming - ESAFORM 2023
Materials Research Proceedings 28 (2023) 807-816

Materials Research Forum LLC
https://doi.org/10.21741/9781644902479-88

overcome this problem, Lode angle parameter ($\bar{\theta}$), which is connected to the third deviatoric stress invariant, has been considered in more recent models (e.g. [8,9]).

In the literature, many attempts have been made to predict forming limits using damage models for incremental metal-forming processes (e.g. [10,11,12,13]). The MMC model is used in one of the authors' earlier studies for flow forming analyses to analyze forming limits and the influence of process parameters in failure. In [14], the MMC model for 6016-T6 aluminum alloy is adopted from the literature to examine the effect of the thickness reduction ratio and the involvement of different roller arrangements on the damage. The strain rate and temperature dependencies for the 4340-steel alloy were included in the plasticity and MMC models in [15], and the implications of process variables including feed rate, roller speed, and offset of the rollers on damage are studied.

The aim of this study is first to calibrate the damage model parameters for Inconel 718 (IN718) through a hybrid experimental-numerical approach. Tensile tests are carried out on four specimen geometries to cover various stress states. The DIC (digital image correlation) technique is used to extract accurate displacement-force values from the experiments. For this material, the plasticity model is constructed using experimental data and FE simulations. The MMC and JC damage models are adopted to model ductile damage. A good correlation between the FE results and the experimental data is achieved with the calibrated models. Then, the failure prediction capabilities of the calibrated damage models are investigated with the simulation of the flow forming process.

Methods

Materials and Tensile Tests.

Tensile tests are performed for IN718 alloy under quasi-static conditions. IN718 is nickel-based precipitation hardenable superalloy where gamma double prime (γ") precipitations are nucleated after the aging treatment to have improved mechanical properties. IN718 has good fatigue and creep strength with high corrosion resistance, and it has been widely used in the hot sections of gas turbines engines such as turbine discs and blades. The chemical composition of the alloy is given in Table 1. For this material, 4 different specimens representing different stress-state conditions are manufactured. The geometries of the specimens are shown in Fig. 1. Specimens are named as smooth tension (ST), notch tension (NT10), plane-strain tension (PST) and in-plane shear (ISS). Specimens are prepared with a thickness of 3 mm. They are spray-painted in white and then patterned in black dots for displacement and strain measurements with the digital image correlation (DIC) method. MTS 100kN Tension-Torsion Fatigue/Static test machine and a high-speed camera are used to obtain the force and displacement data. Displacement-controlled tests are carried out with a strain rate of 1 mm/min. NCORR open-source 2D DIC software is used to process the images captured by the high-speed camera. Force data is obtained from the test machine while the displacements are extracted from the DIC analysis using a virtual extensometer at the center of the specimens spanning vertically with a gauge length of 40, 50, 8.1, and 28 mm for ST, NT10, PST and ISS specimens, respectively.

Material Forming - ESAFORM 2023
Materials Research Proceedings 28 (2023) 807-816

Materials Research Forum LLC
https://doi.org/10.21741/9781644902479-88

Table 1: Chemical composition of IN718 [wt.%].

Element	Ni	Nb	Mo	Ti	Al	Cr	Cu	Si	Fe	Mn	C
Content	54.35	4.96	2.77	0.95	0.56	18.59	0.04	0.32	16.68	0.28	0.5

Plasticity and Damage Models.

The constitutive behavior is governed by the classic J2 plasticity with isotropic hardening. The extended voce rule is used to define the hardening law, as

$$\sigma = \sigma_y + q_1\left(1 - c_1 e^{-b_1 \bar{\varepsilon}_p}\right) + q_2\left(1 - c_2 e^{-b_2 \bar{\varepsilon}_p}\right) \tag{1}$$

where σ_y is the yield stress of IN718 and $\bar{\varepsilon}_p$ is the equivalent plastic strain. q_i, c_i and b_i are material-specific constants.

Two different damage models are chosen to evaluate their applicability to flow forming simulation. The first of these models is the JC model, which is only stress triaxiality and equivalent plastic strain dependent. Due to its simplicity, it has been frequently used in many applications in the literature. In this study, the JC model is utilized independently of strain rate and temperature, as

$$\varepsilon_f(T) = \widehat{D}_1 + \widehat{D}_2 \exp(-\widehat{D}_3 T) \tag{2}$$

where, T is the stress triaxiality and ε_f is the plastic strain value at failure. $\widehat{D}_1$, $\widehat{D}_2$ and $\widehat{D}_3$ are calibration parameters for the JC model. The second model is the MMC model defined as

$$\varepsilon_f(T, \bar{\theta}) = \left\{ \frac{A}{\widehat{C}_2}\left[\widehat{C}_3 + \frac{\sqrt{3}}{2-\sqrt{3}}\left(\widehat{C}_4 - \widehat{C}_3\right)\left(\sec\left(\frac{\bar{\theta}\pi}{6}\right) - 1\right)\right] \times \left[\sqrt{\frac{1+\widehat{C}_1^2}{3}} + \cos\left(\frac{\bar{\theta}\pi}{6}\right) + \widehat{C}_1\left(T + \frac{1}{3}\sin\left(\frac{\bar{\theta}\pi}{6}\right)\right)\right] \right\}^{-\frac{1}{n}} \tag{3}$$

where, T is the stress triaxiality, $\bar{\theta}$ is the Lode angle parameter and ε_f is the plastic strain value at failure. A, n, $\widehat{C}_1$, $\widehat{C}_2$, $\widehat{C}_3$ and $\widehat{C}_4$ are calibration parameters for MMC model. Stress triaxiality and Lode angle parameter are formalized as

$$T = \frac{\sigma_m}{\sigma_{eq}} \text{ and } \bar{\theta} = 1 - \frac{6\theta}{\pi} = 1 - \cos^{-1}\left(\frac{J_3}{2\,\sigma_{eq}^3}\right) \tag{4}$$

where the mean stress is $\sigma_m = \frac{tr(\sigma)}{3} = \frac{\sigma_{11} + \sigma_{22} + \sigma_{33}}{3}$, σ_{eq} is the von-Mises equivalent stress and J_3 is defined as the third deviatoric stress invariants.

The damage accumulation rule is expressed with the following integral

$$D = \int_0^{\varepsilon_f} \frac{d\bar{\varepsilon}_p}{\varepsilon_f} \tag{5}$$

Initially, the material is assumed to be undamaged, $D = 0$, and when D reaches 1 and is interpreted as the material is completely failed.

FE Modelling.

Displacement controlled explicit FE simulations are conducted to calibrate and verify the plasticity and damage model parameters. The MMC model is implemented as a user-defined field subroutine (VUSDFLD) while the inbuilt JC damage model in Abaqus is used. The geometries and mesh of 4 different specimens are shown in Fig. 1. One-fourth of geometries are simulated for

ST, NT and PST utilizing the symmetry planes to reduce the computational cost. For the ISS, analysis is taken with the full model. 8-node linear brick elements (C3D8R) with reduced integration are used and element deletion is utilized to represent material failure. To have boundary conditions consistent with experiments, the parts held and pulled in the machine are modeled as rigid bodies while the section where DIC analysis is conducted is modeled as a deformable body.

Calibrated and validated damage models are then applied to the finite element model of the backward flow forming (FF) process using the dynamic explicit solver of Abaqus. The FE model and mesh are shown in Fig. 2. This model consists of a preform, a mandrel, and 3 rollers rotating and moving in the axial direction. The process is called the backward FF process because the material flows in the direction opposite to the axial movement of the rollers. Mandrel and rollers are modeled as rigid bodies while the preform tube is a deformable body. The rollers are placed around the mandrel with 120 degrees between them with a certain axial offset. This ensures that the thickness of the material is incrementally reduced and a thickness reduction ratio of 40% in total is applied. Both tangential and normal contact are featured in the model with a 0.05 friction coefficient for tangential contact. The preform tube is meshed using hexahedral elements with reduced integration (C3D8R) and enhanced hourglass control is used to prevent mesh distortion. There are 168000 elements in total, with 7 elements in the thickness direction.

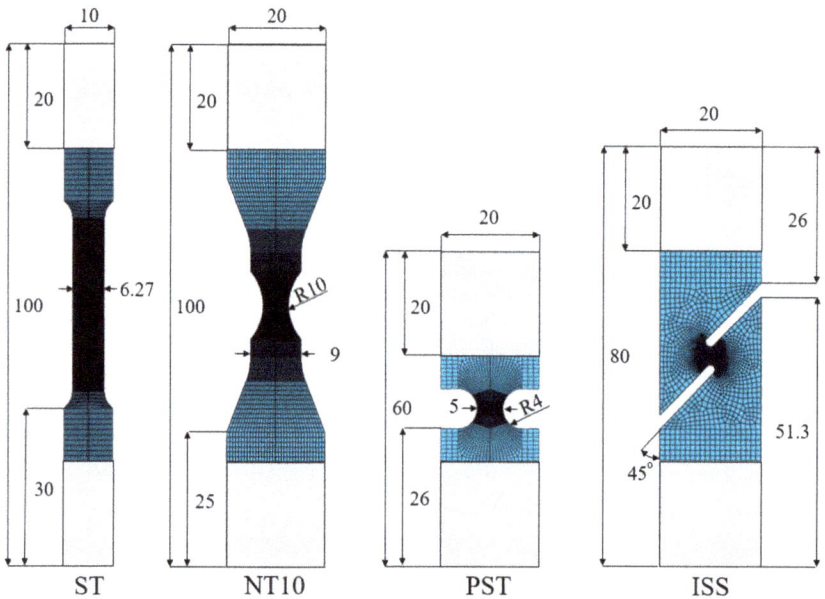

Fig. 1. Dimensions and finite element models of specimens.

Material Forming - ESAFORM 2023   Materials Research Forum LLC
Materials Research Proceedings 28 (2023) 807-816   https://doi.org/10.21741/9781644902479-88

Results and Discussion

Plasticity Model Calibration.

Since an uncoupled modeling approach is used in the current work, hardening parameters can be calibrated with the experimental data independently of the damage parameters. Using the force-displacement data obtained from the experiments, the true stress-strains curves are obtained using up to necking from the ST specimen. The hardening law parameters are then fitted using the MATLAB curve-fitting tool. The yield stress and material constants of the hardening laws are shown in Table 2. Density, Young's modulus and Poisson's ratio are taken as 8.22 g/cm^3, 200 GPa and 0.294, respectively.

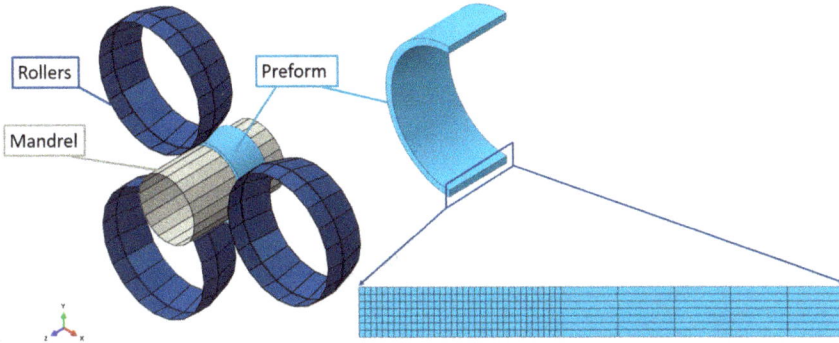

Fig. 2. Finite element model of flow forming process.

Table 2. Calibrated parameters of the hardening law.

σ_y [MPa]	q_1 [MPa]	c_1	b_1	q_2 [MPa]	c_2	b_2
789	499.6	0.1731	106.7	499.6	1.761	4.351

Damage Model calibration.

The MMC damage model parameters are calibrated using a hybrid experimental and numerical approach which has been applied and verified for a wide range of materials and experiments in the literature. In this approach, plastic strain averaged stress triaxiality and Lode angle parameter data are extracted from the FE simulations of the specimens up to the experimentally observed failure point. The place where sudden decreases in force-displacement curves is taken as the fracture initiation and the equivalent plastic strain value at this point is selected as failure strain. The failure strain, averaged T and $\bar{\theta}$ values are taken from the critical elements which have the highest equivalent plastic strain. In ST, NT10 and PST specimens, the critical elements are in the center of the specimens, while the critical points of the ISS are in the middle of the curved region in the gauge section. The failure strain, averaged T and $\bar{\theta}$ values are given in Table 3.

Table 3. Fracture strain, averaged T and $\bar{\theta}$ for all specimens.

Specimen	T_{ave}	$\bar{\theta}_{ave}$	ε_f
ST	0.4040	0.9614	0.6512
NT	0.4849	0.7650	0.5245
PST	0.5494	0.5882	0.4564
ISS	0.1205	0.1546	0.5069

Then, the calibration of the MMC and JC model parameters is simply done using the MATLAB curve fitting tool. Since there are 3 calibration parameters in the JC criterion, using ST, NT10, and PST specimens is sufficient for calibration. For the data given in Table 3, 5 calibration parameters are identified while $\hat{C}_4$ is taken as 1 for the MMC model. Firstly, $\hat{C}_3$ is assumed to be 1 and the other 4 parameters are fitted to the damage criterion. Then, $\hat{C}_3$ is included in the parameter fitting by keeping A as constant. The calibrated damage parameters are given in Table 4.

Table 4. Calibrated parameters of the MMC and JC damage models for IN718.

	$\hat{D}_1$	$\hat{D}_2$	$\hat{D}_3$			
JC	0.04	1.798	2.679			
	A [MPa]	n	$\hat{C}_1$	$\hat{C}_2$	$\hat{C}_3$	$\hat{C}_4$
MMC	1946	2.216	10.39	14420	4.517	1

Ductile Failure Simulations.

The results obtained from FE simulations with the calibrated damage models for 4 different specimens are shown in Fig. 3. Note that experiments are repeated 3 times for each geometry to check variation between specimens. Experimental results are shown with black dashed lines, while the orange and blue solid lines represent JC and MMC simulations, respectively. Due to the changes in manufacturing and micromechanical inhomogeneities, slight variations in failure strain are observed between tests. In the smooth tension test, the variation is found to be higher than normal. Nevertheless, the failure strain value is calibrated based on the average of multiple tests for each specimen geometry. It is clear that the models are able to capture the experimental force-displacement relation for all specimens. In ST, NT10 and PST simulations, JC and MMC models yield almost exactly the same response. However, for ISS specimen, although the failure points predictions are similar, the subsequent failure response is vastly different for the JC model.

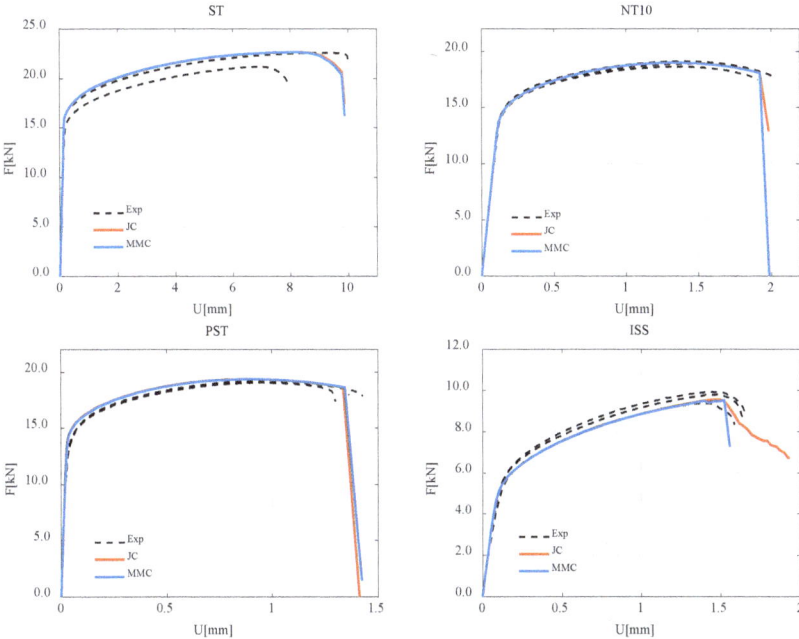

Fig. 3. Experimental comparison of JC and MMC damage models. Force vs. displacement curves for 4 tensile specimens.

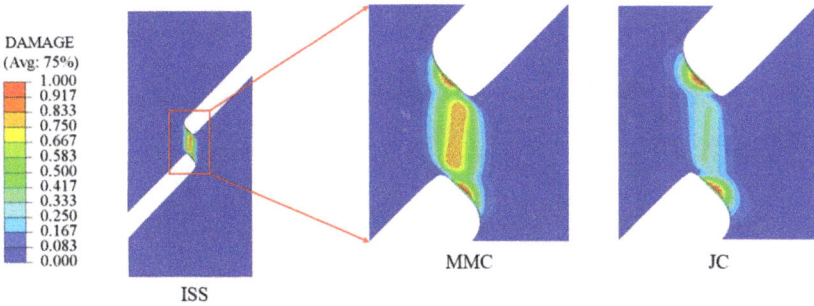

Fig. 4. Comparison JC and MMC damage distribution for ISS at the onset of failure.

In Fig. 4, the damage distributions at the onset of failure are plotted for the ISS specimen using the JC and MMC models. The JC model is known to be insufficient at predicting failure in shear dominant loads due to the lack of Lode parameter effect. One can see that in the gauge section, the JC model predicts lower damage values compared to the MMC model. This results in a gradual decrease of load for the JC model while the MMC model yields a sudden drop after the initiation of failure which is more compatible with the experimental observations.

Material Forming - ESAFORM 2023
Materials Research Proceedings 28 (2023) 807-816

Materials Research Forum LLC
https://doi.org/10.21741/9781644902479-88

Flow Forming Process Simulations.

The verified damage models are applied for the simulation of the flow forming process with two different preform geometries. Geometries differ by their thickness to inner diameter ratios. Preform 2 has double thickness to diameter ratio compared to Preform 1. Simulations are performed at the same feed rate, revolution speed, and the other geometrical parameters are kept constant. Simulation results are depicted in Fig. 5. The highest damage is predicted to be at the inner surfaces where the preform is in contact with the mandrel for both models. Due to the shear dominant nature of the process, the JC model gives significantly smaller damage accumulations compared to the MMC model. At 40% reduction, the damage exceeds 1 with the MMC model while it is much less than 1 for the JC model. With a higher thickness to diameter ratio, the damage is found to increase and spread over a larger area. It is normally expected that the crack would form on the outer surface of the flow formed specimen, which is not captured with the current models. However, referring to a previous study of the authors [15], a more realistic modeling approach with temperature dependent plasticity and damage models is important and would change the damage distribution.

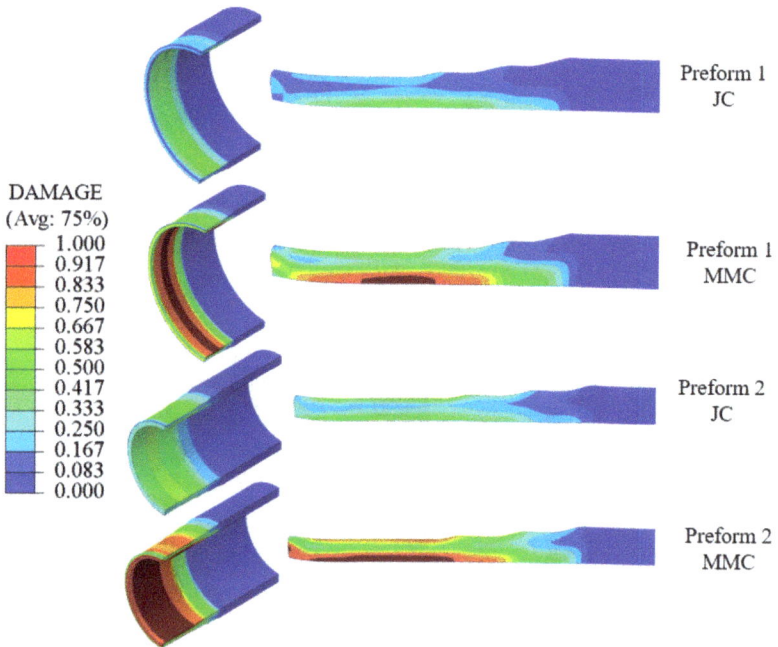

Fig. 5. Distribution of damage over the preform at 2 different geometries.

Material Forming - ESAFORM 2023   Materials Research Forum LLC
Materials Research Proceedings 28 (2023) 807-816   https://doi.org/10.21741/9781644902479-88

Summary
In this work, ductile failure model calibration is performed with experiments using four geometries for IN718 nickel alloy. Calibrated models are then verified through FE element analysis of the specimens and results are found to be in good agreement with experimental data. The MMC model is more capable of predicting experimental force-displacement curves in the shear dominant failure case. Furthermore, the two damage models are implemented in a flow forming simulation to discuss the differences in damage accumulation. Models show a disparity in the distribution of damage at two different preform geometries. It is concluded that the MMC failure criterion is a more appropriate choice for forming processes such as flow forming. It should be noted that the current FE model should be extended to include temperature and strain rate effects to accurately make an experimentally comparative study.

References
[1] A. Karakas, T.O. Fenercioğlu, T. Yalçinkaya, The influence of flow forming on the precipitation characteristics of Al2024 alloys, Mater. Lett. 299 (2021) 130066. https://doi.org/10.1016/j.matlet.2021.130066

[2] A.L. Gurson, Continuum theory of ductile rupture by void nucleation and growth: part I—yield criteria and flow rules for porous, J. Eng. Mater. Technol. 99 (1977) 2-15. https://doi.org/10.1115/1.3443401

[3] J. Lemaitre, Coupled elasto-plasticity and damage constitutive equations, Comput. Methods Appl. Mech. Eng. 51 (1985) 31-49. https://doi.org/10.1016/0045-7825(85)90026-X

[4] T. Yalçinkaya, C. Erdogan, I.T. Tandogan, A. Cocks, Formulation and implementation of a new porous plasticity model, Procedia Struct. Integr. 21 (2019) 46-51. https://doi.org/10.1016/j.prostr.2019.12.085

[5] G.R. Johnson, W.H. Cook, Fracture characteristics of three metals subjected to various strains, strain rates, temperatures and pressures, Eng. Fract. Mech. 21 (1985) 31-48. https://doi.org/10.1016/0013-7944(85)90052-9

[6] Y. Bai, T. Wierzbicki, A new model of metal plasticity and fracture with pressure and Lode dependence, Int. J. Plast. 24 (2008) 1071-1096. https://doi.org/10.1016/j.ijplas.2007.09.004

[7] T. Wierzbicki, Y. Bao, Y.W. Lee, Y. Bai, Calibration and evaluation of seven fracture models, Int. J. Mech. Sci. 47 (2005) 719-743. https://doi.org/10.1016/j.ijmecsci.2005.03.003

[8] Y. Bai, T. Wierzbicki, Application of extended mohr-coulomb criterion to ductile fracture. Int. J. Fract., 161 (2010) 1-20. https://doi.org/10.1007/s10704-009-9422-8

[9] D. Mohr, S.J. Marcadet, Micromechanically-motivated phenomenological Hosford-Coulomb model for predicting ductile fracture initiation at low stress triaxialities, Int. J. Solid Struct. 67-68 (2015) 40-55. https://doi.org/10.1016/j.ijsolstr.2015.02.024

[10] R. Li, Z. Zheng, M. Zhan, H. Zhang, Y. Lei, A comparative study of three forms of an uncoupled damage model as fracture judgment for thin-walled metal sheets, Thin-Walled Struct. 169 (2021) 108321. https://doi.org/10.1016/j.tws.2021.108321

[11] W. Xu, H. Wu, H. Ma, D. Shan, Damage evolution and ductile fracture prediction during tube spinning of titanium alloy, Int. J. Mech. Sci. 135 (2018) 226-239. https://doi.org/10.1016/j.ijmecsci.2017.11.024

[12] H. Wu, W. Xu, D. Shan, B.C. Jin, An extended gtn model for low stress triaxiality and application in spinning forming, J. Mater. Process. Technol. 263 (2019) 112-128. https://doi.org/10.1016/j.jmatprotec.2018.07.032

[13] A.K. Singh, A. Kumar, K.L. Narasimhan, R. Singh, Understanding the deformation and fracture mechanisms in backward flow-forming process of Ti-6Al-4V alloy via a shear modified

continuous damage model, J. Mater. Process. Technol. 292 (2021) 117060. https://doi.org/10.1016/j.jmatprotec.2021.117060

[14] H. Vural, C. Erdoğan, T.O. Fenercioğlu, T. Yalçinkaya, Ductile failure prediction during the flow forming process, Procedia Struct. Integr. 35 (2022) 25-33. https://doi.org/10.1016/j.prostr.2021.12.044

[15] C. Erdogan, H. Vural, T.O. Fenercioglu, T. Yalcinkaya, Effect of process parameters on ductile failure behavior of flow forming process, Procedia Struct. Integr. 42 (2022) 1643-1650. https://doi.org/10.1016/j.prostr.2022.12.207

Material Forming - ESAFORM 2023
Materials Research Proceedings 28 (2023) 817-824

Materials Research Forum LLC
https://doi.org/10.21741/9781644902479-89

Deformation analysis of ZnMgAl coated steel sheet using digital image correlation (DIC) system

LISIECKA-GRACA Paulina[1,a] *, MUSZKA Krzysztof[1,b], KWIECIEŃ Marcin[1,c] and DZIURDZIA Jakub[2,d]

[1]AGH University of Science and Technology, Faculty of Metals Engineering and Industrial Computer Science, 30 Mickiewicza Av., 30-059, Krakow, Poland

[2]ArcelorMittal Tubular Products, 1 Ujastek St., 30-969, Kraków, Poland

[a]graca@agh.edu.pl, [b]muszka@agh.edu.pl, [c]mkwiecie@agh.edu.pl, [d]jakub.dziurdzia@arcelormittal.com

Keywords: Digital Image Correlation (DIC), ZnMgAl Coating, Tubular Products

Abstract. Protective metallic coatings may be applied to the finished tubes and pipes using the hot dip method or may be applied to cold rolled sheet that is then used in the roll forming process of tubes manufacturing. In the latter case applied coating must ensure the proper adhesion and lack of cracks during deformation. In this work, the main objective is to assess the plastic deformability of steel sheets covered with an advanced ZnMgAl coating. The assessment of the susceptibility of the ZnMgAl coatings to plastic deformation was carried out using three-point bending tests, supported by the digital image correlation technique. A standard bending test of a steel sheet with a ZnMgAl coating at different bending angles was also used. The comparison of the results of the tests carried out allowed for the formulation of conclusions for direct use in industrial practice.

Introduction

Optimization and control of the processes of industrial metal forming of steel products requires an understanding of the behavior of the charge material under deformation conditions similar to those, in which it will be formed. Pipes with open-joints are produced from steel sheet in the rolling process, so the accompanying deformations in the bending process are significant [1]. The steel sheet used in the production process should be characterized by good mechanical properties and formability. It should also be remembered that in a pipe manufactured using the open-joint method, the weakest point is the joint/weld. The protective coating can also be used to protect the weld. There are many types of protective coatings on the market today. Many of them are based on zinc. Galvanized steels produced by hot-dip galvanizing find many applications, e.g. in the construction and automotive industries [2-4]. By introducing aluminum and magnesium into a standard zinc coating, a ZnAlMg coating was obtained, which provides excellent corrosion and wear resistance [5]. One of them is the Zn Al3,5Mg3 (Trade name: Magnelis®) coating patented by ArcelorMittal. This coating is characterized by the great corrosion resistance and wide range of possible applications. Compared to standard Zn coatings, Magnelis provides similar level of corrosion resistance protection with much thinner layer of coating. So far, this coating has not been widely studied in terms of its ductility. The coated ZnAlMg steel plate can be used to the production process of the open-joint tubes. This however requires high cracking resistance both of the steel sheet and coating during the roll-forming process. Even a micro damage of the coating may lead to cracking in the most deformed areas and may be the cause of the deteriorations in the in-service corrosion resistance [5-7]. Thus, obtaining the information of the correlation between microstructure and mechanical properties during the roll forming of open-joint tube process is necessary. This knowledge will allow to improve the quality of the coating and to optimize the parameters of the deformation process. Therefore, the main purpose of the presented work was to

Material Forming - ESAFORM 2023
Materials Research Proceedings 28 (2023) 817-824

Materials Research Forum LLC
https://doi.org/10.21741/9781644902479-89

assess the properties of ZnMgAl coated steel sheet used for the production of tubular products where it is subjected to the process of deformation by bending.

Materials

Three types of steel sheets were used in the tests. The base material was hot-rolled steel sheet, the chemical composition of which is shown in Table 1. The base material was E220 steel sheet with a ZnMgAl coating with a total thickness of 1.4 mm. The other two steel sheets of the S235 grade were taken as the reference material - one with a standard Zn coating and the other without this coating. The thickness of both sheets was 1.5 mm.

First, the quality of every coating was investigated using the Nova NanoSEM FEI scanning electron microscope (SEM). Additionally, the thickness and homogeneity of the surface was measured (Fig.1).

Table. 1. Based chemical composition of the steel sheet (% wt.).

Material	C	Mn	Si	P	S	Al	Cu	Sn
E220	0,049	0,20	0,004	0,023	0,015	0,032	0,021	0,002
S235JRH	0,15	0,12	0,030	0,020	0,025	0,01	0,55	-

Fig. 1a and b present the initial microstructure of the cross – section of specimen with standard Zn coating (as received). Analysis of the sheet steel with Zn coating showed good coating adherence to the base sheet. The mean thickness of the Zn coating was approximately 98 μm. During the microstructural analysis the cracks of the Zn coating were not observed.

A similar analysis was performed for the ZnMgAl-coated sample (Fig. 1c, d). In the case of the main research material, the measurement of the coating thickness showed that the ZnMgAl layer is smaller than in the case of the standard Zn layer and is about 38 - 40 μm. Its microstructural analysis showed good adhesive properties and cracks were also not visible.

Based on the microstructural analysis of steel sheets coated with Zn and ZnMgAl it can be concluded that the initial quality of both samples is sufficient for their use in the rolling process.

Material Forming - ESAFORM 2023   Materials Research Forum LLC
Materials Research Proceedings 28 (2023) 817-824   https://doi.org/10.21741/9781644902479-89

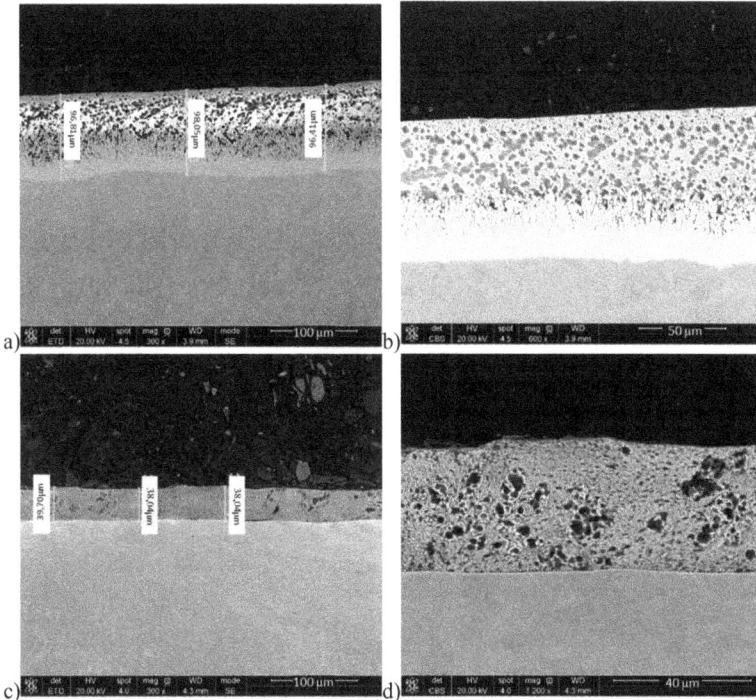

*Fig. 1. The example of the cross – section of the steel sheet with standard Zn coating – a); b)
and ZnMgAl coating – c); d).*

Experimental Procedure

In order to obtain the data for the analysis of deformation susceptibility of steel sheet with ZnMgAl coating, the experimental procedure was divided into two steps. The first stage included the three-point bending test. The radius of the upper tool was equal 5 mm. In this part of work, the three different sheets were deformed: 1) base material (with no coating) 2) sheet with standard Zn coating, and 3) sheet with ZnMgAl coating. The experiments were performed at room temperature with the velocity of the bending load of 1mm/min. The experimental procedure was combined with Digital Image Correlation (DIC) analysis. The DIC system consist of high resolutions CCD 5mpx camera. In order to obtained the accreted results, the DIC analysis was made using two different lenses. In first case the whole surface of the samples were analyzed ale for this the lens of 2.8/50 was used. In the second case only most deformed area was analyzed. and for this measurement the stereo microscope was used as a lens. Postprocessing of the obtained images was made using Istra4D software. The second stage of the investigation was focused only on the steel sheet with ZnMgAl coating. In this stage, the bending test based on the ISO 7438:2016 standard was applied. The bending angle ranging between 90 and 180 deg was used. Similarly, like in the pervious test, the bending test was made at the room temperature, but the angle on the upper tool was equal 30 deg.

Material Forming - ESAFORM 2023

Materials Research Proceedings 28 (2023) 817-824

Materials Research Forum LLC

https://doi.org/10.21741/9781644902479-89

Results and Discussion

In the presented work, the DIC strain measurement was applied to analyze deformation of coated sheets subjected to the three-point bending test. During the test, DIC method was applied to track the strain distribution during the deformation process [8-9]. The upper tool displacement was kept the same for the all analyzed specimens. Results obtained for the one-time step before end of the deformation process with the stress–strain curve are presented in the Fig. 2. DIC strain distribution maps show typical strain accumulation for the three-point bending test localized in the lower part of the specimen. The highest strain accumulation was observed in the sample with standard zinc coating and the lowest value was obtained for the sample with ZnMgAl coating. In the case of sheet with zinc coating the additional strain concentration can be observed in the upper part of the specimen (under the upper tool), what suggests the possibility of the zinc coating crack occurrence in this area.

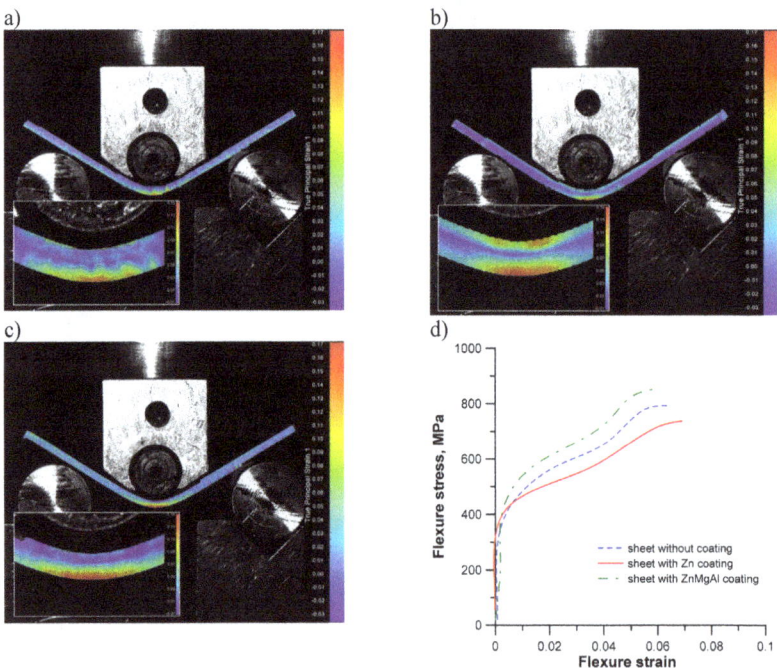

Fig. 2. The true principal strain distributions maps obtained in DIC analysis for sheet plate without the coating – a); with standard zinc coating – b); with ZnMgAl coating – c); and the stress – strain curve obtained in the three – point bending test using two type of the lenses.

The main aim to use the DIC analysis was the estimate the local deformation value on the sample thickness during the bending test in order to related these values with the coating deformation susceptibility in related to the open join tube production process.

Additionally, based on the obtained DIC results, it can be expected that on the sheet with the zinc coating the cracks will appear in the external and internal surfaces of the specimen. For the specimen without coating the cracking of the oxide scale also took place. For confirmation, all

Material Forming - ESAFORM 2023
Materials Research Proceedings 28 (2023) 817-824

Materials Research Forum LLC
https://doi.org/10.21741/9781644902479-89

specimens after the three-point bending test were analyzed using Zeiss optical microscope (Fig. 3). Observations were carried out on the outer and inner surfaces, where the accumulation of deformations was the greatest (Fig. 3a).

The obtained micrographs revealed the flaking of the oxide scale layer - especially on the internal surface of the specimen with no coating (Fig. 3b). In the case of the second specimen with zinc coating clear cracking both at external and internal surface was visible. A digital crack assessment was made and compared with patterns based on the PN-EN ISO 4628-4:2004(U) standard. It was found that the internal cracks can be assigned to the ASTM 4 group and external cracks to the ASTM 2 group. For the last analyzed specimen, with the ZnMgAl coating, no visible cracks were found at any of the analyzed surfaces (internal and external).

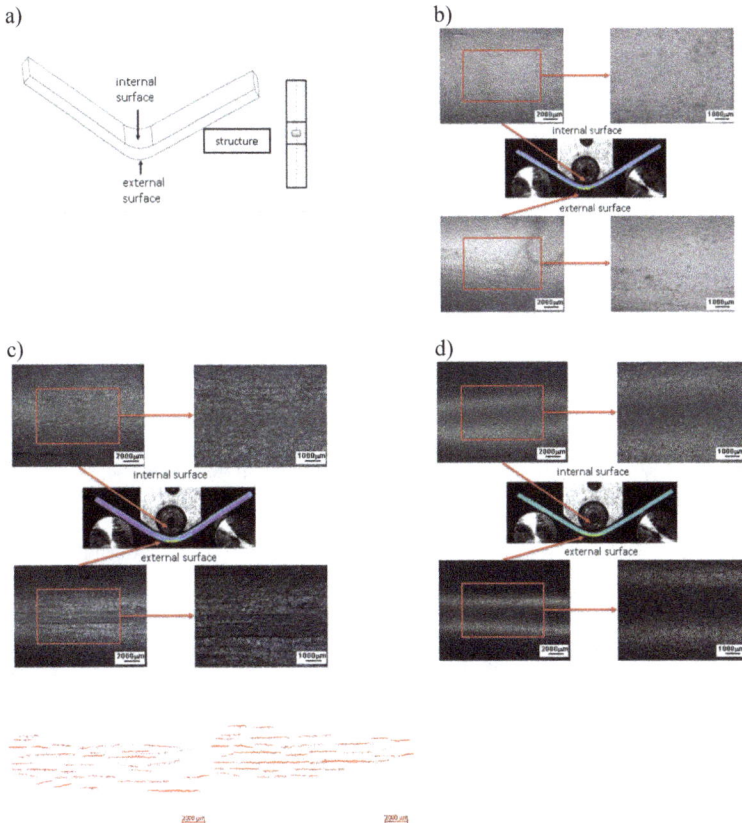

Fig. 3. Position of the optical microscope analysis of surface quality bending sample on the internal and external surface - a). Obtained results for the specimen without the coating - b); with standard zinc coating with comparison to the patterns from PN-EN ISO 4628-4:2004(U) standard - c) and with ZnMgAl coating - d).

Material Forming - ESAFORM 2023   Materials Research Forum LLC
Materials Research Proceedings 28 (2023) 817-824   https://doi.org/10.21741/9781644902479-89

The surface that was observed using DIC system was subjected to the scanning electron microscope observation using FEI Nova Nano SEM. For the analysis the specimen with zinc coating and ZnMgAl coating was used. Fig. 4 shows that in the case of the specimen with standard zinc coating the coating detachment and cracking (marker with the red arrows) can be observed. On the second analyzed specimen the any layer damage was not found.

a)   b)

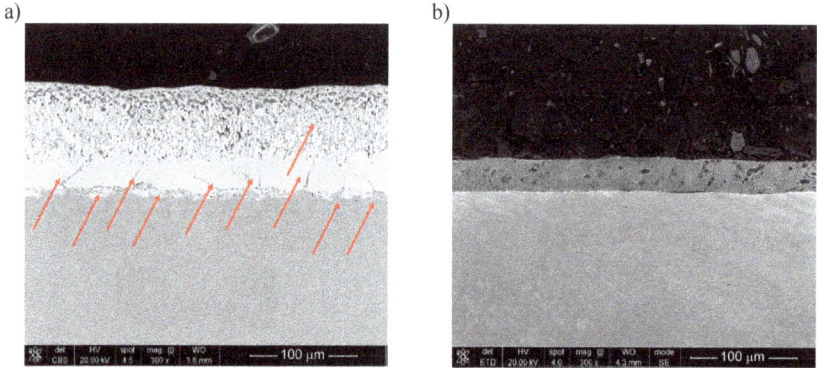

Fig. 4. The SEM image for specimen with Zn coating – a) and ZnMgAl coating – b).

The analysis of the three-point bending test results confirmed that the steel sheet with ZnMgAl coating is suitable to the deformation process.

Additionally, in order to design the roll-forming process, bending test is usually performed to check the springback angle and select the roll forming pass design. In the current study this test was also performed but its aim was more focused on the analysis of the ZnMgAl coating adhesion. The obtained results for the bending test with angle equal to 90deg are presented in the Fig. 5. Additional analysis performed after bending shows that, similarly to the previous test, the cracks were not visible either of the surface (internal and external).

Internal surface   **External surface**

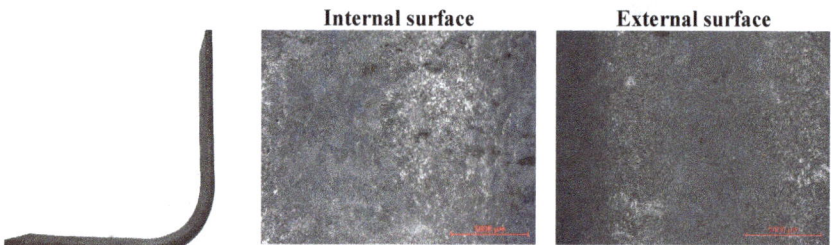

Fig. 5. The shape of the specimen after bending test with microstructure of internal and external surface.

Summary

The tests carried out and the analysis of their results allowed for the assessment of the suitability of the steel sheet with the ZnMgAl coating for the process of forming pipes with an open joint. For this purpose, a standard bend test and a three-point bend test combined with a digital image correlation system were used. In the case of the three-point bending test, the test results of a steel sheet without a protective coating and with a standard zinc coating were taken as the starting point. Based on the DIC analysis, areas of strain concentration were determined where the protective coating could crack. Both bending tests, i.e. standard and three-point bending, showed that the tested material with ZnMgAl coating can be successfully used in the pipe production process. Based on the presented results, the following conclusions can be drawn:

- The microstructural analysis of the sheet steel with coating shows good adherence to the base steel for both analyzed coatings.
- In the case of the sheet plate without the coating strain distribution analysis showed the points localization at maximum strain value. Additional microstructure analysis showed that in this area the flaking of scale layer occurred.
- The DIC analysis of steel sheet with standard zinc coating showed the highest strain concentration both at internal and external surface. Microstructural analysis showed significant cracks. Based on the ISO standard was coating cracking was assigned to the ASTM 4 group for the internal surface and ASTM 2 group for the external surface.
- Based on the DIC analysis of the plate with ZnMgAl coating the strain concentration (beside standard strain localization characterized for the three-point bending test) was not observed. The microstructure analysis confirmed lack of the visible cracks on the surface.

Acknowledgements

The support from the National Centre for Research and Development, Poland through the grant no. POIR.01.01.01.00.0510/18 is greatly appreciated.

References

[1] P. H. Chan, K. Y. Tshai, M. Johnson, S. Li, The flexural properties of composite repaired pipeline: Numerical simulation and experimental validation, Compos. Struct. 133 (2015) 312-321. https://doi.org/10.1016/j.compstruct.2015.07.066

[2] A. Masoud, S. Bekir, K.J. Bart, P. Yutao, Cracking behavior and formability of Zn-Al-Mg coatings, Mater. Des. 212 (2021) 1-15. https://doi.org/10.1016/j.matdes.2021.110215

[3] S.M.A. Shibli, B.N. Meena, R. Remya, A review on recent approaches in the field of hot dip zinc galvanizing process, Surf. Coat. Technol. 262 (2015) 210-215. https://doi.org/10.1016/j.surfcoat.2014.12.054

[4] S.T. Vagge, V.S. Raja, R. Ganesh Narayanan, Effect of deformation on the electrochemical behavior of hot-dip galvanized steel sheets, Appl. Surf. Sci. 253 (2007) 8415-8421. https://doi.org/10.1016/j.apsusc.2007.04.045

[5] C. Yao, H. Lv, T. Zhu, W. Zheng, X. Yuan, W. Gao. Effect of Mg content on microstructure and corrosion behavior of hot dipped Zn-Al-Mg coatings, J. Alloy. Compd. 670 (2016) 239-248. https://doi.org/10.1016/j.jallcom.2016.02.026

[6] S. Thomas, N. Birbilis, M.S. Venkatraman, I.S. Coles, Self-repairing oxides to protect zinc: Review, discussion and prospects, Corros. Sci. 69 (2013) 11-22. https://doi.org/10.1016/j.corsci.2013.01.011

[7] B. Gao, S.W. Li, Y. Hao, G.F. Tu, L. Hu, S.H. Yin, The Corrosion Mechanism of Zn-5% Al-0.3% Mg Coating, Adv. Mat. Res. 189 (2011) 1284-1287. https://doi.org/10.4028/www.scientific.net/AMR.189-193.1284

[8] T.C. Chu, W.F. Ranson, M.A. Sutton, Applications of digital-image-correlation techniques to experimental mechanics, Exp. Mech. 25 (1985) 232. https://doi.org/10.1007/BF02325092

Material Forming - ESAFORM 2023   Materials Research Forum LLC
Materials Research Proceedings 28 (2023) 817-824   https://doi.org/10.21741/9781644902479-89

[9] T. He, L. Liu, A. Makeev, B. Shonkwiler, Characterization of stress–strain behavior of composites using digital image correlation and finite element analysis, Compos. Struct. 140 (2016) 84. https://doi.org/10.1016/j.compstruct.2015.12.018

Material Forming - ESAFORM 2023
Materials Research Proceedings 28 (2023) 825-832

Materials Research Forum LLC
https://doi.org/10.21741/9781644902479-90

Hybrid SPD process of aluminium 6060 for microforming

PRESZ Wojciech[1,a] *, JASIŃSKI Cezary[1,b], MORAWIŃSKI Łukasz[1,c]
and ORŁOWSKA Marta[2,d]

[1]Warsaw University of Technology, Narbutta 85, 02-524 Warsaw, Poland

[2]Military University of Technology, gen. Sylwestra Kaliskiego 2, 00-908 Warsaw, Poland

[a]wojciech.presz@pw.edu.pl, [b]cezary.jasinski@pw.edu.pl, [c]lukasz.morawinski@pw.edu.pl,
[d]marta.orlowska@wat.edu.pl

Keywords: UFG Metals, Aluminium 6060, Hybrid SPD, Microforming

Abstract. The increasing share and expansion of miniature devices requires the mastery of miniature parts production, including metal parts. In metal microforming, due to the conditions of mechanical similarity, the aim is to use materials with the smallest grain. Ultra-fine-grained metals (UFG) fulfill this requirement. These metals can be obtained, inter alia, in SPD processes such as ECAP. The work uses the extension of the SPD based on ECAP with additional metal forming operations necessary to obtain blanks for microforming in the form of a 0.2 mm foil. Three variants of the technological process were performed. This foil was then subjected to a micro-drawing operation aimed at determining the influence of the foil preparation process on the sheet metal microforming process flow.

Introduction

The miniaturization of devices [1], which has been progressing for several decades, imposes the miniaturization of the parts used for their production, which are largely produced by metal forming technology [2]. Reducing the dimensions to about one millimeter causes a change in the dominant physical phenomena. Difficulties in the accurate design of technological processes for miniature parts, combined with the growing demand for them, led to the separation of a relatively new branch of metal forming, which is microforming [3]. This is because of the natural granularity of the polycrystalline materials used [4] and the surface layer treated as an area with a certain thickness, mainly dependent on the structure of the material [5]. The share of both of these features increases as the product becomes smaller, the material of which can no longer be considered uniform throughout its volume. The proposed avoidance of many of the unfavorable scale effects is the concept of using ultrafine grain metals (UFG) [6-8]. Within the presented work, the field of interest was narrowed down to sheet metal microforming processes. As a method of obtaining ultra-fine grain, the ECAP process was adopted on the stand used in previous works [9]. The selected material for the tests is aluminium alloy 6060, which was chosen because of its high (in the group of aluminium alloys) strength properties.

The main purpose of the work was to develop a research method allowing for the assessment of the possibility of using this alloy with the UFG structure obtained by the ECAP method for sheet metal microforming processes. It should be noted that the UFG materials obtained from the ECAP process are in the form of rods that must undergo further deformations [10], so that the final effect is a foil with a UFG structure. The influence of the technological path of preparing foil with a thickness of 0.2 mm made of aluminium alloy 6060 on the possibility of sheet metal microforming was initially examined. The process of free micro-drawing was adopted as the verification process. It was chosen because of the simplicity of implementation, which favours further miniaturization and reduces the number of process parameters in research. Due to the

Material Forming - ESAFORM 2023   Materials Research Forum LLC
Materials Research Proceedings 28 (2023) 825-832   https://doi.org/10.21741/9781644902479-90

difficult formability of the selected material discussed in the literature, it was decided to use heat treatments to improve this feature.

Material Preparation

The extended UFG metal fabrication process with the ECAP main operation was used in the work to prepare 0.2 mm thick 6060 aluminium foil. Three series of material in the form of foil, marked respectively with the symbols CG (material not subjected to ECAP), UFG F (material subjected to ECAP in the delivery condition) and UFG G (material supersaturated before ECAP) were punched with a 4 mm diameter punch. Then, the obtained discs were used to carry out the micro-drawing operation.

The hot-extruded bar PN EN 6060 ø50 was used as the initial material. The bar was divided into billets with dimensions of 8x8x45 mm by milling according to the scheme shown in Fig. 1a. Part of the billets were subjected to supersaturation (annealing for 2 hours at 525°C, cooling in water). These billets were used to prepare a series of UFG G material. Then, a two channel turns mtECAP (2x110°) was performed six times at 150°C. ECAP was performed for a batch of supersaturated material (UFG G) and as delivered (UFG F). The rods obtained in this way were subjected to free upsetting (Fig. 1b) at 150°C, obtaining plates with a thickness of 1 mm. Upsetting was carried out in three subsequent operations with a comparable increase in surface area (~2 times) to enable renewal of lubricating layer. In addition, identical upsetting was applied to the billets that were not subject to the ECAP. The diagram of the upsetting operation is shown in Fig. 1c. The obtained material was mainly subjected to natural ageing, however, the use of elevated temperature in the ECAP and upsetting operations meant that the material stayed in total for almost an hour at a temperature close to 150°C, which could initiate the processes related to artificial ageing.

a)   b)   c)

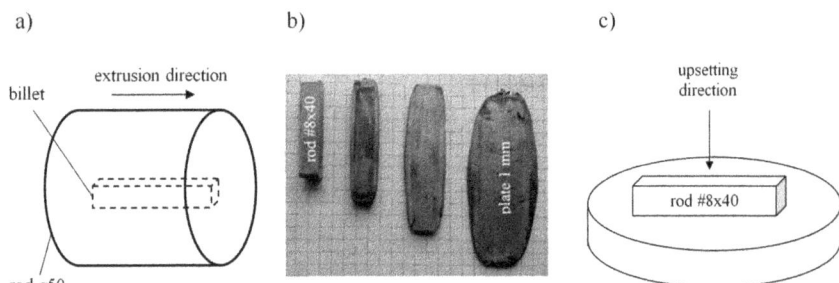

Fig. 1. Scheme of orientation of the taken billets in relation to the ø50 mm rod (a), the result of subsequent upsetting operations at a thickness of ≈ 5, 3 and 1 mm (b) and the scheme of upsetting operations (c).

In the next stage of the process, the obtained plates were cut into 6 mm wide strips, which were rolled to a thickness of 0.2 mm at room temperature to obtain a foil intended for microforming. The method of taking strips from plates are shown in Fig. 2. Full material preparation process is presented in Table 1.

Material Forming - ESAFORM 2023
Materials Research Proceedings 28 (2023) 825-832

Materials Research Forum LLC
https://doi.org/10.21741/9781644902479-90

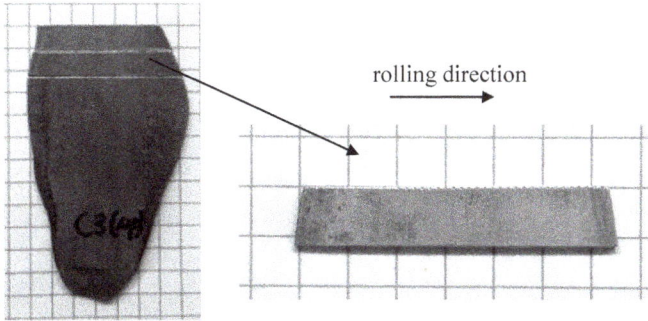

Fig. 2. Preparation of metal strips for the rolling process.

Table 1. Material preparation process (green color mean that operation was performed).

Operation	Cutting from ø50 rod	Supersaturation	mtECAP	Upsetting	Cutting	Rolling
Result	billet (#8x45 mm)	billet (#8x45 mm)	Rod (#8x40 mm)	Plate (1 mm thick)	Strip (6 mm wide)	Foil (0,2 mm thick)
CG						
UFG F						
UFG G						

Material Analysis

In order to verify the effect of supersaturation on the material properties after the ECAP, a uniaxial tensile test for specimen with 10 mm gauge length [9] was performed, in which the yield strength $R_{p0.2}$, tensile strength R_m and elongation A were determined. These parameters were determined for the material immediately after the ECAP process and after additional upsetting. Tensile test for a thin foil was not performed. Fig. 3 presents the obtained results and compares them with the parameters of the material as delivered.

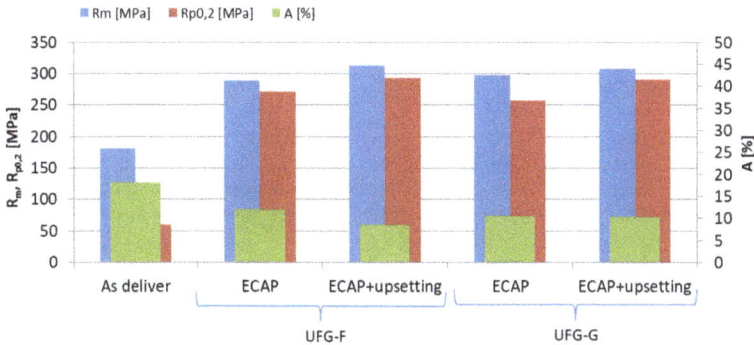

Fig. 3. Comparision of R_m, $R_{p0.2}$ and A for prepared series of material in two stages of the process.

Material Forming - ESAFORM 2023   Materials Research Forum LLC
Materials Research Proceedings 28 (2023) 825-832   https://doi.org/10.21741/9781644902479-90

The obtained 0.2 mm foils were subjected to EBSD structural analysis. Fig. 4 shows the corresponding grain size distribution for CG and UFG foil.

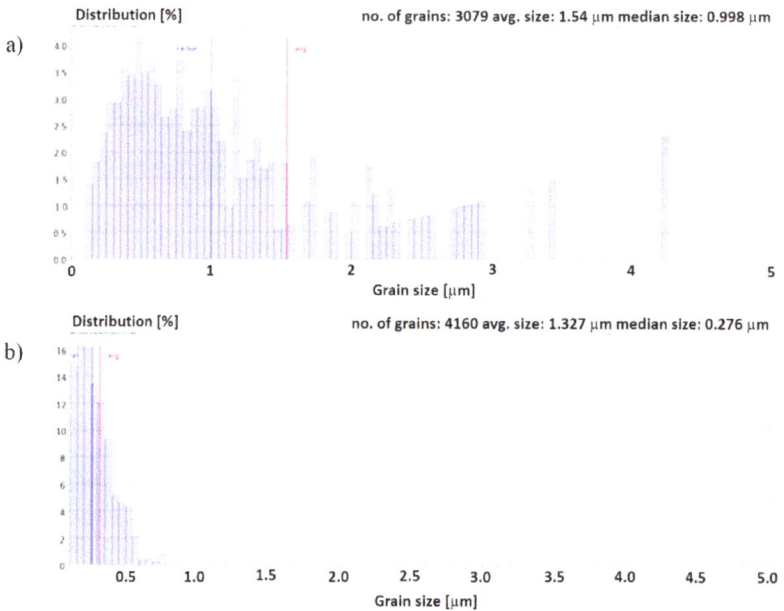

Fig. 4. Grain size distribution of 6060 aluminum foil for CG (a) and UFG-F (b) series.

Collected data indicate a noticeable effect of supersaturation on the properties of the obtained material, however, this effect is different immediately after ECAP and after additional upsetting. The supersaturated material (UFG-G series) shows higher strength and lower plasticity after ECAP compared to the UFG-F series, however, after additional upsetting, this relationship is reversed and the UFG-G series shows lower strength and higher plasticity. This phenomenon may be caused by the influence of increased tamperture occurring during upsetting, which may partially eliminate the strengthening effects. Finally, a positive effect of supersaturation on the formability of the obtained UFG plates was observed.

Structural analysis of foils intended for sheet metal microforming leads to the conclusion that the ECAP process unifies the structure of the material in the entire volume of the obtained foil. The foil made of the CG material is characterized by a much greater grain size variation, despite the fact that the large plastic deformation that took place during subsequent metal forming operations significantly fragmented the grains.

Formability Determination

Formability determination was performed by means of micro-drawing process. Carrying out the drawing process requires a preliminary operation, which is the preparation of the preform. In this case it was a disc with a diameter of 3.8 mm. A micro-blanking process with a blank holder, which in turn required preforms in the form of 10x7x0.2 mm strips. In this process, a precise micro-device placed on the testing machine table was used. The micro-device ensured precise positioning

Material Forming - ESAFORM 2023
Materials Research Proceedings 28 (2023) 825-832

Materials Research Forum LLC
https://doi.org/10.21741/9781644902479-90

of the tools. The micro-blanking operation was additionally used to analyze the technological variants of CG, UFG-F and UFG-G. The courses of forces were recorded (Fig. 6a), the cut strengths and the fill factors of the graph were determined (Table 2). To reduce the complexity of the tools, free micro-drawing was carried out in the system shown in Fig. 5a, consisting of the same device as for micro-blanking. For this kind of process the drawing ratio should be bigger than 0.55 [1] , and 0.58 was chosen.

a)   b)

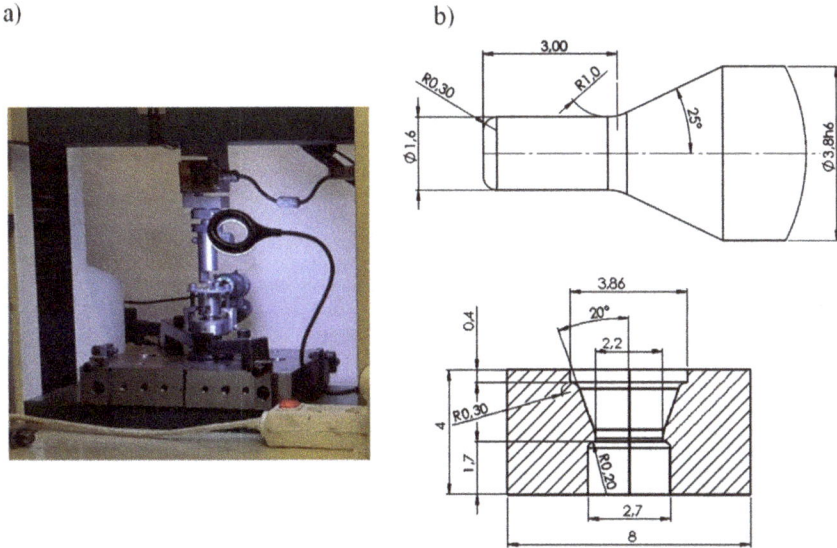

Fig. 5. Setup for micro-drawing (a) and dimensions of the tools used (b).

The dimensions of the tools are shown in Fig. 5b. Lubrication was applied with a lubricant based on MoS2 from the die side and light oil from the punch side. The coefficient of extrusion in the process was 0.54. The following description refers to the force recorded during the process (Fig. 6b).

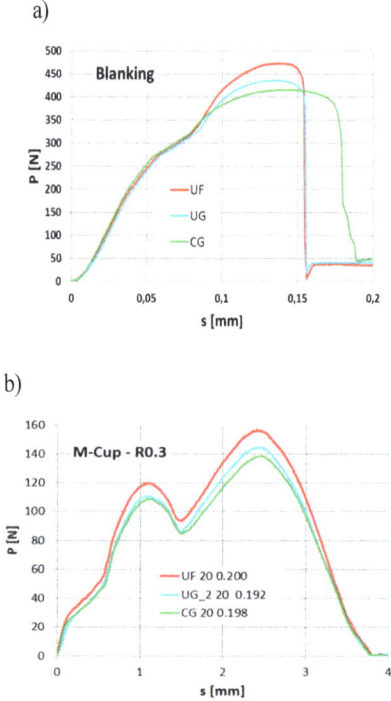

Fig. 6. Course of forces during micro-blanking (a) and micro-drawing (b) of 0.2 mm foil for three series of material.

The disk is placed in the cavity of the matrix (Fig. 5b) enabling its precise positioning. The precisely guided punch presses centrally on the foil, the force of the process increases to the value of the first maximum. After reaching it, the edge of the disc begins to be drawn into the die, sliding on its conical surface. As the punch moves, the force decreases and reaches a minimum value, after which it increases again due to the strengthening of the material and the thickening of the edge of the cup. When the strengthening and thickening of the material is no longer able to compensate for the decreasing perimeter of the edge, the force of the process decreases until its full completion. The threat here is the possible cracking of the bottom, which is the main phenomenon limiting the process.

Table 2. Cut strength R_t and graph fill factor λ for three series of 0.2 mm foil.

	CG	UF	UG
R_t [MPa]	168	185	181
λ	0,67	0,51	0,55

Material Forming - ESAFORM 2023   Materials Research Forum LLC
Materials Research Proceedings 28 (2023) 825-832   https://doi.org/10.21741/9781644902479-90

Despite the slightly higher cut strength determined for the material from the UFG-F series, the effect of supersaturation before the ECAP on the blanking process conditions was not observed. An increase in the micro-drawing process force was observed in the case of the UFG-F material. With regard to the CG and UFG-G materials, slight differences in the force course were observed, requiring further research. The shape and appearance of the cups surface (Fig. 7) as well as the size of the lugs indicating the presence of flat anisotropy in the assessment of the image from the light microscope do not indicate significant differences in individual material series.

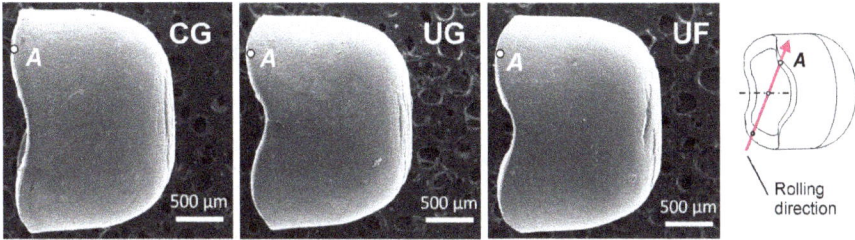

Fig. 7. SEM images of obtained micro cups for three series of material.

Summary
The presented paper proposes a research method for selection of process parameters enabling the use of UFG materials obtained by the ECAP method for use in sheet metal microforming. The applied ECAP has been enriched with the process of obtaining thin foils by initial multiple upsetting with the possibility of increasing the temperature and subsequent multiple rolling at room temperature. The obtained material in the form of a foil with a thickness of 0.2 mm made of aluminium alloy 6060 was then subjected to a micro-drawing operation recognized as a process determining the formability of the prepared materials in the conditions of sheet metal microforming. Necessary to prepare preforms for the micro-drawing process, the micro-blanking process was used to additionally compare the characteristics of foils from different series.

Based on the prepared research, it was found that the obtained material in the form of a foil is characterized by a uniform structure and is consistent throughout its volume without structural defects and cracks. The ECAP process plays a key role in unifying the structure throughout the volume of the foil. Its omission results in a significant decrease in the homogeneity of the material, despite the large deformations to which the material is subjected during upsetting and rolling, which causes significant grain refinement comparable to ECAP.

Despite the noticeable differences in strength properties occurring after the upsetting process for the supersaturated material and the material processed as delivered, no effect of supersaturation of the billets material before the ECAP process on formability was observed in the conditions of the sheet metal microforming.

Some differences in process flow for prepared materials was observed in micro-blanking and micro-drawing (Fig. 6), but for clearly distinguishes the formability there are planed extension of micro-drawing process in the next research.

References
[1] A.B. Frazier, R.O. Warrington, C. Friedrich, The Miniaturization Technologies: Past, Present, and Future, IEEE Trans Ind Electron. 42 (1995) 423-430. https://doi.org/10.1109/41.464603

[2] A. Jäger, S. Habr, K. Tesař, Twinning-detwinning assisted reversible plasticity in thin magnesium wires prepared by one-step direct extrusion, Mater. Des. 110 (2016) 895-902. http://doi.org/10.1016/j.matdes.2016.08.016

[3] M.W. Fu, W.L. Chan, A review on the state-of-the-art microforming technologies, Int. J. Adv. Manuf. Technol. 67 (2013) 2411–2437. https://doi.org/10.1007/s00170-012-4661-7

[4] Z. Jiang, J. Zhao, H. Lu, D. Wei, K. Manabe, X. Zhao, X. Zhang, Influences of temperature and grain size on the material deformability in microforming process, Int. J. Mater. Form. 10 (2017) 753–764. https://doi.org/10.1007/s12289-016-1317-4

[5] U. Engel, Tribology in microforming, Wear 260 (2006) 265-273. https://doi.org/10.1016/j.wear.2005.04.021

[6] W. Presz, A. Rosochowski, The influence of grain size on surface quality of microformed components, 9th Int. Conf. Mater. Form. ESAFORM 2006, Glas UK, 2006, pp. 587-590.

[7] W. Presz, Scale effect in design of the pre-stressed micro-dies for microforming., Comput. Meth. Mater. Sci. 16 (20160 1-8.

[8] W.J. Kim, S.J. Yoo, H.K. Kim, Superplastic microforming of Mg-9Al-1Zn alloy with ultrafine-grained microstructure, Scr. Mater. 59 (2008) 599-602.

[9] M. Ciemiorek, W. Chromiński, C. Jasiński, M. Lewandowska, Microstructural changes and formability of Al–Mg ultrafine-grained aluminum plates processed by multi-turn ECAP and upsetting, Mater. Sci. Eng. A. 831 (2022) 142202. https://doi.org/10.1016/j.msea.2021.142202

[10] M. Lipińska, L. Olejnik, M. Lewandowska, A new hybrid process to produce ultrafine grained aluminium plates, Mater. Sci. Eng. A 714 (2018) 105-116. https://doi.org/10.1016/j.msea.2017.12.096

[11] H. Tschaetsch, Metal Forming Practise, ISBN-10 3-540-33216-2, Springer Berlin Heidelberg New York, 2006, pp. 185-186.

Material Forming - ESAFORM 2023
Materials Research Proceedings 28 (2023) 833-838

Materials Research Forum LLC
https://doi.org/10.21741/9781644902479-91

Influence of intercritical annealing temperature on formability and mechanical properties of medium-manganese-steel in press hardening

TILLY Karl Johann[1,a] *, PLUM Tobias[1,b], BAILLY David[1,c] and HIRT Gerhard[1,d]

[1] Institute of Metal Forming, RWTH Aachen University, Aachen, Germany

[a]karl.tilly@ibf.rwth-aachen.de, [b]tobias.plum@ibf.rwth-aachen.de,
[c]david.bailly@ibf.rwth-aachen.de, [d]gerhard.hirt@ibf.rwth-aachen.de

Keywords: Press Hardening, Medium Manganese Steel, Formability, Hot Stamping

Abstract. Ultra-high-strengh-steel parts produced by press hardening are widely used in the automotive sector for lightweight construction and passenger safety applications. Medium-manganese-steels (MMnS) are currently investigated as an alternative to boron-manganese steels. Their favorable mechanical properties of high strength and high ductility after quenching are often based on rather complex heat treatment strategies resulting in a multiphase microstructure. One possible way such a microstructure can be obtained, is intercritical annealing and subsequent quenching during the press hardening process. For this processing route formability during hot stamping and final properties of the quenched material are dependent on the annealing temperature. For a successful part production, an annealing temperature that satisfies both, in-process properties, as well as final mechanical properties is mandatory. In this paper, the suitability of the intercritical annealing process route for press hardening of MMnS is investigated for a specific MMnS alloy. Formability in hot stamping conditions with respect to different annealing temperatures (range 700°C – 800°C) are examined by performing hot tensile tests, as well as experimental trials using a hot stamping tool. Final properties are analyzed by tensile tests. The formability during hot tensile tests and hot stamping show a strong positive correlation to the annealing temperature, while the values for uniform and ultimate elongation of the quenched material shows a strong negative correlation to the annealing temperature. The tensile strength of the quenched material shows a low sensitivity to the annealing temperature. No annealing temperature that satisfies both, in-process and final properties, could be found for the investigated MMnS alloy.

Introduction

Increasing demands for passenger safety and reduction of CO_2 emissions in the automotive sector led to the increased usage of high strength steels in autobody applications. The main advantage of advanced-high-strength-steels is the lightweight construction potential due to their mechanical properties.[1] Direct press hardening, consisting of the three steps austenitization, hot stamping and tool-quenching, is a widely used process to produce high strength autobody parts from sheet metals.[2] The commonly used manganese-boron steels such as 22MnB5 typically achieve tensile strengths of 1500 MPa und residual formabilities of minimum 5 % in the quenched state with fully martensitic microstructure [3]. In order to improve the performance of parts produced from manganese boron steels strategies such as tailored heating, or the usage of tailored materials such as tailor-welded blanks or cladded material have been investigated. The strategies of tailored processes or properties for press hardening of manganese-boron-steels mostly rely on reducing the strength of sections of the structural component in exchange for an improved ductility. Therefore the overall component performance, for example total energy absorption in a crash, can be

Material Forming - ESAFORM 2023
Materials Research Proceedings 28 (2023) 833-838

Materials Research Forum LLC
https://doi.org/10.21741/9781644902479-91

improved. On the downside to the production process of parts with tailored properties is more complex as for example segmented tools or tailor welded blanks are necessary. [3,4]

As an alternative to the commonly used manganese-boron-steels, steels of the medium-manganese alloy class (MMnS) have been investigated, with the aim of providing a material with high strengths and simultaneously high ductility for press hardening processing. The favorable mechanical properties of MMnS are based on a fine-grained multi-phase microstructure after press hardening that consists of retained austenite and martensite. To achieve this microstructure, it is often necessary to employ multistep heat treatment procedures (for example Quench and Partitioning) before or after the hot stamping operation, wich leads to a more complex processing compared to manganese boron steels [5-9].

In this paper it will be investigated whether a single heat treatment of intercritical annealing is suitable for hot stamping processing of a specific MMnS alloy.

Investigated Alloy and Processing

For the investigations presented in this paper, 2 mm thick sheets of a medium manganese steel alloy were used. The chemical composition of the steel is shown in Tab. 1. The schematic temperature curve of the intercritical annealing and phase fractions before and after quenching are shown in Fig. 1. Before quenching, different fractions of austenite (γ) and primary martensite (α'_p) depend on the annealing temperature. After quenching, a phase transformation of the austenite to retained austenite (γ_r) and secondary martensite (α'_s) as shown is expected, while the primary martensite percentage is anticipated to be stable.

Table 1. Chemical composition (wt. %).

Fe	C	Mn	Si	Al	Cr	P	S	N
bal.	0.30	4.99	1.55	0.004	0.04	0.005	0.003	0.004

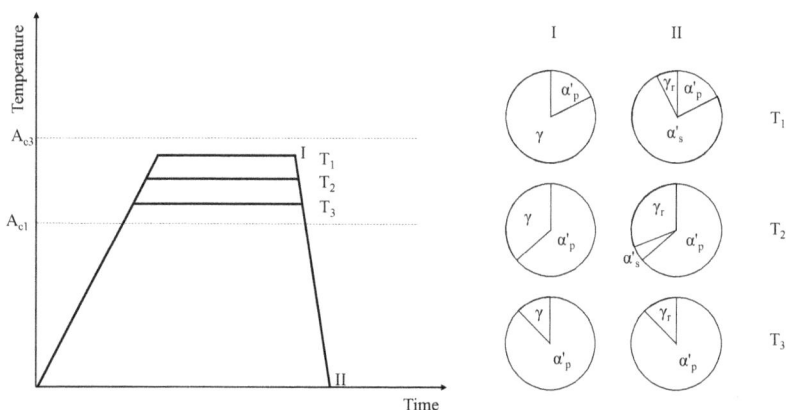

Fig. 1. Schematic illustration of phase fractions before and after quenching from different intercritical annealing temperatures.

Material Forming - ESAFORM 2023   Materials Research Forum LLC
Materials Research Proceedings 28 (2023) 833-838   https://doi.org/10.21741/9781644902479-91

Hot Tensile Tests

In order to verify the expected material behavior in terms of dependence of formability during hot stamping on the blank temperature during prior annealing, hot tensile tests were performed on a TA Instruments DIL805 dilatometer. The annealing temperature was varied in an interval of 700°C to 800°C. All specimens were inductively heated to the respective annealing temperature at a rate of 10 K/s. The soaking time was set to 5 min. After annealing, the specimens were quenched to 600°C and tensile tests were performed at a strain rate of 0.5 1/s. The uniform elongation is taken as an indicator of the materials formability with respect to the annealing temperature.

Fig. 2 shows the measured uniform elongation values at different annealing temperature. As expected, higher annealing temperatures, and therefore higher phase fractions of austenite, result in higher ductility. The highest uniform elongation could be found for an annealing temperature of 800°C (100 % austenite). It is notable that in a relatively small temperature interval from 720°C to 750°C the ductility of the material increases rapidly, indicating a switch between the materials behavior being dominated by the properties of the martensitic phase to being dominated by the properties of the austenitic phase.

A logistic function (Eq. 1) can be fitted to the data points to be able to approximate a continuous correlation between formability and annealing temperature. The formabilities of the single martensitic and austenitic phases are represented by the lower and upper limits of the curve.

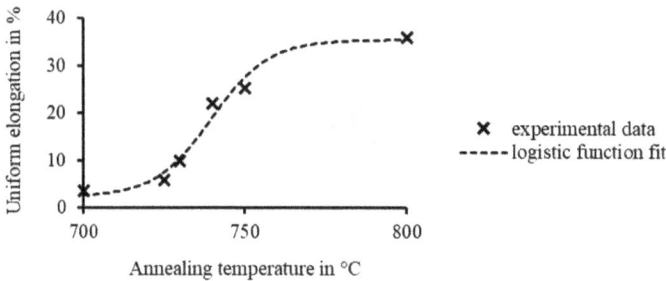

Fig. 2. Dependence of uniform elongation on the annealing temperature during hot tensile testing.

$$A_g = \frac{A_{g,max}}{1+e^{-k(T-T_0)}}$$ (1)

Hot Stamping Experiments

For further investigation of the material's formability at different annealing temperatures hot stamping experiments were performed on a single acting hydraulic deep drawing press using a hot stamping tool resembling a miniaturized b-pillar part. The tool does not have an active water-cooling system. However, as cycle times were larger than 10 minutes, the tool temperature can be assumed to be room temperature. Prior to annealing the blanks are spray-coated with boron nitride for friction reduction and basic scaling protection. The blanks were heated in a convection furnace for 7 minutes at different annealing temperatures (700°C, 730°C, 750°C, 800°C). After annealing the blanks were manually transferred to the pressing tool and formed with a constant tool speed of 35 mm/s (drawing depth of 25 mm). After forming the parts are quenched in the closed tool for 10 s.

Material Forming - ESAFORM 2023   Materials Research Forum LLC
Materials Research Proceedings 28 (2023) 833-838   https://doi.org/10.21741/9781644902479-91

Fig. 3 a) shows the dimensions of the hot stamped T-Shape profile. The notches at the bottom of the part are necessary for aligning the blank in the tool using two pilot rods. As can be seen in Fig. 3 b) only an annealing temperature of 800°C (fully austenitic microstructure) leads to a crack-free part. At annealing temperatures of 700°C, 730°C, and 750°C severe crack form at the upper part of the profile. With increasing annealing temperature additional cracks at the left and right shoulder part of the part start to appear. One possible explanation for cracks occurring at more locations with increasing annealing temperature is the increasing ductility of the material. At 700°C annealing temperature, the crack in the center part of the profile starts to form at very low drawing depths due to the poor ductility of the material. Therefore, almost all deformation for the rest of the drawing process takes place in the cracked region on the sheet. At higher annealing temperatures, the material is more ductile. The crack in the center appears at a higher drawing depth which leads to a more uniform thinning of the material which again leads to additional cracks forming.

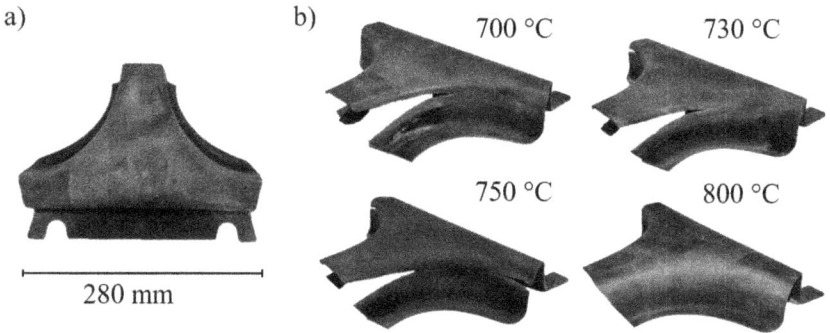

Fig. 3. a) Dimensions of T-shape profile. b) T-shape profiles processed with different annealing temperatures.

Material Properties of Quenched Specimens
In order to assess the mechanical properties of the quenched material with respect to the annealing temperature, further quenching experiments were conducted using the hot stamping tool. For each annealing temperature a rectangular sheet was heated according to the annealing of the T-shape profiles and quenched between the blank holder and the die of the hot stamping tool for 10s. Tensile test specimens were cut from the quenched sheets and quasi-static tensile tests were performed at room temperature. Table 2 shows the mechanical properties in the quenched state with respect to the different annealing temperatures.

Material Forming - ESAFORM 2023
Materials Research Proceedings 28 (2023) 833-838

Materials Research Forum LLC
https://doi.org/10.21741/9781644902479-91

Table 2. Mechanical properties in quenched state with respect to annealing temperature.

Annealing temperature [°C]	$R_{p0,2}$ [MPa]	R_m [MPa]	A_g [%]	A [%]
700	576	1250	21.0	21.2
730	420	1277	6.4	6.4
750	574	1263	2.4	2.4
800	877	1300	0.6	0.6

While yield strength and tensile strength only show a low sensitivity to the annealing temperature, uniform and total elongation show a strong dependence on the annealing temperature. An annealing temperature of 800°C results in a fully austenitic microstructure after annealing and a fully martensitic microstructure after quenching. This microstructure displays a very brittle behavior. With decreasing annealing temperature, and therefore higher phase fraction of retained austenite in the quenched state, the ductility of the material increases. At 730°C annealing temperature the ductility is comparable to standard manganese boron steels. At 700°C annealing temperature, the material displays very favorable mechanical properties in the quenched state.

Summary

This paper investigates the suitability of intercritical annealing and subsequent press hardening for processing high strength parts from medium manganese steel sheets. The choice of a specific annealing temperature influences phase fractions of austenite and martensite in the microstructure during hot stamping and after quenching. For different annealing temperatures formability was investigated by hot tensile tests and hot stamping experiments. Final mechanical properties were investigated by tensile tests. Hot tensile tests show an increasing formability with increasing annealing temperatures. In hot stamping experiments, part production without cracking was only possible for an annealing temperature of 800°C.

While yield strength and tensile strength in quenched material state show very low sensitivity to the annealing temperature, ductility increases heavily with decreasing annealing temperature. Mechanical properties of tensile strengths of 1250 MPa and total elongations of 21.2 % could be achieved at an annealing temperature of 700°C. While for both, in-process and final mechanical properties, an adequate annealing temperature could be found, no investigated annealing temperature satisfies both conditions, as a strong trade-off relationship between favorable final mechanical properties and formability during hot stamping was evident.

As a result, due to the identified trade-off relationship of the intercritical annealing route, it is not yet possible to achieve optimal processing conditions and final properties with the same annealing temperature. In this regard, further research has to be conducted in the processing of the investigated medium manganese steel alloy in press hardening.

Acknowledgement

Funded by the Deutsche Forschungsgemeinschaft (DFG, German Research Foundation) under Germany's Excellence Strategy – EXC-2023 Internet of Production – 390621612.

References

[1] J.H. Schmitt, T. Iung, New developments of advanced high-strength steels for automotive applications, Comptes Rendus Physique 19 (2018) 641-656. https://doi.org/10.1016/j.crhy.2018.11.004

[2] H. Karbasian, A.E. Tekkaya, A review on hot stamping, J. Mater. Process. Technol. 210 (2010) 2103-2118. https://doi.org/10.1016/j.jmatprotec.2010.07.019

[3] D. Rosenstock, J. Banik,; S. Myslowicki, Hot stamping steel grades with increased tensile strength and ductility - MBW-K 1900, tribond 1200 and tribond 1400, IOP Conf. Ser.: Mater. Sci. Eng. 651 (2019) 012040. https://doi.org/10.1088/1757-899X/651/1/012040

[4] M. Merklein, M. Wieland, M. Lechner, S. Bruschi, A. Ghiotti, Hot stamping of boron steel sheets with tailored properties: A review, J. Mater. Process. Technol. 228 (2016) 11-24. https://doi.org/10.1016/j.jmatprotec.2015.09.023

[5] J. Speer, R. Rana, D. Matlock, A. Glover, G. Thomas, E. de Moor, Processing Variants in Medium-Mn Steels, Metals 9 (2019) 771. https://doi.org/10.3390/met9070771

[6] B. De Cooman, J. Speer, Quench and Partitioning Steel: a New AHSS Concept for Automotive Anit-Intrusion Applications, Steel Res. Int. 77 (2006) 634-640. https://doi.org/10.1002/srin.200606441

[7] C. Wang, W. Li, S. Han, L. Zhang, Y. Chang, W. Cao, H. Dong, Warm Stamping Technology of the Medium Manganese Steel, Steel Res. Int. 89 (2018) 1700360. https://doi.org/10.1002/srin.201700360

[8] Y.-K. Lee, J. Han, Current opinion in medium manganese steel, Mater. Sci. Technol. 31 (2015) 843-856. https://doi.org/10.1179/1743284714Y.0000000722

[9] C. Blankart, S. Wesselmecking, U. Krupp, Influence of Quenching and Partitioning Parameters on Phase Transformations and Mechanical Properties of Medium Manganese Steel for Press-Hardening Application, Metals 11 (2021) 1879. https://doi.org/10.3390/met11111879

Material Forming - ESAFORM 2023   Materials Research Forum LLC
Materials Research Proceedings 28 (2023) 839-846   https://doi.org/10.21741/9781644902479-92

Specific behavior of high-manganese steels in the context of temperature increase during dynamic deformation

JABŁOŃSKA Magdalena B.[1,a] *, GRONOSTAJSKI Zbigniew[2,b],
TKOCZ Marek[1,c], JASIAK Katarzyna[2,d], KOWALCZYK Karolina[1,e],
SKWARSKI Mateusz[2,f] and KOSTKA Michał[1,g]

[1]Silesian University of Technology, Faculty of Materials Engineering,
Krasińskiego 8, 40-019 Katowice, Poland

[2]Wrocław University of Technology, Faculty of Mechanical Engineering,
Ignacego Łukasiewicza 5, 50-371 Wrocław, Poland

[a]magdalena.jablonska@polsl.pl, [b]zbigniew.gronostajski@pwr.edu.pl,
[c]marek.tkocz@polsl.pl, [d]katarzyna.jasiak@pwr.edu.pl, [e]karolina.kowalczyk@polsl.pl,
[f]mateusz.skwarski@pwr.edu.pl, [g]michkos976@student.polsl.pl

Keywords: High-Manganese Steel, TWIP, Dynamic Deformation, Microstructure, Temperature, Twinning

Abstract. In recent years, a development of AHSS steels for manufacturing parts for the automotive industry is the observed trend. The high-manganese steels with aluminium and silicon addition, exhibiting twinning induced plasticity (TWIP) effect, are one of the most interesting modern materials, due to their unique combination of both very good strength and great ductility. However, the material behaviour during plastic deformation depends not only on the chemical composition but also on deformation conditions, inter alia, strain rate and temperature. TWIP steels can be used for production of energy-absorbing parts, therefore it is very important to analyse their deformation behaviour at high strain rates. The paper presents the effect of deformation in quasi-static and dynamic conditions on the microstructure of an experimental TWIP steel. The experiments were performed on tensile testing machine and on the flywheel machine. The microstructure was analyzed by optical and scanning transmission electron microscopy. Thanks to the measurements during the quasi-static test and numerical simulations of both tensile tests, the temperature increase was determined in the sample region from which the sections for microstructural studies were taken. It was found that the temperature increase in dynamic conditions can affect the microstructure evolution in the investigated TWIP steel.

Introduction

The autobody components made of non-ferrous materials are nowadays more common than they used to be earlier, however, the most responsible parts in terms of passengers safety are still made of steel [1,2]. In recent years, considerable efforts have been focused on the development of high-manganese steels for the automotive industry. With the specific Mn, Al and Si content, these steels are characterized with the stacking fault energy (SFE) value in the range of ca. 20 to ca. 50 mJ/m^2 which favors twinning as a dominant deformation mechanism. This phenomenon was ultimately confirmed in many studies [2,3]. The concept of a steel with the so called twinning induced plasticity (TWIP) effect was proposed for the first time by Grassel and Fromayer [4]. During the analysis of steel with 15 and 25% of Mn content, the authors established that the unique combination of high mechanical properties and high ductility is determined by the mechanical twins formation in the austenitic grains [5]. The same authors proved that TWIP steel with an advantageous combination of mechanical and plastic properties should contain the amount of manganese within the range of 20% wt. to 30% wt., the aluminium content – from 3% wt. to 5%

Material Forming - ESAFORM 2023
Materials Research Proceedings 28 (2023) 839-846

Materials Research Forum LLC
https://doi.org/10.21741/9781644902479-92

wt. and the carbon content not exceeding 0.6% wt. The TWIP steels are characterized by good work hardening, which is mostly the result of twinning, but also dynamic strain ageing (DSA). Moreover, these steels are extremely sensitive to the strain rate [6]. The studied carried out in [7] proved an influence of the strain rate on the ability to transfer dynamic loads by the TWIP steel, confirming an increase in the impact resistance under the influence of increasing strain rate. The dynamic deformation conditions favors the mechanical twins growth in TWIP steel [8, 9]. However, the high strain rate causes the rapid temperature increase in a deformed region. It was proved that temperature rise above 200°C in numerous MnAl steels causes the increase of the SFE value which promotes the dislocation slip and slows down the mechanical twins growth [10]. There's a limited information in the literature about the effect of heat generated during dynamic deformation on the TWIP steel microstructure evolution [11].

The aim of the study was to determine a temperature increase due to conversion of the deformation work into the heat in experimental TWIP steel samples subjected to tensile tests at two different conditions and to analyse how the heat generated affects the deformation mechanism in the investigated steel.

Research Methodology

The research was carried out on an experimental high-manganese steel with the chemical composition given in Table 1. The steel was smelted in a vacuum induction furnace and cast using the gravity casting technique. 120 mm long ingots with a cross-section of 20 x 40 mm were homogenized in an air furnace at 1200°C and then hot-rolled to a thickness of ca. 2 mm. After hot rolling, the steel was subjected to solution heat treatment (1100°C / 2 h / water quenching).

Table 1. The chemical composition of the investigated steel.

Element content, wt. %									
C	Mn	Al	V	P	S	Ce	La	Nd	N	Fe
0.42	21.10	2.55	0.002	<0.01	0.006	0.011	<0.005	<0.005	43ppm	bal.

The dog-bone samples with the dimensions given in Fig. 1a were cut-out of the obtained strips to perform the quasi-static tensile test (at the strain rate of 0.5 s^{-1} – Fig. 1b) on the Instron tensile testing machine and the dynamic one (at the strain rate of 1000 s^{-1} – Fig. 1c) on the flywheel machine. The latter method is described in [12]. The initial temperature of samples was ca. 20°C. The temperature changes in the sample tested at the lower strain rate were recorded by FLIR T840 thermal imaging camera. Temperature measurement methodology during quasi-static tensile tests was presented in [13].

a)   b)   c)

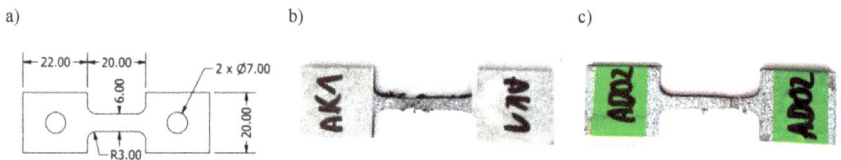

Fig. 1. The sample dimensions used in tensile tests (a) and samples after quasi-static (b) and dynamic (c) tensile tests.

The numerical modelling of quasi-static and dynamic tensile tests were performed in Forge NxT 3.2 – a commercial finite element method software dedicated for metal forming and heat

Material Forming - ESAFORM 2023
Materials Research Proceedings 28 (2023) 839-846

Materials Research Forum LLC
https://doi.org/10.21741/9781644902479-92

treatment simulations. To reduce the computation time, a geometric model of ⅛th of a sample was used (Fig. 2). The fine tetrahedral mesh was generated in the sample volume.

a)   b)

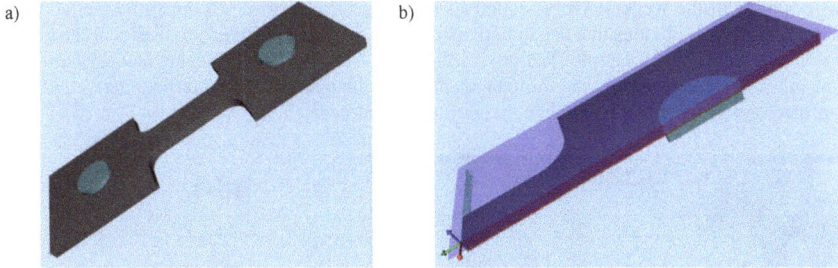

Fig. 2. The geometric model of a sample used in tensile tests with the gauge length of 10 mm, the width of 6 mm and the thickness of 1.4 mm (a) and symmetry planes utilized in simulations (b).

The flow stress of the investigated TWIP steel was defined with the equation:

$$\sigma_p = A \cdot \exp{(m_1 T)} \cdot (\varepsilon + \varepsilon_{b0})^{m_2} \tag{1}$$

where: T – temperature of the sample, ε – equivalent strain, A, m_1, m_2, ε_{b0} – coefficients of the Eq. 1.

The coefficients in Eq. 1 were calculated separately for each strain rate on the basis of the experimentally determined mechanical properties (such as the offset yield strength $R_{p0.2}$, the ultimate tensile strength R_m and the uniform elongation A_g), by means of Cold Rheology Generation Tool included in the Forge NxT 3.2 software, The Eq. 1 coefficient values as well as the mechanical properties obtained in experiments are collected in Table 2. The physical and thermal properties such as density (7400 kg/m³), specific heat (500 J/kgK) and thermal conductivity (17 W/mK) were taken from [14]. Adiabatic conditions on the sample-die interface were assumed.

Table 2. Mechanical properties and coefficients of the flow stress function.

Strain rate $[s^{-1}]$	$R_{p0.2}$ [MPa]	R_m [MPa]	A_g [%]	A	m_1	m_2	ε_{b0}
0.5	245	523	54	1169.71	-0.0009	0.468808	0.0370259
1000	410	650	48	1397.58	-0,0009	0.467435	0.0753929

Microstructural observations were made on microsections parallel to the longitudinal sample axis, in the necked region, by means of the light microscope Olympus GX71 and scanning transmission electron microscope Hitachi HD-2300A. In order to reveal the microstructure, the material was etched in 6% Nital (94 ml of ethyl alcohol, 6 ml of HNO₃ acid).

Results and Discussion
To verify the correctness of the simulation results, calculated temperature distributions and the maximum temperature change in the sample deformed at 0.5 s⁻¹ were compared to the corresponding experimental measurements (Fig. 3). It seems like quite good accuracy of

Material Forming - ESAFORM 2023
Materials Research Proceedings 28 (2023) 839-846

Materials Research Forum LLC
https://doi.org/10.21741/9781644902479-92

simulation results was obtained. In both cases the maximum temperature started to rise intensively on the onset of necking which was predicted in the simulation almost in the same moment as it happened in the experiment. The most notable difference is the necking location. In the real test it has occurred in the weakest cross-section of the gauge length while in the simulation – at the sample center. It is obvious that the numerical model definition has forced prediction of the necked region at the center of a sample. The other solution would be possible if some inconsistency or a flaw in the sample geometry, the finite element mesh, the material model or boundary conditions appeared or was there something like that defined on purpose.

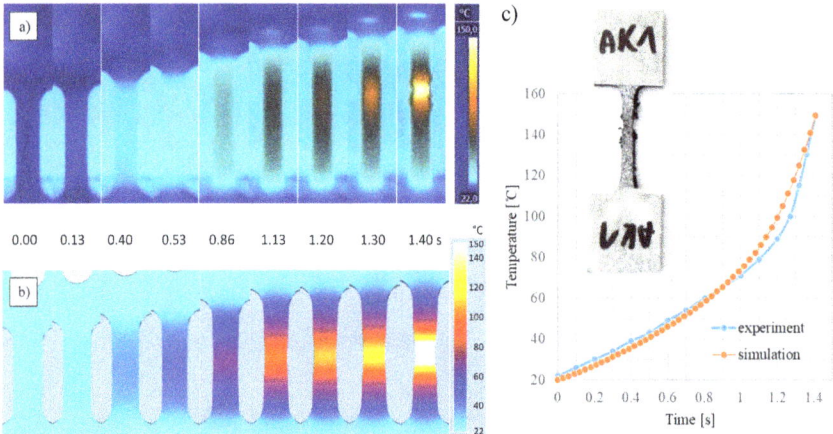

Fig. 3. Comparison of measured (a) and calculated (b) temperature distributions and the maximum temperature (c) in the sample tested at the strain rate of 0.5 s⁻¹.

The mechanical properties, presented in Table 2, as well as the corresponding flow curves (Fig. 4c) indicate that significantly higher stresses are required to deform of the investigated steel under dynamic loading. It is well reflected in the von Mises stress distributions presented in Figs. 4a and 4b. However it is worth to notice that the steel still exhibits the excellent uniform elongation (A_g) at very high strain rates. It is only slightly smaller than the elongation observed in the quasi-static conditions.

Material Forming - ESAFORM 2023
Materials Research Proceedings 28 (2023) 839-846

Materials Research Forum LLC
https://doi.org/10.21741/9781644902479-92

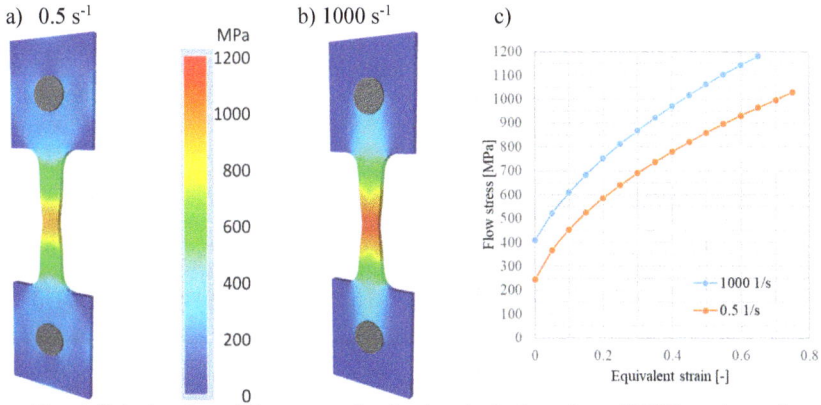

Fig. 4. Calculated von Mises stress distributions in the investigated TWIP steel samples subjected to (a) quasi-static and (b) dynamic tensile tests as well as a comparison of TWIP steel flow curves for the analysed strain rates (c).

A comparison of temperature changes in both selected deformation cases is presented in Fig. 5. As it was expected, the significantly higher flow stress of the investigated steel during deformation at the strain rate of 1000 s⁻¹ caused more intensive temperature increase in the sample. According to the numerical simulation results, the temperature at the center after reaching equivalent strain of ca. 0.8 can exceed 200°C. Similar research results were presented in the work [11]. Such a deformation temperature in some MnAl steels increases stacking fault energy (SFE) and promotes dislocation slip.

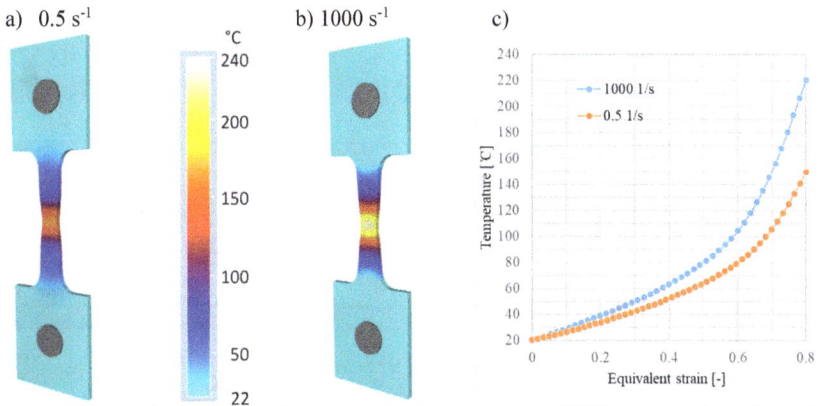

Fig. 5. Calculated temperature distributions in the investigated TWIP steel samples subjected to (a) quasi-static and (b) dynamic tensile tests as well as a comparison of the maximum temperature (c).

Material Forming - ESAFORM 2023
Materials Research Proceedings 28 (2023) 839-846

Materials Research Forum LLC
https://doi.org/10.21741/9781644902479-92

Microstructure Analysis

The microstructure of the studied steel at the initial state is fully austenitic (Fig. 6a). It is characterized by the presence of austenite grains with characteristic coherent annealing twins (Fig. 6b) and tangled dislocations observed near the grain boundaries (Fig. 6c).

After the tensile tests, effects of deformation are clearly visible in the microstructure. The austenite grains are elongated towards the tensile direction (Figs. 6 d,g). In the austenite matrix containing dislocations cells, the generation of mechanical twins occurs in the primary and secondary twinning system (Figs. 6 e,f). At the strain rate of 1000 s^{-1}, the generation of multi-twins as well as nano-twins is observed (Figs. 6 h,i). This leads to the activation of several slip and twinning systems. Moreover, the twin bundles can be observed. The interactions between twins take place and individual micro shear bands appear (Fig. 6h).

The development of micro shear bands results from the evolution of dislocation structure. This phenomenon can be associated with the intensive temperature increase calculated in the sample deformed at the strain rate of 1000°C. It could cause the increase of stacking fault energy and thus, possibly, the *rearrangement of the dislocation structure*.

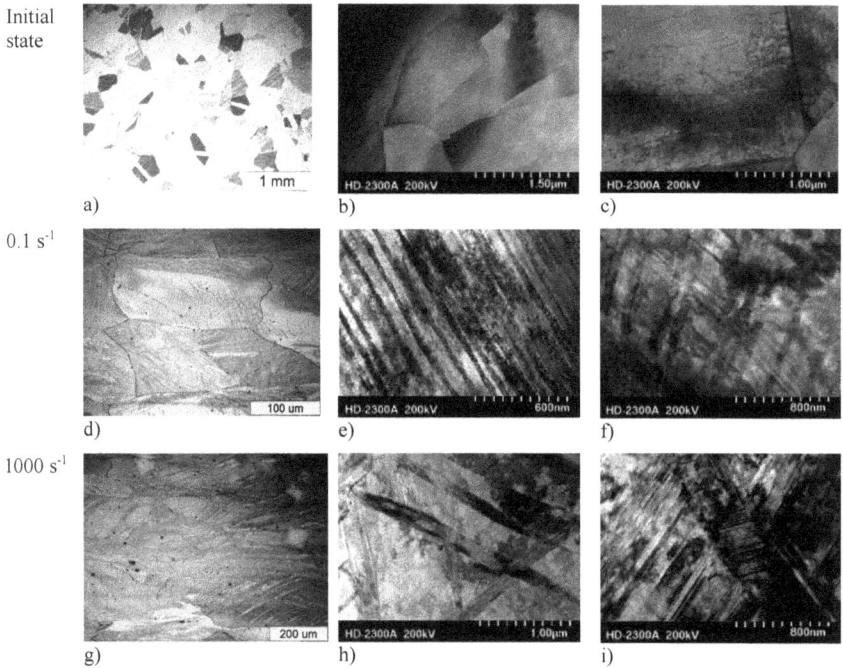

Fig. 6. The microstructure of the investigated steel in the initial state and after deformation at various strain rates; taken from the necked regions where the equivalent strain was ca. 0.8.

Summary
The experimental, high-manganese austenitic steel tested in this work can be classified as a material exhibiting TWIP effect. It was proved that it has good strength, comparable to the currently used steels for elements that increase the safety of vehicles, and very high elongation, even during deformation at very high strain rates. This unique features indicate that TWIP steels are excellent materials for the production of energy-absorbing parts that are designed to withstand severe deformation without fracture during a collision.

The convergence of experimental and simulation results was obtained for the test performed in the quasi-static conditions. This suggests that the prepared simulation model can provide correct results also for dynamic deformation conditions. The conducted numerical simulations made it possible to determine the temperature increase in the sample tested at high strain rate, which was not possible during the experiment.

During deformation at the strain rate of 0.1 s⁻¹, the temperature of the investigated steel reached over 150°C in the necked region of the tensile sample. In these conditions the mechanical twinning is a dominant deformation mechanism. However, at the strain rate of 1000 s⁻¹, the temperature noticeably exceeding 200°C was calculated in the region from which micro-sections for microstructural investigations were taken. The effects of microstructure evolution observed in the sample indicate that such a high temperature probably affected the stacking fault energy of the investigated steel as well as promoted the development of micro-bands and the rearrangement of dislocation structure.

Acknowledgments
The financial support of the National Science Centre, Poland, granted under the project UMO-2019/35/B/ST8/02184 "Effect of the heat generated during deformation at high strain rates on the structure and properties of high manganese steels with twinning as the dominant deformation mechanism", is gratefully acknowledged.

References
[1] D. Frómeta, A. Lara, L. Grifé, T. Dieudonné, P. Dietsch, J. Rehrl, C. Suppan, D. Casellas, J. Calvo, Fracture resistance of advanced high-strength steel sheets for automotive applications, Metall. Mater. Trans. A: Phys. Metall. Mater. Sci. 52 (2021) 840-856. https://doi.org/10.1007/s11661-020-06119-y

[2] K.S Raghavan., A.S Sastri., M.J Marcinkowski., Nature of work-hardening behavior in Hadfields manganese steel, Trans. Metall. Soc. Aime 245 (1969) 1569-1575.

[3] L. Remy, Kinetics of fcc. deformation twinning and its relationship to stress-strain behaviour, Acta Metall. 26 (1978) 443–451. https://doi.org/10.1016/0001-6160(78)90170-0

[4] O. Grässel, G. Frommeyer, Effect of martensitic phase transformation and deformation twinning on mechanical properties of Fe-Mn-Si-Al steels, Mater. Sci. Technol. 14 (1998) 1213-1217. https://doi.org/10.1179/mst.1998.14.12.1213

[5] A. Kozłowska, K. Radwański, K. Matus, L. Samek, A. Grajcar, Mechanical stability of retained austenite in aluminum-containing medium-Mn steel deformed at different temperatures, Arch. Civil Mech. Eng. 21 (2021) 269-275. https://doi.org/10.1007/s43452-021-00177-8

[6] Y.F. Shen, N. Jia, R.D.K. Misra, L. Zuo, Softening behavior by excessive twinning and adiabatic heating at high strain rate in a Fe-20Mn-0.6C TWIP steel, Acta Mater. 103 (2016) 229-242. https://doi.org/10.1016/j.actamat.2015.09.061

[7] A. Śmiglewicz, M.B. Jabłońska, The effect of strain rate on the impact strength of the high-Mn steel, Metalurgia 54 (2015) 631-634. ISSN 0543-5846 UDC – UDK 669.74.15:539.55.375:531.76=111

Material Forming - ESAFORM 2023   Materials Research Forum LLC
Materials Research Proceedings 28 (2023) 839-846   https://doi.org/10.21741/9781644902479-92

[8] A. Śmiglewicz, W. Moćko, K. Rodak, I. Bednarczyk, M.B. Jabłońska, Study of dislocation substructures in high-Mn steels after dynamic deformation tests, Acta Phys. Polon. A 130 (2016) 942-945. https://doi.org/10.12693/APhysPolA.130.942

[9] H.K. Yang, Z.J. Zhang, F.Y. Dong, Q.Q. Duan, Z.F. Zhang, Strain rate effects on tensile deformation behaviors for Fe-22Mn-0.6C-(1.5Al) twinning-induced plasticity steel, Mater. Sci. Eng. A 607 (2014) 551–558. https://doi.org/10.1016/j.msea.2014.04.043

[10] V. Shterner, I.B. Timokhina, H. Beladi, On the work-hardening behaviour of a high manganese TWIP steel at different deformation temperatures, Mater. Sci. Eng. A 669 (2016) 437-446. https://doi.org/10.1016/j.msea.2016.05.104

[11] M.B. Jabłońska, Structure and properties of a high-manganese austenitic steel strengthened in the result of mechanical twinning in processes of dynamic deformation, Monograph, Silesian University of Technology Publisher, ISBN: 978-83-7880-363-8.

[12] Z. Gronostajski, A. Niechajowicz, R. Kuziak, J. Krawczyk, S. Polak, The effect of the strain rate on the stress-strain curve and microstructure of AHSS, J. Mater. Process. Technol. 242 (2017) 246-259. https://doi.org/10.1016/j.jmatprotec.2016.11.023

[13] M.B. Jabłońska, K. Jasiak, K. Kowalczyk, I. Bednarczyk, M. Skwarski, M. Tkocz, Z. Gronostajski, Deformation behaviour of high-manganese steel with addition of niobium under quasi-static tensile loading, Mater. Sci.-Poland 40 (2022) 1-11. https://doi.org/10.2478/msp-2022-0029

[14] V. Garcia-Garcia, Microstructural and mechanical analysis of double pass dissimilar welds of twinning induced plasticity steel to austenitic/duplex stainless steels, Int. J. Press. Vessel. Piping. 198 (2022) 104665. https://doi.org/10.1016/j.ijpvp.2022.104665

Material Forming - ESAFORM 2023
Materials Research Proceedings 28 (2023) 847-854

Materials Research Forum LLC
https://doi.org/10.21741/9781644902479-93

Effect of prestrain on mechanical behavior of aluminum alloys

VINCZE Gabriela[1,2,a] *, BUTUC Marilena C.[1,2,b], WEN Wei[3,c], YÁNEZ Jesús[1,d] and LOPES Diogo[1,e]

[1]Centre of Mechanical Technology and Automation (TEMA), Department of Mechanical Engineering, University of Aveiro, 3810-193 Aveiro, Portugal

[2]LASI - Intelligent Systems Associate Laboratory, Portugal

[3]Department of Engineering, Lancaster University, Lancaster LA1 4YR, UK

[a]gvincze@ua.pt, [b]cbutuc@ua.pt, [c]w.wen2@lancaster.ac.uk, [d]jvallez@ua.pt, [e]diogoaflopes@ua.pt

Keywords: Aluminum Alloys, Mechanical Behavior, Asymmetric Rolling

Abstract. Sheet metal forming involves many times large plastic strain and strain path changes. It is well known that the plastic behavior of metals is strain path sensitive. In monotonic loading, the microstructure as well as the crystallographic texture evolve during deformation leading usually to a gradual material hardening that tends to saturate at large strains. Such evolution can be interrupted if a severe change of strain path occurs. The simplest way to change drastically the strain path is through reverse loading, namely, by loading the material in opposite direction to the previous one. In this case, the material behavior can show one or more characteristics, such as the Bauschinger effect, transient hardening, softening or hardening. This work investigated the effects of the prestrain on the mechanical response of the material subjected to reverse simple shear. The prestrain is produced by rolling, either symmetric or asymmetric, and different amounts of equivalent strain. Three routes of rolling are used, namely, symmetric, asymmetric continuous, and asymmetric reverse [1]. The Bauschinger effect slightly decreases with the prestrain increase for the initial material. After rolling, the Bauschinger effect is insensitive to the rolling route, and it is also insensitive to the amount of the rolling prestrain.

Introduction

Aluminum alloys are very good choices for many applications from the aerospace industry to packaging, due to their advantageous strength-to-weight ratio and their high recyclability without loss of their properties. However, the forming operation of aluminum alloys faces some challenges, due to the low formability at room temperature and high springback. The latter mainly due to the low stiffness compared with steels. It is also well known that a high influence on the mechanical behavior of aluminum alloys subject to plastic deformation is related to their texture. An easy way to change the texture of the material is by rolling. Moreover, asymmetric rolling allows the creation of different texture components by controlling the rolling route. Another important factor during forming operation is the strain path changes. An abrupt change in strain path can cause a Bauschinger effect, a transient behavior, permanent softening, or higher hardening rate compared to monotonic loading. The effects of strain path changes were extensively studied over the years by experimental and numerical methods [2-8]. Moreover, many researchers stated that the Bauschinger effect is associated with the amplitude of springback and is an indispensable parameter that should be considered for an accurate prediction of springback [9,10]. The aim of this study is to analyze the influence of the amount of prestrain in the Bauschinger parameter for a large range of values.

Material Forming - ESAFORM 2023
Materials Research Proceedings 28 (2023) 847-854

Materials Research Forum LLC
https://doi.org/10.21741/9781644902479-93

Material and Experimental methods

Material.

The material used in this study is an aluminum-silicon-magnesium alloy AA6022-T4 with an initial thickness of 2 mm and chemical composition given in Table 1. The material has a strong cube texture typical for recrystallized aluminum sheets as can be seen in Fig. 1. A complete characterization of this material can be found in [1,11].

Table 1. Chemical composition of the material as-received wt. %.

Si	Fe	Cu	Mn	Mg	Cr	Zn	Ti	Others	Al
0.8-1.5	0.05-02	0.01-0.11	0.02-0.1	0.45-0.7	0.1	0.25	0.15	0.15	Balance

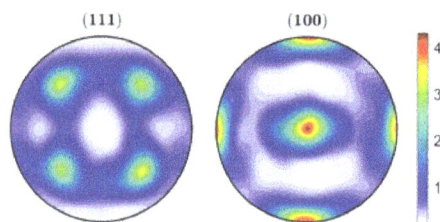

Fig. 1. Pole figures {111},{100} of the initial material.

Asymmetric rolling.

The in-house asymmetric rolling mill presented in Fig. 2 was used for rolling. The diameters of the rolls are 180 mm and the length is 300 mm. Detailed description of this mill can be found in [1]. The asymmetry was introduced by imposing different speeds for the upper and bottom rolls of the mill. For symmetric rolling, the angular speed of both rolls was 15 rpm. For asymmetric rolling, a ratio of 1.36 of the rolls' speeds was produced by the angular speeds of 15 rpm and 11 rpm.

Fig. 2. Asymmetric rolling mill.

The total thickness reduction was 50%, and was obtained in 4 passes, with a thickness reduction of 15% per pass. Regarding asymmetric rolling, two types of strain paths were produced by the rotation of the sheet between two subsequent passes. Namely, (1) asymmetric continuous (ARC) when no rotation of the sheet occurred between two subsequent passes, and (2) asymmetric reverse

Material Forming - ESAFORM 2023   Materials Research Forum LLC
Materials Research Proceedings 28 (2023) 847-854   https://doi.org/10.21741/9781644902479-93

route (ARR) when rotation of 180° around the rolling direction (RD) occurred between two subsequent passes. The rolling route produces a forward or a reverse shear deformation in sheet thickness, corresponding to ARC and ARR respectively. An illustration of the rolling route is presented in Fig. 3.

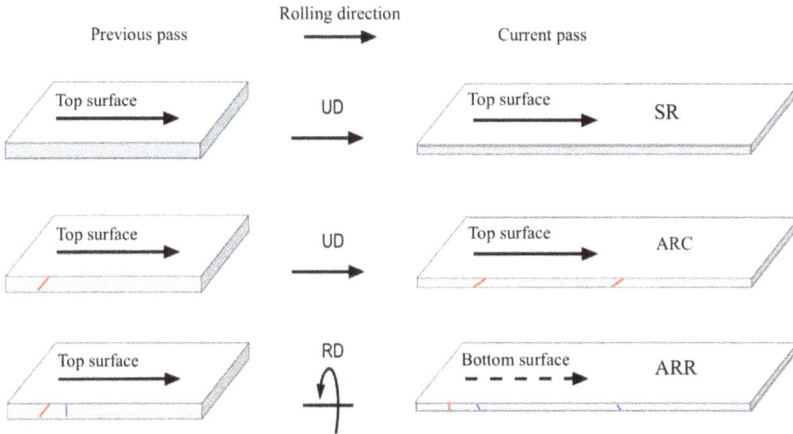

Fig. 3. Rolling routes: SR - symmetric rolling, ARC - asymmetric rolling continuous, ARR - asymmetric rolling reverse.

Simple shear test.

The study of the Bauschinger effect on sheet metals by tension-compression test is difficult due to the buckling that occurs during compression. Thus, to avoid buckling, an easy way is to promote reverse loading in simple shear tests. The as-received material was tested in simple shear and reverse loading at 10%, 20% and 30% shear prestrain. The rolled material was tested in reverse shear for a shear prestrain nearby 10%. The in-house simple shear device was used for these tests. The sample geometry is a rectangle with 34x13 [mm^2], corresponding to length and width, respectively. The thickness is variable according to the rolling. The shear deformation area is 34x3 [mm^2]. The strain measurements were made by Digital Image Correlation using the GOM system and the software ARAMIS 5M. The setup of the simple shear test can be seen in Fig. 4.

Fig. 4. Setup of the simple shear test.

Material Forming - ESAFORM 2023   Materials Research Forum LLC
Materials Research Proceedings 28 (2023) 847-854   https://doi.org/10.21741/9781644902479-93

Bauschinger Parameter

The Bauschinger coefficient was calculated as proposed by Hou et al. [12]

$$\beta = \frac{\tau_{p2}}{\tau_{p1}} \tag{1}$$

Where τ_{p1} and τ_{p2}, denoting the flow stress at the end of prestrain and the yield stress of reloading respectively, are schematically represented in Fig 5.

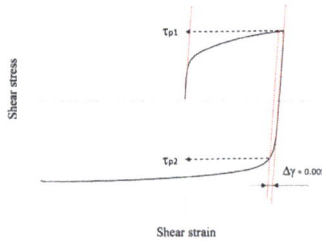

Fig. 5. Schematic representation of Bauschinger parameter calculation.

Thus, if $\beta = 1$, means that Bauschinger effect does not occur. If β decreases, means that the Bauschinger effect is more pronounced and opposite if β increases.

Results and Discussion

Initial material.

The initial material was tested in reverse shear for three amounts of prestrain, namely 10%, 20% and 30 % shear strain. The results are presented in Fig. 6 and are in agreement with data existing in the literature [3]. Namely, it can be noticed the existence of the Bauschinger effect, no plateau and a higher strain hardening that led to an overshooting of the monotonic shear stress-shear strain curve.

It can be observed from Fig. 7 that the Bauschinger coefficient increases with the increase of pre-strain amount for initial material, which means that the Bauschinger effect is reduced with the increase of pre-strain.

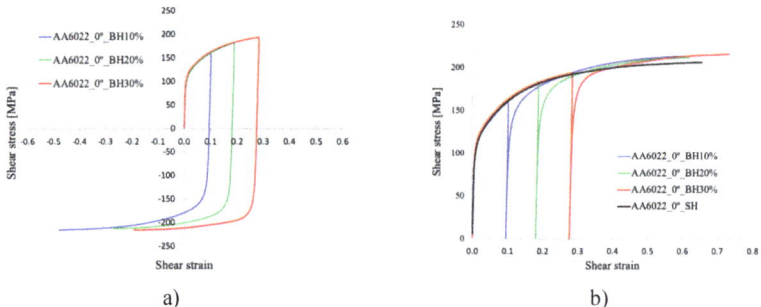

a)   b)

Fig. 6. Simple shear and reverse shear of AA6022-T4: a) reverse shear; b) monotonic and revere shear transformed to the first quadrant.

Material Forming - ESAFORM 2023   Materials Research Forum LLC
Materials Research Proceedings 28 (2023) 847-854   https://doi.org/10.21741/9781644902479-93

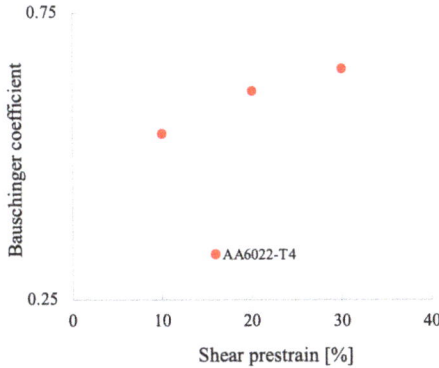

Fig. 7. Evolution of Bauschinger coefficient with the prestrain for material before rolling.

After rolling.

The shear stress-shear strain curves after each pass of rolling are presented in Fig. 8. It can be observed that after each pass the curves are superimposed which means that the material behavior is not affected by the rolling route. After rolling, the material has the same response as the initial material when submitted to reverse loading, showing the Bauschinger effect without plateau. It is worth to mention that the difference in the prestrain deformation is introduced by the difficulty in controlling the test when DIC is used. The Bauschinger coefficient seems to stabilize with the increase of prestrain as can be seen in Fig. 9, where the β is plotted versus equivalent strain corresponding to prestrain, and it is produced by shear or rolling + shear. The values after rolling are very close to the one obtained for the initial material after 30% prestrain. This trend is observed for all three routes. A very small difference in β between the rolling routes can be observed with the lowest value corresponding to ARC and the highest to ARR. The material processed by SR has the lowest variation of β between the rolling passes, only 2%, while for ARC and ARR this variation is about 5% and 4%, respectively. The saturation of the Bauschinger effect observed in figure 9 is related to the dislocation density. In theory, the short-term and long-term Bauschinger effects are caused by the dislocation in the pile-up that can travel backwards upon reversal loading. Their reverse motion is assisted by the back-stress, leading to a drop in the flow stress. Those dislocations also tend to recombine with other dislocations, leading to a lower dislocation hardening level. After a large accumulated strain, more dislocation will be pinned in the forest structures and thus become non-reversible, leading to a less significant Bauschinger effect.

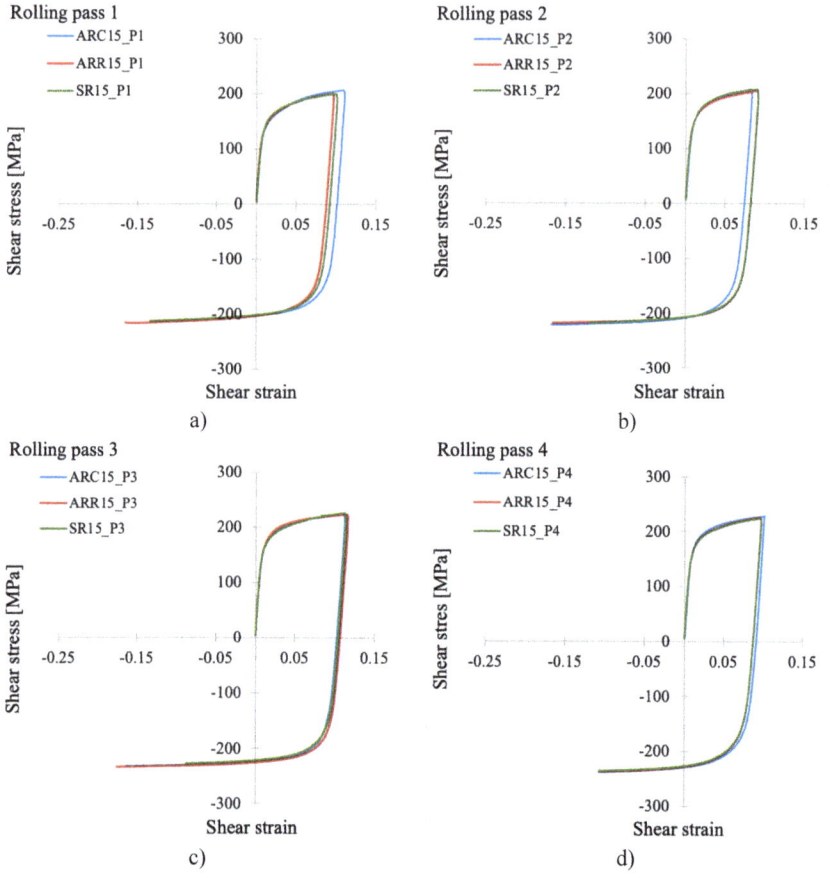

Fig. 8. Shear stress-shear strain after ARC, ARR and SR: a) pass 1; b) pass 2; c) pass 3 and d) pass 4.

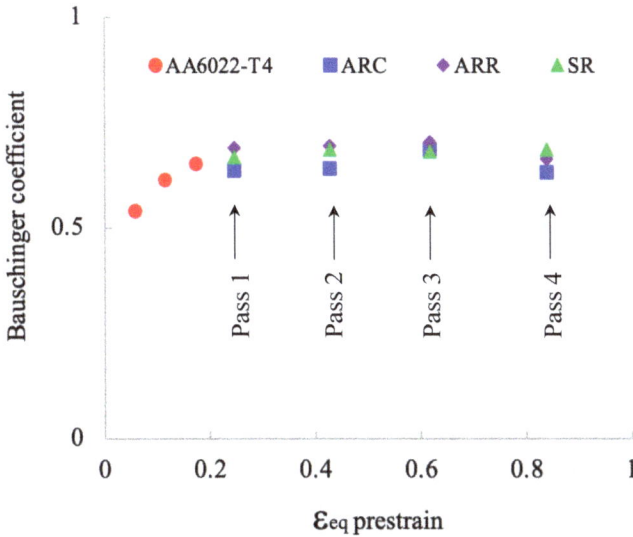

Fig. 9. Evolution of Bauschinger parameter with rolling pass.

Summary

The dependence of the Bauschinger effect on the prestrain quantity was investigated by reverse simple shear test. A large equivalent plastic prestrain was produced by three types of rolling, i.e. symmetric, asymmetric continuous, and asymmetric reverse, rolled in 4 passes. The Bauschinger effect seems to be almost insensitive to the rolling route. The results show an initial increase of the Bauschinger parameter with the increase of prestrain amount followed by a stabilization after approximately 20% equivalent prestrain. These results suggest that the contribution of the Bauschinger effect determined for low prestrain level can be used for the prediction of springback even for large plastic deformation. Nevertheless, it is worth mentioning that other sources of springback should be considered since recent results obtained in V-bending and U-bending tests for the material processed by the same conditions show that the springback after rolling increases.

Acknowledgments

This work was supported by the projects POCI-01-0145-FEDER-032362 (PTDC/EME-ESP/32362/2017) financed by the Operational Program for Competitiveness and Internationalization, in its FEDER/FNR component, and the Portuguese Foundation of Science and Technology (FCT), in its State Budget component (OE), UIDB/00481/2020 and UIDP/00481/2020 - FCT and CENTRO-01-0145-FEDER-022083 - Centro Portugal Regional Operational Program (Centro2020), under the PORTUGAL 2020 Partnership Agreement, through the European Regional Development Fund, and 2022.05783.PTDC financed exclusively by national funds through FCT, I.P. and composed only of one State Budget component (OE).

References

[1] G. Vincze, A.B. Pereira, D.A.F. Lopes, J.M.V. Yánez, M.C. Butuc, Study on Asymmetric Rolling Process Applied to Aluminum Alloy Sheets, Machines 10 (2022) 641. https://doi.org/10.3390/machines10080641

[2] G. Vincze, E.F. Rauch, J.J. Gracio, F. Barlat, A.B. Lopes, A comparison of the mechanical behaviour of an AA1050 and a low carbon steel deformed upon strain reversal, Acta Mater. 53 (2005) 1005-1013. https://doi.org/10.1016/j.actamat.2004.10.046

[3] E.F. Rauch, J.J. Gracio, F. Barlat, Work-hardening model for polycrystalline metals under strain reversal at large strains, Acta Mater. 55 (2007) 2939-2948. https://doi.org/10.1016/j.actamat.2007.01.003

[4] J-H. Schmitt, E.L. Shen, J.L. Raphanel, A parameter for measuring the magnitude of a change of strain path: validation and comparison with experiments on low carbon steel, Int. J. Plast. 10 (1994) 535-551. https://doi.org/10.1016/0749-6419(94)90013-2

[5] C. Teodosiu, Z. Hu, Z., Microstructure in the continuum modeling of plastic anisotropy, in: J.V. Cartensen, T. Leffers, T. Lorentzen, O.B. Pedersen, B.F. Sørensen, G. Winther (Eds.), Proceedings of the Risø International Symposium on Material Science: Modelling of Structure and Mechanics of Materials from Microscale to products, Roskilde, Denmark. Risø National Laboratory, 1998, pp. 149-168.

[6] J. Qin, B. Holmedal, O.S. Hopperstad, Experimental characterization and modeling of aluminum alloy AA3103 for complex single and double strain-path changes, Int. J. Plast. 101 (2018) 156-169. https://doi.org/10.1016/j.ijplas.2018.08.011

[7] M. Butuc, F. Barlat, J. Gracio, G. Vincze, A Theoretical Study of the Effect of the Double Strain Path Change on the Forming Limits of Metal Sheet, Key Eng. Mater. 554-557 (2013) 127-138. https://doi.org/10.4028/www.scientific.net/KEM.554-557.127.

[8] G. Vincze, F. Barlat, E.F. Rauch, C.N. Tome, M.C. Butuc, J.J. Grácio, Experiments and modeling of low carbon steel sheet subjected to double strain path changes, Metall. Mater Trans. A44 (2013) 4475-4479. https://doi.org/10.1007/s11661-013-1895-4

[9] M. Banu, M. Takamura, T. Hama, O. Naidim, C. Teodosiu, A. Makinouchi, Simulation of springback and wrinkling in stamping of a dual phase steel rail-shaped part, J. Mater. Process. Technol. 173 (2006) 178-184. https://doi.org/10.1016/j.jmatprotec.2005.11.023

[10] M.C. Oliveira, J.L. Alves, B.M. Chaparro, L.F. Menzes, Study on the influence of work-hardening modelling in springback prediction, Int. J. Plast. 23 (2007) 516-543. https://doi.org/10.1016/j.ijplas.2006.07.003

[11] G. Vincze, A. Lopes, M.C. Butuc, J. Vallez Yánez, D. Lopes, L. Holz, A. Graça, A.B. Pereira, Numerical and Experimental Analysis of the Anisotropy Evolution in Aluminium Alloys Processed by Asymmetric Rolling, European J. Computat. Mech. 31 (2022) 319-350. https://doi.org/10.13052/ejcm2642-2085.3131

[12] Y. Hou, M-G Lee, L. Jianping, M. Junying, Experimental characterization and modeling of complex anisotropic hardening in quenching and partitioning (Q&P) steel subject to biaxial non-proportional loadings, Int. J. Plast. 156 (2022) 156-169. https://doi.org/10.1016/j.ijplas.2022.103347

Material Forming - ESAFORM 2023
Materials Research Proceedings 28 (2023) 855-865

Materials Research Forum LLC
https://doi.org/10.21741/9781644902479-94

Determination of the forming limit curve of locally annealed aluminum blanks

PICCININNI Antonio[1,a] *, LATTANZI Attilio[2,b], ROSSI Marco[2,c]
and PALUMBO Gianfranco[1,d]

[1]Department of Mechanical Engineering, Mathematics & Management Engineering, Division of Production Technologies, Polytecnic University of Bari, via Orabona 4, 70125 Bari, Italy

[2]Department of Industrial Engineering and Mathematical Sciences, Università Politecnica delle Marche, via Brecce Bianche 12, 60131 Ancona, Italy

[a]antonio.piccininni@poliba.it, [b] a.lattanzi@staff.univpm.it, [c]m.rossi@staff.univpm.it, [d]gianfranco.palumbo@poliba.it

Keywords: Laser Heat Treatment, Aluminum Alloy, FEM, FLC, Bulge Test

Abstract. The research of innovative manufacturing routes to improve the complexity of structural Aluminum (Al) components and reduce the vehicles' weight remains an open question. The adoption of short-term heat treatments to obtain an optimized distribution of properties has shown great potential in enhancing the formability of Al alloys at room temperature. Such a complex approach needs the implementation of methodologies based on numerical simulations able to correctly simulate the forming process. Accordingly, the definition of a proper identification procedure to provide reliable constitutive parameters, possibly employing a limited number of tests, becomes crucial. In this work, we introduce a novel methodology for the evaluation of the Forming Limit Curve (FLC) of laser annealed AA5752-H32 sheet based on the material information from different heterogeneous tests carried out in specific tribological condition, i.e. in absence of friction. Therefore, we investigate an experimental procedure to simultaneously generate an equi-biaxial mechanical state over a wide range of annealing conditions by means of hydraulic bulge test. The results are integrated with full-field data from uniaxial tensile and plane strain tests to provide a testing protocol for the characterization of the failure behavior accounting for the different temperature/time conditions of the sheet metal.

Introduction

The continuous demand for greener and lighter transportations is moving the attention to the improvement of the design criteria, among which the choice of the material for the structural components plays a key role. At the same time, it is widely recognized that Aluminum (Al) alloys are ranked as an ideal candidate to match those requirements [1]. Nevertheless, excellent fire resistance (especially when dealing with the subway applications [2]) combined with high strength-to-weight ratio are partially counterbalanced by the poor formability at room temperature [3]. Several solutions have been investigated over the last decades to overcome such a limitation: the increase of the working temperature – i.e. the theoretical base of the forming operations in warm conditions – has shown great potentialities over a large variety of components [4,5]. A promising and innovative alternative to the warm forming principle is based on splitting the manufacturing process into two separate steps: during the first phase, the material properties are locally modified by means of a short-term heat treatment (for example, by means of a laser heating) so that the blank, once cooled down, can be stamped at room temperature. It has been demonstrated that the optimization of the distribution of the material properties is able to enhance the formability at room temperature [6,7].

Splitting the process into two separate sub-steps automatically implies the need of a robust and systematic approach to design both the steps, especially due to the large number of parameters

Material Forming - ESAFORM 2023   Materials Research Forum LLC
Materials Research Proceedings 28 (2023) 855-865   https://doi.org/10.21741/9781644902479-94

involved in. Moreover, if the numerical approach is indicated as the most reliable methodology to design such a complex process, the proper characterization of the material behavior is a key aspect to be addressed to study in detail the mechanical response in the following forming operation.

While the effect of annealing on the constitutive behavior of the alloy can be investigated by considering, separately, samples of the same material treated at different grades, recently, the application of full-field measurements as the Digital Image Correlation (DIC) has allowed to disentangle the heterogeneous distribution of mechanical properties due to the local annealing [8, 18-20]. In this framework, dealing with heterogenous information makes possible to involve advanced calibration methods based on inverse problems. Generally, they are distinguished in numerical simulation-based methods, such as the Finite Element Model Updating (FEMU), and energy balance-based methods, namely the Virtual Fields Method (VFM) or the Equilibrium Gap Method (EGM) [11]. In the field of metal plasticity, inverse identification methods offer different strategies to produce an enriched calibrations even in the case of complex material models [23]: for instance, for the identification of the hardening behavior including post-necking data [13,14], or to characterize the anisotropic plasticity models [12,15,16] or the thermomechanical behavior of metals [17].

The present work investigates the stretching behavior of a locally laser-annealed AA5752-H32 sheet in a hydraulic bulge test. The aim is to provide heterogenous material data in the biaxial and equi-biaxial stretch conditions spanning over the whole annealing range, that can be coupled with an inverse approach as in [24], or to build a more comprehensive Forming Limit Curve. Different heating strategies have been preliminarily simulated by solving the thermal transient problem and the modification of the material properties have also been numerically predicted. The nodal peak temperatures, being strictly connected to the final level of annealing, were then transferred on the structural model, where the material behavior was modelled according to the hardening law that has been previously calibrated by the same authors [18]. The results, integrated with those coming from the plane strain [19] and from the drawing side, allow to provide a testing methodology to characterize the necking behavior at different material conditions using a limited number of tests.

Material and Methodology
Material.
Bulge tests were simulated considering locally modified AA5754 Al blanks, initially purchased in the H32 state, whose composition is reported in Table 1.

Table 1. Chemical composition of the investigated alloy.

	Al [%]	Cr [%]	Cu [%]	Fe [%]	Mg [%]	Mn [%]	Si [%]	Ti [%]	Zn [%]
AA5754-H32	Bal.	0.024	0.057	0.377	3.057	0.187	0.259	0.017	0.025

Annealing investigation.
The prediction of the material properties at the end of the local heating started from a preliminary characterization step based on the physical simulation of several annealing treatments exploiting the potentialities of the Gleeble system (model 3180). Dog-bone specimens (AA5754-H32) were subjected to a heat treatment composed of a preliminary heating step (rate around 900°C/s), holding at the test temperature and rapid cooling down to room temperature. Being the ends of the specimen clamped by cooled jaws, a parabolic distribution of temperatures occurs during the whole test (equivalent to simultaneously subjecting the same specimen to several heat treatments at the temperatures belonging to the parabolic distribution): hardness tests, carried out once the specimen cools down to room temperature, allowed to easily associate the modification of the material property to the temperature-time history. Hardness measurements were arranged in terms of the Ann variable variable according to Eq. 1: HV_{H32}, HV_{meas} and HV_{H111} refer to the

hardness of the material in the as-received conditions (H32), measured one and in the annealed state (H111), respectively.

$$Ann = \frac{HV_{H32} - HV_{meas}}{HV_{H32} - HV_{H111}} \tag{1}$$

The distribution of the defined variable, as well as its dependency with the peak temperature (T_{peak}), fitted by the sigmoid function, widely used in several applications [14], expressed in Eq. 2, where T^* and λ represents the curve's midpoint and growth rate (or steepness), respectively.

$$Ann = \frac{1}{1 + \exp[-\lambda \cdot (T^* - T_{peak})]} \tag{2}$$

Although Eq. 2 offers a direct analytical description of the percentage of annealing due to the temperature distribution, it does not provide direct information about the different mechanical properties and, therefore, cannot be straightforwardly adopted in structural problems. In this sense, the authors proposed a constitutive model to predict the hardening behavior as function of the T_{peak}. By expressing the annealing level through the general variable T_{peak}, the corresponding flow curve can be predicted using the following logistic expression:

$$\sigma(\varepsilon_p, T_{peak}) = \sigma_{H111}(\varepsilon_p) + \frac{\sigma_{H32}(\varepsilon_p) - \sigma_{H111}(\varepsilon_p)}{1 + \exp\left\{-\left[\beta \tanh\left(\frac{\varepsilon_p}{\varepsilon_0}\right) + c_\beta\right](T^* - T_{peak})\right\}} \tag{3}$$

where ε_p is the accumulated plastic strain, the $\sigma_{H32}(\varepsilon_p)$ and $\sigma_{H111}(\varepsilon_p)$ indicates the flow curves of the material at H32 and fully-annealed (H111) conditions, respectively; β, ε_0, T^* and c_β are the material coefficients regulating the logistic expression. All the coefficients can be inversely calibrated by considering the flow stress curves at different annealing grades. Details about the calibration of the hardening model in Eq. 3 by using a hybrid numerical-experimental procedure on uniaxial tensile data can be found in [18].

Two-steps FE analysis of the heat-treatment and mechanical test.

The adopted approach started from the simulation of the laser heating by solving the thermal transient problem. The investigated heating strategies were based on: (i) a circular track described by the center of the laser spot (indicated by the red square in Fig. 1a) with a radius equal to 30 mm (labelled as AH, standing for *Annular Heating*) and (ii) a linear track, 40 mm long, at two different distances from the center of the blank (20 mm and 30 mm, respectively labelled as LT20 and LT30, where LT stands for *Linear Track*). The simulated strategies were designed to investigate the equi-biaxial stress state over different material conditions (the gradient of properties came from the local laser heating) within the same bulge test.

Material Forming - ESAFORM 2023
Materials Research Proceedings 28 (2023) 855-865

Materials Research Forum LLC
https://doi.org/10.21741/9781644902479-94

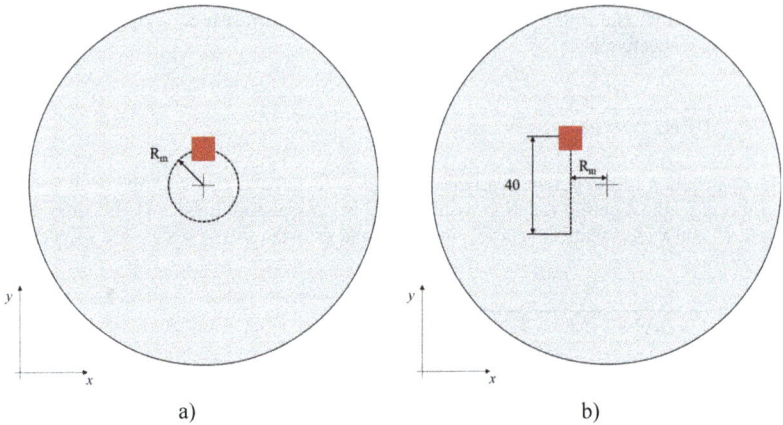

a)   b)

Fig. 1. The investigated heating strategies: a) AH, b) LT. The diameter of the blank is 300 mm, while the gauge area has a diameter of 200 mm (corresponding to the die opening).

Laser heating strategies were simulated implementing a temperature feedback loop (DFLUX user subroutine), setting the heating temperature at 550°C and the feed rate at 5 mm/s. Results from the thermal simulations were processed by means of a Python script to numerically predict the final distribution of the variable *Ann* according to Eq. 2.

In this study, we consider the peak temperature T_{peak} as the common empirical variable to connect the predicted level of annealing with the hardening behavior. Thus, the computed T_{peak} from the transient-thermal analysis is used as input of the structural analysis through the constitutive model in Eq. 3. The forming bulge is replicated numerically by modelling the die as rigid body and the aluminum blank as continuum shell under the hypothesis of plane stress condition due to the small thickness of the sheet metal. Note that the constitutive equations were implemented in Abaqus/Standard® by means of a user material subroutine, employing the well-known Euler Backward integration algorithm for the reconstruction of stress from strain data. Other implicit or semi-implicit algorithms can be also used for the stress update as, for example, the one introduced in [22] for anisotropic plasticity problems.

In this work, we also assumed an isotropic plasticity behavior, while the hardening behavior of the AA5754 was described by using a modified version of the Voce's law, according to the previous experimental investigation reported in [18]:

$$\sigma(\varepsilon_p) = Y(1 + \varepsilon_p) - R \exp(-b\varepsilon_p). \tag{4}$$

All the material coefficients used in this numerical study, including the ones characterizing the logistic formulation in Eq. 3, are listed in Table 2.

Material Forming - ESAFORM 2023
Materials Research Proceedings 28 (2023) 855-865

Materials Research Forum LLC
https://doi.org/10.21741/9781644902479-94

Table 2. Material coefficient of the constitutive equation.

Modified Swift hardening law			
Material state	Y	R	b
As received (H32)	232.74	98.32	20.08
Fully annealed	213.36	148.53	26.99
Logistic hardening model			
β	c_β	T^*	ε_0
0.01095	0.01423	301.35	0.005

Results

Treatment strategies for the heterogeneous bulge test.

Results from the thermal simulations were initially analyzed in terms of maximum temperature distribution: Fig. 2 confirms that the heating temperature close to 550°C was reached and kept along the whole heat treatment in each of the investigated strategies.

R_m=20 mm, L=40 mm R_m=30 mm, L=40 mm R_m=30 mm

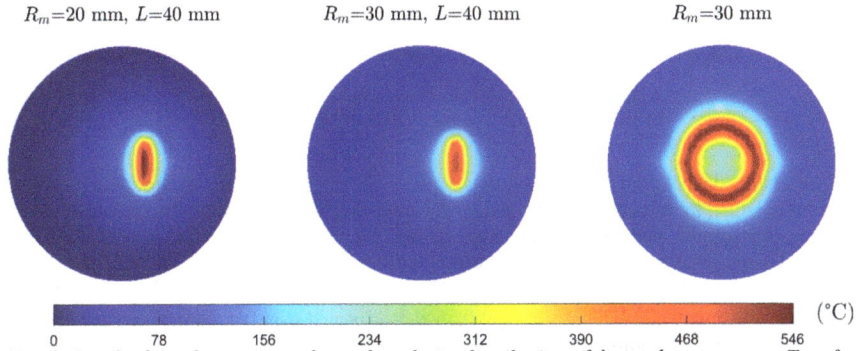

(°C)

0   78   156   234   312   390   468   546

Fig. 2. Results from the transient-thermal analysis: distribution of the peak temperature T_{peak} for the three heat-treatment strategies.

The final distribution of properties, express in terms of the variable *Ann*, was compared among the three strategies along a radial path as shown in Fig. 3 (the *0* radial position corresponds to the center of the circular blank).

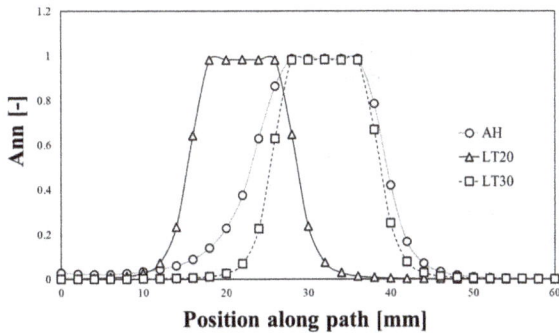

Fig. 3. Predicted distribution of the Ann variable along the radial

Material Forming - ESAFORM 2023
Materials Research Proceedings 28 (2023) 855-865

Materials Research Forum LLC
https://doi.org/10.21741/9781644902479-94

It can be observed that all the geometries show an almost complete annealing (*Ann* close to 1) in the region irradiated by the laser spot (around 10 mm wide in the radial direction). It can also be seen that the annular tracking (AH) modifies a slightly larger portion of the blank close to its center (the transition region is less steep if compared with the *Ann* distribution resulting from the LT30 strategy).

R_m=20 mm, L=40 mm R_m=30 mm, L=40 mm R_m=30 mm

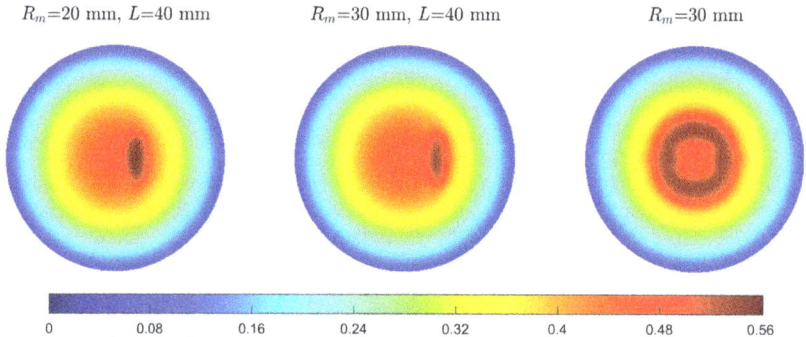

| 0 | 0.08 | 0.16 | 0.24 | 0.32 | 0.4 | 0.48 | 0.56 |

Fig. 4. Maps of accumulated equivalent plastic strain over the bulge test gauge area for the three heat-treatment strategies.

The design of a heterogenous test requires a quantitative method to classify the strain and stress distribution, that, for this study, must consider the annealing temperature and the amount of plastic deformation accumulated during the loading phase. Fig. 4 depicts the maximum plastic strain achieved in our simulated experiments. Note that we impose the same maximum forming pressure for each case study (4.4 MPa) to obtain comparable values of equivalent plastic strain $\bar{\varepsilon}_p$. In all the cases, the maximum $\bar{\varepsilon}_p$ obtained is around 0.56 and is located over the material points identified by the fully annealed condition, although they are quite far from the dome apex. It is worth noticing that, although we imposed the same target forming pressure to all the tests, obtaining similar values of maximum equivalent plastic strain, the macroscopic effect of the annealing treatment on the specimen can be observed on the maximum dome height, which is 60 mm for the LT strategy and 62 mm for the AH one.

To be consistent with previous studies aimed to quantify the heterogeneity of the mechanical states produced by unconventional mechanical test [19,25], the classification and comparison of the three strategies is carried out following two main visualization methods:

- The diagram of major and minor principal strains, also used to infer the FLC;
- The diagram of normalized stress (widely used in the case of anisotropic plasticity), where the stress state for each material point is reported on the yield surface.

It is worth noting that the following diagrams includes only datapoints where the $\bar{\varepsilon}_p > 2\%$: this because low values of plastic deformation – corresponding to low values of dome height – are strongly affected by uncertainties when measured with optical techniques as DIC.

Material Forming - ESAFORM 2023
Materials Research Proceedings 28 (2023) 855-865

Materials Research Forum LLC
https://doi.org/10.21741/9781644902479-94

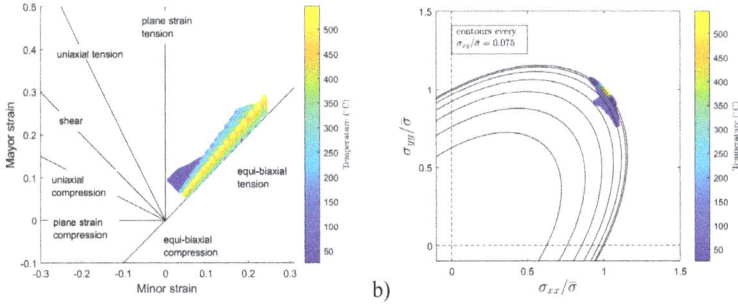

Fig. 5. Distribution of major and minor strains (a) and stress states on the normalized stress plane (b) produced by the bulge test on the specimen exposed to the linear treatment with R_m=20 mm and L=40 mm.

When looking at the results from the LT20, the forming limit diagram shown in Fig. 5a suggests that all the points close to the fully annealed state (yellow point) are characterized by a slope of the strain path slightly higher than that of the equi-biaxial state. On the other hand, material points characterized by lower level of annealing (peak temperature around 400°C) are closer to the bisector of the first quadrant. When moving far from the locally annealed zone, the strain path diverges from the equi-biaxial state more evidently.

The distance between the linear track and the center of the circular blank seems not to have an appreciable effect: in fact, when increasing the distance up to 30 mm, the evolution of the strain paths (Fig. 6a) resembles the distribution shown in Fig. 5a.

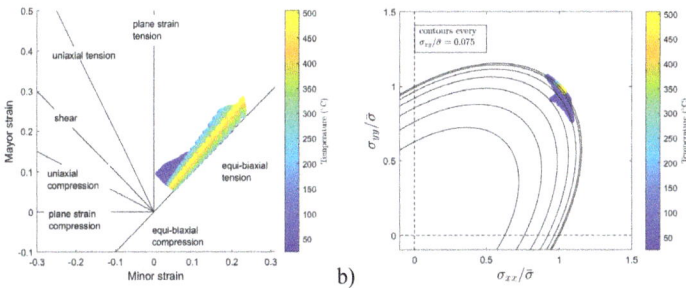

Fig. 6. Distribution of major and minor strains (a) and stress states on the normalized stress plane (b) produced by the bulge test on the specimen exposed to the linear treatment with R_m=30 mm and L=40 mm.

The evolution of the strain paths slightly changes when focusing the attention on the AH strategy. The scatter of the slope in the strain paths from the material point at the higher level of annealing is larger and a wider span of intermediate annealing conditions reaches the almost equi-biaxial state during the forming (Fig. 7, a and b).

Material Forming - ESAFORM 2023   Materials Research Forum LLC
Materials Research Proceedings 28 (2023) 855-865   https://doi.org/10.21741/9781644902479-94

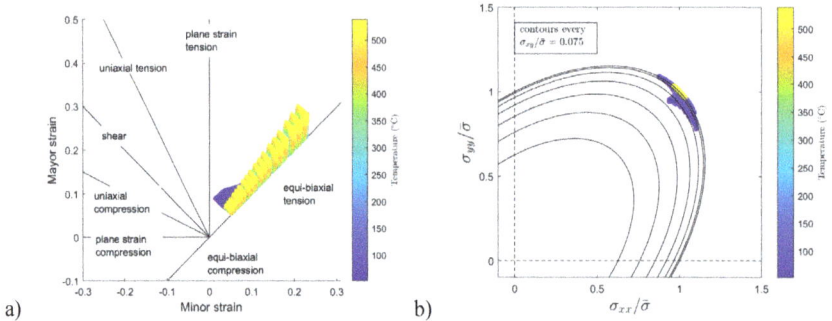

Fig. 7. *Distribution of major and minor strains (a) and stress states on the normalized stress plane (b) produced by the bulge test on the specimen exposed to the circular treatment with* $R_m=30$ *mm.*

The difference between the three investigated geometries are confirmed when looking at the strain paths from a radial path of nodes: for both the LT strategies, the points characterized by the fully annealed conditions, despite slightly diverging from the equi-biaxial state, reached a final strain condition higher than the other points characterized by intermediate levels of annealing (Fig. 8a and Fig. 8b).

On the other hand, when the AH strategy is taken into account, the final strain components (major and minor) are quite close regardless of the reached level of annealing.

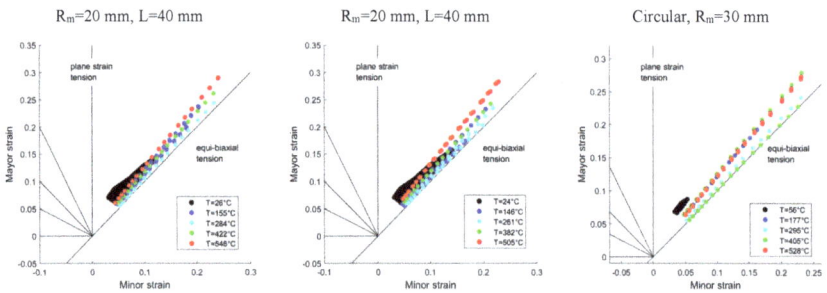

Fig. 8. *Distribution of major and minor strains obtained by sampling measurement points along a horizontal path in the case of five annealing temperatures, for all the heat-treatment configurations.*

FLC curve reconstruction from different heterogenous tests.

The strain path distributions from the locally annealed bulge tests, combined with the strain evolutions from the other two types of tests (tensile test and plane strain tests) allow to cover a wide portion of the forming limit diagrams. The onset of necking, particularly difficult to infer from the numerical simulation of the characterization tests (plane strain and bulge test) can be qualitatively drawn, embracing the limit conditions not only belonging to the material points in the fully annealed conditions but also some characterized by an intermediate level of annealing.

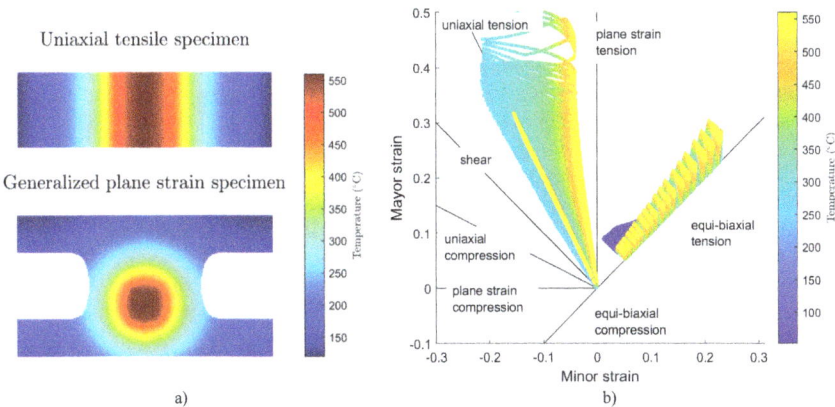

Fig. 9. Integration of different heterogeneous tests with the bulge specimen exposed to circular heat-treatment: a) distribution of the peak temperature for the uniaxial and generalized plane strain specimens, b) combined FLC diagram including all full-field data.

Summary

The present work focuses the attention on the third characterization tests when considering a gradient of properties in a AA5754-H32 blank that has been locally annealed by means of laser heating. The simulation of the bulge test, in combination with the tensile and the plane strain tests, complete the set of characterization methodology to assess the strain behavior of a locally-modified blank. The three investigated laser heating, different in terms of extent of the annealed region (a single linear track against a circular path) seemed not to have a significant influence on the evolution of the slope of the strain paths from points both in annealed conditions and those characterized by an intermediate level of annealing. Nevertheless, when looking at the strain paths along the radial direction, the AH strategy led to a more uniform distribution of strains, since material points at different levels of annealing reached almost the same limit strain (major and minor components). The results from the numerical simulation of the bulge test, combined with those from the plane strain test and tensile tests allowed to cover a wide portion of the forming limit diagram. It must be highlighted that, at this stage, the numerical simulation does not provide direct information about the necking instability and failure behavior due to the absence of a proper modelling of plastic damage. Therefore, the FLD reconstruction here presented should be intended as an investigation about the domain of principal strains that can be achieved by the material points over the whole range of annealing. As a consequence, a qualitative indication of a forming limit diagram could only be drawn.

References

[1] A. Taub, E. De Moor, A. Luo, D.K. Matlock, J.G. Speer, U. Vaidya, Materials for Automotive Lightweighting, Annu. Rev. Mater. Res. 49 (2019) 327–59. https://doi.org/10.1146/annurev-matsci-070218-010134

[2] DIN EN 45545-2. Railway applications - Fire protection on railway vehicles - Part 2: Requirements for fire behaviour of materials and components 2016.

[3] S.S. Hecker, Formability of Aluminum Alloy Sheets, J. Eng. Mater. Technol. 97 (1975) 66-73. https://doi.org/10.1115/1.3443263

[4] G. Palumbo, A. Piccininni, Numerical-experimental investigations on the manufacturing of an aluminium bipolar plate for proton exchange membrane fuel cells by warm hydroforming, Int. J. Adv. Manuf. Technol. 69 (2013) 731–742. https://doi.org/10.1007/s00170-013-5047-1

[5] A. Piccininni, A. Lo Franco, G. Palumbo, Warm Forming Process for an AA5754 Train Window Panel, J. Manuf. Sci. Eng. 144 (2022) 1–12. https://doi.org/10.1115/1.4052583

[6] M. Geiger, M. Merklein, U. Vogt, Aluminum tailored heat treated blanks, Prod. Eng. 3 (2009) 401-410. https://doi.org/10.1007/s11740-009-0179-8

[7] A. Piccininni, G. Palumbo, Design and optimization of the local laser treatment to improve the formability of age hardenable aluminium alloys, Materials 13 (2020). https://doi.org/10.3390/ma13071576

[8] M. Rossi, A. Lattanzi, A. Piccininni, P. Guglielmi, G. Palumbo, Study of Tailor Heat Treated Blanks Using the Fourier-series-based VFM, Procedia Manuf. 47 (2020) 904-909. https://doi.org/10.1016/j.promfg.2020.04.278

[9] M.A. Sutton, J.J. Orteu, H.W. Schreier, Image Correlation for Shape, Motion and Deformation, Springer, 2009.

[10] F. Pierron, M. Grediac, The virtual fields method: Extracting constitutive mechanical parameters from full-field deformation measurements, Springer New York, NY, 2012. https://doi.org/10.1007/978-1-4614-1824-5

[11] S. Avril, M. Bonnet, A.S. Bretelle, M. Grédiac, F. Hild, P. Ienny, F. Latourte, D. Lemosse, S. Pagano, E. Pagnacco, F. Pierron, Overview of identification methods of mechanical parameters based on full-field measurements, Exp. Mech. 48 (2008) 381–402. https://doi.org/10.1007/s11340-008-9148-y

[12] D. Lecompte, S. Cooreman, S. Coppieters, J. Vantomme, H. Sol, D. Debruyne, Parameter identification for anisotropic plasticity model using digital image correlation, Eur. J. Comput. Mech. 18 (2009) 393–418. https://doi.org/10.3166/ejcm.18.393-418

[13] M. Grédiac, F. Pierron, Applying the Virtual Fields Method to the identification of elasto-plastic constitutive parameters, Int. J. Plast. 22 (2006) 602-627. https://doi.org/10.1016/j.ijplas.2005.04.007

[14] M. Rossi, A. Lattanzi, F. Barlat, A general linear method to evaluate the hardening behaviour of metals at large strain with full-field measurements, Strain 54 (2018) 12265. https://doi.org/10.1111/str.12265

[15] M. Rossi, F. Pierron, M. Štamborská, Application of the virtual fields method to large strain anisotropic plasticity, Int. J. Solid. Struct. 97–98 (2016) 322-335. https://doi.org/10.1016/j.ijsolstr.2016.07.015

[16] A. Lattanzi, F. Barlat, F. Pierron, A. Marek, M. Rossi, Inverse identification strategies for the characterization of transformation-based anisotropic plasticity models with the non-linear VFM, Int. J. Mech. Sci. 173 (2020) 105422. https://doi.org/10.1016/j.ijmecsci.2020.105422

[17] J.M.P. Martins, S. Thuillier, A. Andrade-Campos, Calibration of a modified Johnson-Cook model using the Virtual Fields Method and a heterogeneous thermo-mechanical tensile test, Int. J. Mech. Sci. 202–203 (2021) 106511. https://doi.org/10.1016/j.ijmecsci.2021.106511

[18] A. Lattanzi, A. Piccininni, P. Guglielmi, M. Rossi, G. Palumbo, A fast methodology for the accurate characterization and simulation of laser heat treated blanks, Int. J. Mech. Sci. 192 (2021) 106134. https://doi.org/10.1016/j.ijmecsci.2020.106134

[19] A. Piccininni, A. Lattanzi, M. Rossi, Investigation of the plane strain behaviour of a laser-heat treated aluminium alloy Key Engineering Material 926 (2022) 1030-1038. doi:10.4028/p-ic4qyv

[20] A. Piccininni, A. Lattanzi, M. Rossi, G. Palumbo, Investigation of The Anisotropic Behaviour of Laser Heat Treated Aluminium Blanks . Paper presented at ESAFORM 2021. 24th International Conference on Material Forming, Liège, Belgique. (2021). https://doi.org/10.25518/esaform21.4086

[21] F. Grytten, B. Holmedal, O.S. Hopperstad, T. Børvik, Evaluation of identification methods for YLD2004-18p, Int. J. Plast. 24 (2008) 2248-2277. https://doi.org/10.1016/j.ijplas.2007.11.005.

[22] M. Rossi, A. Lattanzi, L. Cortese, D. Amodio, An approximated computational method for fast stress reconstruction in large strain plasticity, Int. J. Numer. Meth. Eng. 121 (2020) 3048-3065. https://doi.org/10.1002/nme.6346

[23] M. Rossi, A. Lattanzi, L. Morichelli, J.M. Martins, S. Thuillier, A. Andrade-Campos, S. Coppieters, Testing methodologies for the calibration of advanced plasticity models for sheet metals: A review, Strain 58 (2022) e12426. https://doi.org/10.1111/str.12426

[24] M. Rossi, A. Lattanzi, F. Barlat, J.H. Kim, Inverse identification of large strain plasticity using the hydraulic bulge-test and full-field measurements, Int. J. Solid. Struct. 242 (2022) 111532. https://doi.org/10.1016/j.ijsolstr.2022.111532

[25] M.G. Oliveira, S. Thuillier, A. Andrade-Campos, Evaluation of heterogeneous mechanical tests for model calibration of sheet metals, J. Strain Anal. Eng. 57 (2021) 208-224. https://doi.org/10.1177/03093247211027061

Friction and wear in metal forming

Material Forming - ESAFORM 2023　　　　　　　　　　　　　　Materials Research Forum LLC
Materials Research Proceedings 28 (2023) 869-878　　　　　https://doi.org/10.21741/9781644902479-95

Galling-free fine blanking of titanium plates by carbon-supersaturated tool steel punch

AIZAWA Tatsuhiko[1,a] *, FUCHIWAKI Kenji[2,b] and DOHDA Kuniaki[3,c]

[1]Shibaura Institute of Technology, 3-15-10 Minami-Rokugo, Ota-City, Tokyo 144-0045, Japan

[2]Hatano Seimitsu Co, 183-7 Hirasawa, Hatano-City, Kanagawa 257-0015, Japan

[3]Northwestern University, Evanstone, IL 60268, USA

[a]taizawa@sic.shibaura-it.ac.jp, [b]kenji-fuchiwaki@hatanoseimitsu.co.jp,
[c]dohda.kuni@northwestern.edu

Keywords: Forging, Fine Blanking, Galling Free, Titanium, Carbon-Supersaturation, Tool Steel Punch, High-Speed Steel Punch, In Situ Solid Lubrication

Abstract. The carbon supersaturated tool steel and high-speed steel punches were utilized to demonstrate that the titanium work materials were forged in high reduction of thickness and fine-blanked without adhesion of their fragments and deposition of their oxide debris particle. This galling free forging and fine blanking came from the in situ formation of the free carbon tribofilm only on the highly stressed interfaces between the punch/die and the titanium work. Under this in situ solid lubrication, the low frictional state was sustained to induce less bulging deformation of work during the forging process. The fine blanking of titanium plates advanced without adhesive wear and fractured regions on their sheared surface. The long life of tools was certified even in fine blanking of titanium and titanium alloy plates.

Introduction

Titanium and titanium alloys have been utilized as typical high strength, light-weight and bio-compatible structural materials in the aerospace industries [1], in the medical applications [2] and in the mechanical part-markets [3]. For manufacturing of those parts and members, various metal forming processes, such as forging, drawing and fine blanking, are utilized to shape their feedstock like bars, plates and billets into products. As reported in [4-5], those processes encountered severe wears by adhesion of fresh titanium fragments and by deposition of titanium oxide debris particles onto the punches and dies. In practical operations during forming the titanium alloy bar to its glass-frames, the cleansing and polishing steps must be included to eliminate the adhesive films and to clean up the debris particles between the successive forging steps under the limited reduction of thickness [4]. In addition, as studied in [5-7], the ceramic dies and their coatings and DLC (Diamond Like Carbon) coatings, were also subjected to highly frictional state with metal sticking during the deep drawing process and in the BOD (Ball-On-Disk) testing in dry, respectively. In particular, TiAlN, TiN and TiCN experienced extremely high friction and seizure against the pure titanium balls.

The pretreatment of SKD11 and AISI420 punches and dies by the low temperature plasma carburizing, played a significant role in dry and cold forging of the pure titanium and β-titanium alloy wires [8-9]. No galling or no adhesion wear occurred in those forging steps even in high reduction of thickness. As explained in detail [10], this anti-galling behavior in metal forming of titanium bars was attributed to in situ formation of free carbon tribofilms onto the highly stressed contact interface of punches and dies to the titanium works during forming process. In addition, this galling-free forging process was characterized by the low friction and low work-hardening even in higher reduction of thickness than 70% [11-12]. This pretreatment to punches and dies are

Material Forming - ESAFORM 2023   Materials Research Forum LLC
Materials Research Proceedings 28 (2023) 869-878   https://doi.org/10.21741/9781644902479-95

attractive to the near-net shaping by forging steps in galling-free and to the fine-blanking with fully burnished surface.

In the present paper, the carbon supersaturated (CS-) tool steel and high-speed steel punches are prepared by the low temperature plasma carburizing. This CS-SKD11 punch and die is utilized to upset the pure titanium bar till the reduction of thickness reaches 70%. The loading - displacement curves are online monitored to describe the compressive and flattening deformation of titanium work. The homogeneous flattening deformation proves that the low frictional state is preserved during forging. The CS-YXR high-speed steel punch is also prepared for fine blanking the pure titanium plates. The normal YXR punch suffers from severe adhesion of titanium fresh surfaces even in a single shot. The punch edges and side surfaces are completely adhered and eroded by the titanium fragments. On the other hand, no galling occurs on the CS-YXR punch edge and side surfaces even after fine blanking the titanium plates repeatedly. This significant difference in galling behavior proves the effectiveness of carbon supersaturation to steel punches and dies for fine blanking the difficult-to-metal-forming work materials.

Methods and Materials
The plasma carburizing system for carbon supersaturation process is first explained with notes on the plasma processing setup and conditions with reference to the previous studies in the literature. The forging and fine blanking setups and systems are also stated with comments on the extension of laboratory-scale methods to the mass production in industries.

Carbon supersaturation process.

The plasma carburizing system was schematically illustrated in Fig. 1a. With reference to the previous studies in [13-14], the hollow cathode device was employed to increase the carbon-ion and CH-radical densities more than 3×10^{17} ions/m^3 and to deepen the carbon supersaturated layer thickness. It was estimated to be 40 μm in case of the plasma carburizing at 673 K for 14.4 ks by 70 Pa.

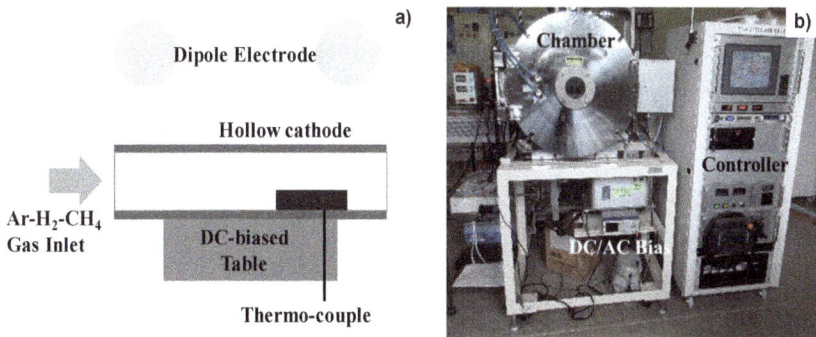

Fig. 1. Low temperature plasma carburizing system. a) Schematic view on the experimental setup, and b) overview on the whole system.

The dipole electrodes were utilized to ignite the RF (Radio-Frequency) plasma while the DC (Direct Current) plasma was also induced by applying the bias voltage to the bottom of hollow cathode. In the following experiments, the RF-voltage and the DC-bias were constant by +220 V and -400 V, respectively. After evacuation down to the base pressure of 0.01 Pa, the argon gas was introduced to a chamber in Fig. 1b at RT to clean the punch and die surfaces. After increasing the process temperature up to 673 K under the argon atmosphere, the hydrogen gas was

Material Forming - ESAFORM 2023
Materials Research Proceedings 28 (2023) 869-878

Materials Research Forum LLC
https://doi.org/10.21741/9781644902479-95

also introduced with the argon gas with the flow rate of 160 mL/min for argon and 20 mL/min, respectively. The total pressure was constant by 70 Pa. After presputtering by the DC-plasmas for 1.8 ks, the methane gas was introduced as a carbon source into argon and hydrogen mixture gas by the flow rate of 20 mL/min. At the specified duration of 14.4 ks, the specimen was cooled down in the chamber under the nitrogen atmosphere before evacuation down to atmospheric pressure. The processing temperature was in situ monitored by the thermocouple, which was embedded into the base plate below the hollow cathode device in Fig. 1a. The total pressure and each flow rate of argon, hydrogen and methane gasses were also measured for process control. The deviation of temperature and pressure in operation was ± 0.1 K and ± 0.05 Pa, respectively. After [8, 11-12], no iron and chromium carbides were synthesized in the plasma-carburized layer. The peak shift of α-ion to the lower 2θ angles in XRD analysis revealed that the α-lattices in the carburized SKD11 layer expanded by themselves through supersaturation of carbon solutes into steel substrates. The maximum surface roughness of punches at 5 mm away from their edge was 0. 35 μm before carburizing and 0.366 μm after carburizing. Hardness of SKD11 punch increased from 700 HV to 950 HV.

Forging process.

The CNC stamping system (ZEN-04; Hoden-Seimitsu, Co., Ltd., Kanagawa, Japan) was used for forging the pure titanium wire with the diameter of 1.0 mm in dry at the room temperature. The load to stroke curves were measured in every reduction of thickness by the load cell, which was embedded into the lower die in Fig. 2a. The overview of this stamping system was depicted in Fig. 2b. No lubricating oils and solid lubricants were utilized in the forging experiments. The punching speed was constant by 1 mm/s.

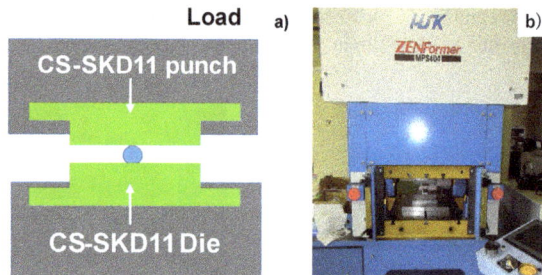

Fig. 2. CNC forging system. a) Schematic view on the forging experimental setup, and b) overview on the whole forging system.

Fine blanking process.

The pure titanium plate with the thickness of 2.0 mm, was blanked with narrow clearance of 4% at the room temperature as illustrated in Fig. 3a. The mechanical stamper (FB 160-FDE; Mori Steel Works Co., Ltd.; Saga, Japan), specially accommodated for fine-blanking process was used for this experiment as shown in Fig. 3b. The punching speed was 5 mm/s.

Material Forming - ESAFORM 2023
Materials Research Proceedings 28 (2023) 869-878

Materials Research Forum LLC
https://doi.org/10.21741/9781644902479-95

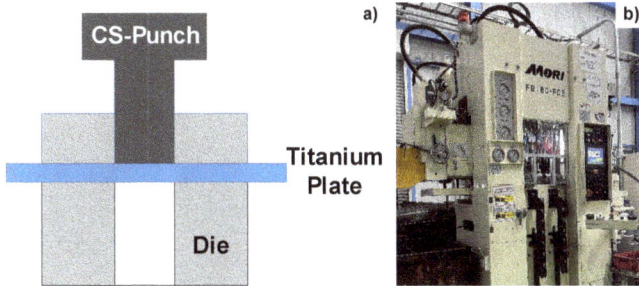

Fig. 3. Fine blanking system. a) Schematic view on the fine-blanking experimental setup, and b) overview on the whole system.

The maximum loading capacity was 1600 kN. The loading sequence for fine blanking was CNC-programmed. The punch with and without carbon-supersaturation were employed to describe the effect of carbon supersaturation on the galling behavior. Each punch was fixed into the punch holder, which was further set up into the upper die set. In the following experiments, FBH9-HMC with the viscosity of 101 m^2/s was utilized as a lubricating oil.

Work and tool materials.

Pure titanium wires and plates were respectively utilized as a work material for upsetting and fine blanking experiments. Their chemical composition consists of hydrogen by 0.0012 mass%, oxygen by 0.097 mass%, nitrogen by 0.007 mass%, iron by 0.042 mass%, carbon by 0.007 mass%, and titanium for balance. A tool steel type SKD11 (or AISI/SAE D3) was employed as a punch material for forging. A high-speed steel type YXR7 was used as a punch for fine blanking. Their chemical compositions were listed in Table 1.

Table 1. Chemical composition of SKD11 and YXR7 punch substrates.

Punch	C	Si	Mn	P	S	Cr	Mo	V	W	Fe
SKD11	1.44	0.3	0.35	0.27	0.06	11.1	0.8	0.2	---	in balance
YXR7	0.8	0.8	0.3	---	---	4.7	5.5	1.3	1.3	in balance

Characterization.

SEM (Scanning Electron Microscopy) - EDX (Energy Dispersive X-ray spectroscopy) was used for microstructure analysis and element mapping on the contact interface of forging and fine-blanking tools to the titanium work materials. Raman spectroscopy (JOEL, Co., Ltd.; Kanagwa, Japan) was also utilized to characterize the binding state of tribofilms.

Results

CNC-stamping was first utilized to describe the upsetting behavior of pure titanium bar with increasing the reduction of thickness. SEM-EDX and Raman spectroscopy were used to analyze the contact interface between the CS-SKD11 punch and the titanium work. Fine blanking system was also utilized to describe the difference of YXR7 punch surface with and without the carbon supersaturation to the fine-blanking punch.

Forging of pure titanium by CS-SKD11 punch.

The pure titanium bar with the diameter of 1.0 mm was upset to the specified reduction of thickness (r), using the CS-SKD11 punch and die. Figure 4 depicts the variation of contact interface width (Wi) with increasing r. When r < 30%, Wi is less than the upset bar width (Wo); e.g., when r = 20%, Wo = 1.15 mm and Wi = 0.70 mm. However, when r > 50%, this Wi approaches to Wo; e.g., when r = 50%, Wo = 1.7 mm and Wi = 1.5 mm. This reveals that the bulging deformation is suppressed with increasing r and the upset titanium bar homogeneously flattens with r.

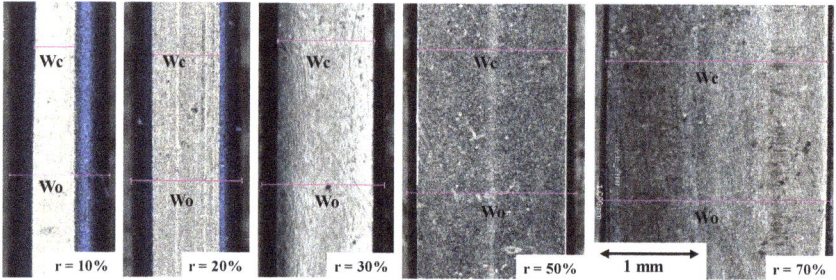

Fig. 4. Variation of the contact interface width at each reduction of thickness by r = 10%, 20%, 30%, 50% and 70% during the upsetting with the use of CS-SKD11 punch and die.

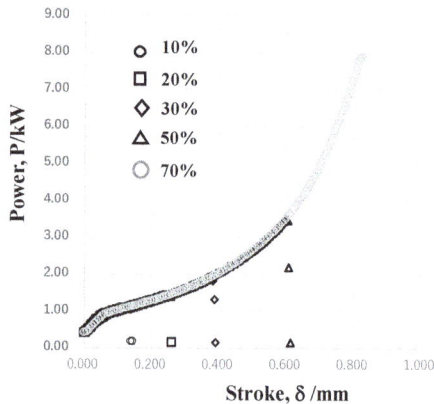

Fig. 5. The measured forging load to stroke relationship at each reduction of thickness for r = 10%, 20%, 30%, 50% and 70%, respectively.

After the empirical relationship between the friction coefficient (μ) and the bulging deformation ratio (= (Wo-Wi)/(2Wo)) in [15], μ is estimated to be 0.05 to 0.1. After the inverse analysis on the friction coefficient using the finite element simulation in [11], μ is estimated to be 0.05. Under this low friction on the interface between the CS-SKD11 punch and the pure titanium works, its thickness is homogeneously upset to r = 70%.

The in-situ measured forging load was plotted against the applied stroke at each reduction of thickness. As depicted in Fig. 5, every load-stroke curve at r = 10%, 20%, 30%, 50% and 70% is edited to one master load-stroke relationship. This implies that the pure titanium work homogeneously deforms with increasing the stroke, irrespective of the reduction of thickness. The forging load abrupt increases in the exponential manner with the stroke when r > 30%. This reveals that the true contact interface area expands by flattening behavior in Fig. 4.

The low friction on the contact interface as well as the homogeneous forging behavior proves that no galling occurs during this forging process with high reduction of thickness. Through the

Material Forming - ESAFORM 2023   Materials Research Forum LLC
Materials Research Proceedings 28 (2023) 869-878   https://doi.org/10.21741/9781644902479-95

precise analysis on the contact interface, the interfacial state is described after upsetting the pure
titanium bars up to r = 70% in twenty shots.

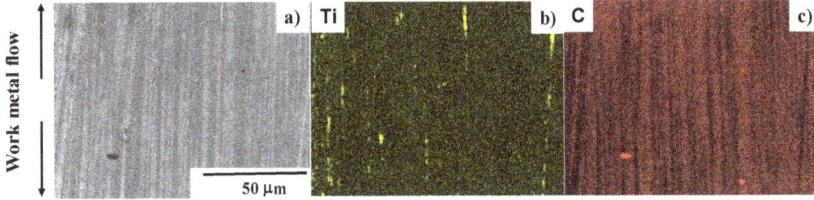

*Fig. 6. SEM image and element mapping on the true contact interface between the CS-punch and
the pure titanium work after continuously forging up to r = 70% in twenty shots.*

As shown in Fig. 6a, many gray stripes are formed on the interface along the direction of work
metal flow. Since the CS-punch surface was cleansed and polished before experiment, no
surfactants and residuals were left on the punch and die surfaces. Hence, these stripes were in situ
formed during the forging steps. SEM-EDX analysis was used to search for the element mapping
in correspondence to these stripes among the main constituent elements of SKD11 such as iron
and chromium, the titanium, the oxygen and the carbon. The iron and chromium maps were the
same as before forging experiments; no correlations to SEM image in Fig. 6a were noticed. As
shown in Fig. 6b, several thin titanium stripes were seen together with the oxygen mapping. There
was no significant correlation to the SEM image. Figure 6c shows the carbon mapping on the
interface; the gray stripes are carbon films, in situ formed onto the contact interface between the
CS-SKD11 punch and the titanium work during the forging process.

This SEM-EDX analysis on the contact interface of CS-SKD11 punch to the titanium work
proves that the in-situ formed free carbon tribofilm works as a solid lubricant medium on the true
contact interface to lower the friction coefficient and to make homogeneous metal flow along the
CS-punch interface. Nearly the same in situ solid lubrication was also observed on the contact
interface between CS-die and the titanium work. These low frictional state of punch and die as
well as the smooth work flow along their interfaces, characterizes the galling-free forging behavior
when using the carbon supersaturated special tools.

To be noticed, no carbon maps in stripes in Fig. 6 were detected outside of the true contact area
on the CS-punch and CS-die surfaces. That is, the applied stresses working on the true contact
interface are necessary to in situ form the free carbon tribofilm. High compressive normal stress
induces the diffusion of carbon solutes in the CS surface zones to the true contact interfaces of
punch and die so that these free carbon agglomerates form the carbon tribofilm only on the contact
interface.

Fine blanking of pure titanium by YXR punch.

In the fine blanking, the punch head is subjected to high normal stresses when punching out the
work materials under the narrow clearance between the punch and die. The punch edge experiences
the high shear stress transients and local metal flow change. The punch side surfaces are also
subjected to shearing flow of work materials. Under these mechanical conditions with more
severity than forging, the plastic distortion of work materials could result in galling damage on the
XYR7 punches when fine-blanking the ductile and high-strength stainless steel and titanium plates.

In the present experiment, a bare YXR7 punch was used as a reference to describe the galling
behavior when fine-blanking the pure titanium plate with the thickness of 2 mm in a single shot
under the lubricating conditions. The CS-YXR7 was also utilized to demonstrate the role of carbon
supersaturation in galling-free fine blanking.

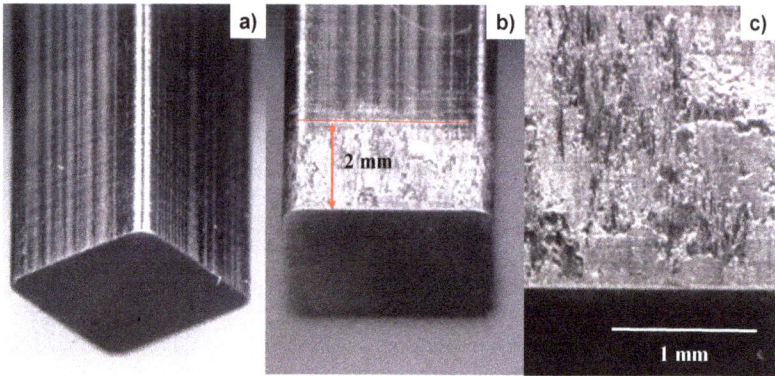

Fig. 7. YXR7 fine-blanking punch before and after fine blanking the pure titanium plate with the thickness of 2.0 mm in a single shot. a) YXR7 punch before fine blanking test, b) YXR7 punch after fine blanking test, and c) adhesion layer on the punch side surfaces.

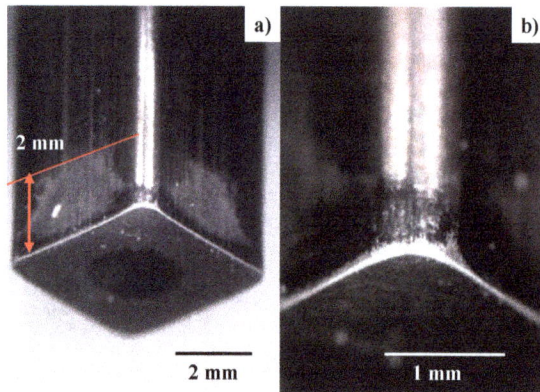

Fig. 8. CS-YXR7 punch after successively fine blanking the austenitic stainless steel AISI304 plates in three shots and the pure titanium plates in two shots. a) CS-YXR7 punch after fine blanking test, and b) punch surfaces around its edge and corner.

Fig. 7a and 7b compare the YXR7 punch surfaces before and after fine blanking the pure titanium plate with the thickness of 2.0 mm. No wears were seen both on the punch head after fine blanking in a single shot. However, its side surfaces with the length around 2 mm were almost covered by the fragments of titanium work. In particular, as seen in Fig. 7c, the titanium fragments adhered and eroded into the YXR7 punch., resulting in total failure of punch. This galling behavior proves that the fresh sheared titanium work is easy to adhere onto the punch side surfaces in the high normal and shear stress conditions under the narrow clearance.

Fine blanking of pure titanium by CS-YXR punch.

CS-YXR7 punch was used to make fine blanking the AISI316 plates with the thickness of 3 mm in three shots and then to punch out the pure titanium plate with the thickness of 2 mm in two shots. Fig. 8a shows the overview of CS-YXR7 after a series of fine blanking shots.

Material Forming - ESAFORM 2023   Materials Research Forum LLC
Materials Research Proceedings 28 (2023) 869-878   https://doi.org/10.21741/9781644902479-95

Fig. 9. The side and bottom surface of titanium blank, punched pout by the CS-YXR7 punch.

In contrast to the severe adhesion of titanium fragments onto the bare YXR7 punch in Fig. 8b and 8c, the CS-XYR7 punch is completely free from the galling of sheared titanium work even after punching the stainless steel and titanium plates in several times. This implies that in situ formed free carbon tribofilms are formed on the highly stressed contact surfaces of punch under the narrow clearance in the similar manner to the in situ solid lubrication as seen in Figure 6 when forging the pure titanium plate. Once the tribofilm is formed on the fine blanking punch, no adhesive wear of metal works occurs even with increasing the number of shots.

This galling-free behavior suggests that the quality of punched-out blank is expected to be improved by this fine blanking process. Fig. 9 depicts the top and side surfaces of punched-out titanium blank. No fractured regions are detected on every four side surface of blank. This proves that CS-YXR7 punch is suitable to high qualification in fine blanking of titanium plates to products.

Discussion

The flash temperature on the highly stressed contact interface between the punches/dies and the work materials was thought as a main process parameter to govern the galling heavier in forging [7, 16]. In fact, the oxidation wear of carbon-based coatings such as diamond and DLC (Diamond Like Carbon) in stamping and forging [17] suggested that the chemical galling had overwhelming influence on the abrasive wear of tool surfaces at hot spots. As compared in Figs. 7 and 8, the active metals such as pure titanium with chemical compatibility to various inorganic materials, induced the adhesive wear onto the steel tool surfaces at RT even under the lubricating conditions. This mechano-chemically adhesive galling under highly stressed condition, accompanied with erosion of titanium into the tools. Under the free carbon tribofilms, in-situ formed onto the carbon supersaturated SKD11 and YXR7 punches, this adhesive galling is completely suppressed to make forging and fine-blanking repeatedly without metallic sticking and seizure. This reveals that inactive tribofilm is needed to be free from the mechano-chemical adhesion galling.

As had been discussed in the dry tribological behavior of carbon-based coatings [6-7], the high friction coefficient, experienced in the TiN, TiCN, TiAlN and other ceramic coatings, much reduced in the silicon-bearing DLC and nano-laminated DLC coatings during the BOD testing. This implies that adhesive galling is effectively suppressed on the bound carbon surface of the amorphous carbon microstructure. In the present in-situ solid lubrication, the free carbon film with nearly the same Raman spectroscopy as the hydrogenated DLC (or a:C:H) works as an inactive buffer interface between the fresh titanium and the steel punch.

Material Forming - ESAFORM 2023   Materials Research Forum LLC
Materials Research Proceedings 28 (2023) 869-878   https://doi.org/10.21741/9781644902479-95

Let us consider the effect of carbon bound state on the frictional behavior. Although the SKD11 houses the iron and chromium carbides by heat treatment and the YXR7 has a low carbon content, severe galling occurs as seen in Fig. 7. The bound carbon in the carbides and the alloying carbon element have no capacity of in situ solid lubrication. The weakly bound carbon in the amorphous carbon film has a little capacity to reduce the friction under mild tribological conditions. The unbound carbon tribofilm has a significant capacity to be free from mechano-chemical adhesion wears in fine blanking. This difference in the bound state of carbons on the contact interface, directly reflects on the galling process of titanium works.

As keenly discussed in [10] and suggested in [18-19], the supersaturated carbon in the steel punches diffuses onto the contact interface to titanium work under high stress gradient, and, forms a free-carbon tribofilm. As seen in Figs. 4 and 8a, no tribofilms were formed outside of the true contact area. In case of forging process, the carbon stripes were formed as a thin tribofilm on the true contact interface of CS-SKD11 punch to titanium work. In case of fine blanking, the carbon film was also formed as a thin tribofilm on the contact interface of CS-YXR7 punch. Remember that high normal stress was applied onto the contact interface when forging the pure titanium wire, and that higher normal stress and high shear stress were combined and applied onto the contact interface when fine-blanking the pure titanium plate. This reveals that tribofilm thickness is dependent on the applied stress level during metal forming processes. To be noticed, the free-carbon tribofilm formed onto the YXR7 punch surface is so tough not to be delaminated even in high and combined stress conditions without galling. Owing to this in situ formation of tough tribofilm, the CS-YXR7 punch is also free from adhesive wearing even with increasing the number of shots as discussed in [20]. This carbon supersaturation into punch and die substrates are necessary to prolong the tool life without severe galling in the fine blanking of titanium plates.

As reported in [21-22], the abrasive and adhesive galling processes were effectively suppressed by the nanotexturing onto the punch side surface when punching out the electrical steel sheets. This was because the debris particles were ejected through the nanotextured grooves to the outside of punching process. This nanotexturing effect can be also accommodated to this fine blanking with the use of CS-steel punches. The titanium and titanium alloys parts with fully burnished surface can be near-net shaped by this fine blanking without significant galling and abrasive wear. The effect of galling-free fine blanking on the product quality and tool life is a next issue to advance this new type tooling for fine blanking.

Summary

The carbon supersaturation by the low temperature plasma carburizing is essential to protect the tool steel and high-speed tool steel punches and dies from adhesion wear during the forging and fine blanking of titanium and titanium alloy works. In particular, the free-carbon tribofilm, in situ formed on the YXR7 punch, minimizes the mechanical and chemical galling damages during the fine-blanking of titanium plates. Under the shearing stress transients in fine-blanking, this in situ solid lubrication plays a role to prevent the punch surfaces from severe galling. This tooling is available in the industries to continuously punch out the complex-shaped blanks of austenitic stainless steels, titanium and titanium alloys with fully burnished surfaces.

Acknowledgment

The authors would like to express their gratitude to S-I. Kurozumi (Nano-Film Coat, llc.) for his help in experiments.

References

[1] G.W. Kuhlman, Forging of titanium alloys, in: Metalworking: Bulk forming, Vol. 14 in ASM Handbook, 2005.

[2] M. Chandrasekaran, Forging of metals and alloys for biomedical applications, Ch. 10 in: Metals for biomedical devices, 2nd Ed. Woodhead Publishing, 2019, pp. 293-310.

Material Forming - ESAFORM 2023   Materials Research Forum LLC
Materials Research Proceedings 28 (2023) 869-878   https://doi.org/10.21741/9781644902479-95

[3] D.S. Fernández, B.P. Wynne, P. Crawforthc, K. Fox, M. Jackson, The effect of forging texture and machining parameters on the fatigue performance of titanium alloy disc components, Int. J. Fatigue 142 (2021) 105949. https://doi.org/10.1016/j.ijfatigue.2020.105949

[4] T. Kihara, Visualization of deforming process of titanium and titanium alloy using high speed camera, Proc. 2019-JSTP Conference 2019, pp. 41-42.

[5] S. Kataoka, M. Murakawa, T. Aizawa, H. Ike, Tribology of dry deep-drawing of various metal sheets wit use of ceramic tools, Surf. Coat. Technol. 178 (2004) 582-590. https://doi.org/10.1016/S0257-8972(03)00930-7

[6] K. Dohda K., T. Aizawa, Tribo-characterization of silicon doped and nano-structured DLC coatings by metal forming simulators, Manuf. Lett. 2 (2014) 82-85. https://doi.org/10.1016/j.mfglet.2014.03.001

[7] K. Dohda, M. Yamamoto, C. Hu, L. Dubar, K.F. Ehman, Galling phenomena in metal forming, Friction 9 (2020) 686-696. https://doi.org/10.1007/s40544-020-0430-z

[8] T. Aizawa, T. Yoshino, Y. Suzuki, T. Shiratori, Free-forging of pure titanium with high reduction of thickness by plasma carburized SKD11 dies, Materials 14 (2021) 2536. https://doi.org/10.3390/ma14102536

[9] T. Aizawa, T. Yoshino, Y. Suzuki, T. Shiratori, Anti-galling cold, dry forging of pure titanium by plasma-carburized Al-SI420J2 dies, Appl. Sci. 11 (2021) 595. https://doi.org/10.3390/app11020595

[10] T. Aizawa, T. Shiratori, In-situ solid lubrication in cold dry forging of titanium by isolated free carbon from carbon-supersaturated dies, J. Friction (2022) (in press).

[11] S. Ishiguro, T. Aizawa, T. Funazuka, T. Shiratori, Green forging of titanium and titanium alloys by using the carbon supersaturated SKD11 dies, Appl. Mech. 3 (2022) 724-739. https://doi.org/10.3390/applmech3030043

[12] T. Aizawa, T. Funazuka, T. Shiratori, Near-net forging of titanium and titanium alloys with low friction and low work hardening by using carbon-supersaturated SKD11 dies, Lubricants 10 (2022) 203. https://doi.org/10.3390/lubricants10090203

[13] T. Aizawa, Low temperature plasma nitriding of austenitic stainless steels, Ch. 3 in Stainless Steels and Alloys, IntechOpen, UK, London, 2019, pp. 31-50.

[14] D.M. Geobel, C. Becatti, I.G. Mikellides, A.L. Ortega, Plasma hollow cathodes, J. App. Phys.130 (2021) 050902. https://doi.org/10.1063/5.0051228

[15] J.J. Hong, W.C. Yeh, Application of response surface methodology to establish friction model of upset forging, Adv. Mech. Eng. 10 (2018) 1-9. https://doi.org/10.1177/1687814018766744

[16] K. Dohda, C. Boher, F. Rezai-Aria, N. Mahayotsanun, Tribology in metal forming at elevated temperatures, Friction 3 (2015) 1-27. https://doi.org/10.1007/s40544-015-0077-3

[17] S. Takeuchi, Treatise on the CVD diamond coatings, Ohm-Sha (2022).

[18] Y. Cao, F. Ernst, G.M. Michal, Colossal carbon supersaturation in austenitic stainless steels carburized at low temperature, Acta Mater. 51 (2003) 4171-4181. https://doi.org/10.1016/S1359-6454(03)00235-0

[19] R. Rementeria, J.D. Poplawsky, E. Urones-Garrote, R. D-Reyes, C. Garcia-Mateo, F.G. Caballero, Carbon supersaturation and clustering in bainitic ferrite at low temperature, Proc. 5th Int. Symp. Steel Science, Nov. 14, 2017, Kyoto, Japan, ISIJ (2017) 29-34.

[20] T. Aizawa, K. Fuchiwaki, Galling-free fine blanking of titanium plates using carbon supersaturated high-speed steel punch, J. Carbon Research (2023) (in press).

[21] T. Aizawa, T. Shiratori, Y. Kira, T. Inohara, Simultaneous nano-texturing onto a CVD-diamond coated piercing punch with femtosecond laser trimming, Appl. Sci. 10 (2020) 2674. https://doi.org/10.3390/app10082674

[22] T. Aizawa, T. Inohara, T. Yoshino, Y. Suzuki, T. Shiratori, Laser treatment of CVD diamond coated punch for ultra-fine piercing of metallic sheets, Ch. 4 in: Engineering Applications of Diamond, Intech Open, UK, London, 2021, pp. 43-65.

Material Forming - ESAFORM 2023
Materials Research Proceedings 28 (2023) 879-890

Materials Research Forum LLC
https://doi.org/10.21741/9781644902479-96

The influence of Particle Hardness on Wear in Sheet Metal Forming

ARINBJARNAR Úlfar[1,a]*, KNOLL Maximilian[1,b], MOGHADAM Marcel[2,c]
and NIELSEN Chris Valentin[1,d]

[1]Department of Civil and Mechanical Engineering, Technical University of Denmark,
Produktionstorvet 425, 2800 Kongens Lyngby, Denmark

[2]FalCom A/S, Laustrupbjerg 7, 2750 Ballerup, Denmark

[a]ulari@dtu.dk, [b]maxkn@dtu.dk, [c]mmoghadam@falcom.net, [d]cvni@dtu.dk

Keywords: Tribological Testing, Particles, Contamination, Lubricant Additive

Abstract. Particles exist in any tribo-system, whether it is closed to the environment or not. These particles originate from various sources, for example: contaminants such as dust or fibres from the environment; wear debris from abrasive/adhesive wear; and particles that are intentionally included in lubricant formulations. The particles affect the tribo-system in which they occur, but it is not always clear how. In this work, three types of chemically inert particles of similar size but different hardness are mixed with an otherwise pure oil and tested tribologically. Three tribological testing methods, pin-on-disc, four-ball, and bending-under-tension, are used to investigate the effect of the particles on friction and wear under sheet metal forming conditions. The hardness of the particles had a large effect on wear development, but little to no effect on the coefficient of friction found by pin-on-disc testing. Including particles of any hardness helped the oil in which they were included resist variation in load, leading to less wear for higher loads compared to the pure oil.

Introduction

Various types of particles are endemic to tribological systems. Some particles are added to a tribo-system on purpose, while others enter the tribo-system accidentally or are unintentionally created through tribological or chemical mechanisms. There are essentially three sources by which the concentration of particles in a running tribo-systems increases over time. One of the sources are particles that are generated through tribological mechanisms such as adhesive or corrosive wear. These can be of different shapes and chemical compositions but would generally include some part of either the workpiece or tool. Another source is the environment, which can introduce particles of material such as dust or fibres from fabrics into the tribo-system. Lastly, depending on the lubricant formulation, the use and aging of the lubricant can lead to bacterial growth, which eventually forms bacterial cells [1]. These cells can be considered solid particles, but often help by improving the wear resistance of the lubricant rather than degrading it. These sources cause an increase in the concentration of particles in a tribo-system, with some particles already existing within the tribo-system, due to e.g., improper cleaning [2] or particles being included in the lubricant formulation.

The presence of particles in a tribo-system is often considered unwanted and has even been used as an indicator of mechanical systems undergoing severe wear [3]. Jiang and Wang [4] investigated the effect of wear particles of different sizes and concentration on wear in a locomotive engine. They applied a roller-on-roller test using a base-oil that included some concentration of particles collected from used oils and filters from running locomotive engines. Their result showed that the concentration of particles and the size of the particles are positively related to the amount of wear that occurs in the tribological test. The particles they used consisted mainly of iron and oxidation products, but also included some copper and aluminium. Abou El

Material Forming - ESAFORM 2023
Materials Research Forum LLC
Materials Research Proceedings 28 (2023) 879-890
https://doi.org/10.21741/9781644902479-96

Naga and Salem [5] investigated how the presence of wear particles in a lubricating oil affected the degradation of the oil. They found that metallic particles can have a catalytic effect and speed up the oxidation of various chemicals in the lubricant composition. An increasing concentration of particles made this effect more pronounced, meaning that the effectiveness of the lubricant was degraded even more quickly. In their paper on the tribology in the wheel-rail contact of locomotives, Olofsson et al. [6] discussed the role that particles play in wear. The role is, independently of where the particles originate, essentially twofold. On one hand, the particles are free to move in a contact interface and therefore lead to three-body abrasive wear. Olofsson et al. note that the hardness of the particles in this case does not have a large effect on the wear rate, provided they are at least 20% harder than the surfaces in contact. This effect is the same as that used by other researchers investigating the use of solid particles in lubricant formulation. On the other hand, the particles can become embedded in a surface and initiate severe two-body abrasive wear.

Many researchers have investigated intentionally adding solid particles to a lubricant to improve the lubricant's tribological performance. Luo et al. [7] added alumina particles with an average size of 78 nm to an oil and found that the particles improved the friction reduction and wear resistance properties of the oil under four-ball and thrust-ring conditions at loads of 147 N and 200 N respectively. The particles they used are harder than the active test components, which had a hardness of between 44 - 66 HRC and were thought to promote rolling over sliding as well as forming a tribo-film that helps resist wear. Padgurskas et al. [8] used iron, copper, and cobalt particles to improve the friction and wear behaviour of a lubricant under four-ball test conditions. They included mixtures of the particles, which showed an even larger improvement in tribological performance compared to single particle lubricants, with the best mixture being equal concentrations of iron and copper. Li et al. [9] investigated the effect of wear particles in a lubricant used in a rolling process and found that the wear particles had an anti-wear and anti-friction effect, at least under the four-ball conditions applied in their work. However, the particle also affected the stability of the lubrication mechanism in a deleterious way, which lead to increased wear after some time. In other work [10], the efficacy of $CaCO_3$ particles of two sizes for improving tribological conditions was evaluated. The particles were added to a base oil, along with a surfactant to ensure proper dispersion, and then tested by applying pin-on-disc testing, four-ball testing, and bending-under-tension testing. The particles were found to drastically improve the tribological performance of the base oil, through what appeared to be physical mechanisms. Peng et al. [11] evaluated the wear resistance of a tribo-system that included SiO_2 particles of different sizes and found that as the size of the particle decreased, the wear resistance and friction reduction improved. At some point, the size of the particles exceeded some critical point and the wear amount, and friction, became higher than for the pure lubricant. An increased concentration of the particles emphasized this behaviour even more.

In summary, depending on the specific properties of particles, such as size and hardness, they can have different effects on the tribo-system in which they are added. Some property of particles in a tribological system means the difference between their acting as abrasives, causing wear, and their acting as ball-bearings to promote rolling over sliding, reducing wear. In this work, three types of particles are added to pure paraffin oil with the aim of investigating the effect of the particle hardness on their role in wear, with a focus on sheet metal forming. Therefore, $CaCO_3$ particles are used as they have been shown to be beneficial to tribo-systems, $BaSO_4$ particles as they are slightly harder than the $CaCO_3$ particles, and SiO_2 particles as they are much harder than the other particles.

Experimental Methods and Materials

Particles. Some of the properties of the particles used in this work, given in supplier datasheets, are shown in Table 1. The size of the particles is similar, with the SiO_2 particles being slightly

Material Forming - ESAFORM 2023   Materials Research Forum LLC
Materials Research Proceedings 28 (2023) 879-890   https://doi.org/10.21741/9781644902479-96

larger but in the same range. The hardness of the particles is given on the Mohs-scale which shows that the $CaCO_3$ particles are softest, the $BaSO_4$ being slightly harder and the SiO_2 particles being significantly harder. All particles used in this work had a nodular shape, i.e., almost spherical.

Table 1. Properties of particles used in this work.

Chemical formula	$CaCO_3$	$BaSO_4$	SiO_2
Commercial name	Polyplex 2	Portaryte® B 10	Silverbond M500
Particle size D10% [µm]	~ 0.4	0.8	2.2
Particle size D50% [µm]	2.0 - 2.6	2.6	5.0
Particle size D90% [µm]	~ 5.4	8.2	11.7
Hardness [Mohs]	3.0	3.5	7.0
Bulk density [g/cm³]	0.8	1.7	0.42
Oil absorption [g/100ml]	17.4	10	23
Source	[12]	[13]	[14]

The lubricant mixtures tested in this work are derived from base mixtures that contain the specific particle, Ph. Eur. grade paraffin oil as the base oil, and Tween60 as surfactant to allow proper and homogeneous dispersion of the particles in the base oil. The base mixtures were prepared in a similar way, involving first dissolving the surfactant in the paraffin oil using a Dispermat CV3-Plus high-speed dissolver, followed by adding the particles and dispersing them using the same high-speed dissolver. The resulting base mixture was then diluted by adding more paraffin oil to arrive at a lubricant mixture of a fixed nominal concentration. The lubricant mixture was then agitated before use to ensure proper dispersal of the particles in the base oil. The nominal concentrations of single particle types tested in this work are 10 wt%, 20 wt% and 40 wt%. They were prepared using a scale that has a resolution of 10 mg, leading to the uncertainty of concentration being less than 1% of the nominal concentration.

Tribological Testing. Pin-on-disc testing involves placing a pin into contact with a rotating disc under some load, as shown in Fig. 1. Also shown in the figure is the specific testing machine used in this work, which is a standard *CSM Instruments Tribometer*. The tribometer has a built-in force transducer that measures friction load, allowing the device to calculate the friction coefficient based on the applied load. In this work the pin-on-disc test was performed according to DIN 50324:1992 for testing of friction. Two loads were applied, 1 N and 10 N, to investigate how the difference in load affects the tendency of the particles for being embedded in surfaces and affecting friction in that way. The disc component used in this testing is made from *Vanadis 4E*, a tool-steel from *Uddeholm* that has been ground and polished to a surface roughness of Ra = 0.06 µm after hardening to 62 HRC. The pin component consists of a ball-holder and a ball, which is made from EN 1.4301 steel, with a hardness of 25 HRC, grade G100 and diameter Ø6 mm. The amplitude of roughness of surfaces that were in contact was therefore much smaller than the particle size for all particle types. A fresh part of the discs and an unused ball were used in each test.

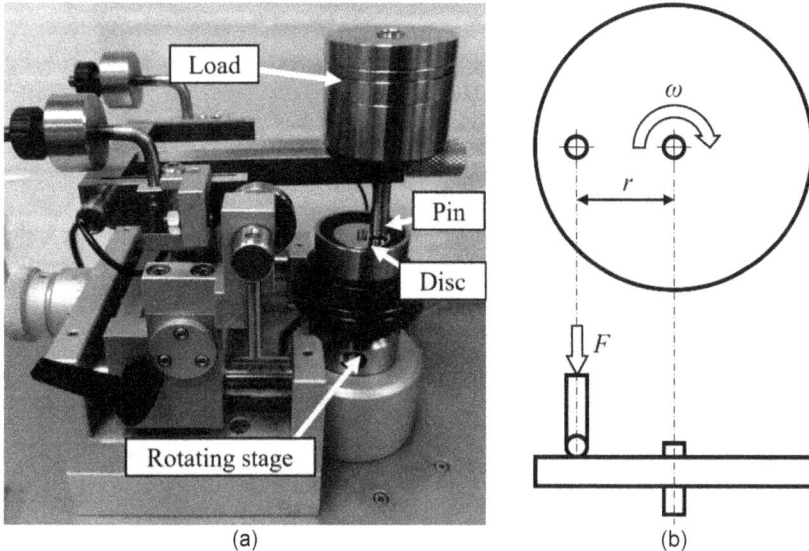

(a)   (b)

Fig. 1. (a) Standard CSM Tribometer with pin-on-disc configuration. (b) Principle of the pin-on-disc test.

Test parameters, besides applied load and track radius, were kept constant through all tests. The sliding speed was set to 50 mm/s, with the total sliding distance being 100 m. The track radii were in the range 8 mm – 14 mm with the rotational speed being adjusted based on the track radii so that the linear sliding speed was constant. The amount of lubricant-particle mixture used in each test, 7 ml – 10 ml, was enough to submerge the disc component to a depth of at least 5 mm. The approximate kinematic viscosity of the mixtures used in this work is on the order of 200 ± 100 cSt, so according to the Hersey-number of the system, the boundary lubrication mechanism is dominant. Each test condition, i.e., particle concentration and type of particle, was repeated three times to account for reproducibility.

Four-ball testing involves establishing sliding contact conditions between three balls that are fixed in place and one ball that rotates under some load, as shown in Fig. 2. Also shown in the figure is the testing machine used in this work, which was custom-built for the Technical University of Denmark. In this work, four-ball testing was performed as a wear test as defined in the ISO 20623:2018 standard, applying a load of 300 ± 5 N or condition C2. The test was performed over a period of $3,600 \pm 1$ seconds at a constant rotational speed of $1,420 \pm 20$ rpm. For each test, a fresh set of 100Cr6 chromium steel bearing balls of hardness 60 HRC - 66 HRC, Ø12.7 mm, and grade G20, were used as test balls. 10 ml of lubricant were used in each test, ensuring that the balls were covered to a depth of at least 5 mm. This test enables the determination of how the presence of particles of different hardness in the lubricant affects the wear resistance properties of the lubricant under boundary lubrication conditions. A further two load levels, 150 N and 450 N were applied for a 10 wt% concentration of particles to see how the load level affects the particle behaviour. Each test condition, i.e., particle concentration, type of particle and load level, was repeated three times to account for reproducibility.

Material Forming - ESAFORM 2023
Materials Research Proceedings 28 (2023) 879-890

Materials Research Forum LLC
https://doi.org/10.21741/9781644902479-96

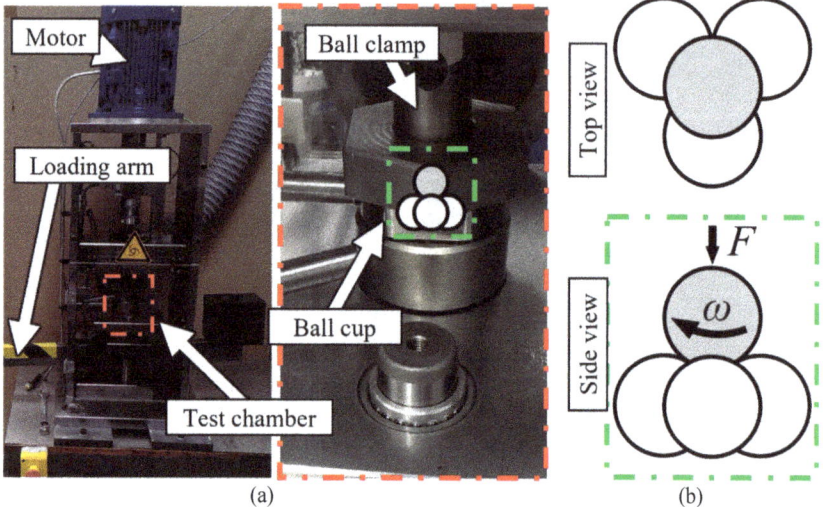

Fig. 2. (a) Four-ball testing machine including close-up of test chamber. (b) Principle of four-ball test.

Bending-under-tension testing was applied as a simulation of a deep drawing process to investigate how the different particles behave in a tribological system that is typically found in industrial sheet metal forming. The principle of the test revolves around bending a strip over a tool-pin under back-tension, as shown in Fig. 3. The difference between the back-tension force, F_b, and the drawing force, F_d, will then be due to the force required to deform the strip and friction. As the force required to deform the strip is nearly constant, then any increase in drawing force will be caused by an increase in friction in the tool-pin/strip interface. The nominal concentration of 20 wt% particles suspended in paraffin oil was applied here. The tool-pins were made from *Vanadis 4E* tool-steel hardened to 62 HRC and polished, while the strip was EN 1.4301 stainless steel with a 2B surface finish and a 30 mm x 1 mm cross-section. The edges of the strip are harder than the middle due to strain-hardening from roller cutting. A back-tension of 180 MPa ± 10 MPa was applied across 50 strokes, the sliding length of which was set to 30 mm. The sliding speed was 30 mm/s, with an idle time of 0.5 s between strokes which meant that the production rate was 40 strokes/min. Each test was repeated so that two data-sets exist for each test set-up.

Material Forming - ESAFORM 2023   Materials Research Forum LLC
Materials Research Proceedings 28 (2023) 879-890   https://doi.org/10.21741/9781644902479-96

(a)

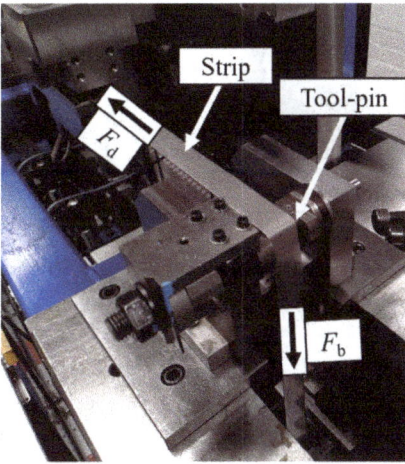

(b)

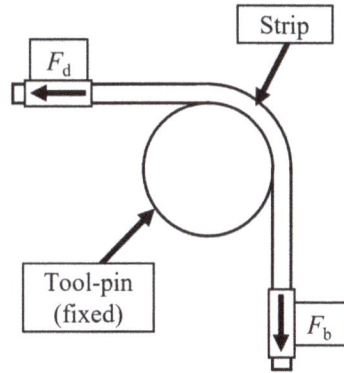

(c)

Fig. 3. Bending-under-tension (BUT) test shown by (a) an overview of the testing machine, (b) a close-up of the testing area, and (c) the principle of the test.

Optical Methods. An Olympus LEXT 4000 laser confocal microscope was used to capture images of wear scars on pin-on-disc discs, four-ball specimens and BUT tool-pins. Captured images were then processed in SPIP, an image processing software. Fig. 4(a) shows a typical disc after pin-on-disc testing, and the location of where images are taken of the wear scars on its surface. Fig. 4(b) shows a typical wear scar found in four-ball testing, along with where its parallel and perpendicular diameters are measured. Reported wear scar diameters are averaged between the two measurements. Fig. 4(c) shows the acquisition strategy employed for acquiring images of the surfaces of tool-pins used in BUT testing. The middle region of wear scars is imaged.

(a)

(b)

(c)

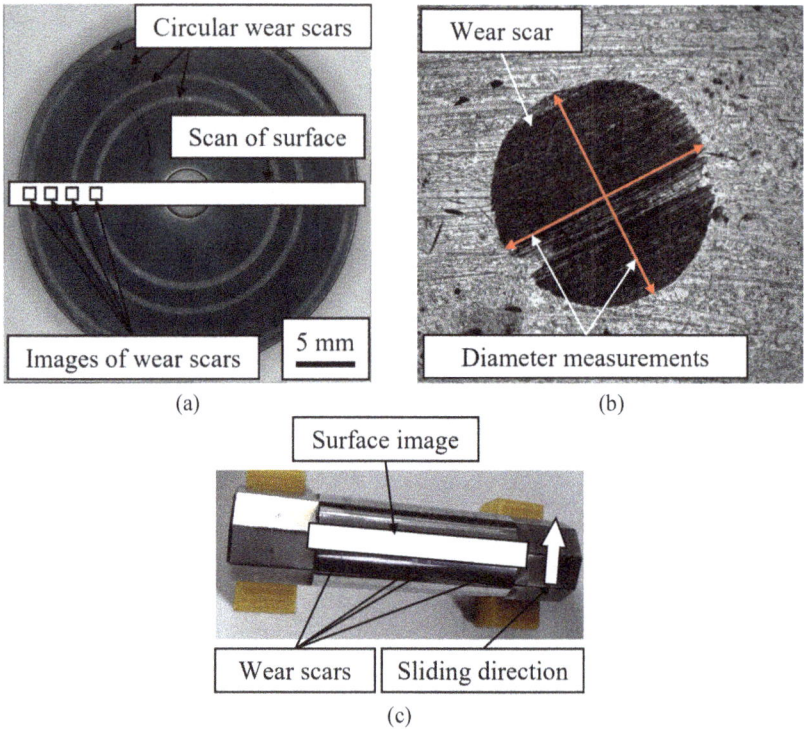

Fig. 4. Explanation of (a) acquisition of images of wear scars from pin-on-disc testing, (b) measuring of wear scars on four-ball specimens, and (c) acquisition of images of wear scars from tool-pin after bending-under-tension testing.

Results

Pin-On-Disc Testing. The influence of the particles on friction was estimated through pin-on-disc testing. The average friction coefficient is determined from the resulting friction profile, and all data-points from each test load are plotted together for comparison and shown in Fig. 5.

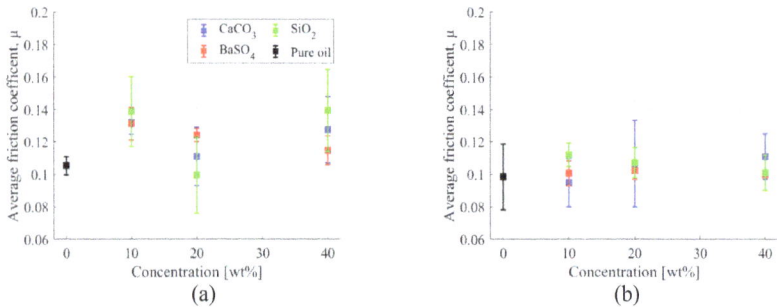

(a)   (b)

Fig. 5. Average friction coefficient found in pin-on-disc tests for (a) 1 N load and (b) 10 N load. Error bars denote standard deviation of data-set.

Material Forming - ESAFORM 2023   Materials Research Forum LLC
Materials Research Proceedings 28 (2023) 879-890   https://doi.org/10.21741/9781644902479-96

No influence by the particle type on friction could be found in the range of loads tested here. Further, compared to the pure oil, friction was not changed much by adding particles to the system. The wear scars that formed on the surface of the discs were, however, very different. Typical examples of the wear scars are shown in Table 2, where wear scar images were taken so that the centre of the disc was to the right of the wear scar. The wear scars from SiO_2 particles were clearly the most pronounced, the size of the wear scar increasing with the concentration of the particles. The $BaSO_4$ particles also showed severe wear scars compared to the pure oil, at least for higher concentrations, which showed very little wear. Applying $CaCO_3$ particles of a small concentration led to a reduction in the size of the wear scar. This effect was emphasized for increasing concentration of $CaCO_3$ to the point that no wear scar could be detected for 20 wt% and 40 wt% concentrations of $CaCO_3$. The applied load did not have much effect on the development of wear when particles were included, although it did increase wear in all cases.

Four-Ball Testing. Two sets of tests were performed using the four-ball configuration, varying either the concentration of particles in the mixture for a constant load, or varying the load for a constant concentration. The results of the former are shown in Fig. 6, whereas the latter is shown in Fig. 7. The wear scar diameter found in the four-ball tests clearly changed as function of particle concentration. For the SiO_2 particles, the wear scar diameter increased compared to the pure paraffin oil and then continued to increase for an increasing concentration of particles. The wear scar for the $BaSO_4$ particles of 10 wt% concentration was smaller than for the pure oil, but then increased with increasing concentration to a similar level as the pure oil. Wear scars found on balls used with $CaCO_3$ particles decreased in size compared to pure oil and were then stable for an increasing concentration of particles.

For changing loads, the behaviour was somewhat different. For the pure oil, as might be expected, the wear scar diameter increased with increasing load as there are no boundary additives included. Including particles prevented this increase, at least somewhat, reducing how much the wear scar diameter increased with increasing load. For a small load, the $CaCO_3$ and $BaSO_4$ particles showed a similar wear scar diameter as the pure paraffin oil. The wear scar diameter increased less with increasing load when these particles were included compared to the pure oil. This is similar to what Ji et al. [15] found for nano particles of $CaCO_3$ under four-ball test conditions in a range of loads using lithium grease as the base lubricant. They explained that for lower loads, the particles behaved based on the ball-bearing mechanism, but that as the load is increased and shearing of the surface becomes more likely, the surface energy of sheared surfaces causes a tribochemical reaction and leads to a boundary layer forming on the worn surface. The SiO_2 particles caused wear for small loads, but then prevented it from growing as the applied load is increased. This is likely due to the particles causing extreme wear in the beginning until the contact pressure is decreased enough that the particles could no longer penetrate the surface of the balls, and instead start to only act in a way that separates the surfaces.

Table 2. Typical wear scars found in pin-on-disc testing using the different particles.

Particle	None		CaCO$_3$		BaSO$_4$		SiO$_2$	
Load	1 N	10N	1 N	10 N	1 N	10 N	1 N	10 N
0 wt%			-	-	-	-	-	-
10 wt%	-	-						
20 wt%	-	-						
40 wt%	-	-						

0.5 mm

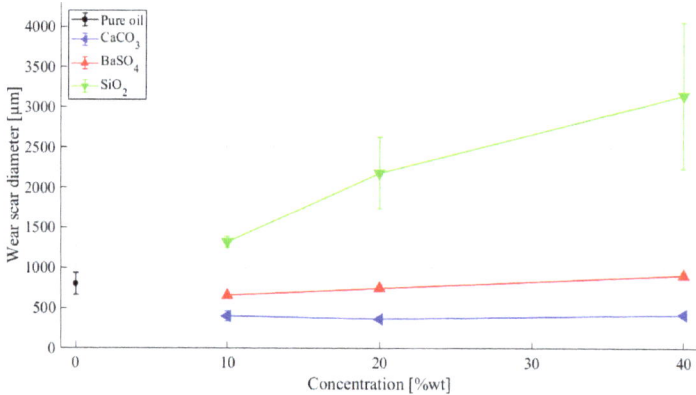

Fig. 6. Wear scar diameters from four-ball testing for different concentrations of particles for load of 300 N. Error bars are plus-minues one standard deviation of measured wear scar diameters.

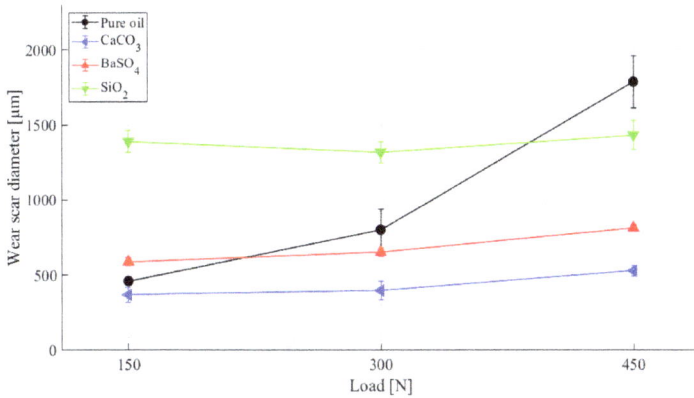

Fig. 7. Wear scar diameter from four-ball testing as a function of applied load for a constant 10 wt% particle concentration. Error bars are plus-minus one standard deviation of measured wear scar diameters.

Bending-Under-Tension Testing. Drawing force, F_d, profiles from BUT testing are shown in Fig. 8. The drawing force was influenced by the presence of particles and the particle type. For the pure oil (PO), pick-up quickly started to occur, leading to a steady increase in the drawing force until the end of the test. Including $BaSO_4$ particles showed a similar development of the force profile, although the force increased less. Including SiO_2 particles led to a quicker increase in force, indicating more severe conditions. The $CaCO_3$ particles showed only a small increase in force across the 1500 mm sliding length, indicating that they are beneficial to the wear resistance properties of the lubricant.

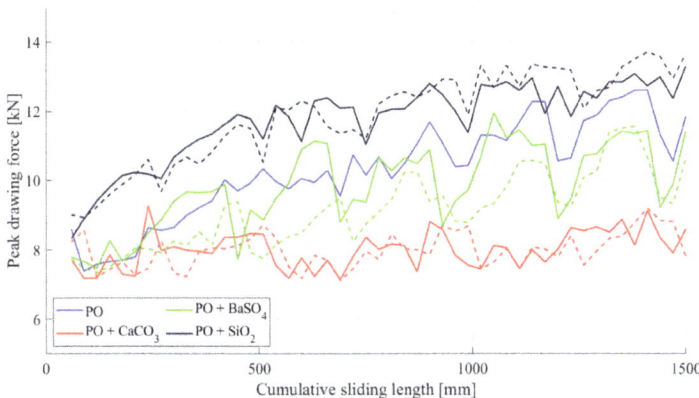

Fig. 8. Drawing force measured in BUT testing. Solid lines and dashed lines of the same colour are repetitions of the same test conditions. The concentration of particles was 20 wt%.

Images of the surfaces of selected tool-pins were captured and are shown in Table 3. The pure oil exhibited a high tendency for pick-up as can be seen on the surface of the tool-pin, which is consistent with the observations of the drawing force. The $CaCO_3$ particles prevented this and led to less wear on the tool-pin surface. The $BaSO_4$ particles helped in reducing wear, but the surface of the tool-pin ended up looking comparable to that of the pure oil. The SiO_2 particles clearly led to much more pick-up, potentially due to localised wear scars causing more favourable conditions for pick-up, or even heating up of the tool-pin near the surface.

Table 3. Surfaces of tool-pins used with different particles of 20 wt% concentration. The drawing direction is from bottom to top of figures. Solid profiles shown in Fig. 8 belong to the surfaces shown here.

Particles	Surface of tool-pins
None	
$CaCO_3$	
$BaSO_4$	
SiO_2	2 mm

Summary

Including particles in a tribo-system, either accidentally or intentionally, has a large effect on the wear development of the tribo-system. The hardness of the particles is a clear factor in how the particles affect the tribo-system, with harder particles increasing wear and softer ones inhibiting it through surface separation without surface penetration. Based on the results of this work, the following conclusions are drawn:
- Inclusion of particles of any hardness, in the force and concentration range tested here, had little to no effect on the coefficient of friction measured in pin-on-disc testing.
- The hardness of particles included in a tribo-system has a large effect on the development of wear in the tribo-system. A higher particle hardness leads to more severe wear.
- Including any particles in a tribo-system makes the system more robust in the face of varying loads. For a pure paraffin oil, wear readily increases with increasing load, but including particles reduced this effect, in the range tested here.
- The particle hardness was important for tribo-systems that include sheet metal forming conditions. High hardness, such as for the SiO_2 particles, lead to increased abrasion and more wear. The $BaSO_4$ did not change conditions much compared to the pure oil, at least not in terms of wear. The $CaCO_3$ particles promoted rolling of the particles as surface penetration was not possible and reduced wear compared to the pure oil.

Acknowledgments

The authors would like to thank A.G. Garcia from the Department of Chemical and Biochemical Engineering at the Technical University of Denmark for help with formulating the lubricant mixtures. Furthermore, U. Arinbjarnar and C.V. Nielsen would like to thank the Danish Council for Independent Research, grant number DFF – 0136-00159A, for the funding of this investigation.

Material Forming - ESAFORM 2023
Materials Research Proceedings 28 (2023) 879-890

Materials Research Forum LLC
https://doi.org/10.21741/9781644902479-96

References

[1] B. Seidel, D. Meyer, Influence of artificial aging on the lubricating ability of water miscible metalworking fluids, Prod. Eng. 13 (2019) 425–435. https://doi.org/10.1007/s11740-019-00891-6

[2] M.R. Sari, A. Haiahem, L. Flamand, Effect of lubricant contamination on gear wear, Tribol. Lett. 27 (2007) 119-126. https://doi.org/10.1007/s11249-007-9215-z.

[3] T. Kjer, Wear rate and concentration of wear particles in lubricating oil, Wear 67 (1981) 217. https://doi.org/10.1016/0043-1648(81)90105-8

[4] Q.Y. Jiang, S.N. Wang, Abrasive wear of locomotive diesel engines and contaminant control, Tribol. Trans. 41 (1998) 605-609.

[5] H.H. Abou El Naga, A.E.M. Salem, Effect of worn metals on the oxidation of lubricating oils, Wear 96 (1984) 267-283. https://doi.org/10.1016/0043-1648(84)90041-3

[6] U. Olofsson, Y. Zhu, S. Abbasi, R. Lewis, S. Lewis, Tribology of the wheel-rail contact-aspects of wear, particle emission and adhesion, Vehicle System Dynamics 51 (2013) 1091-1120. https://doi.org/10.1080/00423114.2013.800215

[7] T. Luo, X. Wei, X. Huang, L. Huang, F. Yang, Tribological properties of Al2O3 nanoparticles as lubricating oil additives, Ceram. Int. 40 (2014) 7143-7149. https://doi.org/10.1016/j.ceramint.2013.12.050

[8] J. Padgurskas, R. Rukuiza, I. Prosyčevas, R. Kreivaitis, Tribological properties of lubricant additives of Fe, Cu and Co nanoparticles, Tribol. Int. 60 (2013) 224-232. https://doi.org/10.1016/j.triboint.2012.10.024

[9] Y. Li, Y. Chen, P. Gao, B. Zhong, F. Ai, L. Li, Y. Xiao, Research on influence of abrasive particles on lubricating performance of the emulsion for cold rolling strip, in: Proceedings of Chinese Materials Conference 2017, 2017, pp. 135-143. https://doi.org/10.1007/978-981-13-0107-0_13

[10] Ú. Arinbjarnar, M. Moghadam, C.V. Nielsen, Application of Calcium Carbonate as Green Lubricant Additive in Sheet Metal Forming, Key Eng. Mater. 926 (2022) 1133-1142. https://doi.org/10.4028/p-x87o62

[11] D.X. Peng, C.H. Chen, Y. Kang, Y.P. Chang, S.Y. Chang, Size effects of SiO2 nanoparticles as oil additives on tribology of lubricant, Ind. Lubric. Tribol. 62 (2010) 111-120. https://doi.org/10.1108/00368791011025656

[12] Calcit, PolyPlex 2, Calcit d.o.o., Stahovica, Slovenia, Jan. 2014. Available online: https://www.calcit.si/assets/PDF/ANG_SPLOSNI_KATALOG_2020_PolyPlex.pdf

[13] Sibelco, Portaryte® B. Sibelco, Maastricht, Netherlands, Mar. 2013. Available online: www.sibelco-specialty-minerals.eu

[14] Sibelco, Silverbond M500. Sibelco, Dessel, Belgium, Oct. 01, 2016. Available online: www.sibelco.eu

[15] X. Ji, Y. Chen, G. Zhao, X. Wang, W. Liu, Tribological properties of CaCO3 nanoparticles as an additive in lithium grease, Tribol. Lett. 41 (2011) 113-119. https://doi.org/10.1007/s11249-010-9688-z

Material Forming - ESAFORM 2023
Materials Research Proceedings 28 (2023) 891-898

Materials Research Forum LLC
https://doi.org/10.21741/9781644902479-97

Surface topography effects on galling of hot dip galvanized sheet metal

VENEMA Jenny[1,a] *, CHEZAN Toni[1,b] and KORVER Frank[1,c]

[1]Tata Steel, Research & Development, PO BOX 10000, 1970 CA IJmuiden, The Netherlands

[a]jenny.venema@tatasteeleurope.com, [b]toni.chezan@tatasteeleurope.com,
[c]frank.korver@tatasteeleurope.com

Keywords: Tribology, Friction, Wear, Zinc Coatings, Galling, Automotive

Abstract. During manufacturing of automotive parts from hot dip galvanized sheet metal, surface asperities of the forming tools can cause breakage of small coating particles. Long and narrow scratches appear on the surface of the form part, a phenomenon known as part surface galling. Laboratory testing using a slider on sheet tests (SOST) are performed in order to investigate surface topography effects on galling. These experiments reveal the dominant effect of tool surface roughness on galling. The tool surface roughness has a large effect on the size of the detached coating particles and the distance before scratch occurrence. If the tool roughness is low enough, no surface scratch formation is observed for the investigated range of sheet surface roughness. At a high tool surface roughness scratches are observed at a very short sliding distance for all tested materials. At an intermediate tool surface roughness, the materials selected for this investigation show measurable differences but no clear trend could be identified.

Introduction

Galling, abrasive surface wear of hot dip galvanized steel (GI) sheet metal can occur during deep drawing, the process commonly used for automotive part manufacturing. Observations of industrial forming processes in the areas with contact with the forming tools revealed two types of contact conditions – normal contact and galling. In normal contact conditions the sheet metal topography is flattened by the tool contact. Such contact areas on the deformed sheet surface can be considerably large, tens of millimeters in the sliding length direction, as the sheet metal slides against the forming tools. In some contact areas unexpected galling behavior in the form of long narrow scratches, in the sheet metal sliding direction [1] was observed. Such scratches are created by tool surface local asperities. The consequence of surface scratch formation might be coating detachment in the form of thin flakes. The flakes might attach to the tool surface resulting in increased tool asperity height and in formation of new tool surface asperities and consequently in acceleration of galling of the following formed part.

In the industry several solution directions exist to prevent galling, namely polishing, hardening and/or coating the forming tool surface and/or the application of extra lubrication. These solution directions indicate already some of the parameters which influence galling, such as tool surface parameters such as roughness and lubrication. Other parameters which influence galling are sheet surface, design (deformation) and process parameters such as pressure, velocity and temperature. In the literature investigations on the mechanisms of galling and the influence of several parameters can be found [2-5]. For example, Van de Heide and Dane [6-7] investigated the influence of lubricant on galling.

In a previous investigation [1] the influence of pressure and tool roughness on galling was investigated with the Slider On Sheet Tester (SOST) on one material with representative roughness. In the current investigations the influence of the tool roughness is investigated for three sheet material roughness levels within the typical intervals found in automotive applications. Both the scratch depth and particle sizes are investigated in great detail.

Material Forming - ESAFORM 2023   Materials Research Forum LLC
Materials Research Proceedings 28 (2023) 891-898   https://doi.org/10.21741/9781644902479-97

Material and Methods

An experimental program was started to evaluate the geometrical surface parameter on galling. The experiments were performed using a SOST capable of providing controlled well-defined contact conditions between the tool surface and metal sheet. Tests were performed on three different GI coated steel sheets and three different tool roughness levels (Table 1). The tribological interactions were studied by Scanning Electron Microscopy (SEM) analysis of particles collected from the tool. The dimensional topography measurements of the tool and sheet surfaces were performed by a confocal microscope.

Table 1. Surface parameters of the sheet and tool material according to ISO 25178.

	Sa [µm]	Sq [µm]	Sz [µm]	Ssk [-]	Sku [-]	Sp [µm]	Sv [µm]
Tool ring 1	0.33	0.43	5.22	-0.43	4.25	2.66	2.55
Tool ring 2	0.74	0.99	11.96	-1.27	8.45	3.39	8.55
Tool ring 3	1.19	1.57	15.30	-0.97	4.89	6.89	8.42
Sheet 1	1.00	1.14	6.64	0.45	2.50	3.85	2.79
Sheet 2	1.17	1.36	7.73	-0.08	2.06	3.92	3.81
Sheet 3	1.59	1.89	12.30	-0.42	2.30	6.62	5.70

A cylindrical tool (Fig.1a) with contact width of 6 mm is pressed with 1.5 bar (resulting in a normal load of 205 N) against the sheet material. The contact situation is therefore a line contact, with a nominal pressure of 190 MPa. The cylinder moves in length direction for 250 mm at a velocity of 50 mm/s, lifted up and brought into a new position to make a new track next to the previous one (Fig.1b) within a few seconds. In this way, every time virgin sheet material is in contact with the tool. The amount of tracks is depended on the tool roughness and availability of material (sheet size). For the smoothest tool (Sa 0.3 µm), between 70 and 75 tracks are tested, resulting in 17.5 to 18.8 m sliding length for the three different materials. This sliding length can represent drawing of approximately close to 200 parts. The tool is rotated (approximately 5°) after each unique parameter setting. In this way a fresh tool material is in contact for each test. A pressure sensitive foil is used to align the tool properly, thereby ensuring homogeneous contact at the start of the test.

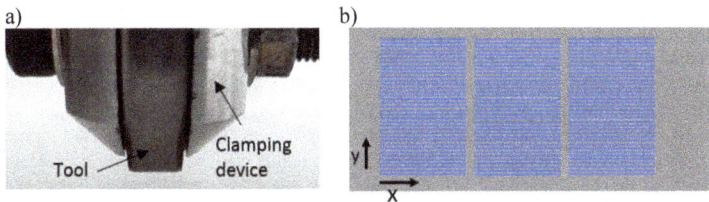

Fig. 1. a) Photograph of cylindrical tool (half) in SOST b) Schematic view of sheet with tracks.

Three different GI coated steel sheets (thickness 0.6 mm) with different surface roughness are tested (Fig. 2). The sheets are cleaned with isopropanol to remove mill applied lubricant on the surface.

Material Forming - ESAFORM 2023
Materials Research Proceedings 28 (2023) 891-898

Materials Research Forum LLC
https://doi.org/10.21741/9781644902479-97

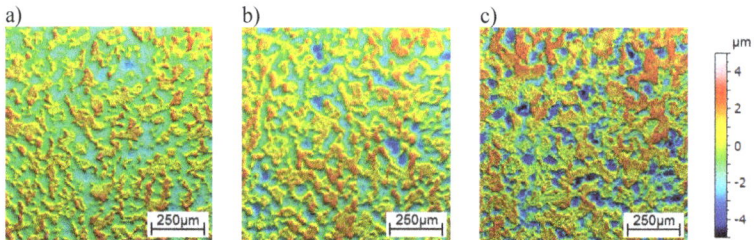

Fig. 2. Confocal measurement sheet surface (size 1x1 mm) a) Sa 1.0 μm b) Sa 1.2 μm
c) Sa 1.6 μm.

The slider (diameter of 43 mm) of tool material DIN 1.2379 has a hardness of 60 ± 2 HRC. Three different tool textures with different roughness (Fig. 3) are tested. The tool surface topography has a defined lay with the grinding direction perpendicular to sliding direction.

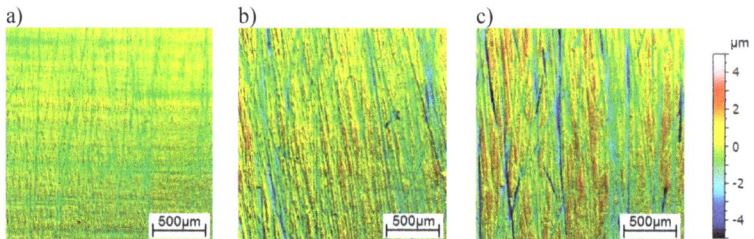

Fig. 3. Confocal measurement tool surface (2x2 mm) a) Sa 0.3 μm b) Sa 0.7 μm c) Sa 1.2 μm.
Sliding direction is horizontal.

Results and Discussion

The appearance of the scratches, ploughing tracks and particles are similar to observations on automotive pressed parts and particles collected from stamping tools [1]. The tool wear mechanism is dynamic. Particles of the GI coating break out during sliding and get trapped in different types of tool surface defects (such as grinding scratches). They stick to the tool surface (Fig.4a & 4b). During repeated sliding the particle build up accumulates. Particle accumulation can cause scratches to sheet material, which can become more severe with sliding distance. However, it is also observed that particles break off resulting in a disappearing scratch or a less severe scratch. The process will continue and the particle accumulation will start again at the same or at another position. Loose particles can also cause 3rd body abrasive wear. These scratches do not necessarily follow the same direction as the sliding direction, but could bend off. This wear mechanism was also observed in a few cases in earlier investigations.

In general, the thickness of individual flakes is lower than the thickness of the GI coating and the GI coating is not spalling off. In case of highest tool roughness the scratch depth becomes close to the thickness of the zinc coating. Cross sectional microscopy is planned to further investigate the fracture of the zinc coating.

After the test the particles were collected with adhesive tape for further investigation in the SEM (Fig. 4c). Mapping of the elements of the particles reveals that it mainly consists of Zn (Fig. 4d) and some coating elements (Al). In general, no tool material elements such as Cr are measured. This indicates that no abrasive tool wear occurs. Also, in most cases almost no Fe is measured, suggesting that the scratches are superficial.

Material Forming - ESAFORM 2023
Materials Research Proceedings 28 (2023) 891-898

Materials Research Forum LLC
https://doi.org/10.21741/9781644902479-97

The tools are ultrasonic cleaned after collection of the particles. After ultrasonic cleaning of the tools, some Zn particles are still attached to the tool surface (Fig. 4b) in which the sliding direction is visible. Remaining particles on the slider after ultrasonic cleaning represents well the industrial case in which grinding is necessary to remove all particles.

Fig. 4. a) Photo of contact surface of tool after test (tool width 6 mm) b) particle remained after ultrasonic cleaning of tool c) SEM BSE collected particles from tool (size 8.5 x 8.5 mm) Tool Sa 0.7 μm Sheet Sa 1.2 μm. SEM /EDX element mapping of Zn (L) (size 8.5 x 8.5 mm) Tool Sa 0.7 μm Sheet Sa 1.2 μm.

Particles collected from tool.

The collected particles are in the majority of the cases flakes, for one material also some other particle morphology is observed. These flakes are very comparable to the flakes collected from industrial tools at OEMs (Fig. 5). The flakes are only a few microns thick, while the width and length are more than 10 times the thickness.

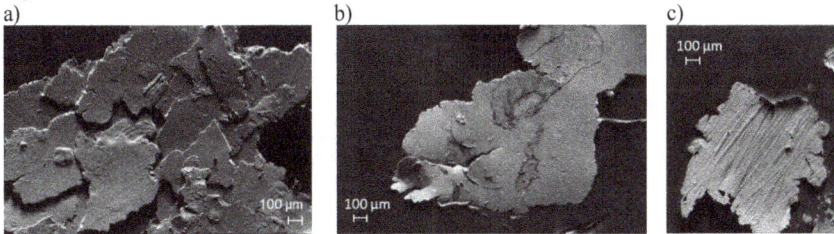

Fig. 5. SEM images a) particles collected from industrial tool b) one particle collected from industrial tool c) one particle collected from SOST sheet Sa 1.2 μm tool Sa 1.2 μm.

The size of the flakes is dependent on the surface roughness, and the average size increases in the majority of cases with increasing tool roughness (Fig. 6).

Fig. 6. SEM analysis of particles after test sheet Sa 1.6 μm with tool roughness a) Sa 0.3 μm b) Sa 0.7 μm c) Sa 1.2 μm.

Material Forming - ESAFORM 2023   Materials Research Forum LLC
Materials Research Proceedings 28 (2023) 891-898   https://doi.org/10.21741/9781644902479-97

Scratches on the sheet surface.

Three dimensional topography measurements are made at specific track numbers (1,10,20,40,60 and last track). A 2D profile cross section is created and the maximum profile depth determined (Fig. 7a). Criterion for a scratch is a larger than 1 μm increase in max profile depth compared to original material. Fig. 7b shows the maximum profile depth versus the sliding distance for sheet material 1 (Sa=1.0 μm) for several tool roughness. In case of low tool roughness (0.3 μm) no scratch occurs. At medium tool roughness (0.7 μm) scratch occurs at 15 m of sliding distance. And for a high tool roughness (1.2 μm) the scratch occurs immediately after start of test. Thus, the tool roughness plays a major effect on the distance to scratch formation.

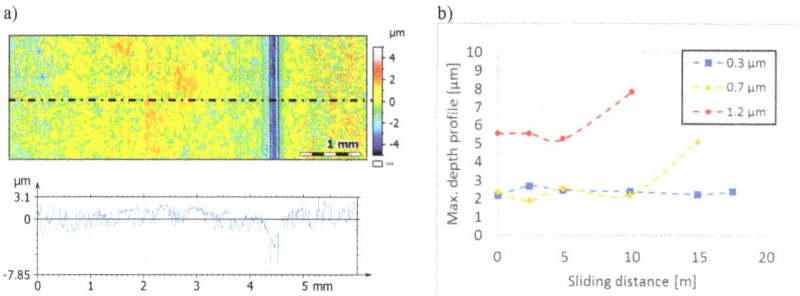

Fig. 7. A) 3D topography measurement sheet Sa 1.0 μm tool Sa 1.2 μm after 10 m sliding. b) max. profile depth versus sliding distance sheet Sa 1.0 μm for the tested tool textures.

For the entire collection of sheet materials no scratches are observed for a low tool roughness of 0.3 μm (Table 2). This shows that the sheet coating can resist relative high contact conditions in case of smooth surface.

In case of intermediate tool roughness the distance before a scratch occurs was different for the three materials considered in this study. A more extended experimental research is necessary to further quantify the sheet coating effect. Also, the number of tests is very limited and further investigations are necessary to determine the range in distance before scratch occurrence for different sheet material and tool roughness combinations. However, trends regarding the increase in tool roughness are observed for each material and each test series (Table 2). An increase in tool roughness results in a decrease in distance to scratch (Table 2) which is probably related to the larger particle size and the presence of more tool surface defects to get stuck. Larger particles can probably be less easily removed out of the contact.

Table 2. Distance [m] to first measured scratch.

Tool roughness Sa	Sheet roughness Sa		
	1.0 [μm]	1.2 [μm]	1.6 [μm]
0.3 [μm]	>17.5	>18.8	>18.5
0.7 [μm]	14.9	4.9	9.9
1.2 [μm]	0.0	2.4	2.4

Material Forming - ESAFORM 2023   Materials Research Forum LLC
Materials Research Proceedings 28 (2023) 891-898   https://doi.org/10.21741/9781644902479-97

Fig. 8 reveals the dynamic nature of galling. For example, at a tool roughness of 0.7 µm a large scratch occurs at track 20, which becomes less severe at track 60. Indicating that a part of the adhered particles broke off from the tool. In case of a roughness of 1.2 µm the position of the scratches also changes, indicating that the buildup and breaking of particles on the tool can occur at several position on the tool.

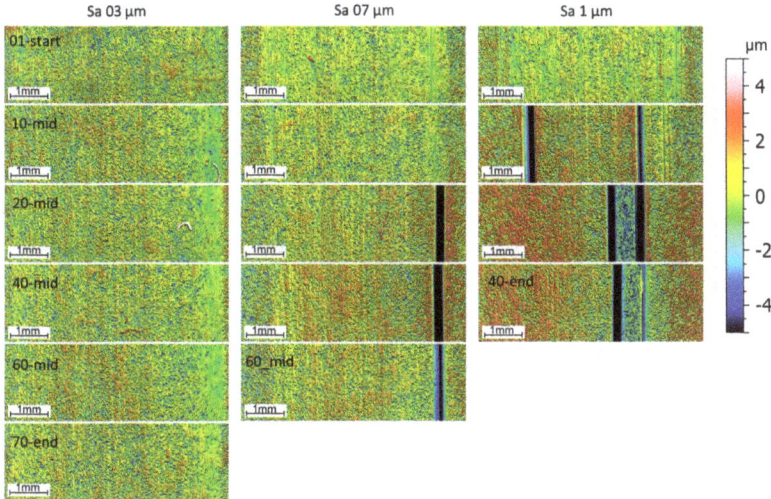

Fig. 8. Confocal measurements for sheet material 2 (Sa 1.2 µm) at positions (# tracks) and for three tool roughness . Sliding in vertical direction.

Coefficient of Friction (COF).

The effect of tool roughness on the COF is also high (Fig.9). With a low tool roughness the friction coefficient is stable and relatively low. Sometimes a certain increase and decrease in COF is related to a scratch, however this certainly is not always the case. The friction coefficient is higher or lower with increasing tool roughness. A different wear regime (cut, wear or ploughing) could cause a different trend in friction coefficient. Recommended is to investigate this effect further with a physical based friction model.

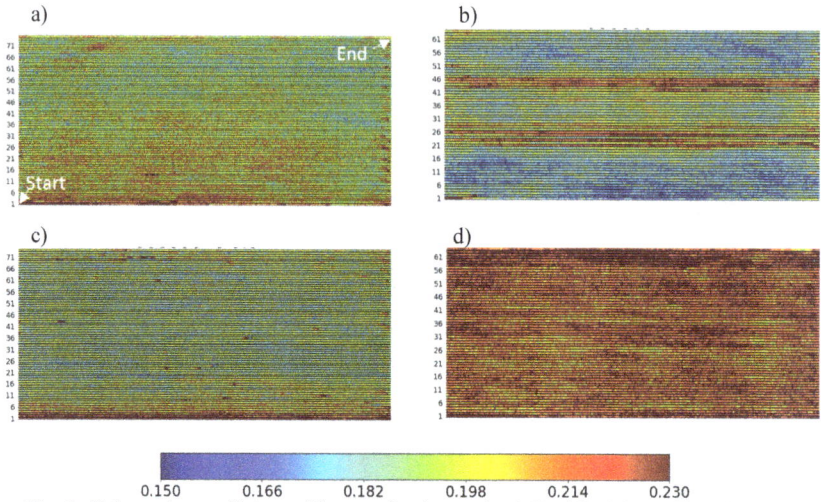

Fig. 9. Colour plot coefficient of friction for the tracks a) Tool Sa 0.3 μm Sheet Sa 1.2 μm
b) Tool Sa 0.7 μm Sheet Sa 1.2 μm c) Tool Sa 0.3 μm Sheet Sa 1.6 μm
d) Tool Sa 0.7 μm Sheet Sa 1.6 μm.

Summary

The experiments reveal the dominant effect of tool surface roughness on galling of the sheet metal coating. It was found that the detached coating flake size increases with increasing tool surface roughness. A too low or too high tool surface roughness resulted in either no surface scratch formation or deep surface scratches after very short sliding distance for all tested sheet materials. The roughness magnitude of the sheet metal surface appeared to influence the galling resistance performance as well for the tests with intermediate tool surface roughness. The material roughness magnitude differences were less clear while compared to the tool surface roughness effect. Further investigations are planned focused on material coating parameters effect on galling performance.

Acknowledgements

The authors would like to thank Ronald van Goethem, Arne Neelen and Marco Appelman for surface measurement and Jan Wörmann for the SEM analysis.

References

[1] J. Venema, F. Korver, T. Chezan, Slider on Sheet Tester Development for Characterizing Galling, ESAFORM 2022, Key Eng. Mater. 926 (2022) 1204–1210. https://doi.org/10.4028/p-uii69m

[2] E. Schedin, Galling mechanisms in sheet forming operations, Wear 179 (1994) 123-128. https://doi.org/10.1016/0043-1648(94)90229-1

[3] H. Kim, J. Sung, F.E. Gondwin, T. Altan, Investigation of galling in forming galvanized advanced high strength steels (AHSSs) using the twist compression test (TCT), J. Mater. Process. Technol. 205 (2008) 459-468. https://doi.org/10.1016/j.jmatprotec.2007.11.281

[4] W. Wang, K. Wang, Y. Zhao, M. Hua, X. Wei, A study on galling initiation in friction coupling stretch bending with advanced high strength hot-dip galvanized sheet, Wear 328-329 (2015) 286–294. http://doi.org/10.1016/j.wear.2015.02.058

Material Forming - ESAFORM 2023   Materials Research Forum LLC
Materials Research Proceedings 28 (2023) 891-898   https://doi.org/10.21741/9781644902479-97

[5] K.G. Budinski, S.T. Budinski, Interpretation of galling tests, Wear 332-333 (2015) 1185-1192. https://doi.org/10.1016/j.wear.2015.01.022

[6] E. van der Heide, A.J. Veld, D.J. Schipper, The effect of lubricant selection on galling in a model wear test, Wear 251 (2001) 973-979.

[7] C.M. Dane, E. van der Heide, Proceedings of the IDDRG '04 Conference, Sindelfingen, Germany, 2004, pp. 182-191.

Material Forming - ESAFORM 2023
Materials Research Proceedings 28 (2023) 899-908

Materials Research Forum LLC
https://doi.org/10.21741/9781644902479-98

Influence of increased die surface roughness on the product quality in rotary swaging

FENERCIOĞLU Tevfik Ozan[1,a] *, SARIYARLIOĞLU Eren Can[1,b],
ADIGÜZEL Erdem[1,c], ŞIMŞEK Ahmet Kürşat[1,d], KOLTAN Umut Kağan [1,e] and
YALÇINKAYA Tuncay [2,f]

[1]Repkon Machine and Tool Industry and Trade Inc. Şile/Istanbul/Turkey

[2]Department of Aerospace Engineering, Middle East Technical University, 06800 Ankara/Turkey

[a]ozan.fenercioglu@repkon.com.tr, [b]eren.sariyarlioglu@repkon.com.tr,
[c]erdem.adiguzel@repkon.com.tr, [d]kursat.simsek@repkon.com.tr, [e]umut.koltan@repkon.com.tr,
[f]yalcinka@metu.edu.tr

Keywords: Metal Forming, Rotary Swaging, Surface Roughness, Finite Element Analysis

Abstract. Rotary swaging is an ascendant forming method for manufacturing axisymmetric parts. High production rate with excellent net shape forming is achieved in recent automation developments. However, precise machine design and tailored process developments are necessary to transfer the high impact type forming loads to workpiece efficiently. The failure of this transfer results in high vibrations of the machine structure and poor product quality, due to the impact loads with high frequencies. The centerpiece of the process development to prevent these disruptive effects is to resolve die specifications such as shape and surface properties. In general forming applications, surface roughness of the dies is perceived as a disruptive element for the product quality and only a small amount is provided to settle lubricants. However, for rotary swaging applications, an optimized surface roughness to increase the load transfer between the die and the workpiece without disrupting the final product surface quality is essential. In this study, for a fixed die shape, the relation between the die surface roughness and the product quality is investigated for macro rotary swaging applications. In particular, the effective transfer of the forming forces to the workpiece is analyzed by using finite element analysis within the scope of surface friction. Consequently, a die set with roughened surface conditions is manufactured by using a novel technique. Real process trials are conducted to validate the results of the analysis.

Introduction

Rotary swaging is an incremental forging operation for tubes and bars. Operation principal of a rotary swaging machine is shown on Figure 1(a). An internal ring with translational sliders is rotated inside a roller cage. Translational sliders hold die assemblies. During the rotation of the internal ring, dies are retracted to an open position due to centrifugal effects and other peripheral supports (support springs). When the die assembly reaches a roller, dies are translated to closed position. Continuous movement of the die assembly results in consecutive blows of dies and cold forging of the workpiece. Hence, the outer diameter of the part decreases with a possible reduction of the cross-section area and with the elongation of the total length. Rod, pipe and tube parts can be shaped by this method and axisymmetric parts can produced. By the addition of a pre-shaped mandrel inside tubular workpiece, it is also possible to form complex shapes inside the workpiece. The mandrel steers the material such that the internal shape of the workpiece reflects the outer shape of the mandrel.

Performance and final mechanical properties of the workpiece are dictated by process parameters and parts (die and mandrel). In particular, dies play the most important role. During

Material Forming - ESAFORM 2023   Materials Research Forum LLC
Materials Research Proceedings 28 (2023) 899-908   https://doi.org/10.21741/9781644902479-98

the process, the workpiece is forced to deform with large loads transferred by dies to reduce the outer diameter. To moderate the disruptive effects of the die impact, dies are designed with a tapered entrance. In this way, shearing of the workpiece is prevented. However, this tapered entrance geometrically results in unwanted axial reaction forces. Each blow of the die pushes back the workpiece in the opposite direction of the elongation. These reaction forces show disruptive effects on overall swaging performance. In the open literature, these disruptive effects are studied in detail under the scope of micro metal forming. Small tubes for needles and concentric bimetallic tubes are manufactured using rotary swaging. Vollertsen called these forces, rejection forces and explained that increased rejection forces result in workpiece buckling [1]. To solve this problem, Böhmermann et al. suggested to increase the effective friction on the forging section between die and workpiece to decrease the reaction forces (Fig. 1 (b)) [2]. By increasing the friction, as the die descends, the end of the tapered section pulls the workpiece and the reaction (rejection) force of the sinking section is supported. Herrmann numerically characterized effects of die friction for macro rotary swaging applications using finite element analysis [3]. 50% reduction is achieved in axial reaction forces by increasing friction coefficient from 0.1 to 0.5 in reduction zone.

Similar investigations were also made for radial forging. Rotary swaging and radial forging are similar processes with different machine configurations. Hence, research results are valid for both applications. Apart from the differences in die motions, another difference is observed in workpiece feeding between both applications. Workpiece feeding is usually force controlled and applied by hydraulic cylinders in radial forging which is replaced by position control systems in modern rotary swaging machines. Hence, when reaction forces reach the limit force of the hydraulic cylinders, feeding fails in radial forging. Semiatin suggested not to use any lubricant between die and workpiece when entrance taper angle is more than 6° [4]. In this way, the effective friction is increased between the die and workpiece. In some applications, forging zone of the die is roughened by hardfacing to increase effective friction and to reduce axial reactions. Recent patents are published where hardfacing (conventional method for surface roughening) is replaced by surface structuring (see e.g. [5]). A wave form is machined on the forging zone to create holding forces to balance reaction forces especially for micro rotary swaging applications.

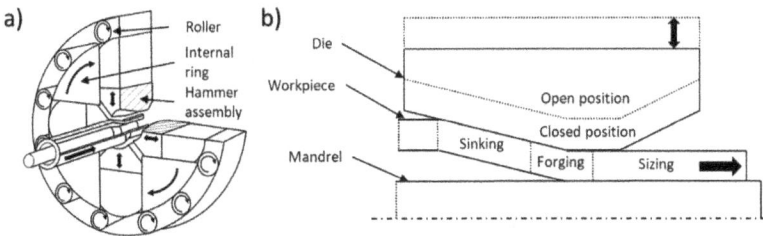

Fig. 1. Schematics of rotary swaging a) Clock-wise rotation of the internal ring rotates die assembly and leads to translational blows when upper surface reaches a roller b) Cross-section of swaging process.

Formerly, effects of the process parameters and parts (die and mandrel) are characterized by process trials. Nevertheless, by the development of reliable metal forming simulations, trial and error methods are evolved to be a final validation tool. Different approaches are found in the literature for the evaluation of rotary swaging and radial swaging. It should be noted that the die-workpiece interactions are similar in rotary swaging and radial forging. Initial studies to foresee the material flow and final mechanical properties are conducted by Grosman and Piela [6]. An

Material Forming - ESAFORM 2023   Materials Research Forum LLC
Materials Research Proceedings 28 (2023) 899-908   https://doi.org/10.21741/9781644902479-98

axisymmetric model is developed where the volumetric change of the cross section during forging sequence is imposed by a radial rigid translation in the simulation. Lahoti and Altan derived an analytical expression to determine process forces and resulting stresses using the upper slab method [7]. This analytical study is further pursued by finite element analysis and experimental validations to develop a die design methodology [8]. As a result of these analysis, the relation between effective friction and forging forces are numerically shown. Also, Tseng et al. put a further step by including thermal effects in analytical expressions [9]. After initial studies of Domblesky et al. on the application of flow formulations for radial forging [10], Yamaguchi et al. developed a simulation and experiment method to predict the grain size of radial forged materials [11]. Recently, Poursina et al., Ameli and Movahhedy, Fan et al, Liu et al. applied displacement formulations for swaging operation using ABAQUS/Explicit and the results showed significant compatibility with the experimental results [12-15]. Load predictions converged to experimental results of Uhlig with an error <5% [16].

Validation of different simulations by experiments ensures the reliability of material process simulations. Displacement formulations with an explicit time scheme will be applied in this study to evaluate the process loads and its relation to surface properties. Although, a solution for high axial reactions by roughening die surface is widely suggested in the literature. A correlation between the surface roughness and mechanical vibrations and reactions is not experimentally shown for macro swaging applications. This study aims to clarify the mechanism behind the surface roughening and to evaluate the disruptive effects of reaction forces. A comparison for different friction conditions is numerically conducted through finite element analysis. Results are validated by product contours and also by position feedbacks of the workpiece pusher system. Moreover, a novel simple method is suggested for surface roughening. In this regard, the current study is unique for correlation of reaction forces on product and process quality and proposes an alternative method to hard facing for the swaging industry in die design.

The paper is organized as follows. Rotary swaging method and the effects of the reaction forces on product quality is briefly presented here, section 2 addresses the finite element analysis of the process cycle. Section 3 validates the numerical analysis through experiments. Then the study is concluded in Section 4.

Finite Element Analysis

ABAQUS/Explicit 2016 solver is used in this study after the quasi-static condition is validated using simulation energy outputs. Kinetic energy of the process is compared to internal energy as suggested in the ABAQUS documentation [17]. Less than 2% kinetic energy is calculated during the process. Hence, the method is found valid and doubts about the stimulation of artificial energy modes are removed. Also, this low amount of kinetic energy let the usage of a mass scaling factor (25). Geometrical simplifications are not applied to better converge to further experiments. Schematic shown on Fig. 1 (b), is represented in Fig. 2 (a) and (b) in association with rotary swaging machine elements and modelling aspects, respectively. Axial motion of the workpiece into dies is given by using workpiece pusher. Movement and forces of the workpiece pusher is recorded during the analysis to evaluate reaction forces. Fig. 2 (b) shows initial and final steps (total simulation duration: 4 seconds) of the simulation. Workpiece is modeled using reduced integration elements with combined stiffness hourglass control (C3D8R, size: 1.2 mm). Hourglass control method is determined by restricting artificial energy to internal energy ratio (<1% achieved). For boundary conditions, both translation and rotation are applied to the workpiece during feeding. Cylindrical kinematic element is used to control the feeding motion. A torsional spring is added to cylindrical connector to prevent the torsion on the workpiece with an extrapolated nonlinear elasticity definition (48kNmm/rad). Elastoplastic properties of the workpiece are constructed from the experiments of Ceschini et al. for 33CrMoV Steel material [17]. Process tools are modeled using rigid body surface elements (R3D4, size: 1.5 mm). Dies rotate around the

Material Forming - ESAFORM 2023   Materials Research Forum LLC
Materials Research Proceedings 28 (2023) 899-908   https://doi.org/10.21741/9781644902479-98

median axis and translates in the radial direction to perform die blows. This kinematic motion is applied to reference points of die surfaces using translator connectors. Penalty contact method with finite sliding is used to model contact interactions between deformable and rigid bodies. Die surface is divided into 3 subsections to separate forging and sizing regions as in Fig. 1 (b). Two models are constructed with forging region having 0.1 and 0.8 friction coefficient values. These two extremum points are selected by considering wall lubricated metal-metal friction (0.1) and non-lubricated (or roughened) metal-metal friction (0.8). Because of the design of the rotary swaging machine, unlike radial forging, a lubricant always operates between workpiece and dies. Hence, suggestions of Semiatin for radial forging in [4] are not possible for rotary swaging process and a roughened surface is the solution.

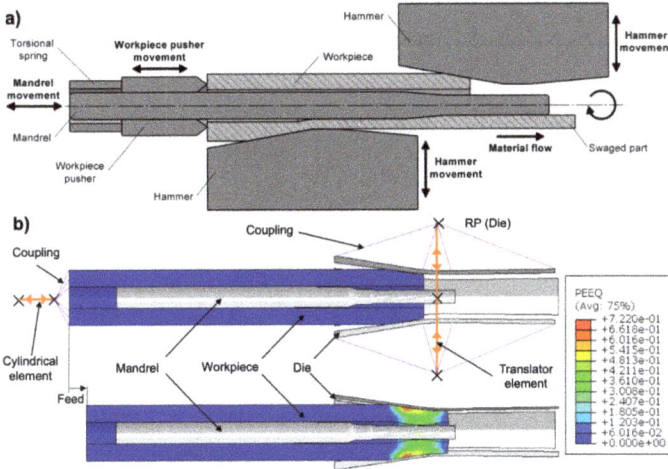

Fig. 2. Rotary swaging analysis model. a) Schematic view of forming section and machine elements, upper and lower die show initial and final position respectively. b) Modelling elements and equivalent plastic strains on deformable body (workpiece) for initial and final steps of the simulation.

Cylindrical element in Fig. 2 represents the workpiece pusher of the rotary swaging machine. Workpiece pusher controls the axial motion of workpiece. Hence, it is exposed to axial reactions. Axial reactions during process simulation are exported from cylindrical element. Fig. 3 shows the load history of the cylindrical connector. When friction coefficient is low (0.1), load profile starts with an increase in each blow. After approximately 2 seconds, a steady load is achieved. This profile is due to entrance tapers of the die and workpiece. Sliding on tapered sections occurs due to low friction. A significant difference is observed for high friction coefficient. During the process, the dies are retracted after the impact. This positive reaction forces prevent the workpiece from escaping back during this retraction. So, upper motion result in a reverse loading with increased friction. The most significant result is the reduction of reaction forces from 12 tons to 7.2 tons when an increased friction is applied between die and workpiece.

Moreover, die torsion force on the workpiece is evaluated for different frictions. Fig. 4 shows relative rotation between the front and rear ends of the cylindrical connector. As can been seen

Material Forming - ESAFORM 2023   Materials Research Forum LLC
Materials Research Proceedings 28 (2023) 899-908   https://doi.org/10.21741/9781644902479-98

from Fig. 2, during consecutive flows, the contact area between dies and workpiece increases in the tangential direction as workpiece translates into the sizing region. This is the reason behind the cumulative character of the Fig. 4. As the process continues, the contact area between the die and the workpiece increases and the workpiece also takes the shape of the die profile. The change of friction coefficient in forging region introduced only a small amount of increase (0.03). This increase is small as expected since the main contribution comes from sizing region where friction coefficient is the same for both models. Also, by using V-shaped dies the increase of disruptive effects due to friction are prevented. Contact lengths are controlled in axial and circumferential directions by the help of this shape. A final evaluation is made for the translation of dies to observe if increased friction creates any disruptive effect on die actuation (shown on Fig. 4). No significant difference is observed when reaction forces on translator connectors which actuate dies during analyses are compared. This is related to the angle of the entrance taper. Motion and forces mainly contribute in the horizontal direction. Hence, deviations in the vertical results are relatively small.

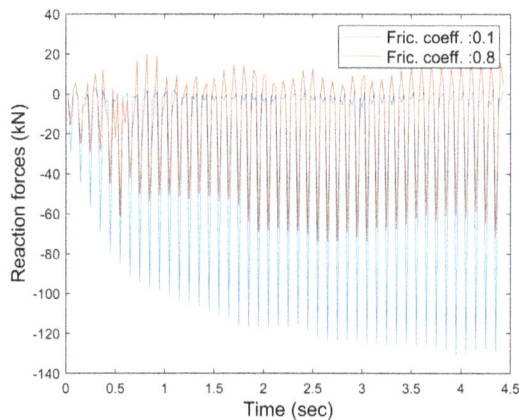

Fig. 3. Reaction forces on workpiece cylindrical connector for different friction coefficients.

The effect of friction coefficient is visualized on Fig. 5 using contact frictional shear pressure outputs. By increasing the surface friction, the contribution of the frictional shearing is increased. During the blow, die pulls the part in the feeding direction which supports the reaction forces created by the tapered section. However, since the die force in vertical direction and contact area are more dependent to workpiece material, contact normal pressures have no significant difference relative to different friction coefficients.

Material Forming - ESAFORM 2023
Materials Research Proceedings 28 (2023) 899-908

Materials Research Forum LLC
https://doi.org/10.21741/9781644902479-98

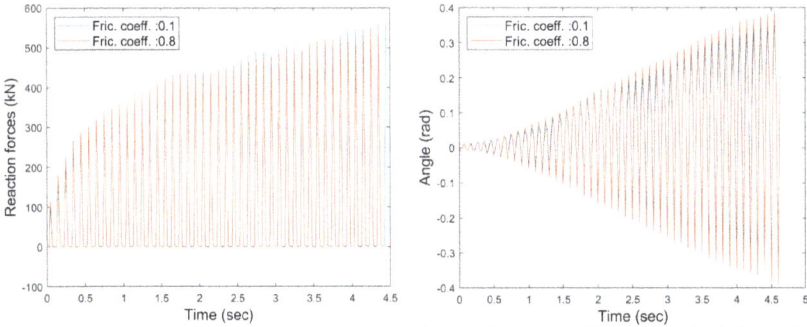

Fig. 4. Reaction forces on die connector for different friction coefficients on the left, Angle between 2 points on cylindrical element for different friction coefficients on the right.

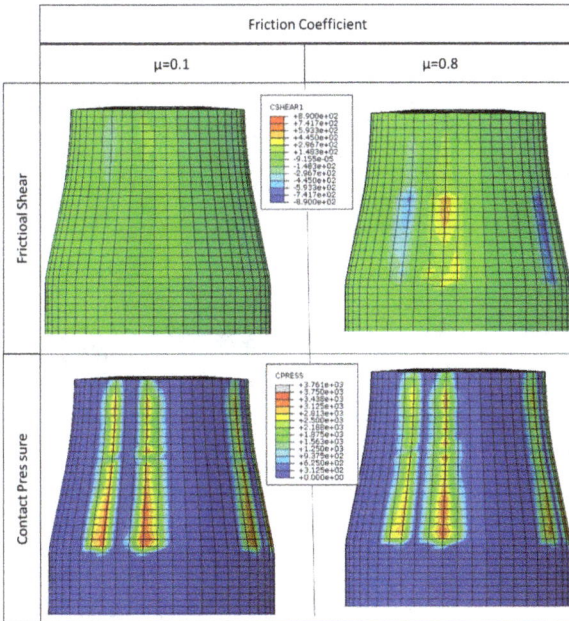

Fig. 5. Frictional shear and contact pressure (in MPa) on the workpiece during forming for different die frictions.

Experimental Validation

In the literature, surface roughening is initially performed using hard facing. Rough areas are coated by hard tungsten layers. During this process, natural pores occur on the surface which creates the necessary roughening. A problem with this method is determined during micro forming studies. During the process, roughening effects disappears as the pores wear out or gets filled by process residues. This problem is especially important for micro forming applications because

Material Forming - ESAFORM 2023
Materials Research Proceedings 28 (2023) 899-908

Materials Research Forum LLC
https://doi.org/10.21741/9781644902479-98

workpiece easily buckles due to reaction forces. It is solved by machining a wave form to swaging dies using electro discharge machining (EDM) [e.g. 5-7]. EDM is an advantageous method to create accurate and reliable shapes but it requires special tool (electrode) development, which is costly. An alternative method is used in this study to validate the effects of the increased friction. A 20W laser engravement machine (JNLINK/LXF-20W, China) is used to engrave the die surface.

1.2379 is used as the die material for this trial. Using fixed engravement parameters (given in Table 1), a sample engravement pattern is studied to increase surface roughness. Surface roughness is determined using Mitutoyo Surftest SJ210 (Japan). An engraved V-type die is shown on Fig. 6. Section A-A is the schematic cross-section of the wave.

Table 1. Engravement parameters.

Speed (mm/s)	Power (W)	Freq. (kHz)	Start TC (US)	Laser Off (US)	End TC (US)	Polygon TC (US)	Distance (mm)	Depth (mm)
100	20	20	300	100	300	100	0.4	30

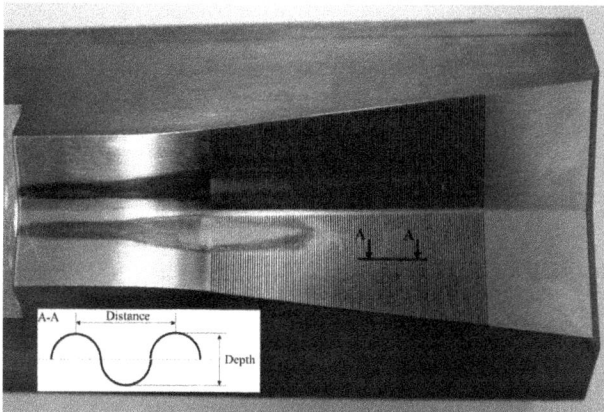

Fig. 6. Engraved swaging die. A-A section shows schematic representation of engravement distance and depth.

V-type dies are composed of 2 sections (left and right) perpendicular to axial direction. Roughness measurements are conducted for these sections separately. Initial and final roughness are presented in Table 2 for 4 dies before and after swaging processes. Initial die geometry is produced using EDM. The difference between the surface roughness parameters for different dies is accepted within tolerances. Surface roughness is measured before and after 3 processes to observe the reliability of the suggested roughening method.

Material Forming - ESAFORM 2023   Materials Research Forum LLC
Materials Research Proceedings 28 (2023) 899-908   https://doi.org/10.21741/9781644902479-98

Table 2. Surface roughness of 4 dies before/after engravement and its deviation after 3 processes.

#	Before engravement		After engravement		After 3 processes	
	R_a (right)	R_a (left)	R_a [R_z] (right)	R_a [R_z](left)	R_a [R_z] (right)	R_a [R_z] (left)
1	0.60	0.29	4.37 [43.7]	4.13 [50.0]	3.81 [43.7]	3.00 [39.2]
2	0.71	0.93	3.58 [45.0]	4.03 [40.1]	4.29 [40.1]	3.55 [30.48]
3	0.50	0.69	5.66 [56.7]	6.66 [68.1]	5.97 [33.8]	6.17 [37.56]
4	0.79	0.48	5.75 [58.8]	7.27 [56.7]	8.91 [53.3]	6.09 [48.66]

After the preparation of engraved dies, process trials are performed to evaluate the effects of surface roughness using REPKON RFFM Series, 4-die rotary swaging machine. The hydraulic cylinder, despite the reaction forces generated on the dies, aims to retain its position. As a result, there is a pressure difference at both ends of the cylinder. This pressure difference is then utilized to calculate the reaction force. Since, there was no infrastructure to measure the friction coefficient of the engraved surface in the production facility, it was decided to conduct a comparative study using engraved and non- engraved die sets to verify the analysis concept. During the test, the mechanism that pushes the workpiece into the die set was operated with position control and the reaction forces on the mechanism were measured. Force measurement results are shown in Fig. 7. The introduction of surface roughness on the die has been found to result in a significant reduction of the reaction forces by half. This validates the application of a roughened die surface concept for macro swaging applications. However, a direct correlation is not established due to lack of friction coefficient determination.

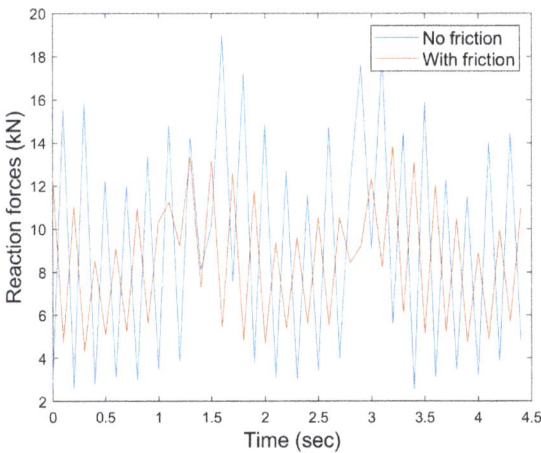

Fig. 7. Workpiece pusher reaction forces during test processes.

Summary
In this study, significant effects of the die surface friction on the rotary swaging process and product quality are determined. Process design requires careful evaluation of benefits and

Material Forming - ESAFORM 2023   Materials Research Forum LLC
Materials Research Proceedings 28 (2023) 899-908   https://doi.org/10.21741/9781644902479-98

drawbacks for roughened die surface application. Main highlights of the study which construct a base for this consideration are shared below:

- Die surface friction results in frictional shearing between die and workpiece. During each blow, die pulls the workpiece in the feeding direction which supports the reaction forces created by the tapered section. Hence, reaction forces and machine vibrations decrease.
- The geometric precision of the product is increased by the high friction area created on the dies. In addition, a more matt structure is obtained on the outer surface of the product.
- Die shape and contact surface play an important role for the contact interactions and alter the flow of the material. During rotary swaging, dies pull the surface in the axial and tangential direction. Tangential component results in undesired torsion. For V-shaped dies, tangential contact area is significantly smaller than axial. Hence, increased surface friction does not change the torsion on the workpiece.
- The main drawbacks of the proposed methods are the wear of die surface and tearing of the workpiece surface as a result of the increased frictional shearing. Sustainability of the proposed technique for each trial requires a further study to determine optimum surface roughness and lubrication to prevent deterioration of the product surface quality and to increase die service life.

References

[1] F. Vollertsen, Ed., Micro Metal Forming, Berlin, Heidelberg: Springer Berlin Heidelberg, 2013. https://doi.org/10.1007/978-3-642-30916-8

[2] F. Böhmermann, H. Hasselbruch, M. Hermann, O. Riemer, A. Mehner, H.-W. Zoch, B. Kuhfuss, Dry rotary swaging–approaches for lubricant free process design, Int. J. Precis. Eng. Manuf.-Green Technol. 2 (2015) 325–331. https://doi.org/10.1007/s40684-015-0039-2

[3] M. Herrmann, Schmierstofffreies Rundkneten / Trockenrundkneten, Dry Rotary Swaging, Mar. 2019.

[4] S.L. Semiatin, Ed., Metalworking: Bulk Forming. ASM International, 2005. https://doi.org/10.31399/asm.hb.v14a.9781627081856

[5] F. Binhack, U.S. Patent No. US 2012/0060577 A1. (2012).

[6] F. Grosman, A. Piela, Metal flow in the deformation gap at primary swaging, J. Mater. Process. Technol. 56 (1996) 404–411. https://doi.org/10.1016/0924-0136(95)01854-9

[7] G.D. Lahoti, T. Altan, Analysis of the Radial Forging Process for Manufacturing Rods and Tubes, J. Eng. Indust. 98 (1976) 265-271. https://doi.org/10.1115/1.3438830

[8] G.D. Lahoti, L. Liuzzi, T. Altan, Design of dies for radial forging of rods and tubes, J. Mech. Work. Technol. 1 (1977) 99–109. https://doi.org/10.1016/0378-3804(77)90016-X

[9] A.A. Tseng, S.X. Tong, T.C. Chen, J. Hashemi, Thermomechanical simulation of a radial forging process, Mater. Des. 15 (1994) 87–98. https://doi.org/10.1016/0261-3069(94)90041-8

[10] J.P. Domblesky, R. Shivpuri, B. Painter, Application of the finite-element method to the radial forging of large diameter tubes, J. Mater. Process. Technol. 49 (1995) 57-74. https://doi.org/10.1016/0924-0136(94)01334-W

[11] M. Yamaguchi, S. Kubota, T. Ohno, T. Nonomura, T. Fukui, Grain Size Prediction of Alloy 718 Billet Forged by Radial Forging Machine Using Numerical and Physical Simulation, Superalloys 718. 625. 706 and Various Derikatives 1 (2001) 300. https://doi.org/10.7449/2001/Superalloys_2001_291_300

[12] B. Ghasemi, H. Alijani, M. Poursina, Prediction of Residual Stresses for a Hollow Product in Cold Radial Forging Process, Int. J. Eng. 28 (2015) 1209–1218.

Material Forming - ESAFORM 2023   Materials Research Forum LLC
Materials Research Proceedings 28 (2023) 899-908   https://doi.org/10.21741/9781644902479-98

[13] A. Ameli, M.R. Movahhedy, A parametric study on residual stresses and forging load in cold radial forging process, Int. J. Adv. Manuf. Technol. 33 (2007) 7-17. https://doi.org/10.1007/s00170-006-0453-2

[14] L. Fan, Z. Wang, H. Wang, 3D finite element modeling and analysis of radial forging processes, J. Manuf. Process. 16 (2014) 329–334. https://doi.org/10.1016/j.jmapro.2014.01.005

[15] L. Liu, L. Fan, Study of Residual Stresses in the Barrel Processed by the Radial Forging, in 2009 Second International Conference on Information and Computing Science, May 2009, vol. 4, pp. 131–134. https://doi.org/10.1109/ICIC.2009.343

[16] A. Uhlig, Investigation of the Motions and the Forces in Radial Swaging, in German, Doctoral dissertation, Technical University Hannover, 1964.

[17] M. Smith, ABAQUS/Standard User's Manual, Version 6.9. Dassault Systèmes Simulia Corp., 2009.

[18] L. Ceschini, A. Morri, A. Morri, S. Messieri, Replacement of Nitrided 33CrMoV Steel with ESR Hot Work Tool Steels for Motorsport Applications: Microstructural and Fatigue Characterization, J. Mater. Eng. Perform. 27 (2018) 3920-3931. https://doi.org/10.1007/s11665-018-3481-9

Material Forming - ESAFORM 2023
Materials Research Proceedings 28 (2023) 909-916

Materials Research Forum LLC
https://doi.org/10.21741/9781644902479-99

Design of a new simulative tribotest for warm forming applications having high contact pressure and surface enlargement

GALDOS Lander[1,a]*, AGIRRE Julen[1,b] and ARANBURU Elixabet[1,c]

[1]Mechanical and Industrial Production Department, Mondragon Unibertsitatea, Loramendi 4,20500 Mondragon, Spain

[a]lgaldos@mondragon.edu, [b]jagirre@mondragon.edu, [c]earanburue@mondragon.edu

Keywords: Tribology, Warm Forming, Fine Blanking

Abstract. Many tribological tests have been developed in the last decades to emulate real cold forging processes at laboratory conditions and to obtain friction coefficients at different process conditions. In this paper a new tribotester is designed and numerically validated to reach extreme contact conditions, high normal contact pressure and surface enlargement, for warm temperature testing of lubricants starting from material in the form of thick sheets or precuts. The numerical simulations prove that this is possible by the use of a combined bending and ironing test.

Introduction

Fine blanking is a manufacturing process capable of producing sheet metal parts with completely smooth cutting surfaces [1]. The process is sometimes known by other names like fine stamping or precision cutting and like standard shear blanking processes, according to DIN 8580, fine blanking falls under the main category of material separation or parting processes, and is defined according to DIN 8588 under the sub-category dividing [2].

The main differences in comparison with the conventional blanking or punching process are the use of a counterpressure which enables to work in a more compressive stress state and decreases the cambering of the final product, the use of very small clearances (0.5-1% of the material thickness instead of 10-20% in conventional shearing) and the introduction of Vee-Rings in the pressure and/or blanking plates. The Vee-Ring is a physical V-shape feature that is located either in the pressure plate, the blanking plate or in both of them, which has the function to hold the punched material outside the blanking line and so prevents lateral flow of the material during the blanking process. In addition, it serves to apply compressive stress to the sheet metal, improving the flow process and delaying the fracture occurrence.

The main advantages of the process are the increased accuracy (smoother sheared surface and better geometrical tolerances) and the need of less production steps to produce high quality functional components that need to be manufactured by metal forming and final machining instead.

The disadvantages of the technology are the high tool wear of the active elements (punches, dies, plates), the high tool costs and the need of specific presses that are only used for fine blanking, as three actions (force application) are needed to run the process. A compound fine blanking die working principle during the blanking process is shown in Fig. 1.

Material Forming - ESAFORM 2023   Materials Research Forum LLC
Materials Research Proceedings 28 (2023) 909-916   https://doi.org/10.21741/9781644902479-99

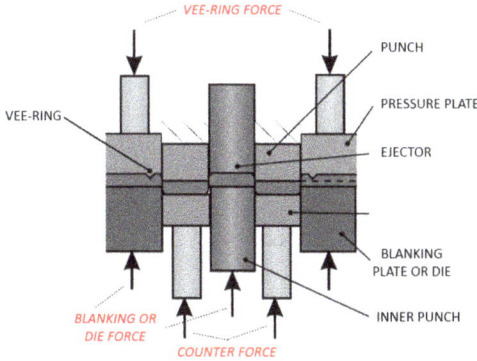

Fig. 1. Compound die set used in fine blanking (recreated from [2]).

Nevertheless, not all the aspects of the fine blanking technology are positive. The main materials that are fine blanked are steels (structural, case hardenable, heat treatable, etc.) which roughly represent the 80-90% of the produced parts all over the world. Other materials like aluminium, cooper and stainless steels are properly processed with the use of proper lubricants that avoid excessive adhesive wear. Anyhow, the elongation capacity of the materials needs to be high enough to guarantee the fine blanking capability of the cut material (complete smooth cutting surface) and this is reached by chemical composition adjustments, a proper thermos-mechanical processing and the soft annealing of the material prior to its processing. For example, in the case of the C45 steel, this annealing heat treatment changes the original microstructure that passes from a ferritic-perlitic typical microstructure to a microstructure that comprises a ferrite matrix with spheroidal cementite embedded in it. This spheroidized microstructure highly increases the fine blanking capability.

Currently, the ultimate tensile strength of fine blanked materials is around 550-600 MPa, both for steel and stainless steels. The current needs for structural lightweighting, especially in the transport industry, are forcing the fine blanking companies to push these material boundaries and to process higher steel grades. Although many attempts have been made by different companies to be able to cut high strength materials, the warm assisted fine blanking seems the most suitable technological approach for this, still causing big challenges to be implemented in serial production.

Warm Assisted Fine Blanking

It is well known that heating of metallic alloys improves their ductility in the absence of blue brittleness or other similar fragility mechanisms. For these reasons, four different research institutes and companies have performed initial fine blanking tests to develop the warm assisted fine blanking process.

Up to the authors best knowledge, researchers of the University of Shanghai were the first ones reporting experimental values in 2020. In their work, aiming to widen the application of the fine-blanking process, the heat-assisted fine-blanking process of AISI 304 stainless steel grade was performed. The cutting of a dog-bone shape component showed that the increase in temperature was beneficial for the final quality of the fine blanked component, being 250°C the optimal temperature for a cutting quality improvement [3].

Later in 2020, Feintool company together with ETH Zurich published an online web publication about their recently developed Thermo-FineBlanking. Unlike the previous work, both the blank

Material Forming - ESAFORM 2023
Materials Research Proceedings 28 (2023) 909-916

Materials Research Forum LLC
https://doi.org/10.21741/9781644902479-99

material and the tool inserts were heated and the aim of the result was to check if conventional steels could be fine blanked without a laborious and expensive heat treatment with a carbide spheroidization level of more than 90%. The cutting force was reduced in a 30% in comparison to the room temperature cutting and fine blanking capability was considerably improved with increased temperature for the 42CrMo4 and 1.4301 (AISI 304) materials [4]. In the same work, authors claimed that new lubricant developments were needed so that a proper tribological condition is obtained with temperatures above 200°C, the limit for conventional fine blanking mineral-oil based lubricants.

Lastly, researchers of the laboratory for Machine Tools and Production Engineering (WZL) of RWTH Aachen University have published three works where they showed preliminary results of the newly developed warm assisted fine blanking process. In Shemet 2021, the authors presented final microhardness values obtained after fine blanking S700MC, 42CrMo4 and 16MnCr5 steels [5]. Later in Esaform 2021, same group presented the cutting forces at different temperatures for the X5CrNi18-10 (AISI 304) stainless steel. In the same paper, final product tolerances and cutting surface quality of different temperatures were also analysed [6]. Finally, in Esaform 2022, the same research group, presented the results obtained when fine blanking a star shape component and the 40MnB4 and 42CrMo4 steels. The process forces were evaluated depending on the sheet metal temperature and a good fine blanking capability of 40MnB4 was confirmed. Moreover, results showed that process forces and product quality were comparable to 42CrMo4 steel [7].

Following the findings of the previous researchers, Mondragon University, together with the Elay company, has recently performed warm forming fine blanking tests using an industrial Mori hydraulic fine blanking press. Different tests were performed by heating the raw material up to 300°C and the results of precedent authors were confirmed by cutting a real component. As stated by Zheng et al. from Shanghai University, the critical point of the industrial application of the process is, among others, the loss of properties of conventional lubricants used in fine blanking and thus, the premature tool wear. For this reason, the main objective of the present paper is to develop a new tribotester that can emulate the fine blanking conditions, extreme contact pressures and surface expansions, at warm temperatures, so that the test results can be used for new lubricants screening and new lubricant formulation optimization.

Tribotesters for Extreme Contact Conditions
Several tribotesters have been developed during the last decades to emulate sheet metal forming and forging processes. Among others the ring test for forging and the strip drawing test for sheet metal forming are the most used tests to estimate the coefficient of friction and for the testing of new lubricants.

Groche et al. compared six different well established friction tests for cold forging operations using one state of the art industrial tribosystem, the contact between the 16MnCrS5 steel and tools made of M2 grade steel with a hardness of 61–63 HRC [8]. The tool surfaces were coated with an AlCrN based coating (Balinit Alcrona Pro) and polished to a roughness of Ra < 0.2 mm. In the same study, the different tribotests were numerically simulated and the basic tribological loads, the contact normal pressure, the surface enlargement, the relative sliding velocity and temperature at the interface were compared.

Due to the individual frictional test setups, it is clearly observed that not all the mentioned tribological loads can be set independently. Each of the tribotest has its own characteristic working window and the Sliding Compression Test seems to be the most flexible option to tune the test and get the desired range of contact conditions. All in all, the tribological conditions found in fine blanking and shearing in general are very aggressive [9-10]. Contact pressure can reach levels as high as 5 GPa and surface enlargements bigger than 50 are locally found in the sheared zone for a fine blanking of 8 mm thick S700MC steel. A summary of the numerical results obtained by the authors is shown in Table 1.

Material Forming - ESAFORM 2023   Materials Research Forum LLC
Materials Research Proceedings 28 (2023) 909-916   https://doi.org/10.21741/9781644902479-99

Table 1. Summary of the numerical results obtained in [8].

Tribological tests	Contact pressure	Surface expansion	Relative sliding velocity	Temperature
Ring Compression	Low	Low	Low	Low
Combined Forward Rod Backward Can Estrusion	Medium	Low	Medium	Medium
Backward Can Extrusion	Medium-High	High	High	High
Backward Can Extrusion with rotation	High	High	Medium	Medium
Upsetting sliding	Medium	Low	Set value	Low
Sliding compression	Medium	Medium	Set value	Medium

For these limitations, also found in cold forging, and to evaluate different lubricants and surface finishings, Hirose et al. developed an Upsetting-Ironing type tribometer for evaluating the tribological performance of lubrication coatings for cold forging [11]. Numerical simulations of the test prove that the test is able to reach a surface enlargement up to 500 and a contact pressure of around 2GPa for a SWRM 10K steel (tensile strength of 480 MPa). During the test a cylindrical billet is first upset for being later deformed by ironing using three ball bearing made of high chromium steel (SUJ2) of diameter 12 mm. Following this idea, a new tribotester is presented in this work for testing thick sheets and to emulate warm fine blanking conditions.

Design of a New Tribostester
The new tribotester that is proposed for fine blanking is shown in Fig. 2. As in the concept presented by Hirose et al. the test is divided in two steps [11]. First, the initial flat blank is bent to obtain an U-shape. Then punch stroke continues and the ironing phase starts. The U-shaped blank, which has been bent with a clearance of 0.5 mm between the punch and the die (per side) is deformed against bearing balls made by SUJ2 steel of diameter 12 mm. At this moment, a build-up of contact pressure and surface enlargement occurs in the desired contact zone. The ball bearings can be exchanged at every new test due to their low cost and the galling can be studied on them, as well as in the tested sample, following the procedure defined by Hirose et al. Chromium tends to adhere in the tested sample and this is used to determine when the galling starts using an SEM microscope.

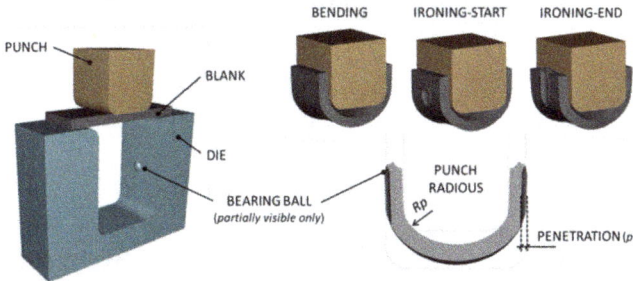

Fig. 2. New tribotester for thick sheet material and extreme tribological conditions.

Material Forming - ESAFORM 2023   Materials Research Forum LLC
Materials Research Proceedings 28 (2023) 909-916   https://doi.org/10.21741/9781644902479-99

The parameters that can be changed during the test are the blank temperature, which can be previously heated in a furnace, the ball penetration (see Fig. 2) and the press speed.

Numerical Modelling of the Test

The new testing concept and set-up has been numerically studied to evaluate the contact conditions reached with different testing conditions, varying the penetration (p) and punch radius (R_p). FORGE NxT 3.2 software has been used for running the numerical simulations and the tools have been considered rigid. The blank material is a S700MC steel which has been modelled using solid tetrahedral elements and a simplified Hensel-Spittel hardening law considered as isotropic (r values close to 1 in different directions). Only one half of the blank has been modelled in the study (see symmetry plane in Fig. 3). Young modulus of 210 GPa and a Poisson coefficient of 0.3 were employed. The material temperature has been fixed to room temperature and no temperature influence has been studied in this work. The different simulation conditions are shown in Table 2 and the numerical model is further detailed in Fig. 3.

Table 2. Different simulation conditions.

Material model – Hensel Spittel (MPa)	$\sigma = 1127 \cdot e^{-0.0009 \cdot T} \cdot (\varepsilon - 0.01354)^{0.0833}$
Friction coefficient (Tresca shear model)	0.05, 0.1, 0.2, 0.3
Penetration (p) - mm	2, 3, 4
Punch radious (R_p) - mm	10, 17.5, 25

The main objectives of varying the selected parameters are:

1. To evaluate the sensitivity of the test to different friction coefficients using different test conditions – a higher force increment when changing the friction coefficient is desirable to perform a screening of different lubricants that are suitable for warm fine blanking
2. To evaluate how the penetration changes the contact pressure and surface enlargement
3. To optimize the punch radius value to avoid excessive curving of the tested sample before the ironing phase

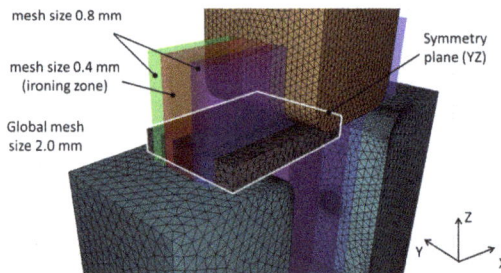

Fig. 3. Numerical model of the bending-ironing tribological test (R10 and p2).

With the purpose of selecting the best tool dimensions of the new tribotester the force evolution curves, surface enlargement variation and the maximum normal contact pressures for each of the simulated case have been evaluated.

A typical force-displacement curve of the punch is shown in Fig. 4a. The curve shows a double dome shape with two force peaks corresponding to the initial bending force and the ironing force respectively. If we solely consider the ironing force, ΔF is defined as the force difference between

Material Forming - ESAFORM 2023   Materials Research Forum LLC
Materials Research Proceedings 28 (2023) 909-916   https://doi.org/10.21741/9781644902479-99

the peak forces taking as a reference the m0.05 friction case. A higher force change means a higher sensitivity to detect different friction coefficients and thus to do a proper evaluation of the new lubricants for fine blanking. Fig. 4b clearly shows that sensitivity of the test increases when the punch radius decreases and the penetration increases. Thus, the best testing parameters in terms of friction sensitivity are R_p=10 mm in combination with p=4 mm.

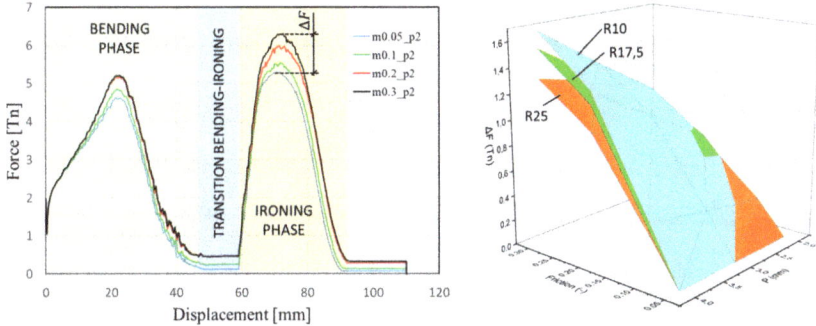

Fig. 4. a) Typical punch Force-Displacement curve for penetration 2 mm and different friction values and b) comparison of ΔF for different configurations.

An example of the numerical results obtained for the field variables maximum surface enlargement and maximum normal contact pressure is shown in Fig. 5a and Fig. 5b respectively. Here, the surface enlargement (named as DSURF) is defined as follows:

$$DSURF = \frac{Surf - Surf_0}{Surf_0} = \frac{Surf}{Surf_0} - 1 \qquad (1)$$

where $Surf$ is the new expanded surface and $Surf_0$ is the initial surface. Consequently, a DSURF value of 3 means that the new surface is four times greater than the original one.

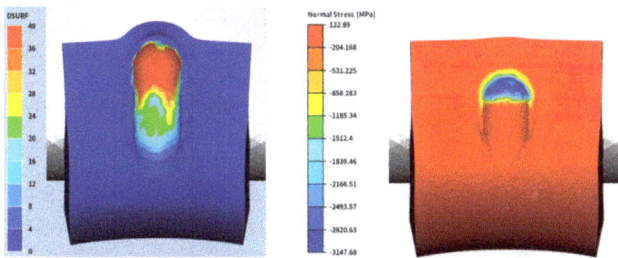

Fig. 5. Example of numerical results for a) Surface enlargement and b) Normal contact pressure.

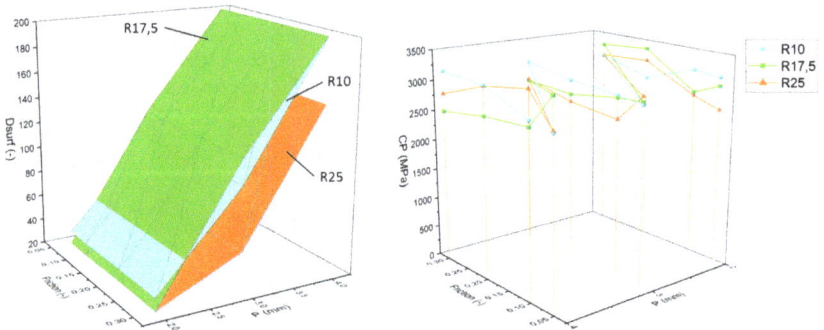

Fig. 6. Influence of parameters on a) Surface enlargement and b) Normal contact pressure.

As it can be observed in Fig. 6a, the maximum surface expansion is mainly dominated by the penetration. Higher penetration results in a higher surface enlargement as already shown by Hirosi et al. where higher amount of upsetting and subsequent ironing achieved higher surface enlargement values. On the contrary, the maximum contact pressure is around 3GPa for all the cases (Fig. 6b). This is probably because the hardening of the material tends to saturate at high strain levels. Nevertheless, both the surface enlargement values and contact pressures are representative of extreme contact conditions and suitable to emulate the fine blanking situation.

Summary

A new tribotester to test lubricants for warm fine blanking has been conceptually defined and numerically validated. The new testing device is simple, can achieve high contact pressures and surface enlargements and uses ball bearings that are cheap and easily exchangeable to analyse wear mechanisms.

The numerical parametric study of the most relevant design dimensions, the ironing penetration (p) and punch radius (R_p), has clearly shown the effect of them on the maximum surface enlargement and contact pressure. The surface enlargement increases with the penetration while the surface enlargement follows a different trend. Similar surface expansion values are obtained for R17.5 and R10 while lower values are observed with R25. This can be probably explained by the fact that using this high punch radius and same sample length, the total extruded length is notably decreased in comparison to the other two cases.

The friction force different between the studied conditions shows that sensitivity to contact condition changes is higher when penetration increases.

Acknowledgments

The authors would like to thank the Spanish Government for the economic support and funding of the HEATFORM European project (SMART Eureka call) and the Elay company for the technical and experimental testing support.

References

[1] R.A. Schmidt, Cold forming and fineblanking, Feintool, Buderus, Wälzholz, Hoesch Hohenlimburg, Germany, 2007.

[2] T. Altan, Metal Forming Handbook /Schuler, Springer-Verlag Berlin Heidelberg, 1998.

[3] Q. Zheng, X. Zhuang, J. Hu, Z. Zhao, Formability of the heat-assisted fine-blanking process for 304 stainless steel plates, Mater. Charact. 166 (2020) 110452. http://doi.org/10.1016/j.matchar.2020.110452

[4] C. Maurer, Development project: Thermo-fineblanking, Feintool, IEPA, 2018. https://blog.feintool.com/en/thermo-fineblanking

[5] I.F. Weiser, A. Feuerhack, T. Bergs, Investigation of the Micro Hardness at the Cut Surface of Fine Blanked Parts with Variation of Sheet Material and Cutting Temperature, Key Eng. Mater. 883 (2021) 269-276.

[6] I.F. Weiser, R. Mannens, A. Feuerhack, T. Bergs, Experimental Investigation of Process Forces and Part Quality for Fine Blanking of Stainless Steel with Inductive Heating, ESAFORM 2021. https://popups.uliege.be/esaform21/index.php?id=2575

[7] I.F. Weiser, T. Herrig, T. Bergs, Fine Blanking Limits of Manganese-Boron-Steel in Fine Blanking Compared to Tempered Steel with Variation of Sheet Metal Temperature, Key Eng. Mater. 926 (2022) 1122-1130. https://doi.org/10.4028/p-7cwhpb

[8] P. Groche, P. Kramer, N. Bay, P. Christiansen, L. Dubar, K. Hayakawa, c. Hu, K. Kitamura, P. Moreau, Friction coefficients in cold forging: A global perspective, CIRP Annals 67 (2018) 261-264. https://doi.org/10.1016/j.cirp.2018.04.106

[9] Y. Abe, R. Yonekawa, K. Sedoguchi, K.I. Mori, Shearing of ultra-high strength steel sheets with step punch, Procedia Manuf. 15 (2018) 597-604. https://doi.org/10.1016/j.promfg.2018.07.283

[10] K.I. Mori, Review of shearing processes of high strength steel sheets, J. Manuf. Mater. Process. 4 (2020) 54. https://doi.org/10.3390/jmmp4020054

[11] M. Hirose, Z.G. Wang, S. Komiyama, An upsetting-ironing type tribo-meter for evaluating tribological performance of lubrication coatings for cold forging, Key Eng. Mater. 535 (2013) 243-246. http://doi.org/10.4028/www.scientific.net/KEM.535-536.243

Incremental and sheet metal forming

Material Forming - ESAFORM 2023
Materials Research Proceedings 28 (2023) 919-928

Materials Research Forum LLC
https://doi.org/10.21741/9781644902479-100

Numerical investigation on the deep drawing of sheet metals with an additively applied coating

MÄRZ Raphaela[1,a *], HAFENECKER Jan[1,b], BARTELS Dominic[2,3,c], SCHMIDT Michael[2,3,d] and MERKLEIN Marion[1,e]

[1]Institute of Manufacturing Technology (LFT), Friedrich-Alexander-Universität Erlangen-Nürnberg, Egerlandstraße 13, 91058 Erlangen, Germany

[2]Institute of Photonic Technologies (LPT), Friedrich-Alexander-Universität Erlangen-Nürnberg, Konrad-Zuse-Straße 3/5, 91052 Erlangen, Germany

[3] Erlangen Graduate School in Advanced Optical Technologies (SAOT), Paul-Gordan-Straße 6, 91052 Erlangen, Germany

[a]raphaela.maerz@fau.de, [b]jan.hafenecker@fau.de, [c]dominic.bartels@lpt.uni-erlangen.de, [d]m.schmidt@blz.org, [e]marion.merklein@fau.de

Keywords: Additive Manufacturing, Forming, Hybrid Parts

Abstract. A promising approach to meet ecological and economical challenges in production industry is the combination of additive manufacturing and forming, which enables the production of hybrid parts. The fusion of the two technologies has the potential to use the benefits of both process classes. One example for the combination of additive manufacturing and forming is a process chain consisting of laser-based directed energy deposition (DED-LB/M) to apply a wear resistant coating on a sheet metal and a subsequent deep drawing process to achieve the final geometry of the component. However, the presence of the local reinforcement influences the subsequent sheet metal forming process. In order to gain more knowledge about the process chain a numerical approach is performed. In this work, the influence of a coating using Bainidur® AM applied by DED-LB/M on the formability of the case hardening steel 16MnCr5 in a deep drawing process is evaluated. For this purpose, a numerical model is built and validated by comparison with experimental results. The finite-element-model serves as the basis for the investigation on the influence of the location of the additively applied coating on the deep drawing process and also enables a deeper understanding of the process.

Introduction

Due to governmental requirements and an increasing environmental awareness, modern production technology has to face new ecological and economical challenges [1]. Therefore, a need for resource efficiency and sustainability in the field of production is coming into focus. In addition to lightweight design [2] another approach to meet the current challenges are resource efficient manufacturing processes. Forming technology is known for its resource efficiency as well as for the high output [3]. On the other hand, a disadvantage of the production technology is its low flexibility [3]. In contrast, the use of additive manufacturing enables a high degree of geometric freedom and a customizability of the components [4]. By combining the two technologies, a production of hybrid parts is possible. Synergy effects such as adapted material properties as well as an extension of the process limits appear and the disadvantages of the two process classes can be compensated [5]. One example for the production of hybrid parts is a process chain consisting of DED-LB/M to apply a wear-resistant coating on a circular blank and a subsequent deep drawing process to achieve the final geometry of the component, which is based on the geometry of a barrel sleeve. The forming process enables a resource efficient production of the component compared to current machining of the barrel sleeve. The use of DED-LB/M before

Material Forming - ESAFORM 2023   Materials Research Forum LLC
Materials Research Proceedings 28 (2023) 919-928   https://doi.org/10.21741/9781644902479-100

the forming process enables a local reinforcement of the blank. Therefore, global energy-intensive heat treatment strategies can be avoided. The authors in [6] show that DED-LB/M can be used to locally strengthen aluminum sheet metal blanks. The formability of the aluminum blanks is limited due to the heat input [6]. Nevertheless, no investigations have yet been carried out into how a high-strength material applied by DED-LB/M affects the formability of a case hardening steel.

Objective and Methodology
The aim of this work is to investigate the influence of an additively applied coating on a deep drawing process. This is necessary to evaluate the formability of hybrid components where a high-strength coating is applied. The methodology used in this work is shown in Fig.1.

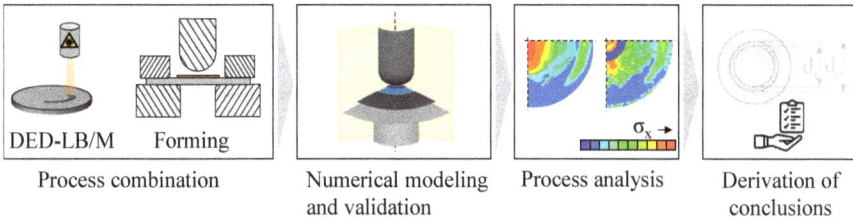

| DED-LB/M   Forming | Numerical modeling and validation | Process analysis | Derivation of conclusions |

Process combination

Fig.1. Methodology.

The basis of the investigation is a process combination of DED-LB/M and a subsequent deep drawing process. In a first step, a FE-model of the forming process is built. In addition to the conventional deep drawing process, the model is extended to include the hybrid approach. The validation is done by comparison with experimental results. In addition to the process force, the geometry of the component is compared in order to evaluate the prediction quality of the FE-model. After the validation, an investigation of the stress state of the uncoated and hybrid components is carried out in order to derive conclusions about the formability of hybrid components and the advantageous placement of the additive coating.

Experimental and Numerical Setup
Material and semi-finished product. To produce the hybrid part a process chain consisting of DED-LB/M and a subsequent deep drawing process is used. The geometry of the semi-finished part is shown in Fig. 2a. A circular blank with a diameter of 105 mm and a thickness of 3.5 mm serves as substrate. As material the case hardening steel 16MnCr5 (1.7131) is used. Before the deep drawing process, a wear-resistant coating is applied on the sheet by DED-LB/M. Therefore, the metal powder Bainidur® AM is used. The applied geometry is shown with its variations in Fig. 2b.

Material Forming - ESAFORM 2023
Materials Research Proceedings 28 (2023) 919-928

Materials Research Forum LLC
https://doi.org/10.21741/9781644902479-100

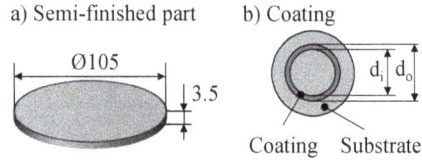

Fig. 2. a) Geometry of the semi-finished part b) Geometry of the coating and its variations.

The coating is applied in a circle and has a height of 0.5 mm. A difference is made between the listed inner and outer diameters in order to determine the influence of the position of the coating on the forming process. To describe the mechanical properties and the hardening behavior of the substrate material and the coating, true stress-true plastic strain curves are determined, which are used for the numerical simulation. The resulting curves are shown in Fig. 3.

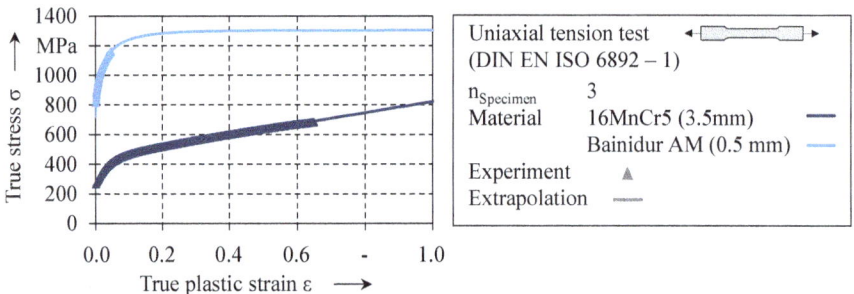

Fig. 3. True stress-true plastic strain curves of the substrate material 16MnCr5 and the coating Bainidur® AM from experimental uniaxial tension tests.

Due to the prevailing stress conditions in deep drawing processes, uniaxial tension tests according to DIN EN ISO 6892-1 [7] at room temperature are carried out. The specimens for the characterization of the substrate material have a measuring length of 50 mm and a width of 12.5 mm. Tensile specimens are also used to characterize the coating. Due to the dimensions of the coating, a smaller specimen geometry with a measuring length of 20 mm and a width of 6 mm is used. The thickness of the specimens corresponds to the thickness of the coating in the experiment. The specimens were first built up as a block through five layers and in the second step separated from the substrate and brought to the final shape by electrical discharge machining. The uniaxial tension tests are carried out using a universal testing machine Zwick Z100 respectively Zwick Z10 for the smaller specimens and a camera-based 3D strain measuring system GOM Aramis. The substrate has a significantly lower initial yield stress of 266 MPa than the coating with a yield stress of 814 MPa. Since higher strains than those determined experimentally

Material Forming - ESAFORM 2023   Materials Research Forum LLC
Materials Research Proceedings 28 (2023) 919-928   https://doi.org/10.21741/9781644902479-100

occur in the real process, the two curves were approximated and extrapolated. Voce generalized [8] is chosen as extrapolation approach for the substrate material 16MnCr5. For the coating, the Hockett-Sherby approach [9] was found to be appropriate.

Laser-based directed energy deposition. Before the forming process, circular blanks are coated using DED-LB/M. Therefore, the experiments are performed on an ERLAS UNIVERSAL 50349 machine (ERLAS GmbH). The machine is equipped with a diode laser with a peak power of 4 kW and a characteristic wavelength of 900 to 1080 nm. The nozzle can be moved using three linear axes. In addition, the cell has a sample holder with two rotary axes, which allows the realization of free-form surfaces. The parameter set used for the coating is based on previous investigations [10].

Deep drawing. The subsequent deep drawing process is carried out on a hydraulic press Lasco TSP 100 So with a maximum force up to 1000 kN. The velocity of the punch is set to 4.7 mm/s and the force of the binder is set to 25 kN. The punch geometry is hemispherical with a diameter of 50 mm. The drawing depth, which is set by hard-stops, amounts to 30 mm. In order to reduce the tribological loads, the blanks are lubricated with KTL N 16. The experiments are carried out at room temperature. After the forming process, the parts are measured optically in order to determine the geometry and the sheet thickness. The measurement is carried out with an Atos Core 300 of the company GOM GmbH.

Numerical setup. The numerical model presented in this paper is built with the finite element software LS-Dyna by Ansys, which allows the simulation of nonlinear physical processes. The generation of the input is done with the help of the graphical pre- and post-processor LS-PrePost V.4.7.0. As solver the MPP Double R11.1 is used. Due to the symmetry of the hybrid part, only a quarter is modelled. Therefore, the Keyword "SPC_SET" is applied to the nodes on the x- and y-edges of the part and the coating. For the tools, namely punch, binder and die, rigid shell elements are used. To avoid unwanted movement, constraints in x- and y-direction are defined for the tools. The numerical setup is shown in Fig. 4.

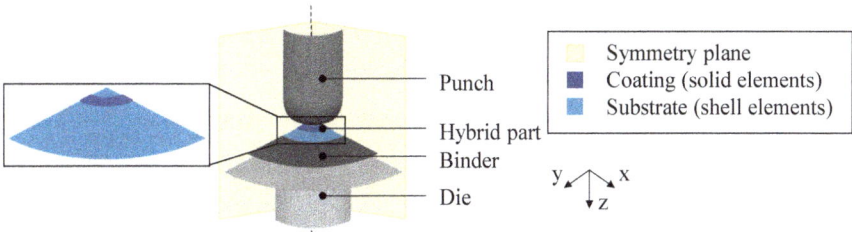

Fig. 4. Setup of numerical model to map the deep drawing process.

The sheet is modeled by fully integrated shell elements. As material model "133-Barlat_YLD2000" is used due to the anisotropic material behavior of the case hardening steel 16MnCr5. In order to map the coating, the material keyword "024-PIECEWISE_LINEAR_PLASTICITY" is applied. In contrast to the substrate material, the coating is modeled with solid elements. To consider the adhesion of the coating to the substrate, the two parts share the same transition nodes. This approach is also chosen in [11]. The contact is modeled with the keyword "CONTACT_FORMING_SURFACE_TO_SURFACE" to avoid penetrations between the tools and the blank.

Material Forming - ESAFORM 2023
Materials Research Proceedings 28 (2023) 919-928

Materials Research Forum LLC
https://doi.org/10.21741/9781644902479-100

Results

Validation. In order to evaluate the prediction quality of the presented FE-model, it is validated by comparison with experimental results. In addition to the hybrid part, the conventional deep drawing process is also considered. In Fig. 5, the comparison of the experimentally and numerically determined maximum process forces is shown.

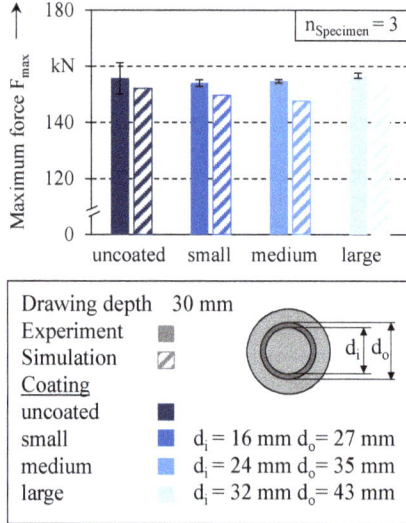

Fig. 5. Comparison of experimental and numerical maximum process force.

Experimentally, a maximum forming force of 155.7 ± 5.1 kN is determined while forming the uncoated blank. In the simulation a maximum forming force of 152.1 kN is reached. With a deviation of 2.6 %, a good agreement between the simulation and the experiment is achieved. Good agreement is also obtained when comparing the numerically and experimentally determined maximum forces of the hybrid components. The largest deviations are evident for the coating with an inner diameter of 24 mm and an outer diameter of 35 mm. Experimentally, a process force of 154.6 ± 0.6 kN is obtained, while the maximum force with 147.5 kN is underestimated in the simulation. However, with a deviation of 4.6 % from the experimentally determined results, a good agreement can be assumed here as well. The influence of the coating on the maximum process force is negligible. The deviations of the variations are within the standard deviation.

In addition to the maximum process forces, the geometry of the part, which was obtained experimentally and calculated numerically, is also compared. The basis for the comparison is the sheet thickness, which was determined at a drawing depth of 30 mm. The sheet thicknesses for the uncoated blank and for the coated blank with an inner diameter of 32 mm and an outer diameter of 43 mm of the coating are shown in Fig. 6. Since only a quarter model was built in the simulation, the resulting thicknesses are mirrored on the y-axis.

Material Forming - ESAFORM 2023   Materials Research Forum LLC
Materials Research Proceedings 28 (2023) 919-928   https://doi.org/10.21741/9781644902479-100

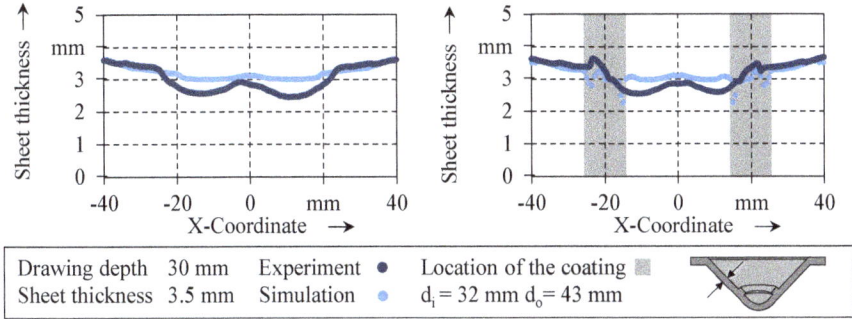

Fig. 6. Experimental and numerical sheet thickness distribution of the a) uncoated part and b) hybrid part.

A good agreement between simulation and experiment is found for the uncoated and the hybrid part in the area of the flange and in the radius area of the die. Deviations are evident in the bottom area of the conventional component. Here, the sheet thickness is overestimated in the simulation for both areas. The reason for this could be the modeling approach of the sheet with shell elements, which is only sufficiently accurate for a sheet with a thickness of 3.5 mm. This upper limit is reached in this case due to the use of a semi-finished part with a sheet thickness of 3.5 mm. Thus, the improvement of the accuracy of the model by using solid elements instead of shell elements has to be verified in future investigations. However, the areas in which the sheet thins out are correctly modeled. Moreover, the determination of the occurring stress states, which is the focus of this contribution, is mainly linked to the process forces, for which only small deviations of less than 5 % are identified. Therefore, it is assumed that the accuracy is adequate for the evaluation of the stress states and the use of the model for further investigations is possible.

Resulting stresses. Different stress states occur during deep drawing. Due to the use of a hemispherical punch, the stress conditions differ from those in conventional deep drawing.

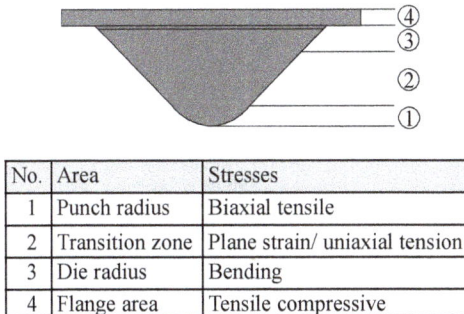

No.	Area	Stresses
1	Punch radius	Biaxial tensile
2	Transition zone	Plane strain/ uniaxial tension
3	Die radius	Bending
4	Flange area	Tensile compressive

Fig. 7. Resulting areas and stress states of a deep drawn part with a hemispherical punch.

Material Forming - ESAFORM 2023
Materials Research Proceedings 28 (2023) 919-928

Materials Research Forum LLC
https://doi.org/10.21741/9781644902479-100

The resulting stress states and the corresponding areas are shown in Fig. 7. The bottom area of the part is characterized by biaxial tensile stresses, similar to conventional deep drawing. The subsequent transition area is initially characterized by a plane strain area. As the punch diameter increases, uniaxial tensile stresses occur. In the area of the die radius, the material undergoes a double bending. In the flange area of the component tensile-compressive stresses occur.

By varying the diameter of the coating, it lies in different areas of the deep-drawn component and thus undergoes different stress states. In the following, the influence of the coating diameter on the resulting stresses is investigated. Since radial tensile and compressive stresses occur predominantly in the formed component, the direction-dependent stresses are considered. The resulting stresses in x-direction at a drawing depth of 30 mm and the interface pressure between the punch and the part are shown in Fig. 8.

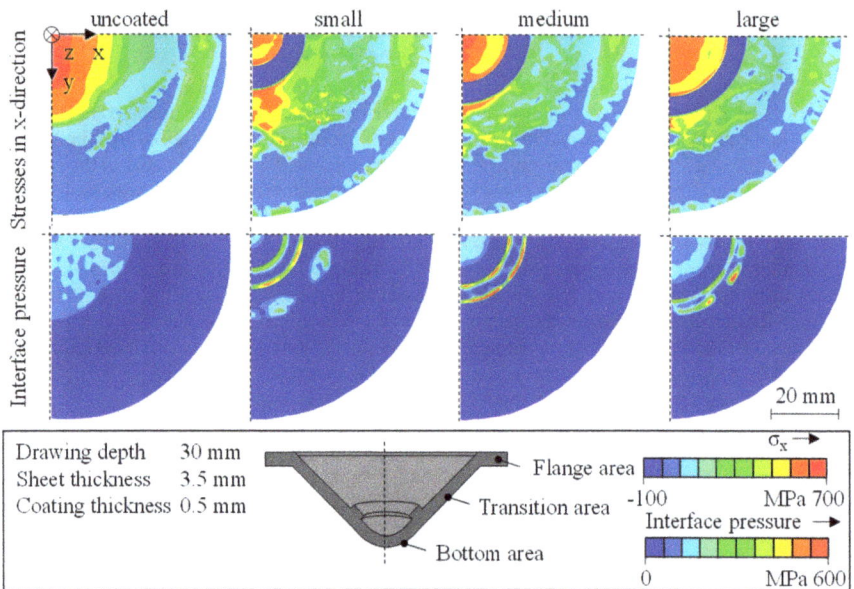

Fig. 8. Comparison of the resulting stresses in x-direction and the interface pressure between the part and the punch in the numerical deep drawing process of the conventional and hybrid parts with coating different diameters.

The maximum tensile stresses occur at the bottom area of the uncoated part. These are more dominant in the y-direction than in the x-direction. The bottom of the component is the most critical area in terms of the formation of cracks due to the maximum stresses in this area. Through the addition of the additively applied coating, the stresses in the bottom area are reduced. The reason for this is the loss of contact with the bottom of the component, as the punch primarily contacts the coating. With an increasing coating diameter, the punch has more contact with the bottom of the component. Therefore, the stresses in the bottom area of the hybrid part increase as well. The use of a coating, however, leads to stress peaks that occur at the inner and outer diameter of the coating, which can be critical regarding a failure of the part at higher drawing depths. A bigger diameter of the coating leads to lower stresses around the coating. In addition to the stresses in x-directions, the major and minor strains of each finite element are investigated in order to evaluate the stress states of the component. The strains were determined numerically and are shown in Fig. 9.

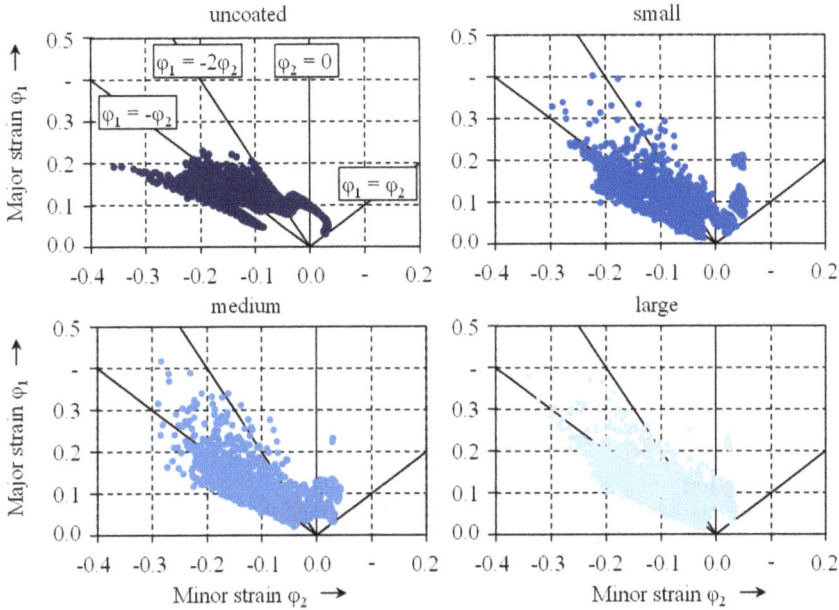

Fig. 9. Major and minor strains from the numerical investigations for the uncoated and hybrid parts.

In the uncoated component, positive major strains and negative minor strains are predominant. This is also observed for the hybrid parts. In the area of the coating positive major and minor strains occur. The major strains are higher at the inner and outer diameter of the coating, especially when using the small coating, and therefore more critical regarding failure. With a bigger diameter of the coating, the load on the coating is more in the area of plane strain with lower major strains. Placing the coating in the plane strain area is therefore more advantageous in terms of formability. Overall, the position of the coating can be used to influence the stress states in the bottom area of the hybrid part. The tensile stresses at the bottom area are reduced through the application of the additive coating. A bigger diameter of the coating leads to reduced stress peaks around the coating, which is due to the plane strain stress condition.

Summary

One approach to meet climate challenges is the combination of different production technologies in order to produce components more resource efficient. An example for this combination is the investigated process chain. In order to evaluate the influence of the additively applied coating on the forming process, a numerical model was built and validated by comparison with experimental

Material Forming - ESAFORM 2023   Materials Research Forum LLC
Materials Research Proceedings 28 (2023) 919-928   https://doi.org/10.21741/9781644902479-100

results. A comparison of the resulting maximum process forces from simulation and experiment showed good agreement. There were significant differences in the distribution of the sheet thickness, which can be explained by the modeling approach of the sheet using shell elements. Since the stress state depends mainly on the process force, the model was nevertheless qualified for these investigations. The use of the coating led to reduced tensile stresses in the bottom area of the hybrid component. An enlargement of the diameter of the coating helped to reduce the stress peaks around the coating. A bigger diameter of the coating also led to lower major strains in the area of plane strain. In further investigations, the modeling approach of the sheet with solid elements should be examined. It would also be of interest to model the heat affected area between the coating and the substrate in order to represent the real process more accurately and to better evaluate the influence of the coating.

Acknowledgement

The authors would like to thank the German Federal Ministry for Economic Affairs and Energy for funding the joint project HyConnect (FKZ: 03LB3010*), within the framework of which the present investigations were carried out. Furthermore, the authors would like to thank the Deutsche Edelstahlwerke Specialty Steel GmbH & Co. KG for providing the powder material and the Schaeffler AG for providing the substrate material for the tests.

References

[1] European Commission, A European Green Deal, 2022. https://ec.europa.eu/info/strategy/priorities-2019-2024/european-green-deal_en

[2] A.E. Tekkaya, N.B. Khalifa, G. Grzancic, R. Hölker, Forming of Lightweight Metal Components: Need for New Technologies, Procedia Eng. 81 (2014) 28-37. https://doi.org/10.1016/j.proeng.2014.09.125

[3] F. Klocke, Manufacturing Processes 4, Springer Berlin Heidelberg, Berlin, Heidelberg, 2013.

[4] I. Gibson, D. Rosen, B. Stucker, Additive Manufacturing Technologies, Springer New York, New York, NY, 2015.

[5] M. Merklein, D. Junker, A. Schaub, F. Neubauer, Hybrid Additive Manufacturing Technologies – An Analysis Regarding Potentials and Applications, Phys. Procedia 83 (2016) 549-559. https://doi.org/10.1016/j.phpro.2016.08.057

[6] M. Bambach, A. Sviridov, A. Weisheit, J. Schleifenbaum, Case Studies on Local Reinforcement of Sheet Metal Components by Laser Additive Manufacturing, Metals 7 (2017) 113. https://doi.org/10.3390/met7040113

[7] DIN EN ISO 6892-1:2020-06, Metallic materials - Tensile testing - Part 1: Method of test at room temperature, Beuth publishing, Berlin.

[8] C. Tome, G.R. Canova, U.F. Kocks, N. Christodoulou, J.J. Jonas, The relation between macroscopic and microscopic strain hardening in F.C.C. polycrystals, Acta Metall. 32 (1984) 1637-1653. https://doi.org/10.1016/0001-6160(84)90222-0

[9] J.E. Hockett, O.D. Sherby, Large strain deformation of polycrystalline metals at low homologous temperatures, J. Mech. Phys. Solid. 23 (1975) 87–98. https://doi.org/10.1016/0022-5096(75)90018-6

[10] M. Kreß, D. Bartels, M. Schmidt, M. Merklein, Material Characterization of Hybrid Components Manufactured by Laser-Based Directed Energy Deposition on Sheet Metal Substrates, KEM 926 (2022) 80--89. https://doi.org/10.4028/p-3vlx01

[11] J. Hafenecker, T. Papke, F. Huber, M. Schmidt, M. Merklein, Modelling of Hybrid Parts Made of Ti-6Al-4V Sheets and Additive Manufactured Structures, in: B.-A. Behrens, A. Brosius, W. Hintze, S. Ihlenfeldt, J.P. Wulfsberg (Eds.), Production at the leading edge of technology,

Material Forming - ESAFORM 2023   Materials Research Forum LLC
Materials Research Proceedings 28 (2023) 919-928   https://doi.org/10.21741/9781644902479-100

Springer Berlin Heidelberg, Berlin, Heidelberg, 2021, pp. 13-22. https://doi.org/10.1007/978-3-662-62138-7_2

Material Forming - ESAFORM 2023
Materials Research Proceedings 28 (2023) 929-936

Materials Research Forum LLC
https://doi.org/10.21741/9781644902479-101

Numerical and experimental investigation of a backward extrusion process for forming geared components from coil

LEICHT Miriam[1,a] *, HENNEBERG Johannes[1,b] and MERKLEIN Marion[1,c]

[1]Institute of Manufacturing Technology (LFT), Friedrich-Alexander-Universität Erlangen-Nürnberg, Egerlandstraße 13, 91058 Erlangen, Germany

[a]miriam.leicht@fau.de, [b]johannes.henneberg@fau.de, [c]marion.merklein@fau.de

Keywords: Sheet-Bulk Metal Forming, Extrusion, Cold Forming

Abstract. Stricter political regulations, increasing ecological awareness of society as well as the pursuit for higher performance of components motivate lightweight construction. Functional integration is one way to realize lightweight design, resulting in increased demands on component geometry. Sheet-bulk metal forming (SBMF) offers the potential to enable an economic and ecological production of functional components through short process chains. SBMF from coil also provides additional advantages regarding high output quantity and short cycle times. However, the industrially application of SBMF from coil is limited due to high tool load and coil-specific challenges like an anisotropic material flow, which negatively affects the part accuracy. In this study, a backward extrusion process from coil for forming functional components with gearing is investigated. Therefore, a numerical process model was built and validated based on experimental results. In order to generate a profound process understanding, a combined numerical-experimental approach was chosen for a fundamental process analysis. The influence of the semi-finished product geometry was investigated by forming rotationally symmetric, pre-cut blanks and coil material. The application of the different sheet geometries was compared based on component- and process-side target quantities. The results indicate an anisotropic material flow as a coil-specific challenge, which leads to a direction-dependent component forming.

Introduction

Climate change and the increasing exhaustion of resources are central challenges of the 21st century [1], which leads to an increasing awareness of sustainability and environmental protection among the population and in politics. In the industrial environment in particular, a trend towards reducing environmental pollution and using resources more efficiently is noticeable. Consequently, higher demands are being placed on efficiency and environmental compatibility of manufacturing technologies and products, which require the development of innovative approaches. One effective approach is the implementation of light-weight design in combination with increased functional integration [2]. However, conventional manufacturing processes are reaching their limits [3]. The innovative process class of sheet bulk forming (SBMF) combines the advantages of sheet metal and bulk forming by applying bulk forming operations to sheet metal semi-finished products [4]. SBMF enables the production of thin-walled components with integrated functional elements by means of a three-dimensional material flow [4]. The use of SBMF-processes is currently limited to the forming of pre-cut blanks as semi-finished products, since they offer flexibility in material supply [5]. However, due to the increasing demand for functionally integrated components, manufacturing processes with short cycle times and high output volumes are gaining in importance. With regard to high output rate, the forming from coil offers the possibility to avoid cycle-time limiting handling systems, because the transport of the components between forming stages is realized by coil [6]. Tajul et al. [7] identified an uncontrolled, anisotropic material flow and unequal forming components as challenges of SBMF from coil. Furthermore, Henneberg et al. [8] investigated the forming of rotationally symmetric pins and cups from coil by backward and

Material Forming - ESAFORM 2023   Materials Research Forum LLC
Materials Research Proceedings 28 (2023) 929-936   https://doi.org/10.21741/9781644902479-101

lateral extrusion, which provided a basic process understanding of SBMF from coil. Additionally, process-, tool- and workpiece-side measures were investigated to reduce the anisotropic material flow and improve the component accuracy, whereby the greatest potential was achieved for local friction adjustment through the use of a modified blankholder surface [8].The investigation of the transferability of these findings for forming of more complex, functional components with secondary forming elements, such as gearing, is currently a subject of research.

Objectives and Methodology

The objective of this investigation is to develop a fundamental process understanding for the forming of functionally integrated, geared SBMF-parts from coil. The industrial use of SBMF from coil is limited due to significant process challenges. Including the coil-specific, anisotropic material flow, which restricts the dimensional accuracy of the component, as well as high tool stress. Against this background, in this research work a backward extrusion process from the coil for the forming of functional components with external gearing is investigated. The special focus is on the evaluation of the manufacturability of accurately formed gearing elements by extrusion from coil. Therefore, the aim of this research is to obtain a profound process understanding and to analyze causes of the forming limits. A process analysis based on a combined numerical-experimental approach is carried out. Thus, a finite element (FE) model of the process is set up and validated by comparison with experimental target values. In order to investigate characteristics of the coil-specific material flow and its causes, the sheet geometry is varied within the process analysis by examining pre-cut blanks and coil material. For this purpose, geometrical and mechanical component properties as well as the process force are analyzed based on experimental and numerical results.

Process Layout

This chapter gives an overview of the process setup for the backward extrusion process as well as the workpiece target geometry. Subsequently, the material used and the lubricant are explained.

Process setup. In this study, a backward extrusion process from coil for the forming of externally geared components is investigated. With regard to a shortening of process chains, which is considered to be an aim of SBMF [9], the forming is carried out in a single punch stroke. In the reference process, a coil with an initial sheet thickness t_0 of 2.0 mm and a width of 30 mm serves as semi-finished product. During forming, the sheet thickness in the forming zone is reduced to a residual sheet thickness of 1 mm with a punch stroke of 1 mm, forming a blind cavity. The resulting workpiece geometry is a pin with abstracted external gearing, shown in Fig. 1a.

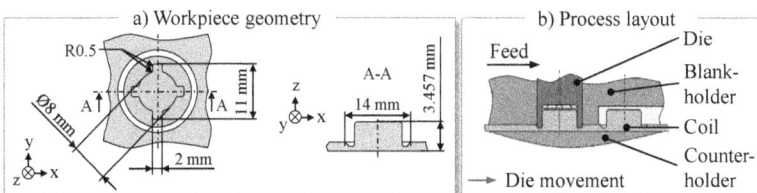

Fig. 1. (a)Workpiece geometry and (b) process layout.

The basic component, the pin, has a diameter of 8 mm and four gearing elements, which are positioned at 90° in and perpendicular to the feed direction. The width of the gearing elements is 2 mm with an outer diameter of 11 mm. The outer diameter of the workpiece has a value of 14 mm. Furthermore, the theoretical pin height with complete material flow into the cavity is 3.4569 mm. To control the material flow into the cavity and to reduce the tool stress, radii are applied at the edge of the blind cavity and at the edges of the pin and gearing. The geometry of the tool parts as well as the process kinematics were designed based on the workpiece geometry. The tool system, shown in Fig. 1b, consists of a die with an internal gearing, a blankholder and a counter-holder.

In the first process step, the blankholder moves in the negative z-direction and fixes the coil with a force of 130 kN by gas pressure springs. Additionally, the blank holder prevents the material flow out of the forming zone as well as the formation of wrinkles. In the next process step, the die moves downwards and the forming process is carried out with a punch stroke of 1 mm. After forming, the die removes, preventing the lifting of the coil by the blankholder. At the end of the process, the blankholder moves to its initial position and the coil is transported out of the forming zone by feed of 20 mm. In the experimental investigation, the active components are integrated into a tooling system, which is installed in the high-speed press BSTA 510-125B2 from Bruderer in order achieve high output quantities. This stroke-controlled press enables the production of 100 components per minute with a maximum nominal force of 510 kN.

Material and lubrication. The workpiece material used is the cold-rolled, mild deep-drawing steel 1.0338 as coil with a nominal sheet thickness of $t_0 = 2$ mm, which is an established material in SBMF [10]. The steel exhibits a purely ferritic microstructure. In addition, the material has a very good formability, which makes it suitable for the production of car body parts. In order to characterize the mechanical material properties, the flow behavior of the deep-drawing steel was experimentally determined in previous research work [11] in the layer compression test according to DIN 50106 [12] up to a true strain of $\varphi = 0.57$. Specimens with a diameter of 10 mm and a height of 14 mm were used in layer compression test, resulting in a compression ratio greater than one. To represent higher true strains, as occur in SBMF [11], the flow curve (Fig. 2) was extrapolated up to a true strain of $\varphi = 4.0$ using the Hockett-Sherby approach.

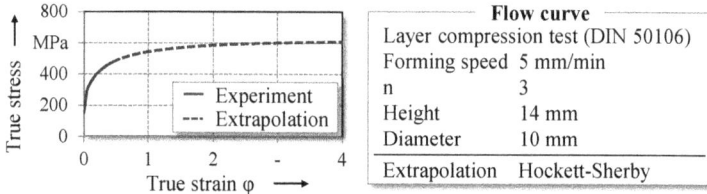

Fig. 2. Flow curve of the workpiece material 1.0338.

The lubricant Beruforge 150 DL from Carl Bechem GmbH, a water-based lubricant with a high solid wax content, was used in the experimental investigation. The suitability of this lubricant for usage in sheet-bulk metal forming has already been demonstrated in [13].

Numerical Process Model and Experimental Validation
In the following, the setup of the simulation model is described. Subsequently, the quality of the numerical model is evaluated by means of an experimental validation.

FE-Model setup. The simulation software Simufact.forming 14.0.1 of Simufact engineering GmbH is used for the numerical investigation of the backward extrusion process. The suitability of this software for SBMF extrusion processes has already been demonstrated in previous research work [11]. Within the analysis of the material flow, the tools are modeled as rigid. To reduce the computational time, the symmetry of the process along the X-axis is applied, resulting in the analysis of a 180° segmented model, as shown in Fig. 3. In addition, a constraint plane was defined in the negative X-direction to represent the influence of the coil.

Material Forming - ESAFORM 2023   Materials Research Forum LLC
Materials Research Proceedings 28 (2023) 929-936   https://doi.org/10.21741/9781644902479-101

Fig. 3. Setup of the numerical part model.

Two strokes are simulated to evaluate the influence of the preceding forming operation. The workpiece was meshed with hexahedral elements with a maximum element size of 0.5 mm according to the recommendations of Tekkaya [14]. This element type allows an accurate map of the three-dimensional material flow. In addition, a cylindrical refinement box (Fig. 3) with a diameter of 14 mm is applied in the forming zone, reducing the element size to 0.125 mm. The tool components are discretized with tetrahedral elements. The blankholder tension is applied by springs with a low stiffness of 10^{-6} N/mm^2 and a constant value of 37.8 MPa. Due to the high local contact pressures in SBMF [4], the friction factor model is used to represent the tribological conditions. A global friction factor of m = 0.1 is chosen, which was identified for the material 1.0338 in previous research [11]. The mechanical properties of the workpiece material are mapped by the determined flow curve (Fig. 2).

Experimental validation. For the evaluation of the quality of the virtual process model, the numerical model is validated according to the recommendations of Tekkaya [14]. This includes the comparison of process- and workpiece-related target values of the experiment with the numerical results. The validation is carried out for the reference process with a coil width of 30 mm and a feed rate of 20 mm, which corresponds to the distance between the centers of the components. In the experiment, six components are formed on one coil section. In order to take the influence of the preceding forming into account, only the last five components are considered for validation. First, the numerical process model is validated based on a comparison of workpiece-related target parameters. Fig. 4 represents a comparison of geometric component parameters.

For the validation, the diameter of the blind cavity is evaluated in X- and Y-direction (Fig. 4a). The determined experimental and numerical values show a good agreement with a deviation of 0.39 % in X-orientation and of 2.30 % in Y-direction. Furthermore, differences in X- and Y-direction can be identified in experiment and simulation, with a larger cavity diameter in Y-orientation, which indicates an anisotropic material flow. Consequently, the numerical model depicts the anisotropic material flow during coil forming. As a further component-side target quantities, the distance of the geared elements is analyzed, shown in Fig. 4b. In experiment and in simulation, this geometric target quantities also exhibits differences in the considered directions with higher values in Y-direction, which indicate an unequal material flow. The deviations between experiment and simulation are 1.2 % in X-direction and 1.0 % in the Y-orientation. In addition, numerically and experimentally an incomplete forming of the gearing is observed, as the theoretical distance of 11.0 mm is not achieved. Furthermore, the determined pin height is analyzed quantitatively (Fig. 4c). With a complete material flow into the cavity, a theoretical pin height of 3.46 mm is obtained. Experimentally (3.02 ± 0.02 mm) and simulatively (3.25 mm), however, the theoretical pin height is not achieved because of the partially uncontrolled material flow. Furthermore, a deviation of 7.1 % between simulation and experiment is determined caused by the complex tribological conditions in SBMF, which cannot be mapped in the simulation [5].

Material Forming - ESAFORM 2023
Materials Research Proceedings 28 (2023) 929-936

Materials Research Forum LLC
https://doi.org/10.21741/9781644902479-101

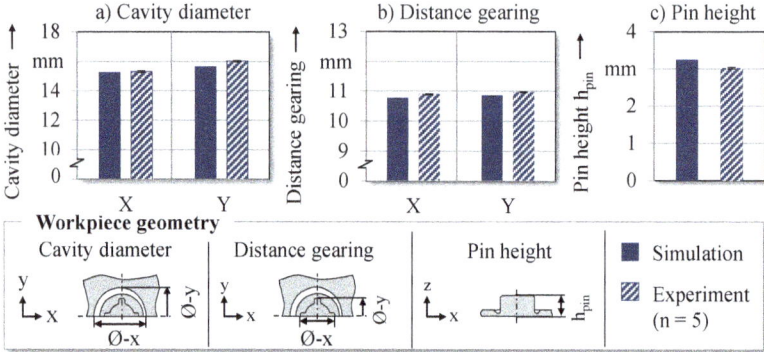

Fig. 4. Validation of the numerical model based on the workpiece geometry: (a) Cavity diameter, (b) distance of the gearing and (c) pin height.

The experimental microhardness distribution is compared with the numerical determined true strain distribution to evaluate the accuracy of the mapping of internal component properties (Fig. 5a). In this regard, a high true strain is considered an indicator of increased strain hardening. The microhardness HV0.05 was determined on sections of the parts with the Fischerscope® HM2000 from Helmut Fischer GmbH according to ISO 14577-1 [15].

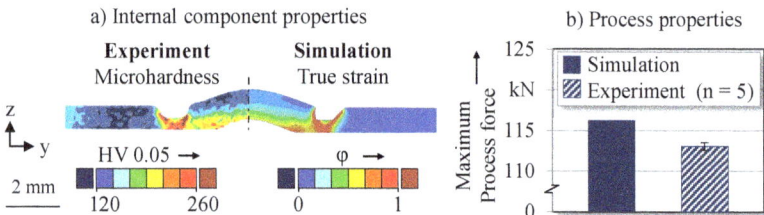

Fig. 5. Comparison of distribution of microhardness and true strain (a) and maximum process (b) force.

The comparison of the microhardness with the true strain distribution shows a high agreement. The maxima of the distributions are achieved in the contact area with the punch. In contrast, low values occur on the upper side of the pin geometry, as the material can flow without flow inhibition in this area. The maximum process force is evaluated as the process-side target variable, which is shown in Fig. 5b. The experimental process force is determined using a piezoelectric load cell 9106A from Kistler AG. In the experiment, a maximum force of 113.08 ± 0.46 kN is required. The maximum numerical force is 116.26 kN. Therefore, the deviation between experiment and simulation is of 2.81 %. In summary, after comparing the part and process-related target values, the numerical process model built up is suitable for realistically representing the forming process.

Material Forming - ESAFORM 2023
Materials Research Proceedings 28 (2023) 929-936

Materials Research Forum LLC
https://doi.org/10.21741/9781644902479-101

Process Analysis

The challenge of SBMF from coil is the anisotropic material flow, as demonstrated in the validation by differences in the process results in the X- and Y-direction. The material flow represents a central difference to the SBMF of pre-cut blanks [8], whereby the sheet geometry is considered as a cause for the occurring anisotropic material flow. In order to generate a profound process understanding of coil-specific challenges as well as extrusion of geared components from coil, the influence of the semi-finished product geometry is investigated in the following. A comparison of forming from coil through single stroke with forming from pre-cut blanks is presented. The selected diameter of the blanks corresponds to the coil width of 30 mm. To analyze the material flow, the numerical radial displacement is depicted in Fig. 6, which provides the displacement of the material elements with respect to their initial position. The numerical results were verified by comparison with the experimental component geometry. A negative radial displacement indicates a displacement in the direction of the component center.

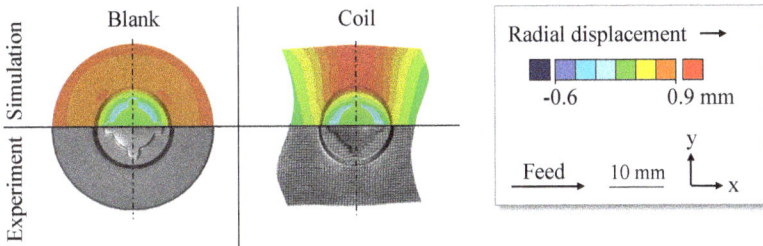

Fig. 6. Influence of the sheet geometry on the radial displacement.

The radial displacement of the circular blank shows a cyclic symmetrical distribution in 90° [5], with an increase in radial displacement outside the bulk forming zone occurring with increasing distance from the centre of the component. Due to the asymmetry of the pin geometry caused by the abstracted gearing, there is also an increased radial displacement locally outwards at an angle of ± 45° to the Y-axis. In contrast to the circular blank, the radial displacement of the coil geometry shows a symmetry with respect to the Y-axis with a high radial displacement perpendicular to the feed direction and a significantly reduced radial displacement in the feed direction. Consequently, an unequal material flow occurs. However, due to the geometry of the coil, the distance from the centre of the part to the outer edge of the coil varies depending on the direction. This leads to a larger contact area between coil and blankholder in X-direction, which reduces the material flow due to the higher friction in this area. In order to analyze the influence of the sheet geometry on the forming of secondary form elements, the form filling of the gearing elements is evaluated by the nominal distance of the abstracted external gearing to the component centre, which is shown in Fig. 7a. The parameter is considered relative to the ideal target values.

When forming the circular blanks, an equal and complete forming of the gearing elements in X- and Y-orientation is achieved with a nominal value of 1.000 ± 0.001. In contrast, the usage of coil results in a difference in nominal distances in the X- and Y-direction, with underfilling of the gearing occurring in both orientations. In Y-direction a value of 0.997 ± 0.001 is determined. In -X-direction a lower nominal distance of 0.993 ± 0.001 occurs. Consequently, the anisotropic material flow leads to an unequal forming and an incomplete form filling of the gearing elements.

Material Forming - ESAFORM 2023
Materials Research Forum LLC
Materials Research Proceedings 28 (2023) 929-936
https://doi.org/10.21741/9781644902479-101

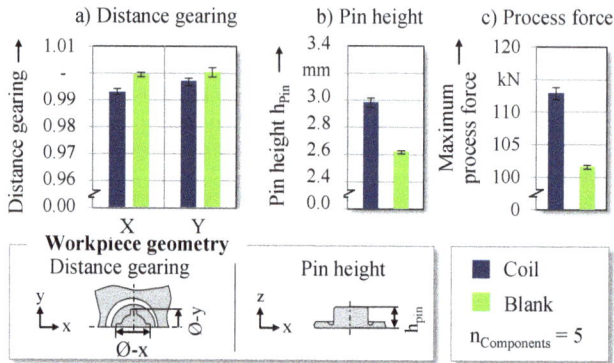

Fig. 7. Influence of the sheet geometry on the distance of the gearing (a), the pin height (b) and the maximum process force (c).

In Fig. 7b, the pin height is considered as further geometric target quantities, which is a parameter for the evaluation of the material utilization. The pin height is 2.62 ± 0.01 mm for the blanks and 2.98 ± 0.03 mm for a coil. Consequently, there is a higher material flow into the cavity when forming the coil. Due to the higher length of the coil in feed direction compared to the diameter of the blank, there is a higher contact between coil and tools as well as a higher material supporting effect, which inhibits the material flow to the outside. Thus, a higher volume of material flows into the die cavity resulting in a greater pin height. The influence of the sheet geometry on the tool stress is evaluated by the maximum process forces (Fig. 7c). When forming the blank, a maximum force of 101.55 ± 0.28 kN is required. For the forming of the coil as sheet geometry, a 10 % higher maximum force results with a value of 112.08 ± 0.84 kN. The increased process force during the forming of the coil is justified by the higher material flow into the bulk forming zone.

Summary

Within this research, a backward extrusion process from coil for the forming of functionally integrated components with geared secondary forming elements is investigated. The focus was on research SMBF- and coil-specific challenges like an anisotropic material flow. In the first step, the process was set up and numerically modeled. Subsequently, the numerical model was validated based on experimental results. In a next step, a combined numerical-experimental approach was used for a fundamental process analysis. The sheet geometry was varied using pre-cut blanks and coils in order to investigate the causes of the anisotropic material flow.

In contrast to the use of circular blanks, the forming of coils leads to an unequal, incomplete forming of the component geometry in and orthogonal to the feed direction. The reason is the non-circular sheet geometry, which causes a higher supporting effect parallel to the feed direction due to the additional material and the higher contact area between coil and tools. Consequently, an anisotropic material flow results. Furthermore, the greater contact between coil and blank holder as well as coil and counter-holder leads to a greater flow of material into the bulk forming zone, resulting in a higher pin height. Therefore, a higher maximum process force is required for the forming of the coil. Despite the higher material flow into the bulk forming zone, the gearing elements were not completely formed, which can be explained by the higher material flow velocity.

In future research work, the influence of the operating mode should be investigated by comparing single stroke and multi-stroke to analyze the influence of previous forming operations.

Material Forming - ESAFORM 2023   Materials Research Forum LLC
Materials Research Proceedings 28 (2023) 929-936   https://doi.org/10.21741/9781644902479-101

Furthermore, the investigation of measures for the target control of the material flow represents major research potential to prevent anisotropic material flow and improve part accuracy.

Acknowledgements

The authors would like to thank the German Research Foundation (DFG), which founded this work within the project DFG ME 2043 99-1 'Fundamental investigations on the production of functional components from coil'. The authors would also like to show their gratitude to Mr. Michael Rindl, who supported the execution of this work within the scope of his project thesis.

References

[1] A. Peterburs, Herausforderungen an die dt. Industrie im 21. Jahrhundert, GRIN Verlag, 2012.

[2] A.E, Tekkaya, N.B. Khalifa, G. Grzancic, R. Hölker, Forming of Lightweight Metal Components: Need for New Technologies, Procedia Eng. 81 (2014) 28–37. https://doi.org/10.1016/j.proeng.2014.09.125

[3] M. Merklein, H. Hagenah, T. Schneider, Blechmassivumformung – Stand der Technik und Ausblick, in: M. Merklein (Eds.), Tagungsband zum 2. Workshop, Blechmassivumform-ung, Meisenbach Verlag, 2013, pp. 1–10.

[4] M. Merklein, J. M. Allwood, B.-A. Behrens, A. Brosius, H. Hagenah, K. Kuzman, K. Mori, A. E. Tekkaya, A. Weckenmann, Bulk forming of sheet metal, CIRP Annals 61 (2012) 725–745. https://doi.org/10.1016/j.cirp.2012.05.007

[5] F. Pilz, J. Henneberg, M. Merklein, Extension of the forming limits of extrusion processes in sheet-bulk metal forming for production of minute functional elements, Manuf. Rev. 7 (2020) 9. https://doi.org/10.1051/mfreview/2020003

[6] A. Birkert, S. Haage, M. Straub, Umformtechnische Herstellung komplexer Karosserieteile - Auslegung von Ziehanlagen, Springer, Berlin Heidelberg, 2013. https://doi.org/10.1007/978-3-642-34670-5

[7] L. Tajul, T. Maeno, K. Mori, Successive Forging of Long Plate Having Inclined Cross-section, Procedia Eng. 81 (2014) 2361–2366. https://doi.org/10.1016/j.proeng.2014.10.334

[8] J. Henneberg, Blechmassivumformung von Funktionsbauteilen aus Bandmaterial, PhD Thesis, FAU, 2022. https://doi.org/10.25593/978-3-96147-580-3

[9] K. Mori, T. Nakano, State-of-the-art of plate forging in Japan, Prod. Eng. Res. Devel. 10 (2016) 81–91. https://doi.org/10.1007/s11740-015-0648-1

[10] D. Gröbel, R. Schulte, P. Hildenbrand, M. Lechner, U. Engel, P. Sieczkarek, S. Wernicke, S. Gies, A.E. Tekkaya, B.A. Behrens, S. Hübner, M. Vucetic, S. Koch, M. Merklein, Manufacturing of functional elements by sheet-bulk metal forming processes, Prod. Eng. Res. Devel. 10 (2016) 63–80. https://doi.org/10.1007/s11740-016-0662-y

[11] D. Gröbel, Herstellung von Nebenformelementen unterschiedlicher Geometrie an Blechen mittels Fließpressverfahren der Blechmassivumformung, PhD Thesis, FAU, 2018. https://doi.org/10.25593/978-3-96147-169-0

[12] DIN Deutsches Institut für Normung e.V., DIN 50106 - Prüfung metallischer Werkstoffe - Druckversuch bei Raumtemperatur, Beuth-Verlag, Berlin, 2016.

[13] H. Vierzigmann, M. Merklein, U. Engel, Friction Conditions in Sheet-Bulk Metal Forming, Procedia Engineering 19 (2011) 377–382. https://doi.org/10.1016/j.proeng.2011.11.128

[14] A.E. Tekkaya, A guide for validation of FE-simulations in bulk metal forming, Arab. J. Sci. Engi. 30 (2005) 113–136.

[15] DIN Deutsches Institut für Normung e.V., DIN EN ISO 14577-1:2015-11: Metallische Werkstoffe - Instrumentierte Eindringprüfung zur Bestimmung der Härte und anderer Werkstoffparameter - Teil 1: Prüfverfahren, Beuth-Verlag, Berlin, 2015.

Material Forming - ESAFORM 2023   Materials Research Forum LLC
Materials Research Proceedings 28 (2023) 937-942   https://doi.org/10.21741/9781644902479-102

Hot stamping of ultra-thin stainless steel for microchannels

GUO Nan[1,a], ZHANG Xianglu[1,b], HOU Zeran[1,c], YANG Daijun[2,d],
MING Pingwen[2,e] and MIN Junying[1,f*]

[1]School of Mechanical Engineering, Tongji University, Shanghai 201804, China

[2]School of Automotive Studies, Tongji University, Shanghai 201804, China

[a]13guonan@tongji.edu.cn, [b]Xianglu_zhang@tongji.edu.cn, [c]zeranhou@tongji.edu.cn,
[d]yangdaijun@tongji.edu.cn, [e]pwming@tongji.edu.cn, [f]junying.min@tongji.edu.cn

Keywords: Stainless Steel, Hot Stamping, Microchannel, Dimensional Accuracy

Abstract. Bipolar plate is one of the core components of a proton exchange membrane fuel cell (PEMFC), in which microchannels with regular distribution separate and distribute the fuel gas at the anode and oxygen/air at the cathode, and remove the reaction products from the cell. Dimensional deviations of microchannels affect assembly accuracy, thereby influencing the efficiency and performance of PEMFC. Ultra-thin stainless steel sheet is the most commonly used material for bipolar plate and stamping is an efficient way to form stainless steel microchannels. However, a challenge faced by the stamping process is how to improve the dimensional accuracy of stainless steel microchannels. Hereby we propose a hot stamping process of ultra-thin stainless steel sheet, which is of high potential to improve the dimensional accuracy of micro-channels. Uniaxial tensile tests are performed at room temperature (RT), 300, 600, and 900°C for an ultra-thin stainless steel 316L (SS316L). Results show that the strength of SS316L at 900 °C decreases significantly compared with that at RT, while the elongation is approximately 44%. Hot stamping process for stainless steel microchannels is developed, in which the ultra-thin sheet is heated by resistance heating. Stainless steel microchannels are hot stamped at 900°C, and the 3D profile and cross-sectional thickness distribution of which are measured. The measurement results show that the dimensional deviations of hot-stamped microchannels are lower than that of cold stamping, in terms of channel depth, rib width, and wall angle. Furthermore, the cross-sectional thickness distribution of the hot-stamped micro-channels has a similar trend as that of the cold stamping, and the thickness at the fillet is not significantly different (avg. + 1 μm) from that of the cold stamping.

Introduction

Fuel cell has a high-energy conversion efficiency of more than 40-50% [1], which is regarded as an ideal technology to utilize hydrogen energy. As the key component of a PEMFC, bipolar plates (BPPs) with fine microchannel features play the role of separating and distributing hydrogen and oxygen/air, providing mechanical support to the fuel cell stack, collecting, and conducting current, etc. Dimensional deviations of microchannel features affect assembly accuracy, thereby influencing the efficiency and performance of PEMFC. With the advantage of high electrical conductivity and mechanical strength as well as low cost, stainless steel is widely utilized to fabricate BPPs, and the stamping process is one of the most used methods to manufacture stainless steel BPPs due to its relatively high productivity and low cost. However, dimensional errors are unavoidable on stainless steel BPPs due to the characteristics of residual stress and springback in stamping process, which contributes to uneven pressure distribution, larger contact resistance, and higher GDL porosity [2]. It is reported that fabricating BPPs with fine flow microchannel features, i.e., flow channels with a channel width, a rib width, and a pitch length of 0.5, 0.5, and 1.5 mm, respectively, is a challenge for the current stamping process [3]. Thus, several advanced forming

Material Forming - ESAFORM 2023   Materials Research Forum LLC
Materials Research Proceedings 28 (2023) 937-942   https://doi.org/10.21741/9781644902479-102

processes have been developed in the literature to increase the dimensional accuracy and the ultimate flow channel features of BPPs. Based on ultra-thin ferritic stainless steel sheets with thickness of 100 and 75 μm, Bong et al. [4] conducted a two-step stamping process to fabricate microchannels, they found that the two-step forming process evidently increase the forming depth and precision of ferritic stainless steel microchannels. Xu et al. [5] revealed that the depth limit of titanium microchannels is increased from 438.1 μm in single-step forming process to 621.1 μm in three-step forming process. Since increasing forming temperature can reduce the strength of materials, thus reducing residual stress and springback and improving the forming accuracy of parts, some heat-assisted forming processes have been proposed for microchannels. By conducting warm stamping process at the temperature up to 300°C, Lakshmi [6] improved the formability of ferritic stainless steel microchannels. Moreover, they found that increasing forming temperature can reduce the springback of microchannels.

In this work, uniaxial tensile tests were performed at room temperature (RT), 300, 600, and 900°C for an ultra-thin stainless steel SS316L. The variation of strength and elongation of ultra-thin SS316L to temperature was discussed to evaluate the properly forming temperature of ultra-thin SS316L. Then a lab-scale hot stamping platform with an on-site resistance heating device was built to verify the hot stamping process of ultra-thin SS316L microchannels. Channel depth, rib width, and wall angle were measured to characterize the dimensional accuracy of ultra-thin SS316L microchannels under hot stamping process.

Experiments

Uniaxial Tensile Test. An ultra-thin austenitic stainless steel SS316L with a thickness of 100 μm was used in this study. Uniaxial tensile tests were performed to obtain the mechanical properties of ultra-thin SS316L at elevated temperatures. Fig. 1a shows the high-temperature uniaxial tensile testing system with a muffle furnace and digital image correlation technic (DIC). The DIC cameras equipped with blue light bandpass filters record the deformation of the specimen through a glass window on the furnace to calculate strain fields. Additionally, two blue LED lights with a wavelength of 450 nm were used to illuminate the specimen from different directions to ensure sufficient image contrast. The geometric dimension of the uniaxial tensile specimen is shown in Fig. 1b according to the ISO 6892-2 standard. All specimens were cut off along the rolling direction by electrical discharge machining to ensure high edge quality. The test temperature of SS316L includes room temperature (RT), 300, 600, and 900°C, and three samples were tested at each temperature to ensure reproducibility.

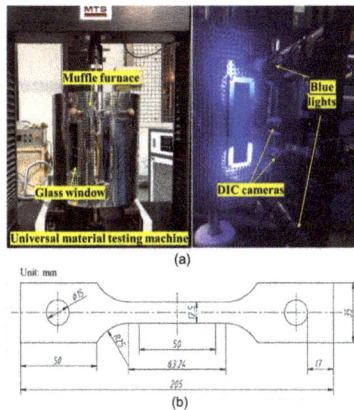

Fig. 1. (a) The high-temperature uniaxial tensile testing system and (b) the geometric dimension of the uniaxial tensile specimen.

Microchannel Hot Stamping Test. The microchannel structure is shown in Fig. 2a, and the cross-sectional geometry of the stamping tool is shown in Fig. 2b, where the dimensions of the channel width, rib width, channel depth, wall angle, and corner radius are also depicted. Fig. 2c illustrates a lab-scale hot stamping platform built in this study, where a pair of electrode clamps is embedded into the system to heat the ultra-thin SS316L sheet by resistance heating. An automatic

Material Forming - ESAFORM 2023

Materials Research Proceedings 28 (2023) 937-942

Materials Research Forum LLC

https://doi.org/10.21741/9781644902479-102

tensioning device composed of compression springs and linear guides is assembled to the clamp to prevent ultra-thin SS316L sheet buckling caused by constrained thermal expansion. During the hot stamping process, the initial distance between the upper and lower tool was 150 mm. The ultra-thin SS316L sheet started to be heated by resistance heating with a power of 22.5 kW at 0 s, and the upper tool started to move downward with a velocity of 140 mm/s at 5 s. The ultra-thin SS316L sheet was heated to the target temperature at 6 s, and meanwhile, the upper tool was down to a distance of 10 mm from the lower tool. Then the power was off and the upper and lower tools were closed at 6.1 s. Then the formed microchannel sample was cooled in the tool from 6.1 s to 7 s. Moreover, a solid lubricant boron nitride (BN) was applied to reduce the friction coefficient between the ultra-thin SS316L sheet and tool.

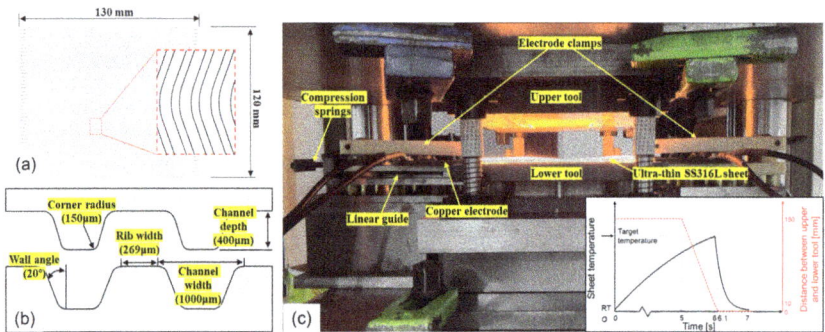

Fig. 2. (a) The designed microchannel structure, (b) the cross-sectional geometry of the stamping tool, and (c) an illustration of the microchannel hot stamping process.

Geometrical and Thickness Measurements. The surface topographies of formed microchannels were measured by a non-contact 3D optical profilometer KEYENCE VR-5000, through which channel width, rib width, and wall angle of the formed microchannels were extracted. Cross-sectional samples were cut along the direction perpendicular to the channels by laser cutting and mounted with resin at room temperature. Then the samples were polished with 800, 1500, and 2500 Grit sandpapers in sequence and observed with an optical microscope to evaluate the thickness of the microchannels.

Results and Discussion

The engineering stress-strain curves of ultra-thin SS316L at different temperatures are shown in Fig. 3a. In the temperature range from RT to 600°C, although the flow stress decreases with the increase of temperature, it can be seen from the stress-strain curves that strain hardening is the dominant. At 900°C, the stress-strain curve of ultra-thin SS316L is more significantly affected by temperature softening, and the strength is reduced to 104 MPa. Fig. 3b represents the variation of ultimate tensile strength and elongation of ultra-thin SS316L with temperature. The ultimate tensile strength (UTS) of ultra-thin SS316L decreases from 620 MPa at room temperature to 104 MPa at 900°C; however, the elongation decreased significantly from room temperature to 600°C, with a minimum value of 26% at 600°C, and with the temperature increasing up to 900°C, the elongation increased to 44%. This indicates that increasing the forming temperature can reduce the strength of the ultra-thin SS316L, thus potentially reduce residual stress and springback to improve the forming accuracy of parts. However, a temperature range, namely 300–600°C, should

Material Forming - ESAFORM 2023
Materials Research Proceedings 28 (2023) 937-942

Materials Research Forum LLC
https://doi.org/10.21741/9781644902479-102

be avoided in developing warm/hot stamping process for ultra-thin SS316L, because the elongation of which is much lower than that at room temperature.

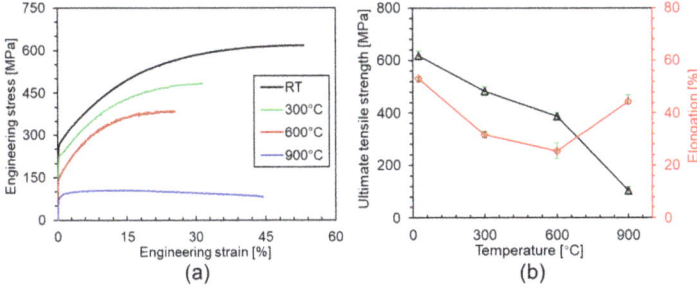

Fig. 3. (a) Engineering stress-strain curves of ultra-thin SS316L at different temperatures and (b) variation of UTS and elongation of ultra-thin SS316L with temperature.

The microchannel features of BPP have significant effects on the distribution of reaction gas, heat and water management, mechanical stability, and consequently on the performance, reliability, and durability of fuel cells [7]. The rib width (a) determines the contact area between the BPP and the MEA, a narrower rib width means a smaller contact area, and a larger contact resistance between the BPP and MEA as well as reduced PEMFC performance. The channel depth (c) and wall angle (β) mainly affect the mass transfer towards MEA and production water removal ability, and a deeper channel depth and a smaller wall angle enhance electrochemical reaction rate and fuel efficiency. Fig. 4 shows a microchannel sample hot-stamped at 900°C, where measurement of microchannel features was conducted within the red dashed wireframe.

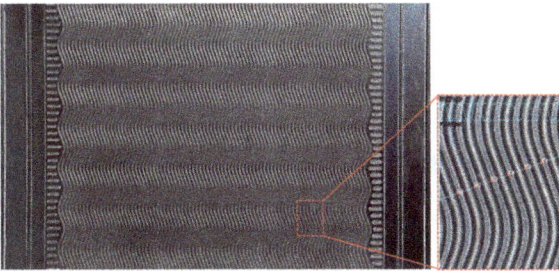

Fig. 4. Hot-stamped microchannel sample.

Measurement results of the microchannel features formed via hot stamping and cold stamping are shown in Fig. 5a. Average c of the hot-stamped microchannel is 394.2 μm, which is closer to the design value (400 μm) compared to that of the cold-stamped microchannel. For a and β, the values of hot-stamped sample are smaller than those of cold-stamped sample, which benefits from the reduced springback under hot stamping process. The dimensional deviations of c, a, and β of the cold-stamped microchannel are larger than that of hot-stamped microchannel. The large dimensional deviations are caused by highly localized stamping force [8]. This indicates that the hot stamping process improves the dimensional accuracy of ultra-thin SS316L microchannels. Fig. 5b represents the thickness distribution of hot-stamped and cold-stamped microchannels, which have a similar trend, i.e., positions 1, 5, and 9 exhibit a larger thickness, and the maximum thinning appears at the fillet where the sheet metal contacts the lower tool (position 2). There is no significant difference between the minimum thicknesses of the hot-stamped and cold-stamped microchannels (average difference at position 2 is 1 μm).

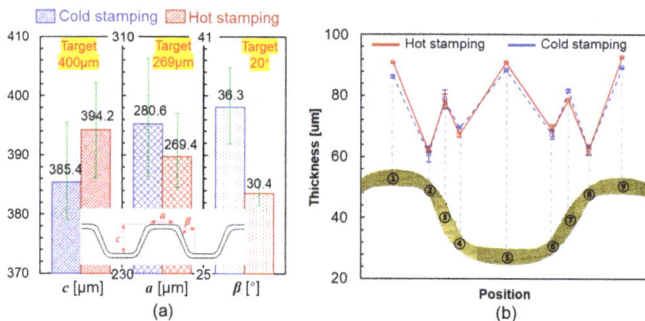

Fig. 5. (a) Comparison of channel depth (c), rib width (a), and wall angle (β) between hot stamping and cold stamping microchannels and (b) cross-sectional thickness distribution of microchannels via hot stamping and cold stamping.

Summary

In this work, the variation of strength and elongation of ultra-thin SS316L at elevated temperatures were investigated by applying high-temperature uniaxial tensile tests. In order to verify the hot stamping process of ultra-thin SS316L microchannels, a lab-scale hot stamping platform with an on-site resistance heating device was built. The dimensional accuracy of hot-stamped ultra-thin SS316L microchannels was characterized, and conclusions can be drawn here:

1) 900°C might be the appropriate hot stamping temperature for ultra-thin SS316L microchannels because of the relative low strength of 104 MPa and an elongation of about 44% of ultra-thin SS316L at 900°C;

2) Compared to cold stamping, hot stamping at 900°C increase the dimensional accuracy of ultra-thin SS316L microchannels in terms of channel depth (from 385.4 μm to 394.2 μm with a target of 400 μm), rib width (from 280.6 μm to 269.4 μm with a target of 269 μm), and wall angle (from 36.3° to 30.4° with a target of 20°).

Acknowledgments

The authors would like to acknowledge the financial support for this research provided through the National Key Research and Development Program of China (No. 2020YFB1505900).

References

[1] K. Sopian, W.R.W. Daud, Challenges and future developments in proton exchange membrane fuel cells, Renewable energy 31(2006) 719-727. https://doi.org/10.1016/j.renene.2005.09.003

[2] L. Peng, Y. Wan, D. Qiu, P. Yi, X. Lai, Dimensional tolerance analysis of proton exchange membrane fuel cells with metallic bipolar plates, J. Power Source. 481 (2021). https://dx.doi.org/10.1016/j.jpowsour.2020.228927

[3] Y. Leng, P. Ming, D. Yang, C. Zhang, Stainless steel bipolar plates for proton exchange membrane fuel cells: Materials, flow channel design and forming processes, J. Power Source. 451 (2020). https://dx.doi.org/10.1016/j.jpowsour.2020.227783

[4] H.J. Bong, J. Lee, J.H. Kim, F. Barlat, M.G Lee, Two-stage forming approach for manufacturing ferritic stainless steel bipolar plates in PEM fuel cell: Experiments and numerical simulations, Int. J. Hydrogen Energ. 42 (2017) 6965-6977. https://dx.doi.org/10.1016/j.ijhydene.2016.12.094

[5] Z. Xu, Z. Li, R. Zhang, T. Jiang, L. Peng, Fabrication of micro channels for titanium PEMFC bipolar plates by multistage forming process, Int. J. Hydrogen Energ. 46 (2021) 11092-110103. https://dx.doi.org/10.1016/j.ijhydene.2020.07.230

[6] R.N. Lakshmi, Forming of ferritic stainless steel bipolar plates, Electronic Theses and Dissertations, 205 (2012). https://scholar.uwindsor.ca/etd/205

[7] J. Wang, Theory and practice of flow field designs for fuel cell scaling-up: A critical review, Appl. Energ. 157 (2015) 640-663. https://doi.org/10.1016/j.apenergy.2015.01.032

[8] C. Turan, Ö.N. Cora, M. Koç, Effect of manufacturing processes on contact resistance characteristics of metallic bipolar plates in PEM fuel cells, Int. J. Hydrogen Energ. 36 (2011) 12370-12380. https://doi.org/10.1016/j.ijhydene.2011.06.091

Material Forming - ESAFORM 2023

Materials Research Proceedings 28 (2023) 943-950

Materials Research Forum LLC

https://doi.org/10.21741/9781644902479-103

Approaches for load path design for stretch forming based on part surface geometry

REITMAIER Lisa-Marie[1,a] *, BAILLY David[1,b] and HIRT Gerhard[1,c]

[1]Institute of Metal Forming, RWTH Aachen University Intzestraße 10, 52072 Aachen, Germany

[a]lisa-marie.reitmaier@ibf.rwth-aachen.de, [b]david.bailly@ibf.rwth-aachen.de, [c]gerhard.hirt@ibf.rwth-aachen.de

Keywords: Stretch Forming, Sheet Metal, Loading Trajectory, Load Path, CAE Methods

Abstract. Stretch forming enables manufacturing large, slightly curved sheet metal parts. If the curvature is multi-directional and complex, parts often cannot be stretch-formed successfully, due to local or inhomogeneous straining that leads to early material failure, buckling or springback. To control strain distribution, the flexibility of load application needs to be improved. This can be achieved through loading at discrete points based on defined cross-sections of part geometries. However this can be applied only by multi grippers, requiring complex plant technology and control. New parameterized approaches for load path generation based on the part surface geometry for single rigid grippers have been developed to improve the flexibility of load paths. The approaches can be mainly distinguished between Surface Transformation method (ST), which transforms a surface into a curve as the base for tangential path creation and the Tool Path Fitting method (TPF). In TPF, paths are created for several cross-sections, similar to the application for multi grippers, and then compiled into one path. In this paper, the load path approaches have been applied and numerically investigated for parts with positively curved translation surfaces. Results of FE simulation of the stretch forming process show the applicability, potential and limitations of the load path design approaches.

Introduction

Stretch forming is an established industrial process for forming automobile body panels such as doors or roofs and aircraft parts [1]. In general, the process is suitable to manufacture large and slightly curved components for various applications. The stretch forming process typically involves a rectangular sheet metal blank clamped at either two opposite sides, less often at all four sides, formed by a relative movement between the grippers and the die. Within the forming process, the grippers either move in only vertical direction (conventional stretch forming) or in both vertical and horizontal directions simultaneously (tangential stretch forming) so that the load is acting tangentially to the die contour. In general, components produced via tangential stretch forming achieve higher geometrical accuracy [1].

A tangential load path is generated based on a representative curve of the part geometry. In the case of a unidimensional-curved component, the curve equals the cross-sectional part contour. For bi-dimensional and more complex curved components, the cross-section varies in transverse direction. In this case, the creation of stretch forming load path is mainly experience based. To apply loading with respect to the overall part surface and to overcome existing process limits, loading at discrete points along transverse curvature according to the longitudinal cross-sectional profiles has been investigated [2,3]. The flexible loading with respect to the complete part geometry surface improves uniform deformation of the sheet metal and contributes to overcome existing process limits for complex shaped parts in stretch forming. To realize these load profiles, multi grippers with separate modules, complex machinery and process control are required. In order to waive such complex technology systems and work with existing rigid gripping modules, Schmitz et al. [4] used a conventional stretch forming operation and presented a method to

Material Forming - ESAFORM 2023
Materials Research Proceedings 28 (2023) 943-950

Materials Research Forum LLC
https://doi.org/10.21741/9781644902479-103

manipulate the strain distribution by selective weakening of the blank with specific hole pattern to homogenize the strain over the part surface. This method involves an optimization algorithm and requires an additional preceding step of laser cutting. In order to avoid complex machinery and additional process steps, alternative parameterized load path design approaches with respect to the overall part surface geometry have been investigated and are presented in this paper.

Approach and Procedure

In this study, different load path design approaches for rigid gripper application based on the surface geometry of bi-dimensional curved components are developed. The approaches can be mainly distinguished between Surface Transformation method (ST), which transforms a surface into a curve as the base for tangential path creation and the Tool Path Fitting method (TPF). In TPF, paths are created for several cross-sections, similar to the application for multi grippers, and then compiled into one path. The part surfaces considered in this work are translation surfaces with positive Gaussian curvature. A translation surface is defined by two curves and is obtained by moving a generator curve, which remains parallel to its initial position, along a guide curve. In this study, the part surfaces and the corresponding stretch forming load paths are created via a developed Python script. Within the scope of this work, the different load path approaches are presented for two different exemplary part geometries. The results are generated through numerical process simulations in LS-Dyna with help of validated stretch forming FE-Models. The criteria for analyzing the different load path approaches are the achieved part geometry and strain distribution.

Material and Methods

Material. The sheet material is 1.4404 austenitic stainless steel in 0.8 mm thickness. The material data is determined via tensile test with samples taken in rolling direction of the sheet. A yield strength of $R_{p0,2}$=279,197 MPa, ultimate tensile strength of R_m =620.043 MPa and ultimate strain A_g=47.697 % were identified.

FE-Model. In order to analyze and compare the forming results after applying different load paths, a FE-Model is established in LS-PrePost 4.3 and the simulations are performed with the explicit solver of LS-Dyna R.11.1.0. The sheet metal blank is 750 x 500 mm in size and discretized with Belytschko-Tsay shell elements with 4 mm. The load is applied on the edge nodes of the short opposite sides of the blank. The sheet is modelled as an elastoplastic deformable body whereas the die is modeled as a rigid body. The die geometry as well as the desired part geometry is limited to bi-dimensional curved translation surfaces with only positive Gaussian curvature. Two-sided stretch forming is conducted.

General approach for load paths determination. To create a conventional load path with only vertical movement, the highest and lowest point of a surface and the final tangential angle α need to be identified (Fig. 1, left). Pre-stretching or post-stretching can be added and involve horizontal movement. A tangential load path (Fig. 1, right) is based on either a cross-sectional curve of the part geometry, or a representative curve that is usually generated based on experience. The path represents an involute of this curve, which can be pictured as wrapping of the sheet metal on the die surface. The involute is defined by the curve and the edge or rather endpoint of the sheet metal. To ensure tensile loads throughout the process a stretching rate of 0.1 % per forming step is applied.

Material Forming - ESAFORM 2023   Materials Research Forum LLC
Materials Research Proceedings 28 (2023) 943-950   https://doi.org/10.21741/9781644902479-103

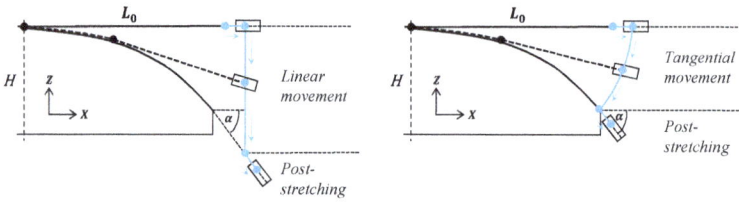

Fig. 1. Schematic drawing of load path determination for a conventional path (left) and a tangential path (right) with an optional pre- or/and post-stretching movement.

Approaches for Load Path Design

The presented approaches for load path design are based on the general approach for tangential stretch forming as stated in the previous section. Therefore, to generate a load path, a defined curve is needed. The Surface Transformation method (ST) transforms a surface into one representative curve as the base for path creation. For the Toolpath Fitting method (TPF), several load paths are created for several cross sections, comparable to the path creation for multi grippers. However, instead of moving multi grippers accordingly, the different paths are compiled into one load path for the rigid single gripper based on different criteria.

Surface Transformation method (ST). In order to transform a multi-curved surface into a single representative curve, all of the surface points are projected onto a plane (cf. Fig. 2). Then, by means of different criteria, a curve is derived from the point cloud. For the first approach, the curve is created based on the lowest points of the surface. Since the first contact point of the sheet metal is always the highest point of a part geometry, respectively the highest point of the lower die, the generated path needs to be corrected by extending the path at the process begin to the same height as the highest point of the part surface geometry.

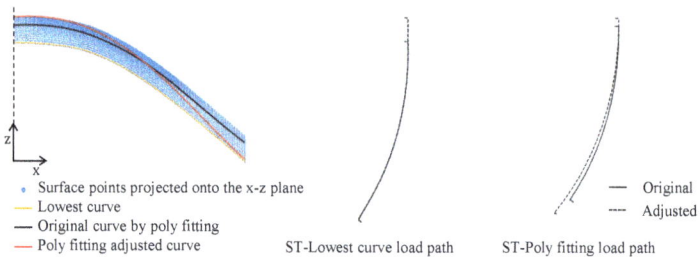

Fig. 2. Projection of coordinate points of a three-dimensional part surface geometry on x-z plane and created representative curves (left) and resulting load paths based on the representative curves (right) generated from two different ST-methods.

In the special case of translation surfaces with only positive Gaussian curvature, the determined lowest curve of the points equals the longitudinal edge curve of the surface. Another load path is based on a curve, which is derived by polynomial fitting through all surface points. The degree of polynomial was set to four. The curve is adapted accordingly in order to include the highest and lowest point of the surface and then the load path is determined based on this curve. Beside the

Material Forming - ESAFORM 2023   Materials Research Forum LLC
Materials Research Proceedings 28 (2023) 943-950   https://doi.org/10.21741/9781644902479-103

highest point, the lowest point needs to be included to be able to fully form the desired part geometry.

Coming to the next Surface Transformation method (Fig.3), the surface is cut in transverse direction (y-direction) by several planes. For every transverse cross-section, the points on the curve (black dotted curve) are fitted into a line (red dotted line). All the lines are projected to x-z plane resulting in one red point per cross-section curve. If a sufficient number of transverse cuts are performed, an averaged curve (connecting the red points in the right figure) can be determined. The highest and the lowest point are integrated in the curve as for the poly fitting method.

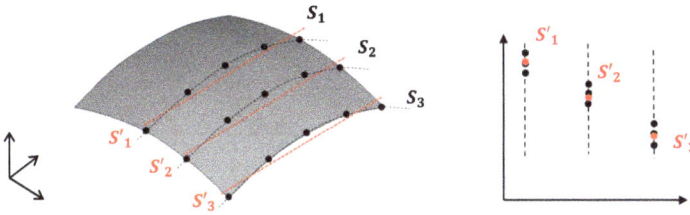

Fig. 3. Simplified sketch of the curve determination by transverse fitting type of Surface Transformation method.

Tool Path Fitting method (TPF). This method (Fig. 4) is based on the generation of load paths determined for each longitudinal cross-sectional curve (black dotted line) of the surface geometry. Within this work, the TPF-method is based on 20 cross-sections. Every load path has the same defined number of steps. So, subsequently for every step, one of the coordinate points of the load paths (J_j, R_i) is picked or fitted into one point. In the first approach, the position that causes the most stretching in the corresponding load step is picked (J_j). In the other approach, this is done by least squares method to find the mean point (T_j).

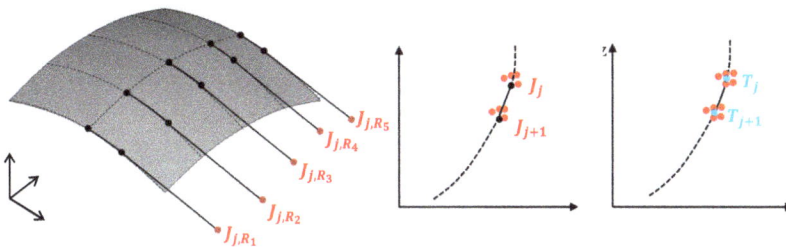

Fig. 4. Simplified sketch of the path determination by Tool Path Fitting method for the lowest point (left) and the mean point (right).

The developed approaches for load path design are summarized in Table 1. The results from the numerical study are presented in the next section.

Material Forming - ESAFORM 2023
Materials Research Proceedings 28 (2023) 943-950

Materials Research Forum LLC
https://doi.org/10.21741/9781644902479-103

Table 1. Summary overview of the developed approaches.

Surface Transformation method (ST)	Tool Path Fitting method (TPF)
Transforms a multi-curved surface into a single representative curve. Therefore, all surface points are projected onto a plane and fitted into a curve by means of different criteria.	Generation of a tangential load path for each specified longitudinal cross-sectional curve of a surface. The paths are compared for every load step and one positions is picked or fitted. The steps are then compiled into one load path.
- ST-Lowest Curve - ST-Poly Fitting - ST-Transverse Fitting	- TPF-Lowest Point - TPF-Mean Point

Results and Discussion

Application of load path design approaches. The presented load path design approaches have been successfully applied for several different translation surface geometries from a defined parameter field. This study focuses on two different exemplary parts shown in Fig. 5 (left), and the corresponding load paths generated via different load path design approaches, shown on the right.

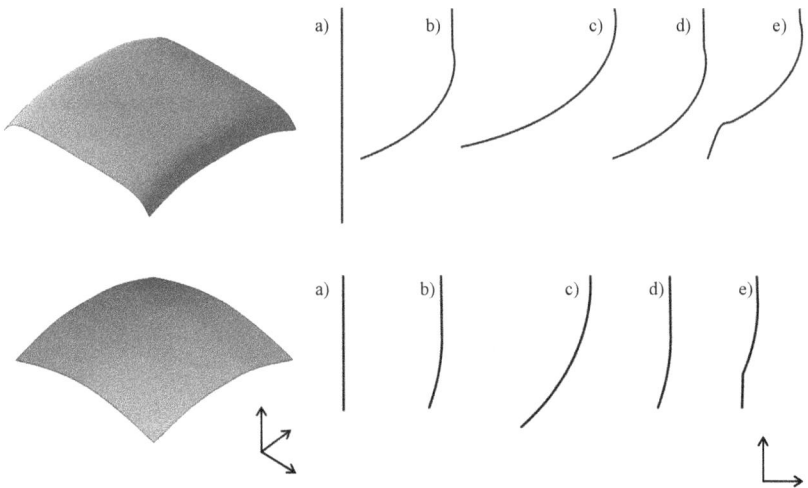

Fig. 5. Visualization of the investigated two different part geometries (left) and load paths a)-e) based on the part surface geometries (right).

Part surface geometry A is relatively flat with a height to length ratio of 0.2 (in both x- and y-direction) and a relatively steep final tangential angle α of 55° towards the edge in x-direction. For part surface geometry B, the height to length ratios in longitudinal and transverse direction are also 0.2 but the final tangential angle α is smaller with 20°. The load paths generated by the different approaches (Fig. 5) are as follows: a) Conventional, b) ST-Lowest Curve, c) ST-Transverse Fitting, d) TPF- Lowest Point, and e) TPF- Mean Point. A generation of a load path by ST-Poly Fitting was not successful, since the generated representative curves were not suitable for

Material Forming - ESAFORM 2023   Materials Research Forum LLC
Materials Research Proceedings 28 (2023) 943-950   https://doi.org/10.21741/9781644902479-103

path creation due to multiple change of curvature. All paths presented above are reduced to the motion without pre- or post-stretching. It can be seen that the conventional path is changing significantly with the tangential angle α or rather the curvature in longitudinal direction. The path with the most tangential movement is the path created by Surface Transformation method with Transverse Fitting (Fig. 5, c). Especially the paths based on the lowest curve of the surface geometry (ST-Lowest Curve (Fig. 5, b)) or the lowest tool position (TPF-Lowest Point (Fig. 5, d)) need a significant linear extension to meet the highest point of the surface geometry. To summarize, besides the ST-Poly Fitting approach, all presented approaches seem to lead to reasonable load paths. The applied load varies with the different load path approaches and therefore, influences the forming result as also discussed by [5].

FE-simulation results. FE-simulations were conducted for the different geometries A and B and the presented stretch forming load paths. Fig. 6 shows the resulting plastic strain distribution for geometries A and B after stretch forming with the above developed load paths.

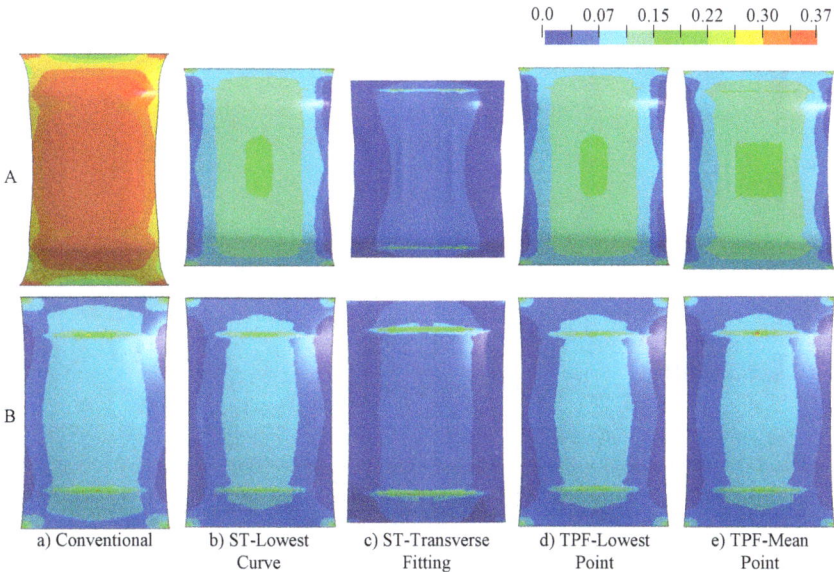

Fig. 6. Effective plastic strain at the final state of deformation under load for geometry A and B.

For the steep curvature of geometry A in longitudinal direction, and the resulting relatively long conventional stretch-forming path, the deformation is expected to cause material failure since the plastic strain of 0.37 exceeds the ultimate strain of the material. Therefore, this part geometry cannot be successfully deformed with the conventional load path.

For load paths b) and d), similar plastic strain distributions can be observed. The path created based on the lowest curve (b) and the choice of the lowest tool position, as in TPF-Lowest Point (d), lead to almost the same resulting path for positively curved translation surfaces. Deviations may be caused by different numbers of data points and consequently different approximations.

As stated before, the load paths based on the ST-Transverse Fitting approach include the most tangential movement. The tangential movement induces less tensile strain and reduces friction

Material Forming - ESAFORM 2023
Materials Research Proceedings 28 (2023) 943-950

Materials Research Forum LLC
https://doi.org/10.21741/9781644902479-103

significantly, since the relative movement between die and blank is lowered to a minimum. Therefore, the formed sheet metal blanks show the smallest overall strains for both geometries. So, local straining which leads to early material failure can be avoided. But especially, while forming bi-dimensionally curved components, buckling can occur due to insufficient stretching in the edge area of the sheet. Fig. 7 shows the cross-sections for the geometries A and B, deformed with the original and an adapted load path based on the ST-Transverse Fitting method.

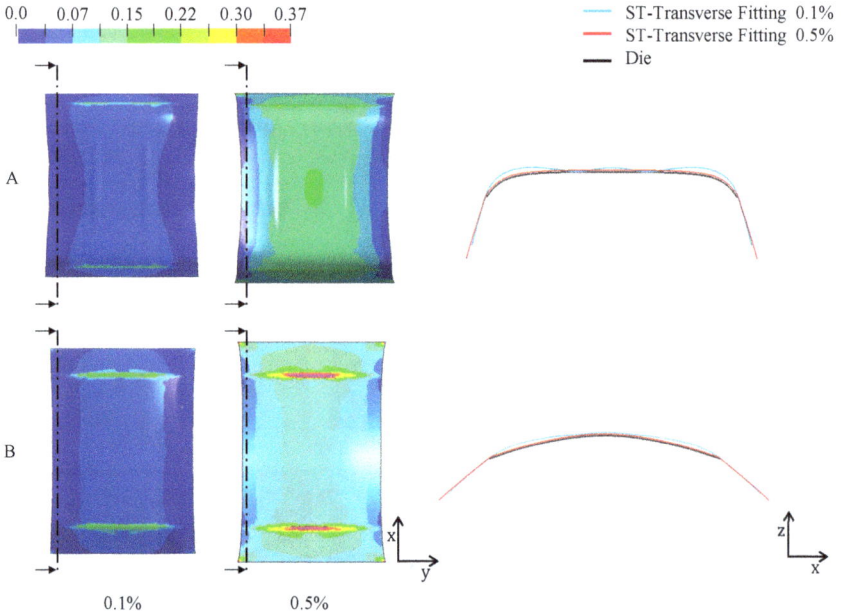

Fig. 7. Effective plastic strain distribution at the final state of deformation under load for geometry A and B for different in-process stretching rates (left) and resulting cross sections (right).

In blue, both cross-sections show deviations from the desired surface geometry denoted in black. Post-stretching is an established measure to apply tensile load at the end of the process. To show the adaptability and therefore flexibility of the load path design approaches, the stretching rate per step was increased from 0.1% to 0.5%. For both geometries, the adaption leads to achievement of the target geometry. The resulting strain distributions are shown in Fig. 7, left. Additional stretching increases the overall strain. The high strains for geometry B for the adapted path can be explained by the simplified die geometries, which are automatically created as shell elements in Python. This leads to element penetration and deformation in the edge area of the simplified die. This example shows that the parameterized path design approaches can be adapted effortlessly and reproducible results can be achieved. For geometry B both simulation results show buckles in the middle area of the blank. The buckles in longitudinal direction occur due to compressive stresses in transvers (y-direction) and cannot be inhibited by additional stretching. The results for a), b) and d) show no center buckles. This could be referred to increased tensile stretching and therefore increased friction between blank and die at the beginning of the process, which prevents the material from flowing in transverse direction. The individual adaption of in process stretching allow for the control of stress and strain distribution throughout the process in the deformed part and can therefore influence the forming result.

Summary

The presented approaches for load path design based on part geometry, despite the ST-Poly Fitting approach, are shown to be applicable for positively curved translation surfaces. They allow for parametrized load path determination and therefore automated process planning for stretch forming double curved surfaces with rigid grippers. Furthermore, the strain distribution within the sheet during stretch forming could be controlled by the parametrized load path design approaches. Since the material properties and the resulting springback are dependent on the stresses and strains in the sheet metal, the parameterized approaches have the potential to specifically control the strain distribution and the parameters in order to achieve the desired forming result. Further research should verify the results of current approaches in experimental tests and the approaches should also be tested for the applicability to more complex shapes. Furthermore, the potential of the load path approaches to control the springback systematically is of high interest and will be investigated. Additionally, it should be investigated if the achievable geometrical part complexity with simple stretch forming plants without multi grippers can be extended with the presented load path approaches.

References

[1] K. Lange, Handbook of metal forming, Society of Manufacturing engineers, 1985.

[2] Z. Yang, Z.-Y. Cai, C.-J. Che, M.-Z. Li, Numerical simulation research on the loading trajectory in stretch forming process based on distributed displacement loading, Int. J. Adv. Manuf. Technol. 82 (2016) 1353–1362. https://doi.org/10.1007/s00170-015-7470-y

[3] Y. Wang, M. Li, H. Liu, J. Xing, Finite Element Simulation of Multi-gripper Flexible Stretch Forming, Procedia Eng. 81 (2014) 2445–2450. https://doi.org/10.1016/j.proeng.2014.10.348

[4] R. Schmitz, M. Winkelmann, D. Bailly, G. Hirt, Homogenisation of the strain distribution in stretch formed parts to improve part properties, AIP Conference Proceedings 1960 (2018) 160025.

[5] D.E. Hardt, W.A. Norfleet, V.M. Valentin, A. Parris, In Process Control of Strain in a Stretch Forming Process, J. Eng. Mater. Tech. 123 (2001) 496-503.

Material Forming - ESAFORM 2023   Materials Research Forum LLC
Materials Research Proceedings 28 (2023) 951-958   https://doi.org/10.21741/9781644902479-104

Modifying mechanical properties of sheet metal materials by work hardening mechanisms induced by selective embossing

HEINZELMANN Pascal[1,a] *, BRIESENICK David[1,b] and LIEWALD Mathias[1,c]

[1]Institute for Metal Forming Technology, University of Stuttgart,
Holzgartenstraße 17, 70174 Stuttgart, Germany

[a]pascal.heinzelmann@ifu.uni-stuttgart.de, [b]david.briesenick@ifu.uni-stuttgart.de,
[c]mathias.liewald@ifu.uni-stuttgart.de

Keywords: Embossing, High Strength Steel, Dual-Phase Steel, Tensile Test

Abstract. The selective modification of mechanical properties of sheet metal materials poses a promising approach for realizing lightweight designs, especially when achieved without additional material input. One method for locally adapting the mechanical properties of sheet metal components is to specifically induce work hardening into the sheet metal material by near-surface embossing. Previous studies have already shown this effect for the sheet metal materials DP500 and DP600. The present paper verifies these findings for embossed high-strength steel DP800 considering different blank thicknesses and embossing depths. During experimental investigations, tensile and bending specimens of different sheet thicknesses were manufactured with definite embossing patterns and subsequently tested with regard to their mechanical properties. To verify the true embossing depth, the specimens were measured optically. As a result of this contribution, it was found that the material properties of high-strength sheet metal materials can be modified for lightweight construction and crash properties by selective embossing. Parameter constellations for increasing the yield strength were found for the materials investigated.

Introduction

Today, automotive traffic accounts for a large amount of global greenhouse gas emissions, which is why both the EU and the United States of America are imposing increasingly stringent restrictions regarding the CO_2 output from motor vehicles [1]. Automotive manufacturers must take this into account during the vehicle development phase, not only concerning driving concepts but also the final vehicle weight. But weight optimization of vehicles is not only a crucial development issue in terms of reducing greenhouse gases but also concerns the availability of metallic raw materials [2]. This is exacerbated in the current situation by Russia's invasion of Ukraine and the tense markets associated with it [3], and the increased costs in the energy sector [4]. Against this background, novel approaches to manufacturing lightweight vehicle structures can significantly contribute to overcoming these challenges.

    Lightweight constructions of car bodies are in most cases based on weight-optimized designs or the alloy-specific adaptation of the sheet metal materials used. In addition, such metallic lightweight structures can also be realized by modifying the component properties via selective work hardening during their forming [5]. In this context, Namoco et al. carried out experiments with aluminium alloys (A6061-T4 and A5052-H34) [6], which focussed on the targeted introduction of work hardening into the sheet metal material using both embossing and reforming

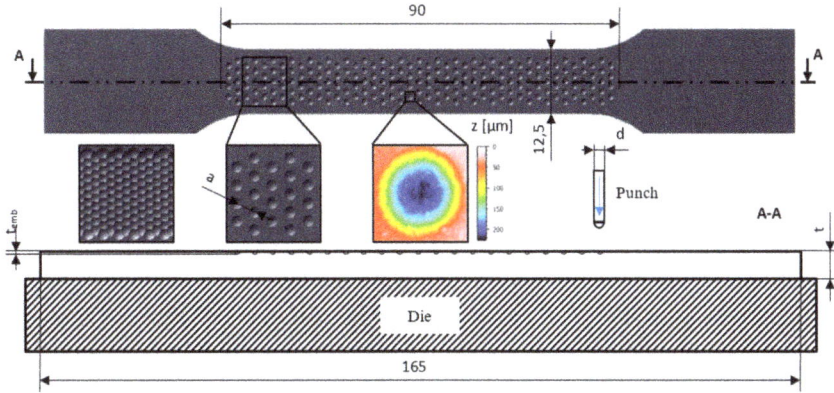

Fig. 1. Schematic of the embossing process with exemplary embossing arrangements.

processes. The results were evaluated based on the obtained values of yield strength, tensile strength and total elongation. Abe et al. examined the formability of embossed deep-drawn components and stated beneficial properties changes due to the embossing process carried out [7]. Walzer et al. investigated DP600 blanks and found that the incorporation of near-surface embossing increased the tensile strength of the sheet metal material [8]. Briesenick et al. carried out bending tests with DP500 blanks embossed in different ways and this way the maximum bending force required was increased by 75 % [9].

Based on these results, the investigations presented in this paper were intended to determine an optimum embossing geometry with regard to the highest possible strength increase of sheet metal structures. The embossing geometry here included the embossing pattern and the embossing depth. Furthermore, the property modifications attainable were evaluated using the example of the two sheet metal materials DP600 and DP800 having different initial yield strengths.

Experimental Method

Previous research work in the field of property modification of sheet materials by means of targeted application of work hardening has predominantly selected DP600 as test material. Materials with higher strengths, however, have not been investigated so far. Due to the existing results with DP600, which serve as a reference, the sheet metal materials DP600 and DP800 were selected for the investigation objectives presented in this paper. In these investigations, the influence of embossing geometries on changes in the mechanical properties yield strength R_p, and elongation at break A was evaluated. In Table *1* the material properties of the unembossed sheet materials are shown.

Table 1. Material properties of the steel blanks used in the tests carried out.

Material	Yield Strength R_p [MPa]	Elongation at Break A [%]
DP600	345.7	26.5
DP800	561.4	18.2

Material Forming - ESAFORM 2023
Materials Research Proceedings 28 (2023) 951-958

Materials Research Forum LLC
https://doi.org/10.21741/9781644902479-104

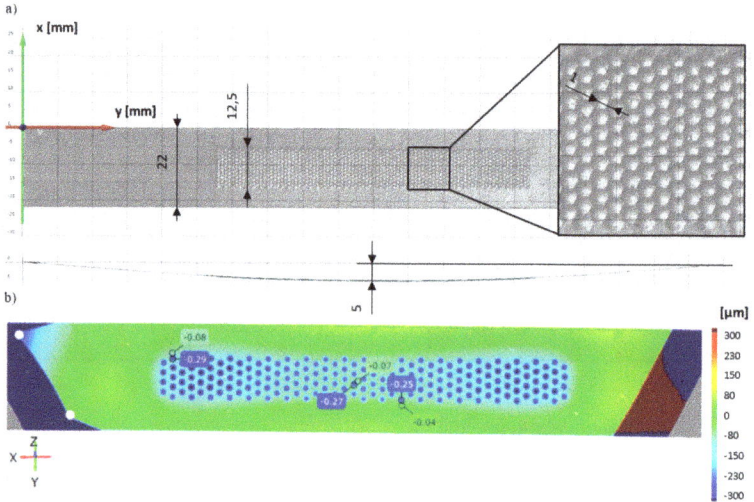

Fig. 2. Measurement of the embossed flat tensile specimens with different embossing geometries.

To evaluate the possible influence of different thicknesses of specimens on the property improvements achievable by means of embossing. The material DP600 was tested with sheet thicknesses t of 1.5 mm and 1.77 mm and DP800 was tested with sheet thicknesses of 1.5 mm and 1.85 mm. The different thicknesses were selected due to the availability of the materials used. To examine the mechanical properties changes, tensile tests were performed with embossed and unembossed specimens. The specimen geometry was determined according to DIN EN ISO 6892-1 [12] with form H and a measuring area of 12.5 mm × 50 mm. The embossing patterns to be investigated were applied beyond the measuring area on a surface of 12,5 mm × 90 mm for all specimens, thus ensuring that failure occurred in the measuring area. A hexagonal arrangement was chosen for the embossing patterns, varying the distance a between neighbouring embossings by 1 mm, 1,5 mm, and 2 mm as well as the embossing depth t_{emb} by 100 μm and 200 μm. In this respect, Fig. 1 shows an embossed flat tensile specimen with the minimal values a of 1 mm and an embossing depth t_{emb} of 100 μm.

A TRUPunch 5000R CNC punching machine was used to emboss the tensile specimen surfaces. This machine can perform up to 1,400 strokes per minute, thus allowing the embossings to be sequentially introduced within relatively short processing times. A modified round punch was used as embossing punch, with the face surface turned into a hemisphere with a diameter d of 4 mm. Due to the springback of the sheet metal material, the punch thereby had to penetrate deeper into the material's surface than the targeted embossing depth. Various test embossing operations were carried out to determine the penetration depth required for this purpose. Furthermore, the embossings were always introduced into the sheet metal material's surface before cutting out the pre-milled specimen geometry. Subsequently, the embossed sheet metal areas were cut out rectangularly (220 mm × 22 mm) and finally milled into the defined specimen shape to compensate for work-hardening caused by the cut-out. Table 2 lists the manufactured specimen variants in detail. Each variant was manufactured in quadruplicate for the reproducibility of the tensile tests.

Material Forming - ESAFORM 2023
Materials Research Proceedings 28 (2023) 951-958

Materials Research Forum LLC
https://doi.org/10.21741/9781644902479-104

Table 2. Specimen designation with details of the material, sheet thickness, embossing distance, and embossing depth.

Specimen Designation	Sheet Metal Material	Sheet Thickness t [mm]	Embossing Distance a [mm]	Embossing Depth t_{emb} [μm]
DP600_T1500_a1000_t100	DP600	1.5	1	100
DP600_T1500_a1500_t100	DP600	1.5	1.5	100
DP600_T1500_a2000_t100	DP600	1.5	2	100
DP600_T1500_a1000_t200	DP600	1.5	1	200
DP600_T1500_a1500_t200	DP600	1.5	1.5	200
DP600_T1500_a2000_t200	DP600	1.5	2	200
DP600_T1770_a1000_t100	DP600	1.77	1	100
DP600_T1770_a1500_t100	DP600	1.77	1.5	100
DP600_T1770_a2000_t100	DP600	1.77	2	100
DP600_T1770_a1000_t200	DP600	1.77	1	200
DP600_T1770_a1500_t200	DP600	1.77	1.5	200
DP600_T1770_a2000_t200	DP600	1.77	2	200
DP800_T1500_a1000_t100	DP800	1.5	1	100
DP800_T1500_a1500_t100	DP800	1.5	1.5	100
DP800_T1500_a2000_t100	DP800	1.5	2	100
DP800_T1500_a1000_t200	DP800	1.5	1	200
DP800_T1500_a1500_t200	DP800	1.5	1.5	200
DP800_T1500_a2000_t200	DP800	1.5	2	200
DP800_T1850_a1000_t100	DP800	1.85	1	100
DP800_T1850_a1500_t100	DP800	1.85	1.5	100
DP800_T1850_a2000_t100	DP800	1.85	2	100
DP800_T1850_a1000_t200	DP800	1.85	1	200
DP800_T1850_a1500_t200	DP800	1.85	1.5	200
DP800_T1850_a2000_t200	DP800	1.85	2	200

To monitor the dimensional accuracy of the produced specimens and their embossings, they were optically measured using a GOM ATOS Compact Scan System, as shown in Fig. 2. Furthermore, the diameters of the embossings on the surface were measured via a Keyence Microscope VHX. Ball segment equations were then used to determine the embossing depths. The Keyence Microscope VHX was used to get fast results to set the right embossing offset. To check all embossing depths over the embossing area the ATOS Compact Scan system was used. Finally, the tensile tests were carried out according to DIN EN ISO 6892-1 [10] using the universal testing machine Roell + Korthaus RKM100.

Material Forming - ESAFORM 2023   Materials Research Forum LLC
Materials Research Proceedings 28 (2023) 951-958   https://doi.org/10.21741/9781644902479-104

Results

For assessing the influence of different embossings on the mechanical properties of sheet metal materials, the two characteristic values yield strength R_p and elongation at break A were determined in each of the tensile tests described above. All characteristic values determined in this respect are summarized in Table 3.

Table 3. Results of the tensile tests.

Experiment	Yield Strength R_p [MPa]	Elongation at Break A [%]
DP600_T1500_a1000_t100	507.35	21.82
DP600_T1500_a1500_t100	500.88	21.95
DP600_T1500_a2000_t100	443.06	22.68
DP600_T1500_a1000_t200	584.34	10.40
DP600_T1500_a1500_t200	545.65	14.48
DP600_T1500_a2000_t200	502.69	19.24
DP600_T1770_a1000_t100	539.41	17.53
DP600_T1770_a1500_t100	486.82	21.23
DP600_T1770_a2000_t100	432.26	24.40
DP600_T1770_a1000_t200	560.92	9.05
DP600_T1770_a1500_t200	528.68	12.67
DP600_T1770_a2000_t200	494.22	17.67
DP800_T1500_a1000_t100	680.81	13.53
DP800_T1500_a1500_t100	655.76	15.44
DP800_T1500_a2000_t100	592.40	16.66
DP800_T1500_a1000_t200	685.00	9.79
DP800_T1500_a1500_t200	661.47	11.81
DP800_T1500_a2000_t200	624.69	14.31
DP800_T1850_a1000_t100	672.87	14.26
DP800_T1850_a1500_t100	649.12	16.69
DP800_T1850_a2000_t100	597.22	17.03
DP800_T1850_a1000_t200	686.07	10.69
DP800_T1850_a1500_t200	673.37	10.88
DP800_T1850_a2000_t200	627.96	13.60

For evaluating these results, the characteristic values achieved must be considered in relation to the respective embossing geometry, i.e., the embossing depth and the embossing distance. Fig. 3 shows these relationships. First, Diagram a) illustrates the yield strengths of those tests performed with specimens having an embossing depth of 100 μm for all test materials and sheet thicknesses investigated. The different embossing distances and respective unembossed reference specimens are shown as coloured bars. Diagram c) shows the same evaluations for the embossing depth of 200 μm. In Fig. 3 b) and d) the evaluation is based on the elongation at break and follows the same scheme then done for the yield strength. The percentage values in the bars of the embossed tests refer to the change in the respective embossing geometry compared to the reference tests of the respective group. For the DP600 material, the increases in yield strength range from 26 % up to 66 %. The range of the DP800 material varies between 5 % and 22 %. When considering the change in elongation at break, a reduction can be observed for all cases. For DP600, this reduction ranges between -66 % and -9 %. For DP800, the reduction occurs in the range of -42 % to -2 %.

An influence of the different plate thicknesses could not be observed for the yield strength. For the elongation at break, a minor influence could be observed.

Considering the influence of the embossing depth on the yield strength, it can be stated that an increase in the embossing depth is accompanied by an increase in the yield strength. This applies more to the lower-strength material DP600 (up to 66 %) than to DP800 (up to 22 %). Thus, some DP600 tests with an embossing depth of 200 μm show similar and partly higher yield strength values than the reference measurements for DP800. An opposite effect can be recognised when considering the influence of the embossing depth on the elongation at break. With an increase in embossing depth, the elongation at break decreases compared to the lower embossing depth. DP600 by up to -66 % and DP800 by up to -45 %. Also, the embossing distance influences the yield strength and the elongation at break. Here, a reduction of the embossing distance leads to an increase in yield strength. This effect is not as high as the effect of the embossing depth, but still considerable. If the yield strength increases due to the reduction of the embossing distance, the elongation at break decreases to a similar extent.

Due to the different patterns, the results of the DP600 material can only be compared with the results of this paper to a limited extent. Qualitatively, it can be said that the results show a similar trend.

Summary

In this paper, the effects of near-surface embossing on yield strength and bearing at fracture were investigated. The dual-phase materials DP600 and DP800 with different plate thicknesses were considered. In general, an increase in the yield strength of the considered DP steel sheets was observed for all tested embossing geometries. In contrast, a reduction was found for the elongation at break for all embossed specimens. The results for the higher-strength material DP800 show property changes due to the introduction of embossing. These changes are overalls smaller in the DP800 material compared to the DP600 material.

When using embossed components, it is therefore important to consider which properties are relevant for the respective application and to what extent the properties should be changed. For this purpose, further research work will be conducted on tests with other materials and embossing geometries. Here, the focus should be when, or in what range, embossing no longer has any influence on higher and highest-strength materials. Moreover, low-strength materials will be considered as well. For a better understanding of the process, additional simulations are needed for providing a more detailed view of the process than just the experimental measurements. Simulations that include the history of the embossing process should also be considered, so the influence of the embossing is also taken into account for further simulative stress analyses. To simulate the whole embossing process, a high computational effort must be made. For faster simulations of embossed parts, the simulation of the processes must be more efficient Possibilities should also be found to transfer the result of the embossing process to simulations of embossed components with little calculation effort.

Material Forming - ESAFORM 2023

Materials Research Forum LLC

Materials Research Proceedings 28 (2023) 951-958

https://doi.org/10.21741/9781644902479-104

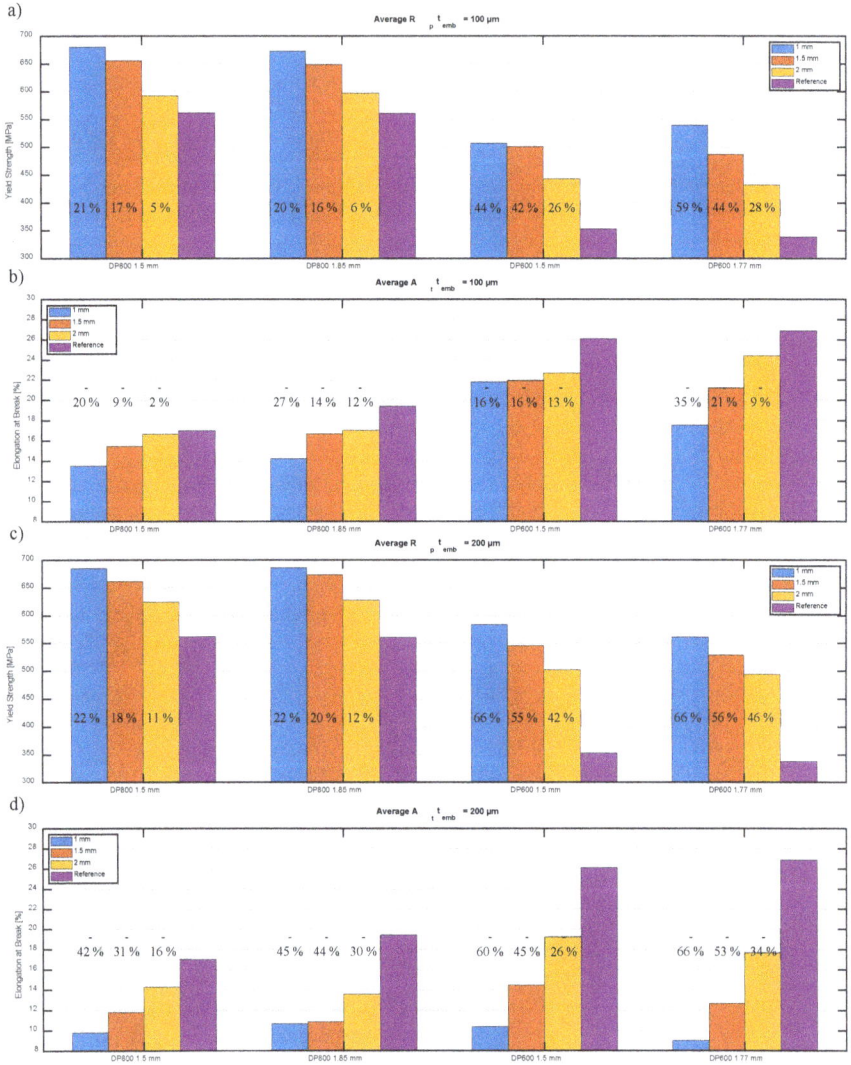

Fig. 3. Evaluation of tensile tests with respect to
yield strength for a) $t_{emb} = 100 \ \mu m$, c) $t_{emb} = 200 \ \mu m$ and
elongation at break for a) $t_{emb} = 100 \ \mu m$, c) $t_{emb} = 200 \ \mu m$.

References

[1] K. Shanmugam, V. Gadhamshetty, P. Yadav, D. Athanassiadis, M. Tysklind, V.K. Upadhyayula, Advanced High-Strength Steel and Carbon Fiber Reinforced Polymer Composite Body in White for Passenger Cars: Environmental Performance and Sustainable Return on Investment under Different Propulsion Modes, ACS Sustainable Chem. Eng. 7 (2019) 4951-4963. https://doi.org/10.1021/acssuschemeng.8b05588

[2] R. Döhrn, Die Lage am Stahlmarkt: Produktion steigt wieder, Probleme bleiben, RWI Konjunkturberichte 72 (2021) 49-58

[3] C. Birkholz, M. Kraus, Die Energiekrise im Standortvergleich: Preiseffekte und Importrisiken. Sonderstudie zum Länderindex Familienunternehmen, München: Stiftung Familienunternehmen, Rep. 978-3-948850-15-9, 2022

[4] O. Holtemöller, S. Kooths, T. Schmidt, T. Wollmershäuser, Gemeinschaftsdiagnose: Energiekrise, Inflation, Rezession und Wohlstandsverlust, Wirtschaftsdienst 102 (2022) 761-765. https://doi.org/10.1007/s10273-022-3291-4

[5] C.S. Namoco, T. Iizuka, R.C. Sagrado, N. Takakura, K. Yamaguchi, Experimental and numerical investigation of restoration behavior of sheet metals subjected to bulging deformation, J. Mater. Process. Technol. 177 (2006) 368–372. https://doi.org/10.1016/j.jmatprotec.2006.03.208

[6] C.S. Namoco, T. Iizuka, K. Narita, N. Takakura, K. Yamaguchi, Effects of embossing and restoration process on the deep drawability of aluminum alloy sheets, J. Mater. Process. Technol. 187-188 (2007) 202–206. https://doi.org/10.1016/j.jmatprotec.2006.11.182

[7] Y. Abe, K. Mori, T. Maeno, S. Ishihara, Y. Kato, Improvement of sheet metal formability by local work-hardening with punch indentation, Prod. Eng. Res. Devel. 13 (2019) 589-597. https://doi.org/10.1007/s11740-019-00910-6

[8] S. Walzer, M. Liewald, Studies on the influence of embossing on the mechanical properties of high-strength sheet metal, AIP Conference Proceedings 2113 (2019) 160006. https://doi.org/10.1063/1.5112703

[9] D. Briesenick, S. Walzer, M. Liewald, Study on the Effect of Embossing on the Bending Properties of High-Strength Sheet Metals, in: Forming the Future: Proceedings of the 13th International Conference on the Technology of Plasticity (Springer eBook Collection), G. Daehn, J. Cao, B. Kinsey, E. Tekkaya, A. Vivek, and Y. Yoshida (Eds.), 1st ed. Cham: Springer International Publishing, Imprint Springer, 2021, pp. 2585-2595.

[10] C. Lesch, N. Kwiaton, F.B. Klose, Advanced High Strength Steels (AHSS) for Automotive Applications – Tailored Properties by Smart Microstructural Adjustments, Steel Res. Int. 88 (2017) 1700210. https://doi.org/10.1002/srin.201700210

[11] F. Ebrahimi, N. Saeidi, M. Raeissi, Microstructural Modifications of Dual-Phase Steels: An Overview of Recent Progress and Challenges, Steel Res. Int. 91 (2020) 2000178. https://doi.org/10.1002/srin.202000178

[12] DIN EN ISO 6892-1:2020-06, Metallische Werkstoffe_-_Zugversuch_-_Teil_1: Prüfverfahren bei Raumtemperatur (ISO_6892-1:2019); Deutsche Fassung EN_ISO_6892-1:2019, Berlin.

Material Forming - ESAFORM 2023
Materials Research Proceedings 28 (2023) 959-968

Materials Research Forum LLC
https://doi.org/10.21741/9781644902479-105

Influence of surface pressure and tool materials on contact heating of aluminum

TRÂN Ricardo[1,a] *, PSYK Verena[1,b], WINTER Sven[1,c] and KRÄUSEL Verena[1,d]

[1]Fraunhofer Institute for Machine Tools and Forming Technology, Reichenhainer Straße 88, 09126 Chemnitz, Germany

[a]ricardo.tran@iwu.fraunhofer.de, [b]verena.psyk@iwu.fraunhofer.de, [c]sven.winter@iwu.fraunhofer.de, [d]verena.kräusel@iwu.fraunhofer.de

Keywords: Heating Technology, Contact Heating, Aluminum, Hot Forming, Tool Technology

Abstract. The implementation of lightweight design concepts can significantly benefit from using highly efficient heating technologies such as contact heating in thermo-mechanical processing of sheet metal components. The investigation of the influence of surface pressure and tool material on the heating time and the heating rate during contact heating is the subject of this publication. A specially manufactured contact heating tool with comprehensive temperature and force measurement was used for studying the effects of different contact plate materials (CuZn39Pb3 and CuCr1Zr), surface pressures (3 MPa - 15 MPa) and variable plate thicknesses (1.0 mm - 5.2 mm) during heating of the aluminum alloy EN AW-7075 up to the solution heat treatment temperature of 475 °C. It was observed that heating time is lower for thinner workpieces. Furthermore, heating times decrease and heating rates increase significantly with increasing surface pressure for a pressure range of 3 MPa - 9 MPa. A further increase in surface pressure is not recommended, because the benefit in terms of further reduction of the heating time is marginal and the strength of the contact plate materials at elevated temperatures is limited. Contact heating using copper plates is significantly faster compared to brass plates and the conventionally used steel plates. Brass plates, however, benefit more from an increase in surface pressure. Both investigated materials allow faster heating than conventional steel plates due to their higher thermal conductivity. Depending on the specific process parameters the heating process can be accelerated to less than one second. Thus, contact heating can be realized within the press cycle.

Introduction

In modern manufacturing, the implementation of lightweight design concepts has been becoming more and more important in order e. g. to save resources in aviation [1], reduce fuel consumption and corresponding CO_2 emission in vehicles with conventional combustion machines [2], increase the range that an electrical vehicle can travel without recharging [3], or improve component function and user-friendlyness e. g. in medical engineering products [4]. These concepts involve design changes e. g. in order to realize integration of functions [5] and the use of typical lightweight materials such as high strength steel [6] or light-metal alloys [7]. These materials usually feature limited formability at room temperature so that forming of the complex geometries typically necessary for lightweight components frequently requires thermo-mechanical processes such as press hardening [8], quenching and partitioning [9], or superplastic forming [10]. In this context, fast and energy-efficient heating strategies are an essential prerequisite. Conventionally, the components are heated in roller hearth furnaces by means of thermal radiation. In practice, these furnaces reach lengths of up to 40 m [8] in order to achieve acceptable cycle times despite of the long furnace residence time. Alternative heating technologies requiring less investment costs and floor space and additionally allowing faster heating are based on resistive [11] and inductive heating [12]. However, resistive heating is efficient only for components with favorable - i. e. high

Material Forming - ESAFORM 2023
Materials Research Proceedings 28 (2023) 959-968

Materials Research Forum LLC
https://doi.org/10.21741/9781644902479-105

- ratio of length and cross section [13] and uniform heating for complex blank geometries is difficult for both - resistive and inductive heating [8]. Direct contact heating is a relatively new approach (suggested in [14]) with high potential to overcome these limitations. In this technology, the contact plates (typically made of metal or ceramics) that are heated by a heat source (e. g. induction coils [14], gas burners [15], cartridge heaters [16]) serve as energy storage and transfer the heat to the workpiece via thermal conduction. Compared to furnace heating, contact heating excels due to extremely high heating rates, which significantly reduces heating time [17]. The mechanism is very similar to the cooling in direct press hardening. From that process it is known that the heat transfer can be controlled by applying a defined contact force (contact pressure) [18, 19]. This suggests that the contact pressure is also a suitable parameter for adapting the heating rate in contact heating processes, which is e. g. relevant for thermo-mechanical processing of high-strength aluminum alloys [20]. Furthermore, it can be expected that different contact plate materials with significantly varying thermal properties (e. g. heat conduction coefficient) affect the heating rate in contact heating. However, neither the influence of surface pressure nor that of different contact plate materials on the heating behavior during contact heating have been investigated yet. Therefore, providing an analysis of these aspects is the aim of this paper.

Experimental Setup and Process Description

For this purpose, the contact plate tool shown in Fig. 1 was developed and mounted to a hydraulic press HS3-1500 by Dunkes with a press force of 1,000 - 15,000 kN. The setup consists of an insulation box that encases exchangeable contact plates, cooling plates and a gas pressure damper plate, that allows applying different surface pressures. This damper plate is coupled with a gas-pressure-based force measurement that enables recording the time-dependent surface pressure between the contact plates and the workpieces.

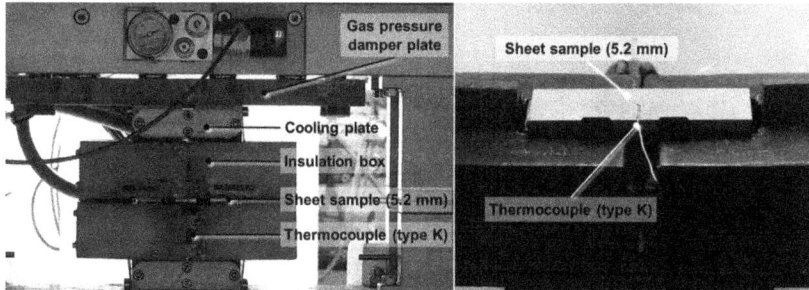

Fig. 1. Experimental setup: front view and view into the open tool with inserted sheet sample.

Material Forming - ESAFORM 2023   Materials Research Forum LLC
Materials Research Proceedings 28 (2023) 959-968   https://doi.org/10.21741/9781644902479-105

This study considers contact plates with a heating surface of 100 mm x 60 mm and a height of 35 mm made of the copper alloy CuCr1Zr and the brass alloy CuZn39Pb3, respectively (see Fig. 2). These materials feature approximately the same heat capacity and different thermal conductivity (see Table 1). Thus, the influence of the thermal conductivity of the contact plate material on the heating process can be investigated.

Fig. 2. Contact plates made of different materials.

Table 1. Thermal material properties of the contact plate materials and the workpiece material at room temperature.

	CuCr1Zr	**CuZn39Pb3**	**EN AW-7075**
heat capacity	383 J/kgK	377 J/kgK	862 J/kgK
thermal conductivity	330 W/mK	123 W/mK	145 W/mK

Both, the upper and the lower contact plate were heated with four individually controlled silicon nitride ceramic heating elements by Bach Resistor Ceramics GmbH. These heating elements feature rectangular geometry in order to facilitate homogeneous temperature distribution. The time-dependent temperatures of the contact plates were recorded using type K thermocouples by OMEGA with a sampling rate of 10 Hz during the tests. Furthermore, the pressure in the gas damping plate was measured and served as basis for determining the surface pressure acting on the contact plates and the workpieces to be heated, respectively.

The specimens considered in this study were made of EN AW-7075 with a width w of 60 mm, a length l of 100 mm and thicknesses s of 1.0 mm, 2.0 mm, 3.0 mm; 4.3 mm and 5.2 mm. In order to record the time-dependent temperature of the workpiece, rectangular grooves were milled into the specimens and thermocouples were inserted in the center of the specimen as shown in Fig. 3. For the sheets with 1.0 mm thickness thermocouples with a diameter d of 0.5 mm were used. For all other sheet thicknesses thermocouples with a diameter of 1 mm were used. The groove width was chosen corresponding to the width of the thermocouple. The groove depth amounts to the sum of half of the sheet thickness and half of the height of the thermocouple. Analogous to the temperature measurement of the contact plates also here type K thermocouples by OMEGA were used and the sampling rate was set to 10 Hz.

Material Forming - ESAFORM 2023   Materials Research Forum LLC
Materials Research Proceedings 28 (2023) 959-968   https://doi.org/10.21741/9781644902479-105

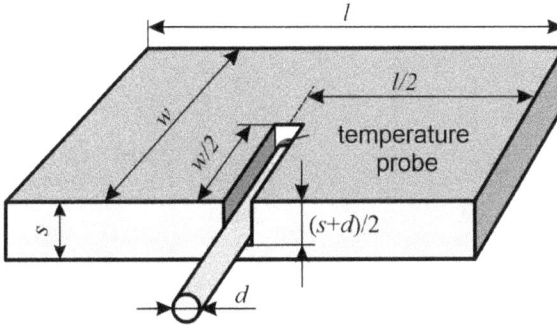

Fig. 3. Principal sketch of the sheet specimen for contact heating tests.

For the contact heating experiments, the contact plates were preheated for 15 minutes in order to achieve homogenous local temperature distribution. As it is known that the temperature rise in the workpiece is very quickly at the beginning of the process, but slows down as it approximates the temperature of the contact plates, the contact plate temperature is usually chosen higher than the target temperature of the workpiece. Here, the contact plates were pre-heated to 500°C in order to achieve the desired specimen temperature of 475°C - i. e. the solution heat treatment temperature of EN AW-7075 - in a reasonable timeframe. Then, a specimen was inserted, the tool was closed within approx. two seconds and a defined surface pressure between the contact plates and the specimen was applied by adjusting the pressure in the gas pressure damper plate correspondingly. For each sample thickness a test series was carried out by varying the surface pressure between 3 MPa and 15 MPa in steps of 3 MPa.

Fig. 4 shows an exemplary process diagram based on the measured time-dependent surface pressure between workpiece and contact plates and corresponding temperatures of workpiece and contact plates for a representative test using copper contact plates. At the moment $t=0$ s, the contact plates are closed, and the sheet specimen is heated from both sides. From this moment, the surface pressure rises approx. linearly up to the target value that is 9 MPa in the regarded case causing rapid heating of the aluminum specimen. At the same time, a temperature reduction of approx. 20 K of the contact plates can be observed as the heat is transferred into the specimen. Correspondingly, an increase of the workpiece temperature occurs. The measured temperature curve is characterized by an initial fast rise that decreases more and more when it approximates the temperature of the contact plates, which is in good agreement with [16]. After reaching the target temperature of 475°C, the test is completed, the specimen is removed from the tool and quenched in water, and the contact plate temperature is re-adjusted for next test of the test series.

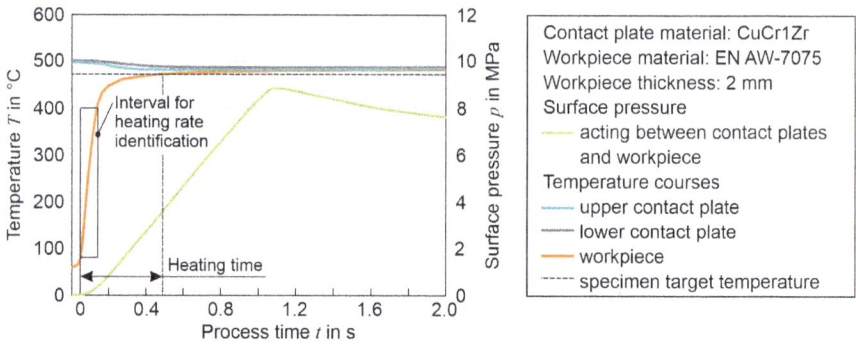

Fig. 4. Time-dependent surface pressure between workpiece and contact plates and corresponding temperatures of workpiece and contact plates.

Based on this measured data, two characteristic values were identified (see also Fig. 4):

1. The heating time t_{heat} was determined as the time interval between closing of the dies ($t=0$ s) and the moment when the specimen temperature reaches the target temperature of 475°C.
2. The maximum heating rate $\dot{T}$ was determined by referring the temperature difference to the corresponding time difference during the fast (and approximately linear) rise of the temperature. This value allows direct comparison to earlier experiments performed with standard contact plate materials presented in [17].

Influence of the Surface Pressure on the Heating of the Workpiece

Fig. 5 shows the heating times at different surface pressures for five different sheet thicknesses (1.0 mm - 5.2 mm). The contact plate material considered here is the copper alloy. The representative curves for all sheet thicknesses indicate that the heating time decreases with increasing surface pressure. However, the curves show that this effect is limited, because at surface pressures of 9.0 MPa and above higher surface pressures have hardly any further positive influence on the heating time, regardless of the sheet thickness to be heated. This suggests that at lower surface pressure the contact conditions are not ideal. This might be attributed to unevenness or roughness of the sheet. With increasing pressure these deficits are reduced and at a surface pressure of 9 MPa all unevenness or roughness is sufficiently leveled so that the contact conditions are almost constant for even higher pressure. Furthermore, the heating time decreases significantly with reduced sheet thickness. For example, the heating times at the maximum sheet thickness of 5.2 mm and a surface pressure of 3.0 MPa are approx. 13.1 s. With a sheet thickness of 1.0 mm, the heating time drops by approx. 86% to 1.8 s for the same surface pressure.

Material Forming - ESAFORM 2023
Materials Research Proceedings 28 (2023) 959-968

Materials Research Forum LLC
https://doi.org/10.21741/9781644902479-105

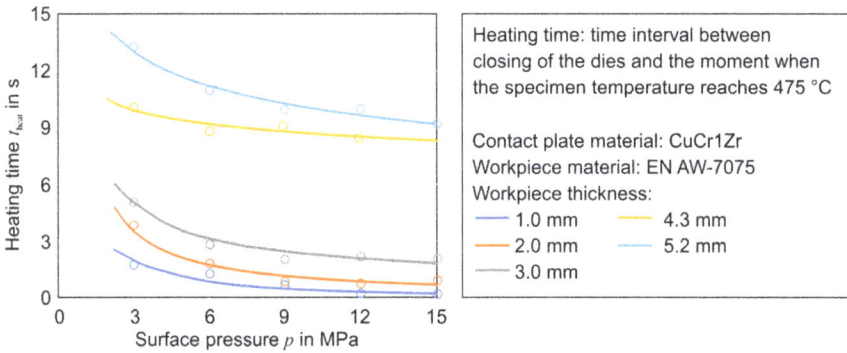

Fig. 5. Heating times for different thicknesses of the aluminum workpiece as function of the surface pressure.

Influence of the Contact Plate Material on The Heating of the Workpiece

In addition to investigating the influence of surface pressure, the material of the contact plates and its influence on the heating rates was studied. For this purpose, additional tests were done using contact plates made of brass alloy. The heating times at varying surface pressures for an exemplary sheet thickness of 2.0 mm are compared to those determined for the copper plates in Fig. 6. In principle, the curves show similar qualitative trends, but the heating times are in general higher in case of the brass plates. This effect can probably be attributed to the lower thermal conductivity of the brass compared to the copper alloy. This means specifically that also in case of heating with brass plates the heating time decreases with increasing surface pressure. The quantitative comparison of the curves determined for the two contact plate materials even shows that in case of heating with brass contact plates the influence of the surface pressure is higher than in case of heating with copper contact plates. For example, the heating under a surface pressure of 3.0 MPa takes more than 10.0 s if brass plates are used, while in case of copper plates such a high heating time was sufficient for heating significantly thicker sheets (see Fig. 5) and heating a sheet with a thickness of 2.0 mm required less than 3.8 seconds. However, also the drop of heating time with increasing surface pressure is more significant if brass plates are used. Already at 6.0 MPa surface pressure, only 4.8 s are required to reach the target temperature, which corresponds to a reduction of 48 %. This percentage reduction is comparable to that of the copper plates (1.9 s at 6.0 MPa). Altogether, the ratio of the heating times for the two contact plate materials (approx. 2.9) corresponds approximately to the inverse ratio of their thermal conductivities (0.37).

Especially if brass plates are used, heating can benefit significantly from a pressure increase, but in addition to the heating time also the temperature dependent stress strain behavior of the contact plate material must be considered when choosing the surface pressure. According to [21], the flow stress of brass alloys similar to the one used here is in the range of 25 MPa - 50 MPa only at a temperature of 550°C, which is only 50 K higher than the contact plate temperature considered here. This means that even though the curves in Fig. 6 suggest that an increase in pressure beyond 15 MPa can still bring a slight reduction of the heating time in case of heating plates made of brass, this advantage must be carefully assessed against the risk of plastic deformation of the contact plates.

Material Forming - ESAFORM 2023   Materials Research Forum LLC
Materials Research Proceedings 28 (2023) 959-968   https://doi.org/10.21741/9781644902479-105

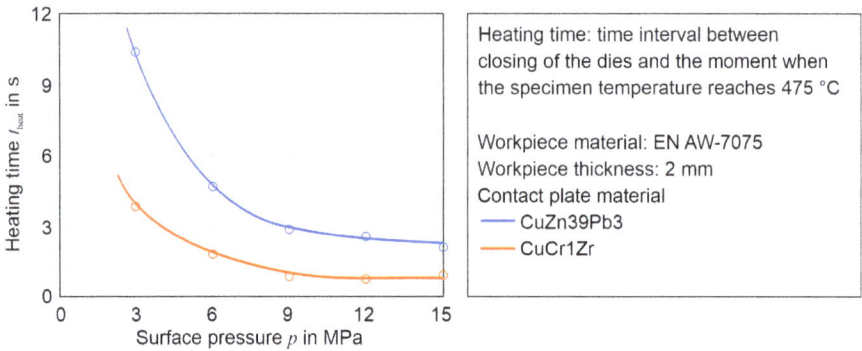

Fig. 6. Heating times for different contact plate materials as function of the surface pressure.

Fig. 7 compares the maximum heating rates $\dot{T}$ of the different contact plate materials for a plate thickness of 2 mm as a function of surface pressure. The heating rates were determined in each case in the range of the linear temperature increase with the maximum slope (see Fig. 4). In addition, the heating rate of the conventionally used contact plate material 1.4828 from [17] was plotted in the diagram for comparison. It can be clearly demonstrated that the maximum heating rates for contact plates made of copper and brass, respectively, are similar at 3 MPa surface pressure. Above 6 MPa surface pressure, however, the values measured for the copper plates increase significantly above the heating rates determined for the brass plates. The maximum heating rates are reached at 12 MPa, but the values for 9 MPa and 15 MPa are at a comparable level. The maximum values amount to approximately 1690 K/s in case of heating plates made of copper and 370 K/s in case of heating plates made of brass. Thus, the maximum heating rate achieved using brass plates is only 23 % of the heating rate achieved using copper plates. At first glance, these results are contrary to those in Figure 6, where the largest difference between the contact plate materials is at 3 MPa and the smallest at 15 MPa. However, the maximum heating rate $\dot{T}$ evaluated here only describes the linear range of the heating process, whereas the heating time t_{heat} covers heating up to 475°C. Heating up to 475°C takes considerably more time and involves strongly non-linear parts of the heating curve (see Fig. 4).

Finally, the heating rate of steel 1.4828 from [17] was plotted. The test conditions were slightly different but still comparable. Especially, the contact plate temperature was slightly lower (480°C instead of 500°C) and the considered workpiece material was a different aluminum alloy with slightly lower heat capacitance but featuring the same thermal conductivity. Moreover, the heating plates and the workpiece were larger in [17], but it can be assumed that this has no significant influence of the heating behavior so the heating rate can be compared. Fig. 7 shows that at a surface pressure of 4.5 MPa, the steel reaches lower heating rates than the predicted heating rates for both materials investigated here. This proves the high potential of the investigated materials in particular the copper alloy for contact heating of aluminum workpieces compared to conventionally used materials as e.g. 1.4828.

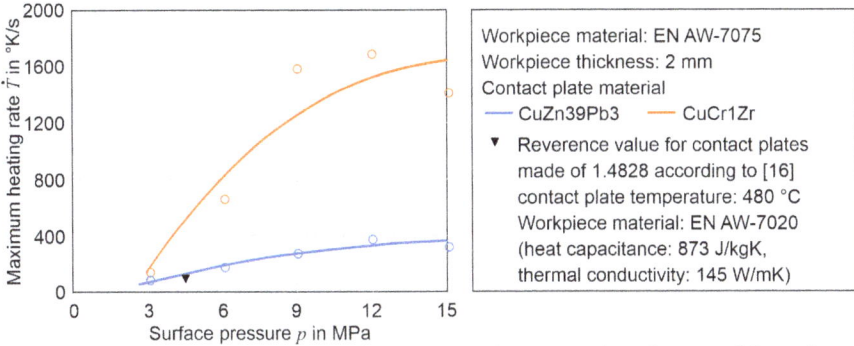

Fig. 7. Maximum heating rates for different contact plate materials as function of the surface pressure.

Summary

In the present study, the effects of two different contact plate materials (copper and brass alloys), five surface pressures and five plate thicknesses on the heating time of the aluminum alloy EN AW-7075 up to the solution heat treatment temperature of 475°C were investigated. A specially manufactured contact heating tool with comprehensive temperature and force measurement was used for this purpose. It was observed for all specimens that fast heating can be realized. Especially at a surface pressure of 9.0 MPa or more, the heating times could be minimized. A further increase in surface pressure is not recommended, because the benefit in terms of further reduction of the heating time is marginal and the strength of the contact plate materials at elevated temperatures relevant for this heating process is limited.

Furthermore, as expected, the heating time decreases significantly with a reduction in sheet thickness. For sheet thicknesses of 1.0 mm and 2.0 mm, heating times of less than 1.0 s could be achieved when heating with the copper plates. The minimum value at 1.0 mm sheet thickness was 0.2 s. The maximum heating rate of 1690 K/s could be achieved at a surface pressure of 12 MPa. This shows the high potential of contact heating especially when heating components with rather small sheet thicknesses. For the highest sheet thicknesses tested here (5.2 mm) and consequently also for the highest volume of the specimen, still a heating rate of 165 K/s could be achieved at the lowest surface pressure. This is significantly more compared to more conventional heating technologies such as furnace heating.

Finally, a comparison of sample heating with brass contact plates was performed. It was found that compared to the heating plates made of copper, especially at low surface pressure significantly higher heating times were necessary, but an increase in surface pressure still allowed reducing the heating time to 2.0 s. Altogether, the following key results were obtained:

- Surface pressures in the range of 9.0 MPa - 12 MPa are recommended for contact heating of aluminum workpieces. Higher pressure brings no further positive effects.
- The lower the sheet thickness, the more positive the contact heating concept. Maximum heating rates of 1690 K/s could be achieved using copper plates.
- Contact heating using copper plates is significantly faster compared to brass plates and the conventionally used steel plates. Heating via brass plates, however, benefits more from an increase in surface pressure.

The presented results prove that contact heating can be realized within the press if suitable surface pressure is applied.

Material Forming - ESAFORM 2023   Materials Research Forum LLC
Materials Research Proceedings 28 (2023) 959-968   https://doi.org/10.21741/9781644902479-105

References

[1] M. Hanna, J. Schwenke, L.-N. Schwede, F. Laukotka, D. Krause, Model-based application of the methodical process for modular lightweight design of aircraft cabins, Procedia CIRP 100 (2021) 637-642. https://doi.org/10.1016/j.procir.2021.05.136

[2] C. Koffler, K. Rohde-Brandenburger, On the calculation of fuel savings through lightweight design in automotive life cycle assessments, Int. J. Life Cycle Assess. 15 (2010) 128-135. https://doi.org/10.1007/s11367-009-0127-z

[3] Q. Liu, Y. Lin, Z. Zong, G. Sun, Q. Li, Lightweight design of carbon twill weave fabric composite body structure for electric vehicle, Compos. Struct. 97 (2013) 231-238. https://doi.org/10.1016/j.compstruct.2012.09.052

[4] C.M. Light, P.H. Chappell, Development of a lightweight and adaptable multiple-axis hand prosthesis, Med. Eng. Phys. 22 (2000) 679-684. https://doi.org/10.1016/s1350-4533(01)00017-0

[5] N. Modler, A. Winkler, A. Filippatos, D. Weck, M. Dannemann, Function-integrative Lightweight Engineering - Design Methods and Applications, Chemie Ingenieur Technik 92 949-959. https://doi.org/10.1002/cite.202000010

[6] Z. Tang, Z. Gu, L. Jia, X. Li, L. Zhu, H. Xu, G. Yu, Research on Lightweight Design and Indirect Hot Stamping Process of the New Ultra-High Strength Steel Seat Bracket, Metals 9 (2019) 833. https://doi.org/10.3390/met9080833

[7] M. Linnemann, V. Psyk, N. Kaden, F. Kersten, M. Schmidtchen, V. Kräusel, M. Dix, U. Prahl, Producing and Processing of Thin Al/Mg/Al Compounds, Eng. Proc. 26 (2022) 8. https://doi.org/10.3390/engproc2022026008

[8] H. Karbasian, A.E. Tekkaya, A review on hot stamping, J. Mater. Process. Technol. 210 (2010) 2103-2118. https://doi.org/10.1016/j.jmatprotec.2010.07.019

[9] S. Winter, M. Werner, R. Haase, V. Psyk, S. Fritsch, M. Böhme, M. Wagner, Processing Q&P steels by hot-metal gas forming: Influence of local cooling rates on the properties and microstructure of a 3rd generation AHSS, J. Mater. Process. Technol. 293 (2021) 117070. https://doi.org/10.1016/j.jmatprotec.2021.117070

[10] R. Trân, F. Reuther, S. Winter, V. Psyk, Process development for a superplastic hot tube gas forming process of titanium (Ti-3Al-2.5V) hollow profiles, Metals 10 (2020). https://doi.org/10.3390/met10091150

[11] K. Mori, S. Maki, Y. Tanaka, Warm and hot stamping of ultra tensile strength steel sheets using resistance heating, CIRP Annals—Manuf. Technol. 54 (2009) 209-212.

[12] R. Kolleck, R. Veit, M. Merklein, J. Lechler, M. Geiger, Investigation on induction heating for hot stamping of boron alloyed steels, CIRP Annals—Manuf. Technol. 58 (2009) 275-278.

[13] R. Kolleck, R. Veit, H. Hofmann, F.J. Lenze, Alternative heating concepts for hot sheet metal forming, 1st International Conference on Hot Sheet Metal Forming of High-Performance Steel, Kassel, Germany, 2008, pp. 239-246.

[14] V. Ploshikhin, A. Prihodovsky, J. Kaiser, R. Bisping, H. Linder, C. Lengsdorf et al., New heating technology for furnace-free press hardening process, Tools and Technologies for Processing Ultra-High Strength Materials, Graz, Austria, 2011.

[15] D. Landgrebe, F. Schieck, J. Schönherr, New Approaches for Improved Efficiency and Flexibility in Process Chaines of Press Hardening, International Mechanical Engineering Congress & Exposition (ASME), Houston/Texas, USA, 13 - 19.11.2015.

[16] J.N. Rasera, K.J. Daun, C.J. Shi, Direct contact heating for hot forming die quenching, Appl. Therm. Eng. 98 (2016) 1165-1173.

[17] R. Trân, L. Kertsch, S. Marx, S. Hebbar, V. Psyk, A. Butz, Towards an efficient Industrial Implementation of W-temper Forming for 7xxx Series Al Alloys, in: G. Daehn, J. Cao, B. Kinsey, E. Tekkaya, A. Vivek, Y. Yoshida (Eds.), Forming the Future, Cham: Springer International Publishing, 2021.

[18] C. Hoff, Untersuchungen der Prozesseinflussgrößen beim Presshärten des höchstfesten Vergütungsstahls 22MnB5, Dissertation, Universität Erlangen-Nürnberg, 2007.

[19] W. Xiao, B. Wang, K. Zheng, J. Zhou, J. Lin, A study of interfacial heat transfer and its effect on quenching when hot stamping AA7075, Arch. Civil Mech. Eng. 18 (2018) 723-730. https://doi.org/10.1016/j.acme.2017.12.001

[20] A.A.M. Smeyers, S. Khosla, Production of formed automotive structural parts from AA7xxx-series aluminium alloys, European Patent EP 2 581 218 B1, 2012.

[21] S. Spigarelli, M. El Mehtedi, M. Cabibbo, F. Gabrielli, D. Ciccarelli, High temperature processing of brass: Constitutive analysis of hot working of Cu-Zn alloys, Mater. Sci. Eng. A 615 (2014) 331-339. http://doi.org/10.1016/j.msea.2014.07.091

Material Forming - ESAFORM 2023
Materials Research Proceedings 28 (2023) 969-976

Materials Research Forum LLC
https://doi.org/10.21741/9781644902479-106

Low-cost tooling concept for customized tube bending by the use of additive manufacturing

TRONVOLL Sigmund A.[1,a*], TREFFEN Helena[1,b], MA Jun[1,c]
and WELO Torgeir[1,d]

[1]Department of Mechanical and Industrial Engineering, Norwegian University of Science and Technology, Trondheim, 7491, Norway

[a]sigmund.tronvoll@ntnu.no, [b]helenatr@stud.ntnu.no, [c]jun.ma@ntnu.no, [d]torgeir.welo@ntnu.no

Keywords: Low-Cost Tooling, Tube Bending, Flexibility, Additive Manufacturing

Abstract. Bending is commonly used in the manufacturing of finished and semi-finished, profile-based products. However, costly profile and bend geometry dependent tooling hampers its applicability for low volume production or prototyping. Additive manufacturing (AM) offers a potential for making tools in low-volume production, which is particularly attractive for customized manufacturing and prototyping with near production intent tooling. In this research, an industrial bending process for tubes and profiles, called rotary drawing bending (RDB), is used as a case. In the RDB process, the mandrel die installed inside the tube or profile blank is of crucial importance to secure the quality of the cross-section of bent shapes. Moreover, this die is normally difficult and expensive to produce. In addition, each mandrel is tailored to a single profile's inner geometry, hence posing an obstacle to acquiring a tooling setup for multiple cross-section dimensions. As a countermeasure, a novel mandrel tooling concept is designed by using a metal rod core with an AM-sleeve fitted to the rod outside, as an easy-to-replicate solution for significantly lowering the costs of offering the capability of forming processes with different cross-section dimensions. Fused filament fabrication (FFF)—a cost-effective AM process—is used for fabricating the mandrel die with various pre-designed and optimized shapes. Using both the AM mandrel in polylactide (PLA) and the conventional metal mandrel in a series of bending experiments of AA6082-T4 Al-alloy tubes, the dimensional and qualitative results of tubes bent with different mandrels is analyzed, discussed, and compared.

Introduction

To meet the increased customer demands for product variety, the manufacturing sector is transforming from mass production to mass customized production, with higher flexibility and on-demand manufacturing [1]. However, for the die-based manufacturing processes such as metal forming, they are increasingly challenged to meet the demands on flexibility [2]. In the metal forming area, the design and fabrication of tools are normally expensive and need a long lead time, which significantly limits the transformation of conventional forming towards the customized forming. This calls for new tooling concepts and methods that allow the time-efficient, cost-effective fabrication of tools for mass customized metal forming.

Additive manufacturing (AM) used for rapid prototyping offers a potential for making tools in low-volume, customized production, which is particularly interested by the industrial metal forming processes. In recent years, some attempts at AM-based tools used in metal forming have been carried out. A typical application of AM as a method for tooling making to realize the metallic dies with optimized cooling channels in hot metal working such as hot stamping [3]. For instance, Komodromos et al. [4] employed a Directed Energy Deposition (DED) to produce the hot stamping tools with integrated cooling channels. After the DED process, the tool surfaces are ball burnished to reduce the surface roughness for more effective heat transfer. Joghan et al. [5] utilized hybrid AM method to make metal laminated forming tools. In this method, laser metal deposition is used

Material Forming - ESAFORM 2023   Materials Research Forum LLC
Materials Research Proceedings 28 (2023) 969-976   https://doi.org/10.21741/9781644902479-106

for bonding the sheets and smoothening the edges, and subsequently milling, roller burnishing, and laser treatment are applied as post-processing for improving the strength and surface finish. Similarly, a layer-laminated manufacturing method and a laser melting process were applied in the manufacturing of the dies with conformal cooling channels for extrusion processes [6]. Chantzis et al. [7] proposed a Design for AM (DfAM) method for hot stamping dies which exploits the benefit of lattice structures for reduced thermal conductivity. It shows that the proposed method can significantly improve the cooling performance of a hot stamping die with printing times reduced by at least 12% compared to traditionally manufactured AM dies. This shows that high end AM systems, as those capable of manufacturing metal parts, has a potential for improving process control and manufacturing rate.

For increased flexibility in low-volume manufacturing, more inexpensive AM-solutions would be preferred, as the associated cost and leadtime is the key parameter, rather than geometric complexity. In this case, fused filament fabrication (abbreviated FFF, also called filament-based extrusion AM) has attracted some scientific interest. Strano et al. [8] demonstrated the potential applicability and versatility of FFF as a rapid tool manufacturing technology for different applications in shearing, bending, deep drawing, and injection molding. Frohn-Sörensen et al. [9] utilized the FFF process to manufacture the tooling to draw a small series of sheet metal parts in combination with the rubber pad forming process. In addition, a variety of common polymer materials (PLA, PA, PETG, PC) were comprehensively compared in compressive and flexural tests to provide an guide for material selection in AM-based polymer tools for metal forming [10].

The body of research on AM-based tooling in metal forming is obviously more extensive than the above-summarized ones. However, most of them focus mainly on cold and hot forming of sheets, and in tooling that will be in compression, and are hence not that affected by FFFs tensile anisotropy [11]. The use of AM in making tools for complex tube bending processes has not been explored earlier, and the process that will be discussed in this paper is rotary draw bending (RDB), which is shown in Fig. 1. This is one of the more popular bending processes, and normally operates with a tooling system including 5-6 complex-shaped tools, in which tooling is one of most important issues affecting the production cost. The tools are also dependent on either or both the bend radius, profile inner- and outer-geometry. Therefore, most workshops keep a limited inventory of tooling variants, which is a significant limitation to their capability in terms of geometric flexibility.

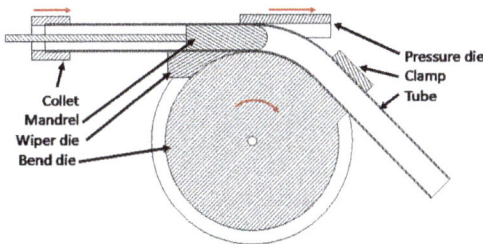

Fig. 1. Conceptual tooling layout of rotary draw bending (RDB).

The aim of this research is to develop a new cost-effective tooling concept for tube rotary draw bending processes by utilizing low-cost FFF manufacturing methods, and initially investigating its applicability to be used for mandrel bending. First, the newly proposed mandrel die concept is introduced. Secondly, using aluminium tubes with different wall thicknesses, a series of bending experiments with and without conventional mandrels and new AM-based mandrels is presented.

Material Forming - ESAFORM 2023   Materials Research Forum LLC
Materials Research Proceedings 28 (2023) 969-976   https://doi.org/10.21741/9781644902479-106

Finally, the dimensional accuracy and qualitative aspects of formed tubes is evaluated to verify the feasibility and capability of the new AM-based mandrel die concept for tube bending.

Low-Cost Tooling Concept for Customized Bending

For both product development and low production purposes using RDB, being able to quickly change the wall thickness of tubes or profiles would be beneficial, but requires change of the mandrel, which is a crucial tool for bending components with a high diameter to thickness ratio. Keeping a catalogue of mandrels suitable for all wall thicknesses could be costly, while acquisition on demand through buying or in-house machining might be preferable. Conventional mandrel shapes, as shown in Fig. 2, does however require multi-step CNC machining, which often comes with a considerable cost and lead time. It is in this domain we believe additive manufacturing could provide a significant benefit in bringing lead time and cost for changing mandrel to a minimum. As a first step in this effort, we are investigating the feasibility of using spherical end mandrels manufactured using standard FFF process equipment and standard PLA filament material.

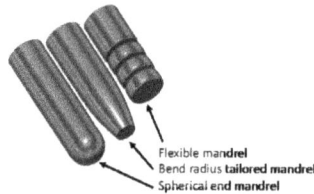

Fig. 2. Mandrel types in tube rotary draw bending.

To provide structural integrity and overcome the inter-layer bonding problems found in FFF parts, the PLA-mandrels used in this study were constructed with a steel core, while adding a bolt-on outer interchangeable PLA sleeve to provide the required geometry, as shown in Fig. 3. As the suggested hybrid mandrel system is believed to be softer than its metal counterpart, we expect that for getting a comparable result, the mandrel extension (shown in Fig. 3) would need to be increased.

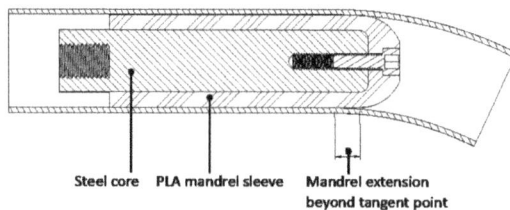

Fig. 3. Hybrid steel/PLA mandrel construction.

Experimental Setup

Using a *Star Technology 800 EVOBEND* tube bending machine, with tooling designed for Ø60 mm profiles and 222 mm bend radius, experiments with tube thicknesses of 3 and 2 mm, and mandrel extension lengths were conducted as shown in Table *1*. The manufactured bends were then evaluated based on the flattening of the cross section radially relative to the bend axis,

Material Forming - ESAFORM 2023
Materials Research Proceedings 28 (2023) 969-976

Materials Research Forum LLC
https://doi.org/10.21741/9781644902479-106

measured using a caliper. For the bend radius and outer tube diameter, experience has shown that 3 mm thickness tubes do not exhibit wrinkling defect if manufactured without mandrel, while 2 mm ones do. The 2mm tube experiments are therefore also evaluated qualitatively on the basis of the mandrel's capability of mitigating the wrinkling defect on 2 mm tubes bent without a mandrel. The steel and AM mandrels, and how they are installed in the tool system is shown in Fig. *4*.

Table 1. Experimental matrix.

Experiment #	Mandrel type	Mandrel extension (e)	Tube thickness	Samples
1	No mandrel		3 mm	2
2	Steel	0 mm	3 mm	2
3	AM	0 mm	3 mm	2
4	AM	5 mm	3 mm	2
5	AM	10 mm	3 mm	2
6	No mandrel		2 mm	2
7	AM	5 mm	2 mm	2
8	AM	10 mm	2 mm	2

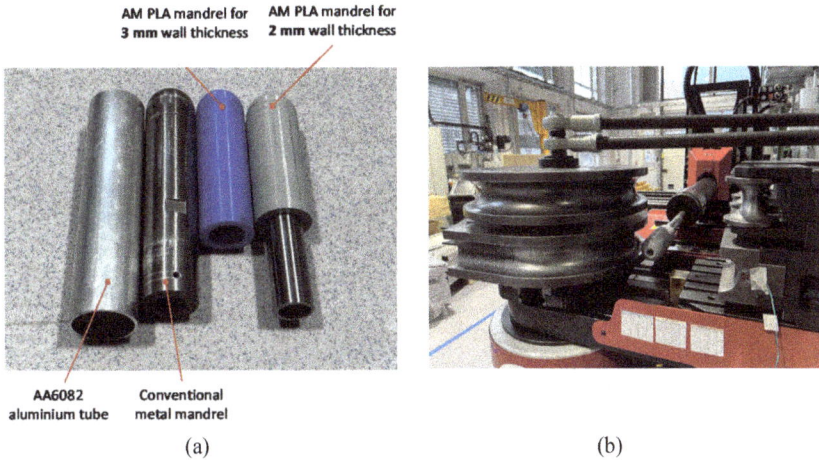

Fig. 4. (a) AM PLA mandrels and conventional metal mandrel, and (b) assembled tooling in the RDB tooling system.

Material Forming - ESAFORM 2023
Materials Research Proceedings 28 (2023) 969-976

Materials Research Forum LLC
https://doi.org/10.21741/9781644902479-106

Results and Discussion

Qualitative and quantitative results for the 3 mm thickness samples are shown in Fig. 5 and in Fig. 6, respectively. Qualitative assessment does not reveal any major difference between the different mandrels. Quantitatively, the AM mandrel provides a significant support to limit tube flattening and reduces the tube flattening with 50% compared to the samples bent without mandrel. Comparing with the steel mandrel, however, the PLA mandrel is seen to provide less support for 0 extension distance, with 1 mm larger cross section flattening (2-2.5 mm in total), suggesting that the PLA mandrel is significantly softer than the metal version. The flattening is reduced to values comparable to those for steel mandrel if the mandrel extension is increased to 10 mm for the PLA mandrel, which results in approximately 1 to 1.5 mm of flattening. Although increased mandrel extension will increase the contact pressure on the mandrel, there is no indications that the PLA mandrel is being degraded significantly, as it shows no cracks or scuffing marks.

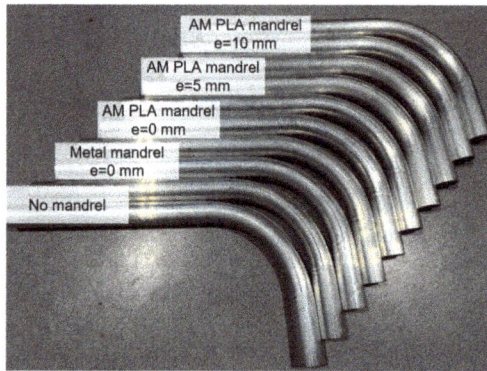

Fig. 5. Qualitative results from tube bending with 2mm wall thickness.

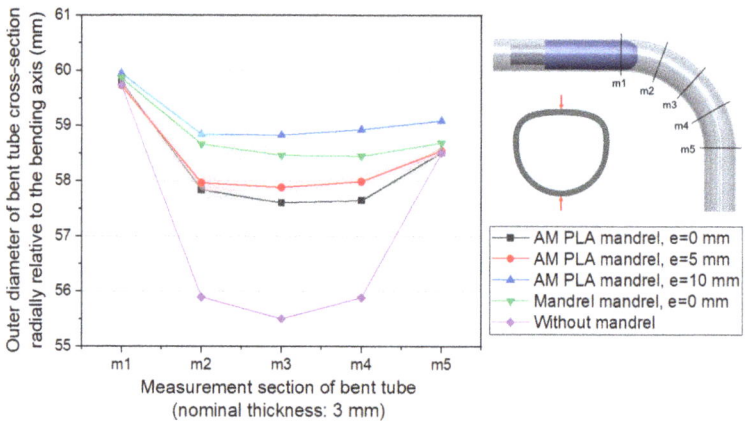

Fig. 6. Radial tube flattening from tube bending with 3 mm wall thickness.

Material Forming - ESAFORM 2023   Materials Research Forum LLC
Materials Research Proceedings 28 (2023) 969-976   https://doi.org/10.21741/9781644902479-106

Qualitative and quantitative results for the 2 mm thick samples are shown in Fig. 5 and in Fig. 6, respectively. As shown in Fig. 5, the lowered wall thickness results in pronounced intrados wrinkling when bending without a mandrel, which is effectively mitigated with the use of the AM mandrel. The samples from bending without a mandrel are left out of the quantitative results due to wrinkling. The 2 mm thickness tubes display a slightly higher flattening than for 3 mm thick tubes, with approximately 3 to 4 mm flattening using an extension distance of 5 mm, which is reduced to approximately 2.5 to 3 mm when increasing the extension distance to 10 mm. As in the case with the mandrel for 3 mm wall thickness tubes, there is in this case also no indications that the PLA mandrel is being degraded significantly, as it shows no cracks or scuffing marks as well.

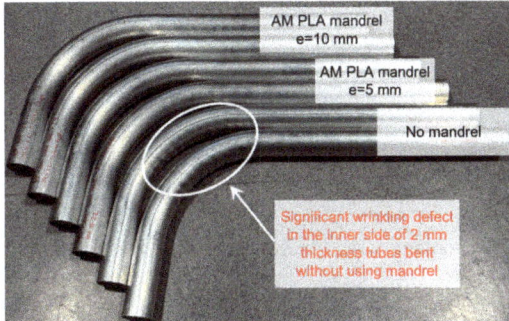

Fig. 7. Qualitative results from tube bending with 2mm wall thickness.

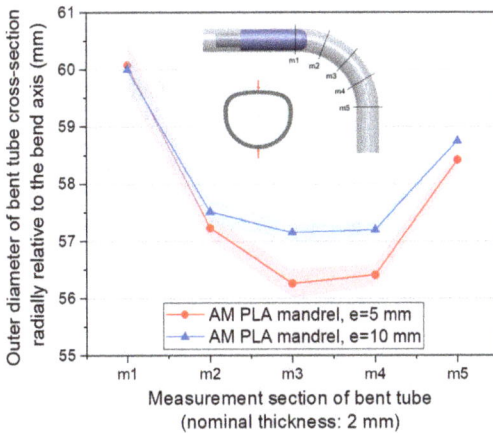

Fig. 8. Radial tube flattening from tube bending with 2 mm wall thickness.

Summary

The proposed tooling concept with a hybrid steel-PLA mandrel is found to provide sufficient support and structural integrity to significantly reduce both the cross section flattening and wrinkling for the investigated cases of tube bending using rotary draw bending of aluminum tubes. With the simplicity of the manufacturing using additive manufacturing, and the low cost associated with the material and manufacturing system for fused filament fabrication, using this type of mandrels provides a significant process improvement over bending without mandrel, and on par with steel mandrels if the mandrel extension length is increased.

As the concept explored in this paper seems promising, replacing other complex tools, as the wiper die could provide additional value. We would also suggest investigating the method's applicability on bending of tubes with higher material strength as steel or titanium. We also believe that further research should target investigating the durability of these AM mandrels, to investigate its applicability for manufacturing larger volumes.

Acknowledgements

The authors gratefully acknowledge the support from NTNU Aluminum Product Innovation Center (NAPIC), and through the *TaFF* project (*Digitally-Controlled, Toolless and Flexible Forming*) in collaboration with *Benteler Automotive Raufoss AS*, funded through *The Research Council of Norway* through grant number 317777.

References

[1] J. Cao, M. Banu, Opportunities and Challenges in Metal Forming for Lightweighting : Review and Future Work, J. Manuf. Sci. Eng. 142 (2020) 1–24. https://doi.org/10.1115/1.4047732

[2] D.Y. Yang, M. Bambach, J. Cao, J.R. Duflou, P. Groche, T. Kuboki, A. Sterzing, A.E. Tekkaya, C.W. Lee, Flexibility in metal forming, CIRP Ann. - Manuf. Technol. 67 (2018) 743-765. https://doi.org/10.1016/j.cirp.2018.05.004

[3] D. Chantzis, X. Liu, D.J. Politis, O. El Fakir, T.Y. Chua, Z. Shi, L. Wang, Review on additive manufacturing of tooling for hot stamping, Int. J. Adv. Manuf. Technol. 109 (2020) 87-107. https://doi.org/10.1007/s00170-020-05622-1

[4] A. Komodromos, F. Kolpak, A.E. Tekkaya, Manufacturing of Integrated Cooling Channels by Directed Energy Deposition for Hot Stamping Tools with Ball Burnished Surfaces, BHM Berg-Und Hüttenmännische Monatshefte. 167 (2022) 428-434. https://doi.org/10.1007/s00501-022-01264-w

[5] H. Dardaei Joghan, M. Hahn, J.T. Sehrt, A.E. Tekkaya, Hybrid additive manufacturing of metal laminated forming tools, CIRP Ann. 00 (2022) 1-4. https://doi.org/10.1016/j.cirp.2022.03.018

[6] R. Hölker, M. Haase, N. Ben Khalifa, A.E. Tekkaya, Hot Extrusion Dies with Conformal Cooling Channels Produced by Additive Manufacturing, Mater. Today Proc. 2 (2015) 4838–4846. https://doi.org/10.1016/J.MATPR.2015.10.028

[7] D. Chantzis, X. Liu, D.J. Politis, Z. Shi, L. Wang, Design for additive manufacturing (DfAM) of hot stamping dies with improved cooling performance under cyclic loading conditions, Addit. Manuf. 37 (2021) 101720. https://doi.org/10.1016/J.ADDMA.2020.101720

[8] M. Strano, K. Rane, M.A. Farid, V. Mussi, V. Zaragoza, M. Monno, Extrusion-based additive manufacturing of forming and molding tools, Int. J. Adv. Manuf. Technol. (2021). https://doi.org/10.1007/s00170-021-07162-8

[9] P. Frohn-sörensen, M. Geueke, T.B. Tuli, M. Manns, B. Engel, P. Manuskript, P. Frohn-sörensen, M. Geueke, T.B. Tuli, M. Manns, B. Engel, 3D printed prototyping tools for flexible sheet metal drawing, (2021) 1-26

Materials Research Forum LLC

https://doi.org/10.21741/9781644902479-106

[10] P. Frohn-Sörensen, M. Geueke, B. Engel, B. Löffler, P. Bickendorf, Compressive and flexural material properties of PC , PLA , PA and PETG for additive tooling in sheet metal forming, (2022). https://doi.org/10.31224/2239

[11] Q. Sun, G.M. Rizvi, C.T. Bellehumeur, P. Gu, Effect of processing conditions on the bonding quality of FDM polymer filaments, Rapid Prototyp. J. 14 (2008) 72-80. https://doi.org/10.1108/13552540810862028

Material Forming - ESAFORM 2023
Materials Research Proceedings 28 (2023) 977-986

Materials Research Forum LLC
https://doi.org/10.21741/9781644902479-107

Experimental investigation of the fluid-structure interaction during deep drawing of fiber metal laminates in the in-situ hybridization process

KRUSE Moritz [1,a,*] and BEN KHALIFA Noomane [1,2,b]

[1]Leuphana University of Lüneburg, Institute for Production Technology and Systems (IPTS), Universitätsallee 1, 21335 Lüneburg, Germany

[2]Helmholtz-Zentrum Hereon, Institute of Material and Process Design, Max-Planck-Straße 1, 21502 Geesthacht, Germany

[a]moritz.kruse@leuphana.de, [b]ben_khalifa@leuphana.de

Keywords: Fiber Metal Laminates, Deep Drawing, In-Situ Hybridization, Fluid-Structure Interaction

Abstract. Matrix accumulations, buckling and tearing of fibers and metal sheets are common defects in the deep drawing of fiber metal laminates. The previously developed in-situ hybridization process is a single-step method for manufacturing three-dimensional fiber metal laminates (FML). During the deep drawing of the FML, a low-viscosity thermoplastic matrix is injected into the dry glass fiber fabric layer using a resin transfer molding process. The concurrent forming and matrix injection results in strong fluid-structure interaction, which is not yet fully understood. To gain a better understanding of this interaction and identify possible adjustments to improve the process, an experimental form-filling investigation was conducted. Using a double dome deep drawing geometry, the forming and infiltration behavior were investigated at different drawing depths with full, partial, and no matrix injection. Surface strain measurements of the metal blanks, thickness measurements of the glass fiber-reinforced polymer layer, and optical analyses of the infiltration quality were used to evaluate the results.

Introduction

Multi-materials allow the engineering of material properties to the specific needs of a given application [1]. Fiber metal laminates (FML) are a multi-material that consists of several layers of fiber-reinforced polymers (FRP) and metal. They were initially developed to address the challenges associated with the use of fiber-reinforced polymers, such as low impact resistance [2]. The material combination offers several advantageous properties, such as improved impact resistance, high strength-to-weight ratio, and little crack propagation [3]. Therefore, FMLs are often used in aircraft body structures, where simple geometries are easy to manufacture [4]. However, their use in other industries, such as the automotive industry, is constrained by the cost and time required to manufacture FML parts with more complex geometries. These parts often require separate forming and bonding steps, as cured FMLs can only be formed to a limited extent [5]. As a result, several investigations have been conducted on one-step manufacturing processes for formed FML, often using half-cured [6, 7] or thermoplastic [8, 9] pre-impregnated fiber materials in a deep drawing process. However, in many cases, strong interactions between fiber, metal and the high-viscosity matrix flow during forming can lead to delamination, inhomogeneous thicknesses of the FRP-layer as well as wrinkling and tearing of the metal sheet and fibers [8,10].

A new manufacturing process for FML was introduced previously [11]. The in-situ hybridization process combines deep drawing with a thermoplastic resin transfer molding process (T-RTM). During deep drawing, a reactive low-viscosity monomeric matrix is injected into the fabric layer (see Fig. 1). After forming, the matrix polymerizes into a thermoplastic and creates

Material Forming - ESAFORM 2023
Materials Research Proceedings 28 (2023) 977-986

Materials Research Forum LLC
https://doi.org/10.21741/9781644902479-107

the interface with the metal sheets. Due to the low matrix viscosity, fibers and metal are in direct contact during the forming process. While a low-viscosity matrix is advantageous for fluid flow and fabric drapability during forming, friction between metal and fabric is higher because of high local normal loads at the roving crossings [12]. Previous investigations have shown that this contact under high normal pressures can reduce the formability of the metal sheet by restricting relative movement between fibers and metal [13]. In deep drawing tests of dry FMLs without matrix injection, tool lubrication led to lower metal and fiber strain [14]. Lower friction between fabric and metal improved the forming further by allowing for better interlayer movement.

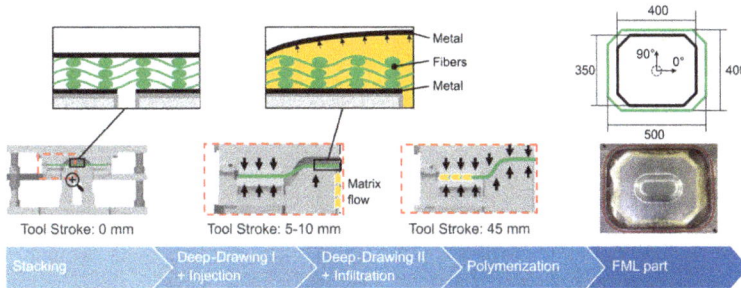

Fig. 1. Combined deep drawing and thermoplastic resin transfer molding (T-RTM) process for the manufacturing of fiber metal laminate parts, as introduced by Mennecart et al. [11].

Double dome geometry parts with a drawing depth of 45 mm could be successfully manufactured without wrinkling or tearing. However, the influence of forming on matrix flow and vice versa has to be investigated to understand fluid-structure interaction in the process and improve infiltration quality and forming. Mennecart et al. [11,15] found that bulging of the metal sheets can occur in regions of high internal pressure due to the matrix injection and low external pressure when no contact with the female die is established. In this work, different amounts of matrix are injected to manufacture parts with different drawing depths, aiming to provide a better understanding of the form-filling behavior. This is necessary to identify the most influential parameters in the process, which are different from those in pure metal deep drawing due to fluid-structure interaction. As a result, possible process improvements are derived.

Materials and Method
The FML specimen consists of a metal layer on the top and bottom, with six layers of a twill 2/2 woven E-glass fabric (280 g/m², Interglas 92125 FK800) with good drapability in between. The metal sheets are made of DC04 (1.0338) with a thickness of 1 mm. The dimensions are shown in Fig. 1. The lower metal sheet has a 19 mm diameter hole in the center for matrix injection. A polytetrafluoroethylene (PTFE) foil (0.025 mm) is used between the top metal sheet and tool for lubrication. After stacking the layers in the tool, deep drawing to 15 % of the final drawing depth is performed to plastically deform the metal sheets and achieve sealing between the sheet and punch due to the increased contact pressure. The thermoplastic matrix (Arkema Elium® 150), with an initial viscosity of 100 mPas, mixed with 2.5 % hardener (dibenzoyl peroxide Perkadox® GB-50X), is injected while deep drawing is continued. Deep drawing is performed with a blank holder force of 190 kN, a punch velocity of 1 mm/s and a tool temperature of 70 °C to accelerate matrix polymerization after forming. When the final drawing depth of 45 mm is reached and the tool is fully closed, the position and blank holder force are held for 20 minutes to allow the matrix to fully polymerize before demolding.

Material Forming - ESAFORM 2023   Materials Research Forum LLC
Materials Research Proceedings 28 (2023) 977-986   https://doi.org/10.21741/9781644902479-107

The fluid-structure interaction and form-filling behavior during the process are investigated under three different conditions: With full infiltration, partial infiltration and without matrix injection. Deep drawing is performed up to 15 %, 50 % and 100 % of the total drawing depth, where the press is halted and the matrix polymerizes. Additionally, three more drawing depths (80 %, 82.5 %, and 85 %) are analyzed for the full infiltration condition. An overview of the experiments is given in Table 1. The target height of 1 mm for the fabric layer corresponds to a fiber volume fraction of 65% and a matrix volume fraction of 35% for the fabric material used in the experiments. The surface area of the metal sheet is 120,000 mm², so the total matrix volume needed for full part infiltration can be calculated to be 42 ml. Approximately 120 ml of matrix are needed to fill the injection channel, so 140 ml are used for the partial infiltration experiments. For the full infiltration experiments, 300 ml are used to ensure that the whole part is infiltrated and any remaining air is flushed out. The formed metal sheets are evaluated with a GOM ARGUS optical surface strain measurement system. The thickness of the glass fiber reinforced polymer (GFRP) layer is measured with a GOM ATOS laser scanning system.

Table 1. Experimental design.

Matrix amount [ml]	Drawing depths [%]
0 (No infiltration)	10, 50, 100
140 (Partial infiltration)	10, 50, 100
300 (Full infiltration)	10, 50, 80, 82.5, 85, 100

In the original process [11], the inner side of the metal sheets is ground in 0° and 90° direction as well as in small circles. A bonding agent (Dynasalan® Glymo by Evonik) is then applied to improve adhesion. For the experiments in this investigation, the inner side of the metal sheet is not ground but treated with a release agent (Henkel Loctite® Frekote HMT2) so that each layer can be investigated separately after manufacturing.

Results

A comparison of parts manufactured with both metal sheet surface treatment methods showed no difference in observed metal strains or fiber draping, which demonstrates the validity of the experiments performed with the release agent. In contrast, previous dry deep drawing experiments without matrix injection demonstrated a strong influence of surface treatment on metal strains [14]. This suggests that the matrix has a lubricating effect on the contact between the fabric and metal, which reduces friction even when the metal sheets are ground.

Full infiltration.

Fig. 2 shows the development of the geometry and strains in the upper metal sheet of the FML during deep drawing. It was previously shown that the strains in the upper metal sheet are more critical regarding tearing than in the lower metal sheet [14]. The vertical distance (major strain) to the forming limit curve (FLC) is used as a measure of how susceptible the metal is to tearing. The applied FLC was obtained as described by Mennecart et al. [13] for DC04 with glass fiber twill fabric interlayer. Because of biaxial tensile strains, the punch radius is the most critical area, even though the die radius has higher major strains.

In the flange, the high blank holder pressure causes high compaction and low permeability of the fabric. As a result, the matrix cannot flow in the flange fast enough during the injection. This causes bulging of the upper metal sheet due to high internal pressure and low external pressure because of the lack of contact with the female die. High friction is present in the flange area between the upper metal sheet and fabric as well as the fabric and lower metal sheet. This hinders the flow of the upper metal sheet and leads to forming similar to a hydro stretch forming process with increased biaxial tensile strains in the bottom and wall area of the part, as can be seen at 50

Material Forming - ESAFORM 2023   Materials Research Forum LLC
Materials Research Proceedings 28 (2023) 977-986   https://doi.org/10.21741/9781644902479-107

% drawing depth. In the lower metal sheet, no bulging can occur in the part bottom because external pressure is established due to punch contact. However, slight bulging does occur in the wall area where punch contact is only established during the compression phase towards the end of the deep drawing. The injection is completed at a drawing depth of 45 %. During the injection, the pressure increases linearly from 1 MPa to a maximum of 2.2 MPa. With increasing drawing depth, the injection pressure increases due to higher flow resistance in the die radius and strain hardening in the bulged metal sheet.

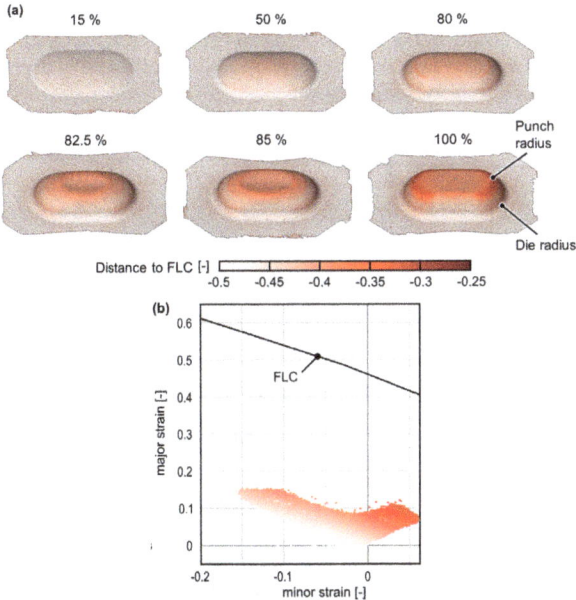

Fig. 2. Development of a part in the in-situ hybridization process with full infiltration (a) Surface strains of the upper metal sheet at different drawing depths (b) Forming limit diagram of the upper metal sheet at 100 % drawing depth (FLC: Forming Limit Curve).

At 80 %, higher strains in the punch radius are present because of metal contact with the fabric, causing the metal to bend over the edge. The wall and bottom area are still bulged. Between 80 % and 82.5 %, the bulge collapses, starting from the center. Comparing the bulge height with the drawing depth shows that die contact is established between 80 % and 82.5 %. When contacting, the die causes the bulge to suddenly collapse inside to achieve a stable condition again. After the brief die contact, there is no contact between the metal sheet and the die at 82.5 % until the die contacts again with further deep drawing between 82.5 % and 85 %. With further closing of the punch and die, the matrix is squeezed to the outside in a donut shape and to the flange with high internal fluid pressure. This is similar to a compression T-RTM process. Because of the bulging and squeezing of the matrix to the outside, high internal fluid pressures lead to high strains in the punch radius of the fully formed upper sheet.

The matrix distribution is shown in detail in Fig. 3a. Because the permeability of the chosen fabric is slightly higher in the 0° direction than in the 90° direction, the flow front initially advances in an oval shape. At 50 % drawing depth, when the injection is complete, the flow front has not reached the edges of the metal sheets yet. Instead, the matrix accumulates in the bulge due to the high blank holder pressure on the fabric. Furthermore, an air cluster in the middle of the bulge and pores can be observed in the matrix layer. One possible reason for this is that air from the injection channel is pushed into the fabric layer before the matrix. Due to the high injection pressure, matrix and air could mix, resulting in air being trapped inside the matrix bulge. Another probable reason for at least some of the air is gas evolution from the matrix during polymerization. Under high temperatures and pressures, as well as with the formation of thick layers, the thermoplastic matrix can produce carbon dioxide as a byproduct during polymerization. Because of the pores and air cluster, the infiltration quality of the fully formed part is poor in some areas. In the final part, air is trapped inside the GFRP layer as small pores and causes insufficiently infiltrated spots in high compression areas like punch and die radius.

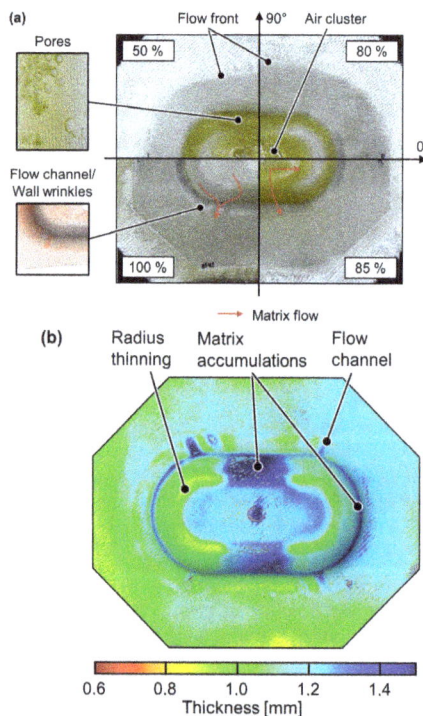

Fig. 3. GFRP layer (a) Matrix flow during the evolution of the geometry (b) Thickness of the cured GFRP layer at 100 % drawing depth.

At 80 %, the flow front has advanced slightly because some matrix is squeezed out of the bulge into the flange. Furthermore, the fabric layer thins out in the punch radius due to high normal compression. When the bulge collapses, the trapped air and matrix are pushed outside and into the flange and wall area. Because of the matrix flow to the outside, the punch radius infiltration is better again at that point. At 85 %, the flange is fully infiltrated. After the collapse, high matrix flow velocities from the part bottom to the flange lead to the development of flow channels in the GFRP that are still present at 100 %. Towards the end of the deep drawing, the part bottom and wall are compressed between the punch and die. The remaining matrix and air in the bulge are squeezed out to the flange with very high fluid pressures. The matrix flow from bottom to flange predominantly occurs over the edges in the 90° direction, as the double-curved geometry in 0° direction causes high biaxial tensile strains. These result in high compactions and lower permeability of the matrix layer with increasing drawing depth. Due to the fluid-structure interaction, flow channels are formed in the die radius next to the double-curved geometry (Fig. 3). The matrix flows predominantly in the 90° direction in the single-curved area. However, wall wrinkles can occur in free-forming zones without punch and die contact due to tangential compressive stresses in the metal sheet. These wrinkles typically occur in the double-curved wall parts in 0° direction. The stress superposition of the internal fluid pressure from the matrix flow and the tangential compressive stresses from the metal sheet causes the development of the flow channels and wall wrinkles in the transition zone between single and double-curved die radii. Wall wrinkling does not occur in the upper metal sheet because die contact is established due to fluid pressure and forming.

The strong fluid-structure interaction causes an uneven thickness distribution in the GFRP layer of the fully formed part, as shown in Fig. 3b. The flange area is relatively uniform and close to the target thickness of 1 mm because the compression in this area is mainly determined by the blank holder pressure. However, the resulting thickness in the bottom and wall area is influenced by the fluid-structure interaction. The punch radius is thinned out due to high fabric compressions from the metal forming process. The flow channels are still visible in the final part thickness. The presence of a flow channel on the top left but not on the bottom left side may indicate that the punch and die were not perfectly aligned, which affected the matrix flow. In the metal strains, no skew is observed. The part bottom is approximately 1.2 mm thick, while the wall area in 90° direction is around 1.5 mm thick due to the matrix flow and resulting matrix accumulations. Another matrix accumulation is observed in the wall area in 0° direction, where the matrix is squeezed out of the punch and die radius into the wall area. In these experiments, the punch was moved using displacement control. However, the bottom thickness suggests, that the tool was not completely closed so no full compression was present in some parts of the wall and bottom. Further compression of approximately 0.2 mm would likely improve the homogeneity of the thickness by reducing the bottom thickness and further compressing the wall area.

Comparison with partial and no infiltration.

In general, no difference in the strains and forming behavior is observed between partial matrix injection and no matrix injection, as shown in Fig. 4. Only very slight bulging in the wall area occurs with partial infiltration, while no bulging takes place during dry deep drawing. At 15 % drawing depth, almost no plastic strains are observed. At 50 %, higher strains than with full matrix injection are present in the punch radius as the upper metal sheet is in contact with the fabric during the entire process. With full infiltration, there are no strain concentrations at 50 % yet but biaxial tensile strains are present in the entire bottom and wall area due to bulging. Overall, the bulging causes higher strains in the final part, in particular in the bottom and punch radius. When no bulging and matrix accumulation takes place, less matrix has to move towards the flange during the compression phase of the process. Therefore, fluid pressure and flow velocity are lower, which prevents the development of flow channel and wall wrinkles in the partial infiltration experiments.

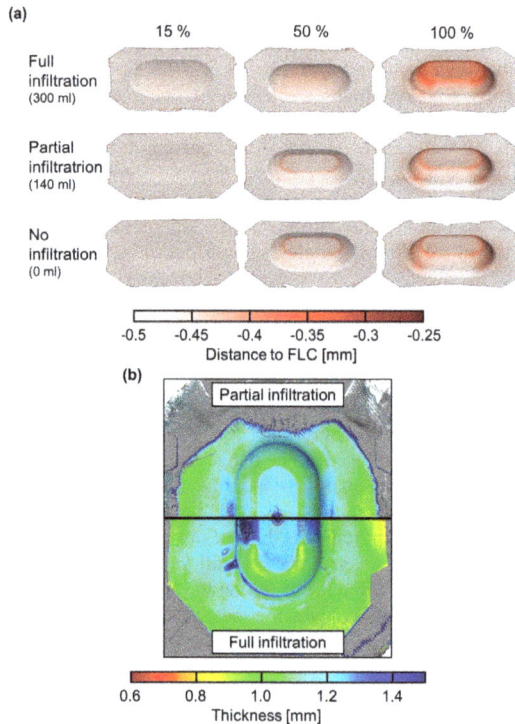

Fig. 4. Comparison with full, partial and without infiltration (a) Development of surface strains in the upper metal sheet at different drawing depths (b) Thickness of the GFRP layer with partial and full infiltration at 100 % drawing depth.

When less matrix is injected, the injection is completed earlier and less bulging is developed on the upper sheet which explains lower injection pressures (1.5 MPa vs 2.2 MPa). With partial infiltration, no air cluster was observed and less matrix accumulates. However, pores are also present at all stages of the process.

When comparing the flow front of the GFRP layer with partial infiltration in Fig. 4b with the flow front in Fig. 3a (full infiltration), a different evolution is observed. With full infiltration, the flow front rather advances in 0° direction because of the fabric's permeability. With partial infiltration, the flow front is further advanced in 90° direction. When comparing the parts at 15 % drawing depth, the flow front with partial infiltration only reaches the die radius, while with full infiltration, parts of the flange have already been infiltrated. With less matrix, the infiltration of the flange area takes place during the later stages of the deep drawing, by squeezing the matrix from part bottom and wall to the flange area. However, the fabric in the punch radius in 0° direction is already compacted further so that the matrix flows in 90° direction when infiltrating the flange. Thus, at small drawing depths and little plastic deformation of the metal sheets, the flow behavior is mainly influenced by the fabric's directional permeabilities. However, with increasing drawing depth, the flow is more influenced by the fabric compaction due to forming.

With partial matrix injection, the thickness distribution of the GFRP layer is much more homogeneous, as shown in Fig. 4b. The part bottom is slightly thinner with partial infiltration and almost no matrix accumulations are present. This could imply that the compression between punch and die might not be necessary with smaller matrix amounts because the compression due to the forming of the upper metal sheet is sufficient to squeeze the matrix into the flange area during deep drawing.

Discussion

Initially, it was expected, that more matrix would be beneficial for the process because it would flush the remaining air out of the fabric layer into the flange and then out of the part. However, this investigation showed that using more matrix is detrimental to the forming and infiltration quality. The use of another matrix system or evacuating the tool might improve the infiltration quality and reduce pores. It should be noted that it was not possible to evaluate the GFRP layer without polymerizing in this investigation. Hence, slight differences might occur in the actual process, such as the flow front not advancing as far as observed, or less pore formation because the matrix accumulations in the bulge are only present for shorter periods of time.

Bulging of the upper metal sheet and matrix accumulations in the early stages of the process should be avoided altogether. Several strategies can be used to avoid or reduce bulging. In general, early infiltration of the whole fabric layer is advantageous because matrix flow at later deep drawing stages is heavily influenced by the part geometry and fabric compression due to forming. Local high fabric compressions lead to inhomogeneous infiltration and possibly dry spots. Ideally, an improved sealing between the punch and lower metal sheet would allow matrix injection prior to deep drawing. Alternatively, deep drawing could be paused at 10–15 % drawing depth to allow for easier infiltration of the whole part during matrix injection. The injection pressure should be reduced so that no bulging of the metal sheet can occur. In that case, the flange is also infiltrated in an RTM process during matrix injection. The injection time would increase significantly because the fabric's permeability in the flange would be very low due to high compactions. During the deep drawing hold, the blank holder force could temporarily be reduced to allow for faster infiltration. Furthermore, a matrix with lower viscosity would increase the flow velocity. By using only as much matrix as is needed to fill the whole part, injection time and fluid-structure interaction can be reduced. A decrease in punch velocity would further reduce internal fluid pressures because the matrix has more time to be distributed. However, the matrix and tool temperature have to be chosen to avoid increased matrix viscosities during forming.

Summary

A form-filling investigation with different matrix amounts was performed to enhance the understanding of geometry evolution and fluid-structure interaction in the in-situ hybridization process. High matrix amounts lead to stronger fluid-structure interaction, particularly bulging of the upper metal sheet during injection, which results in matrix accumulations in the final part. The matrix flow is significantly influenced by the local fabric compaction due to compression from the formed metal sheets. Inversely, high fluid pressures can influence the metal forming, resulting in increased metal strains and the development of wrinkling in areas where stress superpositions from fluid pressure and forming occur. Overall, using smaller amounts of matrix leads to weaker fluid-structure interaction, resulting in more homogeneous thickness distributions and reduced metal strains in the final part. However, pores are still present in all manufactured parts, possibly due to gas development from the matrix or remaining air in the part and tool because no vacuum was applied before the injection. Apart from evacuating the tool, strategies were presented to improve the process by avoiding bulging. Lower injection pressures, temporarily pausing the deep drawing process or reducing the blank holder force during injection could lead to reduced fluid pressures and fluid-structure interaction. The derived process improvements should be investigated in the future. Presumably, more complex geometries and higher drawing depths can be achieved with adjusted process parameters.

Material Forming - ESAFORM 2023   Materials Research Forum LLC
Materials Research Proceedings 28 (2023) 977-986   https://doi.org/10.21741/9781644902479-107

Acknowledgements

The authors would like to thank the German Research Foundation (DFG) for funding the projects BE 5196/4-1 and BE 5196/4-2. The matrix system and hardener were kindly provided by the Arkema Group. The bonding agent was kindly provided by Evonik Industries AG. The authors would like to thank Mr. Marvin Gerdes for the help in performing experiments and Mr. Henrik O. Werner for the help in planning experiments and discussing results.

References

[1] S. Bruschi, J. Cao, M. Merklein, J. Yanagimoto, Forming of metal-based composite parts, CIRP Annals 70 (2021) 567-588. https://doi.org/10.1016/j.cirp.2021.05.009

[2] S. Krishnakumar, Fiber Metal Laminates — The Synthesis of Metals and Composites, Mater. Manuf. Process. 9 (1994) 295–354. https://doi.org/10.1080/10426919408934905

[3] H.E. Etri, M.E. Korkmaz, M.K. Gupta, M. Gunay, J. Xu, A state-of-the-art review on mechanical characteristics of different fiber metal laminates for aerospace and structural applications, Int. J. Adv. Manuf. Technol. 123 (2022) 2965–2991. https://doi.org/10.1007/s00170-022-10277-1

[4] A. Vlot, Glare: History of the development of a new aircraft material, Kluwer Acad. Publ, Dordrecht, 2001. https://doi.org/10.1007/0-306-48398-X

[5] H. Blala, L. Lang, S. Khan, L. Li, A comparative study on the GLARE stamp forming behavior using cured and non-cured preparation followed by hot-pressing, Int. J. Adv. Manuf. Technol. 115 (2021) 1461-1473. https://doi.org/10.1007/s00170-021-07196-y

[6] H. Blala, L. Lang, L. Li, S. Alexandrov, Deep drawing of fiber metal laminates using an innovative material design and manufacturing process, Compos. Commun. 23 (2021) 100590. https://doi.org/10.1016/j.coco.2020.100590

[7] T. Heggemann, W. Homberg, H. Sapli, Combined Curing and Forming of Fiber Metal Laminates, Procedia Manuf. 47 (2020) 36–42. https://doi.org/10.1016/j.promfg.2020.04.118

[8] A. Rajabi, M. Kadkhodayan, M. Manoochehri, R. Farjadfar, Deep-drawing of thermoplastic metal-composite structures: Experimental investigations, statistical analyses and finite element modeling, J. Mater. Process. Technol. 215 (2015) 159-170. https://doi.org/10.1016/j.jmatprotec.2014.08.012

[9] T. Wollmann, M. Hahn, S. Wiedemann, A. Zeiser, J. Jaschinski, N. Modler, N. Ben Khalifa, F. Meißen, C. Paul, Thermoplastic fibre metal laminates: Stiffness properties and forming behaviour by means of deep drawing, Arch. Civ. Mech. Eng. 18 (2018) 442-450. https://doi.org/10.1016/j.acme.2017.09.001

[10] Z. Ding, H. Wang, J. Luo, N. Li, A review on forming technologies of fibre metal laminates, Int. J. Lightweight Mater. Manuf. 4 (2021) 110-126. https://doi.org/10.1016/j.ijlmm.2020.06.006

[11] T. Mennecart, H. Werner, N. Ben Khalifa, K.A. Weidenmann, Developments and Analyses of Alternative Processes for the Manufacturing of Fiber Metal Laminates, in: Volume 2: Materials; Joint MSEC-NAMRC-Manufacturing USA, American Society of Mechanical Engineers, 2018. https://doi.org/10.1115/MSEC2018-6447

[12] M. Kruse, H.O. Werner, H. Chen, T. Mennecart, W.V. Liebig, K.A. Weidenmann, N. Ben Khalifa, Investigation of the friction behavior between dry/infiltrated glass fiber fabric and metal sheet during deep drawing of fiber metal laminates, Prod. Eng. Res. Devel. (2022). https://doi.org/10.1007/s11740-022-01141-y

[13] T. Mennecart, S. Gies, N. Ben Khalifa, A.E. Tekkaya, Analysis of the Influence of Fibers on the Formability of Metal Blanks in Manufacturing Processes for Fiber Metal Laminates, JMMP 3 (2019) 2. https://doi.org/10.3390/jmmp3010002

[14] M. Kruse, J. Lehmann, N. Ben Khalifa, Parameter Investigation for the In-Situ Hybridization Process by Deep Drawing of Dry Fiber-Metal-Laminates, in: WGP 2022, LNPE, 2023. https://doi.org/10.1007/978-3-031-18318-8_13

Material Forming - ESAFORM 2023
Materials Research Forum LLC

Materials Research Proceedings 28 (2023) 977-986
https://doi.org/10.21741/9781644902479-107

[15] T. Mennecart, L. Hiegemann, N.B. Khalifa, Analysis of the forming behaviour of in-situ drawn sandwich sheets, Procedia Eng. 207 (2017) 890-895. https://doi.org/10.1016/j.proeng.2017.10.847

Material Forming - ESAFORM 2023
Materials Research Proceedings 28 (2023) 987-996

Materials Research Forum LLC
https://doi.org/10.21741/9781644902479-108

Modification of the surface integrity of powder metallurgically produced S390 via deep rolling

HERRMANN Peter[1,a], HERRIG Tim[1,b] and BERGS Thomas[1,2,c]

[1]Laboratory for Machine Tools and Production Engineering (WZL) of RWTH Aachen University, Campus-Boulevard 30, 52074 Aachen

[2]Fraunhofer Institute for Production Technology IPT, Steinbachstr. 17, 52074 Aachen

[a]p.herrmann@wzl.rwth-aachen.de, [b]t.herrig@wzl.rwth-aachen.de, [c]t.bergs@wzl.rwth-aachen.de

Keywords: Deep Rolling, Powder-Metallurgical Steel, Surface Integrity, Residual Stress

Abstract. Fine blanking is an economical process for manufacturing sheet metal workpieces with high sheared surface quality. When machining high-strength steels, material fatigue leads to increased punch wear, which reduces the economic efficiency of the process. This fatigue of the cutting edge and lateral punch surface can be counteracted by mechanical surface treatments. Deep rolling has proved particularly useful for such surface modification, as it allows both: machining of the lateral punch surface and the application of the cutting edge rounding required for fine blanking. For the precise design of the fine blanking punch contour especially the macroscopic deformation of the workpiece is decisive. In this paper, the possibility of specifically modifying the surface integrity of hardened and powder metallurgically produced S390 by means of the incremental surface treatment process deep rolling is investigated. By varying the decisive process parameters rolling pressure, ball diameter and step over distance, their influence on surface integrity is determined. The surface integrity is afterwards characterized by macro hardness, surface topography and residual stress state and microstructural images.

Introduction

Due to the limited availability of geo resources, their use has to be made more efficient by reducing demand and by responsible handling. In mechanical engineering, this results in high demands on the functional integration of technical components. From an economic point of view, short and efficient process chains are required, from which the need for fewer and technologically more complex tools derives. Thus, the demand on the dimensional accuracy and the mechanical properties of tools increase [1].

In particular, fine blanking offers a suitable potential for the economical production of functional components with high quality, due to the high sheared surface quality. Fine blanked components often have to withstand significant loads, which is why high-strength materials are increasingly used in fine blanking [2]. The challenge in processing higher-strength materials is to control the increased tool loads [3]. The highest stress is present on the lateral surface of fine blanking punches. This challenge can be met by improving wear resistance of active surfaces of fine blanking punches.

Deep rolling is a promising technology for increasing the service life of alternately loaded fine blanking punches [4]. Deep rolling is an industrially established process that is used to increase the service life of dynamically loaded components [5]. The process principle is based on incremental elastic-plastic deformation of the component edge zone with a hard rolling body. The local forming of the surface-near edge zone leads to a smoothing of the roughness peaks, induction of residual compressive stresses and strain hardening. These edge zone properties have proven a positive effect on the fatigue strength of components [6].

Material Forming - ESAFORM 2023   Materials Research Forum LLC
Materials Research Proceedings 28 (2023) 987-996   https://doi.org/10.21741/9781644902479-108

The cause-and-effect relations between the process kinetics in incremental forming processes, such as deep rolling, and the resulting surface integrity are widely known phenomenologically. However, to extend the application field of deep rolling to include the knowledge-based design of deep rolling processes, material-specific knowledge is required. The overall objective of the studies performed is thereby to reduce abrasive wear on fine blanking punches by means of introducing compressive stresses and surface smoothing. For this reason, specific cause-and-effect relations between the process parameters of deep rolling and the surface integrity of powder metallurgically produced S390 high-speed steel are investigated in this work by means of an experimental parameter study. Powder metallurgically produced S390 high-speed steel is a typical material for fine blanking punches that are conventionally hardened and PVD-coated [7].

Materials and Methods
First, the material of the cylindrical hardened S390 specimens is discussed. Then the experimental setup for the deep rolling within a lathe is presented and the experimental plan is explained. Finally, the measurement methods and the parameters obtained are presented.
The cylindrical specimens (diameter d = 49 mm; length l =60 mm) were set to a hardness of 722 HV30 (61.1 HRC) in preparation for the test. The final specimen geometry was then adjusted by grinding. The material used is the powder-metallurgical S390 from BÖHLER (VOESTALPINE BÖHLER EDELSTAHL GMBH & CO KG, Kapfenberg, Austria), which has the chemical composition given in table 1.

Table 1. Chemical Composition of powder-metallurgical S390 in m-% [8].

C	Si	Mn	P	S	Cr	Mo	Ni	V	W	Co	O
1.63	0.30	0.26	0.018	0.018	4.91	2.28	0.20	5.12	10.09	8.32	0.0041

Due to the chemical composition and the powder-metallurgical microstructure, the compressive strength of the material after heat treatment is in the range of 2,800 to 3,900 MPa. At the same time, the material achieves high toughness: with a hardness of up to 68 HRC, bending strengths of over 4,000 MPa are achieved. The material is characterized by a homogeneous microstructure and, in particular, by the alloying elements tungsten and cobalt, by high wear and high-temperature strength. [8]
Deep Rolling Setup.
The practical test was carried out with a hydrodynamic deep rolling tool from ECOROLL (ECOROLL AG, Celle, Germany) on a conventional lathe. Fig. 1 shows a schematic representation of the test setup and the operating principle of the hydrodynamic deep rolling tool. The lubricant and hydraulic fluid for the hydrodynamic deep rolling tool is the machine's lubricant supply.

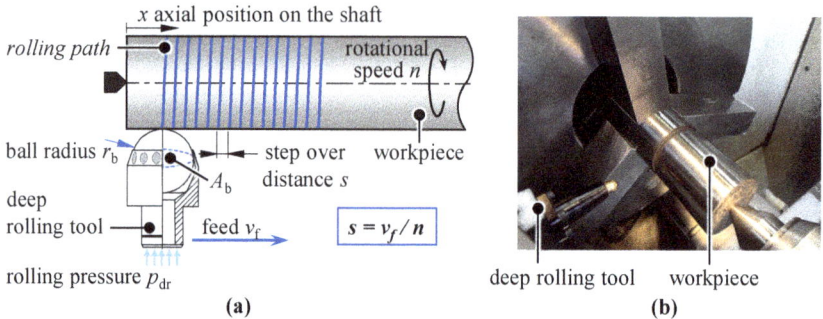

Fig. 1. (a) Test setup including tool and workpiece and properties of the hydrodynamic deep rolling tool from the ECOROLL AG according to [9] and (b) Arrangement of tool and specimen in the lathe.

Fig. 1a shows the schematic test setup. While the workpiece rotates at a constant rotational speed n, the tool moves in axial direction along the workpiece at the feed v_f. From the rotational speed n and feed v_f, the step over distance s results to $s = v_f/n$.

Fig. 1a schematically shows the characteristics of the hydrodynamic deep rolling tool. The rolling ball can be moved dynamically by an infeed during the process. This ensures that the rolling ball and the workpiece are in constant contact during the process. The rolling force F_{dr} is proportional to the rolling pressure p_{dr} and the ball cross-section area A_b and results from the linear relation:

$$F_{dr} = A_b \cdot p_{dr} \cdot (1 - \delta) \text{ with } A_b = d_b^2/4 \cdot \pi. \tag{1}$$

The friction loss δ characterizes the pressure loss between the hydraulic unit and the real pressure applied to the rolling ball. Different loss terms result for different ball diameters d_b and the correspondingly different rolling ball holder:

- $d_b = 3$ mm $\rightarrow \delta_{3mm} = 0.261$
- $d_b = 6$ mm $\rightarrow \delta_{6mm} = 0.236$
- $d_b = 13$ mm $\rightarrow \delta_{13mm} = 0.297$

To determine the friction loss, the rolling force applied at a set pressure was determined for each ball diameter. For this purpose, a test workpiece was fixed on a force measurement platform and the rolling forces for different rolling pressures were determined for each of the three tools ($d_b = 3$ mm, $d_b = 6$ mm and $d_b = 13$ mm). By knowing F_{dr}, p_{dr} and d_{dr}, the friction loss delta could be determined according to Eq. 1, which is specific and constant for each of the three tools.

Test Design.

As a first step it was demonstrated that a significant modification of roughness, hardness and/or residual stress state can be induced by incremental surface treatment. To demonstrate this modification experimentally and to grant a deeper understanding around cause-effect relationships between the deep rolling parameters and the resulting surface integrity, a full factorial experimental design with twofold repetition was used. The process parameters that can be set at the process directly, are the rolling ball diameter d_{dr}, the rolling pressure p_{dr} and step over distance s, which have already been described in the section *Deep Rolling Setup*. The actual parameter values used are shown in Table 2.

Material Forming - ESAFORM 2023   Materials Research Forum LLC
Materials Research Proceedings 28 (2023) 987-996   https://doi.org/10.21741/9781644902479-108

Table 2. Parameters of the executed deep rolling experiments.

Deep Rolling Parameter	Parameter Values
Rolling Pressure p_{dr} /MPa	10; 25; 40
Ball diameter d_{dr} /mm	3, 6, 13
Step over distance s /μm	100; 300; 500

In the case of the ECOROLL tool used, the rolling pressure p_{dr} can be varied continuously in the range from 0 to 40 MPa, with this maximum value being fully utilized by the test plan. The step over distances s are selected in such a way that complete machining of the surface is ensured even with the smallest ball impression, but at the same time no excessive machining time occurs. In contrast to the other two parameter variations, the ball diameter d_{dr} increments are not selected equidistantly, since only the three specified diameters are available as tools here. In total, 27 different test setups result from the full-factorial test plan. In the following, the samples are denoted by p_{dr}/MPa-d_{dr}/mm-s/μm (Ex. p_{dr} =10 MPa; d_{dr} = 3 mm; s = 100 μm → 10-3-100).

Measurement methods and objectives.

After the practical test execution, roughness measurements were carried out according to EN ISO 25178. The arithmetical mean height of the surface (S_a) and the maximum height of the surface (S_z) were determined as 3d-surface roughness parameters. To describe the waviness, the value *WDsm* (Mean period length of the dominant waviness) was used in accordance with VDA 2007. These topography measurements were made on a MAHR LD 260 equipped with a MAHR LP C 45-20-5/90° probe (MAHR GMBH, Göttingen, Germany). Furthermore, HV5 and HV30 hardness tests according to Vickers (EN ISO 6507) were carried out. For statistical validation, these test values were recorded five times each. The test values given below are the average of these five values, each of which was measured on a ZWICKROELL ZHU 50 measuring instrument (ZWICKROELL GMBH & CO. KG, Ulm, Germany). In order not to modify the surface condition, the surfaces were specially prepared for the test. The residual stress states presented were analyzed by means of X-ray diffraction using XSTRESS 3000 Mythen (STRESSTECH GMBH, Rennerod, Germany). The depth increments at which the residual stresses were measured are noted at the corresponding position in the diagram.

Results

In order to derive cause-effect relations between the deep rolling parameters and the surface integrity, the smoothing of roughness peaks, the hardness and the residual stress state of the results were investigated. In the following, the results of these investigations are presented and corresponding interactions are derived.

Roughness Measurements. The roughness was reduced with almost all deep rolling setups. The maximum roughness reduction from S_a = 0.44 μm to S_a = 0.21 μm was achieved with a rolling pressure p_{dr} = 40 MPa, ball diameter d_b = 6 mm and step over distance s = 100 μm. On average, S_a was reduced by 27% to 0.32 μm. Fig. 2 shows these influences deep of the rolling parameters on the roughness S_a and additionally on S_z. A measuring point corresponds in each case to the arithmetic mean of the nine corresponding measured values.

Material Forming - ESAFORM 2023   Materials Research Forum LLC
Materials Research Proceedings 28 (2023) 987-996   https://doi.org/10.21741/9781644902479-108

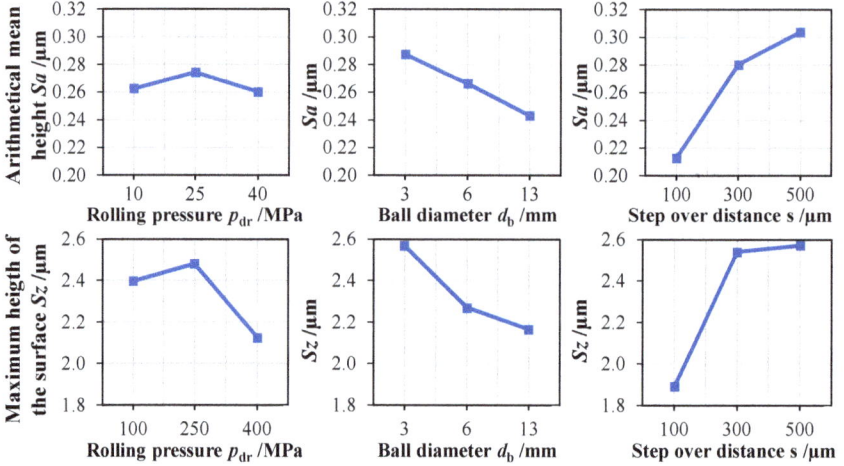

Fig. 2. Main effect plots of the influences of the deep rolling parameters on the arithmetical mean height and the maximum height of the surface.

With respect to S_a, the rolling pressure shows the smallest influence of the three parameters. With increasing rolling pressure p_{dr}, no clear trend in S_a can be identified. There is a clear correlation between the rolling ball diameter d_b and S_a. S_a decreases significantly with increasing d_b. The opposite behavior is found for the dependence on step over distance s. With increasing s, S_a also increases. The results for S_z show similar dependencies as for S_a. With regard to the rolling pressure p_{dr}, no clear trend is established, S_z decreases with increasing ball diameter d_b and increases with larger step over distance s.

In addition to roughness, the waviness of the topography was investigated. In the initial state, the length of the dominant waviness $WDsm$ was determined to 0.363 mm. Fig. 3 shows the measured values for $WDsm$ for all test setups.

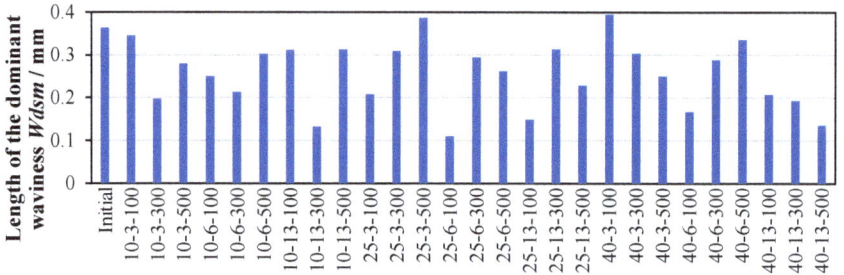

Fig. 3. Length of the dominant waviness Wdsm.

None of the three test parameters shows a correlation with the length of the dominant waviness $WDsm$. In test series carried out with an identical test setup but an initial hardness of 51 HRC, the

Material Forming - ESAFORM 2023
Materials Research Proceedings 28 (2023) 987-996

Materials Research Forum LLC
https://doi.org/10.21741/9781644902479-108

selected step over distance s and $WDsm$ are identical. This expected value is not achieved here. The results rather suggest that there is no correlation between test parameters and $WDsm$.

Hardness tests.

Fig. 4 shows an overall summary of the Vickers hardness tests (HV30). Significant hardness increase was achieved with all deep rolling setups. The maximum hardness increase from 722 HV30 in the initial condition (61.3 HRC) to 882 HV30 (65.7 HRC) was obtained with a rolling pressure p_{dr} = 40 MPa, a ball diameter d_b = 13 mm and a step over distance s = 500 µm. Overall, the hardness values vary in a range of 90 HV30.

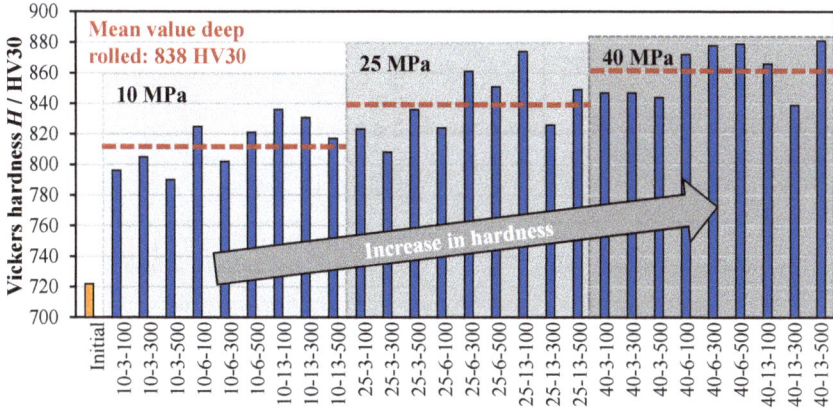

Fig. 4. Overview over the Vickers hardness HV30 for each setup.

On average, a hardness increase of 16% (116 HV30) to 838.6 HV30 was achieved. All hardness values obtained for a rolling pressure p_{dr} = 10 MPa are below this average, whereas all hardness values for p_{dr} = 40 MPa are above this value. The values for p_{dr} = 25 MPa are in between. This tendency can also be seen in the main effect diagrams shown in Fig. 5. The data shown include test values for HV5 as well as HV30. The comparison of the two test forces with each other shows significantly higher HV5 values than HV30 values for a majority of the data points.

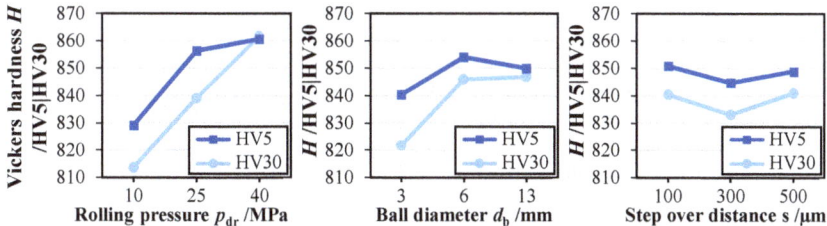

Fig. 5. Main effect plots of the influence of the deep rolling parameters on Vickers hardness HV5 and Vickers hardness HV30.

Material Forming - ESAFORM 2023   Materials Research Forum LLC
Materials Research Proceedings 28 (2023) 987-996   https://doi.org/10.21741/9781644902479-108

Due to the lower depth of penetration of the indenter at the low selected test load (HV5), the area of the edge zone closer to the surface has a higher hardness than the edge zone depth covered by the HV30 test. Furthermore, for the HV30 values there is a correlation between the rolling pressure p_{dr} and the Vicker hardness H. H also increases with the rolling ball diameter d_b, although only marginally between $d_b = 6$ mm and $d_b = 13$ mm; for the HV5 values this trend is even regressive. In contrast, no clear trend emerges with regard to the step over distance s. Similar hardness values are achieved with all three step over distances s. The test values of H and s do not correlate with each other, which is why s is no longer considered in the following analysis.

In Fig 6. interaction diagrams of the Vickers hardness HV30 as a function of rolling pressure p_{dr} and rolling ball diameter d_b and as a function of d_b and p_{dr} are represented.

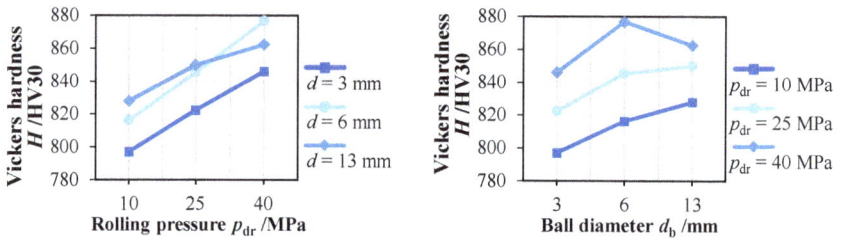

Fig. 6. Interaction diagram of Vickers hardness HV30 as a function of rolling pressure and rolling ball diameter and as a function of rolling ball diameter and rolling pressure.

For each set ball diameter d_b, an increase in rolling pressure is accompanied by a significant increase in hardness. It should be noted that, except for one data point, the largest ball diameter d_b produces the highest hardness at the same pressure. This trend is illustrated in the right hand diagram. The two diagrams show that the differences between the ball diameters of $d_b = 6$ mm and $d_b = 13$ mm are comparatively small. The hardness values for $d_b = 3$ mm are always lower than those of the other two ball diameters at constant pressure. Nevertheless, even with $d_b = 3$ mm, an increase in hardness can be achieved which is significantly above the hardness value obtained on average.

In contrast to the previous diagrams, in Fig. 7 the Vickers hardness H is not given as a function of one of the three process parameters listed in the test design, but as a function of a resulting variable. The rolling forces F_{dr} given in Fig. 7 are determined according to Eq 1.

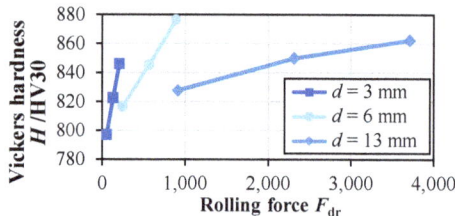

Fig. 7. Interaction diagram of Vickers hardness HV30 as a function of rolling force and rolling ball diameter.

Material Forming - ESAFORM 2023
Materials Research Proceedings 28 (2023) 987-996

Materials Research Forum LLC
https://doi.org/10.21741/9781644902479-108

For each of the three ball diameters d_b, the Vickers hardness H correlates with the rolling force F_{dr}. While the smallest hardness increase also occurs at lowest F_{dr} used, the largest hardness values do not occur at $F_{dr,max} \approx 3{,}711$ N but at $F_{dr} \approx 912$ N and the average ball diameter $d_b = 6$ mm. On average, the smallest increases in hardness are obtained with $d_b = 3$ mm and the largest with $d_b = 13$ mm. Nevertheless, even with $d_b = 3$ mm and the comparatively low rolling force $F_{dr} = 208$ N, a hardness increase up to 846 HV30 can be achieved, which is higher than the 839 HV30 achieved on average.

Residual stress analysis.

In addition to the results presented so far, exemplary residual stress curves were measured. All graphs in Fig. 8 show the residual stresses in the axial direction of the specimens, respectively orthogonal to the rolling path. The three graphs show the residual stress curves for the non-deep-rolled initial condition; a rolling pressure $p_{dr} = 40$ MPa, the rolling ball diameter $d_b = 3$ mm and a step over distance s = 500 µm, as well as $p_{dr} = 10$ MPa, $d_b = 3$ mm and s = 100 µm.

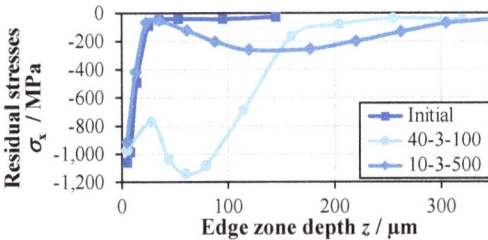

Fig. 8. Depth profile of the residual stress state of selected parameter setups.

In the initial state, a compressive residual stress state is present up to a edge zone depth of $z \approx 25$ µm. At the surface, these residual stresses reach a value of $\sigma_x \approx$ -1,000 MPa. A similar value remains after deep rolling with the rolling pressure $p_{dr} = 10$ MPa. However, here a slight compressive residual stress state is evident at deeper edge zone depths in contrast to the initial condition. For the deep rolling setup with $p_{dr} = 40$ MPa, a similar value of $\sigma_x \approx$ -1,000 MPa is again present at the surface. Here, the maximum residual compressive stresses are, at a depth $z = 60$ µm, at $\sigma_x \approx$ -1,240 MPa. In the range $z > 150$ µm, however, the test setup with lower $p_{dr} = 10$ MPa exhibits greater residual compressive stresses.

Discussion

The surface integrity of powder-metallurgical S390 with an initial hardness of 722 HV30 (61.1 HRC) and an arithmetical height of 0.44 µm can be significantly modified by hydrodynamic deep rolling. With regard to the smoothing of roughness peaks, large rolling ball diameters and a small step over distance have a particularly positive effect. Minimal roughness can be achieved with small rolling ball diameters, but not with large step over distances. Since the influence of rolling pressure on roughness does not show a clear tendency, it can also be assumed that the rolling force, which is proportional to the rolling pressure, has only a subordinate influence in the parameter space considered. It remains unclear to what extent a further reduction in step over distance or larger rolling ball diameters will cause further smoothing.

The results of the waviness analysis show that the spherical ball impression is not introduced into the surface as a distinct path. It is not possible to create a defined path as an impression of the rolling ball at the high initial hardness. Otherwise, the length of the dominant waviness would result directly from the step over distance used. Since this is possible at lower initial hardness, it can be assumed that the tool is deflected by the material accumulation next to the rolling path that occurs during deep rolling.

The increase in hardness was achieved in particular by using large rolling ball diameters and rolling pressures. A more detailed examination reveals that the rolling ball diameter has only a minor influence, since the maximum hardness increase was not achieved with the largest rolling ball diameter and large hardness increases were achieved independently from the ball diameter. Since neither the rolling pressure nor the rolling force are the main influencing parameters, it can be assumed that the surface pressure present in the contact has a decisive influence.

By comparing HV5 and HV30 test results, it could be shown that the near-surface edge zone achieves a higher hardness than deeper material regions. Furthermore, the HV5 and HV30 test values converge strongly with increasing rolling pressure and increasing rolling ball diameter. This circumstance shows that with increasing rolling pressure and increasing rolling ball diameter, deeper edge zone layers (tested with HV30) reach a similarly high hardness as the near-surface edge zone region (tested with HV5). The work hardening is introduced correspondingly deeper into the workpiece.

Furthermore, it was found that despite the comparatively low rolling force of 208 N, high residual compressive stresses of over 1,100 MPa were introduced into the surface-near edge zone. In contrast to the hardness increase, which was already significant at a rolling force of 52 N, the same parameter setup shows a clear but only slight residual compressive stress introduction. Due to the significantly higher hardness values achieved with a larger rolling ball diameter and greater rolling force, it can be assumed that, with appropriate parameter selection, significantly higher residual compressive stresses will also occur.

Summary
In order to achieve the overall objective of increasing the strength of fine blanking punches with regard to abrasive wear, an experimental study was carried out in this paper. The cause-effect relationships determined are used to design a knowledge-based deep rolling process that increases the service life of fine blanking punches by specifically modifying the surface integrity.

Since it can be assumed that low roughness, high hardness values, and the highest possible near-surface compressive residual stresses lead to high strength, several recommendations can be made for the design of a suitable deep rolling process:

- Minimum roughness can be achieved with any rolling ball diameter. The main parameter influencing the resulting roughness is the step over distance, which should be selected as small as possible.
- High hardness increases can be induced in particular with high rolling pressures. In addition, high rolling pressure causes deeper work hardening.
- Larger rolling ball diameters tend to have a positive effect on hardness, but maximum hardness is not achieved at the largest rolling ball diameter selected. It is assumed that in the process the highest possible surface pressure leads to large increases in hardness.
- Significant residual compressive stresses occur even at low rolling forces. It can also be assumed that higher and deeper residual compressive stresses are produced with greater energy input.

For the design of a fine blanking die, the hardness to which the material is to be set before deep rolling must be taken into account. In addition, a suitable cutting edge rounding must be selected, taking into account the macroscopic deformation caused by deep rolling.

Based on the known phenomenological relationships, it can be assumed that increased strength can be achieved by the shown modifications of the surface integrity. Nevertheless, it remains to provide this evidence experimentally. For this purpose, the possible change in surface integrity due to the subsequent coating process must be determined. Subsequently, practical fine blanking tests must be carried out on higher-strength sheets in order to provide practical proof of the strength increase.

Funding

This work was funded by the Deutsche Forschungsgemeinschaft (DFG, German Research Foundation) – project number 423492562

References

[1] T. Bergs, M. Wilms, O. Henrichs, S. Weber, G. Stepien, T. Dannen, M. Prümmer, K. Arntz, Individual process chains in toolmaking through data and model-based forecasts, in: Laboratory for Machine Tools and Production Engineering (WZL) of RWTH Aachen University (Ed.), 30th Aachen Machine Tool Colloquium 2021, Aachen, 2021, pp. 80-102.

[2] U. Aravind, U. Chakkingal, P. Venugopal, A Review of Fine Blanking: Influence of Die Design and Process Parameters on Edge Quality, J. Mater. Eng. Perform. 30 (2021) 1-32. https://doi.org/10.1007/s11665-020-05339-y

[3] Y. Abe, T. Kato, K-i, Mori, S. Nishino, Mechanical clinching of ultra-high strength steel sheets and strength of joints, J. Mater. Process. Technol. 214 (2014) 2112–2118. https://doi.org/10.1016/j.jmatprotec.2014.03.003

[4] M. Krobath, T. Klünsner, W. Ecker, M. Deller, N. Leitner, S. Marsoner, Tensile stresses in fine blanking tools and their relevance to tool fracture behavior, Int. J. Machine Tool. Manuf. 126 (2018) 44–50. https://doi.org/10.1016/j.ijmachtools.2017.12.005

[5] D. Meyer, J. Kämmler, Surface Integrity of AISI 4140 After Deep Rolling with Varied External and Internal Loads, Procedia CIRP 45 (2016) 363-366. https://doi.org/10.1016/j.procir.2016.02.356

[6] X. Wang, X. Xiong, K. Huang, S. Ying, M. Tang, X. Qu, W. Ji, C. Qian, Z. Cai, Effects of Deep Rolling on the Microstructure Modification and Fatigue Life of 35Cr2Ni4MoA Bolt Threads, Metals 12 (2022) 1224. https://doi.org/10.3390/met12071224

[7] K. Bobzin, C. Kalscheuer, M. Carlet, D.C. Hoffmann, T. Bergs, L. Uhlmann, Low-Temperature Physical Vapor Deposition TiAlCrSiN Coated High-Speed Steel: Comparison Between Shot-Peened and Polished Substrate Condition, Adv. Eng. Mater. 24 (2022) 2200099. https://doi.org/10.1002/adem.202200099

[8] voestalpine Böhler Edelstahl GmbH & Co. KG, S390 DE: Microclean, Karpfenberg, 2010.

[9] F. Klocke, Fertigungsverfahren 4, Springer Berlin Heidelberg, Berlin, Heidelberg, 2017.

Material Forming - ESAFORM 2023
Materials Research Proceedings 28 (2023) 987-1006

Materials Research Forum LLC
https://doi.org/10.21741/9781644902479-109

Electrically-assisted deep drawing of 5754 aluminium alloy sheet

DOBRAS Daniel[1,a] *, ZIMNIAK Zbigniew[1,b] and ZWIERZCHOWSKI Maciej[1,c]

[1]Department of Metal Forming, Welding and Metrology, Wrocław University of Science and Technology, 7-9 Ignacego Łukasiewicza Street, 50-371 Wrocław, Poland

[a]daniel.dobras@pwr.edu.pl, [b]zbigniew.zimniak@pwr.edu.pl, [c]maciej.zwierzchowski@pwr.edu.pl

Keywords: Electrically-Assisted Forming, Aluminium Alloys, Deep Drawing

Abstract. The effect of current pulse application on the mechanical behaviour and plasticity of the 5754 aluminium alloy was studied. Tensile and deep drawing tests were conducted. The 5754 aluminium alloy in two different states of hardening was used: H111 and H22. The results show that the application of current pulses can significantly increase the plasticity of the examined alloys in the case of the tensile test. The dynamic recovery process is the main process responsible for the increase in plasticity of the material. However, in the case of the deep drawing process, it was observed that the increase in the material formability is low, and further studies are needed.

Introduction

Aluminium alloys have become very popular in the automotive industry owing to their good specific strength and corrosion resistance. However, the formability of many aluminium alloys at room temperature is low [1,2]. Warm and hot forming methods are applied in metal forming processes because they improve the formability of materials such as aluminium or magnesium alloys [3]. However, these methods have many drawbacks, such as increased adhesion of the die and decreased die strength and lubrication effectiveness [4].

Electrically-Assisted Forming (EAF) is proposed as an alternative method to the warm and hot forming methods. It is commonly known that the application of current pulses during plastic forming of metals can significantly increase their formability, reduce flow stress, and avoid or reduce some of the above-mentioned drawbacks [5–7]. Jeong et al. [8] showed that the application of current pulses during tension can increase the material elongation by over 200% in the case of the 5052-H32 aluminium alloy (AA) [9] and the as-extruded AZ91 magnesium alloy. Because this significant increase of the material plasticity during EAF processes cannot be explained only by the simple Joule heat law, many scientists have tried to explain this phenomenon [10]. There is still no unequivocal proof of existence of non-thermal effects such as the electroplastic effect [11], the magnetoplastic effect [12] or others [13]. The microscale Joule heat theory, more popular in the recent years, has been confirmed by the simulation [14], grain boundary melting [15] or such processes as dynamic recovery and recrystallization [8,16]. However, it does not explain all the observed phenomena [16]. Recently, it has been proved that the application of even low energy current pulses can lead to defect reconfiguration and a change in the dislocation pattern [17–19].

Many industrial metal forming processes have been supported by the application of current pulses. EAF processes such as wire drawing [20,21] or rolling [22,23] have been successfully realized. However, more difficult processes, in terms of EAF, such as deep drawing and press stamping, still need to be developed due to the unsatisfactory results. Only in the case of magnesium alloys the increase of plasticity has been meaningful [24,25]. In the electrically-assisted deep drawing processes, the following problems should still be overcome: excessive heat transfer from the sample to the dies, limited possibility of measuring the temperature or the need to apply higher currents [26–28].

Material Forming - ESAFORM 2023   Materials Research Forum LLC
Materials Research Proceedings 28 (2023) 987-1006   https://doi.org/10.21741/9781644902479-109

Materials and Methods

The tested material was the 5754 aluminium alloy, delivered in two different states of hardening: H111 and H22. The thickness of both as-received sheets was 1 mm. In the present work, two different electrically-assisted forming processes were conducted: the tensile test and the deep drawing test. The tensile samples with a gauge length of 75 mm and a gauge width of 12 mm were prepared by way of milling along the rolling direction of the sheet. The tensile tests were performed at a strain rate of 0.0025 s^{-1} until fracture using an INSTRON 3369 tensile machine.

The deep drawing tests were carried out at the punch speed of 0.4 mm/s until fracture using an Erichsen 142 Sheet Metal Testing Machine. The experimental study proved that a greater contact between the sample and the dies results in more heat transfer. Therefore, to avoid excessive heat transfer from the sample to the dies, the Erichsen machine was working in the sheet holder quick release (SHQR) mode. This means that, when the set draw depth is reached, the blank holder force is released, and the test continues until stopped. For the same reason, the dies were made of stainless steel because of its low thermal conductivity and high electrical resistance. Before the SHQR, the blank holder force was 35 and 40 kN, in the case of the H111 and the H22 state, respectively. A circular punch with a diameter of 30 mm and a radius of 4 mm as well as a matrix with a diameter of 33.5 mm and a radius of 4 mm were used in the deep drawing tests. A special sample, with experimentally fitted dimensions visible in Fig. 1, was used for the tests. The distance of 130 mm (Fig. 1) is the distance between the edges of the electrodes attached to the sample. The central, circular part of the sample with a radius of 32 mm was cut on the sides in order to obtain the highest current density. The dies were isolated from the machine. The sample mounted to the electrodes before the deep drawing process is visible in Fig. 2.

During all the tests, the electric pulses were generated by a self-constructed current pulse generator working at 2.52 V [29] and applied to the sample through copper electrodes attached to it. The applied current was pulsed with a pulse duration of t_d and a period of t_p, and its shape was created by the function generator device. The first pulse of electric current was applied after a time of t_p from the beginning of the tensile test, and from the moment of sheet holder force release in the case of deep drawing. An oscilloscope and a Rogowski coil were used to measure the current flowing through the samples. Each test was repeated three times. One side of the sample was sprayed with black paint, and its temperature was measured by a FLIR T440 infrared thermal imaging camera during the test.

The microstructural observations of the selected specimens were conducted with a VEGA3 TESCAN Scanning Electron Microscope (SEM) operating at 20 kV, and equipped with an Electron Backscattered Diffraction (EBSD) detector (Oxford Instruments). Mechanical grinding and electrolytic polishing using A2 Struers reagent were applied to prepare the specimen surfaces for the EBSD analysis. The applied step size was 0.2 μm. In the case of the non-EA specimen, the single-iteration grain dilation cleanup procedure was performed only one time and affected less than 4% of the measurement pixels.

Material Forming - ESAFORM 2023
Materials Research Proceedings 28 (2023) 987-1006

Materials Research Forum LLC
https://doi.org/10.21741/9781644902479-109

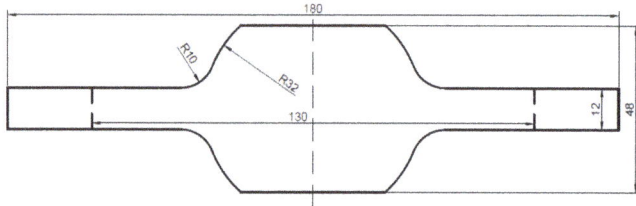

Fig. 1. The shape and dimensions (in mm) of the samples used in the deep drawing tests.

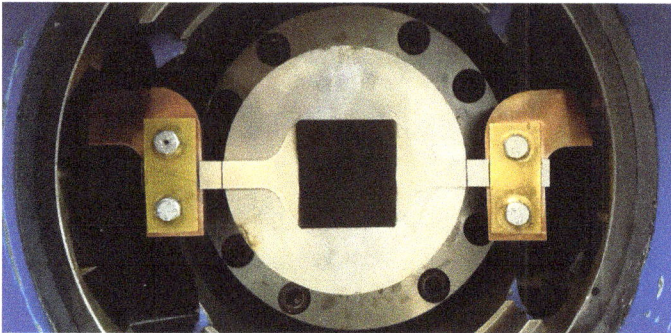

Fig. 2. The setup of the deep drawing sample before the process.

Results and Discussion

One of the most important current parameters in the EAF processes is the current density. Here, the nominal current density is defined as the current measured by the Rogowski coil divided by the cross-sectional area of the sample. The obtained values of the current density, depending on the applied process type and current parameters, are given in Table 1 and 2. Note that the current density in the case of deep drawing is about four times lower than in the case of the tensile test as a result of a four time higher width of the deep drawing samples and no possibility to increase the current. Therefore, as can be seen in Table 1 and 2, the pulses were applied more often in the case of deep drawing.

Table 1. Influence of current parameters on the current density for tensile test.

t_d / t_p [ms / s]	Current density [A/mm^2]	
	5754-H111	5754-H22
400 / 50	190	-
400 / 24	-	200

Table 2. Influence of current parameters on the current density for deep drawing test.

t_d / t_p [ms / s]	Current density [A/mm^2]	
	5754-H111	5754-H22
400 / 2		
400 / 1.2	45	45
500 / 2.5		
500 / 1.5		

Material Forming - ESAFORM 2023   Materials Research Forum LLC
Materials Research Proceedings 28 (2023) 987-1006   https://doi.org/10.21741/9781644902479-109

The engineering stress-strain curves of AA5754-H111 and AA5754-H22 with and without the current pulse application are shown in Fig. 3a-b. It is clearly seen that the application of the current pulses increase the material elongation. As can be seen, when the current pulse is applied, immediately, the engineering stress decreases and the temperature increases. The significant engineering stress drop is, among others, correlated with the sample's thermal expansion. However, before the next pulse application, the temperature drops almost to the room value. In the case of the annealed state (H111), the average value of engineering strain increased from 22.86 to 32.73% (increase to about 143% of the basic value). However, in the case of the hardened alloy (H22), the

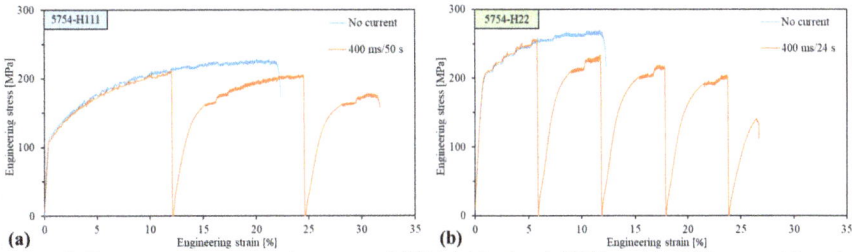

Fig. 3. Engineering stress-strain curves of 5754 AA in the a) H111 and b) H22 state with and without current application.

average value of engineering strain increased from 12.27 to 27.13% (increase to about 221% of the basic value). It is worth mentioning that the ratio of t_d/t_p was 1/60 in the case of the hardened alloy, and it was the best ratio in terms of increasing the material elongation. Nevertheless, in the case of the annealed alloy the similar increase the material elongation occurred for a wide range of t_d/t_p ratios. The average values of the maximal registered temperatures during tensile tests were 364 and 390°C, for AA5754-H111 and AA5754-H22, respectively. However, in both cases, the average temperature during all the tensile tests was under 200°C, which is generally a lower temperature than the warm forming temperature for these alloys.

Fig. 4a-b present the deep drawing force-displacement curves of the 5754 aluminium alloy. The force represents the force at the punch in the deep drawing process and the displacement is the displacement of the punch, which corresponds with the depth of the drawpiece. The SHQR took places 5.4 and 3.4 mm from the beginning of the tests for the annealed and the hardened alloy, respectively. The above-mentioned values were experimentally designated, and the obvious decrease of the force corresponds to the SHQR moment. Unlike the tensile tests, in the deep drawing tests, the application of the current pulses did not lead to a significant increase of the drawpieces' heights. Only in one case, the height of the drawpiece increased noticeably, when the 500 ms/1.5 s current parameters were applied to the 5754-H111 aluminium alloy (Fig. 4a). In this case, the displacement at the punch increased about 8% in comparison with the baseline value (Fig. 5a), but in other cases, the displacement increase or decrease were not greater than 2% (Fig. 5a-b). The average values of the maximal registered temperatures during the deep drawing tests, for the different current parameters, are presented in Fig. 6a-b. The temperature did not exceed 80°C in any case. A picture of an example drawpiece of 5754-H22 AA is visible in Fig. 7. A characteristic fracture is visible on the top edge of the drawpiece.

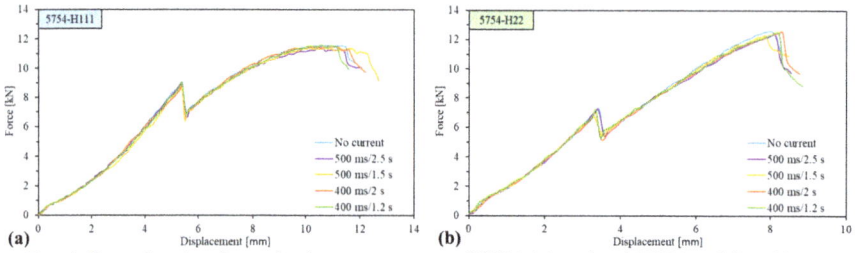

Fig. 4. Deep drawing force-displacement curves of 5754 AA in the a) H111 and b) H22 state with and without current application.

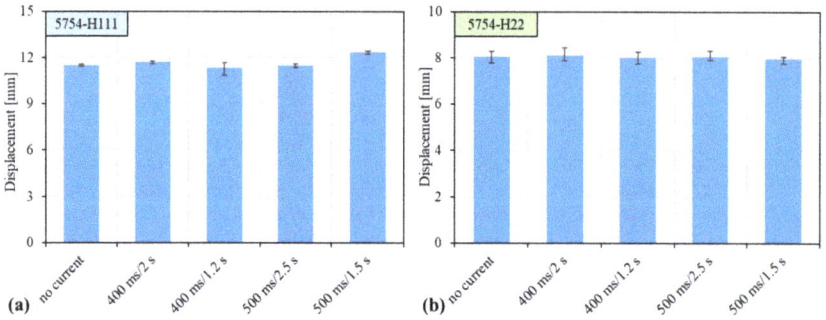

Fig. 5. Influence of the current parameters on the height of the drawpieces of 5754 AA in the a) H111 and b) H22 state.

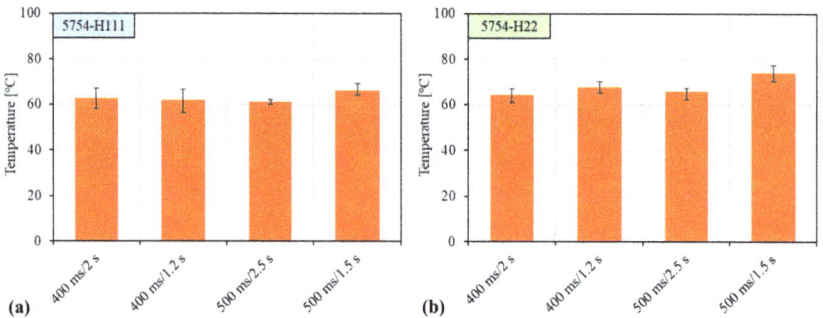

Fig. 6. Influence of the current parameters on the maximal temperature of the drawpieces of 5754 AA in the a) H111 and b) H22 state.

Material Forming - ESAFORM 2023   Materials Research Forum LLC
Materials Research Proceedings 28 (2023) 987-1006   https://doi.org/10.21741/9781644902479-109

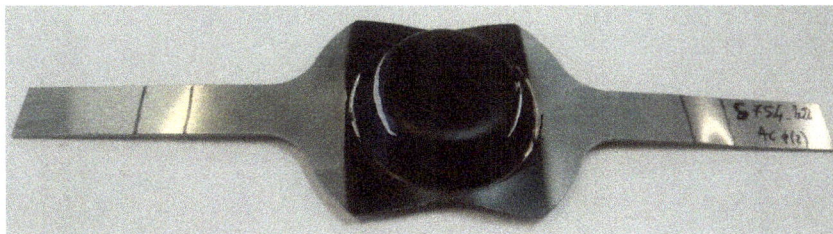

Fig. 7. The example drawpiece of 5754-H22 AA.

Even if the application of the current pulses leads to a significant increase of the material elongation in the tensile tests, it is difficult to reproduce the same conditions and results in the deep drawing process. The current density present during deep drawing is about four times lower than in the case of a tensile, and it is the main reason for the insignificant increase of the material formability. Low current density resulted in low temperature of the tested samples, which did not increase even when the pulses were applied more often. Finally, the dynamic recovery and recrystallization processes cannot take place and increase the material plasticity when the temperature is too low. The solution of the above-mentioned problems will be the application of a large high-current pulse generator, with the possibility of applying high currents. However, this solution increases the process costs and makes it more difficult, yet it will be necessary in potential industrial applications.

The highest increase of the material plasticity, owing to the pulsed current application, occurred in the case of the 5754-H22 AA tensile sample. Therefore, the specimens from the 5754-H22 AA tensile tests were selected for the microstructural analysis. In order to better analyze the effect of the current pulses on the structural changes, two additional tests were conducted. In the first case, the EA tensile test was immediately stopped after the fourth pulse application (Fig. 3b). It happened at 90% elongation of the EA tensile sample. In the second case, the non-EA tensile sample was stretched to a 90% value of its initial elongation and then the test was also stopped. The specimens for the microstructural analysis were cut from the middle of the above-mentioned tensile samples.

The Inverse Pole Figure (IPF) maps of the non-EA and EA specimens are presented in Fig. 8 a) and b), respectively. Due to the high deformation of the crystal lattice, the initial Hit Rate value of the non-EA specimen was only about 60%, which is typical for highly deformed materials. Although the EA specimen was deformed to a greater extent (about two times greater elongation, Fig. 3b), the obtained Hit Rate in this case was more than 80%. It means that the crystal lattice deformation is lower than for the specimen without the current application. If the material has been deformed more, but its lattice is less deformed, then the application of the current pulses and the increase of temperature resulting from it lead to a lattice structure rebuild. The IPF maps represent the above-mentioned case.

Many small grains are visible (Fig. 8b) at grain boundaries of the big and medium size grains in the case of specimen with the pulsed current application. It could mean that the dynamic recrystallization process occurred, especially because, for a while, the temperature during the EA tensile tests reached the value of $0.6T_m$, where T_m is the homologous temperature of aluminium. However, a more detailed analysis is needed to verify this hypothesis.

To analyze the recrystallization process, the Grain Orientation Spread (GOS) parameter was used. Generally, it is assumed that the GOS parameter for recrystallized grains is less than 1.5-2 °. However, this threshold value in the literature ranges from 1 to 3 ° [9,30]. In this study the

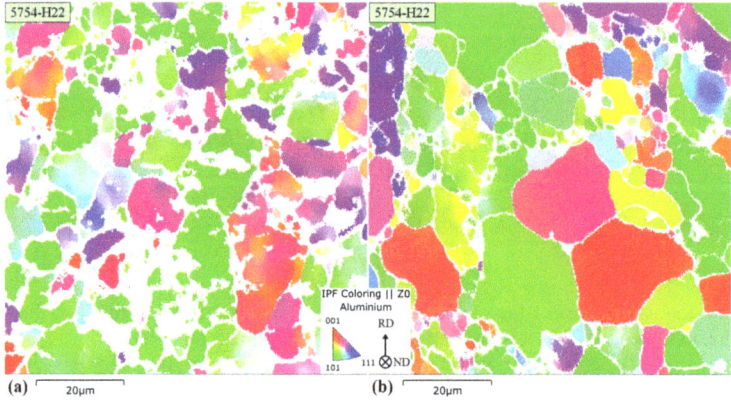

Fig. 8. IPF maps of 5754-H22 AA from a stopped a) non-EA and b) EA tensile test.

recrystallized grains are defined by the GOS less than 1.5 °. The GOS map of non-pulsed specimen is presented in Fig. 9a. It is clearly seen that the most grains are characterized by the GOS higher than 2 °, and only about 13% indexed pixels - by the GOS less than 1.5 °. It could mean that these grains with lower GOS parameter were not deformed or deformed in a small range. Completely different results are shown by the GOS map of the pulsed specimen (Fig. 9b). The effect of temperature resulted in a significant increase in the surface of grains, with the GOS less than 1.5 °. It this case about 62% indexed pixels is characterized by the GOS less than 1.5 °. However, mainly the large grains take small values of the GOS parameter. Therefore, it can be concluded that the intense dynamic recovery process occurred in these large grains. This recovery process led to a structure rebuild of the crystal lattice in grains and thus to the low GOS parameter. On the other hand, many small grains with the GOS less than 1.5 ° are visible at grain boundaries of the bigger grains. It could mean that the dynamic recrystallization process started in these places, however, a more detailed analysis using transmission electron microscopy should be performed to prove it.

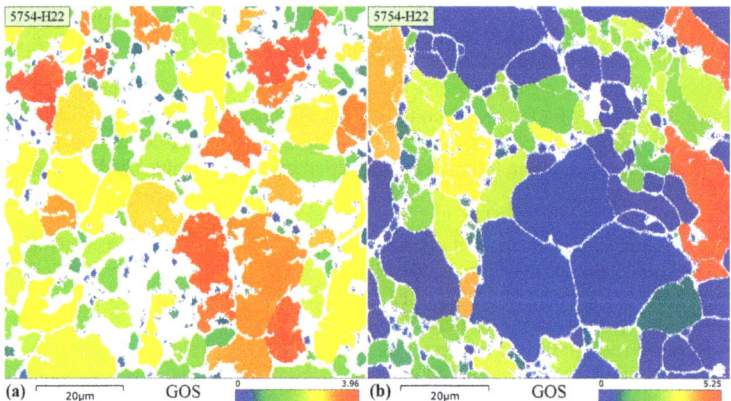

Fig. 9. GOS maps of 5754-H22 AA from a stopped a) non-EA and b) EA tensile test.

Material Forming - ESAFORM 2023   Materials Research Forum LLC
Materials Research Proceedings 28 (2023) 987-1006   https://doi.org/10.21741/9781644902479-109

Summary

In the present work, electrically-assisted tensile tests and deep drawing processes of the 5754 aluminium alloy in different states of hardening were carried out. The main conclusions are:

1. The application of current pulses can lead to a significant increase of the material elongation in the tensile test, especially in the case of the H22 state.

2. The EBSD analysis proved that the intense dynamic recovery is the main factor responsible for the structure rebuild of the crystal lattice and thus the increase of the material plasticity.

3. The application of the sheet holder quick release mode, new in terms of electrically-assisted forming, can lead to a decrease of the excessive heat transfer from sample to the dies.

4. The application of the current pulses did not lead to a significant increase of the material formability in the case of the deep drawing process. The main difficulties in effective implementation of the electrically-assisted deep drawing processes are the necessity of the application of high currents and an excessive heat transfer from the sample to the stamping dies.

References

[1] Z. Gronostajski, K. Jaśkiewicz, P. Kaczyński, M. Skwarski, S. Polak, J. Krawczyk, W. Chorzępa, P. Trzpis, W-Temper forming of B-pillar from 7075 aluminum alloy, CIRP Ann. 71 (2022) 221–224. https://doi.org/10.1016/j.cirp.2022.03.019

[2] H. Deng, Y. Mao, G. Li, X. Zhang, J. Cui, AA5052 failure prediction of electromagnetic flanging process using a combined fracture model, Arch. Civ. Mech. Eng. 22 (2022) 1-17. https://doi.org/10.1007/s43452-022-00390-z

[3] S. Toros, F. Ozturk, I. Kacar, Review of warm forming of aluminum-magnesium alloys, J. Mater. Process. Technol. 207 (2008) 1–12. https://doi.org/10.1016/j.jmatprotec.2008.03.057

[4] W.A. Salandro, J.T. Roth, Formation of 5052 aluminum channels using Electrically-Assisted Manufacturing (EAM), Proc. ASME Int. Manuf. Sci. Eng. Conf. 2009, MSEC2009. 2 (2009) 599-608. https://doi.org/10.1115/MSEC2009-84117

[5] H.R. Dong, X.Q. Li, Y. Li, Y.H. Wang, H.B. Wang, X.Y. Peng, D.S. Li, A review of electrically assisted heat treatment and forming of aluminum alloy sheet, Int. J. Adv. Manuf. Technol. 120 (2022) 7079–7099. https://doi.org/10.1007/s00170-022-08996-6

[6] E. Simonetto, S. Bruschi, A. Ghiotti, Electroplastic effect on AA1050 plastic flow behavior in H24 tempered and fully annealed conditions, Procedia Manuf. 34 (2019) 83-89. https://doi.org/10.1016/j.promfg.2019.06.124

[7] B.J. Ruszkiewicz, T. Grimm, I. Ragai, L. Mears, J.T. Roth, A Review of Electrically-Assisted Manufacturing with Emphasis on Modeling and Understanding of the Electroplastic Effect, J. Manuf. Sci. Eng. Trans. ASME. 139 (2017) 1–15. https://doi.org/10.1115/1.4036716

[8] J.H. Roh, J.J. Seo, S.T. Hong, M.J. Kim, H.N. Han, J.T. Roth, The mechanical behavior of 5052-H32 aluminum alloys under a pulsed electric current, Int. J. Plast. 58 (2014) 84-99. https://doi.org/10.1016/j.ijplas.2014.02.002

[9] H.J. Jeong, M.J. Kim, J.W. Park, C.D. Yim, J.J. Kim, O.D. Kwon, P.P. Madakashira, H.N. Han, Effect of pulsed electric current on dissolution of Mg17Al12 phases in as-extruded AZ91 magnesium alloy, Mater. Sci. Eng. A. 684 (2017) 668-676. https://doi.org/10.1016/j.msea.2016.12.103

[10] N.K. Dimitrov, Y. Liu, M.F. Horstemeyer, Electroplasticity: A review of mechanisms in electro-mechanical coupling of ductile metals, Mech. Adv. Mater. Struct. 0 (2020) 1-12. https://doi.org/10.1080/15376494.2020.1789925

[11] K. Okazaki, M. Kagawa, H. Conrad, A study of the electroplastic effect in metals, Scr. Metall. 12 (1978) 1063-1068. https://doi.org/10.1016/0036-9748(78)90026-1

[12] M.I. Molotskii, Theoretical basis for electro- and magnetoplasticity, Mater. Sci. Eng. A. 287 (2000) 248-258. https://doi.org/10.1016/s0921-5093(00)00782-6

Material Forming - ESAFORM 2023   Materials Research Forum LLC
Materials Research Proceedings 28 (2023) 987-1006   https://doi.org/10.21741/9781644902479-109

[13] B.J. Ruszkiewicz, L. Mears, J.T. Roth, Investigation of Heterogeneous Joule Heating as the Explanation for the Transient Electroplastic Stress Drop in Pulsed Tension of 7075-T6 Aluminum, J. Manuf. Sci. Eng. Trans. ASME. 140 (2018) 1–11. https://doi.org/10.1115/1.4040349

[14] J. Zhao, G.X. Wang, Y. Dong, C. Ye, Multiscale modeling of localized resistive heating in nanocrystalline metals subjected to electropulsing, J. Appl. Phys. 122 (2017). https://doi.org/10.1063/1.4998938

[15] R. Fan, J. Magargee, P. Hu, J. Cao, Influence of grain size and grain boundaries on the thermal and mechanical behavior of 70/30 brass under electrically-assisted deformation, Mater. Sci. Eng. A. 574 (2013) 218–225. https://doi.org/10.1016/j.msea.2013.02.066

[16] M.J. Kim, S. Yoon, S. Park, H.J. Jeong, J.W. Park, K. Kim, J. Jo, T. Heo, S.T. Hong, S.H. Cho, Y.K. Kwon, I.S. Choi, M. Kim, H.N. Han, Elucidating the origin of electroplasticity in metallic materials, Appl. Mater. Today. 21 (2020) 100874. https://doi.org/10.1016/j.apmt.2020.100874

[17] S. Zhao, R. Zhang, Y. Chong, X. Li, A. Abu-Odeh, E. Rothchild, D.C. Chrzan, M. Asta, J.W. Morris, A.M. Minor, Defect reconfiguration in a Ti–Al alloy via electroplasticity, Nat. Mater. 20 (2021) 468-472. https://doi.org/10.1038/s41563-020-00817-z

[18] X. Li, J. Turner, K. Bustillo, A.M. Minor, In situ transmission electron microscopy investigation of electroplasticity in single crystal nickel, Acta Mater. 223 (2022) 117461. https://doi.org/10.1016/j.actamat.2021.117461

[19] X. Zhang, H. Li, M. Zhan, Z. Zheng, J. Gao, G. Shao, Electron force-induced dislocations annihilation and regeneration of a superalloy through electrical in-situ transmission electron microscopy observations, J. Mater. Sci. Technol. 36 (2020) 79-83. https://doi.org/10.1016/j.jmst.2019.08.008

[20] Z. Zimniak, G. Radkiewicz, The electroplastic effect in the cold-drawing of copper wires for the automotive industry, Arch. Civ. Mech. Eng. 8 (2008) 173–179. https://doi.org/10.1016/S1644-9665(12)60204-0

[21] G. Tang, J. Zhang, M. Zheng, J. Zhang, W. Fang, Q. Li, Experimental study of electroplastic effect on stainless steel wire 304L, Mater. Sci. Eng. A. 281 (2000) 263-267. https://doi.org/10.1016/s0921-5093(99)00708-x

[22] Z. Xu, G. Tang, S. Tian, F. Ding, H. Tian, Research of electroplastic rolling of AZ31 Mg alloy strip, J. Mater. Process. Technol. 182 (2007) 128-133. https://doi.org/10.1016/j.jmatprotec.2006.07.019

[23] H.D. Nguyen-Tran, H.S. Oh, S.T. Hong, H.N. Han, J. Cao, S.H. Ahn, D.M. Chun, A review of electrically-assisted manufacturing, Int. J. Precis. Eng. Manuf. - Green Technol. 2 (2015) 365-376. https://doi.org/10.1007/s40684-015-0045-4

[24] H. Xie, X. Dong, Z. Ai, Q. Wang, F. Peng, K. Liu, F. Chen, J. Wang, Experimental investigation on electrically assisted cylindrical deep drawing of AZ31B magnesium alloy sheet, Int. J. Adv. Manuf. Technol. 86 (2016) 1063–1069. https://doi.org/10.1007/s00170-015-8246-0

[25] Z. Lv, Y. Zhou, L. Zhan, Z. Zang, B. Zhou, S. Qin, Electrically assisted deep drawing on high-strength steel sheet, Int. J. Adv. Manuf. Technol. 112 (2021) 763-773. https://doi.org/10.1007/s00170-020-06335-1

[26] G.F. Wang, B. Wang, S.S. Jang, K.F. Zhang, Pulse current auxiliary thermal deep drawing of SiCp/2024Al composite sheet with poor formability, J. Mater. Eng. Perform. 21 (2012) 2062-2066. https://doi.org/10.1007/s11665-012-0165-8

[27] H.G. Park, B.S. Kang, J. Kim, Numerical modeling and experimental verification for high-speed forming of Al5052 with single current pulse, Metals 9 (2019). https://doi.org/10.3390/met9121311

Materials Research Forum LLC
https://doi.org/10.21741/9781644902479-109

[28] Y. Zhao, L. Peng, X. Lai, Influence of the electric pulse on springback during stretch U-bending of Ti6Al4V titanium alloy sheets, J. Mater. Process. Technol. 261 (2018) 12–23. https://doi.org/10.1016/j.jmatprotec.2018.05.030

[29] G. Lesiuk, Z. Zimniak, W. Wiśniewski, J.A.F.O. Correia, Fatigue lifetime improvement in AISI 304 stainless steel due to high-density electropulsing, Procedia Struct. Integr. 5 (2017) 928–934. https://doi.org/10.1016/j.prostr.2017.07.118

[30] M.H. Alvi, S. Cheong, H. Weiland, A.D. Rollett, Recrystallization and texture development in hot rolled 1050 aluminum, Mater. Sci. Forum. 467–470 (2004) 357–362. https://doi.org/10.4028/www.scientific.net/msf.467-470.357

Material Forming - ESAFORM 2023
Materials Research Proceedings 28 (2023) 1007-1014

Materials Research Forum LLC
https://doi.org/10.21741/9781644902479-110

Effect of shear angle in shearing on stretch flangeability of ultra-high strength steel sheets

YAGITA Ryo[1,a] *, KIMURA Shunsuke[1,b] and ABE Yohei[1,c]

[1]Toyohashi University of Technology, 1-1, Hibarigaoka, Tempaku, Toyohashi, Aichi 441-8580 Japan

[a]yagita@plast.me.tut.ac.jp, [b]kimura@plast.me.tut.ac.jp, [c]abe@plast.me.tut.ac.jp

Keywords: Stretch Flangeability, Ultra-High Strength Steel Sheets, Shearing

Abstract. The effect of the shear angle in shearing on the stretch flangeability of 980 MPa ultra-high strength steel sheets was examined. The sheared edges of the ultra-high strength steel sheets sheared at different shear angles were investigated, and then the cracking on the sheared edge after stretch flanging was observed. The length of the fracture surface in the sheared edge was decreased by shearing with the punch with a shear angle, whereas many cracks and burrs at the edge of the fracture surface were caused, and the boundary between the burnished and fracture surfaces became rougher. In stretch flanging, the cracks in the fracture bottom side rapidly increased with the bending length. The stretch flangeability increased at a shear angle, where the number of cracks per unit length was the highest, i.e. this indicates that the large number of cracks in the sheared edge were effective in suppressing penetration cracking to fracture in the stretch flanging.

Introduction

To improve the fuel consumption of automobiles, the reduction in weight is intensively required in the automobile industry. To reduce the weight, the application to automotive parts using high strength and ultra-high strength steel sheets increases. The ultra-high strength steel sheet having a tensile strength above 1000 MPa is effective for automotive parts. However, in cold stamping of ultra-high strength steel sheets, there are many resolving problems such as large stamping load [1], large springback [2], low formability [3], tool failure [4] and hydrogen-induced delayed fracture [5].

Steel sheets are typically cut by shearing, and then are formed by stamping to the parts. When steel sheets are bent into a concave shape in stamping, tensile stress are generated at the concave shape edges. Because the sheared edge of the sheet has large plastic deformation and fracture in shearing, the sheared edge qualities was usually changed. Yagita et al. [6] showed that the sheared edge of the ultra-high strength steel sheets was affected by the shear angle in the punch and the blankholding force. The sheared edge qualities for edge cracking in the hole expansion [7] and the stretch flange abilities [8] were investigated. Sartkulvanich et al. [9] showed that the damage value is related to the blanking clearance in the hole expansion of 590 MPa steel sheet.

On the other hand, the studies have been performed to improve stretch flanging ability through material composition and crystalline structure. Pan et al. [10] produced 780 MPa steel sheet with different compositions and showed that stretch flangeability was affected by the strength difference between ferrite and martensite, and that the addition of Nb could increase ferrite strength and improve stretch flangeability. Gwon et al. [11] showed that in the stretch flanging of TWIP steels, cracking occurs mainly at the grain boundaries between ferrite and martensite, and that grain refinement can improve stretch flangeability. Choi et al. [12] stretch flanged three different types of steel sheets and showed that the alternation of FCC and BCC layers in the steel resulted in lower stretch flangeability due to voids at the phase interfaces. Wang et al. [13] improved the stretch flangeability of TRIP steel by annealing the steel to generate a banded structure of soft ferrite phases, which caused cracks to propagate in the rolling direction.

Material Forming - ESAFORM 2023   Materials Research Forum LLC
Materials Research Proceedings 28 (2023) 1007-1014   https://doi.org/10.21741/9781644902479-110

A schematic illustration of crack initiation in a stretch flanging is shown in Fig. 1. In stretch flanging, a tensile stress is generated at the blank edge, which is deformed in bending. If the roughness of the sheared edge is large, cracks will occur in the product after stretch flanging, triggered by cracks at the blank edge. If the cracks penetrate, the product will be defective, so it is necessary to stretch flange the product to prevent the cracks from penetration. Since the presence of cracks on sheared surfaces increases the potential for crack penetration, it is important to investigate the formation of cracks by shear and their expansion by stretch flanging.

In this paper, the sheared edges of ultra-high-strength blanks sheared at different shear angles were investigated, and the size of the cracks when stretch-flanged was observed.

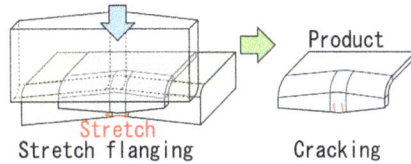

Fig. 1. Crack initiation in stretch flanging.

Materials and Methods
The mechanical properties of the sheet measured by uniaxial tension test are shown in Table 1. The sheet that the nominal tensile strength was 980 MPa was used. Steel sheet was galvanized alloy zinc (GA) steel sheet. The nominal thickness was 1.2 mm.

Table 1. Mechanical properties of steel sheets.

Steel sheets	Thickness [mm]	Galvannealed	Tensile strength [MPa]	Elongation [%]	Reduction in area [%]	n-value [-]
980 MPa	1.20	Yes	1003	14.1	58.5	0.128

The method of measuring cracks in the stretch flanging of a sheared steel sheets is shown in Fig. 2. A 150 mm wide steel sheet was sheared to an appropriate length, and the sheared edges were observed. The sheared steel sheets were then stretch flanged, and cracks were observed at the bending deformed edges.

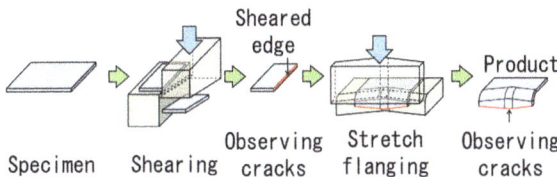

Fig. 2. Method of measuring cracks in stretch flanging of sheared steel sheets.

Shearing conditions are shown in Fig. 3. Each steel sheet was sheared by a die mounted on a servo press (AMADA, SDE-8018). The mean punch speed was 90 mm/s, the shear angles of the punches α, were 0.0, 0.5, and 1.5°, and the clearance between the punch and die was 0.12 mm. The steel sheets were fixed to the die by a blankholder bolted to the die. Counter blocks of urethane rubber (Shore hardness A90) were attached to the under the punch to prevent steel sheets from dropping.

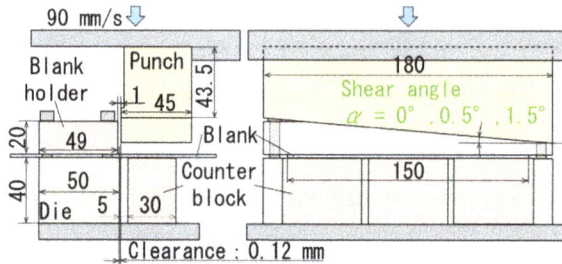

Fig. 3. Shearing conditions.

Stretch flanging conditions are shown in Fig. 4. Each steel sheet was stretch flanged by a die mounted on a servo press (AMADA, SDE-1522). The punch and die had a center angle of 155°, the center was rounded at R50, and the clearance was 1.32 mm. The maximum distance between the sheared edges of the steel sheets and the center of the die was defined as the bend length l. The steel sheets were formed in 5 mm increments from 15 mm to 45 mm.

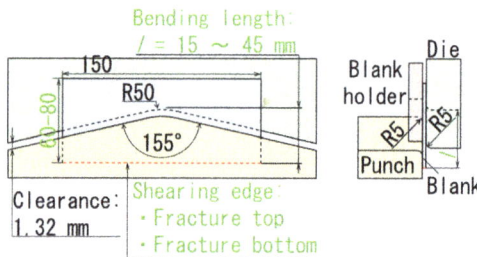

Fig. 4. Stretch flanging conditions.

The steel sheets were mounted on the die by blankholders bolted to the die under two conditions: the fracture top side, where the fracture surface was inside the bend, and the fracture bottom side, where the fracture surface was outside the bend. Anti-rust oil was used as the lubricant.

Results

Sheared Edges of Steel Sheets. The shearing load - punch stroke curve is shown in Fig. 5. The peak shearing load increases with decreasing shear angle, whereas the total stroke decreases.

The sheared edges are shown in Fig. 6(a) to 6(c), and the qualities of sheared edges of sheared sheets are summarized in Fig. 6(d). The sheared edges consisted of the rollover, the burnished surface, and the fracture surface. The ratio of burnished surface was increased by shearing with the punch with shear angle, i.e. the ratio of fracture surface was

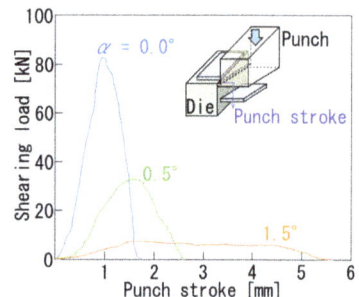

Fig. 5. Shearing load - punch stroke curve.

Material Forming - ESAFORM 2023
Materials Research Proceedings 28 (2023) 1007-1014

Materials Research Forum LLC
https://doi.org/10.21741/9781644902479-110

decreased. In particular, the burnished surface ratio was highest at α =0.5°.

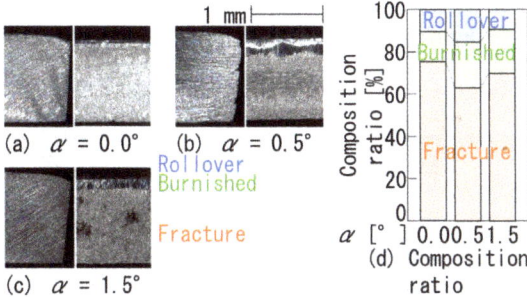

Fig. 6. Sheared edges and quality of sheared edge of sheared sheet.

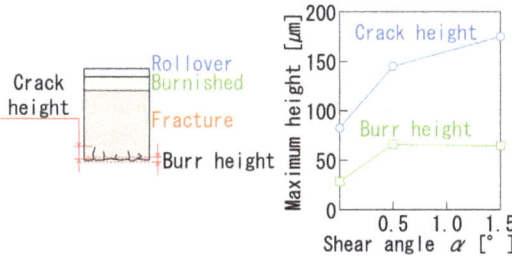

Fig. 7. Maximum crack height and burr height.

In the fracture surface of the sheared edges, cracks that extend into the inside of the fracture surface and small burrs at the edges of the fracture surface were observed. The crack height was defined as the length of the crack from the bottom edge of the fracture surface. The burr height was the length of the burr from the sheet bottom surface. The maximum crack height and burr height sheared at each shear angles are shown in Fig. 7, and the number of cracks per unit width was shown in Fig. 8. The maximum crack height and burr height in the sheared edge with the punch having the shear angle were larger than those in the sheared edge in α = 0.0°. The cracks per unit width in α = 0.5° was the largest.

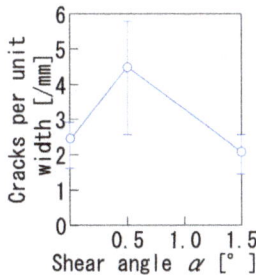

Fig. 8. Number of cracks per unit width.

The boundary edges between the burnished and fractured surfaces of the sheared edge are shown in Fig. 9. The burnished surfaces with the punch having the shear angle were rougher. The boundary edge between the burnished and fractured surfaces was rough with a punch having the shear angle $\alpha = 0.5°$.

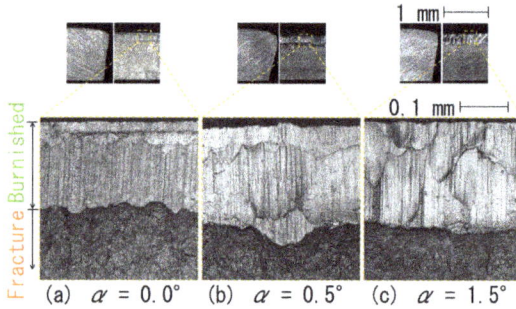

(a) $\alpha = 0.0°$ (b) $\alpha = 0.5°$ (c) $\alpha = 1.5°$

Fig. 9. Boundary between burnished and fractured surfaces.

Cracks after stretch flanging. The stretch flanging ratio with $\alpha = 0.0°$, the fracture bottom and l = 45 mm is shown in Fig. 10. The stretch flanging ratio was calculated by the difference of gage length on the blank before and after stretch flanging. The initial gage length s was 2 mm. The stretch flanging ratio was the largest at the center of the blank, and deformation was concentrated in a range of approximately -15 mm to +15 mm from the center.

The definition of the crack height after stretch flanging is shown in Fig. 11. Cracks were observed at the bending center of the stretch flanging, and the maximum crack height was classified into the following three types;

i) No cracks: crack height less than 100 μm,
ii) Cracking: crack height larger than 100 μm that do not penetrate in the thickness direction,
iii) Penetration: cracks penetrating in the direction of the sheet thickness.

Cracks were further classified into two types with a threshold value of 300 μm (1/4 of the sheet thickness).

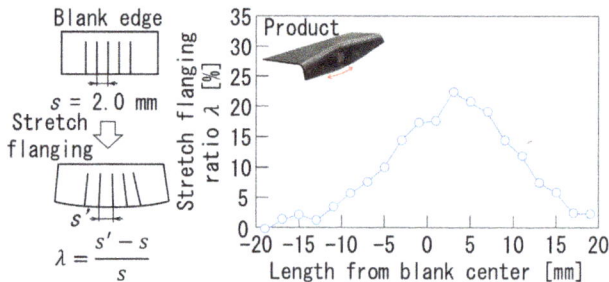

Fig. 10. Stretch flanging ratio with α = 0.0°, fracture bottom side and l = 45 mm.

Material Forming - ESAFORM 2023
Materials Research Proceedings 28 (2023) 1007-1014

Materials Research Forum LLC
https://doi.org/10.21741/9781644902479-110

Pattern	No cracks	Cracking	Penetration
Product Blank edges		>100 μm	
Crack height	○ ≦100 μm	□ <300 μm △ ≧300 μm	× Cracks penetration

Cracking in blank edge

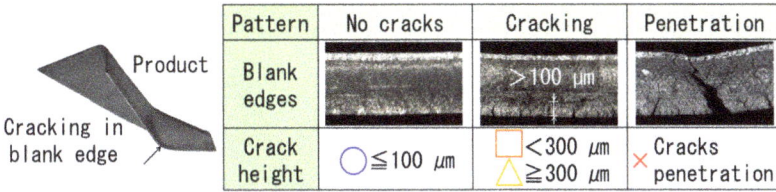

Fig. 11. Definition of crack height after stretch flanging.

The crack heights after stretch flanging on the fracture top and bottom sides are shown in Fig. 12. For both the conditions, as the bending length increased, the crack heights in the sheared edges increased at $\alpha = 0.0°$ and $1.5°$, although the cracks did not penetrate. On the other hand, the increment of crack height at $0.5°$ was small. The crack height after stretch flanging was affected by the sheared edge cut by the punch with the shear angle.

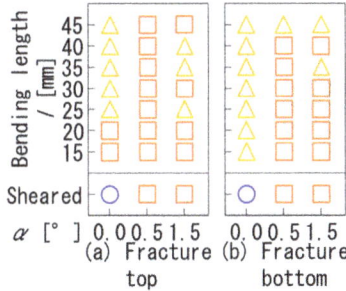

Fig. 12. Crack heights after stretch flanging on the fracture top and bottom sides.

The maximum crack heights for each shear angle after stretch flanging are shown in Fig. 13. A maximum crack height in the sheared edge after stretch flanging was measured in each bending length. On the fracture top side, the maximum crack heights tended to increase as the bending length increased, especially in $\alpha = 0.0°$ and $1.5°$. On the fracture bottom side, the maximum crack height was large in $\alpha = 0.0°$, whereas, the height did not increase significantly. On the other hand, in $\alpha = 0.5°$ and $1.5°$, the maximum crack heights increased with increasing bending length, especially at $l = 45$mm where the crack extended significantly in $\alpha = 1.5°$. It seems that the maximum crack height and the increment of the height in $\alpha = 0.5°$ were small, although the increment was depending on the sheared edge and the bending side.

Material Forming - ESAFORM 2023
Materials Research Proceedings 28 (2023) 1007-1014

Materials Research Forum LLC
https://doi.org/10.21741/9781644902479-110

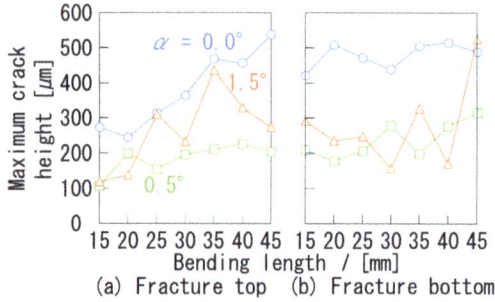

Fig. 13. Maximum crack heights for each shear angle.

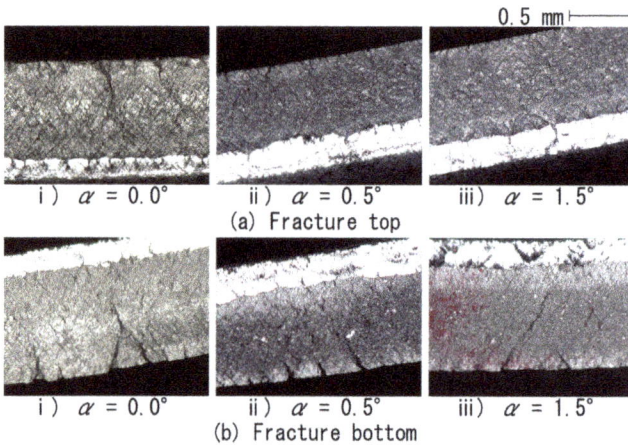

Fig. 14. Edge cracks after stretch flanging with l = 45 mm.

The edge cracks after stretch flanging for each shear angle at l = 45 mm are shown in Fig. 14. Not only in the fracture surface but also in the boundary portion between the burnished and fractured surfaces, the cracks occurred. Although the large cracks with small number are observed in α = 0.0° and 1.5°, the small cracks with large number in fracture surface are observed in α = 0.5°. It seems that a large number of initial cracks as shown in Fig. 8 in the sheared edge distributed the deformation in stretch flanging, and then the occurrence of the large crack was prevented.

Summary

In this paper, the edges of ultra-high-strength blanks sheared with three different shear angles were investigated, and then the size and growth of the cracks in the sheared edge after stretch flanging were observed, with the following results:
1) The ratio of burnished surface was increased by shearing with the punch with shear angle, i.e. the ratio of fracture surface was decreased, and in particular, the burnished surface ratio was highest in the shear angle of 0.5°.
2) The crack height increase after stretch flanging was suppressed in the shear angle of 0.5° in both the fracture top and bottom sides.

Material Forming - ESAFORM 2023   Materials Research Forum LLC
Materials Research Proceedings 28 (2023) 1007-1014   https://doi.org/10.21741/9781644902479-110

3) It seemed that the maximum crack height and the crack increment in the shear angle of 0.5° after stretch flanging were small, although the increment of the maximum crack height was depending on the sheared edge and the bending side.

4) It seemed that a large number of cracks are effective in suppressing penetration cracking caused in stretch flanging, because the stretch flangeability increased in the shear angle of 0.5°, where the number of cracks per unit length were the largest.

References

[1] M.S. Billur, T. Altan, Challenges in forming advanced high strength steels, Proc. New Developments in Sheet Metal Forming (2012) 285-304.

[2] K. Mori, K. Akita, Y. Abe, Springback behaviour in bending of ultra-high-strength steel sheets using CNC servo press. Int. J. Mach. Tools Manuf. 47 (2007) 321-325. https://doi.org/10.1016/j.ijmachtools.2006.03.013

[3] M. Kaupper, M. Merklein, Bendability of advanced high strength steels - A new evaluation procedure, CIRP Annals - Manuf. Technol. 62 (2013) 247-250. https://doi.org/10.1016/j.cirp.2013.03.049

[4] Y. Abe, T. Ohmi, K. Mori, T. Masuda, Improvement of formability in deep drawing of ultra-high strength steel sheets by coating of die, J. Mater. Process. Technol. 214 (2014) 1838-1843. https://doi.org/10.1016/j.jmatprotec.2014.03.023

[5] K. Mori, Y. Abe, K. Sedoguchi, Delayed fracture in cold blanking of ultra-high strength steel sheets, CIRP Annals - Manuf. Technol. 68 (2019) 297-300. https://doi.org/10.1016/j.cirp.2019.04.111

[6] R. Yagita, Y. Abe, Y. Munesada, K. Mori, Deformation Behaviour in Shearing of Ultra-High Strength Steel Sheets under Insufficient Blankholding force, Procedia Manuf. 50 (2020) 26-31. https://doi.org/10.1016/j.promfg.2020.08.006

[7] S. Nasheralahkami, W. Zhou, S. Golovashchenko, Study of Sheared Edge Formability of Ultra-High Strength DP980 Sheet Metal Blanks, J. Manuf. Sci. Eng. 141 (2019) 091009. https://doi.org/10.1115/1.4044098

[8] Y. Abe, R. Yonekawa, K. Sedoguchi, K. Mori, Shearing of ultra-high strength steel sheets with step punch, Procedia Manuf. 15 (2018) 597-604. https://doi.org/10.1016/j.promfg.2018.07.283

[9] P. Sartkulvanich, B. Kroenauer, R. Golle, A. Konieczny, T. Altan, Finite element analysis of the effect of blanked edge quality upon stretch flanging of AHSS, CIRP Annals - Manuf. Technol. 59 (2010) 279-282. https://doi.org/10.1016/j.cirp.2010.03.108

[10] L. Pan, J. Xiong, Z. Zuo, W. Tan, J. Wang, W. Yu, Study of the stretch-flangeability improvement of dual phase steel, Procedia Manuf. 50 (2020) 761-764. https://doi.org/10.1016/j.promfg.2020.08.137

[11] H. Gwon, J.H. Kim, J-K. Kim, D-W. Suh, S-J. Kim, Role of grain size on deformation microstructures and stretch-flangeability of TWIP steel, Mater. Sci. Eng. A 773 (2020) 138861. https://doi.org/10.1016/j.msea.2019.138861

[12] Y.T. Choi, P. Asghari-Rad, J.W. Bae, H.S. Kim, Effect of phase interface on stretch-flangeability of metastable ferrous medium-entropy alloys, Mater. Sci. Eng. A 852 (2022) 143683. https://doi.org/10.1016/j.msea.2022.143683

[13] Y. Wang, Y. Xu, X. Wang, J. Zhang, F. Peng, X. Gu, Y. Wang, W. Zhao, Improving the stretch flangeability of ultra-high strength TRIP-assisted steels by introducing banded structure, Mater. Sci. Eng. A 852 (2022) 143722. https://doi.org/10.1016/j.msea.2022.143722

Material Forming - ESAFORM 2023   Materials Research Forum LLC
Materials Research Proceedings 28 (2023) 1015-1020   https://doi.org/10.21741/9781644902479-111

Effect of reduction ratio in flow forming process on microstructure and mechanical properties of a 6082 Al alloy

MUTLU Mehmet[2,a,*], ÖZSOY Atasan[1,b], FENERCIOĞLU Tevfik Ozan[2,c], KARAKAŞ Aptullah[2,d] and BAYDOĞAN Murat[3,e]

[1]Repkon Production Technologies Contracting Industry & Trade Inc., 34980 Şile, Istanbul, Türkiye

[2]Repkon Machine and Tool Industry and Trade Inc., 34980 Şile, Istanbul, Türkiye

[3]Istanbul Technical University, Department of Metallurgy and Materials Engineering, 34469 Maslak, Istanbul, Türkiye

[a]mehmet.mutlu@repkon.com.tr, [b]atasan.ozsoy@repkon.com.tr, [c]ozan.fenercioglu@repkon.com.tr [d]aptullah.karakas@repkon.com.tr, [e]baydogan@itu.edu.tr

Keywords: 6082 Al Alloy, Age Hardening, Flow Forming, Strain Hardening

Abstract. Flow forming is a cold deformation process in which hollow cylindrical or conical parts with different geometric configurations are produced using tools such as balls, rollers, or flow forming wheels on specialized mandrels. Because it enables the production of parts without any further modifications or with minimal modifications before their use in service, the process is categorized as an NSF technology (net-shape forming), and therefore the flow formed parts can be considered as a final product. The aim of this study is to investigate the microstructure and mechanical properties of a flow formed 6082 Al alloy, which was initially in W-temper condition. Hollow cylindrical preforms were first manufactured by machining, and subsequently solution heat treated and quenched. Then, the parts were flow formed with 3 different reduction ratios (45%, 55% and 65%) prior to aging at 177 °C for 8 h to achieve T8 temper condition. Microstructures of the flow formed parts were examined by an optical microscope, and hardness and tensile tests were conducted. The results revealed that increasing reduction ratio slightly decreases hardness and strength with almost constant ductility.

Introduction

Al-Mg-Si alloys (6xxx series alloys) are commonly used heat treatable Al alloys because they offer a combination of high strength, high corrosion resistance, good weldability and low density. In order to obtain appropriate mechanical strength, 6xxx Al alloys are frequently extruded, shaped, and then artificially aged. During these forming processes, the material is subjected to different levels of stress and deformation, precipitation of the second phases occurs in the microstructure of the deformed material during subsequent aging, and the pre-deformation history significantly affects the precipitation behavior and the aging response [1]. Cold working after solution heat treatment and quenching causes nucleation sites for finer precipitates, which further increases strength [2].

The flow forming is a metal forming method which is used to produce hollow cylindrical parts or conical elements by using tools in the form of balls, rollers, or flow forming wheels on a particular mandrel. In order to extrude the material under pressure in the axial direction, the material must first be brought to the state of plastic flow. As a result, the workpiece's length increases and its diameter decreases [3]. During flow forming, the amount of applied plastic deformation is controlled by various process parameters. The most important parameter among them is reduction ratio, namely, percent deformation through the thickness direction.

Material Forming - ESAFORM 2023
Materials Research Forum LLC
Materials Research Proceedings 28 (2023) 1015-1020
https://doi.org/10.21741/9781644902479-111

Most of the studies concerning flow forming processes include surface roughness prediction depending on the flow forming parameters [4], residual stress estimation [5], forming forces arising from the process variables, and investigation the relationships between the microstructure and the mechanical properties of flow formed materials. In this context, De et al [4] investigated the surface roughness of an H30 aluminum alloy after being flow formed, and established a correlation between the surface roughness and the flow forming parameters. Srivastwa et al [6] measured the flow forming forces depending on the feed rate and the rotational speed of the mandrel during flow forming process of 2014 and 7075 Al alloys in annealed condition. They found that effect of varying feed rate had a higher effect on the resulting forming forces than the effect of the varying rotational speed. They also reported that axial force was the dominant force with respect to radial and circumferential forces. Haghshenas et al [7] investigated the mechanical properties 5052 and 6061 Al alloys after 20 to 60% flow forming. They reported that the yield strength of both alloys increased after the flow forming, with the increment in 5052 Al alloy was 47% higher than that of 6061 alloy due to a higher strain hardening coefficient of 5052 Al alloy. Gao et al [8] investigated the effect of recrystallization and solution annealing heat treatments on the microstructures and mechanical properties of a flow formed 2219 Al alloy. They reported that the flow forming increased strength and decreased ductility. Following the recrystallization annealing at 435°C for 30 min, strength of the flow formed samples decreased first, and increased again with increased recrystallization temperature. For the solution annealing heat treatment conducted after the flow forming, both strength and ductility increased with increasing solution annealing temperature. As seen from the previous works, aluminum alloys were subjected to flow forming process to investigate their response to flow forming in the view point of their surface quality and the final mechanical properties after being flow formed. It is therefore, the main aim of the present work is to investigate the effect of reduction ratio during flow forming process, and subsequent aging on the microstructure and mechanical properties of a 6082 Al alloy.

Experimental Procedure
6082 Al preform tubes were first solution annealed (W-temper condition) at 530 °C for 1 h, and then quenched in 20 vol.% polymer containing water [9]. The flow forming process was performed on solution treated alloy in W-temper condition in three different reduction ratios (45%, 55% and 65%), and finally the tubes were aged at 177°C for 8 h. Fig. 1 shows an image of preform material (left) and flow-formed materials (with increasing reduction ratio from left to right). The preform and the flow formed tubes were hereafter designated as Preform, S45, S55 and S65 samples according to the reduction ratio.

Fig. 1. General view of the preform (left) and flow formed 6082 Al tubes with increasing reduction ratios of 45%, 55%, and 65% from left to right.

Microstructures of the preform and the flow formed samples were examined under an Olympus BX53M optical microscope after being prepared by standard metallographic procedure [10] and etched by 25 s immersion into the Groesbeck's reagent (100 ml water, 4 g NaOH, 4 g KMnO4). Hardness of the samples was measured by an Emco Test DuraScan 20 tester using a Vickers pyramid indenter and 300 g load according to ASTM E384 standard [11]. The hardness measurement was carried out through the thickness of the samples at regular intervals from the outer surface (in contact with the rollers) to the inner surface (in contact with the mandrel). The tensile tests were conducted on an Instron 3382 model universal testing machine using the samples, which were prepared according to ASTM E8/E8M standard [12] in the longitudinal direction to the forming direction.

Results and Discussions
Fig. 2 shows optical micrographs of the flow formed samples. As seen in Fig. 2a, the grains are clearly visible in S45 sample, as being approximately 20 µm length and 5 µm width grains, which were elongated along the flow direction. In the microstructures of S55 and S65 samples (Fig. 2b and c), the reduction ratio was higher, and the grains were not clearly visible due to higher amount of reduction on those samples. Despite all samples were subjected to aging heat treatment at 177°C for 8 h, there was no recrystallization in the microstructures due to a lower temperature of aging, which is insufficient for recrystallization. Considering the fact that recrystallization takes place easier when the reduction rate increases [13], it was seen that even 65% of reduction in the thickness direction did not lead to recrystallization of the samples.

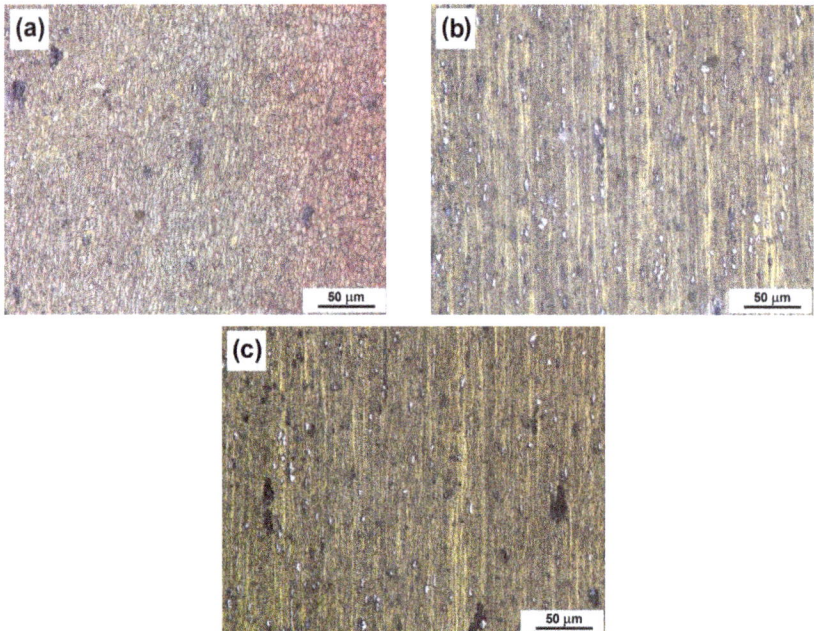

Fig. 2. Optical micrographs of the flow formed samples (a) S45, (b) S55, and (c) S65.

Material Forming - ESAFORM 2023
Materials Research Proceedings 28 (2023) 1015-1020

Materials Research Forum LLC
https://doi.org/10.21741/9781644902479-111

Table 1 lists the hardness of the flow formed samples. It is interesting to note that hardness of the flow formed and the aged samples slightly reduced when the reduction ratio increased. On the other hand, even though hardness of S45 sample close to the inner surface was slightly higher, there was no clear correlation between the hardness values depending on the hardness measurement locations.

Table 1. Hardness measurements of the flow formed and aged samples.

Hardness measurement locations	Hardness, HV0.3		
	S45	S55	S65
1 (close to outer surface)	99	92	93
2	103	96	96
3	98	93	99
4	101	92	93
5 (close to inner surface)	110	93	95

Tensile test results are listed in Table 2, and the stress – strain graphs are presented in Fig. 3. Similar to the hardness variation, strength values of S45 sample are slightly higher than those of S55 and S65 samples. However, the ductility, in terms of elongation, almost remained constant when the reduction ratio increases. Considering that lower reduction ratio results in a higher hardness and strength, and lack of hardness increment with increased reduction ratio, it was concluded that that age hardening mechanism is more effective than strain hardening mechanism in determining the final mechanical properties.

Table 2. Tensile test results of the flow formed and aged samples.

Properties	S45	S55	S65
Yield strength, [MPa]	272	262	268
Ultimate tensile strength, [MPa]	290	275	282
Elongation at fracture, [%]	11.8	11.3	10.8

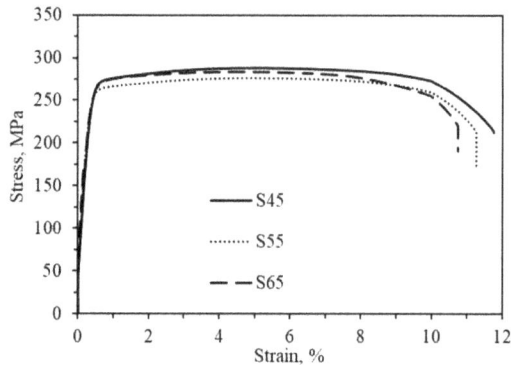

Fig. 3. Stress strain graphs of the flow formed and aged samples.

Summary

The effect of cold deformation by the flow forming on microstructure and mechanical properties of a 6082 Al alloy after the aging was investigated. As the reduction ratio increases from 45% to 65%, the grains are elongated along the flow direction. For lower reduction ratio (S45 sample), the grains are still visible, while, the grains were almost invisible in the samples subjected to reduction ratio (S55 and S65 samples). There was no recrystallization in the microstructures indicating the aging temperature (177°C) after the flow forming is insufficient for recrystallization to take place. Hardness and strength are the highest for the samples subjected to a lower reduction ratio during the flow forming. Increasing deformation slightly reduced hardness and strength, with exhibiting almost constant ductility. The obtained results finally indicate that age hardening mechanism in the present 6082 Al alloy is more effective in determining the final mechanical properties.

References

[1] M. Kolar, K.O Pedersen, , S. Gulbrandsen-Dahl, K. Marthinsen, Combined effect of deformation and artificial aging on mechanical properties of Al–Mg–Si Alloy, Transactions of Nonferrous Metals Society of China 22 (2012) 1824-1830, https://doi.org/10.1016/S1003-6326(11)61393-9

[2] Karakaş, T.O. Fenercioğlu, T. Yalçınkaya, The influence of flow forming on the precipitation characteristics of Al2024 alloys, Mater. Lett. 299 (2021) 130066. https://doi.org/10.1016/j.matlet.2021.130066

[3] Standard Practice for Heat Treatment of Wrought Aluminum Alloys, ASTM B918/B918M-20a, January 2022.

[4] T.N. De, B. Podder, N.B. Hui, C. Mondal, Experimental study and analysis of surface roughness of the flow formed H30 alloy tubes, Materials Today: Proceedings 38 (2021) 3190-3197. https://doi.org/10.1016/j.matpr.2020.09.647

[5] D. Tsivoulas, J. Quinta da Fonseca, M. Tuffs, M. Preuss, Effects of flow forming parameters on the development of residual stresses in Cr–Mo–V steel tubes, Mater. Sci. Eng. A 624 (2015) 193–202. https://doi.org/10.1016/j.msea.2014.11.068

[6] A.K. Srivastwa, P.K. Singh, S. KumarExperimental investigation of flow forming forces in Al7075 and Al2014 – A comparative study, Materials Today: Proceedings 47 (2021) 2715-2719. https://doi.org/10.1016/j.matpr.2021.02.781

[7] M. Haghshenas, J.T. Wood, R.J. Klassen, Investigation of strain-hardening rate on splined mandrel flow forming of 5052 and 6061 aluminum alloys, Mater. Sci. Eng. A 532 (2012) 287-294. https://doi.org/10.1016/j.msea.2011.10.094

Material Forming - ESAFORM 2023   Materials Research Forum LLC
Materials Research Proceedings 28 (2023) 1015-1020   https://doi.org/10.21741/9781644902479-111

[8] P.F. Gao, Z.P. Ren, M. Zhan, L. Xing, Tailoring of the microstructure and mechanical properties of the flow formed aluminum alloy sheet, J. Alloy. Compd. 928 (2022) 167139. https://doi.org/10.1016/j.jallcom.2022.167139

[9] J. Friis, B. Holmedal, Ø. Ryen, E. Nes, O.R. Myhr, Ø. Grong, T. Furu, K. Marthinsen, Work Hardening Behaviour of Heat-Treatable Al-Mg-Si-Alloys, Mater. Sci. Forum 519–521 (2006) 1901–1906. https://doi.org/10.4028/www.scientific.net/msf.519-521.1901

[10] Standard Guide for Preparation of Metallographic Specimens, ASTM E3-11, June 12, 2017.

[11] Standard Test Method for Microindentation Hardness of Materials, ASTM E384-17, June 1, 2017.

[12] Standard Test Methods for Tension Testing of Metallic Materials, ASTM E8/E8M-22, July 19, 2022.

[13] J.D. Verhoeven, Fundamentals of Physical Metallurgy, First ed., Wiley, Michigan, 1975. ISBN 0471906166.

Material Forming - ESAFORM 2023   Materials Research Forum LLC
Materials Research Proceedings 28 (2023) 1021-1028   https://doi.org/10.21741/9781644902479-112

Flow forming and recrystallization behaviour of CuZn30 alloy

MUTLU Mehmet[1,a, *], KARAKAŞ Aptullah[1,b], KUŞDEMIR Hakan[1,c],
KOLTAN Umut Kağan[1,d] and YALÇINKAYA Tuncay [2,e]

[1]Repkon Machine and Tool Industry and Trade Inc., 34980 Şile, Istanbul, Turkey

[2]Department of Aerospace Engineering, Middle East Technical University, 06800 Ankara, Turkey

[a]mehmet.mutlu@repkon.com.tr, [b]aptullah.karakas@repkon.com.tr,
[c]hakan.kusdemir@repkon.com.tr, [d]umut.koltan@repkon.com.tr, [e]yalcinka@metu.edu.tr

Keywords: Flow Forming of CuZn30, Recrystallization of CuZn30

Abstract. CuZn30, which is also called cartridge brass, is an alloy used commonly in the production of large-calibre round cartridge cases. They are usually produced via cupping of a disc and consecutive deep drawing steps to decrease the wall thickness, with an annealing process in between each step to restore formability. In this study the manufacturing of cartridge brass (CuZn30) tubes is conducted through the flow forming process. In order to evaluate the flow forming behaviour, the preforms are manufactured by machining the CuZn30 billets, then the flow forming processes is applied. Thereafter, different temperature ranges (350, 450, and 550°C for 1 h) are applied to flow formed samples in order to determine the proper recrystallization annealing temperature. The obtained microstructures and the mechanical properties are studied and revealed that the flow forming process is successfully realized, and the microstructure of the material is mapped with respect to the subsequent heat treatment temperature for recrystallization. Spherical and new grains are precisely generated after recrystallization annealing at 450 and 550°C, but only partial recrystallization is obtained at 350°C.

Introduction

The alloy CuZn30, also known as cartridge brass, is best known for its use in the production of large-calibre round cartridge cases. In order to reduce the wall thickness in a series of deep drawing steps, they are typically produced by cupping a disc, followed by each deep drawing step being followed by an annealing step to regain formability [1].

Flow forming could also be a suitable alternative method for decreasing the wall thickness of the case. It is an efficient technique for manufacturing precision tubular products and is an incremental forming process in which the wall thickness of a product is decreased by passing rollers over the material once or multiple times [2]. For the flow forming process, in addition to process parameters such as the feed rate, rotation speed, and reduction ratio, which concern the process window, material properties are also a major factor that determines the process outcome. Bylya et al. [3] investigated the influence of the elastic-plastic properties on the formability of the material and pointed out resilience, strain hardening, and tensile area reduction as significant parameters, which affect the material's ability to redistribute its volume along the mandrel. For a material with high strain hardening, the deformed material at the top strengthens and restricts the upward elastic expansion of the material at the bottom. This directs the expansion along the mandrel and allows large elongations under hydrostatic compression. A higher resilience means a higher elastic compression and, therefore, expansion.

Cartridge brass has excellent cold forming and high strain hardening capability. However, its resilience is not as high as for example high strength steel due its lower yield strength. In this work, the flow forming behaviour of CuZn30 is studied considering its plastic properties within the scope

Material Forming - ESAFORM 2023
Materials Research Proceedings 28 (2023) 1021-1028

Materials Research Forum LLC
https://doi.org/10.21741/9781644902479-112

of cartridge case production as well as its recrystallization behaviour to determine the optimum process for the best final product. Heat treatment is required in between the cold forming steps for further forming by recrystallization and is usually done at around 500 to 650°C for the deep drawing processes according to reduction ratios. [4].

Experimental Procedure
CuZn30 alloy is procured in hot extruded condition. Optical emission spectroscopy is employed to confirm the material grade. The material composition matches with the composition of CuZn30 (C26000) according to ASTM B19 as seen in Table 1 [5].

Table 1. Chemical composition of the preform.

	Zn	Pb	Sn	P	Fe	Ni	Al	Bi	Cu
Value [%]	28.67	0.022	0.014	0.004	0.023	0.01	0.0027	< 0.001	71.2066

Fig. 1 Preform (PRE) part on the left and Flow-formed (FF1) on the right.

For the flow forming operations, the materials are machined into preforms in the form of hollow cylinders closed at one end with an outer diameter of 125.20 mm, wall thickness of 5.2 mm and a length of 300 mm. Preform is then flow formed with a feed rate of 110 mm/sec and rotation speed of 160 rpm with 40% thickness reduction ratio at one cycle which are illustrated in Fig.1. In this work, a preform and a flow formed parts are investigated through the microstructural analysis and the mechanical testing. Metallographic specimens from longitudinal and transverse sections for each case is prepared according to ASTM E3 [6]. The specimens are etched with Klemm's III Reagent 3 minutes and Acidic Ferric Chloride 1 minute by immersion, separately. After etching, microstructures are analyzed under an optical microscope, Olympus BX53M. The hardness measurements are conducted with a Vickers micro indentation method using EmcoTest DuraScan – 20 according to ASTM E384 [7]. For longitudinal and transverse sections of specimens, the hardness is measured 10 times, starting from the forming surface in contact with the rollers, along

Material Forming - ESAFORM 2023
Materials Research Proceedings 28 (2023) 1021-1028

Materials Research Forum LLC
https://doi.org/10.21741/9781644902479-112

the diameter at specified intervals. Tensile testing specimens are prepared by machining as shown in Fig. 2 and tested three times according to ASTM E8 [8] using an Instron 3382 model Universal Testing Machine.

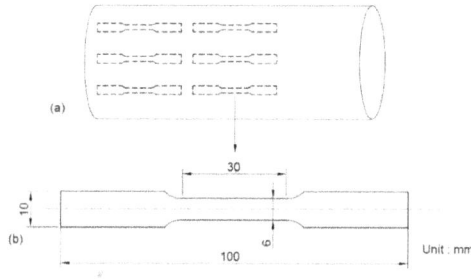

Fig. 2. The cutting position of the specimen for tensile testing (a); and (b) the specimen dimensions for tensile testing.

To investigate the recrystallization behaviour, flow formed samples (abbreviated as FF1) with dimensions around 5x5 cm are cut for microstructure analysis and hardness measurements and tensile testing specimens are machined from the preform and the flow formed parts. After manufacturing the specimens, analysis and testing are carried out, afterwards, the specimens are subjected to recrystallization heat treatment by open atmosphere laboratory furnace. The data given for 40% cold worked CuZn30 (FF1) is used as a reference for determining the recrystallization heat treatment temperature. However, the recrystallization behaviour is expected to vary with the amount of cold work as plastic deformation increases the internal energy of the material. Three heat treatment temperatures, 350, 450 and 550°C, are determined for examination. The specimens are heated with a rate of 10°C/min and held at the respective temperatures for 1 hour followed by air cooling. Microstructure analysis, hardness measurements and tensile tests of the heat-treated specimens are carried out accordingly.

Results and Discussions
The hardness of the preform is measured in the range of 75 – 85 HV. After the flow forming process, the hardness increased to 190 – 235 HV. Near the outer surface where the rollers contact occur, hardness is around 235 HV, and it drops to 190 HV at the inner surface which is in contact with the mandrel. The amount of deformation is higher at the outer surface where the force is applied by the rollers. The hardness values obtained from the longitudinal sections are similar to the transverse section. Microstructure image taken from the preform is presented in Fig. 3. CuZn30 contains approximately 30% Zn. At this concentration, only α phase forms in the microstructure. As illustrated inn Fig. 3, α grains and annealing twins are observed.

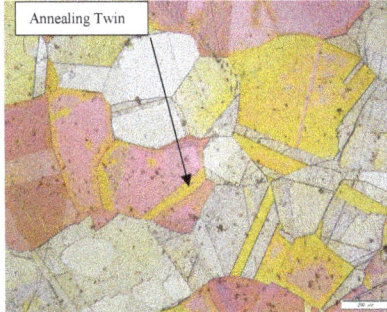

Fig. 3. Microstructure images from transverse cross-sections of PRE-specimen at 100X, Klemm's III Reagent-3min.

The microstructures of flow formed FF1 samples are presented in Fig. 4. The effective hardening mechanism in CuZn30, which consists of a single phase, is deformation hardening. Since α phase has a low stacking fault energy, it is difficult for screw dislocations to perform a cross-slip. Due to this restriction in dislocation movement, the material has high strain hardening properties [9].

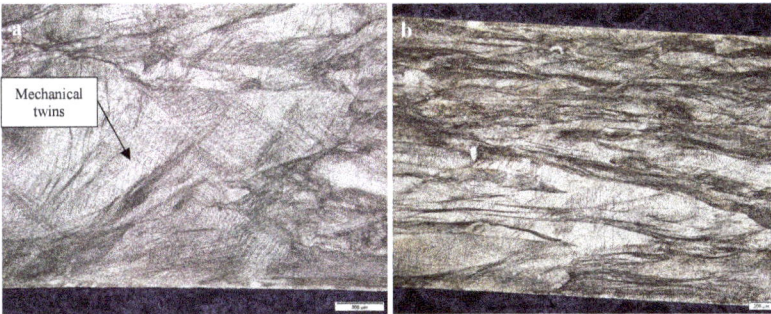

Fig. 4. Microstructure images from transverse cross-section of FF1 specimen at (a) 100X and (b) 50X, Acidic Ferric Chloride-1min.

Since slip systems are limited in materials with low stacking fault energy, mechanical twins are observed in microstructure after the deformation. In flow formed sample (FF1), the mechanical twins are seen as thin parallel lines side by side. These are the local high shear strain zones seen in ductile materials that caused by a high amount of deformation [10]. The ultimate tensile strength (UTS) hardening exponent n of flow-formed (FF1) is calculated as 0.12 and as for the preform (PRE), it is calculated as 0.48. This shows that the preform has a higher strain hardening ability compared to FF1 specimen.

Material Forming - ESAFORM 2023
Materials Research Proceedings 28 (2023) 1021-1028

Materials Research Forum LLC
https://doi.org/10.21741/9781644902479-112

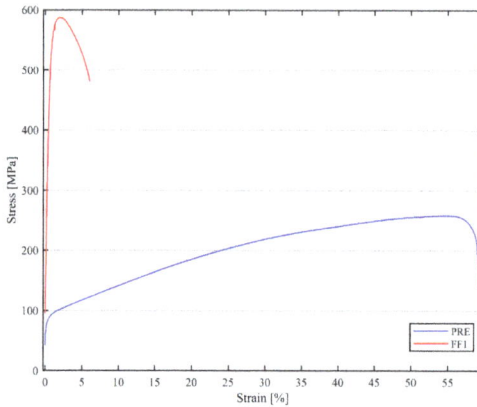

Fig. 5. Stress-strain curves of PRE and FF1 specimens.

The hardness values of heat-treated parts at 350, 450 and 550°C, are measured between 120 – 130 HV, 90 – 110 HV and 75 – 90 HV respectively. The hardness dropped to preform hardness values after heat treating at 550°C. The microstructure of heat-treated samples at 350°C, 450°C and 550°C are given in Fig. 6, 7 and 8 respectively. The microstructure of the transverse, longitudinal and planar sections are similar to each other. As shown in Fig. 6, the mechanical twins did not completely disappear in FF1 after the heat treatment at 350°C. Grain morphology could not occur completely in microstructure, but it could be observed that the number of mechanical twins decreased and partial recrystallization occurred. The recrystallization of FF1 can be completed by increasing the heat treatment time at the application temperature. The microstructures after the heat treatment at 450°C and 550°C are presented in Fig. 7 and 8 respectively. It is observed that recrystallization has taken place in the microstructure. At the higher heat treatment temperature, the grain sizes increased as shown in the figures.

Fig. 6. FF1 heat treated at 350°C taken at 500X, Acidic Ferric Chloride-1min.

Material Forming - ESAFORM 2023
Materials Research Proceedings 28 (2023) 1021-1028

Materials Research Forum LLC
https://doi.org/10.21741/9781644902479-112

Fig. 7. FF1 heat treated at 450 °C taken at 500X, Klemm's III Reagent-3min.

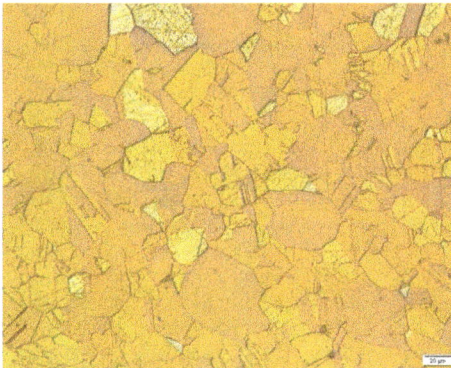

Fig. 8. FF1 heat treated at 550 °C taken at 500X, Klemm's III Reagent-3min.

The tensile test results of the heat-treated samples are shown in Fig. 9 As the heat treatment temperature increases, the strength of the material decreases and its ductility increases. The strength level of FF1 specimen heat treated at 550°C approaches quite close to the strength levels of preform.

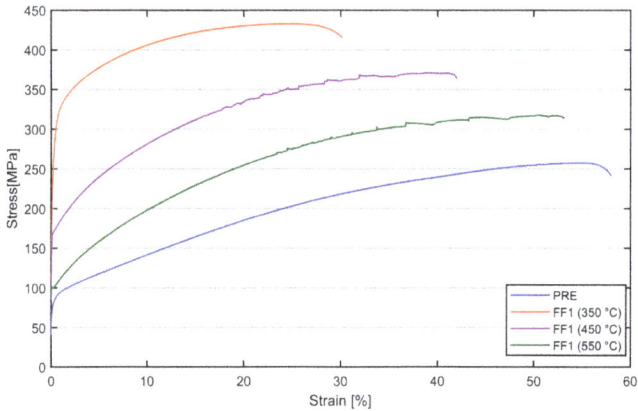

Fig. 9. Stress-strain curves of recrystallization heat treated specimens.

Summary

This paper investigates the formability of cartridge brass using flow forming process as well as the behaviour of recrystallization following cold deformation by heat treatment at three different temperature levels, which are 350°C, 450°C, and 550°C. After the flow forming process, a substantial strain-hardening is obtained, with tensile strength increasing twofold and yield strength increasing fivefold, while elongation decreasing to 5% from 55%. This elongation level makes the material quite brittle, therefore, it is very difficult to use the cartridge brass at this elongation levels because of its low toughness. In order to eliminate brittleness, both recrystallization and annealing procedures are carried out. The annealing processes are applied at three different temperature values. Microstructural analysis and mechanical testing revealed that the recrystallization process succeeds at 450°C and 550°C, while recrystallization could not be completely affected at 350°C. As for tensile testing of the FF1 specimen, which is annealed at 350°C, it is observed that elongation increases which could be related to recovery. The recrystallization mechanism necessitates more temperature in order to lower dislocation density; therefore, only the recovery mechanism occurs at 350°C.

References

[1] A. Doig, Military Metallurgy, 1998, pp. 23 - 29. https://doi.org/10.1201/9781003059400

[2] A. Karakaş, T. Ozan Fenercioğlu, T. Yalçinkaya, The influence of flow forming on the precipitation characteristics of Al2024 alloys, Mater. Lett. 299 (2021) 130066. https://doi.org/10.1016/j.matlet.2021.130066

[3] O. Bylya, T. Khismatullin, P. Blackwell, R. Vasin, The effect of elasto-plastic properties of materials on their formability by flow forming, J. Mater. Process. Technol. 252 (2018) 34-44. https://doi.org/10.1016/j.jmatprotec.2017.09.007

[4] E. El-Danaf, M. Soliman, A. Al-Mutlaq, Correlation of Grain Size, Stacking Fault Energy, and Texture in Cu-Al Alloys Deformed under Simulated Rolling Conditions, Adv. Mater. Sci. Eng. 2015 (2015) 1-12. https://doi.org/10.1155/2015/953130

[5] ASTM, B19-Standard Specification for Cartridge Brass Sheet, Strip, Plate, Bar, and Disks, Annual Book or ASTM Standards, American Society for Testing and Materials, Vol. 02.01

[6] ASTM, E3-11- Standard Guide for Preparation of Metallographic Specimens, Annual Book or ASTM Standards, American Society for Testing and Materials, Vol. 3.01

Material Forming - ESAFORM 2023   Materials Research Forum LLC
Materials Research Proceedings 28 (2023) 1021-1028   https://doi.org/10.21741/9781644902479-112

[7] ASTM, E384-17- Standard Test Method for Microindentation Hardness of Materials, Annual Book or ASTM Standards, American Society for Testing and Materials, Vol. 3.01

[8] ASTM, E8-13. Standard Test Methods of Tension Testing of Metallic Materials, Annual Book or ASTM Standards, American Society for Testing and Materials, Vol. 3.01

[9] G. Xiao, N. Tao, K. Lu, Microstructures and mechanical properties of a Cu–Zn alloy subjected to cryogenic dynamic plastic deformation, Mater. Sci. Eng. A 513-514 (2009) 13-21. https://doi.org/10.1016/j.msea.2009.01.022

[10] S. Cronje, R.E. Kroon, W.D. Roos, J.H. Neethling, Twinning in copper deformed at high strain rates, Bull. Mater. Sci. 36 (2013) 157–162. https://doi.org/10.1007/s12034-013-0445-4

Material Forming - ESAFORM 2023
Materials Research Proceedings 28 (2023) 1029-1035

Materials Research Forum LLC
https://doi.org/10.21741/9781644902479-113

Mechanical and microstructural properties of AISI 4140 after flow-forming process

YAZGAN Elif[1,a] *, MUTLU Mehmet[2,b], AYDIN Güneş[1,c], KARAKAS Aptullah[2,d], FENERCIOGLU Tevfik Ozan [2,e] and BAYDOGAN Murat[3,f]

[1]Repkon Production Technologies Contracting Industry & Trade Inc., 34980 Şile, Istanbul, Türkiye

[2]Repkon Machine and Tool Industry and Trade Inc., 34980 Şile, Istanbul, Türkiye

[3]Istanbul Technical University, Department of Metallurgy and Materials Engineering, 34469 Maslak, Istanbul, Türkiye

[a]elif.yazgan@repkon.com.tr, [b]mehmet.mutlu@repkon.com.tr , [c]gunes.aydin@repkon.com.tr, [d]aptullah.karakas@repkon.com.tr, [e]ozan.fenercioglu@repkon.com.tr, [f]baydogan@itu.edu.tr

Keywords: Flow Forming, Heat Treatment, AISI 4140

Abstract. Flow-forming is a cold deformation process to form dimensionally precise and rotationally symmetrical parts. Strain hardening effect of the flow forming process, and possibility of producing cylindrical part are the advantages especially for aerospace industry. The purpose of this study is to investigate the effect of initial microstructure of an AISI 4140 steel on the microstructure and mechanical properties after being flow formed by 70% as the reduction ratio in the thickness direction. In this context, as-received steel was heat treated to standard quenched and tempered condition, and an additional annealing was also performed. Before and after the flow forming process, the microstructure was examined, hardness and tensile tests were conducted. The results revealed that the additional annealing was beneficial to obtain a crack free material after the flow forming process of a heat treated material.

Introduction

Flow forming is a cold deformation process commonly used in metal forming industry. With this process, dimensionally accurate and rotationally symmetrical parts can easily be produced [1,2]. Cold deformation during flow forming improves hardness and strength properties of the formed parts due to strain hardening mechanism, which occurs as a result of generation of the dislocations within the material and their interactions with each other [3]. In the flow forming process, a relatively short and thick starting material in tube form (preform) is formed into a longer and thinner tube by means of a rotating mandrel inside the preform, and one or more rollers outside the preform. There are basically two flow forming methods depending on the flow direction of the material. When the flow direction of the material is towards the front of the roller, it is called as forward flow forming, and when the flow direction of the material is the rear of the roller, it is called as backward flow forming. Higher dimensional accuracy, higher inner and outer surface quality, improved hardness and strength, and finer and uniform directional grain structure can be achieved by the flow forming process [4].

Flow forming is an active research area both in academic and industrial point of view. In this context, several works have been published so far focusing on Al alloys [5], Mg alloys [6], Ti alloys [7] and steels. Beside experimental works concerning the relationship between input parameters of the flow forming, and final microstructure and mechanical properties of the flow formed alloys, numerical modelling coupled with the experiments was also made. For example, Banerjee et al [8] studied efficiency of two artificial neural network architectures to estimate the final dimensions of large tubes with respect to the input parameters, by using three optimization

Material Forming - ESAFORM 2023   Materials Research Forum LLC
Materials Research Proceedings 28 (2023) 1029-1035   https://doi.org/10.21741/9781644902479-113

techniques, and concluded that BFGS (Broyden–Fletcher–Goldfarb–Shanno) tuned Elman Neural Network (ENN) provided satisfactorily statistical performance with a faster computational time with respect to LM (Levenberg–Marquardt) tuned method. In another work, Roula et al [9] conducted mechanical tests and finite element analysis to model the flow forming behavior and predict the necking behavior during flow forming. On the other hand, Xu et al [10] studied fatigue crack growth rate of a 34CrMo4 steel which was flow formed after hot drawing. They reported that the hot drawn and cold flow formed steel exhibited a higher resistance to the crack growth than the base metal, fatigue crack growth rate increased with increasing stress ratio, and the sample direction had a little effect on fatigue crack growth rate. Karakaş et al [11] investigated the mechanical properties of an annealed 5140 steel tube after being flow formed in comparison to the annealed condition. They reported that after the flow forming, hardness, yield strength and tensile strength were all significantly improved by a factor of 1.2, 2.6 and 1.6 times, respectively, with a corresponding decrease in elongation at fracture by 50%. They also found that hardness through the thickness direction of the flow formed tube significantly decreased from the outer surface to the inner surface.

AISI 4140 alloy is a medium carbon low alloy steel, which is generally used in quenched and tempered condition to meet the strength and hardness specifications in industrial applications. Although the flow forming has a potentiality to modify the microstructure and mechanical properties of AISI 4140 steel, to the best of our knowledge, its mechanical properties have not been studied so far. It is therefore, in this study, it is aimed to investigate the effect of the initial microstructure of an AISI 4140 low alloyed steel on the microstructure and mechanical properties after flow forming.

Methodology

AISI 4140 steel is used as the starting material (the preform) for the flow forming process. Table 1 shows the chemical composition of a hot rolled and annealed AISI 4140 steel use as the preform material in the present study. Preform tubes have the dimensions of 730 mm in length and 5.5 mm in wall thickness. They were subjected to the flow forming process with a 70% reduction ratio in the thickness direction. Flow forming process parameters were as follows; feed rate was 0.5A mm/min, spindle rate was 1.95A rpm, and a cooled emulsion was used as the lubricant (A is factor of company know-how.). Final dimensions of the tubes were 1965 mm in length and 1.6 mm in wall thickness. In the present study, the preform and flow formed parts were shown in Fig. 1.

Table 1. Chemical composition of the AISI 4140 steel used as the preform in this study.

Element	C	Si	Mn	P	S	Cr	Mo	Ni	Al	Cu	Fe
wt.%	0.41	0.23	0.68	0.008	0.013	0.97	0.19	0.12	0,012	0,18	Balance

In order to investigate the effect of initial microstructure, three samples were prepared including as-received (AR), quenched and tempered (QT), and further annealed samples (QTA) in addition to QT condition. For QT condition, the samples were first austenitized at 845°C for 30 min, quenched in oil at room temperature, and tempered at 500°C for 30 min. Additional annealing was carried out at 600°C for 3 h. The microstructures of the preform and the flow formed samples were examined on an Olympus BX53M optical microscope after being prepared by the standard manner [12] and etched with 2% Nital. The hardness was measured on the cross sections of the tubes at specified intervals along the thickness direction by an Emco Test Dura Scan – 20 hardness tester using a Vickers indenter under a load of 300 g according to ASTM E384-17 standard [13].

Material Forming - ESAFORM 2023   Materials Research Forum LLC
Materials Research Proceedings 28 (2023) 1029-1035   https://doi.org/10.21741/9781644902479-113

Fig. 1. Preform tube (Left) and flow formed tube (Right).

The tensile tests were conducted on an Instron 3382 model universal testing machine by using longitudinal samples prepared along the forming direction according to ASTM E8M standard [14]. Three tensile specimens were tested for each condition.

Result and Discussion

Microstructural examinations.

Fig. 2 shows the optical micrographs of the preforms and the flow formed samples representing each condition of the AISI 4140 steel. The microstructure of AR sample preforms has an equiaxed grain structure composed of ferrite and pearlite (Fig. 2a). The flow forming strongly affected this microstructure resulting in severely elongated grains along the forming direction, as expected (Fig. 2b). QT sample has a tempered martensite structure before the flow forming (Fig. 2c), and exhibited a cold worked structure after the flow forming (Fig. 2d). Additional annealing led to a tempered martensite structure (Fig. 2e), which was coarser with respect to that of QT sample. The flow forming of QTA sample resulted in elongated grains (Fig. 2f) as in the case of the previous conditions. The microstructures of all flow formed samples were similar to each other, which are characterized by severely deformed grains along the forming direction without any remarkable discontinuities in the microstructure.

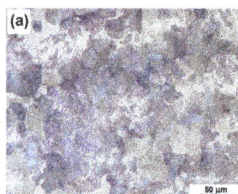

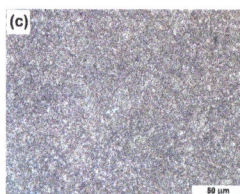

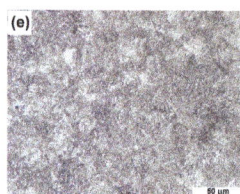

Material Forming - ESAFORM 2023
Materials Research Proceedings 28 (2023) 1029-1035

Materials Research Forum LLC
https://doi.org/10.21741/9781644902479-113

Fig. 2. Optical micrographs of the preform samples (Fig. 2a, c and e) and the flow formed samples (Fig. 2b, d and f). AR samples (Fig. 2a and b), QT samples (Fig. 2c and d), and QTA samples (Fig. 2e and f).

Hardness test results.

Hardness of the samples before and after the flow forming process was listed in Table 2. Among the preform samples, hardness increased with an increasing order for AR, QTA and QT samples. QT samples had the highest hardness, additional annealing slightly reduced hardness, as a result of a higher tempering temperature. Depending on the measurement location, the hardness slightly reduce from the outer surface to the inner surface for AR sample preforms, suggesting that the outer rollers are more effective for hardness increment. However, there was no systematic correlation depending on the measurement locations for QT and QTA samples. After being flow formed, the hardness of all samples increased due to strain hardening. The highest increment (approximately 50%) was observed for AR sample. QT and QTA samples had almost the same hardness, exhibiting a hardness increment by approximately 10-15% after the flow forming process. This indicates that strain hardening mechanism is less effective for heat treated samples than AR sample. No significant variation is observed among the hardness values of the samples from the outer and the inner surface of the flow formed samples.

Table 2. Hardness of the preforms and the flow formed samples.

Distance from the outer surface, [mm]	Hardness, [HV0.3]					
	Preform			Flow formed		
	AR	QT	QTA	AR	QT	QTA
0.15	204	396	354	312	422	435
0.30	212	374	366	305	431	419
0.44	203	388	357	308	417	424
0.60	189	389	379	310	419	413
Average	202	387	364	309	422	423

Tensile test results.

Tensile test results were listed in Table 3, and stress – strain graphs were given in Fig. 3. Among the preform samples, QT samples have the highest strength values with an elongation at fracture of 11.2% as a measure of ductility. Additional annealing reduced strength and ductility. This is a result of the coarser microstructure of QTA sample with respect to QT sample. When the samples were flow formed, regardless of their initial conditions, their strength values were improved with a corresponding decrease in ductility. This shows that hardness and strength values generally

Material Forming - ESAFORM 2023   Materials Research Forum LLC
Materials Research Proceedings 28 (2023) 1029-1035   https://doi.org/10.21741/9781644902479-113

varied similar to each other, except for strength decrement in QTA sample. Although its hardness was equivalent with that of QT sample after the flow forming, strength values of QTA sample were significantly lower than those of QT samples after the flow forming, with slightly higher ductility. On the other hand, for both preform and the flow formed samples, AR samples exhibited lower strength and higher ductility. Similar to the hardness increment upon the flow forming, the strength values increased more in AR samples (approximately 110% for yield strength, and 50% for ultimate tensile strength). On the other hand, for the heat treated samples (QT and QTA samples), this increment was significantly lower (approximately 10-25%). It is also interesting to note that mechanical properties of QTA sample before the flow forming were almost similar to those of AR sample after the flow forming. The tensile test results revealed that initial microstructures of AISI 4140 steel was highly effective in determining the final properties after the flow forming.

Table 3. Tensile test results of the preforms and the flow formed samples.

Properties	Preform			Flow formed		
	AR	QT	QTA	AR	QT	QTA
Yield strength, [MPa]	345	1118	962	738	1342	1059
Ultimate tensile strength, [MPa]	692	1209	1074	1057	1483	1363
Elongation of fracture, %	16.2	11.2	9.3	8.8	5.1	6.6

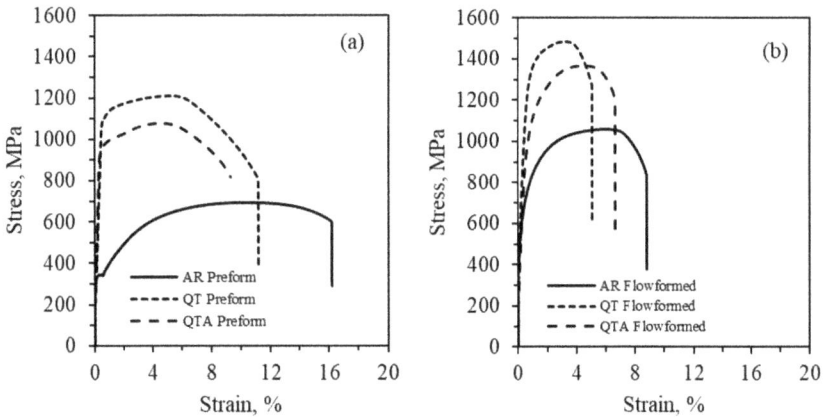

Fig. 3. Stress – strain graphs of (a) the preform and (b) flow formed parts.

Visual examination.
Visual examination of the samples after the flow forming process revealed that AR and QTA samples exhibited no visible defect inside or outside the formed tubes. However, there was a small surface crack inside QT sample as seen in Fig. 4, despite that QT sample preform has slightly higher ductility than QTA sample. The observed crack on QT sample was possibly resulted from its tempered martensite microstructure exhibiting the highest hardness and yield strength. This suggests that additional annealing after quenching and tempering is beneficial to avoid crack

formation during the flow forming, and might be attributed to a coarser tempered martensite microstructure with a lower hardness and yield strength of QTA sample.

Fig. 4. Small surface crack formed inside QT sample.

Summary
In this study, the effect of various initial microstructures of AISI 4140 steel preforms on the microstructure and mechanical properties, and surface quality after being flow formed was investigated. Heat treatments significantly affected the microstructure and mechanical properties of AISI 4140 steel after the flow forming Comparative study of three initial microstructures, namely, as-received (AR), quenched and tempered (QT), additionally tempered after quenching and tempering (QTA) leads to following conclusions:

1. The initial microstructure of AISI 4140 steel significantly affects the microstructure and mechanical properties of the flow formed parts. In this context, AR microstructure exhibited higher increment (approximately 50%) in strength values after the flow forming with a corresponding decrease (approximately 45%) in ductility. On the other hand, the strength increment after the flow forming of QT and QTA samples was lower (10-25%) when compared to that of AR sample.
2. Ductility of all samples in terms of elongation at fracture decreased by the flow forming. However, the highest decrement was observed in QT samples (approximately 55%), while QTA sample exhibited the lowest decrement in ductility (approximately 30%).
3. The highest hardness (387 HV) and strength (1483 MPa) were obtained from QT sample. However, this condition leads to a crack formation on the surface during the flow forming, Additional annealing after quenching and tempering avoided the crack formation during the flow forming. This suggest that a coarser tempered microstructure developed by the additional annealing is beneficial to produce crack free parts for the investigated AISI 4140 steel.

References

[1] A. Karakaş, T.O. Fenercioğlu, T. Yalçinkaya, The influence of flow forming on the precipitation characteristics of Al2024 alloys, Mater. Lett. 299 (2021) 1-4. https://doi.org/10.1016/j.matlet.2021.130066

[2] S. Kalpakjian, S. Rajagopal, Spinning of Tubes: A Review, J. Appl. Metal Work. 2 (1982) 211–223. https://doi.org/10.1007/BF02834039

[3] M.J. Roy, R.J. Klassen, J.T. Wood, Evolution of plastic stra.in during a flow forming process, J. Mater. Process. Technol. 209 (2009) 1018-1025. https://doi.org/10.1016/j.jmatprotec.2008.03.030

[4] D. Marini, D. Cunningham, J.R. Corney, A review of flow forming processes and mechanism, Key Eng. Mat. 651-653 (2015) 750-758. https://doi.org/10.4028/www.scientific.net/KEM.651-653.750

[5] X. Zeng, X.G. Fan, H.W. Li, M. Zhan, S.H. Li, K. Q. Wu, T.W. Ren, Heterogeneous microstructure and mechanical property of thin-walled tubular part with cross inner ribs produced by flow forming, Mat. Sci. Eng. A-Struct. 790 (2020) 139702. https://doi.org/10.1016/j.msea.2020.139702

[6] Q. Xia, J. Long, G. Xiao, S. Yuan, Y. Qin, Deformation mechanism of ZK61 magnesium alloy cylindrical parts with longitudinal inner ribs during hot backward flow forming, J. Mater. Process. Technol. 296 (2021) 117197. https://doi.org/10.1016/j.jmatprotec.2021.117197

[7] Z. Lei, P. Gao, X. Wang, M. Zhan, H. Li, Analysis of anisotropy mechanism in the mechanical property of titanium alloy tube formed through hot flow forming, J. Mater. Sci. Technol. 86 (2021) 77–90. https://doi.org/10.1016/j.jmst.2021.01.038

[8] P. Banerjee, R. Laha, M.K. Dikshit, N.B. Hui, S. Rana,V.K. Pathak, K.K. Saxena, C. Prakash, D. Buddhi, A study on the performance of various predictive models based on artificial neural network for backwardmetal flow forming process, Int. J. Interact. Des. Manuf. (IJIDeM) (2022). https://doi.org/10.1007/s12008-022-01079-6

[9] A.M. Roula, K. Mocellin, H. Traphöner, A.E. Tekkaya, P-O Bouchard, Influence of mechanical characterization on the prediction of necking issues during sheet flow forming process, J. Mater. Process. Technol. 306 (2022) 117620. https://doi.org/10.1016/j.jmatprotec.2022.117620

[10] X. Xu, C. Lu, Y. Li, X. Ma, W. Jin, Fatigue Crack Growth Characteristics of 34CrMo4 Steel for Gas Cylinders by Cold Flow Forming after Hot Drawing, Metals 11 (2021) 133. https://doi.org/10.3390/met11010133

[11] A. Karakaş, A.C. Kocabıçak, S. Yalçınkaya, Y.Şahin, Flow Forming Process for Annealed AISI 5140 Alloy Steel Tubes. in: M. Abdel Wahab (Ed.), Proceedings of the 8th International Conference on Fracture, Fatigue and Wear, FFW 2020 2020, Lecture Notes in Mechanical Engineering. Springer, Singapore. https://doi.org/10.1007/978-981-15-9893-7_10

[12] Standard Guide for Preparation of Metallographic Specimens, ASTM E3-11, June 12, 2017.

[13] Standard Test Method for Microindentation Hardness of Materials, ASTM E384-17, June 1, 2017.

[14] Standard Test Methods for Tension Testing of Metallic Materials, ASTM E8/E8M-22, July 19, 2022.

Material Forming - ESAFORM 2023   Materials Research Forum LLC
Materials Research Proceedings 28 (2023) 1037-1046   https://doi.org/10.21741/9781644902479-114

Simulation of incremental sheet metal forming for making U-channel in two light-weight alloys

BHUSHAN Bharat[1,a], RAMKUMAR Janakarajan[1,b]* and DIXIT Uday Shanker[2,c]

[1]Department of Mechanical Engineering, Indian Institute of Technology Kanpur,
Kanpur, Uttar Pradesh 208016, India

[2]Department of Mechanical Engineering, Indian Institute of Technology Guwahati,
Guwahati, Assam 781039, India

[a]bharatb@iitk.ac.in, [b]jrkumar@iitk.ac.in, [c]uday@iitg.ac.in

Keywords: Incremental Sheet Metal Forming, Finite Element Method, U-Channel, Shell Element

Abstract. Incremental sheet metal forming (ISMF) is a viable method for fabricating complicated three-dimensional structures from sheet metal. It is characterized by localized deformation and is effective for both rapid prototyping and low-volume manufacturing. ISMF technology is suitable for quick product development time with affordable tooling and for deforming difficult-to-form materials. Simulation of the process reduces costly hit-and-trial attempts for manufacturing an accurate product. This article presents the simulation for producing a U-channel made of aluminum (Al 6061-T6) and titanium (Ti-6Al-4V) alloys. A finite element method (FEM) package, ABAQUS®, has been used for simulation using shell elements. The effect of various parameters on the forming forces is discussed using two different tools, flat and hemispherical-end.

Introduction

For the past few decades, numerical simulations have been playing a vital role in understanding the underlying physics of various manufacturing processes. In processes where the machine operation and raw material costs are very high, simulation results assist in reducing the expensive hit-and-trial attempts. Sheet metal forming is a metal forming process where a sheet of metal is shaped into a product by deforming it plastically. Incremental sheet metal forming (ISMF) is one of the advanced and innovative techniques in sheet metal forming processes. This die-less process has gained popularity thanks to its high flexibility and cost-effectiveness for small-batch production. This process can form difficult-to-form materials such as titanium, aluminium, and magnesium [1]. Because of its capability to produce high geometric accuracy and surface finish, ISMF finds many applications in the automotive, aerospace, and medical sectors. Numerical modelling of the ISMF is an active area of research [2]. Modelling and simulation can provide all the information about the distribution of stresses and strains, which is crucial for finding out spring back and the possibility for defects. Many advanced finite element methods (FEM) software packages such as ABAQUS®, LS-DYNA, and PAM-STAMP are used to simulate ISMF. Among these packages, ABAQUS® /Explicit is a general-purpose package and hence, has wider availability in organizations. One can model the ISMF process using explicit or implicit dynamic methods [3].

A single-point incremental forming is an ISMF process wherein a tool presses the clamped sheet from the top without having any support from the bottom. The tool can be of different shapes, which is plunged into the sheet incrementally; path planning of the tool is also important. During the deformation, a very high reaction force is developed on the tool, depending on the sheet thickness and material. Arens et al. [4] developed a mathematical relation for the approximate prediction of axial force in ISMF. In the force prediction model, the axial force was the function

Material Forming - ESAFORM 2023   Materials Research Forum LLC
Materials Research Proceedings 28 (2023) 1037-1046   https://doi.org/10.21741/9781644902479-114

of the tensile strength (not the yield strength) with other parameters. No effect of strain hardening was included. Chang et al. [5] highlighted that due to localized and step-wise deformation in the ISMF process, the axial forming force is drastically reduced compared to conventional sheet forming processes. They noted that elastic deformation of the sheet was the major cause of the axial force fluctuation.

Petek et al. [6] demonstrated that the forming force in ISMF is very small as compared to deep drawing and is independent of the size of the product. Forming angle, tool diameter, and step-depth were the major influencing factors for the force distribution during the ISMF process. The authors also observed that the rotation of the tool has no effect on the force, but it affects the surface quality of the product. Asghar et al. [7] observed a good agreement between the forces obtained from FEM and the experimental study. The axial force is greater than the radial force on the deep-drawn component; comparatively, the tangential component of force is much smaller. Hence, geometrical accuracy is mainly influenced by axial and radial forces.

ISMF process provides good formability, low surface roughness, and high geometric accuracy. Formability improvement is one of the major advantages offered by ISMF. Shamsari et al. [8] carried out a single and two-stage ISMF for 70° wall angle truncated cone. Two-stage forming provided 26% improvement in forming depth compared to a single stage; sheet thinning was also reduced by around 45%. The effect of a hemispherical-end and flat-end tool on deformation behavior was also studied. A straight tool path has been used for each simulation, with different step depths as per the number of stages. Najm et al. [9] observed that a flat-end tool with a small corner radius offered the best thickness distribution stability. Ziran et al. [10] found that as compared to hemispherical-end tools, flat-end tools provided greater profile accuracy and formability.

The primary objective of this work is to carry out FE simulations for the forming of Al 6061-T6 and titanium (Ti-6Al-4V) alloys. Although simulations of ISMF have been performed by several researchers; however, many research articles provide only partial information about the results. This work attempts to understand the efficacy of simulations for proper process planning of ISMF. Alloys chosen for simulations are used in the automobile, aerospace, and medical sectors. Simulations were performed for making U-channels and evaluating the effects of step depth and coefficient of friction between the tool and workpiece.

Problem Definition
The deformation in the ISMF process is due to the horizontal and vertical linear movement of the tool along the defined tool path over the workpiece. Due to the motion of the tool, plastic deformation takes place, resulting in the development of stresses, reaction forces, and strains in the workpiece. Fig. 1 shows a forming tool starting from its initial position and moving forward in a horizontal direction at a certain depth. After one pass, more depth is provided depending on step-size. Then the tool moves backward and reaches its original position. This process is continued till the desired form is produced. During the deformation of the workpiece, high reaction forces act on the tool, and thinning of the sheet also takes place. Fig. 1 shows the schematic of the process along with the tool path. Here a spherical-end tool is shown, but the simulations were carried out with a flat-end tool too.

Material Forming - ESAFORM 2023
Materials Research Proceedings 28 (2023) 1037-1046

Materials Research Forum LLC
https://doi.org/10.21741/9781644902479-114

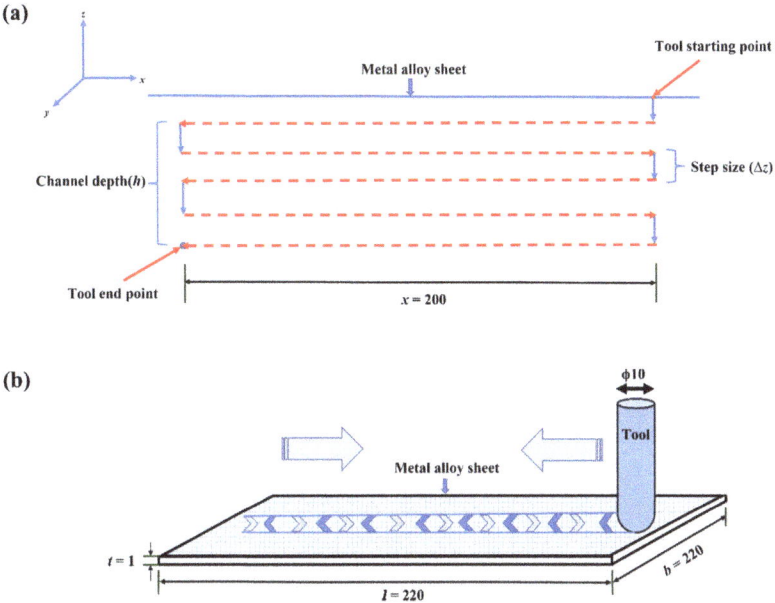

Fig. 1. Forming of U-channel: (a) tool path, (b) total imparted depth h = 5 mm with step size Δz = 1 mm. (All dimensions shown in the figure are in mm).

Numerical Simulation

Simulations were performed in ABAQUS®/Explicit FE software (SIMULIA™ by Dassault Systems®, France). Explicit analysis has been chosen because of its computational efficiency. Considering that strains are small and the sheet is already hardened, strain hardening has been neglected. The sheet blank of aluminum alloy Al 6061-T6 of thickness 1 mm was modeled. Considering the low thickness of the sheet, shell elements were chosen. The use of shell elements provided much less computational time than that required with solid elements [11]. The plasticity model used for numerical simulations for both materials is the 'Perfectly plastic isotropic hardening model based on von Mises yield criterion'. The S4R four-node square shell element was used to mesh the sheet. S4R is a 4-node double curved thin or thick shell with reduced integration, hourglass control, and finite membrane strains. For various parameters, the approximate element sizes of 1 mm, 2 mm, and 4 mm were taken. Simulations were performed to obtain different output parameters such as stress, strain, reaction force on the tool, the thickness of the sheet at different values of coefficient of friction between tool and workpiece, number of stages, and two different types of the tool.

Fig. 2 shows the hemispherical-end tool and the flat-end tool with a diameter of 10 mm. A corner radius of 2.5 mm was given to the flat-end tool. As the analytical rigid property was given to both tools, no material properties were assigned to them. Additionally, no rotational speed was given to the tool, as the effect of rotation was manifested in the form of friction.

Material Forming - ESAFORM 2023   Materials Research Forum LLC
Materials Research Proceedings 28 (2023) 1037-1046   https://doi.org/10.21741/9781644902479-114

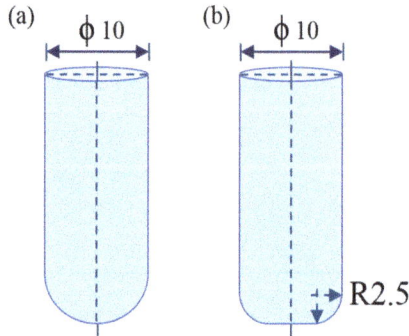

Fig. 2. Incremental sheet metal forming tools: (a) hemispherical-end, and (b) flat-end with chamfered edges. (All dimensions are in mm).

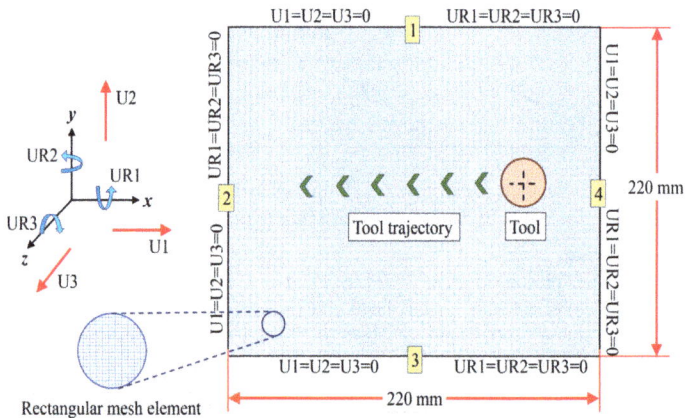

Fig. 3. Domain with boundary conditions. U1, U1, and U3 represent displacement, and UR1, UR2, and UR3 represent rotation about the x, y, and z axes, respectively. The sheet thickness is kept at 1 mm, and the edge length of the square mesh element is 2 mm.

Fig. 3 shows that the sheet was square in shape, with cross-sectional dimensions of 220 mm × 220 mm and a thickness of 1 mm. The sheet was modeled using a deformable, shell, planer type base feature. All edges of the blank sheet were fixed by applying an ENCASTRE (U1=U2=U3=UR1=UR2=UR3=0) boundary condition. In this type of displacement boundary condition sheet was clamped with all its edges. The translation and rotational motion along the x, y, and z axis were restricted. The boundary condition provided for the tool path was of amplitude type. In this type of boundary condition, the field variable varied with time. Co-ordinates of tool path were the field variables. Surface-to-surface explicit type contact property was given for dynamic explicit analysis. A master-slave algorithm was used to describe the frictional contact between tool and work. The tool moves from one edge to the other edge and repeated its path as per the number of stages provided.

Material Forming - ESAFORM 2023
Materials Research Proceedings 28 (2023) 1037-1046

Materials Research Forum LLC
https://doi.org/10.21741/9781644902479-114

The mesh sensitivity analysis was performed. Square S4R shell elements of 1 mm, 2 mm, and 4 mm size were taken by considering the effect of friction. Reaction force was estimated with different mesh sizes. CPU time increased drastically as mesh size decreased. Simulation time was about 20 minutes, 85 minutes, and 225 minutes for the element sizes of 4 mm, 2 mm, and 1 mm, respectively, for one problem. The material properties of the different materials are listed in Table 1.

Table 1. Physical and mechanical properties of Aluminum 6061- T6 and Titanium Ti-6Al-4V [12, 13].

Properties	Aluminum	Titanium
Density [g/cc]	2.7	4.43
Tensile Yield Strength [MPa]	276	880
Ultimate Tensile Strength [MPa]	310	950
Poisson's Ratio	0.33	0.34
Modulus of Elasticity [MPa]	68900	113800

Results and Discussion

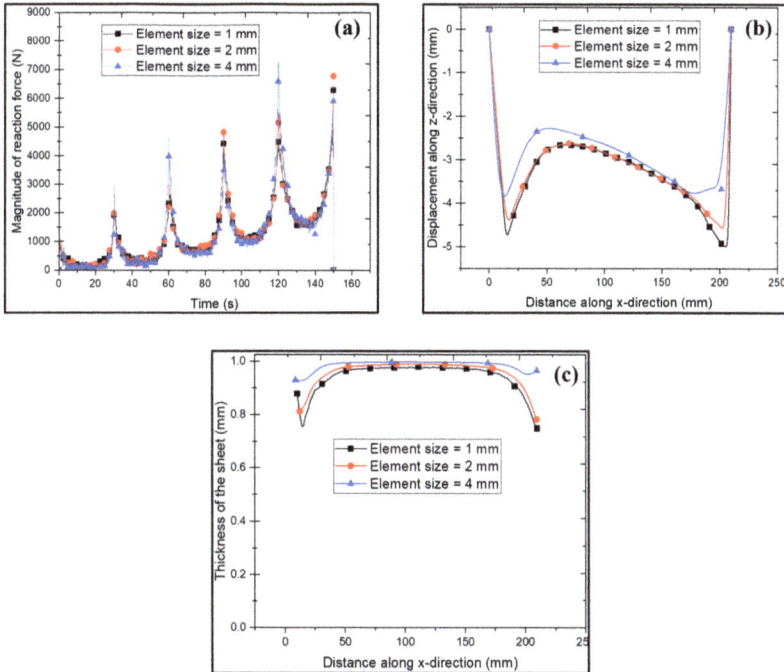

Fig. 4. Element size effect on (a) reaction force developed on the tool, (b) displacement achieved in the z-direction, and (c) thickness of the sheet after deformation.

Material Forming - ESAFORM 2023   Materials Research Forum LLC
Materials Research Proceedings 28 (2023) 1037-1046   https://doi.org/10.21741/9781644902479-114

Mesh sensitivity analysis.

For simulation, mesh size plays a vital role. In Fig. 4, an aluminium alloy has been used to investigate the mesh sensitivity. Smaller element size becomes too costly in terms of simulation time. Mesh sensitivity affects the magnitude of the reaction force developed on the tool, as shown in Fig 4 (a). Element size effect on the magnitude of reaction force developed was insignificant, but it affects the other process parameters significantly. Fig 4 (b) shows that with an element size of 1 mm, the spring back effect is high compared to the element size of 2- and 4-mm. Fig. 4 (c) showed that thickness distribution with larger element size was better, and low thickness reduction was observed. Considering a trade-off between accuracy and computational time, a 2 mm element size was chosen for subsequent parametric study. Fig. 5 shows the von Mises stress developed in aluminium alloy sheet.

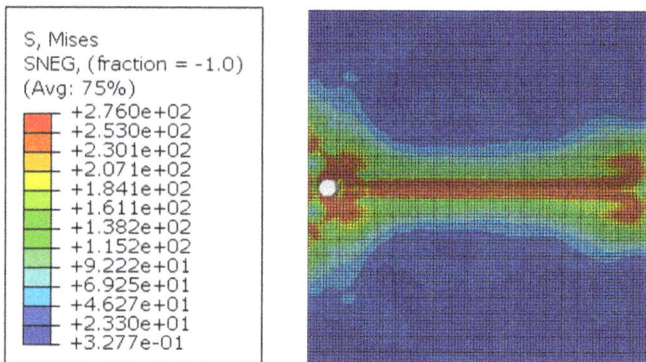

S, Mises
SNEG, (fraction = -1.0)
(Avg: 75%)
+2.760e+02
+2.530e+02
+2.301e+02
+2.071e+02
+1.841e+02
+1.611e+02
+1.382e+02
+1.152e+02
+9.222e+01
+6.925e+01
+4.627e+01
+2.330e+01
+3.277e-01

Fig. 5. Numerical simulation response of deformation with aluminium alloy.

Effect of different parameters on the reaction force.

In ISMF, the forming forces depend upon the tool path, and the feasibility of forming a reliable component also depends upon it. In the present case, the straight tool path was chosen for simulation, and forming forces were determined for different process parameters. One of the input process parameters was the coefficient of friction. The penalty-based condition with a hard contact surface was given between the tool and the workpiece.

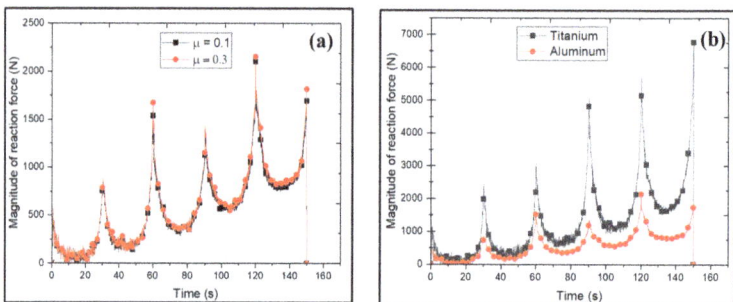

Fig. 6. Time-history plots of resultant reaction force with (a) different coefficients of friction values and (b) different materials.

Material Forming - ESAFORM 2023
Materials Research Proceedings 28 (2023) 1037-1046

Materials Research Forum LLC
https://doi.org/10.21741/9781644902479-114

Force was measured by taking every 0.1 s unit of time in field output request in the step module of ABAQUS. It was observed that the friction coefficient has very little influence on the reaction forces acting on the tool Fig 6 (a). A tool with a hemispherical-end was used by taking $\mu=0.1$ and $\mu=0.3$ with a 2 mm mesh size, i.e., (approximate global size). The observed magnitude of force was the same for these friction coefficient values. These results inferred that the coefficient of friction had an insignificant effect on the reaction force developed on the tool in any direction. When the tool stopped and reversed back during its motion along the assigned path and simultaneously moved in the z-direction in accordance with step size, it resulted in a sudden increase in reaction force. Thus, a sharp spike in reaction force was observed at the edges.

Fig. 6 (b) showed that the magnitude of the reaction force on the tool depends upon the type of material used. The developed reaction force with titanium alloy was more compared to aluminum alloy. It was due to the high tensile strength of the titanium alloy. For the same amount of deformation, by keeping all other parameters the same, force developed with titanium should obviously be high.

For making channels by ISMF process, different types of tools can be used, such as hemispherical-end tool, flat-end tool, and flat-end tool with corner radius. Simulation results suggest that the magnitude of reaction force required for making a channel is relatively less with the hemispherical-ended tool, as shown in Fig. 7.

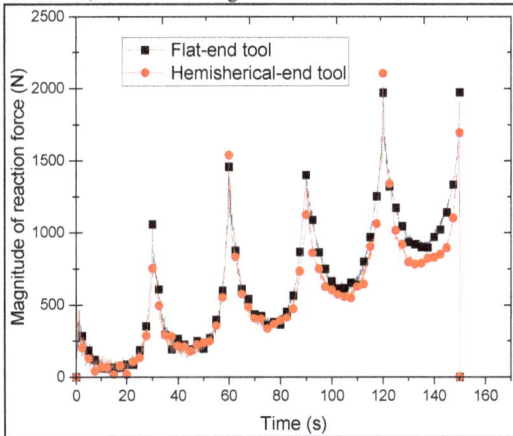

Fig. 7. Time-history plots of magnitude of resultant reaction force with different types of tools.

Effect of the step size.

The effect of step depth on different response parameters was investigated. Forming tool was treated here as a rigid body with no deformation. Reaction force developed on the tool, displacement achieved in the z-direction, and final thickness of the sheet were analyzed for two different step sizes. Fig 8 (a) showed that the reaction force developed was high with a large step size.

Sheet thinning is the major factor in sheet forming. Here, it was observed that with different step sizes, there was not much difference in the final thickness of the sheet. The spring back effect with a small step was more compared to a high step size, and the depth achieved in the z-direction was more with a large step size Fig. 8 (b), (c).

Materials Research Forum LLC

https://doi.org/10.21741/9781644902479-114

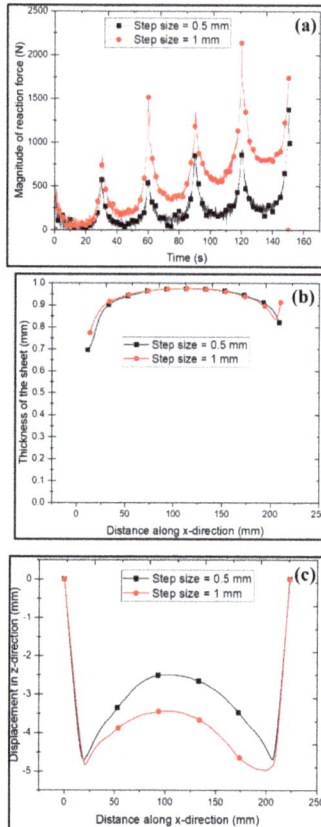

Fig. 8. Effect of step size on (a) reaction force developed on the tool, (b) thickness of the sheet after deformation, and (c) displacement achieved in the z-direction.

Experimental validation was carried out qualitatively from published literature. The trend of various parameters is observed to be in good agreement with numerical simulation data. It was observed that reaction force increases with an increase in step size [14]. From this reference, it is noted that the trend of force is in good agreement as it was done in the simulation in this article. Formability was higher with a large step size, but the thickness distribution was better with a low step size [15]. Strain hardening has a negligible effect on the force developed. There is less effect of the coefficient of friction on the formability and reaction force [16].

Summary

In recent years, the ISMF process has been intensively investigated using the FEM. In this work, numerical simulations were performed for ISMF in the U-channel model. Aluminum and titanium alloys were deformed with flat-ended and hemispherical-ended tools. To obtain a depth of 5 mm, simulations were performed with various step sizes. Based on the analysis of simulation results, the following conclusions can be drawn:

Material Forming - ESAFORM 2023
Materials Research Proceedings 28 (2023) 1037-1046

Materials Research Forum LLC
https://doi.org/10.21741/9781644902479-114

1. The size of the element has a significant effect on the resulting values of some output parameters. However, decreasing the element size makes the simulation computationally intensive, time-consuming, and cost-ineffective.

2. No significant difference was observed with different coefficient of friction values on reaction force.

3. It is observed that the reaction force developed on the tool was higher with high-strength alloy and the flat-end tool experienced a high magnitude of reaction force compared to a hemispherical end tool.

4. Step size affects the reaction force devolved on the tool and formability (measured in the form of depth achieved) but has little influence on the thickness reduction of the sheet.

Future work aims to validate the results with in-house experiments and study formability.

References

[1] I. Bagudanch, G. Centeno, C. Vallellano, M.L. Garcia-Romeu, Forming force in single point incremental forming under different bending conditions, Procedia Eng. 63 (2013) 354-360. https://doi.org/10.1016/j.proeng.2013.08.207

[2] M. Tisza, Numerical modelling and simulation in sheet metal forming, J. Mater. Process. Technol. 151 (2004) 58-62. https://doi.org/ 10.1016/j.jmatprotec.2004.04.009

[3] L. Taylor, J. Cao, A.P. Karafillis, M.C. Boyce, Numerical simulations of sheet metal forming, J. Mater. Process. Technol. 50 (1995) 168-179. https://doi.org/10.1016/0924-0136(94)01378-E

[4] R. Aerens, P. Eyckens, A. van Bael, J.R. Duflou, Force prediction for single point incremental forming deduced from experimental and FEM observations, Int. J. Adv. Manuf. Technol. 46 (2010) 969–982. https://doi.org/ 10.1007/s00170-009-2160-2

[5] Z. Chang, J. Chen, Analytical model and experimental validation of surface roughness for incremental sheet metal forming parts, Int. J. Mach. Tools Manuf. 146 (2019) 103453. https://doi.org/10.1016/j.ijmachtools.2019.103453

[6] A. Petek, K. Kuzman, J. Kopač, Deformations and forces analysis of single point incremental sheet metal forming, Arch. Mater. Sci. Eng. 35 (2009) 107-116. http://www.amse.acmsse.h2.pl/vol35_2/3527.pdf

[7] J. Asghar, R. Lingam, E. Shibin, N.V. Reddy, Tool path design for enhancement of accuracy in single-point incremental forming, Proceedings of the Institute of Mechanical Engineers Part: B J. Eng. Manuf. 228 (2014) 1027-1035. https://doi.org/10.1177/0954405413512812

[8] M. Shamsari, M.J. Mirnia, M. Elyasi, H. Baseri, Formability improvement in single point incremental forming of truncated cone using a two-stage hybrid deformation strategy, Int. J. Adv. Manuf. Technol. 94 (2018) 2357-2368. https://doi.org/10.1007/s00170-017-1031-5

[9] S.M. Najm, I. Paniti, Study on effecting parameters of flat and hemispherical end tools in spif of aluminium foils, Tehnicki Vjesnik 27 (2020) 1844-1849. https://doi.org/10.17559/TV-20190513181910

[10] X. Ziran, L. Gao, G. Hussain, Z. Cui, The performance of flat end and hemispherical end tools in single-point incremental forming, Int. J. Adv. Manuf. Technol. 46 (2010) 1113-1118. https://doi.org/10.1007/s00170-009-2179-4

[11] H.Y. Shahare, A.K. Dubey, P. Kumar, H. Yu, A. Pesin, D. Pustovoytov, P. Tandon, A comparative investigation of conventional and hammering-assisted incremental sheet forming processes for aa1050 h14 sheets, Metals 11 (2021). https://doi.org/10.3390/met11111862

[12] Information on https://asm.matweb.com/search/SpecificMaterial.asp?bassnum=ma6061t6

[13] Information on https://asm.matweb.com/search/SpecificMaterial.asp?bassnum=mtp641

[14] J. Duflou, Y. Tunçkol, A. Szekeres, P. Vanherck, Experimental study on force measurements for single point incremental forming, J. Mater. Process. Technol. 189 (2007) 65-72. https://doi.org/10.1016/j.jmatprotec.2007.01.005

[15] H.B. Lu, Y.le Li, Z.B. Liu, S. Liu, P.A. Meehan, Study on step depth for part accuracy improvement in incremental sheet forming process, Adv. Mater. Res. 939 (2014) 274-280. https://doi.org/10.4028/www.scientific.net/AMR.939.274

[16] B. Lu, Y. Fang, D.K. Xu, J. Chen, H. Ou, N.H. Moser, J. Cao, Mechanism investigation of friction-related effects in single point incremental forming using a developed oblique roller-ball tool, Int. J. Mach. Tool. Manuf. 85 (2014) 14-29. https://doi.org/10.1016/j.ijmachtools.2014.04.007

Material Forming - ESAFORM 2023
Materials Research Proceedings 28 (2023) 1047-1056

Materials Research Forum LLC
https://doi.org/10.21741/9781644902479-115

Flange wrinkling in incremental shape rolling

ESSA Abdelrahman[1,a] *, ABEYRATHNA Buddhika[1,b],
ROLFE Bernard[2,c], and WEISS Matthias[1,d]

[1]Institute for Frontier Materials, Deakin University, Waurn Ponds, Pigdons Rd., Geelong, VIC 3216, Australia

[2]School of Mechanical Engineering, Deakin University, Waurn Ponds, Pigdons Rd., Geelong, VIC 3216, Australia

[a]aessa@deakin.edu.au, [b]buddhika.abeyrathna@deakin.edu.au, [c]bernard.rolfe@deakin.edu.au, [d]matthias.weiss@deakin.edu.au

Keywords: Incremental Shape Rolling, Flexible Roll Forming, Finite Element Analysis, Wrinkling

Abstract. Automotive structural components from Advanced High Strength Steels (AHSS) can be manufactured with Flexible Roll Forming (FRF). Flange wrinkling is a common shape defect in FRF, this restricts the application of FRF in the automotive industry. The new Incremental Shape Rolling process (ISR) showed that a high tensile transverse strain is developed in the flange and that assists the plastic deformation in the flange. Hence, wrinkle severity can be significantly reduced when weight-optimized components are formed. In this study, the ISR process is applied to a variable width profile from DP600. This investigated profile is a modified automotive component. The forming strains and wrinkling severity in ISR are compared with those obtained from the FRF case. The ISR and the FRF experimental trials are performed on a prototype FRF facility and then used to validate the numerical models which are applied to analyse the deformation behaviour in both processes. The ISR results show a significant reduction in wrinkling. This is due to the high tensile transverse strain that is developed in the ISR flange which facilitates the plastic deformation and hence the flange is stably compressed to the required shape.

Nomenclatures

d_y	Increment size
f	Flange length
MAE	Mean absolute error
n	Number of points considered for the wrinkling evaluation
R	Radius of the pre-cut blank
R_n	Roll nose radius
r_i	Inner radius of the part
t	Material thickness
XYZ	Global coordinate system
xyz	Local coordinate system
x_i	X coordinate of the deformed flange edge
$\hat{x}_i$	X coordinate of the ideal flange edge
ε_{th}	Theoretical required longitudinal strain
φ	Bend angle

Material Forming - ESAFORM 2023
Materials Research Proceedings 28 (2023) 1047-1056

Materials Research Forum LLC
https://doi.org/10.21741/9781644902479-115

Introduction

Advanced High Strength Steels AHSS are increasingly used in the automotive industry for structural components due to their high strength-to-weight ratio [1, 2]. Roll Forming (RF) is a common forming process for the manufacturing of lightweight automotive structural components from high-strength and low-ductility materials [3]. However, RF is limited to the forming of simple components that have a constant cross-section [4]. Flexible roll forming was therefore developed to enable the forming of complex components with cross-sections that vary along the length of the profile [5]. Flange wrinkling is one of the common shape defects in FRF [6] and occurs if the required longitudinal compressive strain in the flange cannot be achieved by stable compression [7]. Limited practical solutions have been proposed to eliminate flange wrinkling, this restricts the shape complexity achievable with FRF from AHSS [6, 8].

Recently, a prototype flexible roll forming facility has been introduced [9, 10] and special forming tooling developed that supports the flange during the deformation and reduces flange wrinkling. However, wrinkling is still observed when flexible roll forming high-strength sheet metal [9] and this limits the widespread application of FRF in the automotive industry.

Incremental Sheet Forming (ISF) processes such as metal spinning and single-or-two point incremental forming are characterized by localized deformation which improves the formability and shape quality [11]. The axial tensile deformation that is formed during the forward paths in metal spinning balances the compressive hoop stress and this allows the forming of wrinkle-free spun [12, 13]. However, due to the limited working range of the ISF milling machines, forming long automotive structural sections is not feasible [14]. In addition, ISF of high-strength materials needs warm forming techniques to reduce the material strength [15].

The new Incremental Shape Rolling process has been recently developed and applied for the forming of long and straight U-profile components [16]. Similar to ISF, a high transverse tensile strain is developed in the flange and this promises to facilitate the plastic deformation in the flange when ISR is applied to weight-optimized components from AHSS.

In this study, the new ISR process is applied to manufacture the critical region of a simplified automotive component with a changing width cross-section. The same component shape is formed with the conventional FRF process on the same prototype forming facility [9] and the material deformation and shape defects are compared between both processes. Abaqus implicit is then used to further analyse the deformation behaviour in both processes.

Profile Geometry and Material

The investigated component has a cross-section that varies in width along its length, see Fig. 1a. The component has a flange length of $f = 18$ mm, an inner radius $r_i = 5$ mm and a material thickness of $t = 1.5$ mm. Wrinkling is only expected to occur in the compression side of the component, where the flange is longitudinally compressed. On the other side, the flange is longitudinally stretched. Therefore, only the compression side of the component is considered in this study and due to shape symmetry around the X-Y plane, only one-quarter of the component is investigated (the highlighted quarter). The pre-cut blank corresponding to the investigated quarter of the component is shown in Fig. 1b. Standard tensile tests were carried out on an Instron 5967 with a 30 kN load cell according to ASTM E8/E8M [17]. The tensile tests were done along the rolling direction and at a test speed of 0.025 mm s^{-1}. The averaged true stress–true strain curve of the DP600 sheet is shown in Fig. 1c. The mechanical properties are obtained by fitting the Hollomon's power law of the tested material and are shown in Table 1.

Material Forming - ESAFORM 2023   Materials Research Forum LLC
Materials Research Proceedings 28 (2023) 1047-1056   https://doi.org/10.21741/9781644902479-115

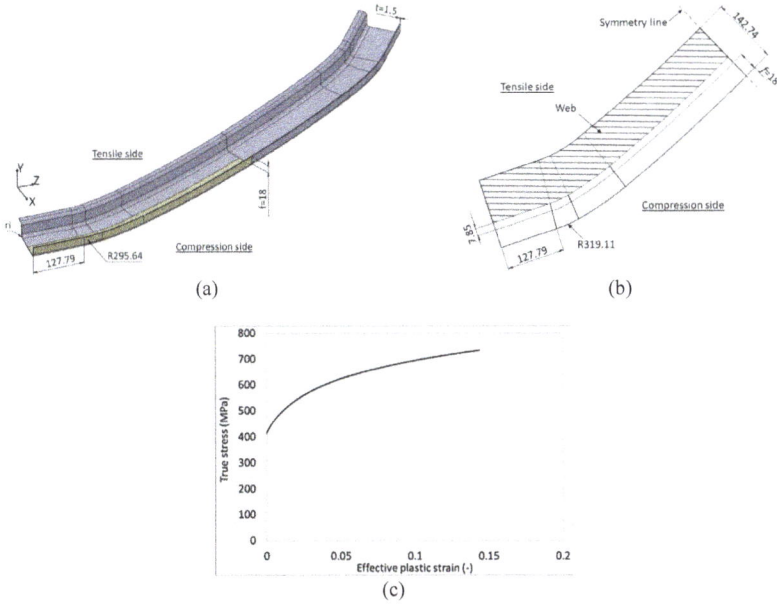

Fig. 1. Schematic showing (a) the full variable width profile and (b) the pre-cut blank corresponding to the investigated quarter of the component and (c) the average true stress-effective plastic strain curve of the DP600 sheet (dimensions in mm).

Table 1. Tensile data of the DP600 sheet.

Material	Thickness t (mm)	0.2% Yield strength (MPa)	Ultimate tensile strength (MPa)	Strain hardening	Strength coefficient (MPa)
DP600	1.5	414.5	733.5	0.11	897

Experimental Trials

Incremental Shape Rolling.

In ISR, the pre-cut blank is held between the clamping dies with the force applied by six hydraulic cylinders. A cylindrical forming roll with a roll nose radius, R_n=1 mm was used, see Fig. 2, and an initial clearance between the forming roll and the top die set to the sheet thickness, t. The hexapod which carries the forming roll has six degrees of freedom. During forming the roll is first moved up incrementally in Y-direction, d_y, and then moved in the X-Z plane to follow the top die contour while keeping the initial clearance, t. The roll rotates around the Y-axis to keep its axis perpendicular to the roll path. At the same time, the clamping dies move linearly in the longitudinal Z-direction. This roll path is repeated several times until the blank is fully formed over the top die.

Material Forming - ESAFORM 2023
Materials Research Proceedings 28 (2023) 1047-1056

Materials Research Forum LLC
https://doi.org/10.21741/9781644902479-115

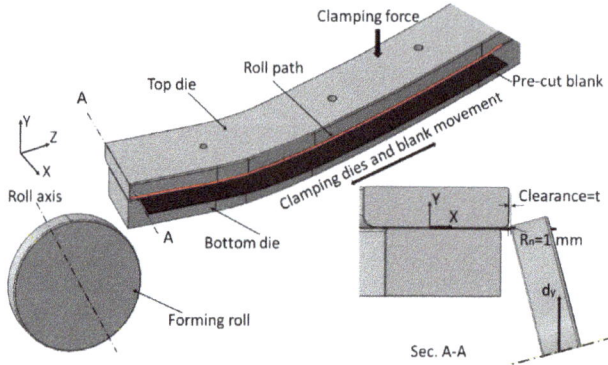

Fig. 2. Schematic drawing of the ISR process of the variable width profile.

The same ISR forming sequence as applied for the ISR of straight components in previous work [16] was implemented in this investigation, where the flange incrementally wraps over the roll nose radius in each pass, and in this way, the flange is formed upwards until the pre-cut is fully formed to the desired shape. In this study, the smallest possible increment size d_y=1 mm was used, as it is reported that the transverse tensile strain increases with decreasing the increment size, d_y [16] which provides the best possible flange quality.

Flexible Roll Forming.

The flexible roll forming trials were done on Deakin's FRF facility, where the blank is held between the clamping dies by six-hydraulic cylinders. One bottom roll (B1) is used to incrementally bend the pre-cut blank into shape with 14 forming passes. This represents the simplest forming and tool approach. More complex forming tools using finger rolls have been applied in previous studies to significantly reduce wrinkling and improve part shape [18]. In this study, a higher winkling tendency was desired to provide a clear difference in forming modes between the ISR and the FRF approach. The corresponding bend angle increment in FRF is 11°, 7°, 10°, 5°, 8°, 5°, 6°, 6°, 6°, 6°, 5°, 5°, 5°, 5°. The left side of Fig. 3 shows the schematic of the relative contact between the roll and the flange during forming, while the FRF bend sequence is shown on the right side of Fig. 3.

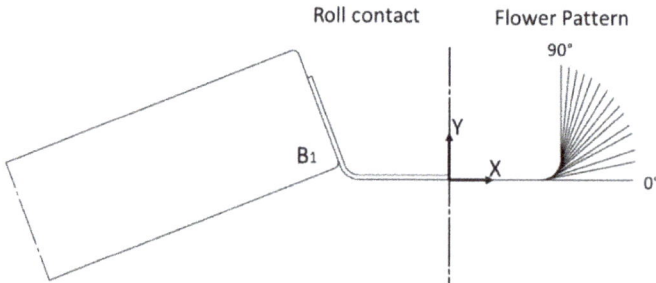

Fig. 3. Schematic of the roll set used in the FRF process and the implemented flower pattern.

Material Forming - ESAFORM 2023
Materials Research Proceedings 28 (2023) 1047-1056

Materials Research Forum LLC
https://doi.org/10.21741/9781644902479-115

Shape Analysis and Strain Measurement.
A 3D laser scanner "CreaForm HandyScan 700" [19] has been used to scan the flange after the final forming pass and before the part was released from the clamping dies. The data was then imported into the Geomagic Qualify software [20] to evaluate the flange quality and to compare it to the ideal shape of the flange. For this, a X-Z plane section cut was created along the length of the flange and located at a distance of 1 mm under the flange edge, see Fig. 4.

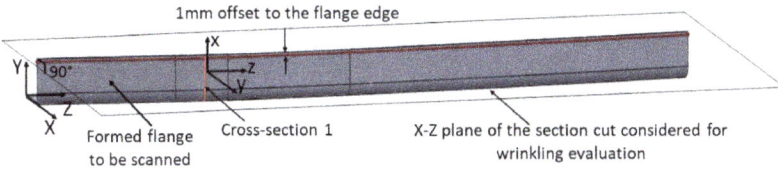

Fig. 4. Formed flange and the path considered for wrinkling evaluation.

Eq. 1 was used to calculate the mean absolute error (MAE) of the flange edge [9] to quantitatively evaluate flange wrinkling. The wrinkles were observed at a Z-coordinate of 70 to 320 mm measured from the lead end of the component and the MAE was therefore calculated in this region, see Fig. 5. Note that, the lead end is the component end where the forming pass starts while the tail end is where the forming pass ends.

$$MAE = \frac{1}{n} \sum_{i=1}^{n} |x_i - \hat{x}_i|. \tag{1}$$

Where, $\hat{x}_i$ is the corresponding X coordinate of the ideal flange edge, xi is the X coordinate of the deformed flange edge and n is the number of points considered for the wrinkling evaluation.

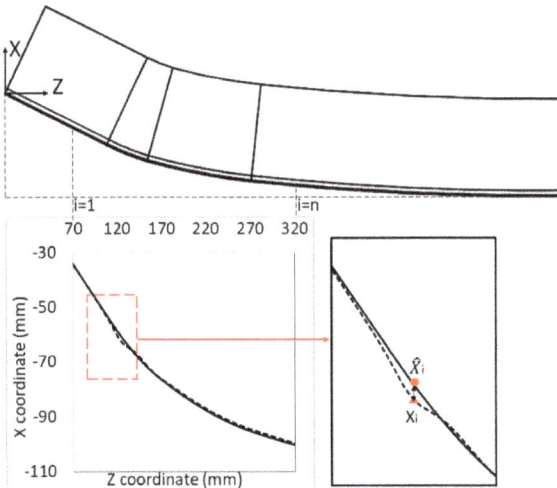

Fig. 5. Flange edge considered for the wrinkling analysis.

Material Forming - ESAFORM 2023   Materials Research Forum LLC
Materials Research Proceedings 28 (2023) 1047-1056   https://doi.org/10.21741/9781644902479-115

The AutoGrid strain measuring system [21] was used to measure the true major and minor strains after the final forming pass and after the component had been removed from the clamping dies. For this, a 2 mm × 2 mm grid was printed on the pre-cut blank and pictures of the formed grid were taken with a camera. The strain components were evaluated in cross-section 1 (S1) in the middle of the critical radius, R=295.64 mm (see Fig. 1a and 6). The local coordinate system (xyz) which follows the formed material was used to plot the strain components; where PExx is the transverse strain and PEzz is the longitudinal strain, while (XYZ) is the global coordinate system, see Fig. 4.

Eq. 2 was used to calculate the theoretical longitudinal strain, ε_{th}, required to form the flange. The R319.11 mm is the most critical radius in the compression side of the pre-cut blank (see Fig. 1a) and this results in the highest longitudinal compression strain in the flange, and severe wrinkling is expected to occur in this critical radius [9]. Based on Eq. 2 the theoretical longitudinal strain that is needed to form the flange in this critical radius is ε_{th} = -0.054.

$$\varepsilon_{th} = ln\frac{R - f(1-cos\varphi)}{R}.$$ (2)

Where ε_{th} is the theoretical required longitudinal strain, R is the radius of the pre-cut blank, f is the flange length and φ is the required bend angle of 90°.

Finite Element Analysis
The finite element simulation of the ISR process and FRF process have been performed with Abaqus Implicit. The clamping dies and the forming roll were modelled as rigid bodies, while the pre-cut blank has been discretized with reduced integration, hexahedral, linear brick elements (C3D8R). Four elements through the blank thickness were used with an element size of 1 mm and 2.5 mm in the X and the Z directions, respectively. A "frictionless contact" was assumed between the forming roll and the blank surfaces to avoid convergence issues [22]. To minimize the penetration of the rigid bodies into the blank surfaces at the constraint locations, the "hard contact condition" was applied [23]. The true stress-effective plastic strain curve shown in Fig. 1c was used together with isotropic hardening and the von Mises yield criteria to define the plastic material behaviour of the DP600 sheet [24, 25]. The elastic properties were defined with Poisson's ratio, ν, which is assumed to be 0.3 [9] and a Young's modulus of 200 GPa [26]. For the analysis of the forming results, the same procedure as used in the experimental trials is applied.

Results
Evaluation of Flange Wrinkling.
Fig. 6a compares the formed flange shape with the ideal shape of the flange after the final forming pass for both FRF and ISR. For a clear presentation, a 10 mm X-offset is applied between each forming case. The experimental results show that a clear wrinkle is formed in the FRF flange while the final shape of the ISR flange is close to the ideal shape. The FEA results underestimate the shape error and only indicate the formation of a buckle in the FRF flange, while there is no wrinkling predicted for the ISR process which correlates with the experimental results. Fig. 6b shows the calculated MAE for each forming case. The FEA results of the MAE accurately predict the experimental trend.

Material Forming - ESAFORM 2023   Materials Research Forum LLC
Materials Research Proceedings 28 (2023) 1047-1056   https://doi.org/10.21741/9781644902479-115

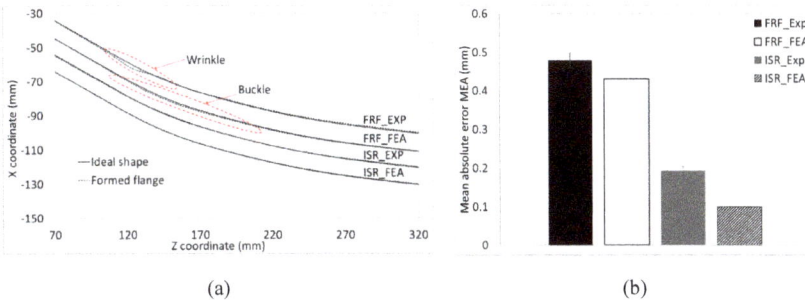

(a)   (b)

Fig. 6. (a) Comparison of the flange edge shape after the final forming pass and (b) MAE of the flange edge.

Material Deformation in the Flange.

The distribution of the transverse and the longitudinal residual strain along Section 1, i.e., in the middle length of the critical radius (see Fig. 4) was measured after the final forming pass and is given in Fig. 7a and b, respectively. The Autogrid system gives a good measurement along the flange length, while a strain measurement was not possible in the profile radius area zone, r_i. The experimental results and the FEA results suggest that a high transverse and longitudinal strain develops in the ISR flange, while there is only a very small level of transverse tensile strain and longitudinal strain in the FRF flange, see Fig. 7. The longitudinal compressive strain that is formed in the ISR flange reaches the theoretically required longitudinal, ε_{th}, in the flange edge, see Fig. 7b. In contrast, the longitudinal compressive strain that develops in the FRF flange is significantly lower than the theoretically required strain, ε_{th}.

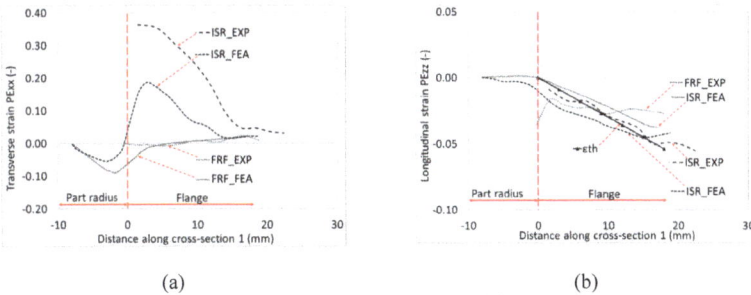

(a)   (b)

Fig. 7. Residual strain along cross-section 1 after the final forming pass (a) transverse direction and (b) longitudinal direction.

Discussion

The Effect of the Forming Method on Wrinkling Severity.

In ISR, the flange is wrapped over the roll nose radius to a larger extent in the early forming passes compared to that in the later forming passes [16]. Thus, the formed transverse tensile strain in ISR reaches its maximum level above the part radius, r_i, and then drops back to reach a minimum at the flange edge, see Fig. 7a.

To produce a complex part contour without wrinkling, the flange needs to compress longitudinally [7]. Eq. 2 was used to calculate the theoretical longitudinal strain, ε_{th}, required in the flange to produce the investigated component. As can be seen in Fig. 7a, ISR results in a high tensile transverse strain. This high transverse strain assists in forming the longitudinal compressive strain that is required to form the investigated flange. As shown in Fig. 7b, the formed longitudinal compressive strain in the flange reaches the theoretically required longitudinal strain. This indicates that the material is stably compressed in the critical zone without wrinkling.

In FRF, bending is the major deformation mechanism in the transverse direction and therefore the transverse tensile strain is low. This results in a small compressive longitudinal strain that is lower than that theoretically required. This means that in FRF the edge length remains longer than required and that therefore the compressive longitudinal stress is released in form of wrinkling rather than stable compression. This explains the severe flange wrinkling that occurred when FRF the component with the simple tooling, see Fig. 6a.

Both the experimental result and the FEA result of the FRF case suggest the formation of a buckle after pass 5, which forms into a wrinkle after pass 7 in the experimental trials. The longitudinal membrane stress, S_{zz}, is investigated along the selected node path in the flange edge (see Fig. 8a) for both the FRF (Fig. 8b) and the ISR case (Fig. 8c). For both processes, the maximum longitudinal compressive stress occurred in the critical radius, R295.64, see Fig. 1a. In FRF, the longitudinal compressive stress increases with the forming passes and reaches the critical buckling initiation stress of 527 MPa [9] after pass 5 (Fig. 8b), which is where the initial buckle was observed. Note that as soon as the buckle initiates the stress will release, i.e., there will be no further stress increase. The residual longitudinal stress formed in ISR is lower than the critical buckling stress and decreases with the forming passes (Fig. 8c). This suggests that the transverse tensile stress that is generated in the flange reduces the longitudinal compressive stress and therefore prevents the development of a wrinkle.

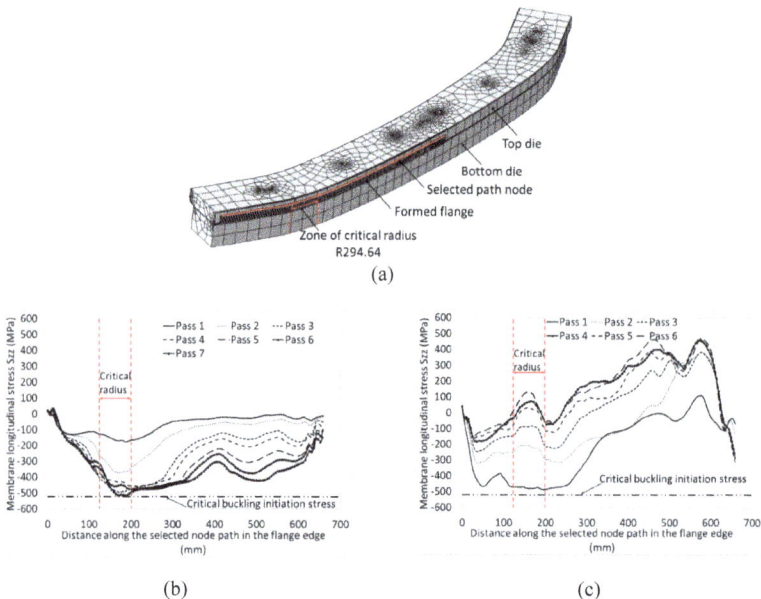

Fig. 8. (a) Node path considered for analysis of longitudinal stress, (b) evaluation of membrane longitudinal stress at the selected node path for FRF and (c) for ISR.

Summary

In this study, the incremental shape rolling of an automotive component with variable width was investigated and the deformation mechanisms and shape quality compared with those obtained from FRF. The major outcomes of this study are:

- ISR enables the manufacture of weight-optimized automotive components from high-strength steel by eliminating the wrinkling severity.
- In ISR, the transverse tensile strain results from the material wrapping over the roll radius.
- The transverse tensile strain that is developed in the ISR flange assists in the forming of the longitudinal compressive strain that is required to form the part contour. In FRF, the transverse and longitudinal strains are low. This leads to wrinkling instead of stable longitudinal compression in the flange.
- ISR represents a promising alternative to conventional sheet forming processes where excessive wrinkling leads to issues.

Acknowledgements

The authors acknowledge data M Sheet Metal Solutions GmbH for the development and manufacture of the 3D roll forming prototyping facility. The authors would like to thank Deakin University Postgraduate Research Scholarships (DUPRS) for their financial support.

References

[1] N. Baluch, Z.M. Udin, C.S. Abdullah, Advanced high strength steel in auto industry: an overview, Eng. Appl. Sci. Res. 4 (2014) 686-689. https://doi.org/10.48084/etasr.444

[2] G. Sun, M. Deng, G. Zheng, Q. Li, Design for cost performance of crashworthy structures made of high strength steel, Thin-Walled Struct. 138 (2019) 458-472. https://doi.org/10.1016/j.tws.2018.07.014

[3] A.D. Deole, M.R. Barnett, M. Weiss, The numerical prediction of ductile fracture of martensitic steel in roll forming, Int. J. Solids. Struct. 144-145 (2018) 20-31. https://doi.org/10.1016/j.ijsolstr.2018.04.011

[4] C. Jiao-Jiao, C. Jian-Guo, Z. Qiu-Fang, L. Jiang, Y. Ning, Z. Rong-guo, A novel approach to springback control of high-strength steel in cold roll forming, Int. J. Adv. Manuf. Technol. 107 (2020) 1793-1804. https://doi.org/10.1007/s00170-020-05154-8

[5] Y. Yan, H. Wang, Q. Li, B. Qian, K. Mpofu, Simulation and experimental verification of flexible roll forming of steel sheets, Int. J. Adv. Manuf. Technol. 72 (2014) 209-220. https://doi.org/10.1007/s00170-014-5667-0

[6] M.M. Kasaei, H.M. Naeini, G.H. Liaghat, C.M.A. Silva, M.B. Silva, P.A.F. Martins, Revisiting the wrinkling limits in flexible roll forming, J. Strain. Anal. Eng. Des. 50 (2015) 528-541. https://doi.org/10.1177/0309324715590956

[7] M.M. Kasaei, H.M. Naeini, B. Abbaszadeh, M. Mohammadi, M. Ghodsi, M. Kiuchi, R. Zolghadr, G. Liaghat, R.A. Tafti, M. S. Tehrani, Flange wrinkling in flexible roll forming Process, Procedia Eng. 81 (2014) 245-250. https://doi.org/10.1016/j.proeng.2014.09.158

[8] P. Groche, A. Zettler, S. Berner, G. Schneider, Development and verification of a one-step-model for the design of flexible roll formed parts, Int. J. Mater. Form. 4 (2010) 371-377. https://doi.org/10.1007/s12289-010-0998-3

[9] B. Abeyrathna, S. Ghanei, B. Rolfe, R. Taube, M. Weiss, Optimising part quality in the flexible roll forming of an automotive component, Int. J. Adv. Manuf. Technol. 118 (2021) 3361-3373. https://doi.org/10.1007/s00170-021-08176-y

[10] S. Ghanei, B. Abeyrathna, B. Rolfe, M. Weiss. Analysis of material behaviour and shape defect compensation in the flexible roll forming of advanced high strength steel. In: IOP Conf Ser Mater Sci Eng.vol. 651. IOP Conf Ser Mater Sci Eng; 2019. https://doi.org/10.1088/1757-899x/651/1/012064.

Material Forming - ESAFORM 2023   Materials Research Forum LLC
Materials Research Proceedings 28 (2023) 1047-1056   https://doi.org/10.21741/9781644902479-115

[11] S. Gatea, H. Ou, G. McCartney, Review on the influence of process parameters in incremental sheet forming, Int. J. Adv. Manuf. Technol. 87 (2016) 479-499. https://doi.org/10.1007/s00170-016-8426-6

[12] O. Music, J.M. Allwood, K. Kawai, A review of the mechanics of metal spinning, J. Mater. Process. Technol. 210 (2010) 3-23. https://doi.org/10.1016/j.jmatprotec.2009.08.021

[13] Z. Jia, L. Li, Z. R. Han, Z. J. Fan, B. M. Liu, Experimental study on wrinkle suppressing in multi-pass drawing spinning of 304 stainless steel cylinder, Int. J. Adv. Manuf. Technol. 100 (2018) 111-116. https://doi.org/10.1007/s00170-018-2712-4

[14] A.K. Behera, R.A. de Sousa, G. Ingarao, V. Oleksik, Single point incremental forming: An assessment of the progress and technology trends from 2005 to 2015, J. Manuf. Process. 27 (2017) 37-62. https://doi.org/10.1016/j.jmapro.2017.03.014

[15] A. Kumar, V. Gulati, P. Kumar, H. Singh, Forming force in incremental sheet forming: a comparative analysis of the state of the art, J. Braz. Soc. Mech. Sci. Eng. 41 (2019). https://doi.org/10.1007/s40430-019-1755-2

[16] A. Essa, B. Abeyrathna, B. Rolfe, M. Weiss, Prototyping of straight section components using incremental shape rolling, Int. J. Adv. Manuf. Tech. 121 (2022) 3883-3901. https://doi.org/10.1007/s00170-022-09600-7

[17] ASTM Standard, Standard Test Methods for Tension Testing of Metallic Materials, in, ASTM International, 2016.

[18] A. Sreenivas, B. Abeyrathna, B. Rolfe, M. Weiss, Longitudinal strain and wrinkling analysis of variable depth flexible roll forming, J. Manuf. Process. 81 (2022) 414-432. https://doi.org/10.1016/j.jmapro.2022.06.063

[19] Information on https://www.creaform3d.com/en;

[20] Information on http://www.geomagic.com/en/;

[21] Information on https://www.vialux.de/en/;

[22] M. Weiss, B. Abeyrathna, D.S. Gangoda, J. Mendiguren, H. Wolfkamp, Bending behaviour and oil canning in roll forming a steel channel, Int. J. Adv. Manuf. Technol. 91 (2017) 2875-2884. https://doi.org/10.1007/s00170-016-9892-6

[23] Y.Y. Woo, S.W. Han, I.Y. Oh, Y.H. Moon, Shape defects in the flexible roll forming of automotive parts, Int J Automot. Technol. 20 (2019) 227-236. https://doi.org/10.1007/s12239-019-0022-y

[24] B. Abeyrathna, B. Rolfe, L. Pan, R. Ge, M. Weiss, Flexible roll forming of an automotive component with variable depth, Adv. Mater. Process. Technol. 2 (2016) 527-538. https://doi.org/10.1080/2374068x.2016.1247234

[25] B. Abeyrathna, B. Rolfe, J. Harrasser, A. Sedlmaier, G. Rui, L. Pan, M. Weiss. Prototyping of automotive components with variable width and depth. In: 36th IDDRG Conference.vol. 36th IDDRG Conference; 2017. https://doi.org/10.1088/1742-6596/896/1/012092

[26] B. Abeyrathna, B. Rolfe, M. Weiss, The effect of process and geometric parameters on longitudinal edge strain and product defects in cold roll forming, Int. J. Adv. Manuf. Technol. 92 (2017) 743-754. https://doi.org/10.1007/s00170-017-0164-x

Innovative joining by forming technologies

Material Forming - ESAFORM 2023
Materials Research Proceedings 28 (2023) 1059-1066

Materials Research Forum LLC
https://doi.org/10.21741/9781644902479-116

Clinching with divided punch to prevent critical neck thicknesses

FALK Tobias[1,a] *, KROPP Thomas[1,b] and DROSSEL Welf-Guntram[1,c]

[1] Fraunhofer Institute for Machine Tools and Forming Technology IWU, Nöthnitzer Straße 44, 01187 Dresden, Germany

[a]tobias.falk@iwu.fraunhofer.de, [b]thomas.kropp@iwu.fraunhofer.de, [c] welf-guntram.drossel@iwu.fraunhofer.de

Keywords: Clinching, Divided Punch, Critical Neck Thickness, Crack Prevention

Abstract. Clinching thin sheet metal into a thicker part can result in low neck thicknesses or even neck cracks. How these critical low values for neck thickness can be counteracted is described in this paper, using a two-part punch, which is devided in an inner and outer unit. At the beginning, both move parallel in the direction of the die until the outer punch stops at a defined position and only the inner one moves further down and forms the clinch point, with its characteristic contour and the geometric values of interlock and neck thickness. Due to the large punch diameter at the beginning of the process, more material initially flows into the neck area than in conventional clinching, so that a greater neck thickness can be achieved. Numerical simulation was used to create the concept and verify it experimentally. The greater neck thickness has a positive effect on the shear tensile strength and can also be transferred to the typical joining direction for clinching (thick into thin material).

Introduction

Mechanical joining technology is widely used to join different sheet materials with different thicknesses and tensile strengths. In contrast to welding, for example, the sheet metal partners can be of different types, as is the case when steel is joined into aluminum. Depending on the specific joining process, there is a recommended arrangement of the sheets (see Fig. 1). For example, in self-pierce riveting, the thinner part, which is more difficult to form, should be positioned on the punch side for optimum joint formation [1]. This is different for clinching, where the harder material should also be on the punch side, but it is recommended that the thicker sheet should be joined into the thinner part [2].

Fig. 1. Cross sections of three different mechanical joining processes (materials and blank thicknesses in the figure): Self-pierce riveting with semi-tubular rivet (left); Self-pierce riveting with solid rivet (middle); Clinching (right).

Clinching

In general, clinching is a simple way to join two or more sheets mechanical. It is frequently used because this joining process does not require a rivet or any other auxiliary joining part and is therefore easier to implement in terms of equipment and has weight advantages over other joining processes. On the other hand, clinching cannot compete with other mechanical joining processes regarding to joint strength. [3]

Material Forming - ESAFORM 2023   Materials Research Forum LLC
Materials Research Proceedings 28 (2023) 1059-1066   https://doi.org/10.21741/9781644902479-116

There are several process variants for clinching, such as clinching with cutting component [4], clinching with preformed hole [5], clinching without point elevation [6] or clinching with flat die [7]. Since these and other special variants are not used in this paper, their description will not be discussed in detail.

In current automotive production, rigid and radial opening clinching dies are mostly used. The rigid dies are characterized by their simple, rotationally symmetrical design. The advantage of radial opening dies is that the material flow is positively influenced by flexible, movable segments, thus the formation of interlock is supported [8].

The actual joining process is similar for both presented die variants. First, the sheets are positioned between the blankholder and die (Fig. 2, I), the moving punch presses the sheet partners toward the die contour (Fig. 2, II), and the joining point is finally formed by radial material flow (Fig. 2, III). This results in a strength-relevant interlock with a sufficiently large neck thickness. [2]

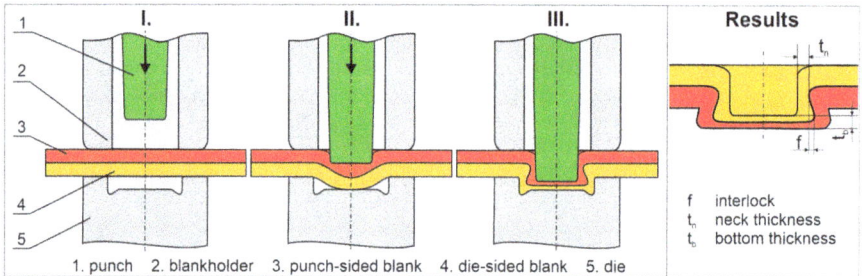

Fig. 2. Three steps of a clinching process (left); characteristic results of a clinching joint (right).

Due to the relatively simple joining process (no intended material cutting as, for example, in the case of semi-tubular self-piercing riveting when piercing the punch-side blank), the clinching process can be simulated easily with the help of numerical computation. Particularly when rigid, rotationally symmetrical dies are used, but also certainly in the case of radial opening dies, it is even possible to use two-dimensional simulations [9], which significantly reduces the complexity and calculation time. Thus, in addition to the pure joining point contour and force-displacement data, critical neck thicknesses and cracks can also be identified and predicted by numerical simulations, as shown in Fig. 3.

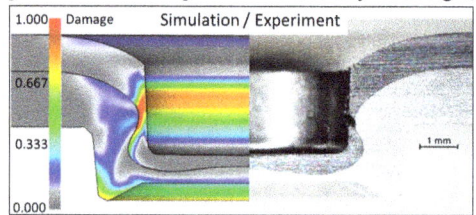

Fig. 3. Comparison of numerical simulation (left) and experimental result (right): HC340LA t = 1.0 mm in EN AW- 5754 t = 1.5 mm) [9].

Sampling

For this study of increasing the neck thickness, a joint with unfavorable sheet metal arrangement and small thickness in the neck region is used. As described above, this tends to be the case for clinching when joining thin into thick material. A steel joint of S350GD $t_1 = 2.0$ mm into S350GD $t_2 = 3.0$ mm is selected (as presented in [10]). As can be seen in Fig. 4(left), a bottom thickness of 1.06 mm results in a clinch point that has significantly lower values for neck thickness than interlock, with average $t_n = 0.25$ mm and $f = 0.42$ mm. A maximum joining force of 80 kN was required for this joint. The rather high force value for clinching can be explained by the large thicknesses and the die diameter of 10 mm.

Material Forming - ESAFORM 2023
Materials Research Proceedings 28 (2023) 1059-1066

Materials Research Forum LLC
https://doi.org/10.21741/9781644902479-116

Furthermore, the cross-section for the typical joining direction is shown on the right in Fig. 4. With an identical joining force (80 kN), almost the same bottom thickness of $t_b = 1.05$ mm is achieved. The neck thickness, averaging $t_n = 0.72$ mm, is significantly greater. The thicker steel sheet on the punch side almost triples the neck thickness but the the interlock ($f = 0.30$ mm) is decreasing. However, the loss of 29 % interlock is not as extreme as the increase of the neck thickness.

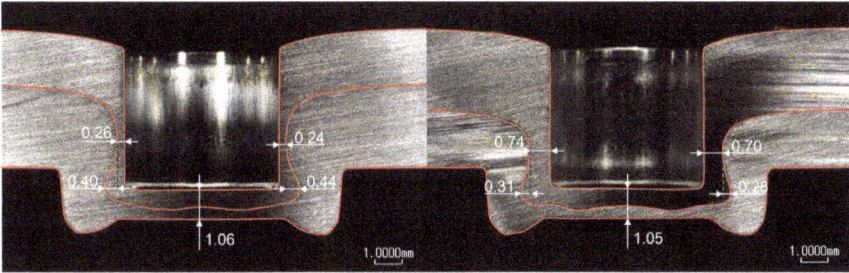

Fig. 4. Cross-sections of S350GD t = 2.0 mm in S350GD t = 3.0 mm (left) and opposite joining direction (right); red line added afterwards for better detection of the contour.

Numerical Simulation

In addition to the available information of the geometric dimensions of the tools, the spring rate of the blankholder and the joining speed, the friction conditions as well as the material properties are important for an exact modeling of the numerical simulation. Due to numerous past projects in which clinch simulations were considered [9, 11, 12], the gained expertise is used and the shear approach is applied for the friction. The selected friction factors have been determined on the base of the already performed numerical computations of the clinching process.

For the integration of the material behavior into the simulations, compression tests [13] are first carried out (Fig. 5, left). Since significantly higher degrees of deformation are achieved during clinching than can be determined in the compression tests, an extrapolation of the data is necessary [14]. For this purpose, individual factors are obtained for different approximation approaches with the experimental data in the initial range and the different results are compared with each other. The least squared error can then be used to find the extrapolation approach that most appropriately reflects the material behavior. For the used S350GD the approximation approach by Orowan is the best fit for the experimental data. The extrapolated and in the simulation used flow curve is shown in Fig. 5 (right).

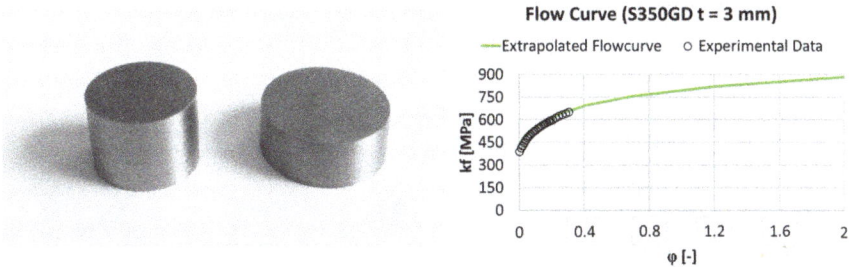

Fig. 5. Original and compressed cylinder sample of the compression test (left); extrapolated flow curve of S350GD t = 3.0 mm with approximation approach by Orowan.

Material Forming - ESAFORM 2023   Materials Research Forum LLC
Materials Research Proceedings 28 (2023) 1059-1066   https://doi.org/10.21741/9781644902479-116

Since only rotationally symmetric tools are used for the experiments, the more simple 2D simulations are used as mentioned above. Using the FEA software DEFORM, a simulation model was built, which can be seen in Fig. 6. Only the two blanks are modeled as elastic-plastic, the tools punch, blankholder and die are defined as rigid for simplification.

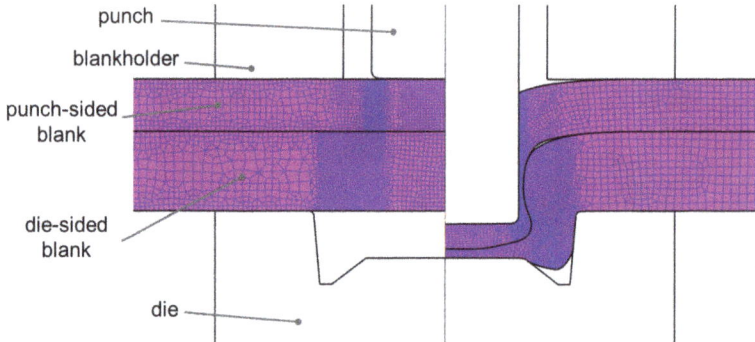

Fig. 6. Numeric simulation model of the clinching process build with the software DEFORM.

Verification

To ensure that the simulation can be used in further investigations, a comparison must be made between the experiment and the numerical calculation. Only if the results of both match sufficiently (geometry and load-stroke-curve), further use is reasonable.

Unlike in Fig. 4, in the following graph (Fig. 7) the colored contour is the result of the simulation. The results of the numerical simulation were generated using Deform V12 software from SFTC. In it, the minimum bottom thickness t_b achieved in the experiments was defined as a stopping criterion in the simulation. With these, areas of high deformation such as interlock and neck areas are generated finer meshes than at the edge of the simulated blanks. For the implementation of friction, a constant shear friction value per contact pair (such as punch/upper blank or die/bottom blank) was determined in parameter studies.

The difference between experimental cross-section and calculated clinch contour is hardly visible. The values of the characteristic geometrical parameters generated by the simulation are also at the same level as those of the experiments. Therefore, the simulation can be considered to have a high accuracy and prediction and can be used in further investigations instead of costly experiments.

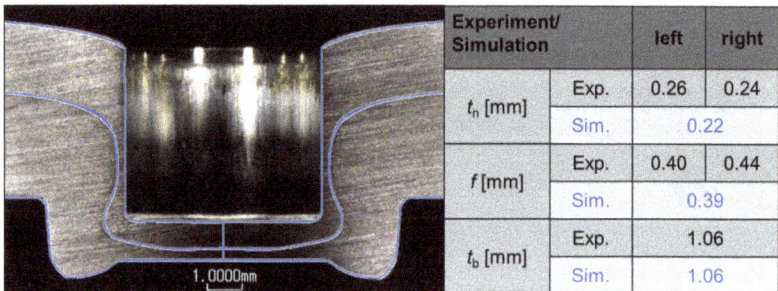

Experiment/ Simulation		left	right
t_n [mm]	Exp.	0.26	0.24
	Sim.	0.22	
f [mm]	Exp.	0.40	0.44
	Sim.	0.39	
t_b [mm]	Exp.	1.06	
	Sim.	1.06	

Fig. 7. Comparison of experiment and simulation: cross-section and simulated contour of the clinch joint (left); characteristic geometrical values (right).

Material Forming - ESAFORM 2023
Materials Research Proceedings 28 (2023) 1059-1066

Materials Research Forum LLC
https://doi.org/10.21741/9781644902479-116

Divided Punch

One approach to bring joints with small neck thicknesses into a non-critical state is to allow more material to flow into the neck area at the beginning of the clinching process. The idea developed here is to use a larger punch diameter which displaces more volume into this neck area. The joining point is then formed by a smaller punch. This does not necessarily have to involve a punch change; the division of the punch into two different diameters is already used in shear clinching [15] (Fig. 8, numbers in the figure are part of the patent [16], where all components are explained by numbers).

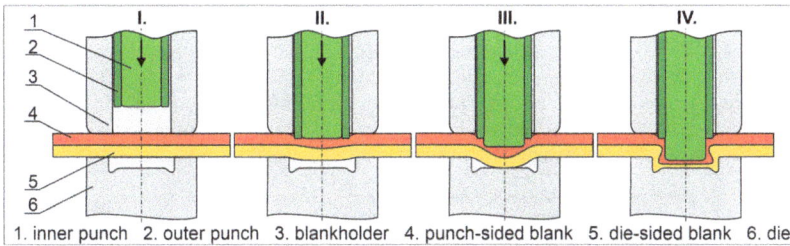

Fig. 8. Division of the punch for clinching with cutting components [16].

Described in more detail, clinching with a divided punch can be structured into four phases. First, just as in the conventional method, the sheets to be joined are positioned between the die and the blankholder (Fig. 9, I.). Then the larger outer punch and the smaller inner punch move simultaneously until a defined stroke is reached (Fig. 9, II.). At this point, the outer punch stops, and only the inner punch continues to move towards the die (Fig. 9, III.) and, as in the conventional case, forms the joining point with its characteristic contour (Fig. 9, IV.).

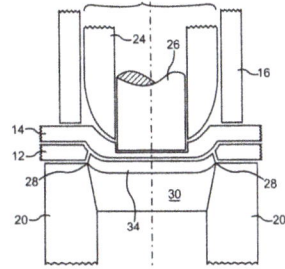

1. inner punch   2. outer punch   3. blankholder   4. punch-sided blank   5. die-sided blank   6. die

Fig. 9. Four steps of clinching with divided punch.

The validated simulations and the concept of punch division are now being used to first develop simulative tool combinations that generate an increase of the neck thickness. In the case of the outer punch, it is not only the pure geometry that is important, but also at which stroke it should stop moving.

In order to show whether a compression of the material at the end of the process has the same effect as pre-compression, a stepped punch is also considered. This stepped punch has a smaller diameter which, after a defined height, has a shoulder where the diameter becomes larger. This punch shape is one-piece and therefore much simpler in design than a tool division.

Fig. 10 shows the results of the simulations and at the same time also those of the experiments. The left subgraph is already known from Fig. 7 and is only included here for better comparison. First of all, it is noticeable that for all three punch variants the simulation accurately predicts the actual experimental result. This is the case for the pure contours as well as for the measured values for neck thickness and interlock.

The differences between the two conventional punch designs (left and center) are marginal, there is no increase of neck thickness by using the stepped punch. The situation is different with the divided tool (the punch contour is based on the one of the stepped punch): Due to the large outer punch, so much material flows into the neck area at the start of the process that the average neck thickness could be increased from 0.25 mm to 0.39 mm (+77 %). Since experience shows that neck thickness and interlock are always in opposite trends, it is not surprising that the interlock decreases with the divided punch. Compared to the significant increase in neck thickness, however,

Material Forming - ESAFORM 2023

Materials Research Proceedings 28 (2023) 1059-1066

Materials Research Forum LLC

https://doi.org/10.21741/9781644902479-116

the loss of only 17 % (from 0.42 mm to 0.35 mm) is significantly lower.

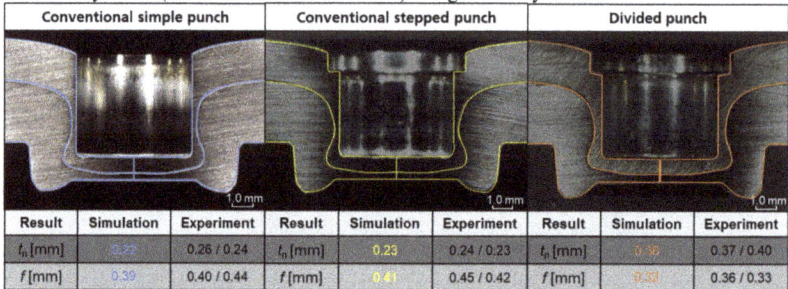

Conventional simple punch			Conventional stepped punch			Divided punch		
Result	**Simulation**	**Experiment**	**Result**	**Simulation**	**Experiment**	**Result**	**Simulation**	**Experiment**
t_n [mm]	0.22	0.26 / 0.24	t_n [mm]	0.23	0.24 / 0.23	t_n [mm]	0.16	0.37 / 0.40
f [mm]	0.39	0.40 / 0.44	f [mm]	0.41	0.45 / 0.42	f [mm]	0.33	0.36 / 0.33

Fig. 10. Comparison of experiments and simulations (colored contours): conventional simple punch (left); conventional stepped punch (middle); divided punch (right).

The declared goal of increasing neck thickness by using a two-part punch appears to be successful. The concept developed is appropriate for the present case. How and whether the larger neck thickness with simultaneously smaller interlock affects the joint strength will be determined in the next step.

Strength Analysis

Since the use of the stepped punch does not have any advantage in the present case, this punch variant is not included in the joint strength investigations. Fig. 11 shows results from shear tensile tests. As discussed in the last chapters, the sheet pairing S350GD 2.0 mm in S350GD 3.0 mm as well as the reverse joining direction can be found in it. In addition, a distinction is made in each case between the conventional simple punch and the divided die.

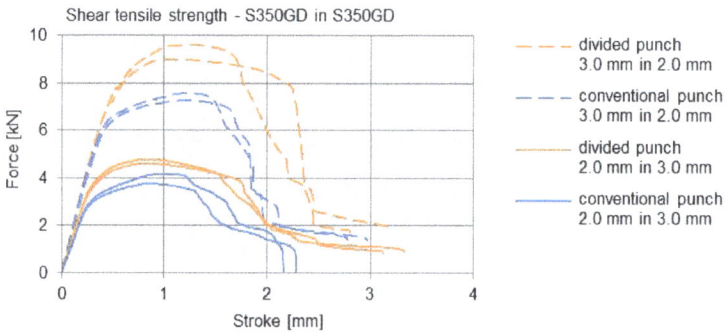

Fig. 11. Results of shear tensile strength: S350GD (2 mm in 3 mm) vs. (3 mm in 2 mm) and conventional simple punch vs. divided punch with pre-stamping.

It can be seen, that for both joining directions (thick in thin and thin in thick) the use of the divided punch has a positive effect on the joint strength under shear tensile load. This is the case both for the maximum forces and for the energy absorption capacity (in connection with the extended strokes). The fact that the use of the thicker blank on the punch side achieves fundamentally higher values than the thinner blank reflects experience. For the untypical blank arrangement for clinching (2 mm in 3 mm), the average maximum force for the variant with pre-stamping can be increased by 19% with the larger punch compared to the simple tool. For reversed blank positioning, even an increase of 25% on average is achieved.

Further Investigations

For the outer punch, not only the diameter is essential, but also how far it moves with the inner punch has an influence on the joint contour. For the sheet pairing S350GD $t = 2$ mm in S350GD $t = 3$ mm investigated in this paper, a small sensitivity analysis is therefore carried out with the help of simulation in order to be able to evaluate the effect on neck thickness and interlock.

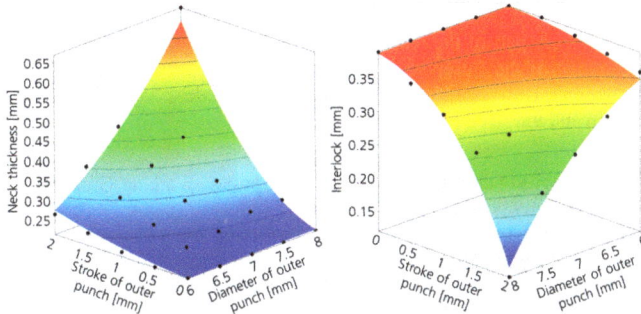

Fig. 12. Results of a sensitivity analysis to show the influence of stroke and diameter of the outer punch on neck thickness (left) and interlock (right).

Fig. 12 shows that the larger the diameter and stroke of the outer punch, the greater the neck thickness. Here, the stroke has a greater influence than the diameter. On the other hand, increasing punch stroke and diameter of the outer tool have a negative effect on the interlock. Depending on the initial situation and requirements, it is therefore also necessary to find a tool setup for the espective compound. The punch division can help to achieve a balanced and target-oriented relationship between neck thickness and interlock.

In further investigations, other joint components should be tested and whether the punch division has a positive effect on the joint contours and strengths. Both different sheet thicknesses and materials (steel, aluminum) in different strengths are of interest here and will be part of the project in the future work. For these other materials, it is also planned to test not only the shear tensile strength but also in head tensile direction, in each case for the reference experiments as well as the optimized with punch division tests. If there is an upper limit of material strengths that can be used with respect to the divided punch, these future investigations will show.

In addition to the rigid dies used here, further studies should analyze the influence when using more flexible designs such as dies with radially opening segments.

The sensitivity analyses presented above could be extended to consider all tools and kinematics and thus provide a fully comprehensive picture of neck thickness increase using punch division. Of course, not only the stroke and diameter of the outer punch influence the relationship between neck thickness and interlock, so the interaction of all parameters should be investigated in more detail.

The obvious disadvantage of the punch division presented here is the need for two decoupled drives. The subject of further research should therefore be whether there are alternative possibilities for pre-stamping. It is conceivable, for example, to use spring units instead of several drives, to carry out the stamping by using a modified blankholder with an integrated stamping ring. This simple tool geometry naturally leads to a loss of flexibility in terms of a freely adjustable punch stroke for pre-stamping.

Summary

In this paper, the issue of small neck thicknesses in clinching was first described and demonstrated using the joint S350GD 2.0 mm in S350GD 3.0 mm. By using a validated simulation model, the

Material Forming - ESAFORM 2023   Materials Research Forum LLC
Materials Research Proceedings 28 (2023) 1059-1066   https://doi.org/10.21741/9781644902479-116

concept of punch division and the positive effect on a neck thickness increase were first simulatively proven. This could be verified by experiments, the comparison to the simulation was convincing and consequently the neck thickness could be increased in these real investigations as well. This led to significantly higher shear tensile strengths for the above-mentioned joint, but also for the joining direction more typical of clinching (thick material joined into thin material).

Acknowledgement

The presented results are part of the research project "Extending the process limits of clinching through pre-stamping" of the European Research Association for Sheet Metal Working (EFB) funded by the program for "Industrial Research" (IGF) of the Federal Ministry of Economic Affairs and Climate Action and the AiF (22329 BR).

References

[1] Technical Bulletin DVS 3410, Self-pierce riveting – basics, Working Group V10 „Mechanical Joining" by the joint committee for DVS and EFB, 2021.

[2] Technical Bulletin DVS 3420, Clinching – basics, Working Group V10 „Mechanical Joining" by the joint committee for DVS and EFB, 2021.

[3] X. He, Clinching for sheet materials, Sci. Technol. Adv. Mater. 18 (2017) 381-405. https://doi.org/10.1080/14686996.2017.1320930

[4] M. Merklein, G. Meschut, M. Müller, R. Hörhold, Grundlegende Untersuchungen zur Verbindung von pressgehärtetem Stahl und Aluminium mittels Schneidclinchen, 2013.

[5] Company information of Eckold GmbH & Co. KG, St. Andreasberg, https://www.eckold.de/de/technologien/clinchen/, accessed 11.11.2022.

[6] Company information of TOX Pressotechnik GmbH & Co. KG, Weingarten https://de.tox-pressotechnik.com/verfahren/clinchen/clinchpunkt-formen/uebersicht-clinchpunkt-formen/, accessed 11.11.2022.

[7] S. Dietrich, Grundlagenuntersuchungen zu neuen matrizenlosen Umformfügeverfahren, Dissertation, Verl. Wiss. Scripten 2007.

[8] M.M. Eshtayeh, M. Hrairi, Recent and future development of the application of finite element analysis in clinching process, Int. J. Adv. Manuf. Technol. 84 (2016) 2589-2608. https://doi.org/10.1007/s00170-015-7781-z

[9] W.-G. Drossel, et al. Unerring planning of clinching processes through the use of mathematical methods, Key Eng. Mater. 611 (2014).

[10] W.-G. Drossel, et al. Mechanisch gefügte Stahlstrukturen in Fahrzeugbau und Bauwesen, FOSTA Bericht P1172, 2020.

[11] M. Götz, F. Leichsenring, T. Kropp, P. Müller, Data mining and machine learning methods applied to A numerical clinching model, Comput. Model. Eng. Sci. 117 (2018) 387-423. http://doi.org/10.31614/cmes.2018.04112

[12] N. Vancraeynest, M. Jäckel, S. Coppieters, W.G. Drossel, Data-Based and Analytical Models for Strength Prediction of Mechanical Joints, Key Eng. Mater. 926 (2022) 1556-1563. https://doi.org/10.4028/p-a33855

[13] S. Coppieters, et al. Large Strain Flow Curve Identification for sheet metal: Process informed method selection, Forming Technology Forum (2019).

[14] V. Thoms, W. Voelkner, D. Süße, Methoden zur Kennwertermittlung für Blechwerkstoffe, EFB Forschungsbericht 187, 2002.

[15] D. Han, R. Hörhold, M. Müller, S. Wiesenmayer, M. Merklein, G. Meschut, Shear-clinching of multi-element specimens of aluminum alloy and ultra-high-strength steel, Key Eng. Mater. 767 (2018) 389-396. https://doi.org/10.4028/www.scientific.net/KEM.767.389

[16] S. Busse, et. al. Joining device for connecting parts to be joined in a stamping process comprises a stamping unit having moving concentric stamps, disclosure document, DE 10 2006 028 568 A1, 2006.

Material Forming - ESAFORM 2023
Materials Research Proceedings 28 (2023) 1067-1074

Materials Research Forum LLC
https://doi.org/10.21741/9781644902479-117

An ANN based approach for the friction stir welding process intrinsic uncertainty

QUARTO Mariangela[1,a]*, BOCCHI Sara[1,b], GIARDINI Claudio[1,c]
and D'URSO Gianluca[1,d]

[1]Department of Management, Information and Production Engineering, University of Bergamo, Via Pasubio 7/b, 24044 Dalmine (BG), Italy

[a]mariangela.quarto@unibg.it, [b]sara.bocchi@unibg.it, [c]claudio.giardini@unibg.it, [d]gianluca.d-urso@unibg.it

Keywords: Artificial Neural Network, Process Variability, Friction Stir Welding

Abstract. Friction Stir Welding is a solid-state bonding process that during last years has caught the researcher's attention for the mechanical characteristics of the welded joints that are quite similar to the properties of the base material. The Friction Stir Welding is affected by several process parameters leading to intrinsic variability in the process. The present paper would introduce a new approach for predicting the surface hardness in different areas of the welded parts. Specifically, this method is based on the hypothesis that multiple Artificial Neural Networks, characterized by the same architecture but different weights, can be used for forecasting both the punctual value of the local hardness and its confidence interval, resulting in taking into account the intrinsic variability of the process.

Introduction

Friction Stir Welding (FSW) is one of the most significant technological developments in metal joining of the last thirty years and it is also considered a sustainable technology thanks to its energy efficiency. Indeed, the energy involved is on average 30% - 40% lower than conventional welds. In addition, the emission of greenhouse gases in FSW is about 30% lower than in traditional welds.

Friction Stir Welding is a welding technology patented by The Welding Institute (TWI) in 1991 [1]. This solid state mechanical joining technique is used to make joints between similar or dissimilar materials that are difficult to be welded with conventional fusion techniques, such as sintered materials, magnesium, copper, Inconel, titanium, metal matrix composites and thermoplastics. In particular, it is used for high-strength aluminum alloys, which are difficult to join using traditional techniques due to their typical microstructural alteration during aging hardening.

From a structural point of view, the FSW introduces welded joints without overlapping edges and allows the elimination of connecting parts, such as rivets, with a consequent reduction in the weight and costs of the structures. This process, still relatively young, has produced great interest in the industrial world so as to achieve enormous development, especially in naval applications, but also in the aerospace, railway and automotive industries. The difficulty of the traditional welding processes is linked to the problems developed due to the high temperatures during processing. Conversely, in Friction Stir Welding, which is a friction process by plastic deformation, fusion temperature is never reached during processing, therefore, the formation of the problems that occur in conventional welds is avoided.

In general, welds made using FSW have a different microstructure with respect to the joints welded with fusion processes because the maximum temperature reached in the heat-affected zone is significantly lower than the melting temperature [2]. By basing the microstructural characterization of the joints on the size of the grains and precipitates, it is possible to divide the structure of the FSW joint into four characteristic zones: the nugget, the thermo-mechanically

Material Forming - ESAFORM 2023

Materials Research Proceedings 28 (2023) 1067-1074

Materials Research Forum LLC

https://doi.org/10.21741/9781644902479-117

affected zone, the heat affected zone and the base material. The nugget is the area of completely recrystallized material close to the welding line; near to the nugget, the thermo-mechanically affected zone (TMAZ) is detectable. In the TMAZ the material undergoes both plastic deformation and heating, whilst only alterations of the microstructure and mechanical properties due to the welding thermal cycle are present in the heat affected zone (HAZ) [3].

For these FSWed joints characteristics, it is evident that the process parameters greatly influence the flow of the plastically deformed material [4]. In particular, there is a strong relationship between the microstructure and the quality of the joints is mainly due to the closed connection between the mechanical properties of the FSWed joints and the microstructural variations of the above-mentioned different zones of the welds. Therefore, particular attention must be paid in the choice of parameters to ensure adequate processing conditions, in order to avoid the formation of potential defects or weakness points in the identified welding areas [2].

Due to the large number of variables that must be considered simultaneously, it is not easy to optimize and predict with a good level of accuracy the process performance and the weld quality as a function of both the process conditions and the weld zones considered. For this reason, optimizing process performance by considering each process parameter individually is not an applicable approach. To overcome this problem, various artificial intelligence techniques have recently been used, among which the Artificial Neural Network stands out [5,6].

An Artificial Neural Network (ANN) is a mathematical model capable of reproducing the reasoning mode of the human brain, simulating its processing capabilities through the connection of neurons, which are the elementary computational elements of any neural network. The ANN, once it has been adequately trained, is able to solve even non-linear problems easily and quickly. To do this, the ANN is composed by an oriented network formed by different layers: the input layer, which collects the external data, the hidden layers, which process the information through the activation functions and the weights of each connection, and the outputs layer.

The training of the neural network is based on the use of a dataset thanks to which the network is able to determine the weights of the interconnections and of the activation function. These weights define the reliability of the prediction and can vary as a function of trainings; in fact, repeating the training defines different weights configurations which generate different results in the prediction. Consequently, it is evident that if the same weights are always considered, the ANN produces a constant predictive response with respect to a specific set of inputs. However, this aspect is a distorted version of reality. Indeed, it is known that every process is characterized by an intrinsic variability caused by external conditions that can influence systems in different and, often unpredictable, ways. The possibility of considering this uncertainty in the analysis and prediction of complex systems is a topic of considerable importance [7]. For all these reasons, the main objective of this work is to develop a model capable of predicting not only the punctual mechanical characteristics reached in the different areas of the weld starting from the welding parameters, but also of being able to evaluate the intrinsic variability of the process.

The proposed approach aims at forecasting the welding properties considering the process' natural variability by means of a group of Artificial Neural Network (ANN) having the same architecture, but different weights. The hypothesis of this work is that predicting the output using this approach allows to define an upper and a lower limit in which the proper results are placed. This idea was applied to the FSW process for the prediction of the hardness value reached by an aluminum alloy (AA 2024-T3) as a function of the main welding parameters (considered as input nodes of the ANN).

Material Forming - ESAFORM 2023
Materials Research Proceedings 28 (2023) 1067-1074

Materials Research Forum LLC
https://doi.org/10.21741/9781644902479-117

Methodology

Experimental tests.

AA2024-T3 sheets were friction stir welded perpendicularly to the rolling direction. The FSW process was performed using a CNC machine tool with a tool characterized by a smooth plane shoulder (16-mm diameter) and a frustum of cone pin shape (maximum and minimum diameters equal to 6 and 4 mm, height equal to 3.9 mm). The tool was realized from a drawn bar of carbon steel C40 without any heat treatment. Table 1 reports the welding parameters considered for the experimental campaign. The difference in the feed rate combined to 1200 rpm and 1500 rpm is linked to the presence of macro and micro defects in the joints welded with 1500 rpm-40 mm/min and 1500 rpm-100 mm/min. These defects led the authors to eliminate these combinations of parameters from further evaluation, considering them not suitable for the material considered.

Table 1. Parameters set-up for experimental tests.

Rotational speed (rpm)	Feed rate (mm/min)
1200	40
1200	70
1200	100
1500	70

All the other process parameters, such as the clamping system for the sheets, the tool inclination (fixed at $3°$), and the tool penetration depth into the sheets (fixed at 3.99 mm), were kept constant.

Rockwell B hardness tests (HRB) were performed for all the welding conditions, following a 5mm spaced grid in the central zone of the top of the specimens, according to ISO 6508. For each sample, three profiles of indentations were carried out moving from the joint axis to the base material.

Developed approach.

One of the main issues in the FSWed joints is represented by the desire to obtain welds having mechanical properties as close as possible to those of the base material. In this context, one of the most important characteristics is the surface hardness of the joints.

Due to the importance of the prediction of these kind of properties without the need to execute experimental tests, the prediction models have become very useful in the scientific research. Nevertheless, these forecasting models are characterized by a limitation in the prevision since they supply a deterministic constant result for the same inputs, not considering the intrinsic variability of the process. This is a huge limitation typical also of the trained artificial neural networks (ANN). Indeed, the hypothesis behind the development of this models is that, even if the same architecture is considered, the ANN can be characterized by different connection weights and activation functions, obtainable through different training iterations. This means that same architecture, but different weights, generate different, even if still reliable, forecasts. This means that it might be possible to develop a model able to predict not only the required punctual value but also its confidence interval. This kind of model has been developed following the flow-chart reported in Fig. 1.

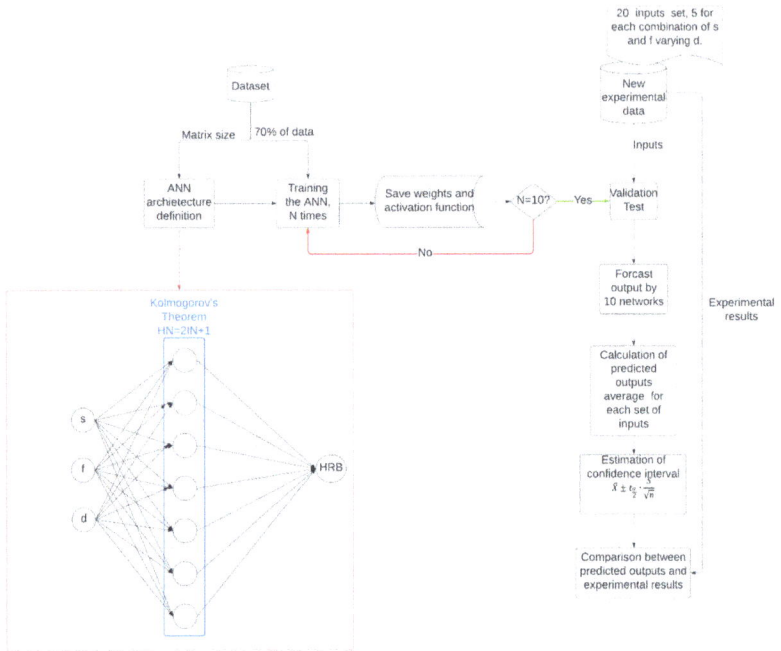

Fig. 1. Algorithm flow-chart.

The following stages are considered:

1. ANN architecture definition. The inputs and the outputs layers were defined according to the welding parameters and the final characteristic to be evaluated. Specifically, the input nodes (IN) were three and represented respectively the rotational speed of the tool (s), the feed rate of the tool (f) and the distance from the welding line (d) where the hardness, which represented the output node (ON), has to be known. A single hidden layer was introduced since the literature supports the idea of introducing few hidden layers to avoid unnecessary complexity of the network. The number of hidden nodes (HN) was selected based on one of the most applied heuristic techniques found in the literature: the Kolmogorov's Theorem, which demonstrates that a network with a single hidden layer can compute any arbitrary function of its input [8]. This theorem allows to estimate the size of the hidden layer as reported in Eq.1.

$$HN = 2 \cdot IN + 1 \tag{1}$$

2. ANN training. The dataset used for the ANN training contained 470 inputs combinations and the relative outputs, considered as target values. The dataset dimension fit with the requirements defined by Alwosheel A. et al. [9]. The dataset was based on the data of experiments performed to evaluate the surface hardness of the welded joints. Through a Matlab code, the ANN was trained to identify the weights, defining N=10. N indicated the number of ANN, with fixed architecture, considered for the multiple ANN approach trained.

Material Forming - ESAFORM 2023   Materials Research Forum LLC
Materials Research Proceedings 28 (2023) 1067-1074   https://doi.org/10.21741/9781644902479-117

This allows to obtain 10 ANNs with the same size, but with the layers differentiated from the weights of the connections and the activation functions. Levenberg–Marquardt algorithm was chosen as the training function for all the ANNs.

3. Forecasting tests. A new group of experimental data were considered for the verification of the applicability of the algorithm. Specifically, 20 combinations of input parameters were selected, and the relative experimental results were collected, ensuring to consider inputs always included in the training interval of the network. These sets of inputs were introduced in the group of the trained ANNs obtaining 10 predictions for each combination, for which the average value was calculated. At this point, the idea of confidence interval was introduced to verify the reliability of the previsions and to forecast the intrinsic variability of the process.

4. Confidence interval definition. Since the standard deviation of the data distribution was unknown, it was necessary to estimate the interval of the population average (μ) using, in addition to the sample mean ($\bar{X}$), the sample standard deviation (S), allowing to estimate the population standard deviation (σ). This can be done assuming that the random variable was normally distributed and, as a consequence, that the confidence interval can be defined by introducing the t-distribution. The t-distribution is similar to the normal distribution (symmetrical, bell-shaped, with mean and median equal to 0), but with a higher probability to fall into its queues than in the central part of the bell. Moreover, increasing the number of degrees of freedom (n), the random variable t tends to the normal distribution. Thus, considering α as the confidence level and n as the sample dimension (represented by the number of repetitions of inputs combination introduced into the dataset), the confidence interval is estimated as reported in Eq. 2. Considering the degrees of freedom equal to 3 and an α equal to 5%, from the statistical table, $t_{\frac{\alpha}{2}} = 4.3027$. The confidence interval was estimated based on the training dataset.

$$\bar{X} - t_{\frac{\alpha}{2}} \cdot \frac{S}{\sqrt{n}} \leq \mu \leq \bar{X} + t_{\frac{\alpha}{2}} \cdot \frac{S}{\sqrt{n}} \tag{2}$$

In this way, the estimated upper and lower bounds allowed to predict and define the interval where the results, in this case the surface hardness, can be placed if the experiments will be repeat more than once. Specifically, from the practical point of view, it was possible to predict HRB with a certain level of variability reproducing the intrinsic variability of the process, linked to external and uncontrollable factors.

Results and Discussion
The validation was performed by comparing the experimental and forecasted data and verifying their placement considering the predicted confidence interval. 20 combinations of inputs, indicating different welding conditions and hardness locations, were selected from the experimental tests (not used for ANN training): five hardness measurements for each combination of welding parameters by randomly selecting the distance from the welding line (d). Table 2 reports the combination used for the validation and the related results obtained from the experiments (performed as reported in previous section) and the average of previsions. It is important to underline that the data used for the validation are not related to a unique welding joint, thus resulting in a nonlinear distribution of the obtained hardness values in the typical FSW W-shape. This choice was made because the main aim of this step was to demonstrate the validity of the developed method considering the totality of the dataset and not limiting it to a determined couple of inputs and their related output. Each ANN responds with a slightly different result from the others and this ANN behavior allows to assimilate these simulation repetitions to the intrinsic variability of the process, such as possible to observe by the standard deviation and the dispersion

of the data is coherent with a stable and feasible process. Furthermore, the error in prevision is very low showing a maximum error equal to 6.25%. As it is possible to observe more clearly in Figure 2, all the average values of the predictions are contained in the estimated confidence interval, the same for the ten predicted values. The validation process allows to demonstrate the reliability of the developed approach not only in the prediction performed by a single ANN, but also in the identification of a confidence interval in which the process results can vary due to the intrinsic variability.

Table 2. Inputs combinations and ANN previsions.

INPUTS			TARGET	PREDICTED OUTPUT		
S	f	d	HRB exp	MEAN	Std. Deviation	%Error
1200	40	75	79	79.5	1.4	0.58%
1200	40	60	79	78.9	0.5	-0.12%
1200	40	10	75	70.3	1.2	-6.25%
1200	40	-2.5	70	67.9	2.3	-2.93%
1200	40	-35	77	77.4	1.0	0.56%
1200	70	50	80	78.4	0.7	-2.05%
1200	70	20	72	74.9	0.8	4.05%
1200	70	2.5	75	71.6	1.2	-4.50%
1200	70	-17.5	74	75.1	0.8	1.45%
1200	70	-45	80	78.4	1.0	-2.03%
1200	100	20	73	75.5	0.6	3.46%
1200	100	2.5	76	73.0	0.6	-3.98%
1200	100	0	76	73.0	0.6	-3.93%
1200	100	-5	71	73.4	0.7	3.31%
1200	100	70	79	78.8	0.5	-0.29%
1500	70	25	79	77.0	1.0	-2.53%
1500	70	10	78	74.5	0.9	-4.46%
1500	70	0	76	73.4	1.3	-3.46%
1500	70	-10	77	74.4	0.8	-3.36%
1500	70	-12.5	79	77.4	1.2	-2.06%

Material Forming - ESAFORM 2023
Materials Research Proceedings 28 (2023) 1067-1074

Materials Research Forum LLC
https://doi.org/10.21741/9781644902479-117

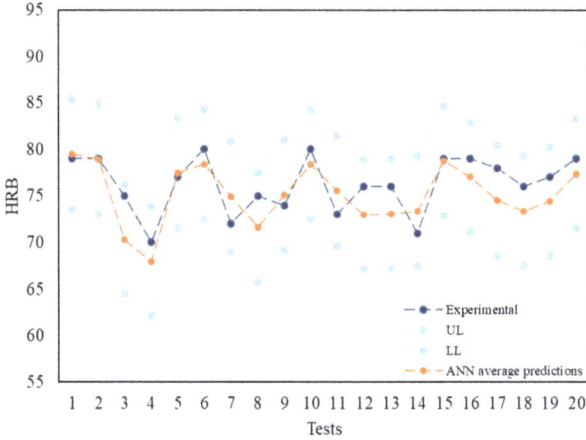

Fig. 2. Validation results.

Summary

A new approach for the prevision of Rockwell B hardness values as a function of the distance from the center of the welding line of joints obtained by the FSW process was developed. The novelty of the model regarded the possibility to predict the results considering the intrinsic variability of the process linked to uncontrollable factors of the process.

This means that, through the proposed approach, it was possible to correctly predict the surface hardness obtainable after a welding process, performed with the welding parameters used for the training. Furthermore, the approach was also able to show the confidence interval of the previsions.

The present approach identifies a new method for the use of ANN in manufacturing processes. In this way, with the same input, it is possible to obtain different outputs, placed in a confidence interval acceptable from the statistical point of view thanks to the different distribution of the weights of the connections, reflecting more a real process that is always characterized by a slight variability in the results.

References

[1] W.M. Thomas, E.D. Nicholas, J.C. Needham, M.G. Murch, P. Templesmith, C.J. Dawes, GB Patent application no. 9125978.8, 1991. Avalible: https://patentimages.storage.googleapis.com/66/60/ad/5f784d0b8653b7/US5460317.pdf (accessed 28 February 2019).

[2] P. Kah, R. Rajan, J. Martikainen, R. Suoranta, Investigation of weld defects in friction-stir welding and fusion welding of aluminium alloys, Int. J. Mech. Mater. Eng. 10 (2015) 1-10. https://doi.org/10.1186/s40712-015-0053-8

[3] M. Jariyaboon, A.J.J. Davenport, R. Ambat, B.J.J. Connolly, S.W.W. Williams, D.A.A. Price, R. Ambat, A.J.J. Davenport, The effect of welding parameters on the corrosion behaviour of friction stir welded AA2024–T351, Corros. Sci. 49 (2007) 877-909. https://doi.org/10.1016/j.corsci.2006.05.038

[4] M. Aissani, S. Gachi, F. Boubenider, Y. Benkedda, Design and optimization of friction stir welding tool, Mater. Manuf. Process. 25 (2010) 1199-1205. https://doi.org/10.1080/10426910903536733

Material Forming - ESAFORM 2023   Materials Research Forum LLC
Materials Research Proceedings 28 (2023) 1067-1074   https://doi.org/10.21741/9781644902479-117

[5] K. Deepandurai, R. Parameshwaran, Multiresponse Optimization of FSW Parameters for Cast AA7075/SiCp Composite, Mater. Manuf. Process. 31 (2016) 1333-1341. https://doi.org/10.1080/10426914.2015.1117628

[6] K. Mallieswaran, R. Padmanabhan, V. Balasubramanian, Friction stir welding parameters optimization for tailored welded blank sheets of AA1100 with AA6061 dissimilar alloy using response surface methodology, Adv. Mater. Process. Technol. 4 (2018) 142-157. https://doi.org/10.1080/2374068X.2017.1410690

[7] W.L. Oberkampf, J.C. Helton, C.A. Joslyn, S.F. Wojtkiewicz, S. Ferson, Challenge problems: Uncertainty in system response given uncertain parameters, in: Reliab. Eng. Syst. Saf., Elsevier, 2004: pp. 11–19. https://doi.org/10.1016/j.ress.2004.03.002

[8] R. Hecht-Nielsen, Kolmogorov's Mapping Neural Network Existence Theorem, in: SOS Printing, 1987.

[9] A. Alwosheel, S. van Cranenburgh, C.G. Chorus, Is your dataset big enough? Sample size requirements when using artificial neural networks for discrete choice analysis, J. Choice Model. 28 (2018) 167-182. https://doi.org/10.1016/J.JOCM.2018.07.002

Material Forming - ESAFORM 2023   Materials Research Forum LLC
Materials Research Proceedings 28 (2023) 1075-1082   https://doi.org/10.21741/9781644902479-118

Modelling of the weld seam in the forming simulation of friction stir welded tailored blanks

BACHMANN Maximilian[1,a] *, RIEDMÜLLER Kim[1,b], LIEWALD Mathias[1,c]
and MERTEN Mathias[2,d]

[1]Institute for Metal Forming Technology, Holzgartenstraße 17, 70174 Stuttgart, Germany

[2]DYNAmore GmbH, Stralauer Platz 34, 10243 Berlin, Germany

[a]maximilian.bachmann@ifu.uni-stuttgart.de, [b]kim.riedmueller@ifu.uni-stuttgart.de,
[c]mail@ifu.uni-stuttgart.de [d]mathias.merten@dynamore.de

Keywords: Tailor Welded Blanks, Material Modelling, Sheet Forming

Abstract. In former papers methods to join different aluminium or different steel plates having same thicknesses are presented. These blanks are often joined by friction stir welding using flat tools. In order to increase the lightweight potential, the Materials Testing Institute of the University of Stuttgart developed a modified friction stir welding process to join aluminium and steel plates of different thicknesses. The process differs from the conventional method in stir welding tool used, which consists of a stepped welding pin and enables combined lap-and-butt joints to be produced. In this paper, a suitable material model is presented to describe this weld seam, allowing the forming behavior of hybrid sheet metal compounds to be realistically simulated.

Introduction

In car body construction, lightweight design approaches are becoming more common, in which weight of components is reduced by substituting one material by a comparatively lighter material (lower density). On the one hand, the weight reduced this way can extend the range of vehicles, which is particularly important in terms of the future or acceptance of electromobility, as charging cycles per kilometre driven play a significant role. On the other hand, lightweight components contribute to regulatory compliance as they lead to a reduction in the CO_2 footprint of vehicles by lowering the energy consumption to drive the vehicle and by requiring less material to be used in production [1]. One challenge arising in the substitution of known materials consists in the required adaptation of the component geometries and the joining processes used [2]. When combining high-strength aluminium alloys or even combinations of aluminium and steel alloys lightweight construction potentials increases. In particular, joining processes established in car body construction, such laser beam and resistance spot welding, show limitations regarding the combination of such materials.

Sheet metal components made of different high-strength aluminium alloys and different sheet thicknesses can be used in the same applications as conventional materials, providing better properties like higher strength and lower density. However, joining of such high-strength aluminium alloys (e.g. 6000-series alloys) by established joining processes such as laser beam welding proves to be difficult due to the materials' tendency to hot cracking during or after welding [3]. In addition, the high heat input of the laser beam welding process causes a reduction of the strength of aluminium alloy, resulting finally in a reduced strength of the joint (weld seam) [4]. For resistance spot welding, high wear of electrodes poses a problem when welding such materials. These problems in joining high-strength aluminium alloys were circumvented in 1991 with the friction stir welding process patented by Thomas et al [5]. In this process, the process energy required is introduced by means of a rotating welding tool, which is pressed into and moved along the joint gap [6]. Since the melting temperature is principally not reached during friction stir

Material Forming - ESAFORM 2023   Materials Research Forum LLC
Materials Research Proceedings 28 (2023) 1075-1082   https://doi.org/10.21741/9781644902479-118

welding due to the method of energy input, the materials to be joined remain in the solid phase during the entire process. Joining of different sheet materials with different sheet thicknesses allows application-specific lightweight solutions to be realized. When joining different sheet materials with different sheet thicknesses, however, the welding tool in general has to contact the joint area flatly. Furthermore, for the joining of aluminium and steel, a joining method is required that minimises the appearance of intermetallic phases, as these reduce the strength of the weld seam, and at the same time the surface of the steel is sufficiently activated to form the joint with the aluminium. To ensure this, the Materials Testing Institute of the University of Stuttgart (MPA) developed a variation of the friction stir welding process to join different aluminium or aluminium and steel blanks of different thicknesses. In this process a combined overlap and butt joint is used as shown in Fig. 1. Here, the thicker metal sheet is provided with a milled step in sheet thickness into which the thinner sheet is inserted before the stir welding process [7].

Fig. 1. Combined lap and butt joint; left: with welding tool, right: connection point [1].

Design of the pin of the welding tool shown in Fig. 1 is shaped in a way that both the front face and the surface of the steel sheet are activated. Additional the tool shoulder only rests on the aluminium and therefore does not rub on the steel as with conventional butt joints. The tool live, especially of the tool shoulder significantly is increased. The presented friction stir welding process can therefore be used both to join high-strength aluminium alloys and to produce mixed aluminium-steel joints [8,9]. The use of mixed aluminium-steel joints has a considerable potential for lightweight materials, as load-adapted components can be produced in which higher stressed sections can be provided with a steel sheet and the remainder with an aluminium sheet. In addition to the possibility of joining different materials of different sheet thicknesses, the developed welding process serves as solution annealing. Components of different materials, sheet thicknesses and material grades manufactured with friction stir welding are known as Tailor Welded Blank (TWB).

Previous studies [10] show that TWBs made of aluminium and steel having the same sheet thickness of the base materials show relatively good forming properties as cups with drawing depths similar to the base materials could be drawn. However, [10] considered only 5000-series alloys on the aluminium side in his investigations. Singar [11] investigated the forming capacity of aluminium-steel TWBs made of HX340LAD (0.8mm) and AA 6014 T4 (1.2mm) joined by the cold metal transfer welding process. However, the investigations shown, that the values achieved for elongation at break (approx. 7%) and forming limits from the forming limit diagram required further work to produce TWBs suitable for industrial use.

By using the friction stir welding process presented in this paper hybrid TWBs consisting of DX54 (1.0 mm) and AA 6016 T4 (2.0 mm) can be produced showing a new potential for forming operations [12]. For these hybrid TWBs a simulative mapping approach is developed with focus on the question how the hybrid TWB has to be modelled in order to describe the forming process realistically. First the material characterization in form of tensile tests and the determination of the forming limit curves is carried out for the base materials and the weld seam. With the results from these characterizations a simulative approach of TWBs is set up by mapping the properties of the three components aluminium, steel and weld seam in separate parts in order to map the forming behaviour and thus increase acceptance of aluminium-steel TWBs in the industry.

Material Forming - ESAFORM 2023   Materials Research Forum LLC
Materials Research Proceedings 28 (2023) 1075-1082   https://doi.org/10.21741/9781644902479-118

Material Characterization

For deriving the material model for the forming simulation, the mechanical properties and the forming limit curves (FLC) of both the base materials (DX54 and AA6016 T4) and the weld seam of the TWBs were determined. Concretely, tensile tests were performed at the DYNAmore GmbH to obtain yield curves and Nakajima tests were carried out at the Institute for Metal Forming Technology (IFU) for determining the FLCs. The weld seam was furthermore characterized by the area of the combined lap and butt joint. The yield curves measured via the tensile tests are shown in Fig. 2. Here, for the base materials specimens were tested with 0°, 45° and 90° to rolling direction and for the weld seam, specimens were tested with 0° and 90° to the welding orientation. Each characterization test was performed using 5 samples. The extrapolation of the data from the tensile test was done using Hockett-Sherby. In Fig.2 the yield curves for AL6016 T4 and DX54 in 0° are presented as these are the rolling directions with the lowest loading limits. In this context, Fig. 2 shows the flow curve of the weld seam specimen taken 0° to the welding orientation which is above the flow curve of the base material AA6016 T4, but below the flow curve of the base material DX54. In contrast, the yield curve of the weld seam taken 90° to the welding orientation is below the yield curve of both base materials. Considering these properties, it can be seen weld seams should be positioned in 0° to the welding orientation in order to obtain components with a correspondingly strength.

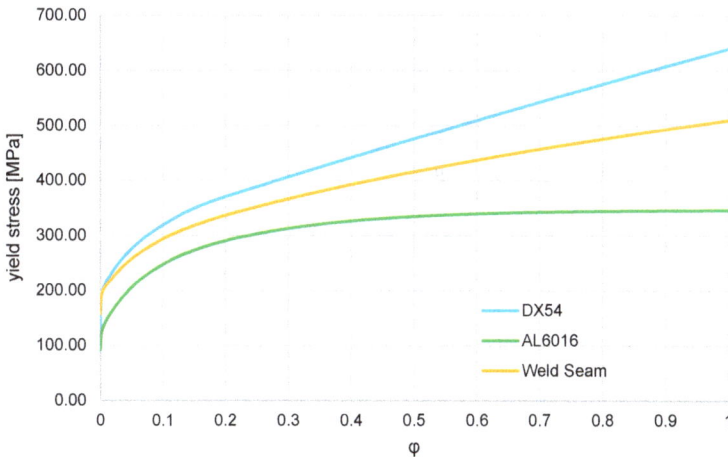

Fig. 2. Yield curves of AA6016 T4, DX54 and weld seam.

The forming limit curves of the TWBs were determined on the basis of 5 support points (waist of 30 mm, 60 mm, 100 mm, 140 mm and 200 mm) according to ISO 12004 [13] using a workshop press of the IFU (see Fig. 3, left and middle). For each of the 5 support points, Nakajima specimens were produced with a welding orientation of 0° and 90°. Each support point was determined using 4 valid samples. The smaller sheet thickness of the aluminium sheet of the TWB was compensated by means of a shim plate, thus ensuring a uniform force application during the testing procedure. The phenomenon of weld seam migration during forming in the direction of the aluminium could be minimized by increasing the pressure of the gas springs of the blank holder.

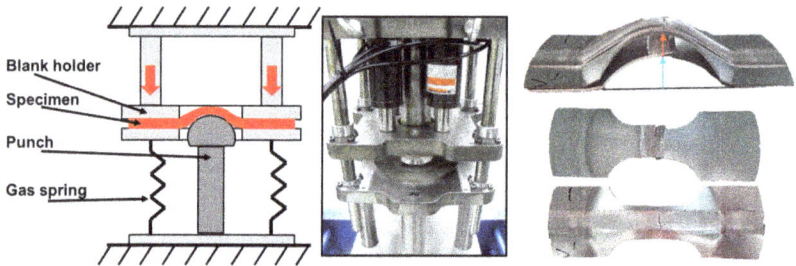

Fig. 3. Principal of the Nakajima test used (left), Nakajima testing device at IFU (middle), TWBs with a welding orientation of 0° and 90° (right).

Depending on the weld direction in the specimens, different results were observed with regard to crack initiation. For the specimens with 0° welding orientation, the crack forms, as usual for Nakajima tests of monolithic specimen, starting from the weld in the middle in the direction of the sheet edge (Fig. 3, right bottom). In this case, the load applied by the punch is absorbed by the weld seam and the base materials. Here, the GOM aramis system could be used to determine the major and minor strains before cracking. Determining the crack initiation for specimen with a welding orientation of 90° proved more difficult. Here, the failure was also initiated in the weld seam, but could not be identified as a local crack on the upper side of the sheet. The applied force of the punch is absorbed only from the weld seam which forces the weld seam to open up from the bottom side of the sheet. Therefore, no crack could be detected on the upper side of the sheet. On the upper side of the sheet only the separation of the weld seam was detected, (see Fig.3 right, middle). In this case, the crack initiation moment could be determined from the history curve of the major and minor strain, since they showed a rapid decrease after separation of the sheets due to the force of the punch. This moment was defined as the crack initiation moment for TWB with welding orientations of 90°.

The different forming behavior of TWBs considering the welding orientation is shown in Fig. 3 (right, top), illustrating the lateral plan view of Nakajima specimens with a waist of 30 mm. The drawing depth at 90° welding orientation is given in blue and the drawing depth at 0° welding orientation is shown in red. Such differences in the forming capability of the TWB could be observed for all waists and are shown in the forming limit diagram for both cases (0° and 90° orientation of the welding seam) in Fig. 4.

Comparison of the two FLCs reveals that TWBs show different forming capability depending on the weld orientation for each loading condition (uniaxial, plain strain, biaxial). These results correspond to those from the tensile tests already presented, as the forming capability of TWBs with a welding orientation of 90° is below the forming capability of 0° at the critical plane strain load.

Material Forming - ESAFORM 2023   Materials Research Forum LLC
Materials Research Proceedings 28 (2023) 1075-1082   https://doi.org/10.21741/9781644902479-118

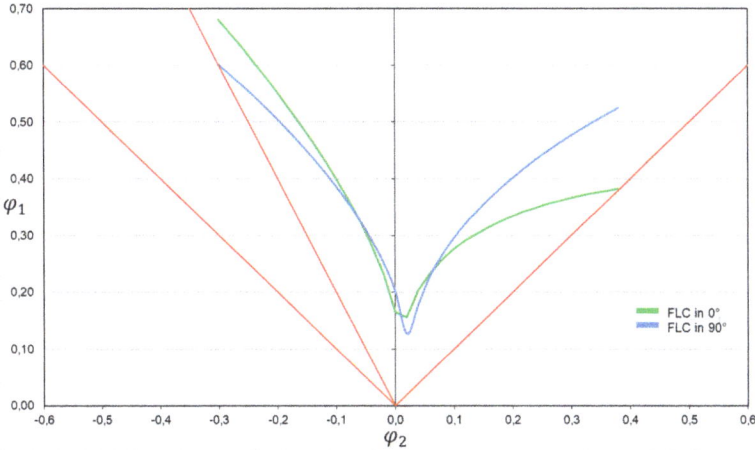

Fig. 4. Forming limit diagram for DX54/AL6016 TWBs in 0° and 90° welding orientation.

Numerical Simulation Setup

The mechanical properties determined for the base materials DX54 and AA6016 T4 as well as the weld seam were subsequently used to set up material models for the simulation of a Nakajima test performed. As described before the failure behavior varies due to the welding orientation and in this approach the welding orientation of 0° was further investigated. The simulations aimed to model the forming behavior of the TWB specimen with different waists and welding orientation of 0° during the Nakajima test as accurately as possible. For simulation setup and performing, the simulation software DYNAFORM and LS-PREPOST were used. Previous simulation work in this field was mainly concerned with the modeling of TWBs consisting of the same material using the software ABAQUS [14], TWBs consisting of different aluminium alloys [15] or TWBs produced by laser welding [16]. In laser-welded TWBs, the width of the weld seam is only a few millimeters compared to friction stir welded TWBs. Moreover, the higher process temperatures of laser welding as opposed to friction stir welding lead to significantly differing forming behavior, especially of the weld seam but also of aluminium sheets as part of a TWB. To account for these characteristics, a new modeling approach was developed for the numerical study on the forming behavior of friction stir welded TWBs presented in this paper, distinguishing it from previous work on laser welded TWBs. The distinction from [14] and [15] is made by considering different materials (aluminium and steel) with different thicknesses. In contrast to [17] this modelling approach on the one hand considers anisotropic material behavior and on the other hand describes the weld seam not only by node definitions, but as an independent material with its specific forming behavior.

In the model approach, the welded blank was divided into three parts (the aluminium side, the weld seam and the steel side) as shown in Fig. 5 a). The weld seam was designed with a width of 20 mm, which corresponds to the area of the combined lap and butt joint of the real TWB. The interface from AA6016 T4 to the weld seam and from the weld seam to DX54 were modeled using the weld function of DYNAFORM. Normally, failure values are assigned to these weld functions to implement the failure behavior at this location. However, as shown during the Nakajima tests, the TWB considered did not fail on the interfaces between the base materials and the weld seam, but within the weld seam. For this reason, the weld function was assigned with 'no failure'. Each of the three parts was assigned with its determined mechanical properties using the material model

Material Forming - ESAFORM 2023   Materials Research Forum LLC
Materials Research Proceedings 28 (2023) 1075-1082   https://doi.org/10.21741/9781644902479-118

MAT_036-3-Parameter_Barlat. The entire welded blank was meshed with shell elements having a size of 1 mm, which means the weld seam has been represented with 20 elements in width. Fig. 5 b) shows the entire simulation setup with punch, blank holder, die and TWB in lateral plan view.

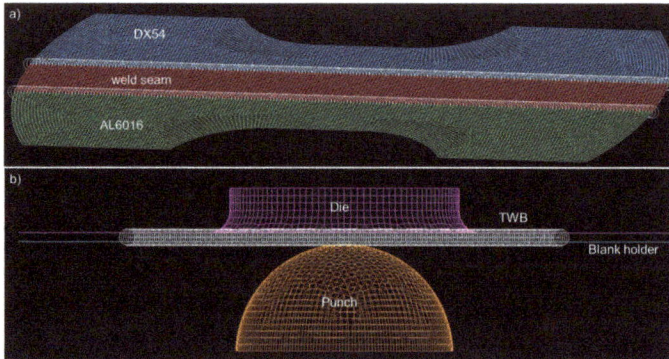

Fig. 5. a) Separation of the blank in 3 parts; b) Simulation setup of Nakajima test considered.

Validation Of Simulation Model

For validation of the simulation results, the strain values measured in the corresponding Nakajima test were compared with the strain values calculated from the simulation. Furthermore, by importing the forming limit curve for the welding orientation of $0°$, the crack initiation moment was determined based on the simulation results and compared with the crack initiation observed in the experiment. In this context, Figure 6 shows the major strain distribution of a Nakajima specimen with a waist of 60 mm calculated by simulation on the left side and measured in the experiment on the right side. On the right side the measurement results from the experiment of three samples are shown. The average value for the major strain at crack initiation moment is $\varphi_1 = 0,283$ in contrast, the simulative determined main strain is $\varphi_1 = 0,31$. Comparison of these strain results revealed the strain distribution could be calculated quite accurately with the established simulation model for TWB consisting of DX54 and AA 6016 T4, but the strain values at crack initiation were overestimated relative to the strain values measured in the experiment. In Fig. 6 the location of crack initiation is highlighted with black arrows. The simulation differs slightly from the experiment with respect to the predicted location of crack initiation. The simulation result indicates this allocation slightly closer to the transition between the weld seam and the base material. The reason for these deviations in the simulation is probably the weld seam exhibits a different mixture of aluminium and steel in the transition area to the base materials compared to the weld seam center. This was not considered in the characterization tests of the weld seam carried out so far.

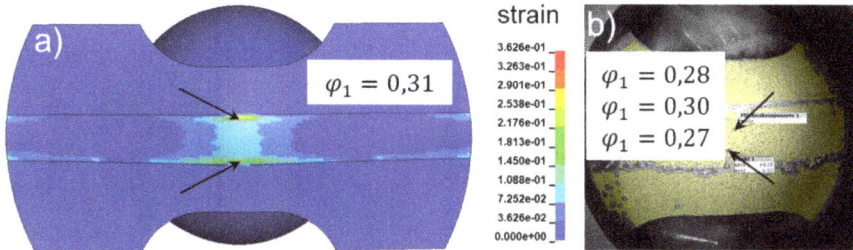

Fig. 6. Strain values before crack initiation moment a) major strain in simulation, b) major strain in experiment.

Summary

In this paper, a modelling approach for the numerical forming simulation of TWBs with a special focus on the consideration of the weld seam is presented. First, the base materials DX54 and AA6016 T4 as well as the weld seam of the TWB considered were characterized. Based on these tests, a simulation model of one of the Nakajima tests was set up and validated using the strains numerically calculated and experimentally measured. The simulation results revealed that the forming behavior of the TWBs could be described quite accurately by mapping the properties of the three components aluminium, weld and steel separately, but the behavior of the weld seam still needs to be represented in more detail as the predicted location of crack initiation varies from the experiments.

In the future, in addition to the strain values, the force curves from simulation and experiment will be compared to base validation on 2 values. Furthermore, a simulative representation of the weld seam with shell and volume elements will be aimed in order to implement hardness measurements of the weld seam in the simulation and thus to consider the transition of the weld seam to the base materials in more detail. Possible variations of the properties of the weld seam will thus considered in further investigations by implementing the hardness profile of the weld seam and linking it to a prior damage. In addition, other connection methods will be considered to map the crack initiation at the correct location. Moreover, simulation of the Nakajima test with welding orientation of 90° will be considered in the future as they have a different failure behavior which was described in this work.

Acknowledgement

The work of this paper was carried out within the framework of the project CO2-HyChain, which is funded by the Technologietransfer-Programm Leichtbau (TTP LB) of the Federal Ministry of Economic Affairs and Climate Action.

References

[1] A. Birkert, S. Haage, M. Straub, Umformtechnischer Herstellung komplexer Karosseriebauteile – Auslegung von Ziehanlagen. Springer Verlag Berlin Heidelberg, 2013.

[2] F. Henning, E. Moeller, Handbuch Leichtbau – Methoden, Werkstoffe, Fertigung. Carl Hanser Verlag München Wien, 2020.

[3] M.W. Mahoney, C. Rhodes, J.G. Flinott, W.H. Bingel, RR. Spurling, Properties of friction stir-welded 7075 T651 aluminium, Metall. Mater. Trans. A 29 (1998) 1955-1964. https://doi.org/10.1007/s11661-998-0021-5

[4] A.K. Lakshminarayanan, V. Balasubramanian, K. Elangovan, Effect of welding processes on tensile properties of AA 6061 aluminium alloy joints, Int. J. Adv. Manuf. Technol. 40 (2009) 286-296. https://doi.org/10.1007/s00170-007-1325-0

[5] W.M. Thomas, E. Nicholas, J. Needham, M. Murch, P. Temple-Smith, C.J. Dawes, Great Britain Patent Application No. 9125978.8 (1991).

[6] H.B. Schmidt, J.H. Hattel, Thermal Modelling of friction stir welding, Scirpta Mater. 58 (2009) 332-337. https://doi.org/10.1016/j.scriptamat.2007.10.008

[7] M. Werz, Experimentelle und numerische Untersuchungen des Rührreibschweißens von Aluminium- und Aluminium-Stahl-Verbindungen zur Verbesserung der mechanischen Eigenschaften, Materialprüfungsanstalt (MPA) Universität Stuttgart, Dissertation, 2020.

[8] T. Wanatabe, H. Takayama, A. Yanagiswa, Joining of aluminium to steel by friction stir welding, J. Mater. Process. Technol. 178 (2006) 342-349. http://doi.org/10.1016%2Fj.jmatprotec.2006.04.117

[9] H. Uzun, C. Donne, A. Arganotto, T. Ghidini, Friction stir welding of dissimilar AL 6013-T4 to X5CrNi18-10 stainless steel, Mater. Des. 26 (2005) 41-46. https://doi.org/10.1016/j.matdes.2004.04.002

[10] T. Tanaka, T. Hirata, N. Shinomiya, N. Shirakawa, Analyses of material flow in sheet forming of friction-stir welds on alloys of mild steel and aluminium, J. Mater. Process. Technol. 22 (2015) 115-134. https://doi.org/10.1016/J.JMATPROTEC.2015.06.030

[11] O. Singar, M. Merklein, Study on the formability characteristics of the weld seam of Aluminium Steel Tailor Hybrid Blanks, Key Eng. Mater. 549 (2013) 302-310. https://doi.org/10.4028/www.scientific.net/KEM.549.302

[12] F. Panzer, M. Schneider, M. Werz, S. Weihe, Friction stir welded and deep drawn multi-matieral tailor welded blanks, Materials Testing, Carl Hanser Verlag, München, 2019.

[13] DIN Deutsches Institut für Normung e.V., Metallische Werkstoffe – Bestimmung der Grenzformänderungskurve für Bleche und Bänder – Teil 2: Bestimmung von Grenzformänderungskurven im Labor, Beuth Verlag GmbH, Berlin, 2021.

[14] W. Lee, K.-H. Chung, D. Kim, J. Kim, C. Kim, K. Okamoto R.H. Wagoner, K. Chung, Experimental and numerical study on formability of friction stir welded TWB sheets based on hemispherical dome stretch tests, Int. J. Plast. 25 (2009) 1626–1654. http://doi.org/10.1016/j.ijplas.2008.08.005

[15] C. Leitao, B.K. Zhang, R. Padmanabhan, D.M. Rodrigues, Influence of weld geometry and mismatch on formability of aluminium tailor welded blanks: numerical and experimental analysis, Sci. Technol. Weld. Join. 16 (2011) 662–668. https://doi.org/10.1179/1362171811Y.0000000055

[16] X. Qiu, The study on Numerical Simulation of Tailor Welded Blanks in Square Cup Stamping, Adv. Mater. Res. 189-193 (2011) 3932-3935. https://doi.org/10.4028/www.scientific.net/AMR.189-193.3932

[17] O. Singar, D. Banabic, Numerical Simulation of Tailored Hybrid Blanks, Proceedings of Romanian Academy, Sereis A, Volume 22, Number2/2021, pp. 177-184.

Material Forming - ESAFORM 2023   Materials Research Forum LLC
Materials Research Proceedings 28 (2023) 1083-1090   https://doi.org/10.21741/9781644902479-119

Comparative investigation of partial cooling methods for induction heating of hybrid bulk components for hot forming

INCE Caner-Veli[1,a] *, KATZ Fabian[1] and RAATZ Annika[1,b]

[1]Leibniz University Hannover, Institute of Assembly Technology, An der Universität 2, 30823 Garbsen, Germany

[a]ince@match.uni-hannover.de, [b]raatz@match.uni-hannover.de

Keywords: Partial Cooling, Form Variable Handling, Function Integrated Handling, Tailored Forming

Abstract. The novel Tailored Forming process chain enables the combination of crucial properties of different materials by manufacturing hybrid components. Thereby, the limitations of monolithic components are surpassed. However, manufacturing hybrid bulk components introduces new challenges for hot forming. For example, when combining steel and aluminium, the main challenge is establishing and maintaining a temperature gradient in the component to match the differing flow stresses of the materials for a successful forging. For establishing the gradient, a particular heating strategy, including inductive heating of the steel and parallel partial cooling of the aluminium, is necessary. After reaching the target temperature, the heated component has to be transferred to the forging die by a robot while maintaining the essential temperature gradient. Therefore, a portable spray nozzle cooling system attached to the robot's end effector was designed in former work. This paper aims to validate spray nozzles for establishing a temperature gradient in a hybrid workpiece with a particular heating strategy compared to a currently used immersion cooling. For the validation, the nozzles will cool a hybrid steel aluminium shaft, whereby the nozzles' operation parameters influence on the temperature gradient will be investigated. Finally, the performance of the nozzles will be compared against the currently used immersion cooling.

Introduction

Nowadays, the research and development of increasingly high-performance and lightweight materials are becoming essential for resource efficiency. With these challenges, conventional mono-material components reach their limits to an increasing degree. To improve these limitations, hybrid components are developed that allow the use of different material properties by combining them. For example, a shaft that carries a high local mechanical load consists of a material that withstands the load and has a high inherent weight. If instead, only the loaded areas consist of the high-strength material, and the remaining part uses a lighter and, therefore, lower-strength material, the shaft would still be suitable for its application and, at the same time, reduce its weight.

Conventional hybrid manufacturing processes merge the different materials in a near-net shape at the end of the process chain, which restricts the geometry and the quality of the joining [1]. To overcome this limitation, a novel process chain for producing solid hybrid components that pairs the materials in a semi-finished workpiece and then forges them together, called "Tailored Forming" has been developed [2]. The early joining process results in better mechanical properties in the joining zone as well as a wider range of possible final geometries for the hybrid components [3]. However, the early joining leads to a new challenge of establishing and maintaining a temperature gradient in the workpiece during the whole process chain. The combined materials have deviating yield stresses, which must be equalized before forging. In the Tailored-Forming-Process steel-steel, steel-aluminium and titan-aluminium pairings are investigated. Steel and titan require forging temperatures that exceed the melting temperature of aluminium. To prevent

Material Forming - ESAFORM 2023   Materials Research Forum LLC
Materials Research Proceedings 28 (2023) 1083-1090   https://doi.org/10.21741/9781644902479-119

melting of the aluminium, which destroys the component, special heating strategies were developed for several hybrid components [4, 5]. The heating takes place locally using induction heating. In this way, only the material requiring higher temperatures for forming is heated. Simultaneously, the aluminium is actively cooled down in order to prevent exceeding the melting temperature. As a result, this treatment establishes a step-like temperature gradient in the component. Unfortunately, the current heating and cooling strategies can only be used stationary, resulting in the temperature gradient not being maintained during a transfer from the induction coil to the forming machine. As a result, equalization processes occur in the component, which can lead to critical temperatures in the joining zone, resulting in a renewed threat of melting.

In order to eliminate the risk of melting during transfer, a handling system was developed with an integrated cooling device [6]. The handling system consists of a form variable gripper that can handle hot forged objects and spray nozzles as a cooling unit. The form variable gripper allows handling diverse geometries without retooling time, whereby the handling gets more versatile as conventional two-finger grippers adjusted for only one geometry. The nozzles are attached to the gripper by an adaptable mounting system, whereby they can be aligned so that only a predefined area is cooled. Utilizing this handling system enables local cooling during the transfer phase, eliminating the risk of melting. The nozzles' cooling parameters must be selected to prevent melting and avoid an undercooling of the steel side.

This paper aims to validate the cooling system to prove the cooling effect needed for the sufficient forging of hybrid bulk components. For this purpose, the cooling system is initially considered stationary, i.e. without handling equipment integration and compared with the currently utilized immersion bath cooling. An example process from the Tailored-Forming-Process is used for the validation. For this purpose, immersion cooling and spray cooling are presented first to introduce and compare the methods. Based on this knowledge, the chosen methodology is described in more detail, and the experimental setup is explained. Afterwards, the results are examined and evaluated in detail. Finally, the paper concludes with a summary and an outlook.

Cooling Methods

Active cooling through controlled heat dissipation is described as quenching and is defined as cooling occurring more rapidly than in static air [7]. Such quenching techniques as immersion cooling and spray cooling can be utilized to establish the necessary temperature gradient. This chapter discusses the primary cooling techniques used for the particular heating strategy.

Immersion cooling. Immersion cooling means that a workpiece is immersed in a liquid bath and thereby cooled. Water, oil and polymers serve as coolants for immersion cooling. Depending on the coolant, the cooling rate can be influenced. For example, the cooling rate of water is higher than that of quenching oil. Further parameters with which the cooling rate can be influenced are the temperature and the flow velocity of the coolant. During immersion cooling, different phases may occur depending on the workpiece temperature, resulting in different cooling rates. A vapor layer forms on the workpiece at a high surface temperature, which has a heat-insulating effect and thus

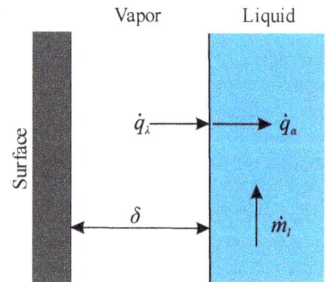

Fig. 1. Leidenfrost scheme demonstrating formation of a vapor layer.

leads to a low cooling rate. This phenomena is called Leidenfrost-Effect [8] and demonstrated in Fig. 1.

The surface temperature is higher than the boiling point of the surrounding liquid. The vapor layer formed separates the liquid from the object and has the thickness δ and the corresponding

Material Forming - ESAFORM 2023   Materials Research Forum LLC
Materials Research Proceedings 28 (2023) 1083-1090   https://doi.org/10.21741/9781644902479-119

coefficient of thermal conductivity, leading to a heat flux $\dot{q}$. In the fluid media, thermal conduction occurs with a different heat transfer coefficient (HTC), for which heat flux is given as a function. Each phase has a specific range for its coefficients, whereby gas coefficients are smaller than coefficients of liquids, which explains the insulating effect of the vapor [9]. The heat flows from the object into the vapor, and then into the liquid, generating free convection $\dot{m}$ in both phases. As soon as the temperature decreases beneath the liquid's boiling temperature, which is also called the Leidenfrost temperature, the vapor layer breaks down, and the heat dissipation increases again. The induction of a flow into the cooling medium influences the Leidenfrost temperature so that the vapor film forms at higher temperatures.

Spray cooling. Another cooling technique is spray cooling, where a nozzle atomizes a liquid coolant and sprays it onto the workpiece to be cooled. This cooling technique serves, for example, for locally varying cooling rates [10]. Atomization can be performed with single-substance or dual-substance nozzles. In the case of single-substance nozzles, the liquid coolant is atomized directly, and in the case of dual-substance nozzles, it is atomized with pressurized air. Compared with single-substance nozzles, dual-substance nozzles achieve higher heat transfer coefficients (HTC) [11]. The impingement density is the quantity of water or liquid that hits the surface of the workpiece per time and area. Furthermore, an increase in the coolant pressure improves the impingement density, positively affecting the cooling rate and, thus, the HTC [12, 13]. This correlation can be utilized in spray cooling to allow flexible adjustment of the cooling rate by varying the pressure. Another feature is that the heat transfer is highest at the center of the spray and decreases radially from it. However, by overlapping the spray cones, the dissipated heat and, consequently, the heat transfer coefficient can be increased [12]. Thus, the radially decreasing intensity of the heat transfer can be counteracted.

Methods

The previously presented cooling methods will now be compared, utilizing a Tailored-Forming-Process, concerning their cooling performance. Initially, the process and its current heating strategy will be presented, followed by a description of the experimental setup and execution.

Investigated Process. To validate the cooling process to establish and maintain the temperature gradient during heating and transport of the manufacturing process of a hybrid shaft developed by Behrens et al. is used [14]. The shaft consists of steel and aluminium (Fig. 2). Friction welding merges the two materials into a semi-finished workpiece. Induction heating of the steel side then prepares the workpiece for the forging process, while the aluminium side remains in a cooling immersion bath with water. This heating strategy achieves the required steel side temperatures of up to 900°C while the aluminium side remains below the melting point. During the following handling into the forging machine, the aluminium is not cooled, whereby the temperature gradient disappears. Therefore, a spray nozzle cooling unit integrated into the handling equipment was developed that maintains the temperature gradient while handling [6]. To confirm the cooling performance of the nozzles, they will be evaluated in stationary use instead of immersion cooling to establish the temperature gradient. If the same temperature gradient as immersion cooling is achieved, the functionality is verified and the handling equipment integrated configuration can be evaluated.

Material Forming - ESAFORM 2023
Materials Research Proceedings 28 (2023) 1083-1090

Materials Research Forum LLC
https://doi.org/10.21741/9781644902479-119

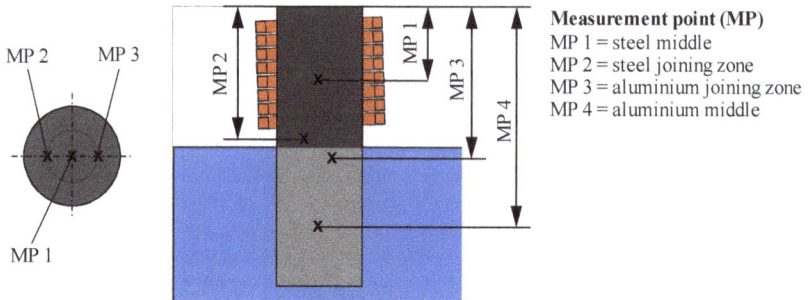

Fig. 2. Scheme of immersion cooled shaft workpiece during inductive heating with marked measurement points (MP).

Experimental Setup. For the validation, the workpiece's steel side is inductively heated, while the the nozzles cool the aluminium. Thermocouples attached to the workpiece measure the temperatures near the joining zone (MP 2 & MP 3) and the center of the respective material volume (MP 1 & MP 4)(Fig. 2) with 10 Hz. Drilled holes permit the thermocouples to be located at the defined positions to measure the temperature. The thermocouples for MS 1 - MS 3 are inserted on the steel end face, while MS 4 passes through the lateral surface of the aluminium. A workpiece can only be used once because the heating and cooling alter the material properties, whereby a second heating would lead to falsified results. Fig. 3 shows the experimental setup, where the workpiece is located in the induction coil. The nozzles are arranged around the aluminium part instead of the immersion bath. Further, the cooling system records the pressurized air and water mass flows. The adjustment of these mass flows occurs through pressure regulation. A valve sets the air pressure, while pressurized air adjusts the water pressure through a tank, where the pressurized air flows in and the water out.

The induction coil heats the workpiece for 28 seconds with a particular power profile, which is designed for immersion cooling, while the nozzles cool the aluminium. The inductive heating unit has a max. power of 44 kW, while the workpiece is heated for 9 s with 55%, then 9 s with 35% and finally 10 s with 20% power to establish the step-like temperature gradient. In order to determine the parameter's influence on the temperature gradient, different parameters for the air and water mass flow are tested. The

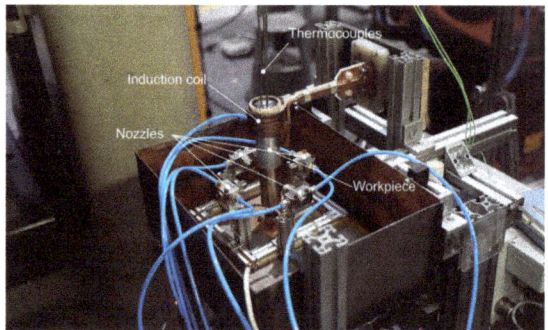

Fig. 3. Experimental setup of the heating unit with spray nozzles.

experimental design includes three parameter sets shown in Tab. 1, each repeated three times in a randomized order. While the pressure parameters are set, the mass flow of the fluids is measured. The air pressure influences the water mass flow antipropotional and is considered insofar as keeping the water pressure constant while varying the air pressure. The original handling process

Material Forming - ESAFORM 2023   Materials Research Forum LLC
Materials Research Proceedings 28 (2023) 1083-1090   https://doi.org/10.21741/9781644902479-119

specifies the pressure limits of each fluid because the utilized robot and surrounding components are not sealed against water and could be damaged. The chosen parameters enable a safe system operation for the tested stationary as well as for the prospective gripper integrated operation.

Table 1. Investigated parameter sets (pressure) and resulting mass flow depending on the parameter.

| | Pressure [bar] | | Mass flow | |
Set	Water	Air	Water [l/min]	Air [m³/h]
1	0.7	1.1	1.6	8.58
2	0.7	1.35	1.02	10.03
3	0.7	1.6	0.7	11.2

Results

In order to compare the cooling performance of the spray cooling with that of the currently utilized immersion bath cooling, the heating process described above is carried out, and the temperatures at the defined measuring points (MS 1 - MS 4) are recorded. The temperature data at the individual points are already available from earlier tests for the immersion bath and serve as a reference. A step-like temperature gradient must be established in the component for spray cooling to be confirmed as a suitable cooling method that should be visible in the temperature curves of the individual measuring points. Furthermore, the aluminium must not exceed its melting point during heating, which would damage the component. Ideally, spray cooling should achieve the same temperature distribution in the component as immersion cooling.

Fig. 4 shows the temperature curve of the workpiece during heating, considering the different parameter sets. All three graphs show the same course, which indicates a good and equal connection in the joining zone, increasing the measurements' significance. If there were strongly deviating joining zone properties of the materials in the individual samples, this would lead to different heat conduction behavior, which would be reflected in the temperature curves. Further, in the curves of MS 2 and MS 3 in all three plots, there is a brief rise in the temperature after finishing the heating of 30 s. This is caused by removing the workpiece's thermocouples towards the end of the heating period. The two thermocouples are inserted in a drilling that leads straight through the steel to the joining zone. Consequently, the sensors pass through the hot steel when they are removed.

One essential factor for a successful forming process is the step-like temperature gradient. For this purpose, Fig. 5 summarizes the temperatures achieved for the corresponding parameter sets. The maximum temperatures during the process were determined for every repetition of each parameter set and measurement point. The first parameter set achieved the best cooling performance in the tests, while the highest air pressure (parameter set 3) resulted in the worst cooling performance. The lowest amount of water is sprayed through the nozzles with high air pressure, which explains the inferior cooling performance of parameter set 3. However, it should be noted that in parameter set 1 at measuring point 4, a strongly deviating maximum temperature occurred. Thermocouple displacement occurred by the spray was observed depending on the orientation of the drilling to the nozzles. The workpiece is manually placed in the induction coil, which causes the rotational orientation of the cylindrical workpiece could vary. Due to the possible displacement of the thermocouple, it experiences a different electric field through the induction coil, which can lead to a significant deviation [15]. Since the parameter sets investigated were restricted by external factors, the temperature gradient may be a local optimum, which would have to be investigated further. Nevertheless, it can already be seen that there is a correlation between the amount of water, the air pressure and the cooling power.

Currently, the process utilizes immersion for the thermal preparation, whereby the component reaches a temperature gradient sufficient for successful forming. Here, the temperature gradient serves as a comparison to demonstrate the applicability of spray cooling. Immersion cooling achieves a lower cooling performance, which the Leidenfrost effect can explain. In the experiments, the aluminium reaches temperatures of approx 110 °C, causing the water to boil and form a vapor phase. The steel also heats up and exceeds the 900 °C limit, which is ideal for steel forming. In the case of spray cooling, the steel side reaches approx 800 °C. The influence on the formability requires further investigation.

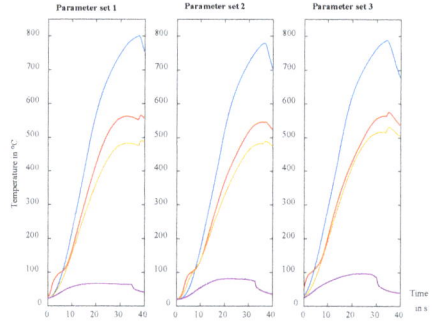

Fig. 4. Temperature in different measurement point (MS) during inductive heating and simultaneous spray cooling.

The cooling of the hybrid workpiece with spray nozzles compared to immersion bath cooling is more intensive. The parameters investigated also show that the cooling performance of the spray nozzles can be variably adjusted, which permits even more precise adjustment of the temperature gradient. Accordingly, it can be assumed that the spray nozzles are suitable for setting the temperature gradient.The results indicate that the necessary cooling performance can be achieved with the nozzles integrated into the handling equipment.

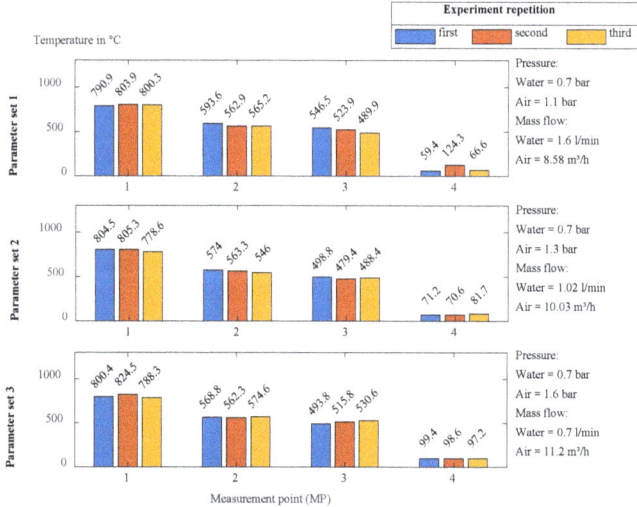

Fig. 5. Bar plot of the experimental results showing the maximal measured temperatures for each parameter set and repetition.

Material Forming - ESAFORM 2023   Materials Research Forum LLC
Materials Research Proceedings 28 (2023) 1083-1090   https://doi.org/10.21741/9781644902479-119

Summary

This paper aimed to validate spray nozzles for establishing a temperature gradient in a hybrid workpiece with a particular heating strategy compared to a currently used immersion bath. The nozzles have the advantage of being able to be integrated into the handling equipment and permit cooling during transport. With immersion cooling, no cooling can be performed during handling, which means that the temperature gradient due to equalization processes disappears.

For the validation, one manufacturing process for hybrid steel aluminium shafts was utilized. The workpiece attached with thermocouples is inductively heated at the steel side, while the aluminium side is cooled with the nozzles. The nozzles operate with compressed air and water, with the pressures of the two fluids as parameters. Three-parameter pairings were made, where the air pressure varies and the water pressure remains constant. The tests were subsequently performed with the corresponding settings. The results show that the mass flows of the two fluids primarily depend on the pressure of the compressed air. Furthermore, parameter combinations with a greater mass flow of water achieve a more effective cooling performance.

Finally, the performance of the nozzles with the best parameter set was compared against the currently used immersion bath cooling. As a result of the comparison, the semi-finished workpiece reached lower temperatures in both materials when cooled with nozzles. Thus, the nozzles can be used for this application, as the nozzle parameters allow a variable cooling performance, which can be further adjusted to approximate the immersion bath cooling.

Outlook

In the future, the parameter range of nozzle cooling must be investigated in more detail since only a small range could be investigated in this work. In these experiments, the forming should also be carried out so that the influence of the cooling capacity on the forming result can be tested simultaneously. Thereby, the effect of 100 °C lower temperature in the steel side on the forming can be investigated, since it is unclear to what extent this precisely influences the forming.

Another aspect to be examined is the nozzle's integration into the handling system and the robotized operation. The robot will be suited with the handling system to grasp the shaft, and the grasping position is essential because that surface cannot be cooled, whereby local heat spots can emerge and possibly damage the workpiece.

Acknowledgement

The results presented in this paper were obtained within the Collaborative Research Centre 1153 'Process chain to produce hybrid high-performance components by Tailored Forming' - 252662854 in subproject C07. The authors would like to thank the German Research Foundation (DFG) for the financial and organizational support of this project.

References

[1] P. Groche, S. Wohletz, M. Brenneis, C. Pabst, F. Resch, Joining by forming - a review on joint mechanisms, applications and future trends, J. Mater. Process. Technol. 214 (2014) 1972-1994. https://doi.org/10.1016/j.jmatprotec.2013.12.022

[2] J. Uhe, B.A. Behrens, Manufacturing of Hybrid Solid Components by Tailored Forming, in: J.P. Wulfsberg, W. Hintze, B.A. Behrens (Eds.), Production at the leading edge of technology, Springer Vieweg, Berlin, Heidelberg, 2019. https://doi.org/10.1007/978-3-662-60417-5_20

[3] B.-A. Behrens, K.-G. Kosch, Production of strong steel-aluminum composites by formation of intermetallic phases in compound forging, Steel Res. Int. 82 (2011) 1261-1265. https://doi.org/10.1002/srin.201100055

[4] B.-A. Behrens, H. Wester, S. Schäfer, C. Büdenbender, Modelling of an induction heating process and resulting material distribution of a hybrid semi-finished product after impact extrusion, ESAFORM 2021, Apr.2021. https://doi.org/10.25518/esaform21.574

[5] C.-V. Ince, A. Chugreeva, C. Böhm, F. Aldakheel, J. Uhe, P. Wriggers, B.-A. Behrens, A. Raatz, A design concept of active cooling for tailored forming workpieces during induction heating, Prod. Eng. 15 (2021) 177-186. https://doi.org/10.1007/s11740-021-01027-5

[6] C.-V. Ince, J. Geggier, C. Bruns, A. Raatz, Development of a form-flexible handling technology with active cooling for hybrid components in forging processes, Procedia CIRP 97 (2021) 27-32. https://doi.org/10.1016/j.procir.2020.05.200

[7] A. Buczek, T. Telejko, Investigation of heat transfer coefficient during quenching in various cooling agents, Int. J. Heat Fluid Flow 44 (2013) 358-364. https://doi.org/10.1016/j.ijheatfluidflow.2013.07.004

[8] M. Shirota, M.A.J. van Limbeek, C. Sun, A. Prosperetti, D. Lohse, Dynamic Leidenfrost Effect: Relevant Time and Length Scales, Phys. Rev. Lett. 116 (2016). https://doi.org/10.1103/PhysRevLett.116.064501

[9] P. Stephan, B1 fundamentals of heat transfer, in: VDI Heat Atlas, Springer Berlin Heidelberg, Berlin, Heidelberg, 2010, pp. 15–30. https://doi.org/10.1007/978-3-540-77877-6_115

[10] S. Schüttenberg, J. Lütjens, M. Hunkel, U. Fritsching, Adapted spray quenching for distortion control, Mat.-wiss. u. Werkstofftech. 43 (2012) 99-104. https://doi.org/10.1002/mawe.201100895

[11] I. Stewart, J.D. Massingham, J.J. Hagers, Heat transfer coefficient effects on spray cooling, Iron Steel Eng. 73 (1996) 17-23.

[12] S. Herbst, K.-F. Steinke, H.J. Maier, A. Milenin, F. Nürnberger, Determination of heat transfer coefficients for complex spray cooling arrangements, in: International Conference on Distortion Engineering, Bremen, 2015, pp. 247–256.

[13] A. Pola, M. Gelfi, G.M. La Vecchia, Simulation and validation of spray quenching applied to heavy forgings, J. Mater. Process. Technol. 213 (2013) 2247-2253. https://doi.org/10.1016/j.jmatprotec.2013.06.019

[14] B.-A. Behrens, J. Uhe, T. Petersen, F. Nürnberger, C. Kahra, I. Ross, R. Laeger, Contact Geometry Modification of Friction-Welded Semi-Finished Products to Improve the Bonding of Hybrid Components, Metals 11 (2021) 115. https://doi.org/10.3390/met11010115

[15] A. Smalcerz, R. Przylucki, Impact of Electromagnetic Field upon Temperature Measurement of Induction Heated Charges, Int. J. Thermophys. 34 (2013) 667-679. https://doi.org/10.1007/s10765-013-1423-1

Material Forming - ESAFORM 2023
Materials Research Proceedings 28 (2023) 1091-1100

Materials Research Forum LLC
https://doi.org/10.21741/9781644902479-120

FSW process mechanics and resulting properties for dissimilar Al-Ti T-joints

RANA Harikrishna [1,a*], BUFFA Gianluca [1,b], FRATINI Livan [1,c]

[1]Department of Engineering, University of Palermo, Viale Delle Scienze, 90128 Palermo, Italy

[a]hairkrishnasinh.rana@unipa.it, [b]gianluca.buffa@unipa.it, [c]livan.fratini@unipa.it

Keywords: FSW, T-Joint, Dissimilar, Skin, Stringer, Titanium, Aluminum, Material Flow

Abstract. Emergent manufacturing demands for superior performance but lightweight structures have pinpointed the development of multi-material and hybrid structures specifically in the aerospace and automotive industrial sectors. Friction stir welding (FSW), a solid-state joining technique has been proven very effective to produce joints between materials possessing extremely diverse thermal and mechanical properties. The present research aims to investigate the feasibility of Al-Ti skin-stringer joints with different plate geometries and placements. The effect of different approaches on material flow, grain morphology, intermittent phases, joint resistance, and microhardness are discussed in depth.

Introduction

Multi-material components are enticing recognition for their unique application base catering to automotive, aerospace, and transportation industries. Multi-material components not only present the unique benefits of demonstrating the properties of the individual materials but their advantages as a union too. The joining between titanium(Ti) and aluminum(Al) can be proposed for numerous applications in aeronautic and automotive sectors, wherein the higher strength-to-weight ratio is crucial in terms of body weight reduction and fuel savings. Conventional thermo-fusion joining techniques struggle with metallurgical challenges owing to a substantial disparity in physical properties, inadequate mutual solubility, and formation of brittle intermetallics (IMCs) [1,2].

The solid-state welding process was adopted into Al-Ti dissimilar joining and had been proven to be an efficacious technology by a few works of literature [3,4]. However, the formation of brittle IMCs like $TiAl_3$, $TiAl_2$, $TiAl$, Ti_3Al, Ti_2Al_5, and Ti_2Al_5 was difficult to control during the welding. Alternatively, FSW has been proven to be an efficient process to eradicate the formation of IMCs up to a certain extent as compared to other solid-state welding processes [3,5,6]. This technique has been efficient to produce extremely arduous dissimilar welding joints owing to its less heat input feature. Moreover, with an innovative approach to in-process cooling Patel et al. [3] have reported a considerable decrease in the quantity of IMCs, which increased joint efficiency. To date, numerous investigations featuring butt and lap configuration joints with different process parameters have been reported [7-9]. According to these studies, the IMC layer thickness can be downed up to 2µm with a proper set of FSW parameters. Moreover, from the literature mentioned, the ideal range for revolutionary pitch (RP) during FSW with butt configuration joints was 0.05-0.10 which delivered 80-110% joint efficiency [7,9,10]. Apart from common process parameters like tool rotation speed (TRS), tool traverse speed (TTS), tool tilt angle, etc. Li et al.[8] experimented with the tool offset and reported the turbulent material flow in the nugget zone (NZ) during FSW with an increase in tool offset and a decrease in RP. However, vortex material flow is desirable during FSW as it distributes the fine Ti particle in Al substrates and makes a uniform composite structure delivering enhanced properties[3]. A few investigations have reported the reduction in the IMCs by using a thin interlayer during the FSW of Al-Ti butt configuration joints. The interlayer alloy having the higher affinity to the substrate withholds the formation of brittle IMCs, ensuing enhanced properties of the prepared joints [11]. Yet, Al-Ti "T" configuration joints produced by FSW have not been reported in the literature (except one reported by own research group) as per author's best knowledge owing to the complexities involved.

The literature survey discussed to the point indicates a lack of literature available in the field of Al-Ti "T" configuration FSW joints despite having a huge potential and promising sectors to explore. The objective of the present investigation is to produce successful Al-Ti "T" configuration FSW joints experimenting with parameters including substrate plate position swap and geometrical features. The influence of these variables on the mechanical and microstructural features like joint strength, microhardness, and IMCs distribution is studied in detail. This investigation is an extension of the work carried out by authors elaborating on the influences of several FSW parameters like TRS, TTS, and material positioning as skin/stringer[12].

Materials and Procedure

Materials and Process.

Substrates of aluminum alloy 6156 (Si 1 wt.%, Mg 0.9 wt.%, Cu 0.9 wt.%, Fe 0.1 wt.%, Mn 0.55 wt.%, Cr 0.125 wt.%, Zn 0.4 wt.%, Al – bal) and commercially pure titanium grade 2 with plate dimensions $140 \times 90 \times 3$ mm and $140 \times 90 \times 2$ mm respectively were used for experiments. A fully automatic FSW machine (Make: ESAB, Model: LEGIO 3ST) was used for conducting the experiments. An experimental setup with fixture and tool is displayed in Fig.s 1(a) and (b). Each experiment was conducted with position control mode and using a push roller on the leading edge of the Tungsten-Rhenium conical tool to ensure uniform contact conditions and plunge. TRS and TTS were chosen from the best welding conditions derived from the authors' recent publication [12]. The experiments were repeated thrice with each processing condition to ensure the credibility of the results. To understand the effect of slots in the skin, specimens were prepared with three different slot depths of 0 mm, 0.5mm and 1 mm longitudinally. Table 1 summarises each parametric condition with its unique experiment ID.

Table 1. Parametric sets with relative specimen IDs.

Specimen ID	A	B	C	D	E	F	G	H	I
Skin Material	Titanium Grade 2			Al 6156					
Skin Slot Depth [mm]	0	0.5	1	0	0.5	1	0	0.5	1
Tool Feed, F [mm min⁻¹]	40			50			70		
Tool Rotation Speed [rev. min⁻¹]	400			800			1000		
Tool Tilt Angle	2.5°								

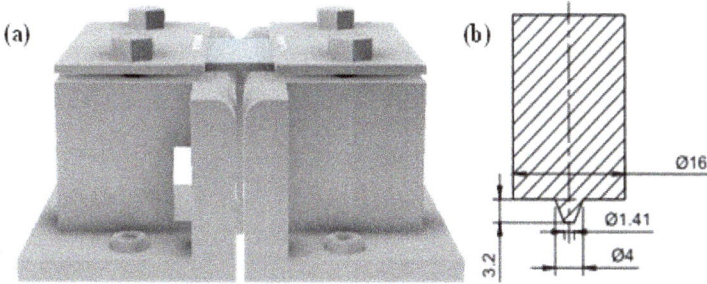

Fig. 1. Experimental setup (a) T-joint custom-designed fixture (b) WC-Re FSW tool geometry.

Microstructural Characterization

The specimens for microstructural examinations were obtained by sectioning at the center of the weld. They were further mounted, ground, and polished finally with 0.5μm diamond paste. Each joint specimen was etched with Keller's reagent (HF 2% vol. + HCL 3% vol. + HNO$_3$ 5% vol. + H$_2$O 90% vol.) for the AA 6156 section and Kroll's reagent (HF 3% vol. + HNO$_3$ 5% vol. + H$_2$O 90% vol.) for Ti Grade 2 section. Further, microstructural characterization was carried out through optical microscopy (OM) (OLYMPUS, Model-Inverted Metallurgical Microscope GX51), scanning electron microscopy (SEM), and energy dispersive spectroscopy (EDS) (Make: Zeiss, Model: Ultra-55 SEM).

Hardness and Tensile Testing

Each joint specimen was examined for microhardness using Eseway 4302 Vickers hardness tester (according to ASTM E-384). A square-based pyramid diamond indenter (136° intersects) was employed for tests with load and dwell time of 5 kg and 15 seconds, respectively. A specially designed fixture as displayed in Fig. 2 was used for conducting T- pull tests on a conventional tensile testing machine with a velocity of 2 mm/minute. The fractured specimens were further examined by SEM fractographic analysis.

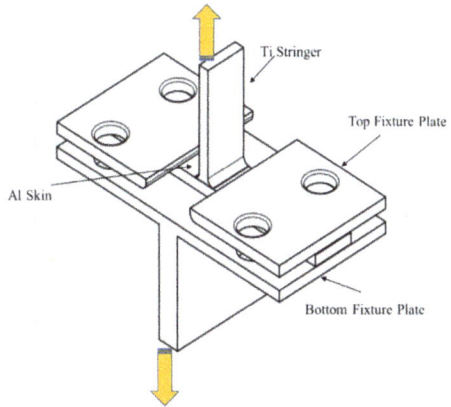

Fig. 2. Schematic of T-pull test fixture.

Material Forming - ESAFORM 2023

Materials Research Proceedings 28 (2023) 1091-1100

Materials Research Forum LLC

https://doi.org/10.21741/9781644902479-120

Results & Discussion

Macro- and Microstructural Analysis

Macrostructural insights.

Macrostructural images of T-joint specimens prepared with Al and Ti as skin and stringer alternatively are displayed in Fig. 4 and Fig. 5. The macrostructure was characterized by very distinctive featured flow arms of different materials and dimensions. These arms are generated by distinctive material flow patterns recorded with varied slot sizes and welding conditions. The size of the flow arm greatly influences the resulting properties owing to the skin/stringer material interlocking. During FSW of T-joints, as soon as the rotating tool pin penetrates the stringer material, it pushes the material underneath the tool to get extruded in the opposite direction, being constrained by the surrounding skin material. Simultaneously, the skin material flows downwards to the sides of the stringer material. As a result, engendered flow arms can be noticed in the vicinity of the thermo-mechanically affected zone (TMAZ) area as schematically represented in Figs 6 (a) and (b). Chen et al [13] have reported and corroborated the matching phenomenon of "extruded hook" during the FSW of Al-Ti lap joint configurations. The specific thermal contribution (STC) ensuing from distinct RP majorly contributes to the size of this flow arm, too.

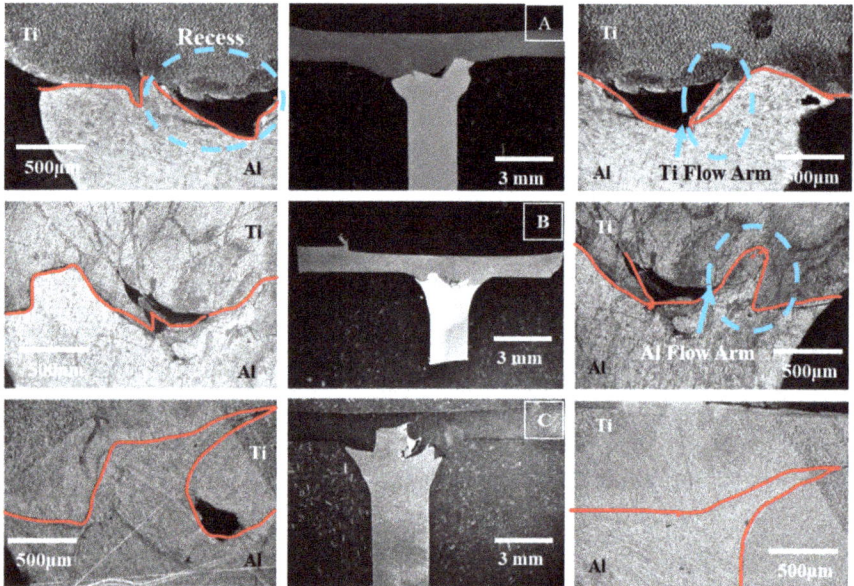

Fig. 4. Macrostructures of Ti skin Al stringer T-joints prepared with the skin-slot depth of (A) 0 mm (B) 0.5 mm (C) 1 mm.

As indicated earlier in the introduction section, FSW parameters were finalized using the author's previous publication. Ti skin/Al stringer samples were prepared with TRS and TTS of 400 rev. min⁻¹ and 40 mm min⁻¹. However, the microstructure of this sample was characterized by a recess defect in the NZ. Such defects incurred owing to a huge property mismatch between skin and stringer materials, making it extremely challenging to get a synchronized material in the NZ. The flow stress at room temperature for Al is ¼ of the Ti, making Al soften in the advanced stage

Material Forming - ESAFORM 2023   Materials Research Forum LLC
Materials Research Proceedings 28 (2023) 1091-1100   https://doi.org/10.21741/9781644902479-120

of the FSW stirring, while Ti resists the deformation. Although such mismatch can be minimized at elevated temperatures with a low RP value parametric set, it is not possible to eradicate. Hence, to minimize the flow resistance of the Ti skin material a novel strategy proposing a longitudinal slot in the skin was experimented as schematically exhibited in Fig. 6(c). Therefrom, the samples with 2 longitudinal slots with depths of 0.5 mm and 1 mm were milled prior to the FSW. The slot width was kept equal to the stringer plate thickness. It was believed that the slot shall not only reduce the volume of the Ti material to be deformed and stirred but increase the flow arm size as indicated schematically in Fig. 6(c). The micrographs of the Ti skin samples welded with the 0.5 mm and 1 mm slot depth are displayed in Fig. 4(B) and (C) respectively. As the proposed strategy, the tool pin not only minimized the recess defect size in NZ up to a great extent but allowed the Al flow arm to penetrate the Ti skin material. Still, it was not possible to omit the recess/pocket in the NZ completely. Moreover, the small Ti flow arms engendered by tool-driven extrusion were also identified in these samples indicating the greater softening of the Ti skin. The flow arm interlocking was also visible in these samples.

Nevertheless, the material flow that happened in the Al skin samples was quite different than in Ti skin and it was possible to extract a few defect-free samples at different processing conditions. As far as the Al skin samples are concerned, these samples were characterized by the longer Ti flow arm penetrating the Al skin as displayed in Fig. 5. As discussed in the preceding work, with a decrease and increase in the RP and STC values respectively the flow arm length was reportedly increased delivering the best joint strength. Therefrom, to further increase the flow arm size, a slot design was proposed. As the proposed strategy, the Ti flow arm size was drastically increased with the greater tool-driven extrusion. However, at the same time the larger extent of the extrusion created a larger void to fill in between two Ti flow arms ensuing into the recess/pocket in each sample prepared with slot (see Fig. 5 (E), (F), (H) and (I)). As mentioned earlier, the profile of this stringer arm is quite important for an interlocking mechanism during the T-pull test. It was observed that, while on one hand its vertical alignment (perpendicular to the skin), creates the least disturbance to material momentum while stirring, on the other hand, the horizontal alignment barricades the material flow toward the center and results in recess/pocket defects[9]. It is worth noticing that the flow arm length grows in the vertical direction with slot depth, but once it crosses a threshold value it starts bending towards the center of the weld as schematically represented in Fig. 7. This can be explained as when the slot depth is increased the shoulder rotates closer to the extruded stringer arm, engendering tremendous pressure on the stirring material, which bends the flow arm in the plane parallel to the shoulder face as visible in specimens F and I. This Ti flow arm boundary barricades the skin material to flow toward the center of NZ leaving back unfilled voids. Hence, as far as the slot depth is concerned the joints prepared with 0.5 mm slot exhibited longer and vertically aligned flow arms compared to others.

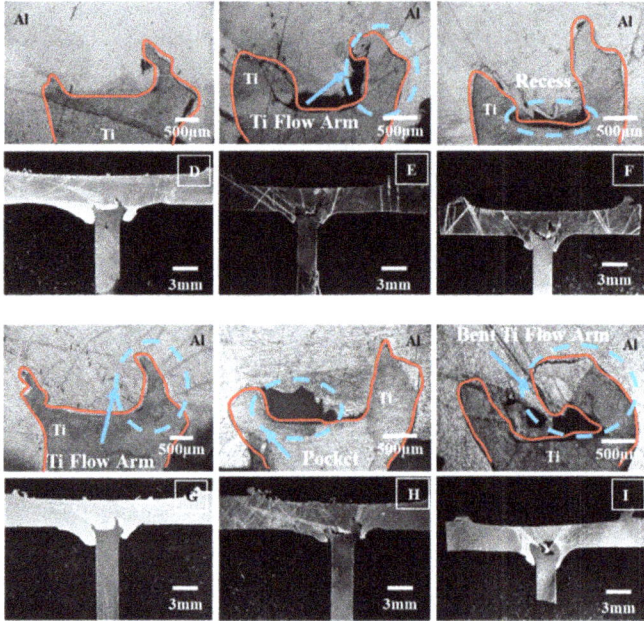

Fig. 5. Macrostructures of Al skin Ti stringer T-joints prepared with slot depth, TRS, and TTS of (D) 0 mm, 800 rev. min^{-1},50 mm min^{-1} (E) 0.5 mm, 800 rev. min^{-1},50 mm min^{-1} (F) 1 mm, 800 rev. min^{-1},50 mm min^{-1} (D) 0 mm, 1000 rev. min^{-1}, 70 mm min^{-1} (E) 0.5 mm, 1000 rev. min^{-1}, 70 mm min^{-1} (F) 1 mm, 1000 rev. min^{-1}, 70 mm min^{-1}.

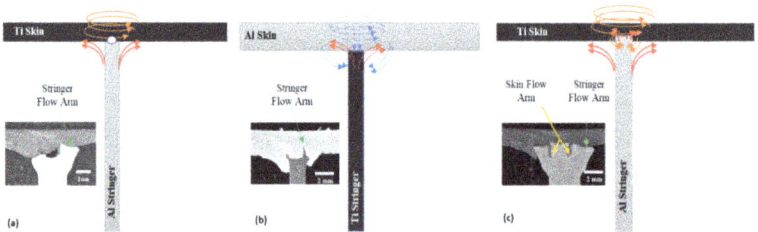

Fig. 6. Schematic exhibits for proposed material flow phenomenon in (a) Ti skin Al stringer (b) Al skin Ti stringer (c) proposed Ti skin with slot Al stringer.

Apart, defects like skin thinning and kissing bonds were also observed in the few samples (viz., B, C, H, I). Kissing bonds are reportedly formed either when the substrate materials are extremely asymmetric or they are inadequately heated and stirred. This eventually promotes the formation of residual oxide layers (e.g. TiO$_2$ in the present investigation) detached from the other substrate [14]. These layers wane the material attachment and raise the localized stresses during weld solidification. Especially for T-joint configuration, as the tool pin rotation plane is parallel to the joint interface, it becomes difficult to slash these oxide layers [12].

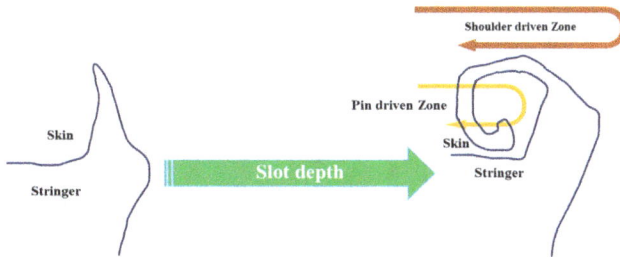

Fig. 7. Schematic representation of change in stringer flow arm profile concerning slot depth.

Microstructural insights.

The SEM and EDS mapping were carried out for the samples with no skin slot and maximum slot depth viz., G and F respectively. The mapping indicated the composite structure near the extruded Ti flow arm for both samples as displayed in Fig. 8. During FSW, when the extruded Ti stringer flow arm reaches the threshold length value, it starts interacting with the rotating shoulder. Further, the Ti flow arm experiences grinding led by the shoulder, leaving fine Ti particles in the vicinity of the flow arm. Rostami et al.[7] have reported the identical occurrence of the composite structure close to the Ti hook profile generated during FSW of Al-Ti lap joints.

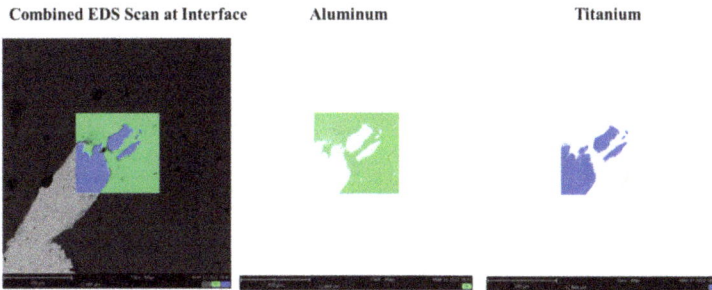

Fig. 8. SEM-EDS mapping of the composite structure near the Ti Flow Arm.

Moreover, the obtained results during the SEM-EDS analysis indicated a very thin and intermittent layer of $TiAl_3$ intermetallic close to the Al-Ti interface as exhibited in Fig. 9. It is noteworthy that the IMC layer thickness ranged from ~ 1 μm to 5 μm, which may be believed to be very lean considering the observed thickness for FSW butt and lap configuration joints [10, 13]. However, a thicker IMC layer was identified in the vicinity of the extruded Ti flow arm. This variable layer thickness at different locations may be understood by the successive events taking place while FSW. On the one side, when the rotating plunges adjacent to the Al-Ti interface, owing to softening of the Al skin led by high frictional heat, the Ti face is pressed downward and extruded as a flow arm leaving an unfilled void. This void remains empty until the tool progress along the joint line and stirring Al enters at the back of the tool.

Material Forming - ESAFORM 2023
Materials Research Proceedings 28 (2023) 1091-1100

Materials Research Forum LLC
https://doi.org/10.21741/9781644902479-120

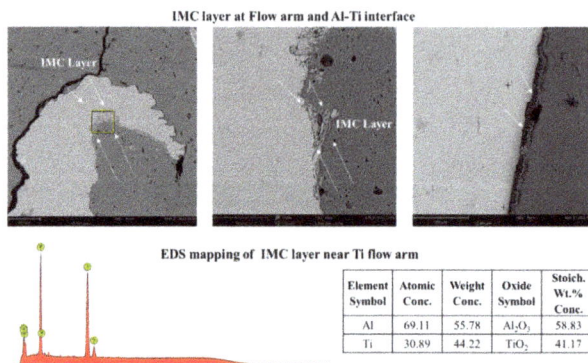

Fig. 9. SEM/EDS mapping of Intermetallic Layer at Al-Ti interface and near the extruded Ti flow arm.

However, by the time Al fills the void completely, the temperatures of both Al and Ti are slightly dropped to a level at which the occurrence of Al-Ti diffusion is very minimal. While of the other side, when the extruded Ti flow arm penetrates Al, the temperatures of both substrates are elevated, leading to the Al-Ti diffusion and formation of a thick layer of IMC. The prevalent IMC phase identified in the NZ was $TiAl_3$. The mechanical properties of the prepared FSW joints are greatly influenced by the location, size, and amount of such IMCs.

Mechanical Properties

The joint strengths observed with T-pull test of all the samples are graphically displayed in Fig. 10. Concerning Ti skin samples, they were characterized by a moderate joint strength, which can be attributed to defects like porosity, recess, etc., and lack of appropriate flow arm interlocking phenomenon. Although the slots allowed the Al flow arm to enter into the Ti skin, there was a lack of proper interlocking. Still, a slight improvement in the joint strength can be observed for the samples prepared with the skin slots. While concerning the Al skin samples, the highest joint strengths of 146 MPa and 144 MPa were observed for sample G and sample D, respectively which were prepared without any skin slot. Although T-joints created with skin slots demonstrated longer extruded flow arms, they were characterized by recess and unfilled cavity defects as described earlier in section 3.1. Alternatively, quite comparable strength of 107 MPa was recorded with the sample I which was prepared with 1 mm slot depth. While comparing the performance of the slot depth in terms of joint strength, the deeper skin slot of 1 mm exhibited considerably superior results which can be attributed to the longer flow arm accompanied by the superior flow arm interlocking with skin in the NZ.

As far as the hardness results are concerned, the substrate materials Ti grade 2 and AA 6156 exhibited a hardness of 115 HV and 40 HV respectively. The hardness values recorded in the NZ of sample G were almost 50 % higher than the unprocessed substrate. The increased hardness values can be attributed to the Al-Ti composite structure, grain refinement, and IMCs present in the NZ of the welded samples [3,15,16]. Such rise in the hardness values can be backed by several strengthening mechanisms viz., shear lag mechanism and dislocation strengthening mechanism. According to the shear lag mechanism, as NZ is characterized by Ti particulate composite structure, during testing when the specimen is loaded the load is transferred from the Al matrix to very hard Ti particulates. The shear stress is generated at the Al-Ti interface due to this load shift, which eventually hinders the dislocation motion and ensures higher properties. As per dislocation strengthening, owing to a large mismatch of coefficient of thermal expansion between two substrates, geometrically indispensable dislocations are engendered at the Al/Ti interface and in the vicinity to the Ti particulates and IMCs, during the cooling period post-thermos-plastic deformation led by FSW. These dislocations impede crack growth and thereby increase joint strength and hardness. In the nutshell, the T-joints prepared with FSW not only exhibited the metallurgical bonding but the mechanically interlocked bonding too which promises an effective approach for the production of dissimilar material T-joints.

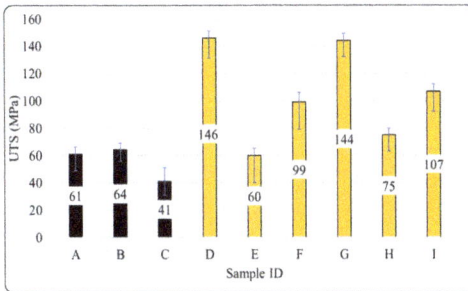

Fig. 10. Joint strength for Ti and Al skin samples.

Summary

This novel investigation describes the successful fabrication of dissimilar T-joints using Ti Gr 2 and AA6156 substrates, wherein the effects of mutual sheet position and innovative skin slot design on macrostructural, microstructural, and mechanical properties of the joints have been discussed in depth. The obtained conclusions are as under:

• AA 6156 substrate skin samples displayed substantially better properties than Ti substrate skin samples.

• The dissimilar material T-joint was characterized by interesting material flow patterns. Particularly, the development and geometry of the extruded flow arm were found to be crucial for mechanical interlocking resulting in improved joint strength. The length of this flow arm is directly proportional to the amount of heat input during the process and skin-slot depth. However, too large values of any of them can bend the flow arm which hinders the appropriate material flow leaving behind defects like cavities, recess, etc, and eventually reducing the mechanical properties.

• The samples produced without skin-slot exhibited superior joint strength as compared to skin-slot samples, owing to the formation of the defects like cavities and recesses in skin-slot samples. However, samples joined with a maximum slot depth of 1 mm exhibited superior joint strength as compared to 0.5 mm slot depth, owing to better mechanical interlocking of the flow arm with Al skin substrate.

• The hardness values of prepared T-joints were improved by 50% and 20% as compared to the unprocessed Al skin substrate and Ti stringer. Such an increase can be attributed to the Ti-particulate composite structure, grain refinement, and IMCs-led strengthening mechanisms.

References

[1] W. Vaidya, M. Horstmann, V. Ventzke, B. Petrovski, M. Koçak, R. Kocik, G. Tempus, Improving interfacial properties of a laser beam welded dissimilar joint of aluminium AA6056 and titanium Ti6Al4V for aeronautical applications, J. Mater. Sci. 45 (2010) 6242-6254. https://doi.org/10.1007/s10853-010-4719-6

Material Forming - ESAFORM 2023   Materials Research Forum LLC
Materials Research Proceedings 28 (2023) 1091-1100   https://doi.org/10.21741/9781644902479-120

[2] W. Shouzheng, L. Yajiang, W. Juan, L. Kun, Research on cracking initiation and propagation near Ti/Al interface during TIG welding of titanium to aluminum, Kovove Mater. 52 (2014) 85-91. http://doi.org/10.4149/km2014285

[3] P. Patel, H. Rana, V. Badheka, V. Patel, W. Li, Effect of active heating and cooling on microstructure and mechanical properties of friction stir–welded dissimilar aluminium alloy and titanium butt joints, Weld World 64 (2020) 365-378. https://doi.org/10.1007/s40194-019-00838-6

[4] Y.-C. Kim, A. Fuji, Factors dominating joint characteristics in Ti–Al friction welds, Sci. Technol. Weld. Join. 7 (2002) 149-154.

[5] L. Fratini, G. Buffa, L.L. Monaco, Improved FE model for simulation of friction stir welding of different materials, Sci. Technol. Weld. Join. 15(3) (2010) 199-207.

[6] A. Simar, M.-N. Avettand-Fènoël, State of the art about dissimilar metal friction stir welding, Sci. Technol. Weld. Join. 22(5) (2017) 389-403. https://doi.org/10.1080/13621718.2016.1251712

[7] H. Rostami, S. Nourouzi, H.J. Aval, Analysis of welding parameters effects on microstructural and mechanical properties of Ti6Al4V and AA5052 dissimilar joint, J. Mech. Sci 32 (2018) 3371-3377. https://doi.org/10.1007/s12206-018-0640-8

[8] B. Li, Z. Zhang, Y. Shen, W. Hu, L. Luo, Dissimilar friction stir welding of Ti–6Al–4V alloy and aluminum alloy employing a modified butt joint configuration: Influences of process variables on the weld interfaces and tensile properties, Mater. Des 53 (2014) 838-848. http://doi.org/10.1016/j.matdes.2013.07.019

[9] A. Kar, S. Suwas, S.V. Kailas, Multi-length scale characterization of microstructure evolution and its consequence on mechanical properties in dissimilar friction stir welding of titanium to aluminum, Metall. Mater. Trans. A 50 (2019) 5153-5173.

[10] Z. Ma, X. Sun, S. Ji, Y. Wang, Y. Yue, Influences of ultrasonic on friction stir welding of Al/Ti dissimilar alloys under different welding conditions, Int. J. Adv. Manuf. Technol. 112 (2021) 2573-2582. https://doi.org/10.1007/s00170-020-06481-6

[11] A. Kar, S.K. Choudhury, S. Suwas, S.V. Kailas, Effect of niobium interlayer in dissimilar friction stir welding of aluminum to titanium, Mater. Charact. 145 (2018) 402-412. http://doi.org/10.1016/j.matchar.2018.09.007

[12] H. Rana, D. Campanella, G. Buffa, L. Fratini, Dissimilar titanium-aluminum skin-stringer joints by FSW: process mechanics and performance, Mater. Manuf. Process. 38 (2023) 471-484. https://doi.org/10.1080/10426914.2022.2116044

[13] Y. Chen, C. Liu, G. Liu, Study on the joining of titanium and aluminum dissimilar alloys by friction stir welding, Open Mater. Sci. 5 (2011). http://doi.org/10.2174/1874088X01105010256

[14] Y. Su, W. Li, F. Gao, A. Vairis, Effect of FSW process on anisotropic of titanium alloy T-joint, Mater. Manuf. Process. 37 (2022) 25-33. https://doi.org/10.1080/10426914.2021.1942911

[15] H. Rana, V. Badheka, A. Kumar, A. Satyaprasad, Strategical parametric investigation on manufacturing of Al–Mg–Zn–Cu alloy surface composites using FSP, Mater. Manuf. Process. 33 (2018) 534-545. https://doi.org/10.1080/10426914.2017.1364752

[16] E. Lertora, C. Mandolfino, C. Gambaro, Effect of welding parameters on AA8090 Al-Li alloy FSW T-joints, Key Eng. Mater. 554c(2013) 985-995. http://doi.org/10.4028/www.scientific.net/KEM.554-557.985

Material Forming - ESAFORM 2023
Materials Research Proceedings 28 (2023) 1101-1110

Materials Research Forum LLC
https://doi.org/10.21741/9781644902479-121

Influence of process parameters on mechanical properties of lamination stacks produced by interlocking

MARTIN Daniel Michael[1,a]*, KUBIK Christian[1,b], KRÜGER Martin[1,c] and GROCHE Peter[1,d]

[1]Technical University of Darmstadt, Institute for Production Engineering and Forming Machines, Otto-Berndt-Straße 2, 64287 Darmstadt, Germany

[a]daniel.martin@ptu.tu-darmstadt.de, [b]kubik@ptu.tu-darmstadt.de, [c]martin@apo-wetzlar.de, [d]peter.groche@ptu.tu-darmstadt.de

Keywords: Interlocking, Lamination Stacks, Electrical Steel

Abstract. Interlocking is mainly used in the manufacture of lamination stacks for the cores for electrical energy converters. The process involves the embossing of nubs, which are subsequently stacked and joined by an interference fit. For an optimal design of interlocked joints, sufficient knowledge of the relationships between various influencing parameters in the manufacture and the resulting stack properties is essential. In addition to the design parameters nub geometry, size and embossing depth, the process parameters embossing clearance, counterpunch and blankholder force influence the joint strength. In addition to these fixed parameters, continuously varying uncertainty such as wear, which can lead to a rounding of the tool edges, affects relevant stack properties like the joint strength. The mechanical loads in the area of the tool edge which are responsible for this rounding mainly act immediately before material separation and have to be considered in the embossing process. However, the influence of punch edge radii on the joint properties during the interlocking process has not yet been investigated. In order to describe the interdependencies between different influencing parameters on the achievable mechanical strength, experimental investigations are carried out. Therefore, cylindrical nubs are interlocked with varying embossing depth, clearance and edge radii. The joint strength is determined via tensile tests. Key findings are that the joint strength of nubs with increasing abrasive wear on the cutting edge of the embossing tool are compensated by a higher embossing depth. The minimum embossing depth required for stacking and the embossing depth at which the optimum strength is achieved depend on the embossing edge radius and thus on the current abrasive wear state.

Introduction

Interlocking is a process for manufacturing iron cores in the stators and rotors of electric motors. These are assembled from thin sheets that are electrically insulated from each other in order to minimise iron losses. To build up a lamination stack, interlocking nubs are introduced into the sheets by an embossing operation and then pressed into each other, creating a force-fit joint. For electrical insulation, the sheets are coated in advance. Various coating materials are used for this purpose, which are classified according to their chemical composition, insulating capacity and the area of application [1]. Common coating types are organic coatings of type C3 and the completely or predominantly inorganic C5 coatings. C3 coatings can contribute to an improvement in punchability compared to uncoated materials. The same applies for C5 coatings depending on the fillers and additives. Therefore, both are in principle suitable for use in punch-stacked components [2]. Since the working principle of interlocking is based on friction and is therefore dependent on the tribological conditions, the choice of the coating material potentially influences the joint quality.

Fig. 1(a) shows the schematic sequence of the punch-stacking process chain consisting of punching, embossing and stacking. For use in electric drives, both the magnetic and the mechanical

Material Forming - ESAFORM 2023
Materials Research Proceedings 28 (2023) 1101-1110

Materials Research Forum LLC
https://doi.org/10.21741/9781644902479-121

properties (joint strength) have to be taken into account when designing interlocked lamination stacks. In order to achieve a high torque density and low power losses, the magnetic properties must be optimised and additional losses caused by the manufacturing process minimised [3].

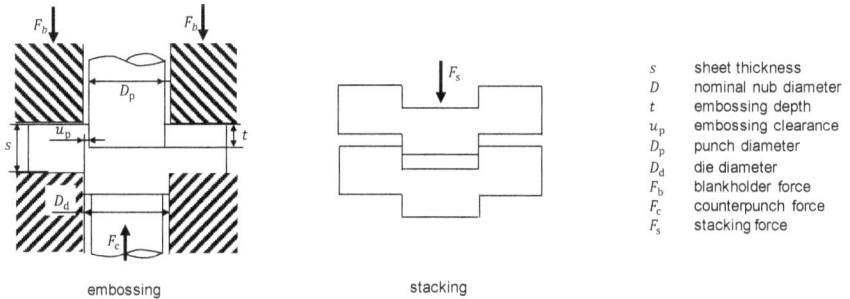

	s	sheet thickness
	D	nominal nub diameter
	t	embossing depth
	u_p	embossing clearance
	D_p	punch diameter
	D_d	die diameter
	F_b	blankholder force
	F_c	counterpunch force
	F_s	stacking force

Fig. 1. Schematic illustration of the interlocking process chain.

From a mechanical point of view, the joint strength between the sheet metal lamellae plays a decisive role. To ensure a reliable connection of the lamination stacks in the subsequent assembly steps, a minimum strength is required for robust process handling. The resulting strength of the punch-stacked joint depends on a complex interaction of a multitude of parameters. The geometry, position and number of nubs as well as the material properties (sheet thickness, alloy combination, coating material) are determined during product design of the electric drive. Parameters such as the embossing clearance, the blankholder and counterpunch force as well as the stacking force are determined by the tool design and can be varied within a specific range. The tool parameter embossing clearance and the process parameter embossing depth have been shown to have the greatest influence on the joint strength in previous investigations [4,5]. As an uncontrollable parameter, wear causes alterations in the tool configuration and thus directly affects the mechanical properties of the embossed nubs. Since the embossing process is a punching process that is stopped just before cracks are initialized into the material, wear on the edge of the embossing tool plays an important role.

From the literature it is known that the so-called punch phase in which the blanking tool separates the products from the sheet metal strip causes abrasive wear on the cutting edge of the tool [6]. This abrasive wear is influenced by the acting contact normal stresses, which correlate with the maximal punch force F_{max}. The maximal punch force is defined as the cutting force $F_{max} = l \cdot s \cdot k_s$, with the cutting line l, the sheet thickness s and the shear strength shear strength k_s, according to Lange [7]. The shear strength depends on a various number of variables such as tool parameters (clearance, tool wear, surface conditions of the active elements), material parameters (tensile strength, elongation at break, alloy composition) and other process parameters (lubrication, temperature, cutting speed) [8].

Due to high accuracy requirements in the cutting step and the influence of the embossing gap on the joint strength, very small clearances between punch and die are selected for both the cutting step and the embossing step in interlocking, which leads to high contact normal stresses and promotes the occurrence of abrasive wear. As mentioned above the embossing step corresponds to a shear cutting process in which the slug is not sheared off completely until it breaks, but remains connected to the base sheet due to a sufficiently small immersion depth of the punch. Because the maximum force occurs shortly before the material fails by tearing off the slug, it can be assumed that only abrasive wear in the form of rounding of the punch edge has to be considered. Further wear phenomena such as adhesions, scratches and chipping are found on the lateral surface and can be neglected during the embossing process. Literature in area of shear cutting demonstrates that abrasive wear directly affects the properties of the manufactured part.

Former studies prove that the occurring tool wear plays a significant role in the shear cutting of electrical sheet and can also have an effect on the magnetic properties [9,10] in addition to the influence on the mechanical properties and the cutting contour [10,11]. Although this correlation for punching tools is present in recent studies, abrasive wear on embossing tools and its effect on the stack properties have not yet been investigated.

An established method of monitoring the wear condition of punching tools is the measuring of force signals during the process [12]. Previous studies have shown that it is possible to estimate the current punch edge radius [13] and the resulting workpiece characteristics [14] based on time series, for example using machine learning algorithms. This methodology can potentially be used to monitor the current wear condition during punch stacking and to adapt to it by appropriate countermeasures. The hypothesis put forward in this paper is that by monitoring the wear condition in-situ, the optimum range in the process window in terms of joint strength can be determined in real time and abrasive wear can be counteracted by adjusting the process parameters. In order to lay the foundation for this, the correlations between the parameters embossing depth, embossing clearance and punch edge radius and the resulting joint strength are to be investigated experimentally in this paper.

Approach

In order to investigate the influence of the main process parameters and the tool wear state on the joint strength, experimental investigations are carried out. For the experiments electrical steel sheets of grade M270-50A according to DIN EN 10106 [15] are used with two different coating materials (C3 and C5). The sheet thickness is 0.5 mm and the specimen have a size of 35x35 mm containing one cylindrical nub per layer. The nubs have a circular cross section with a diameter of 6 mm. The single sheets where stacked using a punch with the same diameter. For the investigation the parameters blankholder force and counterpunch force are kept constant. The embossing depth is varied in the range 60 % to 125 % in relation to the sheet thickness. The embossing gap is 2 % or 1 % of the sheet thickness. Table 1 gives an overview over the experimental parameters.

Table 1. Dimensions and parameters used in the experimental investigation.

material and design parameters		constant process and tool parameters		varied properties and parameters	
sheet thickness s	0.5 mm	stroke height	35 mm	coating type	C3, C5
specimen size	35 x 35 mm	stroke speed	15 strokes/min	embossing depth t	60 % - 125 %
material	M270-50A	blankholder force F_b	3,000 N	embossing clearance u_p	2 %, 1 %
punch diameter D_p	6 mm	counterpunch force F_c	800 N	punch edge radius	sharp (R00), 0.1 mm (R01), 0.2 mm (R02), 0.3 mm (R03)
		stacking force F_s	5,654 N		

Fig. 2(a) shows the test tool used for the embossing step and the test set-up for stacking the nubs. The tool used for embossing is a test tool for simple shear cutting operations with a closed cutting contour. A round punch with a diameter of 6mm is used. To investigate the influence of tool wear in the form of abrasion, punches with an artificially rounded embossing punch edge (indicated in Fig. 2(a)) with different radii (see Table 1) are used for embossing in addition to a punch with a sharp edge. The tool is equipped with a piezoelectrical force washer (Kistler 9015A) to acquire the embossing force and an eddy current sensor (µEpsilon EU8) to acquire the motion sequence of the tool.

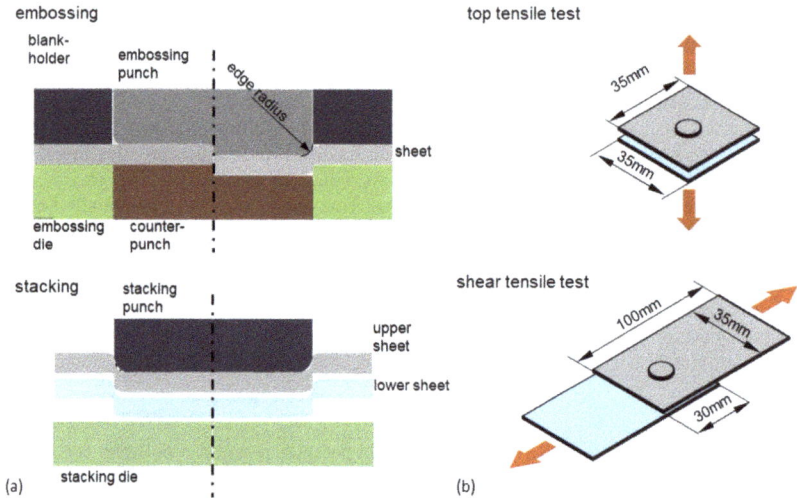

Fig. 2. Experimental setup for embossing, stacking and tensile testing.

The samples are stacked on a tensile-pressure testing machine, whereby the lower sheet is placed on the convex side of the nub on a flat pressure plate and a punch with the same geometry as the embossing punch presses the nub of the upper sheet into the lower nub. The punch movement is force-controlled at 500 N/s until a defined upper force limit is reached, for which a value of 5,654 N was empirically determined, which corresponds to a compressive stress of 200 MPa in relation to the nub cross-section. In order to prevent the lamination stack from sticking to the punch due to radial clamping forces caused by elastic deformation, the stacking punch has an edge radius that is 0.1 mm larger than that of the embossing punch.

Fig. 3 shows the curves of the punch force over time for both materials as a function for different embossing depths using a sharp punch (R00) and for a slightly rounded punch with an edge radius of 0.1 mm (R01). The graphs are synchronised so that the times of impact of the blankholder coincide. Since the embossing depth is varied by adjusting the bottom dead centre of the press, the punch speed changes with the embossing depth, therefore the time at which the punch hits the sheet also changes, as can be seen from the rapid increase at between 0.5 s and 0.6 s. With the rounded edge, the punch consistently hits the sheet a little later, as the length of the punch is slightly shorter with the artificially worn rounded-edge punches, but the hold-down is set the same. In the embossing phase, different characteristic progressions can be seen depending on the embossing depth. At high embossing depths, the typical force curve of shear cutting results with an abrupt drop in force after reaching the critical breaking stress. Using a sharp punch, this can be observed at embossing depths of 100 %. In this case, fracture initiation already takes place at embossing depths in the range of 80 % to 85 %. Below this, the curve of the punch force is decisively dominated by the punch travel and the corresponding counterpunch and blankholder forces.

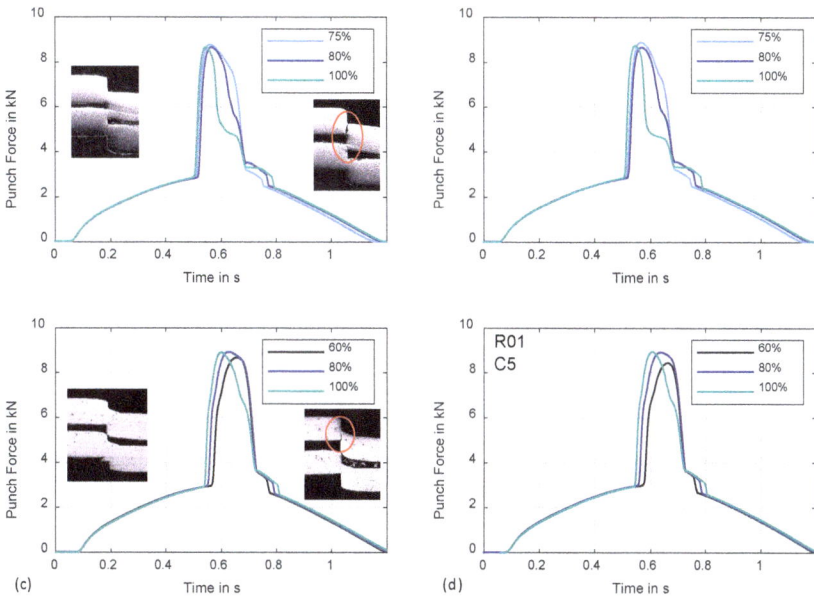

Fig. 3. Punch force curves for different embossing depths (in relation to the sheet thickness) with sharp (R00) embossing tool, (a) and (b), and rounded tool (R01) with edge radius 0.1 mm, (c) and (d); (a) and (c) show the curves for the material with C3 coating, (b) and (d) the material with C5 coating.

When using a worn punch, the material breaks at higher embossing depths. In this state, tensile and compressive stresses are superimposed in the forming zone, local stress peaks are reduced, and the material allows greater plastic deformations. As a result, the embossed nubs do not tear off at embossing depths of 90 % and below.

Fig. 4 demonstrates how the punch force changes with different edge radii at the same embossing depth (100 %). It can be seen that the maximum force increases with increasing edge radius. This is consistent with investigations of force signals during shear cutting [16]. Since the edge radius increases, the effective clearance decreases depending on the depth the tool penetrates the die. As a result of a smaller clearance, the maximum cutting force is reduced. This effect is counteracted by an increasing radius, due to the greater load which must be applied to the tool system to induce cracks in the material. According to Kubik et al., the effect of superimposed tensile and compressive stresses dominates, resulting in an increased maximal force [16].

Furthermore, it can be observed that in the case of a slight fillet, the shape of the curve changes significantly. There is a sharp drop in the force and a kink in the curve immediately before reaching the bottom dead centre (which is passed through at approximately 0.63 s), which indicates that the tensile strength of the material has been reached locally and damage and possibly initial cracking occurs in the area of the tool edges, so that a further increase in the embossing depth can be expected to cause the punching nub to tear off. It is noticeable that the curves for R01 and R02 are very close together. Only a further increase of the edge radius to 0.3mm causes the critical

Material Forming - ESAFORM 2023
Materials Research Proceedings 28 (2023) 1101-1110

Materials Research Forum LLC
https://doi.org/10.21741/9781644902479-121

embossing depth to shift upwards until breakage occurs to such an extent that cracking does not yet occur at 100 % embossing depth. Due to these characteristic changes in the force curve, it is possible to predict the state of wear, expressed by the edge radius for the case at hand, using known methods [6].

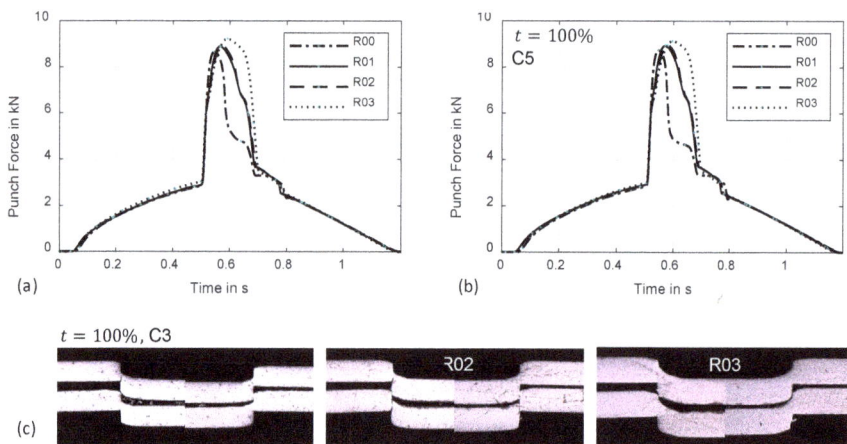

Fig. 4. Punch force curves for different edge radii of the embossing tool for the materials with C3 coating (a) and C5 coating (b) at a stamping depth of 100%.

The joint strength was determined by quasi-static tensile tests, as illustrated in Fig. 2(b) with five repetitions per process parameter set whereby the focus is on the maximum force achieved in the load direction until separation of the stack. For the case R01 shear as well as top tensile tests were performed and evaluated. The results for stamping depths of 80 %, 90 % and 100 % show that the shear strength is significantly higher than the pull-off strength in top tensile test in all cases (see Fig. 5), which can be explained by the fact that in the direction of loading in the sheet plane, in addition to the force-fit connection, they are also joined by a form-fit force component. Nevertheless, both tests show the same qualitative correlation when varying the embossing depth. In the following, only the results from the top tensile tests are used to simplify the investigation of the influence of edge radius, embossing depth and embossing clearance.

Results

In order to find out how the relationship between embossing depth and joint strength changes with increasing tool wear, investigations are carried out with one sharp and three rounded embossing tools. Nubs with different embossing depths are introduced into the samples and then two sheets are pressed together. In the following, the maximum forces achieved in the top tensile test are used to evaluate the joint strength.

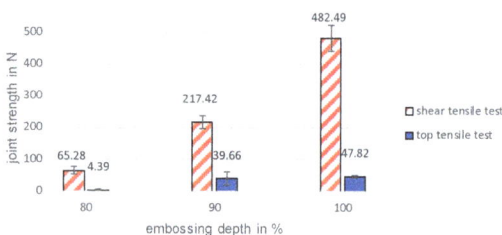

Fig. 5. Comparison of shear and top tensile test for an embossing radius of 0.1 mm.

Material Forming - ESAFORM 2023   Materials Research Forum LLC
Materials Research Proceedings 28 (2023) 1101-1110   https://doi.org/10.21741/9781644902479-121

The results are intended to provide the basis for making a real-time prediction of the joint strength by monitoring the wear condition on the basis of force signals. In addition, it will be investigated whether a targeted rounding of the tool edge can have a positive effect on the joint strength.

Fig. 6 shows the results of the tensile tests for the materials with C3 and C5 coating. Using a sharp tool (R00), a stacking with the present configuration is possible from an embossing depth of 70% (with C3) or 75% (with C5). Below this, no permanent joint can be created. The strength values for R00 of 13.75N at 70% and 14.09N at 75% are at an almost constant, low level in this range. Further increase of the embossing depth to 80% leads to a slightly decreased joint strength. For C5, the lowest embossing depth with which a successful package was produced is 75%. The strength values of 9.45N and 9.69N for 75% and 80% are also very close to each other, whereas the maximum strength value of 12.5 N is measured at 85%. The maximum achievable embossing depth before the nub breaks off is 80% for the C3 material and 85% for C5.

In the tests with rounded tool edges, three main observations can be made:

1. Both the minimum embossing depth at which a laminations stack can be joined and the maximum embossing depth before nub breakage occurs increase significantly compared to a sharp tool. The increase in the minimum embossing depth is due to the fact that, because of the rounded tool edge, there is a significant radius on the inside of the dimple as well as on the underside of the sheet in the rollover zone of the nub, which increases with increasing tool edge radius. This reduces the remaining actual contact area on which the contact normal stress can act in the radial direction. The maximum stamping depth increases because with a rounded punch, the stress peaks at the sharp edges are lowered, causing cracking to occur later. For C5, in all cases with a rounded tool, regardless of the edge radius, a joint can only be made from an embossing depth of 90%. With the largest radius of 0.3mm (R03), nubs with a maximum embossing depth of 125% can be successfully stacked for both C3 and C5.

2. After reaching an optimum, the strength decreases again with a further increase in the embossing depth. This could indicate that if the embossing depth is increased too much, the crack formation is already initiated at least locally and the thus pre-damaged nub no longer has the necessary strength in the subsequent stacking process to build up a press-fit with the same joint strength.

3. The maximum achievable strength with optimal choice of embossing depth increases significantly compared to the case with sharp embossing tool. It can be clearly seen that the maximum achievable joint strength in all cases with a rounded tool edge is above the strength with a sharp tool, regardless of the radius. The optimum is achieved at 100% embossing depth with a slight rounded tool with a radius of 0.1mm and is 47.82N with C3 coating or 44.42N with C5 coating. With further increase, the maximum strength decreases again and tends to occur at higher embossing depths.

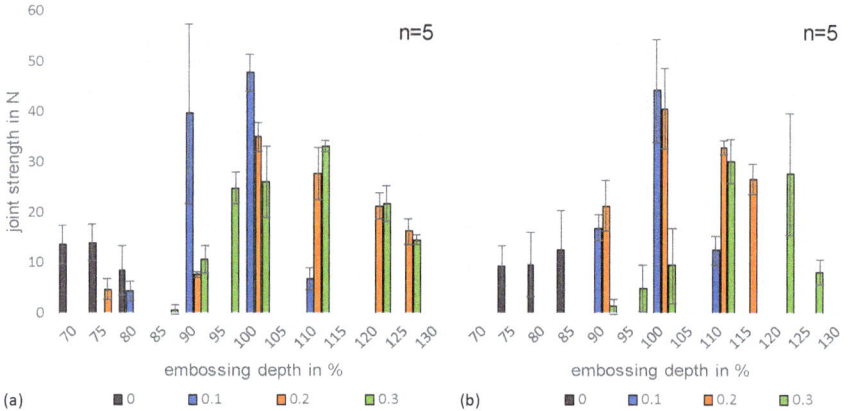

Fig. 6. Results of top tensile tests for different embossing depths and embossing tool radii at samples with C3 coating (a) and coating of type C5 (b).

Based on the tests carried out, the process window for the present joint configuration with a cylindrical nub of 6mm diameter can be determined empirically using an embossing clearance of 2%. In Fig. 7 this is shown as a function of the punch edge radius and the embossing depth. On this basis, both an upper and a lower limit can be read off for the respective radius. These differ only slightly for the two materials examined with different coatings.

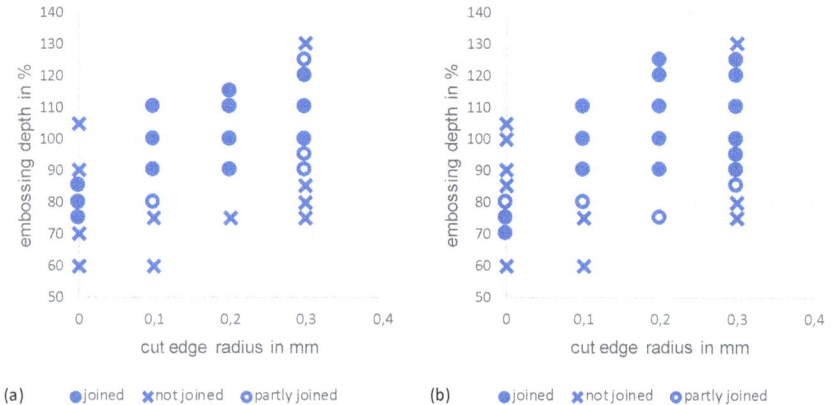

Fig. 7. Process windows for the investigated interlocking setup for test stacks of C3 (a) and C5 (b) coated sheets.

For achieving high-quality lamination a small clearance between the embossing punch and die is favorable [4]. Therefore, the influence of the embossing clearance is further investigated. Fig. 8 shows that the joint strength increases when the embossing clearance decreases from 2 % to 1 %, which is in accordance with studies by Lin et al. [4], where the same relationship between the embossing clearance and the joint strength was established with a sharp punch of 2mm diameter. In addition, the results indicate that the maximum embossing depth at which successful stacking

Material Forming - ESAFORM 2023
Materials Research Proceedings 28 (2023) 1101-1110

Materials Research Forum LLC
https://doi.org/10.21741/9781644902479-121

is possible is about the same as with a smaller gap, although higher local stresses are expected in the embossing step. Furthermore, the lower process limit for the embossing depth is lower with a smaller embossing gap, which leads to a significantly larger process window. This is valid for both C3 and C5 coated material.

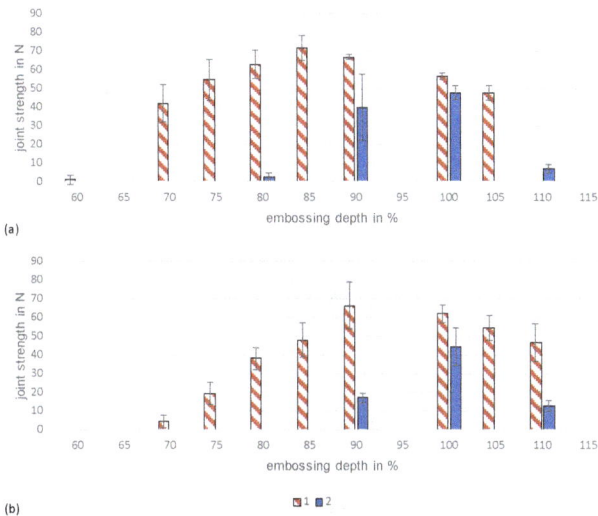

Fig. 8. Joint strength determined in top tensile tests for different embossing clearances 1 % and 2 % at various embossing depth with (a) C3 and (b) C5 coating.

Summary

In this work the influence of tool wear expressed by a rounding of the embossing tool edge on the joint strength of interlocked lamination stacks with cylindrical nubs was investigated.

It can be found that the achievable embossing depth increases when using an embossing punch with a rounded tool edge. Coincidently, the minimum embossing depth required for stacking and the embossing depth at which the optimum strength is achieved depend on the punch edge radius. If the embossing depth is regarded as a control variable, this knowledge can be used to react to wear occurring on the tool edge by adjusting the process parameter embossing depth. Another approach, which is obvious in order to design the interlocking die with regard to optimum mechanical joint strength, is to design the tools for stamping from the outset in such a way that an edge radius is provided in order to achieve the optimum of the achievable joint strength.

The comparison of the two different coating types C3 and C5, which are commonly used for blanking of electrical sheets shows no clear difference in the suitability for interlocked lamination stacks.

Acknowledgement

The presented results are part of the research projects "Production-induced properties in interlocking of lamination stacks for stator cores" of the European Research Association for Sheet Metal Working (EFB). The authors would like to thank the German Federation of Industrial Research Associations (AiF) and the German Federal Ministry for Economic Affairs and Climate Action (Bundesministerium für Wirtschaft und Klimaschutz, BMWK) for funding within the framework of project no. IGF 22036 N. Further thanks go to Dirk A. Molitor for the constructive technical discussions during the preparation of this paper. Furthermore, the authors would like to

Material Forming - ESAFORM 2023   Materials Research Forum LLC
Materials Research Proceedings 28 (2023) 1101-1110   https://doi.org/10.21741/9781644902479-121

thank Bruderer AG for providing the high-speed press BSTA 810-145 on which the tests were carried out.

References

[1] DIN EN 10342:2005-09, Magnetic materials - Classification of surface insulations of electrical steel sheet, strip and laminations; German version EN 10342:2005. https://doi.org/10.31030/9643607

[2] M. Lindenmo, A. Coombs, D. Snell, Advantages, properties and types of coatings on non-oriented electrical steels, J. Magnet. Magnet. Mater. 215-216 (2000) 79-82. https://doi.org/10.1016/S0304-8853(00)00071-8

[3] E. Lamprecht, Der Einfluss der Fertigungsverfahren auf die Wirbelstromverluste von Stator-Einzelzahnblechpaketen für den Einsatz in Hybrid- und Elektrofahrzeugen, Fertigungstechnik - Erlangen vol. 247, PhD thesis, 2014. ISBN 9783875253627

[4] H.-S. Lin, H.-C. Fu, L.-H. Liu et al., Stacking with cylindrical spots in lamination of stamped electrical steel sheets, Procedia Eng. 207 (2017) 992-997. https://doi.org/10.1016/j.proeng.2017.10.864

[5] D.M. Martin, F. Bäcker, B. Deusinger, C. Kubik, P. Gehringer, J. Schröder, P. Groche, Design Guidelines for interlocked stator cores made of CoFe sheets, IOP Conf. Ser.: Mater. Sci. Eng. 1238 (2022) 012036. https://doi.org/10.1088/1757-899X/1238/1/012036

[6] C. Kubik, M. Becker, D.A. Molitor, P. Groche, Towards a systematical approach for wear detection in sheet metal forming using machine learning, Prod. Eng. Res. Devel. 17 (2022) 21-36. https://doi.org/10.1007/s11740-022-01150-x

[7] K. Lange, Handbook of Metal Forming, J. Appl. Metalwork. 4 (1986) 188. https://doi.org/10.1007/BF02834383

[8] C. Kubik, D.A. Molitor, M. Rojahn, P. Groche, Towards a real-time tool state detection in sheet metal forming processes validated by wear classification during blanking, In: IOP Conference Series: Mater. Sci. Eng. 1238 (2022). https://doi.org/10.1088/1757-899X/1238/1/012067

[9] M. Regnet, A. Kremser, M. Reinlein, P. Szary, U. Abele, Influence of Cutting Tool Wear on Core Losses and Magnetizing Demand of Electrical Steel Sheets, 9th International Electric Drives Production Conference (EDPC), 2019, pp. 1-6. https://doi.org/10.1109/EDPC48408.2019.9011866

[10] K. Senda, M. Ishida, Y. Nakasu, M. Yagi, Influence of shearing process on domain structure and magnetic properties of non-oriented electrical steel, J. Magnet. Magnet. Mater. 304 (2006) e513-e515. https://doi.org/10.1016/j.jmmm.2006.02.139

[11] H.A. Weiss, N. Leuning, S. Steentjes, K. Hameyer, T. Andorfer, S. Jenner, W. Volk, Influence of shear cutting parameters on the electromagnetic properties of non-oriented electrical steel sheets, Journal of Magnetism and Magnetic Materials, 2017. https://doi.org/10.1016/j.jmmm.2016.08.002

[12] P. Groche, J. Hohmann, D. Übelacker, Overview and comparison of different sensor positions and measuring methods for the process force measurement in stamping operations, Measurement 135 (2019) 122-130. https://doi.org/10.1016/j.measurement.2018.11.058

[13] W. Klingenberg, T.W. de Boer, Condition-based maintenance in punching/blanking of sheet metal, Int. J. Mach. Tool. Manuf. 48 (2008) 589-598. https://doi.org/10.1016/j.ijmachtools.2007.08.013

[14] R. Hambli, Prediction of burr height formation in blanking processes using neural network, Int. J. Mech. Sci. 44 (2002) 2089-2102. https://doi.org/10.1016/S0020-7403(02)00168-6

[15] DIN EN 10106:2016-03, Cold rolled non-oriented electrical steel strip and sheet delivered in the fully processed state, German version EN10106:2015. https://doi.org/10.31030/2315581

[16] C. Kubik, J. Hohmann, P. Groche, Exploitation of force displacement curves in blanking – Feature Engineering beyond defect detection, Int. J. Adv. Manuf. Technol. 113 (2021) 261-278.

Material Forming - ESAFORM 2023
Materials Research Proceedings 28 (2023) 1111-1118

Materials Research Forum LLC
https://doi.org/10.21741/9781644902479-122

Versatile self-piercing riveting with a tumbling superimposed punch

WITUSCHEK Simon[1,a] *, ELBEL Leonie[1,b] and LECHNER Michael[1,c]

[1]Institue of Manufacturing Technology, Friedrich-Alexander-Universität Erlangen-Nürnberg, Egerlandstr. 13, 91058 Erlangen, Germany

[a]Simon.Wituschek@fau.de, [b]Leonie.Elbel@fau.de, [c]Michael.Lechner@fau.de

Keywords: Joining, Multi-Material-System, Versatile Joining, Self-Piercing Riveting

Abstract. Increasing resource efficiency is a major challenge and affects almost every aspect of social and economic life. The mobility sector in particular is responsible for a large share of primary energy consumption and is increasingly in the focus of public interest. One possibility to adress these challenges is to reduce the vehicle weight by means of lightweight construction technologies such as multi-material systems. These assemblies consist of workpieces with different mechanical and geometrical properties, which poses a major challenge for joining technology. Mechanical joining processes such as semi-tubular self-piercing riveting are often used in the production of these assemblies, but due to their process characteristics, they are rigid and can only react to changing process variables to a limited extent. One way to increase the versatility of self-piercing riveting is to superimpose a tumbling kinematics on the punch. During tumbling, an angular offset of the punch axis to the tool axis is set and the contact area between punch and workpiece is reduced. In this work, investigations were carried out to determine how the tumbling strategy, consisting of the parameters tumbling angle, tumbling onset and tumbling kinematics, affects the material flow of the rivet element. For this purpose, experimental tests are conducted with the typical materials of conventional multi-material systems and the geometric joint formations are determined by means of macrographs.

Key findings
- Analysis and identification of the significant influencing parameters of the tumbling strategy on the process combination and their interactions
- Evaluation of possibilities of a targeted material flow control to influence the geometric joint formation

Introduction

The energy and climate crisis demands a more efficient use of resources and technology. A major part of this ongoing change has to take place in the mobility sector [1]. Stricter limits for CO_2 emissions were imposed recently [2]. One way of increasing efficiency in automotive engineering is weight reduction [3]. Especially the battery weight in electric vehicles can be partially reduced this way [3]. Since the body in withe is an important part of the vehicle's total weight, efforts are often focused on replacing conventional steel used in body design with lightweight materials such as aluminum, fiber-reinforced plastics or high strength steel [4]. Reducing the overall mass of a passenger car with an internal combustion engine by 100 kg can save about 0,5 l of fuel per 100 km [5]. In addition to reducing weight and increasing efficiency, the mechanical strength and crash behaviour of a vehicle can also be improved by the selective use of materials [4]. This approach is generally called multi-material design [3]. The spectrum of materials used in this strategy includes light metals, such as titanium or aluminum, as well as plastics and composites [6], up to ceramics [7]. Due to the different chemical, mechanical and thermal properties of these materials, joining technology is one of the greatest challenges of multi-material design [6]. While conventional

Material Forming - ESAFORM 2023   Materials Research Forum LLC
Materials Research Proceedings 28 (2023) 1111-1118   https://doi.org/10.21741/9781644902479-122

thermal joining processes reach their limits when it comes to joining dissimilar materials [8], mechanical joining technologies such as semi-tubular self-piercing riveting (SPR) are becoming increasingly important [9]. Since SPR is a rather rigid process with limited flexibility, it cannot comply with the increasing demand for rapid response to individual product configurations [3]. One way to improve the variability of SPR is to superimpose the process with a tumbling kinematics on the punch. The punch is tilted by a certain angle, which reduces the contact area between the tool and the workpiece [10], as shown in Fig. 1b). As a result, the joining forces are reduced, and the process limits are increased [11], due to a higher number of parameters. Furthermore, the flexibility and versatility of the process can be enhanced [3].

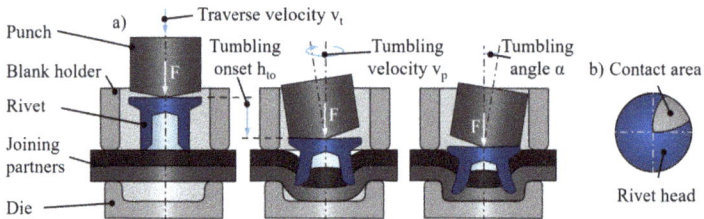

Fig. 1. a) Components and process parameters [13] of semi-tubular self-piercing riveting with tumbling punch and b) resulting contact area.

At present, the potential of this process combination cannot be exploited to its full extent, since no holistic understanding about the correlation of the process parameters with the geometrical joint formation of the SPR joints exists. To improve the process knowledge, an extensive series of experiments was planned and executed, analyzing the effects of various parameters and tumbling strategy combinations based on macrographs. Therefore, the tumbling angle α, the tumbling speed v_t, the tumbling kinematics, and the tumbling onset h_{to} were varied. These parameters are further illustrated in Fig. 1a) and are explained in more detail below. The results of the analysis provide a basis for the efficient use of the process combination to achieve a higher versatility. In particular, conclusions regarding the process and material flow control in semi-tubular self-piercing riveting are drawn.

Tumbling Self-Piercing Riveting Process

For the investigation of the process combination of a tumbling superimposed semi-tubular self-piercing riveting process, the tool shown in Fig. 2 is utilized. The tool design provides the ability to investigate a variety of parameters of the joining and tumbling strategy [3]. The core of the tool setup is the combination of a rotating and a linear axis, which enables predominantly rotating and linear kinematic models to be executed and a free path planning of the contact surface between punch and rivet head can be implemented. Furthermore, the adjustment mechanism for the tumbling angle and the punch movement is excluded from the force path of the tool, thus enabling highly dynamic movements. The tool is installed in a conventional universal testing machine of the type Walter-Bai, which performs the tool stroke in the z-direction. As materials for the investigations, the steel HCT590X with a sheet thickness of $t_0 = 1.5$ mm and the aluminium alloy EN-AW6014 with a sheet thickness of $t_0 = 2.0$ mm are selected. Due to their mechanical and geometric properties, these two materials and sheet thicknesses represent common multi-material systems and their challenges for joining technology. The die type used is a flat die with a diameter of 8.5 mm, a depth of 1.7 mm and a draft angle of 5°. The rivet element is a C-rivet with a shaft diameter of 5.3 mm and a rivet height of 5.0 mm from industrial applications.

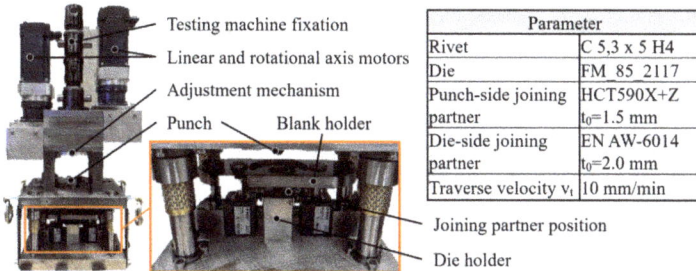

Parameter	
Rivet	C 5,3 x 5 H4
Die	FM_85_2117
Punch-side joining partner	HCT590X+Z t_0=1.5 mm
Die-side joining partner	EN AW-6014 t_0=2.0 mm
Traverse velocity v_t	10 mm/min

Fig. 2. Semi-tubular self-piercing riveting tool with detail views and tool parameters.

The investigations are carried out to identify the influences of individual parameters and their interactions with each other. For this purpose, the four process parameters of the tumbling strategy consisting of the tumbling angle α, the tumbling velocity v_t, the tumbling kinematics and the tumbling onset h_{to} are varied and evaluated. An overview of the variations of the process parameters are shown in Fig. 3. The tumbling angle is investigated in three stages. As a lower limit, the tumbling angle $\alpha = 1°$ is selected, since at an angle of $\alpha = 0°$ no tumbling angle is applied and the process is similar to a conventional semi-tubular SPR process. As a result, no correlations can be identified between the tumbling and the joining process. The upper limit of the investigated tumbling angles is set at $\alpha = 5°$. Previous studies have shown that in certain circumstances cracks in the rivet shaft of multi-material systems occur at angles higher than $\alpha = 5°$ [3]. For a finer gradation, the tumbling angle $\alpha = 3°$ is also investigated. Furthermore, the tumbling kinematics is varied as a process variable. A distinction can be made between predominantly rotating and predominantly linear movements. For the rotating kinematics models, circular and spiral kinematics are applied. The two models differ mainly in the adjustment of the maximum tumbling angle, which is approached in the first revolution in case of circular kinematics and is built up over the entire joining process in case of spiral kinematics. As a predominantly linear kinematics model, the kinematic shown in Fig. 3d) is investigated, which has fundamentally different motion characteristics. The motion has a significantly higher proportion in the radial direction and thus affects the material flow during the joining process.

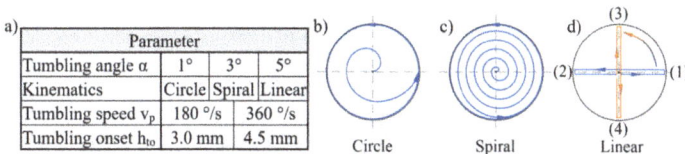

Parameter			
Tumbling angle α	1°	3°	5°
Kinematics	Circle	Spiral	Linear
Tumbling speed v_p	180 °/s		360 °/s
Tumbling onset h_{to}	3.0 mm		4.5 mm

Fig. 3. a) Overview of the process parameter variation and schematic illustration of b) circular c) spiral and d) linear kinematics.

As a further process parameter, the influence of the tumbling velocity on the joining process is investigated. This parameter is varied in the steps 180 °/s and 360 °/s. Since the traverse speed, which generates the z-stroke of the punch, is constant at $v_t = 10$ mm/min, an increase in the movement speed enhances the distance moved by the contact surface during the joining process, which can be described in terms of the number of revolutions in case of a rotating movement pattern. For the investigations with linear kinematics, the same path speeds are applied as for the rotating models. The fourth parameter of the tumbling strategy is the tumbling onset. As shown in Fig. 1a), the tumbling onset corresponds to the stroke of the punch from the first contact with the rivet to the start of the tumbling motion. It is selected based on the process phases of conventional semi-tubular SPR with $h_{to} = 3.0$ mm and $h_{to} = 4.5$ mm. A tumbling motion of the punch before the cutting phase of the punch-side joining partner is completed causes a large angular misalignment of the rivet in the joint. Therefore, the influence of an onset of the punch movement during spreading with $h_{to} = 3.0$ mm and setting with $h_{to} = 4.5$ mm is examined. The process phases are identified during sampling with a tumbling angle $\alpha = 0°$ using force signals [12].

In the investigation, a total of three tests per parameter combination are conducted in order to identify statistical and process uncertainties. In total, the test setup comprises 108 tests. The geometric joint formations, shown in Fig. 4, are determined to evaluate the influences of the individual parameters on the joint and their interactions. For this purpose, macrographs are prepared and the standard geometric parameters relevant for the joint quality are determined. These consist of the undercut, the rivet head end position and the residual sheet thickness [13].

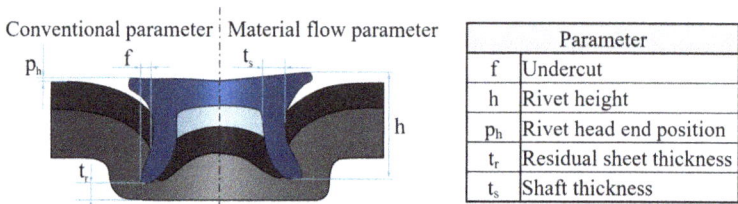

Parameter	
f	Undercut
h	Rivet height
p_h	Rivet head end position
t_r	Residual sheet thickness
t_s	Shaft thickness

Fig. 4. Geometric joint parameter for conventional and material flow parameter.

The investigations are also intended to identify and determine the material flow components. For this reason, the maximum rivet shaft thickness is measured in order to be able to identify radial material flow components in the joint. Furthermore, the rivet height is determined, which also provides insights into the effects of the individual parameters on the joint.

Evaluation of the Influencing Parameters

To determine the influence of the parameter variation, the rivet head geometry was analysed using 3D images. Fig. 5 shows three profile geometries of rivet heads, which differ in the tumbling angle. The other parameters remain constant. Therefore, the circular kinematics, the tumbling onset of $h_{to} = 3.0$ mm, and the tumbling velocity of $v_p = 180 °/s$ were chosen since the influence of the tumbling angle can be visualized best with this configuration. The colours of the rivet heads indicate how far the rivet extends above the surface of the punch-side joining partner. It is evident that the rivet head sits significantly lower when the tumbling angle is increased. Accordingly, by using a greater tumbling angle it is possible to directly regulate the rivet head end position for example to obtain a plane surface of rivet head and joining partner. The planarity improves with increasing angle and the conical shape caused by the punch is reduced.

Material Forming - ESAFORM 2023
Materials Research Proceedings 28 (2023) 1111-1118

Materials Research Forum LLC
https://doi.org/10.21741/9781644902479-122

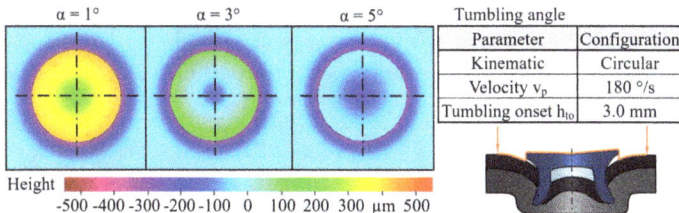

Fig. 5. Analysis of the rivet head geometry with varying tumbling angle.

Thus, it can be determined, the rivet head end position decreases when the tumbling angle is increased. In order to identify process characteristics about the material flow, the residual sheet thickness is examined. Fig. 6 shows the correlation of the residual sheet thickness with the kinematics, the tumbling angle, and the tumbling onset. Since the tumbling velocity has shown hardly any influence on the residual sheet thickness, both velocities are included in the average values plotted. It can be stated that the tumbling onset has a major influence on the residual sheet thickness when circular or spiral kinematics are applied. The linear kinematics leads to significantly more inhomogeneous results and is therefore statistically unstable. If tumbling starts after a 3.0 mm stroke, the residual sheet thickness increases with the tumbling angle. The difference from $\alpha = 1°$ to $\alpha = 3°$ is significantly greater compared to the variation from 3° to 5°. However, if the tumbling starts at $h_{to} = 4.5$ mm, the residual sheet thickness remains almost constant and is barely affected by other parameters. This means that the effect of the tumbling angle on the residual sheet thickness is a characteristic attribute of the tumbling punch, which only occurs if the effective tumbling process is sufficiently applied. It can be seen that not only the onset but also the duration of the tumbling has a significant influence on the geometric joint formation.

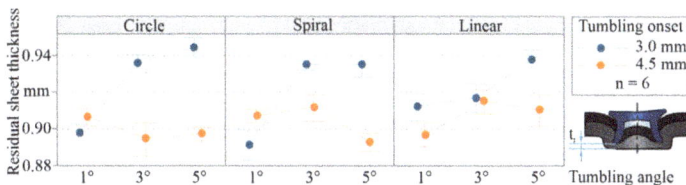

Fig. 6. Average residual sheet thickness depending on kinematics, tumbling angle and onset.

The fact that the residual sheet thickens as the tumbling angle increases contradicts decreasing rivet head end position. Hence, the tumbling punch must result in a deformation of the rivet. Therefore, the rivet height and the thickening of the rivet shaft were investigated to identify the direction of the material flow. Fig. 7 shows the correlation of the rivet height with the kinematics, the tumbling angle, the tumbling velocity, and a constant tumbling onset. It is apparent that the rivet is significantly compressed as a large tumbling angle is selected. The residual sheet thickness can therefore increase as the change in the rivet height is greater than the change in the rivet head end position. The figure also indicates that the deformation of the rivet is significantly less when linear kinematics are used. This shows that the effects of the tumbling punch are not well utilized by linear kinematics. Additionally, it can be seen that increasing the tumbling velocity leads to an enhanced rivet height change, especially with a large tumbling angle. To understand the effect of the velocity on the joint, process limits can be examined. If the tumbling velocity is increased infinitely, the tumbling process is equal to the conventional SPR. Accordingly, a very high velocity neutralizes the effects of the tumbling punch. A similar case applies to a tumbling velocity approaching zero. Hence, there must be an interval, in which the tumbling velocity ideally amplifies the effects of the tumbling punch however, detailed investigations with additionally varying velocities of the traverse are required therefore.

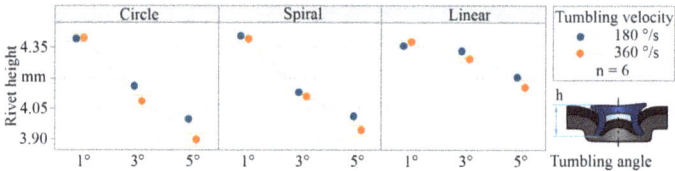

Fig. 7. Average rivet height depending on kinematics, tumbling angle, and tumbling velocity.

A reduction in rivet height leads to a partial radial material flow. The thickening of the rivet shaft increases the clamping effect of the joint and improves the force fit. Fig. 8 can be used to estimate the material flow. The outlines of rivet geometries extracted from the macrographs are shown. The reduction of the rivet height as well as the thickening of the rivet shaft by increasing the tumbling angle can be clearly seen in Fig. 8a). A significant part of the material flows radially inwards, as there is a flow constraint to the outside. The thickening of the rivet shaft also influences the undercut of the joint. For instance, the analysis showed that the undercut does not increase any further above a tumbling angle of $\alpha = 3°$, since the shaft thickening pushes the lowest and innermost point of the punch-side joining partner, at which the undercut is measured, further outward. This effect can be observed particulary well, when taking the average undercut for circular kinematics with 3.0 mm tumbling onset as an example. When the tumbling angle is increased from $\alpha = 1°$ to $\alpha = 3°$, the undercut changes from $f_{1°} = 0.32$ mm to $f_{3°} = 0.37$ mm. However, if the angle is further increased to 5°, the measured undercut even decreases slightly, resulting in an average value of $f_{5°} = 0.35$ mm.

In Fig. 8c) the influence of the kinematics is shown. While the outline is barely distinguishable for circular and spiral kinematics, the outline of the linear kinematis stands out from the other two kinematics, especially on the left side. As shown before, the rivet height remains greater with this kinematics. As a result, the radial material flow is reduced and the rivet shaft barely thickens. Also, the rivet head end position remains higher and the rivet foot is pushed outwards significantly less compared to the other kinematics. Thus, it can be confirmed that the effects of the tumbling punch vary in intensity due to the selection of the kinematics.

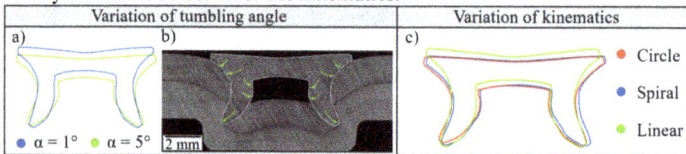

Fig. 8. Comparison of rivet outlines depending on a) tumbling angle and c) kinematics to determine b) material flow behaviour using 3.0 mm tumbling onset and a velocity of 180 °/s.

Summary

In this work, a semi-tubular self-piercing riveting process with superimposed tumbling kinematics on the punch was investigated. This involved deliberately varying the tumbling strategy in order to examine the effects of the process parameters tumbling angle, kinematics, tumbling onset and tumbling velocity on the geometric joint formation. In addition to the conventional geometric parameters such as rivet head end position, residual sheet thickness and undercut, the rivet height and the thickness of the rivet shaft were analysed to obtain findings on the material flow. It can be stated that the rivet geometry can be significantly influenced by the selected parameters, especially by the tumbling angle. When varying the tumbling kinematics, the circular and spiral kinematics show similar results. The linear kinematics weakens the effects of the tumbling punch and partly causes statistically uncertain results. Accordingly, the following findings refer primarily to the predominantly rotational kinematics.

The analysis of tumbling onset showed that the characteristics of tumbling are less identifiable when onset is delayed. This is particularly important for the residual sheet thickness since it can only be affected by the tumbling angle when the tumbling onset is in an early process stage. This understanding improves the flexibility of the process, as different sheet thickness combinations can thus be joined without having to exchange the die or the rivet. In general, a low rivet head end position is targeted in semi-tubular self-piercing riveting, ensuring that the surface of the rivet head and the punch-sided joining partner form a plane surface. It can be identified that the rivet head end position can be improved by increasing the tumbling angle. Furthermore, it was determined that an early tumbling onset in combination with a large tumbling angle and a high tumbling velocity cause a reduction of the rivet height. This leads to a thickening of the rivet shaft and thus to radial material flow. As a result, the force fit of the joint can be strengthened and the load-bearing capacity affected.

The results of this work provide a solid basis for the process control of tumbling semi-tubular self-piercing riveting. Nevertheless, some aspects offer further research potential. For example, due to the thickening of the rivet shaft the undercut can no longer be increased above a certain tumbling angle. However, since the undercut is decisive for the joint strength, the correlation between these two parameters could be investigated in more detail. Furthermore, there is potential for varying the kinematics, as only one configuration per kinematics was examined here. To separate the effects of tumbling onset and tumbling duration more precisely, it would also be interesting to conduct experiments with a tumbling motion only in the respective process phase. In addition, it is advisable to analyse the significance of the radial material flow for the load-bearing capacity and the joint strength of the connection in more detail.

Acknowledgment

Funded by the Deutsche Forschungsgemeinschaft (DFG, German Research Foundation) - TRR 285-Project-ID 418701707.

References

[1] European Commission, European Green Deal, Brussels, 2021.

[2] Regulation (EU) 2019/631 of the european parliament and of the council: of 17 April 2019, 2022. Available online: https://eur-lex.europa.eu/legal-content/EN/TXT/HTML/?uri=CELEX:32019R0631&from=EN. (accessed 23 November 2022).

[3] B. Bader, E. Türck, T. Vietor, Multi Material Design, A current overview of the used potential in automotive industries, in: Lehnert, Dröder (Eds.), Technologies for economical and functional lightweight design, Springer Berlin Heidelberg, Berlin, Heidelberg, 2019, pp. 3-13.

[4] G. Meschut, V. Janzen, T. Olfermann, Innovative and Highly Productive Joining Technologies for Multi-Material Lightweight Car Body Structures, J. Mater. Eng. Perform. 23 (2014) 1515-1523. https://doi.org/10.1007/s11665-014-0962-3

[5] Federal Ministry for Economics Affairs and Climate, Lightweighting. BMWI. Available: https://www.bmwk.de/Redaktion/EN/Dossier/lightweighting.html, 2022. (accessed 10 November 2022).

[6] S. Kleemann, D. Inkermann, B. Bader, E. Türck, T. Vietor, A Semi-Formal Approach to Structure and Access Knowledge for Multi-Material-Design, 21st International Conference On Eengineering Design, Vancouver (2017) pp. 289-298.

[7] H.E. Friedrich, Leichtbau in der Fahrzeugtechnik, Springer Fachmedien Wiesbaden, Wiesbaden, 2013.

[8] K. Martinsen, S.J. Hu, B.E. Carlson, Joining of dissimilar materials, CIRP Annals 64 (2015) 679-699. https://doi.org/10.1016/j.cirp.2015.05.006

[9] F. Kappe, S. Wituschek, M. Bobbert, G. Meschut, Determining the properties of multi-range semi-tubular self-piercing riveted joints. Prod. Eng. Res. Devel. 16 (2022) 363-378. https://doi.org/10.1007/s11740-022-01105-2

[10]M. Plančak, D. Vilotić, M. Stefanović, D. Movrin, Igor Kačmarčik, 2012. Orbital Forging - A plausible alternative for bulk metal forming, J. Trend. Develop. Machin. Associat. Technol. 16 (2022) 35-38.

[11]M. Merklein, D. Gröbel, M. Löffler, T. Schneider, P. Hildenbrand, Sheet-bulk metal forming - forming of functional components from sheet metals, MATEC Web of Conferences 21 (2015) 1001. https://doi.org/10.1051/matecconf/20152101001

[12]S. Wituschek, M, Lechner, Investigation of the influence of the tumbling angle on a tumbling self-piercing riveting process, Proceedings of the Institution of Mechanical Engineers Part L-J. Mater. Des. Appl. 236 (2022) 1151-1345. https://doi.org/10.1177/14644207221080068

[13]G. Meschut, M. Merklein, A. Brosius, D. Drummer, L. Fratini, U. Füssel, M. Gude, W. Homberg, P.A.F. Martins, M. Bobbert, M. Lechner, R. Kupfer, B. Gröger, D. Han, J. Kalich, F. Kappe, T. Kleffel, D. Köhler, C.-M. Kuball, J. Popp, D. Römisch, J. Troschitz, C. Wischer, S. Wituschek, M. Wolf, Review on mechanical joining by plastic deformation, J. Adv. Join. Process. 5 (2022) 100-113. https://doi.org/10.1016/j.jajp.2022.100113

Lionel Fourment MS on optimization and inverse analysis in forming

Material Forming - ESAFORM 2023
Materials Research Proceedings 28 (2023) 1121-1130

Materials Research Forum LLC
https://doi.org/10.21741/9781644902479-123

On the comparison of heterogeneous mechanical tests for sheet metal characterization

GONÇALVES Mafalda[1,a *], OLIVEIRA Miguel Guimarães[1,2,b],
THUILLIER Sandrine[2,c] and ANDRADE-CAMPOS António[1,d]

[1]Centre for Mechanical Technology and Automation (TEMA), LASI – Intelligent Systems Associate Laboratory, Department of Mechanical Engineering, University of Aveiro, Portugal

[2]Univ. Bretagne Sud, UMR CNRS, IRDL, F-56100 Lorient, France

[a]mafalda.goncalves@ua.pt, [b]oliveiramiguel@ua.pt, [c]sandrine.thuillier@univ-ub.fr, [d]gilac@ua.pt

Keywords: Heterogeneous Tests, KPIs, Material Behavior, Sheet Metal

Abstract. The characterization of sheet metal behavior is of utmost importance for the accurate virtualization of sheet metal forming processes. Newly proposed mechanical testing approaches are overcoming the use of standard mechanical tests. Test configurations with more complex geometries present richer mechanical fields and, therefore, provide a higher quantity of valuable information about the material behavior in a more efficient manner. To extract that information, full-field measurement techniques such as Digital Image Correlation are being used. Although several test designs have already been proposed, the choice of the best one to calibrate a chosen mechanical model is still an issue. This work aims at proposing Key Performance Indicators (KPIs) that are able to rank mechanical tests by their potential to enhance the material behavior characterization process. These metrics evaluate quantitatively the quality and the importance of the data that each test can provide. The potential of three test designs to characterize accurately sheet metal mechanical behavior is analyzed using the proposed KPIs. From a uniaxial tensile loading test up to rupture, the numerical mechanical information is extracted, and the performance of each test is evaluated and compared.

Introduction

Sheet metal forming processes play a major role in the development of mechanical parts for several industries such as automotive and aircraft. The virtualization of these processes has gained wide popularity in the transition to a more sustainable industry. Reduced costs, time, and material waste associated with the experimental task can be achieved. However, an accurate digitalization of the forming process is only possible when the numerical model can represent reality. For instance, the material constitutive model chosen to define the material behavior has to be well calibrated, having a huge repercussion on the obtained results. A classical model calibration procedure demands a huge quantity of data from several standard mechanical tests. The time and costs associated with this task led to the emergence of new ways of mechanical testing. The so-called heterogeneous testing consists in test configurations that have the potential to provide a significant amount of data with just a single test. Whether they present more complex boundary conditions or geometries, heterogeneities in the mechanical fields are induced, leading to a more informative test.

Several innovative designs have already been proposed to overcome the standardized mechanical testing. The design process of these tests has been led mainly by two approaches searching for: (i) the heterogeneity of the mechanical fields that are induced on the specimen [1–7] and (ii) the quality of the material model parameters that are indentified [8–11]. A more comprehensive review of the mechanical tests commonly used for material characterization and model calibration can be found in [14].

The use of heterogeneous mechanical tests is only achievable due to the emergence of full-field measurement techniques, such as Digital Image Correlation [15]. As an optical technique, it has

Material Forming - ESAFORM 2023   Materials Research Forum LLC
Materials Research Proceedings 28 (2023) 1121-1130   https://doi.org/10.21741/9781644902479-123

the ability to extract information about the strain fields at each material point. Therefore, a larger quantity of information can be extracted from one heterogeneous test using full-fields techniques. With this data, it is possible to calibrate material constitutive models more efficiently using inverse methodologies such as the Finite Element Model Updating (FEMU) technique [16] and the Virtual Fields Method (VFM) [17]. The quality of the identification depends on the quality of the mechanical information that can be retrieved from the specimen. For instance, the heterogeneity of the mechanical fields, the magnitude and distribution of the equivalent plastic strain and the sensitivity of the induced fields to the parameters to be identified are aspects that have a huge influence on the identification quality.

Several designs of tests have been proposed with the aim of trying to characterize material behavior more accurately and cost-effectively. However, it is still unclear how to choose the best test geometry to calibrate a chosen material model. The choice of the most suitable test is not straightforward, so appropriate metrics should be established to evaluate and compare the performance of each test in the model calibration and material identification procedures.

This work aims at filling this gap and establishing Key Performance Indicators (KPIs) that analyze the potential of each test design to provide valuable data for the material characterization and model calibration procedures. Three heterogeneous tests, deformed up to rupture under a uniaxial load, were chosen to be analyzed. The investigation is in a first step purely numerical. Based on the mechanical fields, the potential of each test was evaluated using the proposed KPIs.

Heterogeneous Tests Analysis

Key Performance Indicators. The aim of this work consists in evaluating and comparing the potential of each test to provide relevant information for the material identification and model calibration procedures. For that, some scalar metrics are here proposed. The first metric consists in a mechanical indicator proposed by Souto et. al. [12] that can be defined as

$$I_{\mathrm{T}} = w_{r1} \frac{\mathrm{Std}(\varepsilon_2/\varepsilon_1)}{w_{a1}} + w_{r2} \frac{(\varepsilon_2/\varepsilon_1)_{\mathrm{R}}}{w_{a2}} + w_{r3} \frac{\mathrm{Std}(\bar{\varepsilon}^{\mathrm{p}})}{w_{a3}} + w_{r4} \frac{\bar{\varepsilon}^{\mathrm{p}}_{\max}}{w_{a4}} + w_{r5} \frac{\mathrm{Av}(\bar{\varepsilon}^{\mathrm{p}})}{w_{a5}}, \qquad (1)$$

where ε_1 and ε_2 are the principal major and minor strains in the sheet plane, respectively. The equivalent plastic strain, $\bar{\varepsilon}^{\mathrm{p}}$, and its maximum value, $\bar{\varepsilon}^{\mathrm{p}}_{\max}$, are also considered as evaluation criteria for the test. This indicator evaluates different features: the strain state standard deviation, the strain state range, the standard deviation, the maximum, and the average value of the equivalent plastic strain. The importance of each term is adjusted and normalized using absolute values (w_{ai}, with $i \in [1,5]$) and relative weights (w_{ri}, with $i \in [1,5]$). The chosen values for these parameters can be seen in Table 1. A more detailed description of the indicator as well as the established values for the normalization are described in [12], however, a higher indicator value means a more informative and richer test in terms of strain heterogeneity.

Table 1. Absolute and relative weights for the adjustment and normalization of the indicator terms.

w_{a1}	w_{a2}	w_{a3}	w_{a4}	w_{a5}	w_{r1}	w_{r2}	w_{r3}	w_{r4}	w_{r5}
1	4	0.25	1	1	0.3	0.03	0.17	0.4	0.1

In a similar way, the scalar indicator proposed by Barroqueiro et. al. [7] is used to evaluate the stress states heterogeneity, in this case, considering only tension, compression, and shear. It can be stated as

$$id = \prod_{s=1}^{3} \left[\frac{3}{\sum_{e=1}^{n} X_e} \sum_{e=1}^{n} (\delta_e^s \, Z_e \, X_e) \right]. \qquad (2)$$

Material Forming - ESAFORM 2023   Materials Research Forum LLC
Materials Research Proceedings 28 (2023) 1121-1130   https://doi.org/10.21741/9781644902479-123

The index s relates to the stress state, compression, shear, and tension, respectively. The term Z_e stands for the penalization of stress concentrations and unstressed material while the parameter δ_e^s is responsible for identifying the stress state in each element e. In this case, the term X_e corresponds to the volume of each finite element. The ideal solution would present the same amount of material in the three stress states (tension, compression, and shear) without stress concentrations or unstressed material. The multiplicative behavior of the indicator leads to values close to zero most the elements are subjected to tensile loading, being only a few subjected to shear and compression. More information on the computation of the indicator can be found in [7].

Another metric, proposed by Oliveira et. al. [18], evaluates the sensitivity of the test to anisotropy. It was derived from Mohr's circle equations, and it was introduced for a plane stress state. Based on the principal angle's formulation, it considers the maximum principal stress in absolute value, and the range of tensile orientations typically used to calibrate the material's anisotropic behavior. Represented by γ and denominated by rotation angle, it refers to the principal direction associated with the maximum principal stress in absolute value and, it ranges from 0° and 90°, being given by

$$\gamma = \begin{cases} 45 & \text{if } \sigma_{xx} = \sigma_{yy} \text{ and } \sigma_{xy} \neq 0 \\ 45(1-q) + q|\beta| & \text{otherwise} \end{cases} \qquad (3)$$

where β is the principal angle and q is an integer that ranges between -1 and 1, that can be defined as

$$q = \frac{\sigma_{xx} - \sigma_{yy}}{|\sigma_{xx} - \sigma_{yy}|} \frac{|\sigma_1| - |\sigma_2|}{||\sigma_1| - |\sigma_2||} \qquad (4)$$

where σ_1 and σ_2 are the principal major and minor stresses, respectively. Since a rotation angle value is related to each material point, an average metric needs to be defined to characterize the sensitivity to anisotropy of each test. Therefore, it is proposed to use the standard deviation measure of the rotation angle to evaluate each test.

Specimen designs. In this work, three specimen designs proposed in the literature were chosen to be analyzed. To improve the quality of the information that can be extracted from a single mechanical test, their geometries were designed from different approaches. The first design was proposed by Rossi et. al. [19] and is usually referenced as Notched. Jones et. al. [2] developed the second specimen geometry, referred as D, via an iterative geometry design process aided by engineering intuition. The last specimen was obtained from a topology-based design methodology [20], designed with the goal of presenting the most heterogeneous displacement field. This will be referred to as TopOpt. The selected geometries are depicted in Fig. 1 along with their dimensions. The three specimen geometries were considered to be machined at 45° with respect to the rolling direction (along the x-direction) in order to enhance the heterogeneity of the mechanical fields.

Material behavior. The material considered in this work is a dual-phase steel (DP600) [21] with a thickness of 0.8 mm. The elastic behavior is considered isotropic, being modeled by Hooke's law. The plastic behavior is anisotropic and is defined by the Yld2000-2d anisotropy yield criterion [22]. The isotropic hardening is characterized by the Swift Law. The material parameters that describe the proposed behavior are in Table 2.

Material Forming - ESAFORM 2023
Materials Research Proceedings 28 (2023) 1121-1130

Materials Research Forum LLC
https://doi.org/10.21741/9781644902479-123

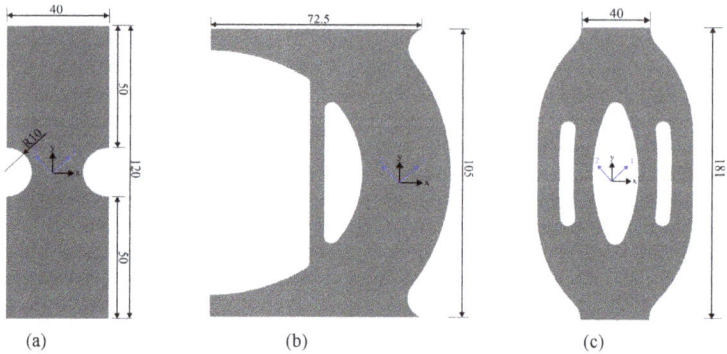

Fig. 1. Specimen designs selected to be analysed in this work: (a) Notched, (b) D
and (c) TopOpt.

Table 2. Elastic and constitutive model parameters for the Swift's law and the Yld2000-2d yield
function of the DP600 [21].

E[GPa]	ν	K[MPa]	ε_0	n				
210	0.3	979.46	0.00535	0.194				
α_1	α_2	α_3	α_4	α_5	α_6	α_7	α_8	a
1.011	0.964	1.191	0.995	1.010	1.018	0.977	0.935	6

Numerical simulation. Finite element simulations were carried out using Abaqus/Standard to submit each test configuration to a uniaxial tensile loading. Four-node shell elements were used with reduced integration and hourglass control. An element size of 0.5 mm was used for each specimen. The numerical simulations were performed with automatic time stepping and a maximum increment size of 0.02. The material behavior was described with the aid of a UMMDP (User Material Model Driver for Plasticity). Regarding the boundary conditions of the test, the bottom edge of the specimen was constrained in all degrees of freedom while the displacement at the top edge was constrained in x- and z- directions. A displacement was applied in y- direction. Each mechanical test is performed up to rupture, being the stopping condition established through the Forming Limit Curve (FLC), which is represented in the major and minor strains diagrams of each test. Therefore, the values of the applied displacement that led to rupture were 2.45 mm, 6.38 mm, and 12.45 mm, for the Notched, D and TopOpt, respectively.

Results and Discussion
In this section, the numerical information extracted is used to evaluate the potential of each test. Fig. 2 exhibits the principal stress and strain diagrams and the distribution of the equivalent plastic strain in the specimen just before rupture. Each line stands for the information of one test configuration. The principal strains and stresses diagrams provide huge information on the heterogeneity of the strain and stress fields induced on the specimens as well as the stress and strain state ranges. The equivalent plastic strain distribution on the specimen allows us to understand the areas where a higher magnitude of the variable is achieved. The material points in the elastic regime are represented in grey since these do not provide much information for the characterization of the plastic behavior. By contrast, the color associated with each material point in the plastic regime corresponds to the magnitude of its equivalent plastic strain. All the diagrams present the same scale between tests for easier comparison. `

Material Forming - ESAFORM 2023

Materials Research Proceedings 28 (2023) 1121-1130

Materials Research Forum LLC

https://doi.org/10.21741/9781644902479-123

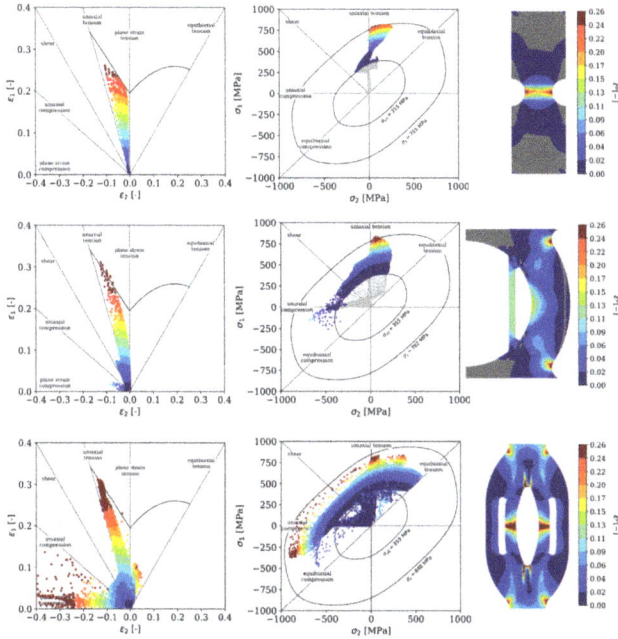

Fig. 2. Principal strains and stresses diagrams and equivalent plastic strain distribution for each test design at the moment just before rupture. Both the initial surface and the one associated with the maximum yield stress are plotted in the principal stress diagrams.

It can be noted that the Notched specimen is the one that presents the smallest stress state range, being limited to uniaxial tension. Similarly, the strain states are mainly located between uniaxial tension and plane strain tension. Regarding the equivalent plastic strain, the higher values of this variable are located in the center part of the specimen. The other material points in the plastic regime are under low values of plastic strain.

In contrast, both D and TopOpt specimens present strain states ranging from plane strain compression and plane strain tension. Although both specimens present material points from uniaxial compression to plane strain compression, the distribution is much denser in the TopOpt specimen. Regarding the principal stresses diagram, there are material points between equibiaxial compression and equibiaxial tension in both specimens, D and TopOpt. However, the latter is the one that presents the largest stress state range. When compared to the Notched specimen, both the D and TopOpt specimens present higher values of plastic strains as well as a larger part of the material points in the plastic regime. In the case of the D specimen, the distribution is more spread, being mainly placed along the curve part of the specimen. The TopOpt specimen presents all the material points in the plastic regime at the moment just before rupture, being the one that achieves higher values of plastic strain. However, these are concentrated in small areas and near the specimen boundaries. This makes the extraction of the information difficult using full-field measurement techniques. It is worth noting that there are material points with plastic strains that are located inside the initial yield surface. This is due to the non-monotonic behavior of some material points. During the test, they enter the plastic regime, but as of a certain point, their strain

Material Forming - ESAFORM 2023   Materials Research Forum LLC
Materials Research Proceedings 28 (2023) 1121-1130   https://doi.org/10.21741/9781644902479-123

paths change direction, leading to a decrease in the induced stresses. Since the equivalent plastic strain consists in a cumulative value, at the moment of rupture, these points are located inside the yield surface but present plastic strains.

With the aim of analyzing quantitatively the information given by the diagrams presented in Fig. 2, the mechanical indicator proposed by [12] has been computed. In Fig. 3, the values of the mechanical indicator for each test configuration are presented as well as its terms individually. All the terms, except for the ones involving the standard deviation computation, were normalized taking into account the size and the number of elements for a fair comparison of the designs.

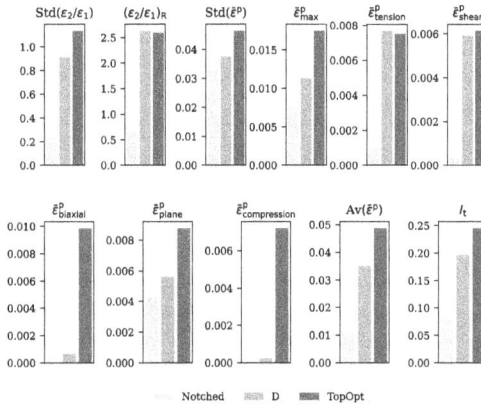

Fig. 3. Values of the mechanical indicator proposed by Souto et. al. [12] and its terms computed individually for all the test designs.

It can be noted that the TopOpt specimen presents the highest overall value of the indicator, followed by the D and the Notched specimens. The higher the indicator value, the more informative the test design is. Regarding the equivalent plastic strain, as has already been referred to previously, the TopOpt presents the highest values and the largest distribution of this variable. It is also the one that presents the most material points in the plastic regime under the strain states evaluated with this indicator. The D specimen also presents an interesting diversity of strain states and equivalent plastic strain distribution. Concerning the strain states range, both specimens present similar performance. Fig. 4 depicts the elements distribution of the D and TopOpt specimens over the strain state range that is limited between -15 and 1. The TopOpt specimen presents a more even distribution of the elements over the presented range although two peaks located around uniaxial tension and between plane strain and uniaxial compression can be noted. In the D specimen, the majority of the material points are between uniaxial tension and shear. Therefore, the standard deviation is higher in the TopOpt specimen than in the D one. However, the second term that evaluates the difference between the maximum and minimum values of the principal strains' ratio is equal for both specimens since these values correspond to the imposed limits. The final value is different due to the normalization of the term concerning the size and number of the elements of each specimen. In previous works [12,23], it was proposed a ranking of the indicator value for well-known standard tests and for new geometries that have been proposed since then. Despite the modification of the strain state range for the computation of the indicator in this work, an updated ranking can be seen in Fig. 5. Also, in the work developed by Thoby et. al [24], it was made a comparison of several specimen designs and this mechanical indicator was used to compare them. Although some terms have been adapted, a similar ranking was obtained.

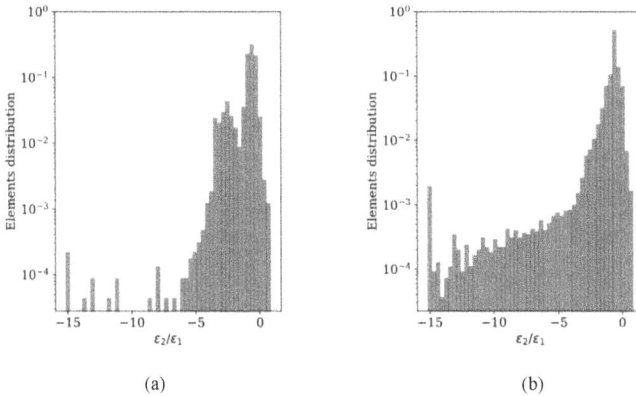

(a)　　　　　　　　　　(b)

Fig. 4. Elements distribution over the strain state range induced in the (a) D and (b) TopOpt specimens.

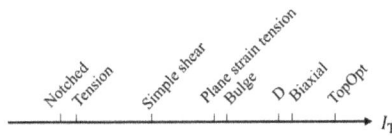

Fig. 5. Ranking for the tests evaluated in [8] and for the Notched, D and TopOpt test designs.

The other mechanical indicator proposed to evaluate the performance of the specimens is the one proposed by [7]. Fig.6 provides the value of the indicator as well as the terms corresponding to each stress state analyzed. Overall, the TopOpt specimen is the one that presents the highest heterogeneity of stress states (tension, compression, and shear). It can be noted that the stress states under which all the specimens present more material points is tension, followed by shear. In compression, there is a significant difference between the TopOpt and the others. The TopOpt specimen was designed considering this indicator, so it is expected that this design presents the highest value.

The rotation angle provides information on the sensitivity of the test to anisotropy. Fig. 7 represents the distribution of the material points over the range of rotation angle values (0° to 90°) for each test configuration. The more dispersed the distribution, the higher the sensitivity to the anisotropic behavior it presents. Therefore, to evaluate quantitatively the information given by the diagrams, it is represented the standard deviation value of the rotation angle of each test design in Table 3. Since the material orientation is 45° in relation to the loading direction, it is expected that most of the material points present rotation angle values in the same order as the material orientation, being the mean value of the rotation angle around 45° for all the test designs. Although the distribution is well dispersed between 0° and 90°, the Notched specimen is the test with the lowest standard deviation value, presenting the majority of the points in the elastic regime with rotation angle values around 45°. However, there is a significant spread of material points in the plastic regime with rotation angle values between -15° and 75°. In contrast, the D specimen presents most of the material points in the plastic regime between rotation angle values of 30° and 60°. The rotation angle standard deviation of the D specimen is considerably higher than the

Materials Research Forum LLC
https://doi.org/10.21741/9781644902479-123

Notched one, pointing out the difference between both distributions. The TopOpt specimen presents the best range of rotation angle values, presenting material points in the plastic regime between 0° and 90°, covering the whole range. Although this leads to a lower density of points over the range, it presents the highest standard deviation value of all the tests, allowing us to notice a good sensitivity to anisotropy.

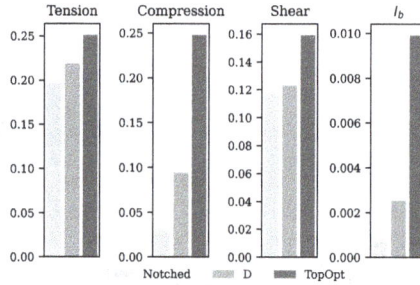

Fig. 6. Values of the performance indicator proposed by Barroqueiro et. al [7] and its terms for all test designs.

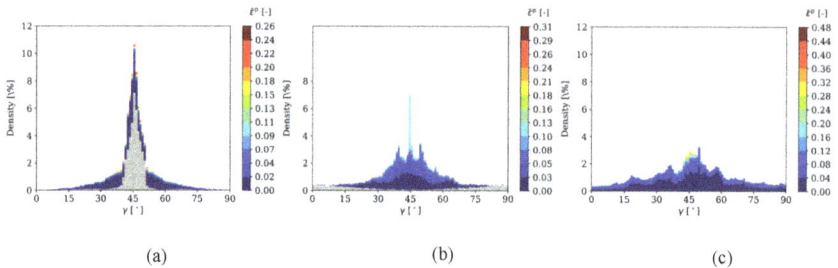

(a) (b) (c)

Fig. 7. Rotation angle values for each test configuration: (a) Notched, (b) D and (c) TopOpt.

Table 3. Mean and standard deviation values of the rotation angle distribution for each test design.

	Notched	D	TopOpt
Av(γ) [°]	45.33	44.92	45.70
Std(γ)	10.16	17.38	19.70

Summary
This work aimed at proposing a set of KPIs to evaluate the performance of several test designs in providing the most informative quantity of data for the material behavior characterization and model calibration procedures.

Three specimen designs were subjected numerically to a uniaxial loading test up to rupture. Based on the obtained mechanical fields, the potential of each test design was analyzed and compared with the others. Firstly, the information given by the principal strains and stresses diagrams was quantified by the computation of the mechanical indicator proposed by Souto et. al. [12]. The obtained values allow us to conclude that the TopOpt specimen presented the larger strain state range and most interesting equivalent plastic strain distribution, followed closely by the D specimen. The stress states heterogeneity present in the specimen was also evaluated by the

Material Forming - ESAFORM 2023   Materials Research Forum LLC
Materials Research Proceedings 28 (2023) 1121-1130   https://doi.org/10.21741/9781644902479-123

mechanical indicator proposed by Barroqueiro et. al. [7]. Due to the tensile conditions of the test, all the specimens presented the majority of their material points under tension, being the TopOpt specimen the one with higher heterogeneity of stress states and, therefore, with the higher indicator value. Based on the rotation angle values distribution, it can be concluded that the TopOpt and the D specimens are the ones that present the higher sensitivity to anisotropy due to the higher standard deviation values. Regarding the TopOpt specimen, due to its complex geometry, it may suffer from buckling when tested. However, this situation may lead to a better performance as long as this behavior can be captured by the full-field measurement technique.

This work already consists in a step closer to a more straightforward approach to choose the most informative heterogeneous mechanical test. At this stage, these KPIs evaluate each test based on the diversity of mechanical phenomena and strain and stress states that are covered. However, there is still a need for metrics that take into account the inverse identification quality, for example, the sensitivity of each test to the model parameters and, also, the full-field measurement technique.

Acknowledgments

This project has received funding from the Research Fund for Coal and Steel under grant agreement No 888153. The authors also acknowledge the financial support under the projects UIDB/00481/2020 and UIDP/00481/2020 – FCT – Fundação para a Ciência e Tecnologia; and CENTRO-01-0145-FEDER-022083 – Centro Portugal Regional Operational Programme (Centro2020), under the PORTUGAL 2020 Partnership Agreement through the European Regional Development Fund. M. Gonçalves is grateful to the FCT for the Ph.D. grant Ref. UI/BD/151257/2021.

Disclaimer

The results reflect only the authors' view, and the European Commission is not responsible for any use that may be made of the information it contains.

References

[1] J.H. Kim, F. Barlat, F. Pierron, M.G. Lee, Determination of Anisotropic Plastic Constitutive Parameters Using the Virtual Fields Method, Exp. Mech. 54 (2014) 1189-204. https://doi.org/10.1007/s11340-014-9879-x

[2] E.M.C. Jones, J.D. Carroll, K.N. Karlson, S.L.B. Kramer, R.B. Lehoucq, P.L. Reu, D.Z. Turner, Parameter covariance and non-uniqueness in material model calibration using the Virtual Fields Method, Comput. Mater. Sci. 152 (2018) 268-290. https://doi.org/10.1016/j.commatsci.2018.05.037

[3] T. Pottier, P. Vacher, F. Toussaint, H. Louche, T. Coudert, Out-of-plane Testing Procedure for Inverse Identification Purpose: Application in Sheet Metal Plasticity, Exp. Mech. 52 (2012) 951–963. https://doi.org/10.1007/s11340-011-9555-3

[4] N. Souto, A. Andrade-Campos, S. Thuillier, Mechanical design of a heterogeneous test for material parameters identification, Int. J. Mater. Form. 10 (2016) 353-367. https://doi.org/10.1007/s12289-016-1284-9

[5] M. Conde, Y. Zhang, J. Henriques, S. Coppieters, A. Andrade-Campos, Design and validation of a heterogeneous interior notched specimen for inverse material parameter identification, Finite Elem. Anal. Des. 214 (2023) 103866.

[6] M. Gonçalves, A. Andrade-Campos, B. Barroqueiro, On the design of mechanical heterogeneous specimens using multilevel topology optimization, Adv. Eng. Softw. 175 (2023) 103314.

[7] B. Barroqueiro, A. Andrade-Campos, J. Dias-de-Oliveira, R.A.F. Valente, Design of mechanical heterogeneous specimens using topology optimization, Int. J. Mech. Sci. 181 (2020) 105764. https://doi.org/10.1016/j.ijmecsci.2020.105764

[8] P. Wang, F. Pierron, M. Rossi, P. Lava, O.T. Thomsen, Optimised experimental characterisation of polymeric foam material using DIC and the virtual fields method, Strain 52 (2016) 59-79.

https://doi.org/10.1111/str.12170

[9] L. Chamoin, C. Jailin, M. Diaz, L. Quesada, Coupling between topology optimization and digital image correlation for the design of specimen dedicated to selected material parameters identification, Int. J. Solids Struct. 193-194 (2020) 270-286. https://doi.org/10.1016/j.ijsolstr.2020.02.032

[10] X. Gu, F. Pierron, Towards the design of a new standard for composite stiffness identification, Compos. Part A Appl. Sci. Manuf. 91 (2016) 448-460. https://doi.org/10.1016/j.compositesa.2016.03.026

[11] M. Rossi, M. Badaloni, P. Lava, D. Debruyne, F. Pierron, A procedure for specimen optimization applied to material testing in plasticity with the virtual fields method, AIP Conf. Proc. 1769 (2016) 200016. https://doi.org/10.1063/1.4963634

[12] N. Souto, S. Thuillier, A. Andrade-Campos, Design of an indicator to characterize and classify mechanical tests for sheet metals, Int. J. Mech. Sci. 101-102 (2015) 252-271. https://doi.org/10.1016/j.ijmecsci.2015.07.026

[13] M. Conde, A. Andrade-Campos, M.G. Oliveira, J.M.P. Martins, Design of heterogeneous interior notched specimens for material mechanical characterization, Esaform 2021, Liège, Belgique, 2021.

[14] F. Pierron, M. Grédiac, Towards Material Testing 2.0. A review of test design for identification of constitutive parameters from full-field measurements, Strain 57 2021) 1-22. https://doi.org/10.1111/str.12370

[15] M. Grédiac, The use of full-field measurement methods in composite material characterization: Interest and limitations, Compos. Part A Appl. Sci. Manuf. 35 (2004) 751-761. https://doi.org/10.1016/j.compositesa.2004.01.019

[16] S. Avril, M. Bonnet, A.S. Bretelle, M. Grédiac, F. Hild, P. Ienny, F. Latourte, D. Lemosse, S. Pagano, E. Pagnacco, F. Pierron, Overview of identification methods of mechanical parameters based on full-field measurements, Exp. Mech. 48 (2008) 381-402. https://doi.org/10.1007/s11340-008-9148-y

[17] F. Pierron, M. Grédiac, The virtual fields method: Extracting constitutive mechanical parameters from full-field deformation measurements, Springer Verlag, 2012.

[18] M.G. Oliveira, S. Thuillier, A. Andrade-Campos, Evaluation of heterogeneous mechanical tests for model calibration of sheet metals, J. Strain. Anal. Eng. Des. 57 (2022) 208-224. https://doi.org/10.1177/03093247211027061

[19] M. Rossi, F. Pierron, M. Štamborská, Application of the virtual fields method to large strain anisotropic plasticity, Int. J. Solids. Struct. 97-98 (2016) 322-335. https://doi.org/10.1016/j.ijsolstr.2016.07.015

[20] M. Gonçalves, A. Andrade-Campos, S. Thuillier, On the topology design of a mechanical heterogeneous specimen using geometric and material nonlinearities, IOP Conf. Ser. Mater. Sci. Eng. 1238 (2022) 012055. https://doi.org/10.1088/1757-899X/1238/1/012055

[21] F. Ozturk, S. Toros, S. Kilic, Effects of Anisotropic Yield Functions on Prediction of Forming Limit Diagrams of DP600 Advanced High Strength Steel, Procedia Eng. 81 (2014) 760-765. https://doi.org/10.1016/j.proeng.2014.10.073

[22] F. Barlat, J.C. Brem, J.W. Yoon, K. Chung, R.E. Dick, D.J. Lege, F. Pourboghrat, S.-H. Choi, E. Chu, Plane stress yield function for aluminum alloy sheets - Part 1: Theory, Int. J. Plast. 19 (2003) 1297-1319. https://doi.org/10.1016/S0749-6419(02)00019-0

[23] J. Aquino, A. Andrade-Campos, J.M.P. Martins, S. Thuillier, Design of heterogeneous mechanical tests: Numerical methodology and experimental validation, Strain 55 (2019) 1-18. https://doi.org/10.1111/str.12313

[24] J.D. Thoby, T. Fourest, B. Langrand, D. Notta-Cuvier, E. Markiewicz, Robustness of specimen design criteria for identification of anisotropic mechanical behaviour from heterogeneous mechanical fields, Comput. Mater. Sci. 207 (2022) 111260. https://doi.org/10.1016/j.commatsci.2022.111260

Material Forming - ESAFORM 2023   Materials Research Forum LLC
Materials Research Proceedings 28 (2023) 1131-1142   https://doi.org/10.21741/9781644902479-124

On the inverse identification of sheet metal mechanical behaviour using a heterogeneous Arcan virtual experiment

HENRIQUES Joao[1,2,a] *, ANDRADE-CAMPOS António[1,2,b] and XAVIER José [2,3,c]

[1]Centre for Mechanical Technology and Automation (TEMA), Department of Mechanical Engineering, University of Aveiro, Campus Universitário de Santiago, 3810-193 Aveiro, Portugal

[2]Intelligent Systems Associate Laboratory (LASI), 4800-058 Guimarães, Portugal

[3]Research and Development Unit for Mechanical and Industrial Engineering (UNIDEMI), Department of Mechanical and Industrial Engineering, NOVA School of Science and Technology, NOVA University Lisbon, 2825-149 Lisbon, Portugal

[a]joaodiogofh@ua.pt, [b]gilac@ua.pt, [c]jmc.xavier@fct.unl.pt

Keywords: Sheet Metal Forming, Anisotropic Plasticity, Arcan Test, Heterogeneous Test Evaluation, Inverse Identification, Digital Image Correlation, Synthetic Images

Abstract. Modelling and simulation are critical stages of product development in modern industry. Simulation tools in solid mechanics use constitutive models and their parameters to describe the behaviour of materials. Nowadays, with the use of heterogeneous test configurations and full-field measurements, it is possible to measure a combination of multiple strain states, allowing for the identification of multiple parameters from a single test with reduced cost and time. This work aims to investigate the potential for obtaining heterogeneous states of strain\stress with the Arcan test configuration. A finite element model was developed using a specimen with a smooth arc section in which the loading and material directions varied, producing tensile, shear, or mixed mode responses. The most heterogeneous test configuration was selected using a heterogeneous criterion and the numerical results were used to generate synthetic speckle pattern images and further processed by digital image correlation (DIC). The DIC results were used as input for the identification procedure through the virtual fields method (VFM) for the simultaneous calibration of the Swift hardening law and the Hill'48 anisotropic yield criterion. The identified solution was compared with the ground truth material parameters. The results show the potential of combining the Arcan test with the VFM to simultaneously identify material parameters for anisotropic plasticity models of sheet metals.

Introduction

Computer-aided engineering systems are a powerful tool used in modern industry to optimise costs and time consumption in the design of new products. Nowadays, in metal forming technology, the development of sheet metal parts tends to be more virtual through the use of numerical simulation. Sheet metal anisotropy is a critical property that greatly influences the accuracy of the numerical results [1]. This anisotropy is a result of the rolling process that produces preferential orientations in the material texture [2], which induces differences in the yield stress along different orientation angles from the rolling direction (RD). Therefore, the accuracy of the simulation is heavily reliant on the calibration of the model that describes the behaviour of the materials during the forming process including the anisotropy. The improved accuracy of complex constitutive models is related to increased flexibility in the mathematical formulation, which translates to a greater number of constitutive parameters to calibrate, resulting in increased calibration complexity [3]. Classically, the calibration procedure was done using standard homogeneous tests [4]. However, as models become more accurate and complex, their calibration has become a difficult task, increasing the number of experimental tests required to accurately calibrate such models.

Material Forming - ESAFORM 2023   Materials Research Forum LLC
Materials Research Proceedings 28 (2023) 1131-1142   https://doi.org/10.21741/9781644902479-124

Recent developments in optical full-field measurement techniques, such as digital image correlation (DIC) [5], have been changing the approach to the calibration procedure of material constitutive models. The increased amount of measurable kinematic data has enabled the transition from simple statically determinate tests to innovative tests with complex geometries that produce heterogeneous states of stress and strain, allowing for the collection of more data and thus reducing the experimental effort [6]. However, there is no closed-form formulation relating local kinematic data to constitutive parameters. As a result, in recent years, there has been an increased interest in the development of robust inverse identification techniques, most notably the virtual fields method (VFM) [7] and the finite element model updating (FEMU) method [8]. These approaches have altered the way mechanical tests are conducted, eventually leading to the development of new standards as cameras gradually replace strain gauges and extensometers in both academia and industry. Nonetheless, the ability to reduce the number of experimental tests required to simultaneously identify material parameters is highly dependent on the test configuration used, which implies that optimised test configurations are required to fully benefit from this new paradigm [6,9].

The mechanical testing scientific community has made significant efforts in recent years to find methodologies for developing optimised specimens [6]. There has been an increase in the development of heterogeneous tests through different methodologies, for instance, using empirical knowledge [10,11] or numerical optimisation procedures [9,12,13]. The specimen design by empirical knowledge is tied to the author's previous experience. However, this approach frequently results in unoptimised specimen shapes, especially when using complex constitutive models, since the relationship between the constitutive parameters and strain measurements can be quite complex [6]. A more logical approach that permits the use of various design variables is the use of numerical optimisation procedures. The design variables can range from the strain states [4,9], identification quality [14] or full-identification chain quality [15,16]. For a more comprehensive review of the state-of-the-art of heterogeneous specimen design, the reader is referred to the work of Pierron and Grédiac [6].

The Arcan test is an interesting test configuration since it allows varying the loading direction in a standard uniaxial tensile testing machine. In the framework of sheet metal plasticity, the majority of the previously mentioned authors focused on optimising the shape of the specimen or using geometries with holes or notches. However, changing the loading direction can also be used to increase strain heterogeneity, which in the case of the Arcan test can result in tension, shear or mixed mode responses. Wang et al. [15,16] designed orthotropic foam tests using the Arcan fixture by varying the loading direction and the material orientation. The authors were able to successfully calibrate the orthotropic linear elastic constitutive model with a single test when using the optimised design parameters. Although some authors used the Arcan test in sheet metal plasticity [17,18], it is seldom used in heterogeneous test design for the calibration of plastic constitutive models. Nonetheless, the Arcan test has the potential to provide interesting heterogeneous test configurations when used for test design in sheet metal plasticity.

The goal of this work is to evaluate the heterogeneity of stress/strain states for different Arcan test configurations using a specimen with a smooth arc section [18]. A finite element model was developed in which the loading and material directions varied, producing tensile, shear, or mixed mode responses. The reference constitutive parameters for the DP600 steel were used [19] to select the most promising test configuration using a heterogeneous criterion [4,12]. To get closer to real experimental data, the FEA results for the most heterogeneous test configuration were used to generate synthetic speckle pattern images, which were further processed by DIC. The parameters for the Hill'48 yield criteria and the Swift hardening law were simultaneously identified using the VFM and compared to the ground truth parameters.

Material Forming - ESAFORM 2023
Materials Research Proceedings 28 (2023) 1131-1142

Materials Research Forum LLC
https://doi.org/10.21741/9781644902479-124

Methodology

Numerical model.

In this work, a finite element model of the Arcan test was developed using a specimen with a smooth arc section [18] under quasi-static loading conditions. The load direction angle (α) and the RD angle (θ) with the x-axis were varied, producing tensile, shear, or mixed mode responses. The angles θ and α assumed the values of 0°, 45° or 90°, totalling nine possible test configuration combinations. The finite element model was implemented in ABAQUS/Standard software [20] assuming plane stress conditions and a thickness of 0.8 mm for the specimen. A mesh convergence study was performed and a total of 2596 four-node plane stress elements (CPS4R) were used with displacement-driven loading conditions. The forming limit curve (FLC) is used as the rupture criteria and the test was conducted until the failure of the specimen with a total of 40 load steps. Fig. 1a depicts the specimen geometry and dimensions (in mm) and Fig. 1b shows the FEA mesh and boundary conditions, including the variation of the material and load direction angles.

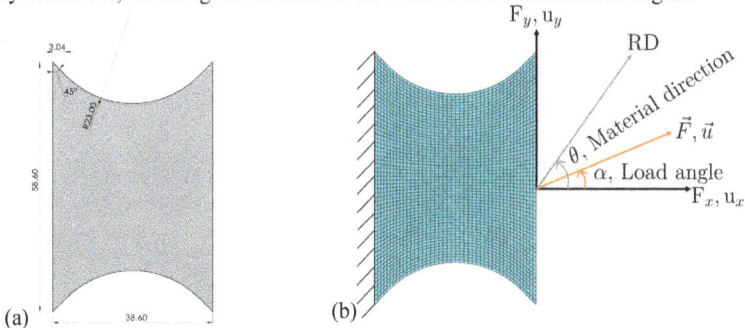

Fig. 1. (a) Geometry and dimensions [mm] of the specimen with smooth arc section [18], (b) Finite element mesh and boundary conditions, showing the variation of the material and load direction angles.

Material and constitutive model.

A sheet metal of DP600 dual-phase steel with 0.8 mm thickness was considered in this work. The elastoplastic behaviour of the material was modelled using phenomenological models, namely isotropic elasticity described by Hooke's law ($E = 210$ GPa and $\nu = 0.3$) and anisotropic plastic behaviour described by the Hill'48 yield criteria [21] and Swift hardening law. The reference parameters of the DP600 steel [19] are presented in Table 2. In this work, the commonly used condition $G+H=1$ is assumed [3], which deduces that the yield stress in the rolling direction corresponds to the yield stress (σ_y). A user-defined material subroutine, Unified Material Model Driver for Plasticity (UMMDp) [22], was used to model the material behaviour.

Test evaluation.

The selection of the most heterogeneous test configuration is not an obvious task. Therefore, two heterogeneity indicators were used to evaluate the richness of strain/stress in each of the test configurations used. The first indicator was initially proposed by Souto et al. [4], and is based on the maximization of five quantities that the ideal test should exhibit and can be written as:

$$IT_1 = w_{r1}\frac{\text{Std}(\varepsilon_2/\varepsilon_1)}{w_{a1}} + w_{r2}\frac{(\varepsilon_2/\varepsilon_1)_R}{w_{a2}} + w_{r3}\frac{\text{Std}(\bar{\varepsilon}^P)}{w_{a3}} + w_{r4}\frac{\bar{\varepsilon}^P_{\max}}{w_{a4}} + w_{r5}\frac{\text{Av}(\bar{\varepsilon}^P)}{w_{a5}}, \quad (1)$$

where ε_1 and ε_2 are the major and minor principal strains, respectively, $\bar{\varepsilon}^P$ is the equivalent plastic strain, $\bar{\varepsilon}^P_{\max}$ is the averaged maximum value of the equivalent plastic strain for each stress/strain state and w_r, w_a are relative and absolute weighting factors, respectively. Since the ratio between

Material Forming - ESAFORM 2023   Materials Research Forum LLC
Materials Research Proceedings 28 (2023) 1131-1142   https://doi.org/10.21741/9781644902479-124

the minor and major principal strains represents various strain states, the first two quantities presume that its distribution should be as wide as possible. The final three quantities are related to the equivalent plastic strain, which should be as heterogeneous and widely distributed as possible, as well as have a high global plastic strain level. The quantities are further normalised and adjusted in terms of importance according to the different weights. The weights used in this work were the recommended from the original work by Souto *et al.* [4] and are presented in Table 1.

Table 1. Absolute and relative weighting factors used for the definition of IT_1[4].

w_{a1}	w_{a2}	w_{a3}	w_{a4}	w_{a5}	w_{r1}	w_{r2}	w_{r3}	w_{r4}	w_{r5}
1	4	0.25	1	1	0.3	0.03	0.17	0.4	0.1

The second heterogeneity indicator is the rotation angle (γ) which evaluates the sensitivity of the specimen to anisotropy and is based on the principal angle formulation [23]. This indicator varies between 0° and 90° and represents the principal direction for the maximum principal stress in absolute value and can be written as:

$$\gamma = \begin{cases} 45 & \text{if } \sigma_{11} = \sigma_{22} \text{ and } \sigma_{12} \neq 0 \\ 45(1-q) + q|\beta| & \text{otherwise} \end{cases}, \tag{2}$$

where σ_{11}, σ_{22}, σ_{12} are the components of the Cauchy stress tensor in the material coordinate system and β is the principal angle in degrees and q is an integer that assumes the value of 1 or -1 that can be calculated from:

$$q = \frac{\sigma_{11} - \sigma_{22}}{|\sigma_{11} - \sigma_{22}|} \frac{|\sigma_1| - |\sigma_2|}{||\sigma_1| - |\sigma_2||}, \tag{3}$$

where σ_1, σ_2 are the major and minor principal stresses, respectively. The formulation is written in such a way that it can represent most of the conceivable stress/strain states in Mohr's circle [23], except for being undefined in the purely biaxial stress state ($\sigma_{11} = \sigma_{22}$ and $\sigma_{12} = 0$).

Synthetic images and digital image correlation.

The numerical results of the most heterogeneous test configuration were used to synthetically deform a speckle pattern image using the FE Deformation module from the MatchID software [24]. The synthetic speckle pattern images were then further processed with the subset-based 2D-DIC using the MatchID software [24]. The goal of this approach is to simulate a real experiment in which the results are measured by using the DIC filter, where spatial averaging is used to reduce the inherent experimental noise. This approach makes it possible to capture the full uncertainty propagation through the identification chain and leads to more realistic identification results that can also be applied to test design [6].

Furthermore, the choice of DIC settings has a significant impact on measurement accuracy [8]. Therefore, the MatchID Performance Analysis module [24] was used to analyse several combinations of these settings in order to find a good balance between spatial resolution and accuracy. The hardware and DIC settings used here are equal to those used in a previous work (see [25]).

Material Forming - ESAFORM 2023
Materials Research Forum LLC
Materials Research Proceedings 28 (2023) 1131-1142
https://doi.org/10.21741/9781644902479-124

Virtual fields method.

The material constitutive parameters were identified using the VFM, which is based on the principle of virtual work. The objective function for the VFM can be written in static conditions while ignoring body forces as:

$$R = \left(-\int_V \boldsymbol{\sigma}\,(\boldsymbol{\chi}, \boldsymbol{\varepsilon}) : \mathrm{grad}\,\mathbf{u}^* \, dV + \int_{\partial V} \mathbf{T} \cdot \mathbf{u}^* \, dS \right)^2 \approx 0, \tag{4}$$

where R is the residual, $\boldsymbol{\sigma}$ is the Cauchy stress tensor which is a function of the material parameter set χ and the strain tensor $\boldsymbol{\varepsilon}$, and $\mathbf{u}^*$ is the virtual displacement vector. The internal virtual work is then calculated for the volume V of the given solid. $\mathbf{T}$ is the traction vector acting on the boundary of the solid and ∂V is the boundary where the loading is applied.

The idea behind the use of VFM for the identification of constitutive parameters is to find the material parameters set that minimises the gap between the internal and external virtual works. Another important aspect of VFM is the selection of appropriate virtual fields $\mathbf{u}^*$, which can be done either manually or automatically. The manual selection of the virtual fields is done by user experience. These must, however, meet the following requirements: (i) $\mathbf{u}^*$ must be kinematically admissible, (ii) $\mathbf{u}^*$ should be constant at the boundary ∂V, in order to use the global force rather than its distribution, which is unknown in experimental tests and (iii) $\mathbf{u}^*$ must be collinear with $\mathbf{T}$ to eliminate components of the loading forces that are unknown [3]. For more details on automatic strategies for the selection of virtual fields see [26,27].

Results and Discussion

Test evaluation. The IT_1 heterogeneity criterion was calculated using the numerical results of each test configuration. To simplify the evaluation, only the results from the last increment before failure were used, since these are a good indication of the stress and strain states observed in a specimen under monotonic loading [4,23]. Each term of the IT_1 criterion was also evaluated separately to see the relative advantages of each test configuration. Fig. 2 depicts the results computed for the IT_1 criterion as well as each of its terms as a function of the load angle (α) for each RD angle (θ).

The results indicate that the shear test configurations (α=90°) are the most heterogeneous, owing to the higher standard deviation of the ratio between the minor and major principal strains, as well as the higher average and maximum value of the equivalent plastic strain. Tensile test configurations (α=0°), on the other hand, provide the least amount of information about the mechanical behaviour of the material, which is not surprising given the relatively simple specimen geometry used, which promotes primarily the tensile stress/strain state localised in a narrow area of the specimen. According to the results, the mixed-mode response test configuration (α=45°) provides more information about the material's mechanical behaviour than the tensile test configurations, but it is still less heterogeneous than the shear test configurations. Furthermore, it can be seen that the shear test configuration with the 90° RD (θ=90°) gives the most information regarding the mechanical behaviour of the material in accordance with the criteria being used. This is primarily due to the larger equivalent plastic strain, average and standard deviation values.

Material Forming - ESAFORM 2023
Materials Research Proceedings 28 (2023) 1131-1142

Materials Research Forum LLC
https://doi.org/10.21741/9781644902479-124

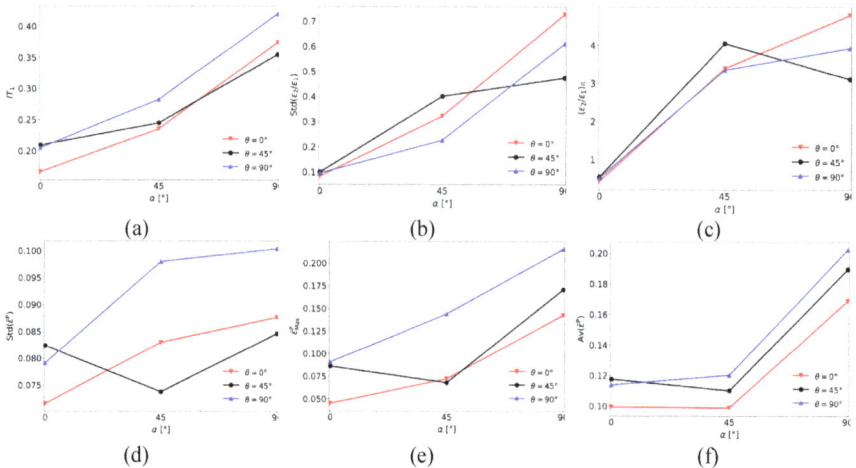

Fig. 2. Mechanical test heterogeneity indicator results for all test configurations: (a) IT_1, (b) $Std(\varepsilon_2/\varepsilon_1)$, (c) $(\varepsilon_2/\varepsilon_1)_R$, (d) $Std(\bar{\varepsilon}^P)$, (e) $\bar{\varepsilon}^P_{max}$ and (f) $Av(\bar{\varepsilon}^P)$.

A qualitative analysis was also carried out by investigating the major and minor principal stresses and strains as well as the rotation angle criterion for the most heterogeneous test configuration of each assessed load angle (see Fig. 3).

According to the principal strains and stresses diagrams, the shear load test with θ=90° exhibits the greatest variety of stress/strain states, ranging from uniaxial compression to pure shear to uniaxial tension. The other test configurations, on the other hand, mostly provide the uniaxial tension state, with the uniaxial tensile test configuration approaching plane strain tension. These qualitative results support the previous analysis using the IT_1 criterion, which indicates that the shear load test configuration is the most heterogeneous, with a higher standard deviation in the ratio between the principal strains.

It is also interesting to locate the material points of higher equivalent plastic strain in both strain and stress diagrams, the rotation angle histogram and the contour represented in the specimen geometry inside the rotation angle diagram. For instance, in Fig. 3a, the points with the highest equivalent plastic strain are in the specimen's centre and are primarily in the uniaxial tension state, with rotation angles ranging from 30° to 60°. According to the rotation angle histogram results, the material points for the loading angles of 45° and 90° (Figs. 3b and 3c) are primarily on the right side (between 45° and 90°), with some points also on the left side of the histogram, but with low levels of equivalent plastic strain, and in the case of the test with a loading angle of 45°, they still are in the elastic regime. On the other hand, the rotation angle histogram for the tensile load test (Fig. 3a) shows a wide distribution of points, with the highest equivalent plastic strain occurring between 30° and 60°, which indicates that this test may be the most sensitive to anisotropy out of the three tests investigated in this qualitative analysis.

Materials Research Forum LLC
https://doi.org/10.21741/9781644902479-124

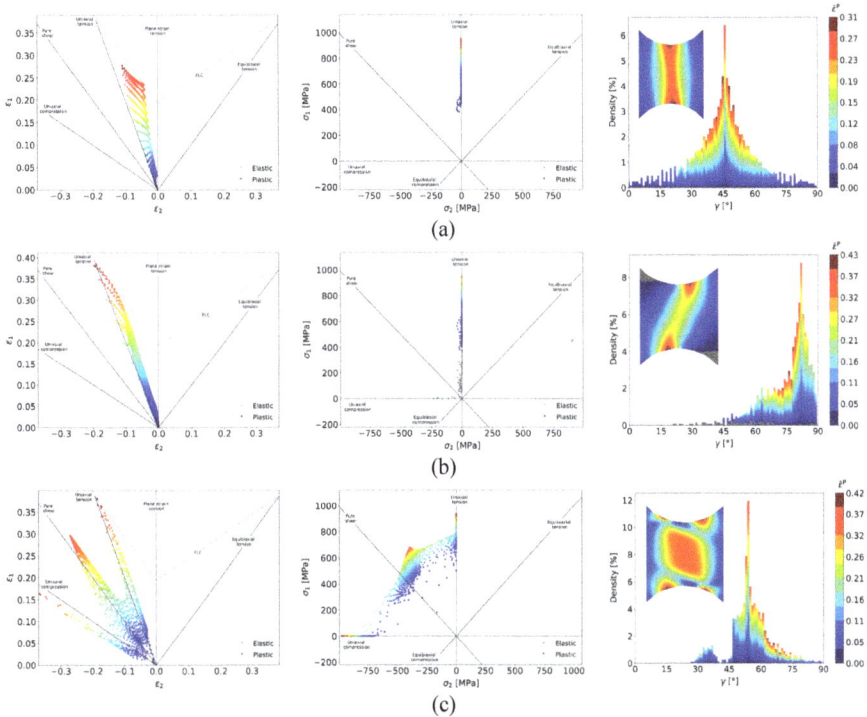

Fig. 3. Numerical results regarding principal strains diagram (left), the principal stresses diagram (center) and rotation angle (right) for the most heterogeneous test configuration of each load angle: (a) α = 0° and θ = 45°, (b) α = 45° and θ = 90° and (c) α = 90° and θ = 90°.

While it can be difficult to quantitively rank all the test configurations, the test with a loading angle of 45° can generally be considered to be the least interesting one. The shear load test appears to outperform the tensile load test in the heterogeneity of strain/stress states. However, the opposite can be said about the rotation angle distribution. Nonetheless, the shear load test with the 90° RD was chosen for the identification process due to significantly higher performance in the stress/strain states and IT_1 criterion results, despite having a narrower distribution of the rotation angle.

Inverse identification.

The numerical results of the shear load test with the 90° RD were used to generate synthetic speckle images, which were then processed by DIC. The DIC results were then used for the inverse identification with VFM with one uniform virtual field, using the VFM module from the MatchID software [24]. Four identification runs were carried out, each with a unique initial set of parameters that were generated using the Latin hypercube sampling (LHS) method. Table 2 lists the reference parameters for the DP600 steel [19], the lower and upper bounds for each parameter, the initial set of parameters used for each identification run, the residual value, the identification results and the absolute relative error.

Table 2. Reference [19] and initial set of parameters, lower and upper bounds, identification results, final residual value and absolute relative error of each of the identification runs.

	F	G	H	N	K [MPa]	ε_0	n
Ref. parameters	0.3748	0.5291	0.4709	1.1125	979.46	0.00535	0.194
Lower bound	0.2624	-	0.3296	0.7787	685.62	0.00370	0.136
Upper bound	0.4872	-	0.6122	1.4462	1273.30	0.00700	0.252
Run 1 - Final residual = 0.6632							
Initial parameters	0.3467	-	0.4356	1.3628	759.08	0.00580	0.238
Id. parameters	0.4185	0.6697	0.3303	1.2970	1054.00	0.00534	0.199
|Relative error| [%]	11.66	26.57	29.86	16.58	7.61	0.19	2.58
Run 2 - Final residual = 0.7142							
Initial parameters	0.4591	-	0.5769	1.0290	1052.92	0.00490	0.179
Id. parameters	0.3925	0.6699	0.3301	1.25	1036	0.00559	0.201
|Relative error| [%]	4.72	26.61	29.90	12.36	5.77	4.49	3.61
Run 3 - Final residual = 0.4432							
Initial parameters	0.2905	-	0.5062	0.8621	906.00	0.00410	0.209
Id. parameters	0.4872	0.6704	0.3296	1.4280	1086.00	0.00371	0.1875
|Relative error| [%]	29.99	26.71	30.01	28.36	10.88	30.65	3.35
Run 4 - Final residual = 0.8209							
Initial parameters	0.4029	-	0.3649	1.1959	1199.84	0.00660	0.150
Id. parameters	0.4872	0.6704	0.3296	1.2430	1043.00	0.00700	0.209
|Relative error| [%]	29.99	26.71	30.01	11.73	6.49	30.84	7.73

The results show that the identified constitutive parameters related to the swift hardening law have a lower absolute relative error than the results for the Hill'48 anisotropy yield criterion. This was expected, given that the qualitative analysis conducted previously indicated that this test configuration contained richer information regarding the stress/strain states than the material anisotropy. Though the difference in residual between all the identification runs was relatively low, identification run 3 had the lowest final residual of all the runs that were performed. However, some parameters seem to have higher absolute errors when compared to the reference parameters. This result is the consequence of the noise and uncertainty propagation throughout the whole identification process when using DIC and experimental images. The results could also be improved by using more virtual fields, as previous research has shown that the number of virtual fields has a significant impact on the performance of VFM identification, with the number of virtual fields increasing the accuracy of the identification [28]. Fig. 4 shows the evolution of the residual between internal and external virtual works during the identification procedure, as well as a comparison of the external virtual work and the final calibrated internal virtual work for each identification run.

The evolution of the residual during the identification process (Fig. 4a) shows that the initial parameters set have a significant impact on the initial and final calibrated residual values, and consequently, on the identification results. The difference in the final residuals between the identification runs, on the other hand, is small, as can be seen in the difference between the external and virtual works (Fig. 4b), where all the curves are nearly coincident. The small difference in

Material Forming - ESAFORM 2023   Materials Research Forum LLC
Materials Research Proceedings 28 (2023) 1131-1142   https://doi.org/10.21741/9781644902479-124

residuals with the variation of the identified parameters also suggests the need for more virtual fields to be used in the identification process, as using just one virtual field could filter out crucial kinematic data about the mechanical behaviour related to a particular constitutive parameter.

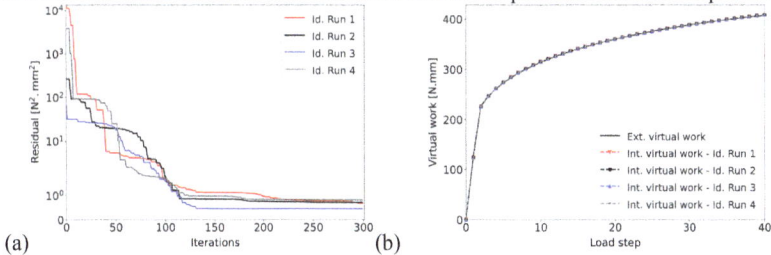

Fig. 4. Identification results in terms of: (a) evolution of the residual during the identitication process and (b) final difference between external and internal virtual works.

In Fig. 5, the Swift hardening law curves and the Hill'48 yield surfaces are compared as part of cross-validation between the reference parameters and the identification results. The numerical results for this test configuration are also plotted in the normalised stress space (see Fig. 5b), with the yield stress calculated from the reference parameters.

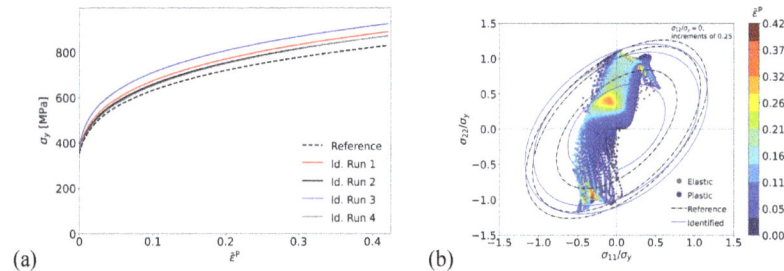

Fig. 5. Results obtained with the identified parameter sets for: (a) Swift law hardening curve for all identification runs and (b) comparison of the reference yield surface with the identified yield surface for the identification run 3 for different levels of σ_{12}/σ_y.

The identified Swift hardening law curves present some divergences from the hardening curve with the reference parameters. These differences become more pronounced as the equivalent plastic strain increases. These results seem to be a consequence of the overall uncertainty propagation in the identification process, in which the global minimum of the residual between the external and internal virtual works may differ from the ground truth parameters due to experimental image noise and the DIC technique's different filtering and spatial resolution. Nevertheless, increasing the number of virtual fields used, even if they are uniform, may improve the accuracy of the parameters identified. Moreover, when comparing the reference and the identification run with the lower residual value (identification run 3), the differences appear to be larger where there is a lack of information in the stress space and lower where there is a higher concentration of points in the stress space. These results are consistent with the previous qualitative analysis of the rotation angle, which revealed a narrower distribution of the rotation angle criterion for this test configuration. A heterogeneous distribution of material points across the whole stress space would almost certainly improve the accuracy of the identified parameters regarding the anisotropic yield criterion.

Summary

This work investigated the use of different indicators to evaluate the heterogeneity of the mechanical behaviour of distinct test configurations when using the Arcan test. The numerical results from the most interesting test configuration were used to generate synthetic speckle images and processed through DIC and further used in the inverse identification with VFM. The main conclusions that can be drawn from this work are as follows:

- Even with a simple specimen geometry, the Arcan test provided interesting heterogeneous test configurations, demonstrating the potential of its use for heterogeneous test design in the context of metal plasticity.
- The IT_1 criterion proved to be a good quantitative evaluation of the diversity of the material's stress/strain states, which appears to have a significant influence on the inverse identification of the hardening parameters. It is, however, inappropriate for quantifying the identifiability of anisotropy parameters.
- The rotation angle parameter seems to be a suitable qualitative criterion for the evaluation of the sensitivity of mechanical tests to anisotropy.
- When using experimental images and DIC, the propagation of noise and uncertainty throughout the identification process has a significant impact on the identified parameters, resulting in deviations from the ground-truth parameters.

In future works, other test configurations could be used for the inverse identification of the isotropic hardening and anisotropic yielding parameters by finding a good balance between the IT_1 and the rotation angle criterions. Additionally, a greater number of uniform virtual fields could be used in the inverse identification of the constitutive parameters, as could automatic virtual field selection strategies such as sensitivity-based virtual fields. Other specimen geometries could also be used to increase the heterogeneity of stress/strain states of the material, which might become even more important when using more complex constitutive models, such as Yld2000-2D.

Acknowledgements

J. Henriques is grateful to the Portuguese Foundation for Science and Technology (FCT) for the Ph.D. grant 2021.05692.BD. This project has received funding from the Research Fund for Coal and Steel under grant agreement No 888153. The authors gratefully acknowledge the financial support of the FCT under the project PTDC/EMEAPL/29713/2017 by UE/FEDER through the programs CENTRO 2020 and COMPETE 2020, and UID/EMS/ 00481/2013-FCT under CENTRO-01-0145-FEDER-022083. The authors also acknowledge the FCT (FCT - MCTES) for its financial support via the projects UIDB/00667/2020 (UNIDEMI).

Disclaimer

The results reflect only the authors' view, and the European Commission is not responsible for any use that may be made of the information it contains.

References

[1] A. Lattanzi, F. Barlat, F. Pierron, A. Marek, M. Rossi, Inverse identification strategies for the characterization of transformation-based anisotropic plasticity models with the non-linear VFM, Int. J. Mech. Sci. 173 (2020) 105422. https://doi.org/10.1016/j.ijmecsci.2020.105422

[2] P.D. Wu, S.R. MacEwen, D.J. Lloyd, M. Jain, P. Tugcu, K.W. Neale, On pre-straining and the evolution of material anisotropy in sheet metals, Int. J. Plast. 21 (2005) 723-739. https://doi.org/10.1016/j.ijplas.2004.05.007

[3] J.M.P. Martins, A. Andrade-Campos, S. Thuillier, Calibration of anisotropic plasticity models using a biaxial test and the virtual fields method, Int. J. Solids Struct. 172 (2019) 21-37. https://doi.org/10.1016/j.ijsolstr.2019.05.019

[4] N. Souto, S. Thuillier, A. Andrade-Campos, Design of an indicator to characterize and classify mechanical tests for sheet metals, Int. J. Mech. Sci. 101-102 (2015) 252-271. https://doi.org/10.1016/j.ijmecsci.2015.07.026

[5] M. Sutton, J. Orteu, H. Schreier, Image correlation for shape, motion and deformation measurements: basic concepts, theory and applications, first ed., Springer, New York, 2009. https://doi.org/10.1007/978-0-387-78747-3

[6] F. Pierron, M. Grédiac, Towards Material Testing 2.0. A review of test design for identification of constitutive parameters from full-field measurements, Strain 57 (2021) 12370. https://doi.org/10.1111/str.12370

[7] F. Pierron, M. Grédiac, The virtual fields method. Extracting constitutive mechanical parameters from full-field deformation measurements, first ed., Springer, New York, 2012. https://doi.org/10.1007/978-1-4614-1824-5

[8] J. Henriques, J. Xavier, A. Andrade-Campos, Identification of Orthotropic Elastic Properties of Wood by a Synthetic Image Approach Based on Digital Image Correlation, Materials 15 (2022) 625. https://doi.org/10.3390/ma15020625

[9] J. Aquino, A. Andrade-Campos, J.M.P. Martins, S. Thuillier, Design of heterogeneous mechanical tests: Numerical methodology and experimental validation, Strain 55 (2019) 12313. https://doi.org/ 10.1111/str.12313

[10] J. Fu, W. Xie, L. Qi, An Identification Method for Anisotropic Plastic Constitutive Parameters of Sheet Metals, Procedia Manuf. 47 (2020) 812-815. https://doi.org/10.1016/j.promfg.2020.04.251

[11] T. Pottier, P. Vacher, F. Toussaint, H. Louche, T. Coudert, Out-of-plane Testing Procedure for Inverse Identification Purpose: Application in Sheet Metal Plasticity, Exp. Mech. 52 (2012) 951-963. https://doi.org/10.1007/s11340-011-9555-3

[12] M. Conde, Y. Zhang, J. Henriques, S. Coppieters, A. Andrade-Campos, Design and validation of a heterogeneous interior notched specimen for inverse material parameter identification, Finite. Elem. Anal. Des. 214 (2023) 103866. https://doi.org/10.1016/j.finel.2022.103866

[13] M. Gonçalves, A. Andrade-Campos, B. Barroqueiro, On the design of mechanical heterogeneous specimens using multilevel topology optimization, Adv. Eng. Softw. 175 (2023) 103314. https://doi.org/10.1016/j.advengsoft.2022.103314

[14] L. Chamoin, C. Jailin, M. Diaz, L. Quesada, Coupling between topology optimization and digital image correlation for the design of specimen dedicated to selected material parameters identification, Int. J. Solids. Struct. 193-194 (2020) 270-286. https://doi.org/10.1016/j.ijsolstr.2020.02.032

[15] P. Wang, F. Pierron, O.T. Thomsen, Identification of Material Parameters of PVC Foams using Digital Image Correlation and the Virtual Fields Method, Exp. Mech. 53 (2013) 1001-1015. https://doi.org/10.1007/s11340-012-9703-4

[16] P. Wang, F. Pierron, M. Rossi, P. Lava, O.T. Thomsen, Optimised Experimental Characterisation of Polymeric Foam Material Using DIC and the Virtual Fields Method, Strain 52 (2016) 59-79. https://doi.org/10.1111/str.12170

[17] M. Shifa, Strength of Aluminum Alloys Under Static Mixed-Mode I/II Loading Conditions, J. Test. Eval. 46 (2018) 294-304. https://doi.org/10.1520/JTE20160475

[18] A. Kumar, M.K. Singha, V. Tiwari, Structural response of metal sheets under combined shear and tension, Structures 26 (2020) 915-933. https://doi.org/10.1016/j.istruc.2020.05.006

Material Forming - ESAFORM 2023   Materials Research Forum LLC
Materials Research Proceedings 28 (2023) 1131-1142   https://doi.org/10.21741/9781644902479-124

[19] F. Ozturk, S. Toros, S. Kilic, Effects of Anisotropic Yield Functions on Prediction of Forming Limit Diagrams of DP600 Advanced High Strength Steel, Procedia Eng. 81 (2014) 760-765. https://doi.org/10.1016/j.proeng.2014.10.073

[20] Dassault Systèmes. Abaqus 2017 documentation, 2017

[21] R. Hill, A theory of the yielding and plastic flow of anisotropic metals, Proc. R. Soc. Lond. A 193 (1948) 281-297. https://doi.org/10.1098/rspa.1948.0045

[22] H. Takizawa, T. Kuwabara, K. Oide, J. Yoshida, Development of the subroutine library 'UMMDp' for anisotropic yield functions commonly applicable to commercial FEM codes, J. Phys.: Conf. Ser. 734 (2016) 032028. https://doi.org/10.1088/1742-6596/734/3/032028

[23] M. Guimarães Oliveira, S. Thuillier, A. Andrade-Campos, Analysis of Heterogeneous Tests for Sheet Metal Mechanical Behavior, Procedia Manuf. 47 (2020) 831-838. https://doi.org/10.1016/j.promfg.2020.04.259

[24] MatchID: Metrology beyond colors. MatchID version 2022.2, 2022.

[25] J. Henriques, M. Conde, A. Andrade-Campos, J. Xavier, Identification of Swift Law Parameters Using FEMU by a Synthetic Image DIC-Based Approach, Key Eng. Mater. 926 (2022) 2211-2221. https://doi.org/ 10.4028/p-33un7m

[26] S. Avril, M. Grédiac, F. Pierron, Sensitivity of the virtual fields method to noisy data, Comput. Mech. 34 (2004) 439-452. https://doi.org/10.1007/s00466-004-0589-6

[27] A. Marek, F.M. Davis, F. Pierron, Sensitivity-based virtual fields for the non-linear virtual fields method, Comput. Mech. 60 (2017) 409-43. https://doi.org/10.1007/s00466-017-1411-6

[28] J.M.P. Martins, S. Thuillier, A. Andrade-Campos, Calibration of Anisotropic Plasticity Models with an Optimized Heterogeneous Test and the Virtual Fields Method, Residual Stress, Thermomechanics & Infrared Imaging and Inverse Problems 6 (2020) 25-32. https://doi.org/10.1007/978-3-030-30098-2_5

Material Forming - ESAFORM 2023   Materials Research Forum LLC
Materials Research Proceedings 28 (2023) 1143-1154   https://doi.org/10.21741/9781644902479-125

On the constraints and consistency in implicit constitutive modelling using ANNs and indirect training

LOURENÇO Rúben[1,2,a], CUETO Elías[3,b], GEORGIEVA Pétia[4,c]
and ANDRADE-CAMPOS António[1,2,d] *

[1] Dept. of Mechanical Engineering, Centre for Mechanical Technology and Automation, University of Aveiro, Portugal

[2] LASI, Intelligent Systems Associate Laboratory, Guimarães, Portugal

[3] Aragon Institute of Engineering Research (I3A), University of Zaragoza, Spain

[4] Institute of Electronics and Informatics Engineering of Aveiro (IEETA); Dept. of Electronics, Telecommunications and Informatics, University of Aveiro, Portugal

[a]rubenl@ua.pt, [b]ecueto@unizar.es, [c]petia@ua.pt, [d]gilac@ua.pt

Keywords: Constitutive Model, Elastoplasticity, Neural Networks, Indirect Training, Constrained Optimization

Abstract. The training of an Artificial Neural Network (ANN) for implicit constitutive modelling mostly relies on labelled data pairs, however, some variables cannot be physically measured in real experiments. As such, the training should preferably be carried out indirectly, making use of experimentally measurable variables. The unconstrained training of an ANN's parameters often leads to spurious responses that do not comply with the physics of the problem. Applying constraints during training ensures not only the physical meaning of the ANN predictions but also potentially increases the convergence to a global minimum, while improving the model's performance. An ANN material model is trained using a novel indirect approach, where the local and global equilibrium conditions are ensured employing the Virtual Fields Method (VFM). An example of physical constraint is analyzed and applied during the training process.

Introduction

ANNs are powerful function approximators that can be used to implicitly learn constitutive relations directly from data, without having to postulate a mathematical formulation [1–3]. Several successful applications of ANNs for implicit modelling of material behavior have been reported in the literature (e.g., [4–6] among others). Most of the approaches rely on training the ANNs with paired data, usually stress-strain, from numerically generated datasets. Nevertheless, in a real experiment, certain variables, such as stresses, are not measurable and, therefore, the training should preferably be carried out indirectly making use of experimentally measurable variables only.

Although a standard ANN could be able to learn the constitutive behavior of a material, given enough data, it usually works as a black-box model in which its structure is not easily interpretable and there is no guarantee that its predictions are usable, as they can violate fundamental laws of mechanics and thermodynamics [3–5]. Thus, it is necessary to enforce physics-based constraints when using ANNs for implicit constitutive modelling. The incorporation of this knowledge into the network allows it (i) to learn the structure of the underlying constitutive relations, (ii) reduce its sensitivity to noise and (iii) increase its performance regarding inputs outside the training domain [7]. Physics-based constraints act as a regularization agent for ANNs, reducing the space of admissible solutions and allowing the network to learn with smaller datasets, as it already does not have to learn those relationships from data [6,7]. These constraints can be enforced using

Material Forming - ESAFORM 2023
Materials Research Proceedings 28 (2023) 1143-1154

Materials Research Forum LLC
https://doi.org/10.21741/9781644902479-125

custom ANN architectures [3], model constraints (e.g., weight constraints) [7] or penalty/regularization terms [5,8]. Some examples of constraints are shown in Table 1.

In the present work two single layer perceptron models are used to model the linear elastic response of a virtual material. A novel indirect training methodology employing the sensitivity-based Virtual Field Method (VFM) [17] is used to train both models and study the application of constraints during training. Here, the applied constraint was the positive definiteness of the tangent stiffness, which in the absence of plastic deformation, degenerates into the elastic stiffness matrix **D**.

Table 1. Examples of material model constraints to enforce during ANN training.

	Formulation	Description
1st law of thermodynamics	$\gamma_{loc} = \boldsymbol{\sigma} : \dot{\boldsymbol{\epsilon}} + \theta\dot{\eta} - \dot{E}$	The work done by stress must either be stored as recoverable internal energy in the solid or dissipated as heat [8,9]
2nd law of thermodynamics	$\gamma_{loc} \geq 0$	For a sample of material subjected to a cycle of deformation, starting and ending with identical strain and internal energy, the total work must be positive or zero [8,9]
Drucker's postulate	$\Delta\sigma_{ij}\Delta\varepsilon_{ij} \geq 0$	The work done by the tractions through the displacements is positive or zero [12]
Symmetric positive definiteness of tangent stiffness	$\dfrac{\partial\boldsymbol{\sigma}}{\partial\boldsymbol{\varepsilon}} = \mathbf{H} > 0$	The tangent stiffness matrix is symmetric positive definite, ensuring that the strain energy is weakly convex [13]
Time consistency	$\lim_{\Delta\varepsilon\to 0} \Delta\boldsymbol{\sigma} = 0$	A consistent material law maps a state of zero strain onto a state of zero stress [13]

The Drucker's stability criterion bears no physical meaning. Not all materials are stable in this sense, however, issues will arise for materials that do not respect it, when used to solve boundary problems [9,17].

Mechanical Test and Dataset Creation
A heterogeneous test was used to generate synthetic data to train the ANN models. The geometry consists of 3×3 mm^2 plate with thickness $t = 0.1$ mm. The initial mesh, geometry and boundary conditions are depicted in Fig. 1. Symmetry boundary conditions are applied to the boundaries at $x = 0$ and $y = 0$ and a surface traction is applied to the boundary at $x = 3$ mm. The traction follows a non-uniform distribution with a single component along the x-direction, varying linearly in the y-direction, according to: $f_x(y) = my + b$, where m and b, respectively, control the slope and intercept of the distribution.

$$\sigma \cdot \mathbf{n} = \begin{Bmatrix} f_x(y) \\ 0 \end{Bmatrix}$$

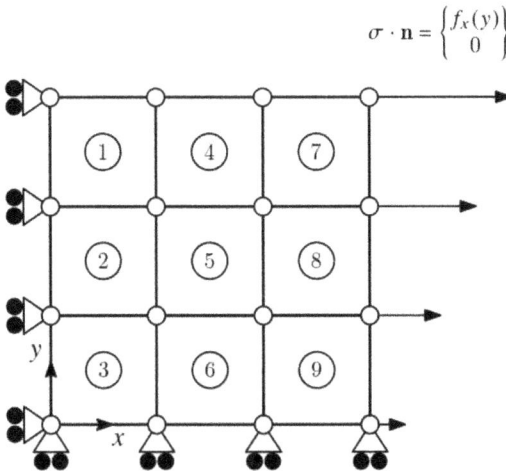

Fig. 1. Heterogeneous test: initial geometry, mesh and boundary conditions, adapted from [14].

The numerical simulations were conducted using the commercial finite element code Abaqus. The model was built with CPS4R elements (bilinear reduced integration plane stress). The elastic parameters were defined as $E = 210$ GPa and $\nu = 0.3$. To generate a training dataset, various simulations were performed with the time period set to 1 and a fixed time increment $\Delta t = 0.001$. For each time step, the strain tensor at the centroid was extracted for all the elements and the resultant force computed from the equilibrium of the internal forces, such that:

$$Fl = \sum_{i=1}^{n} \sigma_i A_i t \tag{1}$$

where F is the global force, l the length of the solid, A is the element's area and t is the thickness. The training and validation datasets were generated for different load distribution parameters, depicted in Table 2. Prior to training and for each mechanical test, the dataset was organized into batches of 9 elements per time increment and shuffled before being split into training (67%) and test data (33%). The input features were scaled in order to have zero mean and unit variance.

Table 2. Loading parameters for the generation of the training and validation datasets.

Training set	$m = \{60,80,100,120\}$ [N /mm]		$b = \{60,80,100\}$ [N]
	m		b
Validation set	50		50
	70		90
	90		65

Material Forming - ESAFORM 2023
Materials Research Proceedings 28 (2023) 1143-1154

Materials Research Forum LLC
https://doi.org/10.21741/9781644902479-125

Neural Network Model

A single layer perceptron with linear activation was chosen to predict the elastic response. The inputs were the components of the strain tensor at a given time t and the outputs were the components of the corresponding stress tensor, as depicted in Fig. 2.

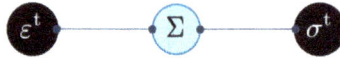

Fig. 2. Single layer perceptron with linear activation.

The model outputs the stress directly according to the following relationship during the forward pass:

$$\sigma = \mathbf{W}\epsilon + \mathbf{b} \iff \begin{Bmatrix} \sigma_{xx} \\ \sigma_{yy} \\ \tau_{xy} \end{Bmatrix} = \begin{bmatrix} w_{11} & w_{12} & w_{13} \\ w_{21} & w_{22} & w_{23} \\ w_{31} & w_{32} & w_{33} \end{bmatrix} \begin{Bmatrix} \epsilon_{xx} \\ \epsilon_{yy} \\ \epsilon_{xy} \end{Bmatrix} + \begin{Bmatrix} b_{11} \\ b_{21} \\ b_{31} \end{Bmatrix} \tag{1}$$

where w_{ij} are the layer weights and b_{ij} the biases. An analogy can be established between the terms of the weight matrix $\mathbf{W}$ and the elasticity matrix $\mathbf{D}$ in the Hooke's law, for an isotropic material under plane stress:

$$\sigma = \mathbf{D}\epsilon \iff \begin{Bmatrix} \sigma_{xx} \\ \sigma_{yy} \\ \tau_{xy} \end{Bmatrix} = \frac{E}{(1+v)(1-2v)} \begin{bmatrix} 1-v & v & 0 \\ v & 1-v & 0 \\ 0 & 0 & \frac{1-2v}{2} \end{bmatrix} \begin{Bmatrix} \epsilon_{xx} \\ \epsilon_{yy} \\ \epsilon_{xy} \end{Bmatrix} \tag{2}$$

with $\mathbf{D}$ defined in terms of the elastic constants E, the Young's modulus and v, the Poisson's ratio. For the material chosen for this work, the elasticity matrix $\mathbf{D}$ is the following:

$$\mathbf{D} = \begin{bmatrix} 230769.231 & 69230.769 & 0 \\ 69230.769 & 230769.231 & 0 \\ 0 & 0 & 80769.231 \end{bmatrix} \tag{3}$$

The architecture presented above was used to train two models in order to learn the elastic response of the material. One of the models was trained following an unconstrained optimization approach, while the other model was trained using a constrained optimization, in order to study how adding constraints would influence the resulting stress predictions.

Indirect Training for Linear Model

Implicit constitutive modelling using ANNs relies on paired data, usually strain and stress tensors, in order to learn the material behavior. However, variables such as stress cannot be obtained from experiments [15]. Therefore, the training must be carried out indirectly, using only measurable data. The VFM, first introduced by Grédiac [16], is a state-of-the-art method employed in the identification of constitutive parameters, known by its computational efficiency and does not resort to FEM for any forward calculations [17]. The key elements behind the VFM are the Principle of Virtual Work (PVW) and the choice of virtual fields. According to the PVW, the internal virtual work must be equal to the external virtual work performed by the external forces and is written by [18]:

$$- \int_V \boldsymbol{\sigma} : \boldsymbol{\varepsilon}^* \, dV + \int_{\partial V} \mathbf{T} \cdot \mathbf{u}^* \, dS = 0 \tag{4}$$

where $\boldsymbol{\varepsilon}^*$ is the virtual strain, $\mathbf{u}^*$ is the virtual displacement, V is the volume of the solid and T is the traction vector. The virtual entities work can be defined independently of the measured displacements/strains. Any number of virtual fields can be used; however, the following conditions should be honored [5,6]: the functions defining the virtual fields should be piece-wise differentiable and kinematically admissible, in order satisfy the displacement boundary conditions. The virtual fields can be manually defined, though the choice is tied to the user's own experience and intuition. Moreover, manually defined virtual fields do not show a temporal evolution [17]. Nonetheless, systematic procedures to automatically define these virtual entities exist, namely: the stiffness-based and the sensitivity-based virtual fields [4,7].

In the present work, sensitivity-based virtual fields were employed to indirectly train a single-layer perceptron for implicit constitutive modelling, following the workflow presented in Fig. 3.

The key concept of the sensitivity-based virtual fields is to apply a perturbation to each of the model's parameters in order to obtain a stress sensitivity and its temporal evolution [17]. The stress field is the only quantity that depends directly on the constitutive parameters in the VFM, so the stress sensitivity maps highlight areas that strongly depend on a given parameter [20]. In the context of this work, the single-layer perceptron from Fig. 2 is taken as a representation of the constitutive model. The spatial sensitivity of stress to each model parameter is computed as:

$$\delta\boldsymbol{\sigma}^{(i)}(\boldsymbol{\epsilon}, \mathbf{W}, t) = \boldsymbol{\sigma}(\boldsymbol{\epsilon}, \mathbf{W}, t) - \boldsymbol{\sigma}(\boldsymbol{\epsilon}, \mathbf{W} + \delta w_i, t) , \; \delta w_i = -0.1 w_i \tag{5}$$

with i being the i-th term of the perceptron's weight matrix $\mathbf{W}$ and t the time step. The virtual displacements $\mathbf{u}^*$ are found starting from the following system of equations applied to a virtual mesh:

$$\delta\boldsymbol{\sigma}^{(i)} = \mathbf{B}\mathbf{u}^{*(i)} \tag{6}$$

where $\mathbf{B}$ is the global strain-displacement matrix, used to map the virtual displacements at the nodes to every virtual strain. If the displacement is prescribed at the boundaries, the traction is unknown and, as such, the displacements are set to zero. If, on the other hand, the traction distribution is unknown and only the resultant force is known, a constant virtual displacement is set at the boundaries. By applying these constraints, a modified global strain-displacement matrix $\overline{\mathbf{B}}$ is obtained and the virtual displacements $\mathbf{u}^*$ can finally be computed as follows:

$$\mathbf{u}^{*(i)} = \text{pinv}(\overline{\mathbf{B}}) \, \delta\boldsymbol{\sigma}^{(i)} \tag{7}$$

with $\text{pinv}(\overline{\mathbf{B}})$ being the pseudo-inverse of the modified strain-displacement matrix. The virtual strains $\boldsymbol{\varepsilon}^*$ are then computed as:

$$\boldsymbol{\varepsilon}^{*(i)} = \mathbf{B}\mathbf{u}^{*(i)} \tag{8}$$

The training process is carried out, with the virtual fields $\mathbf{u}^{*(i)}$ and $\boldsymbol{\varepsilon}^{*(i)}$ being updated multiple times in order to evaluate the loss resulting from the application of the PVW and update the weight matrix $\mathbf{W}$ accordingly. The loss to be minimized is the following:

$$\mathcal{L}(\mathbf{W}, \mathbf{b}, \boldsymbol{\epsilon}) = \sum_{i=1}^{n_{\text{VFs}}} \left[\frac{1}{(\alpha^{(i)})^2} \sum_{t=1}^{n_t} \left(\sum_{j=1}^{n_{\text{pts}}} \boldsymbol{\sigma}^j(\mathbf{W}, \mathbf{b}, \boldsymbol{\epsilon}, t) \cdot \boldsymbol{\varepsilon}^{*j^{(i)}}(t) \cdot S^j - W_{\text{ext}}^*(t) \right)^2 \right] \tag{9}$$

with S^j being the surface area of the j-th measurement point and $\alpha^{(i)}$ the scaling factor of the i-th virtual field, employed to guarantee similar magnitudes between different virtual fields. In the present work, the scaling factor was defined based on the mean of the 30% highest internal virtual work values.

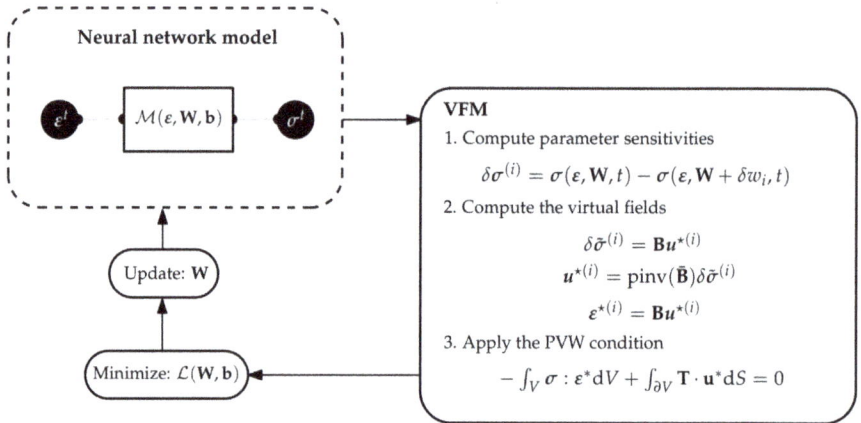

Fig. 3. Coupled ANN-VFM method for implicit constitutive modelling

Unconstrained Training Analysis

The Adam algorithm was used to optimize the network weights, with an initial learning rate set to 0.1, scheduled to be reduced using a multiplier of 0.2 if no improvement in the training loss was registered after 3 epochs. For the unconstrained training model, the network was set to train during a maximum of 10000 epochs. However, an early-stopping criterion was triggered when no further improvement was observed in the test loss, after 2886 epochs. The learning curves are plotted in Fig. 4, showing a sharp decline during training with a low value in the order of 10^{-3} achieved at the end and almost no gap being observed between the train and test curves, indicating the model did not overfit the data.

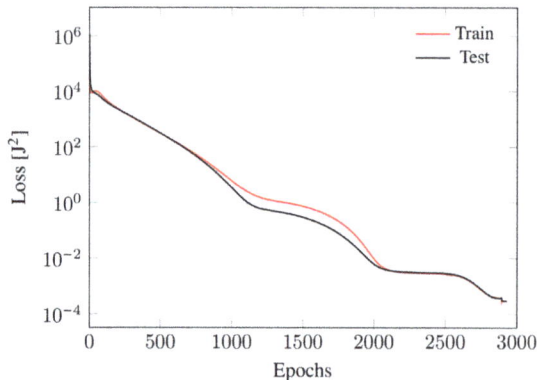

Fig. 4. Learning curves for the unconstrained training model.

Materials Research Forum LLC
https://doi.org/10.21741/9781644902479-125

During training, the layer weights should evolve in such a manner that, at the end of the process, the unscaled matrix **W** approximates **D** as much as possible, with the bias vector **b** being equal to or almost zero. Due to the scaling of the inputs, the parameters hold no significance, so the inverse scaling is needed here to compare both entities. An overview of both scaled and unscaled model parameters obtained after training is shown in Table 3.

Table 3. Scaled and unscaled model parameters for the unconstrained model after training.

	Layer weights **W**			Biases **b**		
Scaled	$\begin{bmatrix} 76.6645 \\ 23.1111 \\ 0.6532 \end{bmatrix}$	$\begin{matrix} 0.01328 \\ 27.4231 \\ 0.3366 \end{matrix}$	$\begin{matrix} 0.0159 \\ 0.0667 \\ 5.1392 \end{matrix}$	$\{111$	0.0997	$5.80\}^{\mathrm{T}}$
Unscaled	$\begin{bmatrix} 209891.4619 \\ 63273.3868 \\ 1788.5317 \end{bmatrix}$	$\begin{matrix} 101.9638 \\ 210502.5075 \\ 2584.3104 \end{matrix}$	$\begin{matrix} 234.9259 \\ 981.5443 \\ 75627.5458 \end{matrix}$	$\{0.0305$	-0.0369	$-0.286\}^{\mathrm{T}}$
Rel. error [%]	$\begin{bmatrix} 9.047 \\ 8.605 \\ \text{N/A} \end{bmatrix}$	$\begin{matrix} 99.853 \\ 8.782 \\ \text{N/A} \end{matrix}$	$\begin{matrix} \text{N/A} \\ \text{N/A} \\ 6.366 \end{matrix}$		N/A	

Examining the unscaled parameters, one can observe that although the single layer perceptron was able to naturally learn nonnegative terms, it was not able to replicate the symmetry of the elasticity matrix, thus failing to guarantee the isotropy of the material. Furthermore, there are significant errors between the unscaled parameters and the terms of the elasticity matrix **D**. Nevertheless, the indices of the most dominant terms (w_{11}, w_{22}, w_{33}) correlate well with those from **D**. The unscaled bias vector holds very small terms in comparison, however not close to zero. From here, we conclude that the model is not able to fully predict the linear elastic response. This is confirmed by observing the plots comparing the real and predicted elastic curves corresponding to two example elements, shown on Fig. 5. The differences stated above, caused the single layer perceptron predictions to achieve correlations ranging from 0.85 to 0.999 and mean absolute errors (MAE) from 0.266 to 5.173.

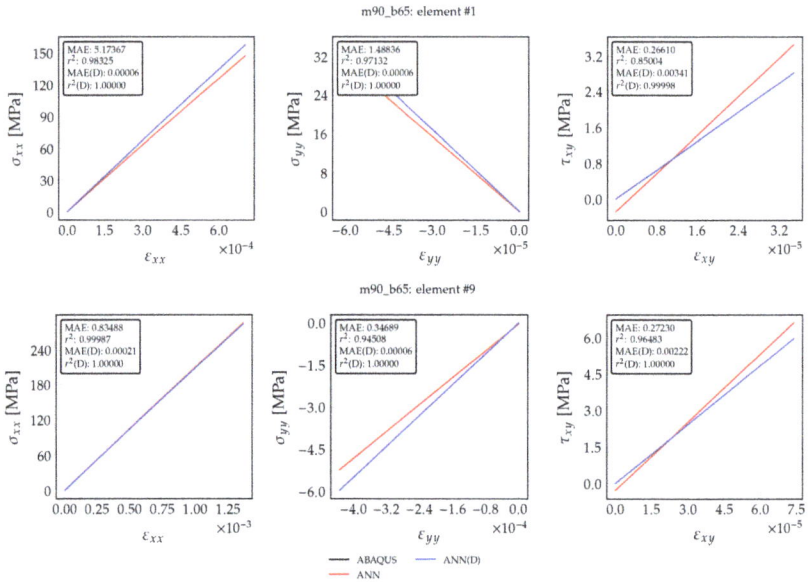

Fig. 5. Validation results for the unconstrained training model for elements 1 and 9.

Constrained Training Analysis

For the constrained training model, the same training parameters were used, with the only difference being that the following set of constraints were applied to the entries of the parameter matrix $\mathbf{W}$:

$$w_{ij} > 0 \wedge w_{13} = w_{31} = 0 \wedge w_{23} = w_{32} = 0 \tag{10}$$

meaning all the terms of $\mathbf{W}$ are forced to be nonnegative and some terms set to be zero, matching the indices of the terms with zero values in $\mathbf{D}$. The training progressed with the early-stopping criterion being triggered when no further improvement was observed in the test loss, after 5381 epochs. The resulting learning curves are plotted in Fig. 6, showing a sharp decline during training with a lower value of loss being achieved at the end (order of 10^{-9}) when compared to the unconstrained model. Similarly, almost no gap is observed between the train and test curves, indicating the model did not overfit the data.

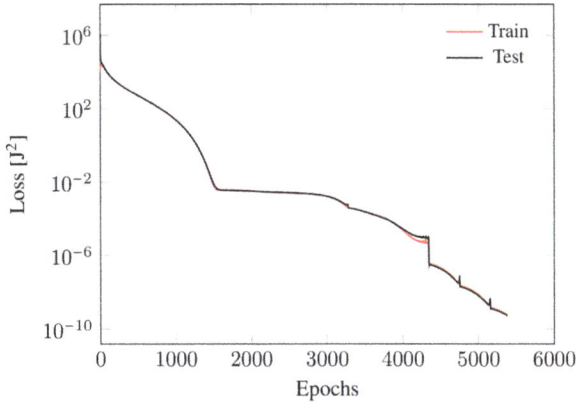

Fig. 6. Learning curves for the constrained training model.

An overview of both scaled and unscaled model parameters obtained after training is shown in Table 4. Examining the unscaled parameters, one can observe that enforcing the constraints defined in (10 was enough for the single layer perceptron to naturally replicate the symmetry of the elasticity matrix, thus guaranteeing the isotropy of the material. Although no constraints were applied to the biases, the unscaled bias vector holds values very close to zero, as it was expected. The model perfectly predicts the elastic response, as it is confirmed by the plots comparing the real and predicted elastic curves corresponding to two example elements, shown on Fig. 7. There are some differences between the unscaled parameters and the terms of the elasticity matrix $\mathbf{D}$, however these are not significantly high to cause de model to output badly predicted curves. The constraints allowed the model to achieve lower mean absolute errors and the highest possible correlations r^2.

Table 4. Scaled and unscaled model parameters for the constrained model after training.

	Layer weights $\mathbf{W}$	Biases $\mathbf{b}$
Scaled	$\begin{bmatrix} 78.4168 & 8.2999 & 0 \\ 23.5103 & 27.8005 & 0 \\ 0 & 0 & 4.9949 \end{bmatrix}$	$\{104.3700 \quad 0 \quad 5.4179\}^T$
Unscaled	$\begin{bmatrix} 230643.294 & 68815.410 & 0 \\ 69149.634 & 230497.092 & 0 \\ 0 & 0 & 80669.215 \end{bmatrix}$	$\{-0.0043 \quad -0.0001 \quad -0.0007\}^T$
Rel. error [%]	$\begin{bmatrix} 0.054 & 0.600 & N/A \\ 0.117 & 0.118 & N/A \\ N/A & N/A & 0.124 \end{bmatrix}$	N/A

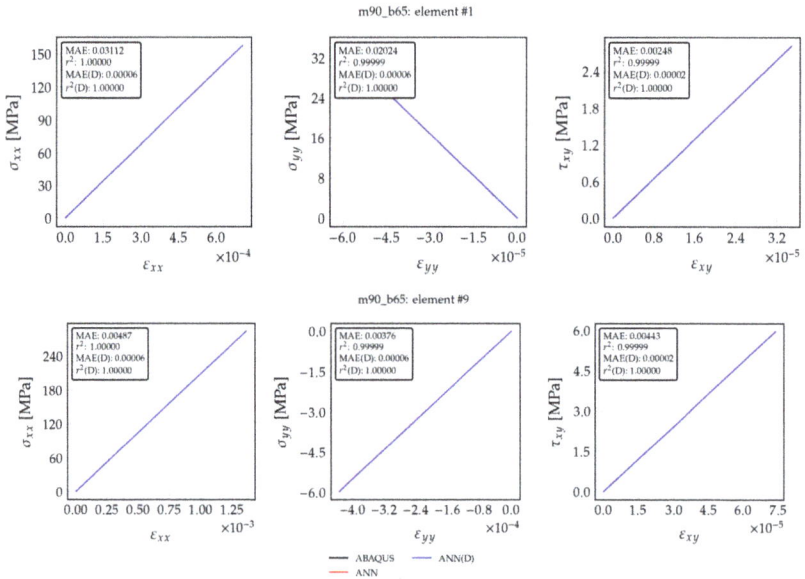

Fig. 7. Validation results for the constrained training model for elements 1 and 9.

Summary

A single layer perceptron model was trained without stress labels employing a new indirect training methodology based on the Virtual Fields Method, which makes use of the Principle of Virtual Work and sensitivity-based virtual fields to guarantee the equilibrium between the external and virtual work. The sensitivity-based virtual fields are a more robust approach than the manually defined virtual fields, providing a systematic way of generating virtual fields that are not static and do not dependent on the user's intuition.

Two different perceptron models with the same architecture were trained in order to learn the linear elastic response of a virtual material and to study the influence of the application of constraints during training. Considering a single layer perceptron with 3 inputs and 3 outputs, an analogy could be established between the mathematical description of the forward pass and the Hooke's law under plane stress, with the perceptron's weight matrix $\mathbf{W}$ resembling the elasticity matrix $\mathbf{D}$. There are several possible ways to constrain the space of admissible solutions of a neural network model. In the present work, by adapting the perceptron's model to the real material model, its parameters gain a real meaning, thus making it easier to control their evolution during training and ensure the predictions are physically admissible. As such, the constraints were applied directly to the perceptron's parameters in order to guarantee nonnegative terms in $\mathbf{W}$ and force some of its terms to be zero, matching those in the elasticity matrix $\mathbf{D}$. It was shown that by applying these constraints, the single layer perceptron naturally learned the symmetry of the elasticity matrix $\mathbf{D}$, respecting the isotropy of the material, and replicated its terms without significant errors, being able to predict a linear elastic response that matches the real one.

Material Forming - ESAFORM 2023
Materials Research Proceedings 28 (2023) 1143-1154

Materials Research Forum LLC
https://doi.org/10.21741/9781644902479-125

Disclaimer
The results reflect only the authors' view, and the European Commission is not responsible for any use that may be made of the information it contains.

Acknowledgements
Rúben Lourenço acknowledges the Portuguese Foundation for Science and Technology (FCT) for the financial support provided through the grant 2020.05279.BD, co-financed by the European Social Fund, through the Regional Operational Programme CENTRO 2020. The authors also gratefully acknowledge the financial support of FCT under the projects UID/EMS/00481/2013-FCT under CENTRO-01-0145-FEDER-022083. This project also supported by the Research Fund for Coal and Steel under grant agreement No 888153.

References
[1] Y. C. Lin, J. Zhang, J. Zhong, Application of neural networks to predict the elevated temperature flow behavior of a low alloy steel, Comput. Mater. Sci. 43 (2008) 752-758. https://doi.org/10.1016/j.commatsci.2008.01.039
[2] K. Hornik, M. Stinchcombe, H. White, Multilayer feedforward networks are universal approximators, Neural Networks 2 (1989) 359-366. https://doi.org/10.1016/0893-6080(89)90020-8
[3] K. Linka, E. Kuhl, A new family of Constitutive Artificial Neural Networks towards automated model discovery, Arxiv (2022). https://doi.org/10.48550/arxiv.2210.02202
[4] J. Ghaboussi, D.A. Pecknold, M. Zhang, R.M. Haj-Ali, Autoprogressive training of neural network constitutive models, Int. J. Numer. Methods Eng. 42 (1998) 105-126. https://doi.org/10.1002/(SICI)1097-0207(19980515)42:1<105::AID-NME356>3.0.CO;2-V
[5] A. Zhang, D. Mohr, Using neural networks to represent von Mises plasticity with isotropic hardening, Int. J. Plast. 132 (2020) 102732. https://doi.org/10.1016/j.ijplas.2020.102732
[6] F. Masi, I. Stefanou, P. Vannucci, V. Maffi-Berthier, Thermodynamics-based Artificial Neural Networks for constitutive modeling, J. Mech. Phys. Solids 147 (2021) 104277. https://doi.org/10.1016/j.jmps.2020.104277
[7] F. As'ad, P. Avery, C. Farhat, A mechanics-informed artificial neural network approach in data-driven constitutive modeling, Int. J. Nume.r Methods Eng. 123 (2022) 2738-2759. https://doi.org/10.1002/NME.6957
[8] L. Borkowski, C. Sorini, A. Chattopadhyay, Recurrent neural network-based multiaxial plasticity model with regularization for physics-informed constraints, Comput. Struct. 258 (2022) 106678. https://doi.org/10.1016/J.COMPSTRUC.2021.106678
[9] M. Raissi, P. Perdikaris, G.E. Karniadakis, Physics-informed neural networks: A deep learning framework for solving forward and inverse problems involving nonlinear partial differential equations, J. Comput. Phys. 378 (2019) 686-707. https://doi.org/10.1016/j.jcp.2018.10.045
[10] E. Cueto, F. Chinesta, Thermodynamics of learning physical phenomena, 2022.
[11] A. Danoun, E. Prulière, Y. Chemisky, Thermodynamically consistent Recurrent Neural Networks to predict non linear behaviors of dissipative materials subjected to non-proportional loading paths, Mech. Mater. (2022) 104436. https://doi.org/10.1016/j.mechmat.2022.104436
[12] A. Bower, Appl. Mech. Solids, CRC Press, 2009.
[13] K. Xu, D.Z. Huang, E. Darve, Learning constitutive relations using symmetric positive definite neural networks, J. Comput. Phys. 428 (2021) 110072. https://doi.org/10.1016/j.jcp.2020.110072
[14] J.M.P. Martins, A. Andrade-Campos, S. Thuillier, Comparison of inverse identification strategies for constitutive mechanical models using full-field measurements, Int. J. Mech. Sci. 145 (2018) 330–345. https://doi.org/10.1016/j.ijmecsci.2018.07.013

[15] X. Liu, F. Tao, H. Du, W. Yu, K. Xu, Learning nonlinear constitutive laws using neural network models based on indirectly measurable data, J. Appl. Mech. Trans. ASME 87 (2020) 1-8. https://doi.org/10.1115/1.4047036

[16] M. Grédiac, Principe des travaux virtuels et identification, Comptes Rendus de l'Académie des Sciences 309 (1989) 1–5.

[17] A. Marek, F.M. Davis, F. Pierron, Sensitivity-based virtual fields for the non-linear virtual fields method, Comput. Mech. 60 (2017) 409–431. https://doi.org/10.1007/s00466-017-1411-6

[18] F. Pierron, M. Grédiac, The virtual fields method: extracting constitutive mechanical parameters from full-field deformation measurements. Springer Science & Business Media, 2012.

[19] M. Grediac, F. Pierront, S. Avrilt, E. Toussaint, The Virtual Fields Method for Extracting Constitutive Parameters From Full-Field Measurements: a Review, Strain 42 (2006) 233-253. https://doi.org/10.1111/j.1475-1305.2006.tb01504.x

[20] A. Marek, F.M. Davis, M. Rossi, F. Pierron, Extension of the sensitivity-based virtual fields to large deformation anisotropic plasticity, Int. J. Mater. Forming 12 (2019) 457-476. https://doi.org/10.1007/S12289-018-1428-1/FIGURES/13

[21] T.B. Stoughton, J.W. Yoon, Review of Drucker's postulate and the issue of plastic stability in metal forming, Int. J. Plast. 22 (2006) 391–433. https://doi.org/10.1016/J.IJPLAS.2005.03.002

Material Forming - ESAFORM 2023
Materials Research Proceedings 28 (2023) 1155-1166

Materials Research Forum LLC
https://doi.org/10.21741/9781644902479-126

Forming process optimisation for variable geometries by machine learning – Convergence analysis and assessment

ZIMMERLING Clemens[1,a] * and KÄRGER Luise[1,b]

[1]Karlsruhe Institute of Technology, Institute of Vehicle System Technology, Karlsruhe, Germany

[a] clemens.zimmerling@kit.edu, [b] luise.kaerger@kit.edu

Keywords: Forming Optimisation, Surrogate, Machine Learning, Variable Geometries

Abstract. For optimum operation, modern production systems require a careful adjustment of the employed manufacturing processes. Physics-based process simulations can effectively support this process optimisation; however, their considerable computation times are often a significant barrier. One option to reduce the computational load is surrogate-based optimisation (SBO). Although SBO generally helps improve convergence, it can turn out unwieldy when the optimisation task varies, e.g. due to frequent component adaptations for customisation. In order to solve such variable optimisation tasks, this work studies how recent advances in machine learning (ML) can enhance and extend current surrogate capabilities. More specifically, an ML-algorithm interacts with generic samples of component geometries in a forming simulation environment and learns to optimise a forming process for variable geometries. The considered example of this work is blank holder optimisation in textile forming. After training, the algorithm is able to give useful recommendations even for new, non-generic geometries. While the prior work considered initial recommendations only, this work studies the convergence behaviour upon component-specific algorithm refinement (optimisation) at the example of two geometries. The convergence of the new pre-trained ML-approach is compared to classical SBO and a genetic algorithm (GA). The results show that initial recommendations indeed converge to the process optimum and that the speed of convergence outperforms the GA and compares roughly to SBO. It is concluded that – once pretrained –the new ML-approach is more efficient on variable optimisation tasks than classical SBO.

Introduction

Most modern production systems are complex systems and require a careful optimisation during production ramp-up. In current practice, this often involves resource-intensive trial-error campaigns combined with expert-judgment based on experience from prior parts. However, shrinking lot sizes, ever shorter development cycles and increasing product diversity severely challenge such empirical approaches and call efficient process optimisation tasks.

Thus, it is found that recurring optimisation tasks for ever-changing geometries or materials, respectively, are a significant economical barrier [1]. This holds all the more when processing delicate materials, such as textiles used for continuous-fibre reinforced plastics (CoFRP). They are usually processed in elaborate, multi-step processes and most often comprise a forming process of a textile. The wide range of adjustable process parameters and the complex, non-linear material behaviour require place high demands on a suitable process configuration and pose a challenging development task.

Material Forming - ESAFORM 2023   Materials Research Forum LLC
Materials Research Proceedings 28 (2023) 1155-1166   https://doi.org/10.21741/9781644902479-126

To reduce the cost of an experimental process development, numerical simulations have gained attention over the last decades [2]. They allow for detailed analyses of complex processes and help concentrate costly experiments on the most promising variants. Also, their inherently digital nature allows a combination with optimisation algorithms. However, they usually involve significant computational efforts and especially repetitive simulations, e.g. iterative optimisation, quickly renders them impracticable in practice.

One option to reduce the numerical effort in such cases is surrogate-based optimisation (SBO) [3]. Surrogates are numerically efficient, data-driven approximations of expensive simulations based on input-output-observations. Once sufficiently trained, optimisation can be done on the surrogate in short time. Overall, SBO results in significant optimisation speed-ups. However, current SBO-approaches are mostly application-specific and fall short on reusability in new scenarios. Even subtle problem variations, e.g. geometry variations in manufacturing, instantly invalidate the surrogate and require resampling of data and reconstructing the surrogate. Thus, demand for generalised models arose.

At the same time, developments in Machine Learning (ML) have achieved remarkable results in complex tasks and may open up new avenues for advanced surrogates . The overarching concept is to sample process observations for a range of generic geometries and analyse it with ML-techniques [4-6]. Recurring patterns in the data may then guide a process optimisation of a new component. Owing to their reconfigurability and ease of evaluation, physics-based numerical process simulations are used for data sampling. Prior work has shown that such models can issue useful process recommendations for new components [4]. However, although the recommendations are useful, they are not strictly optimal but show some deviations to the true optimum. Thus, this work studies whether or not the initial ML-recommendations converge to the true optimum upon component-specific refinement.

Optimisation Methodology
Optimisation Approaches.

Formally, a forming simulation can be seen as a function $\varphi: P \xrightarrow{G} Q$ which maps variable process parameters $p \in P$ to a part quality descriptor $q \in Q$ for a given component geometry $g \in G$. Process optimisation then amounts to searching the optimal process parameters $p^* = \arg\min q(p)$ which yield the best quality[1]. Finding this optimum is a profound task, though, and this work compares three different workflows as shown in Fig. 1.

One approach is to directly couple optimisation algorithms with the simulation φ, e.g. genetic algorithms as Fig. 1a) shows. For a given geometry g, they determine the process parameter optimum p^* by iterative evaluation, variation and combination of parameter combinations. However, φ is generally costly to evaluate and thus, iterating p^* until convergence quickly become prohibitively computation-intensive.

In order to increase efficiency, surrogate-based optimisation (SBO) constructs a numerically efficient substitute function $\mu_{\mathrm{srg}}: P \rightarrow Q$, the "surrogate". The surrogate seeks approximates φ from a set of pre-sampled observations and allows to do the optimisation in short time on μ_{srg} instead of φ, cf. Fig. 1b) [3]. The obtained candidate solution p^*_{srg} is in turn validated in a simulation run and this new observation is fed back to the database. The procedure then repeats until convergence.

Numerous case studies across disciplines have reported substantial optimisation speed-ups by SBO, see e.g. [7]. However, current SBO-strategies provide mostly application-specific, one-off models and struggle with unforeseen task variations. This impairs reusability in new scenarios:

---

[1] In manufacturing the part quality q is often expressed by the extent of defects like cracks or wrinkles which are sought to be minimised, not maximised.

Material Forming - ESAFORM 2023   Materials Research Forum LLC
Materials Research Proceedings 28 (2023) 1155-1166   https://doi.org/10.21741/9781644902479-126

Even a subtle problem variation, e.g. a change of material or geometry, instantly invalidates the surrogate and requires resampling of data and reconstructing the surrogate [4]. Thus, demand for generalised models has been identified early on [8].

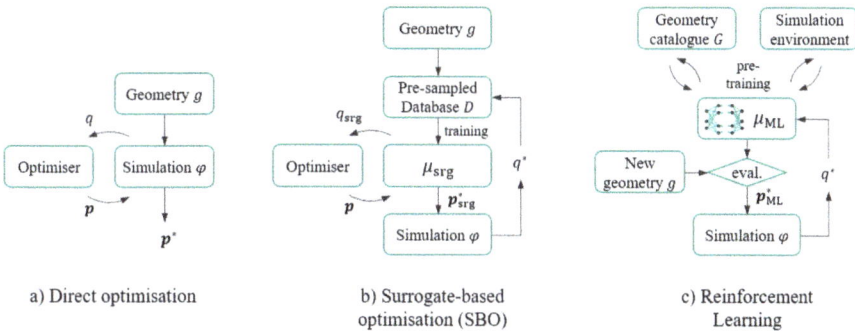

Fig. 1. Workflows of a) direct, b) surrogate-based optimisation (SBO) and c) Reinforcement Learning on multiple geometries.

As a remedy, prior work of the authors [4] and [9] suggests to give up on classical surrogate models μ_{srg} in favour of a more generalised function $\mu: G \rightarrow P$. While classical surrogate models are in most cases used for fix geometries or consider parametric geometries only, μ is meant to learn the underlying part-process-relations from a whole set of non-parametric generic training geometries. After training, it shall be able to give parameter recommendations even for new geometries which are not part of the training geometries. A detailed description of the implementation is given in the prior publications [4] and [9] and thus, only a brief glimpse is provided in this work.

Reinforcement Learning for process optimisation.

The overarching idea is that the ML model function interacts with a set of generic geometries (geometry catalogue G) in a simulation environment. The workflow is visualised in Fig. 2a).

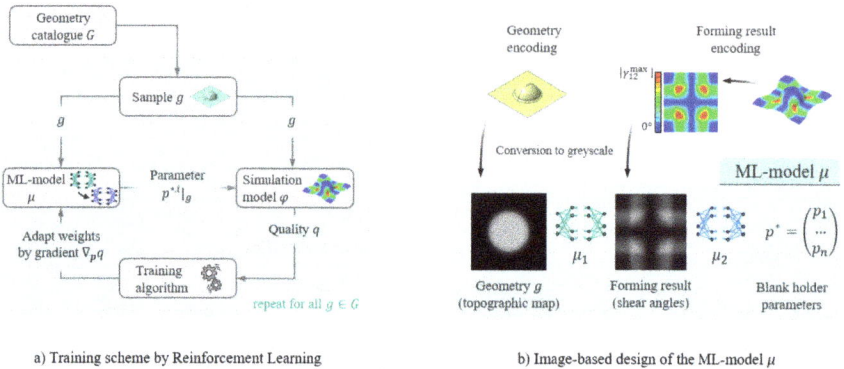

a) Training scheme by Reinforcement Learning

b) Image-based design of the ML-model μ

Fig. 2. a) Training scheme via Reinforcement Learning and b) image-based design of the ML-model $\boldsymbol{\mu}$ [4,9].

In each training iteration i, a geometry g is drawn from the geometry catalogue. Then the ML-model μ analyses g and issues an estimation of optimal parameters $p^*|_g$. This parameter recommendation is then evaluated in a simulation run φ to determine the resulting part quality q. If $p^*|_g$ indeed improves the part quality, then the ML-model is encouraged to give similar recommendations for similar geometries in the future. Formally, this means that the gradient $\nabla_p q$ is determined and used to adapt μ's parameter recommendations in direction of increasing quality q. The procedure iterates until the forming quality across all geometries in the catalogue seizes to improve.

In the context of material forming, image-based approaches for μ – as illustrated in Fig. 2b) – have been proposed [4-6]. Compared to conventional geometry parameters, e.g. length, width, fillet radii or angles, image-based geometry descriptions have proven robust and versatile when applicability to variable geometries is key. In this work and the prior works of the authors, μ consists of two nested functions, μ_1 and μ_2, both of which are image-processing neural networks, so-called convolutional neural networks. The functions serve different purposes as Fig. 2b) visualises: μ_1 interprets the geometry and issues an estimation where and to what extent material strains are likely to occur. More specifically, the shear strain is evaluated as in-plane shear is the dominant deformation mechanism in engineering fabrics. Then, μ_2 interprets this strain-estimation and devises a parameter recommendation p^*.

Simulation Model and Optimisation Task
This work picks up on the models from the prior works [4,9] and is based on experimentally validated simulation approaches. It considers forming of cuboid geometries with the aid of pressure pads, cf. Fig. 3a). The cuboids can be varied in their dimensions w_1 and w_2 (width and length) as sub-image b) shows. For all geometries, the pads can be positioned freely around the perimeter of the geometries in order to control the draw-in of the fabric during forming.

Material Forming - ESAFORM 2023
Materials Research Proceedings 28 (2023) 1155-1166

Materials Research Forum LLC
https://doi.org/10.21741/9781644902479-126

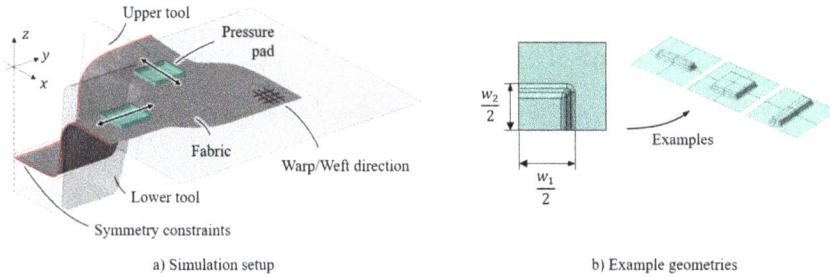

a) Simulation setup

b) Example geometries

Fig. 3. a) Simulation setup for fabric forming with pressure pads; b) sample geometries in the geometry catalogue. Images are adapted from the prior work [4].

The cuboids are deliberately severe and show a strong tendency towards wrinkling. The overall quality goal during forming is to obtain a wrinkle-free forming result. As a direct measure of wrinkling, the fabric curvature κ is evaluated according to [11]. More specifically, the 99.5 % percentile κ_{qnt} of a Weibull fit to the fabric curvature distribution quantifies the quality, cf. Fig. 4a). The contour plot in Fig. 4b) shows κ_{qnt} for a single geometry as a function of the position of the pressure pads $p_{1,2}$ along with a top-view-visualisation of the pressure pads. The plot has been established by a fine grid-sampling of all possible pad positions for a single geometry. Note that these grid-samples are *by no means* involved in algorithm training but serve only for visualisation of κ_{qnt} (objective function) from an 'omniscient' perspective. The yellow marker illustrates how a specific pad position relates to the contour plot and vice-versa. Note that the plot is normalised so as to facilitate comparability with the following plots.

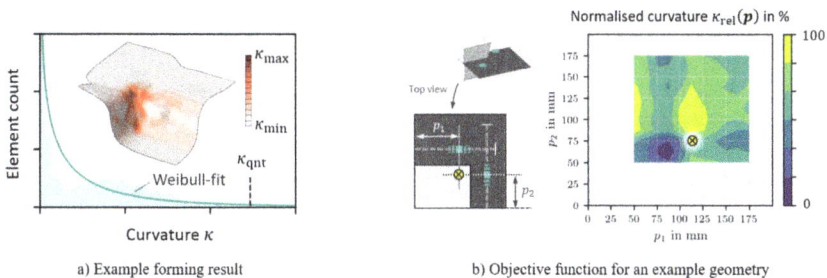

a) Example forming result

b) Objective function for an example geometry

Fig. 4. a) Exemplary forming result with wrinkling, histogram of the fabric curvature plus Weibull fit; b) Contour plot of the obtained curvature (quality) depending on the pad positions $p_{1,2}$. Images adapted from prior work [4].

Clearly, different pad positions significantly alter the obtained curvature and a distinct optimum, i.e. minimal curvature, can be observed. All geometries show such an optimum; however, its location varies and μ shall estimate its position for each geometry in the catalogue.

Material Forming - ESAFORM 2023   Materials Research Forum LLC
Materials Research Proceedings 28 (2023) 1155-1166   https://doi.org/10.21741/9781644902479-126

Results and Discussion

In the prior work [9], μ is trained on a set of box-geometries shown in Fig. 3b). The results show that it gives useful estimations not just for new box-geometries but also – to a certain degree – for 'non-box' geometries. See [9] for further details on the results. Two of these test geometries shall be further analysed in this work. They are shown in Fig. 5 on the left: A rotational symmetric shell with conical tips (g_1) and a double-dome geometry (g_2).

The contour plots in Fig. 5 show the forming quality (curvature) for all pad positions obtained by a grid-sampling approach. Note that they are not a subset of the box-geometries but, as they are doubly-symmetric and near-convex, they will still show a similar forming behaviour: Both contour plots feature a minimum similar to the box-geometry in Fig. 4b). Some qualitative differences can be observed, though: For g_1, the optimum is not a sharp point but a plateau-like region in the bottom-left of the plot. In contrast, the optimum for g_2 is comparably distinct and surrounded by two, connected maxima in a funnel-like shape while the majority of the plot is of approximately constant, mediocre quality.

Additionally, the plots feature three types of markers, namely two large and 30 small markers. The blue large marker visualises the best quality obtained during grid-sampling which is assumed to be the 'true optimum' p^*. The yellow large marker denotes μ's initial process recommendation $p_{ML}^{*,0}$ immediately after training on box-geometries. If both the blue and the yellow marker coincided, that would imply that μ had made a perfect estimation of the process optimum. However, such a perfect estimation is highly unlikely since μ makes inference for a new geometry on the basis of generic box-geometries. Thus, the two markers are close to each other but still lie some distance apart. More specifically, an overestimation of $\approx 18\ \%\ (g_1) \approx 13\ \%\ (g_2)$ relative to the parameter range $25\ \text{mm} \leq p_{1,2} \leq 200\ \text{mm}$ is observed.

However, the question arises, whether the yellow marker approaches the blue marker (true optimum) upon component-specific continuation of training. That is, having been pretrained on generic boxes, μ interacts now only with geometry g_1 or g_2, respectively. Thereby μ can iteratively refine on these specific geometries. The evolution of μ's recommendation $p_{ML}^{*,i}$, is visualised by the smaller markers whose hue denotes their order of appearance, i.e. refinement iteration $i_{rfn} = 1$ (black) to $i_{rfn} = 30$ (white).

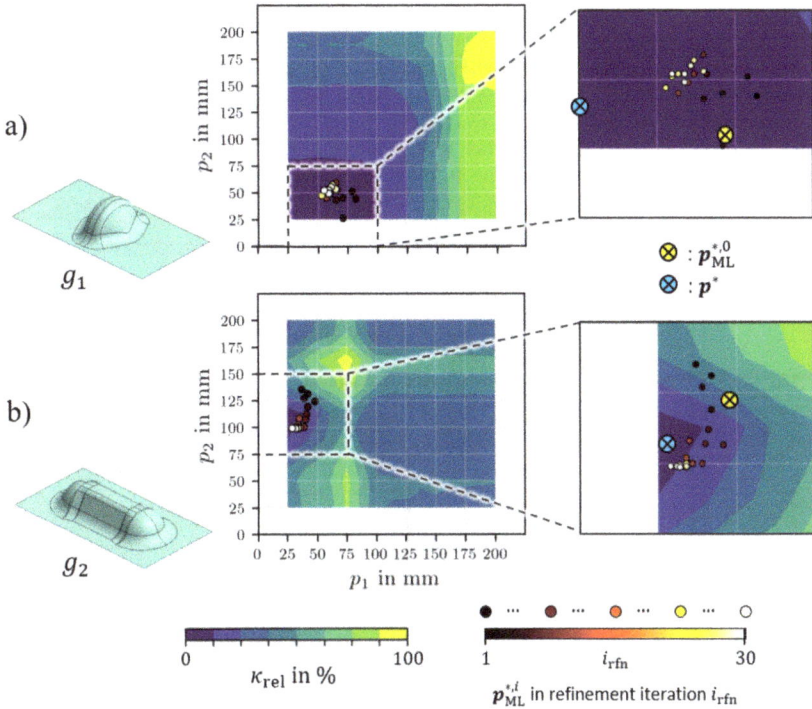

*Fig. 5. Contour plots with **μ**'s process recommendations during component-specific continuation of training for a) geometry **g_1** (rotational symmetric shell) and b) **g_2** (double-dome).*

However, the question arises, whether the yellow marker approaches the blue marker (true optimum) upon component-specific continuation of training. That is, after the training on generic box-geometries, μ interacts now only with geometry g_1 or g_2, respectively. Thereby μ is meant to iteratively refine its recommendations on these specific geometries. The evolution of μ's recommendation $p_{ML}^{*,i}$, is visualised by the smaller markers, while their hue denotes their order of appearance, i.e. refinement iteration $i_{rfn} = 1$ (black) to $i_{rfn} = 30$ (white).

The plots reveal a disparate refinement behaviour: For geometry g_1 in subplot a), the refined markers appear somewhat incoherently scattered around the initial recommendation. At most, a light tendency to the top left can be observed. In contrast, for g_2 (double-dome) in subplot b), the markers do accumulate and show a coherent evolution: At first, the markers move to the top left. But, since this deteriorates the results, they reverse and gradually move downwards before concentrating near the minimum.

These observations can be explained when examining the objective functions in the contour plots: For g_1 in subplot a), the plateau-characteristic makes it difficult to identify a gradient for an improved pad position. Accordingly, the recommendations (markers) appear erratic. In contrast, a distinct optimum is observable for g_2 and – after a few iterations for 'orientation' – the markers move in direction of improved part quality. For some reason however, they only approach but do not reach the observed optimum (blue marker). This may be due to numerical noise when evaluating the forming simulation.

Ultimately, Fig. 6 illustrates the optimisation progress for both geometries by means of the objective function. The diagrams support the above line of thought: Sub-image a) shows that the forming quality κ_{rel} stays practically constant for g_1 through all refinement iterations, while it improves (declines) for g_2 after correction of an initial peak. Overall, the results indicate that– like a regular surrogate – μ can indeed refine its initial recommendations with new evidence.

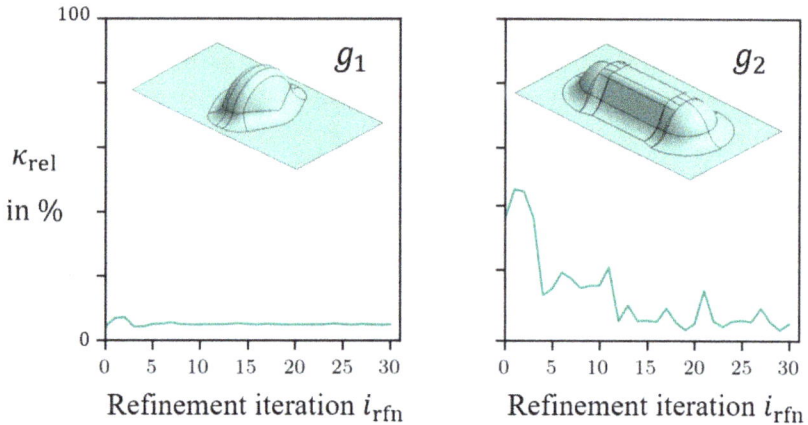

Fig. 6. Evolution of the optimisation objective κ_{rel} during component-specific refinement of μ on g_1 and g_2.

Summary

Having seen that the RL-based optimisation method indeed shows optimisation behaviour, the three optimisation approaches from Fig. 1 – direct optimisation with a genetic algorithm (GA), SBO and RL – can be compared to each other. Each method is applied to the geometries g_1 and g_2 for inspection of the optimisation behaviour. The GA stems from the publicly available optimisation toolbox 'Dakota' [12] and the SBO reference method is described in detail in [13].

Note that the optimisation approaches employ different iterative schemes: The GA utilises a generation-based principle with each generation comprising $n_{idv} = 20$ individuals. It is set to terminate after $n_g = 15$ generations, i.e. after a total of $n_g \cdot n_{idv} = 15 \cdot 20 = 300$ simulations. This limit prevents excessive computation times and was determined empirically in a prior analysis on a fast – but much simpler – substitute model, cf. [4]. The limit is set such that at least one individual becomes (near-)optimal; yet it does not necessarily imply full convergence of the whole population around the optimum. For prevention of a potential quality-loss, the best found solution from the previous generation is (unaltered) carried over to the next generation (so-called 'elitist' selection). Both SBO and RL employ $n_i = 30$ refinement iterations (simulations). However, SBO requires a component-specific sampling – in this work Latin Hypercube Sampling – which accounts for additional $n_s = 20$ a-priori simulations before optimisation start. Also note, that RL needs no specific sampling due to its pretraining on box-geometries. Thus, it can directly start refining its recommendations for optimisation.

Geometry g_1.

For a direct comparison, Fig. 7 plots the progress of each optimiser on the objective κ_{rel} in one diagram along with a detail view for closer inspection (grey shade). Besides the sheer sequence of κ_{rel} (thin dashed line), the plots also show their lower envelope (bold solid line). Essentially, it gives for each iteration the 'so-far-best' solution and visualises, how fast each optimiser reduces the objective function. Overall, all graphs show a successful optimisation behaviour as they all approach the minimum ($\kappa_{rel} = 0\,\%$).

However, although all approaches improve the solution over the iterations, the gain during additional optimisation iterations is limited as the first solutions are already near-optimal ($\kappa_{rel} \approx 1...5\,\%$). Thus, they offer little room for improvement with further iterations. This phenomenon can be explained by inspection of the contour plot, cf. Fig.5a): Since the optimum is not a sharp point but a comparably large plateau, the optimisers find it right at the beginning and thus have no room for improvement during iterations.

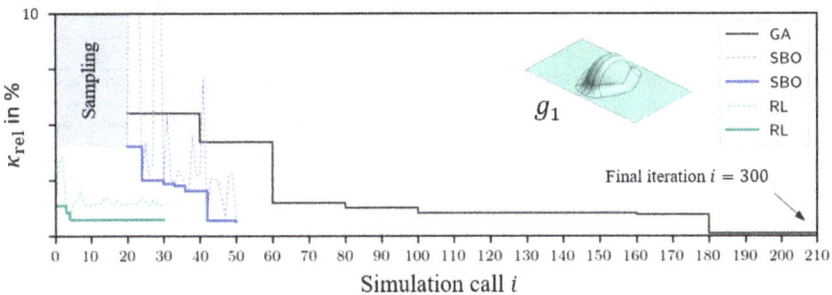

Fig. 7. Juxtaposition of κ_{rel} during optimisation on g_1 for the GA, SBO and the RL-approach. The plots of GA and SBO are offset to account for the simulations until a first iteration occurs. Note that κ_{rel} is near-optimal directly from the beginning ($\kappa_{rel} \approx 1...5\,\%$). In accord, the plot is magnified with a y-axis focus on the lower 10%.

Despite the limited gain, the GA finds the overall-best solution ($\kappa_{rel} = 0.2\,\%$), while both RL and SBO ($\kappa_{rel} = 0.82\,\%$) remain slightly inferior ($\kappa_{rel} = 0.78\,\%$). Note however, that RL and SBO require far fewer iterations to reach their final value, i.e. they converge faster. Also note that all algorithms involve elements of randomness. Thus, re-running them will probably lead to slightly different graphs and optimisation results.

Geometry g_2.

In a similar manner, Fig. 8 shows graphs for each optimisation approach on g_2 (double-dome) The overall behaviour of the algorithms stays the same as before: A successful optimisation can be observed. However, this time a notable effect of additional iterations is observed. Due to the majority of mediocre process responses (plateaus), all methods start at $\kappa_{rel} \approx 40...50\,\%$ in their first iteration, but then the progresses differ.

Material Forming - ESAFORM 2023   Materials Research Forum LLC
Materials Research Proceedings 28 (2023) 1155-1166   https://doi.org/10.21741/9781644902479-126

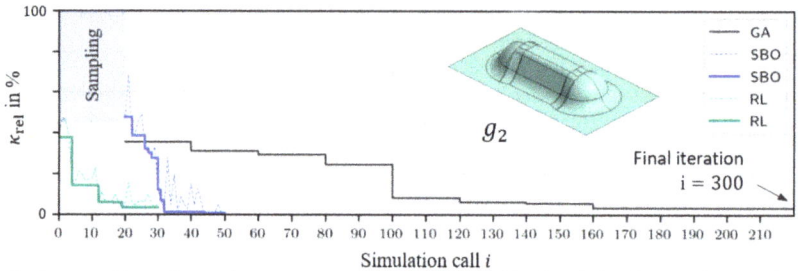

Fig. 8. Juxtaposition of κ_{rel} during optimisation on g_2 (double-dome) for the GA, SBO and the RL-approach. The plots of GA and SBO are offset to account for the number of simulations until a first iteration.

Overall, the plot resembles the previous: The graphs decline in a monotonous manner to the minimum (κ_{rel} = 0%) and RL and SBO outperform the GA. However, this time all graphs show a substantial improvement of the objective function κ_{rel} during optimisation. All optimisers find a solution within a 5 %-range around the optimum, while SBO finds the overall best-solution with $\kappa_{rel} \approx 2$ %. The GA reaches the 5 %-range after 160 simulations and RL after 18 iterations.

Speed of convergence. Ultimately, the speed of convergence shall be briefly assessed. To this end, the convergence metric

$$C = \int_{i=0}^{i_{max}} \kappa_{rel}(i)\, di = \sum_{i=0}^{i_{max}} \kappa_{rel}(i)\, \Delta i = \sum_{i=0}^{i_{max}} \kappa_{rel}(i) \tag{1}$$

is used. Essentially, it is the area under the (solid) graphs until each optimiser has reached its final value in iteration i_{max}. Due to the integer-abscissa, $di = \Delta i = 1$ holds which simplifies the integral to a sum. Overall, a smaller value of C implies faster convergence. If an approach finds a good solution early on, the following summands and thus C becomes smaller. This reflects the desired behaviour during optimisation of expensive functions, i.e. rapid minimisation with only few function evaluations. Table 1 below summarises the convergence metrics C and numbers of simulations i_{max} each optimisation approach required until its final value.

Table 1: Convergence comparison of GA, SBO and RL according to the convergence metric $\mathbf{C}$ and the required number of iterations $\mathbf{i_{max}}$.

Geometry	GA		SBO		RL	
	C	i_{max}	C	i_{max}	C	i_{max}
g_1	4.39	180	0.79	42	0.31	4
g_2	37.13	160	13.50	32	4.02	19
sum	-	340	-	74	-	23

For both geometries, g_1 and g_2, SBO and the new RL-based method outperforms the GA by a large margin: While the GA requires 160 or 180 simulations, respectively, SBO and RL require less than 50 simulations to reach an approximately equal forming result. Equally large differences can be observed for the C. When comparing SBO and RL directly, RL appears more efficient than SBO. This is mainly because the RL-based approach does not require a component-specific sampling due to its pretraining on generic (box-)geometries. It must be noted however, that the computational effort for this pretraining may considerable.

Material Forming - ESAFORM 2023
Materials Research Proceedings 28 (2023) 1155-1166

Materials Research Forum LLC
https://doi.org/10.21741/9781644902479-126

The advantage of the pretrained RL-model becomes even more evident when optimising multiple geometries: Suppose, both geometries g_1 and g_2 need to be manufactured and their processes must be optimised. Optimising both geometries with SBO requires in total 74 simulations of which $2 \cdot n_s = 40$ simulations are required just for sampling. In contrast, the RL-model can do without component-specific sampling due to its pretraining on boxes and takes only 23 simulations in total. Although the exact numbers will certainly vary in different applications, the non-necessity of the a-priori sampling substantially cuts the computational effort.

Summary

This work studies the convergence behaviour of a previously proposed RL-based approach for process optimisation of variable geometries. The approach centres around an ML-model which interacts iteratively with a catalogue of generic geometries in a simulation environment. It thereby learns which geometry requires which optimal process configuration. The approach is studied at an example from prior work, namely optimisation of pressure pad positions during textile forming. Prior work has shown that – after the (pre-)training on generic geometries – the RL-approach indeed gives useful, near-optimal process recommendations [4,9]. This work picks up on these results and investigates, whether the recommendations converge to the true optimum upon component-specific refinement and if so, how fast.

The results are twofold: First, the results on two demonstrator geometries hint that the proposed RL-model indeed converges to the true process optimum. Second, a comparison to two conventional optimisation approaches – a genetic algorithm (GA) and a 'classical' surrogate-based approach (SBO) – shows that RL and SBO outperform the GA and show a similar convergence during refinement iterations. However, SBO needs a component-specific a-priori sampling while the RL-based method is pre-trained on generic geometries and can directly start refining its recommendations. Thus – once trained – it speeds up the optimisation process similar to a classical surrogate but beneficially cuts the component-specific sampling effort. However, this comes at the cost of substantial numerical efforts for algorithm pretraining.

The results of this work and the prior work show that it is indeed possible to extract usable part-process-relations from generic samples and use them during process optimisation for new components. While this work outlines their principal potential, the developed techniques yet need to be advanced to more complex and application-centred use-cases from industrial practice. Regarding textile-forming, this means the integration of more complex process scenarios, e.g. more complex geometries, additional process parameters and variable material properties. In the wider sense, the methods can also be tested on other processes inside and outside material forming.

Acknowledgment

This work is part of the IGF research project *OptiFeed* (21949 N) of the research association *Forschungskuratorium Textil e.V.*, Reinhardtstraße 14-16, 10117 Berlin, funded via the *AiF* within the program for supporting the Industrial Collective Research (IGF) from funds of the Federal Ministry of Economic Affairs and Climate Protection (BMWK). It is also part of the Young Investigator Group (YIG) *Green Mobility*, generously funded by *Vector Stiftung*.

References

[1] N. Shamsaei, A. Yadollahi, L. Bian, S.M. Thompson, An overview of direct laser deposition for additive manufacturing - part 2, Additive Manuf. 8 (2015) 12-35. https://doi.org/10.1016/j.addma.2015.07.002

[2] L. Kärger, A. Bernath, F. Fritz, S. Galkin, D. Magagnato, A. Oeckerath, A. Schön, F. Henning, Development and validation of a cae chain for unidirectional fibre reinforced composite components, Compos. Struct. 132 (2015) 350-358. https://doi.org/10.1016/j.compstruct.2015.05.047

Material Forming - ESAFORM 2023
Materials Research Proceedings 28 (2023) 1155-1166

Materials Research Forum LLC
https://doi.org/10.21741/9781644902479-126

[3] S. Koziel, L. Leifsson, Surrogate-Based Modeling and Optimization: Applications in Engineering. Springer Science+Business Media, New York, 2013. https://doi.org/10.1007/978-1-4614-7551-4

[4] C. Zimmerling, C. Poppe, O. Stein, L. Kärger, Optimisation of manufacturing process parameters for variable component geometries using reinforcement learning, Mat. Des. 214 (2022) 110423. https://doi.org/10.1016/j.matdes.2022.110423

[5] H.R. Attar, H. Zhou, A. Foster, N. Li, Rapid feasibility assessment of components to be formed through hot stamping: A deep learning approach, J. Manuf. Process. 68 (2021) 1650-1671. https://doi.org/10.1016/j.jmapro.2021.06.011

[6] H. Zhou, Q. Xu, Z. Nie, N. Li, A study on using image-based machine learning methods to develop surrogate models of stamp forming simulations, J. Manuf. Sci. Eng. 144 (2022) 021012. https://doi.org/10.1115/1.4051604

[7] R.T. Haftka, D. Villanueva, A. Chaudhuri, Parallel surrogate-assisted global optimization with expensive functions – a survey, Struct. Multidisc. Optimization 54 (2016) 3-13. https://doi.org/10.1007/s00158-016-1432-3

[8] H. El Kadi, Modeling the mechanical behavior of fiber-reinforced polymeric composite materials using artificial neural networks - a review, Compos. Struct. 73 (2006) 1-23. https://doi.org/10.1016/j.compstruct.2005.01.020

[9] C. Zimmerling, C. Poppe, L. Kärger, Estimating optimum process parameters in textile draping of variable part geometries - A reinforcement learning approach, Procedia manuf. 47 (2020) 847-854. https://doi.org/10.1016/j.promfg.2020.04.263

[10] C. Zimmerling, D. Trippe, B. Fengler, L. Kärger, An approach for rapid prediction of textile draping results for variable composite component geometries using deep neural networks, AIP Conference Proceedings 2113, ESAFORM 2019 in Vitoria-Gasteiz/Spain (2019), Art. 020007. https://doi.org/10.1063/1.5112512

[11] S. Haanappel, Forming of UD fibre reinforced thermoplastics: A critical evaluation of intra-ply shear. PhD-Thesis, Universiteit Twente, 2013. https://doi.org/10.3990/1.9789036535014

[12] Sandia National Laboraties. DAKOTA user's manual - version 6.15, www.dakota.sandia.gov/sites/default/files/docs/6.15/Users-6.15.0.pdf (2021), last accessed: 31.08.2022.

[13] C. Zimmerling, P. Schindler, J. Seuffert, L. Kärger: Deep neural networks as surrogate models for time-efficient manufacturing process optimisation. PoPuPS of ULiège Library, ESAFORM 2021 in Liège/Belgium (2021). https://doi.org/10.25518/esaform21.3882

Material Forming - ESAFORM 2023
Materials Research Proceedings 28 (2023) 1167-1174

Materials Research Forum LLC
https://doi.org/10.21741/9781644902479-127

Identification of the large strain flow curve of high strength steel via the torsion test and FEMU

VANCRAEYNEST Niels[1,a]*, COOREMAN Steven[2,b] and COPPIETERS Sam[1,c]

[1]Elooi lab, Department of Materials Engineering, KU Leuven, KU Leuven-Gent Campus, Gebroeders De Smetstraat 1, 9000 Ghent, Belgium

[2] Applications & Solutions department, ArcelorMittal Global R&D Gent / OCAS NV, Belgium

[a]niels.vancraeynest@kuleuven.be, [b]steven.cooreman@arcelormittal.com, [c]sam.coppieters@kuleuven.be

Keywords: Large Strain Flow Curve, Torsion Test, Inverse Identification

Abstract. A torsion specimen is removed from the as-received thick high strength steel sheet (S700MC with a nominal thickness of 12 mm). The test material has a maximum uniform tensile strain of about 12 %. The torsion test is conducted up to fracture. The experimentally acquired torque-angle curve is then used to inversely identify the large strain flow curve up to an equivalent plastic strain of approximately 1. The identification strategy is based on the Finite Element Model Updating (FEMU) approach.

Introduction

Nowadays, numerical simulations are widely used in industry, hence generally considered to be a common engineering tool. However, the quality of simulations highly depends on the accuracy of the inputs. In metal forming simulations, it is well-known that the accuracy of the digital representation of the plastic material behaviour, i.e. the so-called material model, is of crucial importance for the predictive accuracy of the simulation. Commercially available finite element codes are still confined to phenomenological material models. In the case of metal plasticity, such models mostly rely on the concept of a yield surface, a hardening law and the associated flow rule. Experimental data is used to calibrate the governing material model parameters. Obviously, the type, quality and the amount of data determines the calibration accuracy. For example, the work hardening behaviour, also referred to as the flow curve, is conventionally calibrated using a standard tensile test in the Rolling Direction (RD). Since the majority of the metal forming processes generate plastic deformations beyond the maximum uniform tensile strain, the standard tensile test is of limited usefulness to determine the work hardening at large plastic strains, i.e. the large strain flow curve. In the case of sheet metal, several dedicated experiments enable to determine the large strain flow curve. Coppieters et al. [1] recently provided a comprehensive review on methods to determine the large strain flow curve of thin sheet metal. However, the majority of these methods cannot be applied to thick steel sheets. Zhang et al. [2] showed that it is possible to inversely extract the large strain flow curve from the diffuse neck in a tensile test. Alternatively, a torsion test can be used to probe large plastic strains without pronounced necking phenomena [3]. As opposed to a compression test, there is no influence of friction in the torsion test. To acquire the flow curve, the experimentally acquired torque-angle curve can be analytically converted to an equivalent stress-strain curve. The rotation angle directly yields the shear strain which is then converted to the equivalent plastic strain. However, for large plastic deformations, the validity of this analytical approach is violated since the relation between shear strain and equivalent strain does not remain linear [4,5]. The analytical post-processing is further complicated by the inhomogeneous stress and strain distribution within the torsion specimen. Indeed, the stress and strain radially increases in the cross section of the specimen. Thin-walled tubes circumvent

Material Forming - ESAFORM 2023   Materials Research Forum LLC
Materials Research Proceedings 28 (2023) 1167-1174   https://doi.org/10.21741/9781644902479-127

this problem, yet often buckle due to eccentricity of the inner and outer diameter or an unfavourable ratio between the wall thickness and radius [6]. Consequently, this work aims at extracting the post-necking work hardening behaviour of thick high-strength steel through a torsion test on a cylindrical bar. According to Petrov et al. [7], the analytical methods lack accuracy and it is recommended to inversely post-process the experimental data acquired during a torsion test. For example, Gavrus et al. [8] proposed a FEMU method to inversely identify the large strain flow curve. A similar approach is followed in this paper to identify a suitable phenomenological hardening law enabling to describe the pre- and post-necking hardening behaviour of a S700MC sheet with a nominal thickness of 12 mm. The paper is structured as follows. In the first section the experimental details of the torsion experiment are described. The second section embarks on the numerical counterpart of the torsion test. The third section introduces the FEMU method. In section four, the results are presented and discussed.

Experimental

The torsion test is conducted on a custom-made tension-torsion machine (see Fig 1). Consequently, one grip (twisting head) is able to rotate and the other grip (weighing head) is restricted to only translational movement along the rotation axis. The rotation is induced by a servo motor and the angle of rotation of the twisting head is measured with an internal encoder. At the weighing head, gripping the other side of the specimen, the twisting moment is measured using a piezoelectric load cell with a moment and tensile capacity of 100 Nm and 5 kN, respectively. To obtain a pure torsion test, the weighing grip is free to move along the rotation axis.

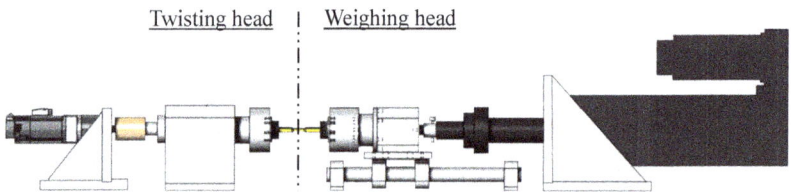

Twisting head | Weighing head

Fig. 1. Schematic overview of the tension-torsion machine. Left: Twisting head controlled by a servo motor; Right: Weighing head capable of measuring torque and tensile force with the possibility to connect a linear actuator to induce tension/compression.

The test material is a high strength low alloy (HSLA) steel S700MC with a nominal thickness of 12 mm. The 'M' and the 'C' indicate that the grade is made by a Thermomechanical Controlled Process (TMCP) and can be cold formed, respectively. Standard tensile tests were conducted and reported in [9], showing that the material exhibits a maximum uniform tensile strain of 0.12. The torsion samples are manufactured from the steel sheet in the RD. This is schematically visualized in the left panel of Fig. 2. First, rectangular samples were removed by waterjet cutting. Through precision turning the final torsion samples were obtained as shown in the right panel of Fig. 2.

Material Forming - ESAFORM 2023   Materials Research Forum LLC
Materials Research Proceedings 28 (2023) 1167-1174   https://doi.org/10.21741/9781644902479-127

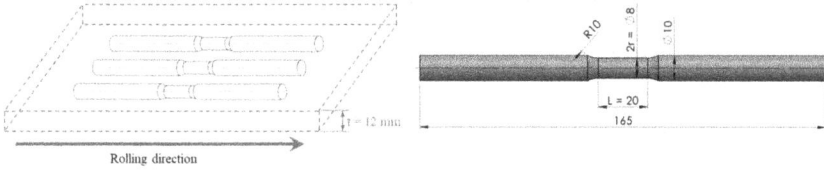

Fig. 2. The orientation of the torsion samples in the steel sheet (left) and geometry of the torsion sample after precision turning (right).

The torsion test is conducted under quasi-static conditions with a nominal strain rate in the order of $\dot{\varepsilon}_{eq} \approx 10^{-3}\frac{1}{s}$. For small rotational angles, the shear strain γ can be converted to the equivalent strain [4]:

$$\dot{\varepsilon}_{eq} = \frac{\dot{\gamma}}{\sqrt{3}} = \frac{r \cdot \dot{\theta}}{L \cdot \sqrt{3}} \tag{1}$$

With r the radius and L the length of the gauge section. $\dot{\theta}$ is the rotational speed of the twisting head. For the dimensions shown in the right panel of Fig. 2, Eq. (1) yields a maximum rotational speed $\dot{\theta}$ of 30 °/min to ensure quasi-static conditions. Three experiments were conducted using a constant speed of rotation *(20 °/min)* up to fracture of the sample. The results can be seen in Fig. 3. The three repetitions are shown by the dashed grey lines. The solid black line is the average of the three experiments. All experiments easily reached *600 °* of rotation prior to fracture. Identical flat fracture surfaces perpendicular to the rotation axis were obtained in the three experiments.

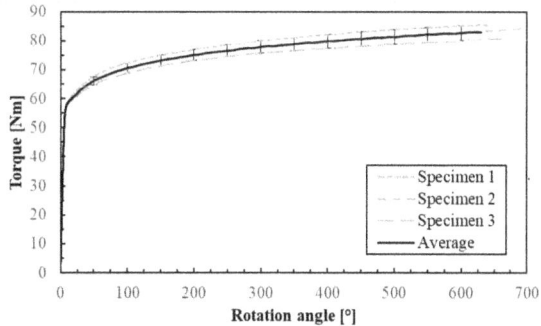

Fig. 3. Experimentally aquired Torque-rotation angle curves.

Numerical
The quasi-static torsion test is simulated using Abaqus/Standard. The torsion sample shown in the right panel of Fig. 2 is modelled in 3D using brick elements. The gripping areas are kinematically coupled onto which the boundary conditions are applied. The reference point referring to the twisting head is constrained in all degrees of freedom but the rotation angle. The reference point associated with the weighing head fixes all degrees of freedom but the translation along the rotation axis. The specimen is meshed with hexahedral elements (C3D8R) with an average size of 0.3 mm in the gauge section. The material is assumed to be elastically and plastically isotropic. Young's modulus and Poisson's ratio are assumed to be equal to 210GPa and 0.3, respectively. The plastic material behaviour is modelled using the von Mises yield criterion and a phenomenological hardening law.

It has been shown in [9] that the pre-necking hardening behaviour of S700MC can be accurately described by Swift's hardening law. Thus, Swift's hardening law can be fitted to pre-necking hardening data acquired with a uniaxial tensile test (UTT). As opposed to Swift's hardening law, Voce's hardening law enables to predict saturation of the work hardening. Given that the large strain flow curve is a composite curve, several researchers proposed to combine existing hardening laws. An example of such approach is the p-model proposed by Coppieters and Kuwabara [10]. The p-model essentially combines Swift's and Voce's hardening and was recently successfully adopted to capture the large strain flow curve of 5182-O aluminium alloy [11].

Table 1. Selected phenomenological hardening laws.

Hardening law	Model description
Swift [12]	$\sigma_{eq} = K\left(\varepsilon_0 + \varepsilon_{eq}^{pl}\right)^n$
Voce [13]	$\sigma_{eq} = C\left[1 - m e^{-B\varepsilon_{eq}^{pl}}\right]$
p-model [10]	$\sigma_{eq} = \begin{cases} K\left(\varepsilon_0 + \varepsilon_{eq}^{pl}\right)^n, & \varepsilon_{eq}^{pl} \leq \varepsilon_{max} \\ K(\varepsilon_0 + \varepsilon_{max})^n + \dfrac{Kn(\varepsilon_0 + \varepsilon_{max})^{n-1}}{p}\left[1 - e^{-p\left(\varepsilon_{eq}^{pl} - \varepsilon_{max}\right)}\right], & \varepsilon_{eq}^{pl} > \varepsilon_{max} \end{cases}$

Table 1 summarises the hardening laws considered in this work. The model parameters will be inversely identified using FEMU. Swift's and Voce' hardening law both involve three unknowns. The p-model was constructed under the assumption that the large strain flow curve can be extracted from the diffuse neck in UTT. The pre-necking data is then readily available, hence the parameters $K, \varepsilon_0, \varepsilon_{max}$ and n are considered to be known and only the post-necking parameter p is subjected to inverse identification. To increase the flexibility of the p-model in the large strain range probed in the torsion test, however, p and ε_{max} are considered to be the unknown hardening parameters in this work. For the p-model, K, ε_0 and n are fixed and equal to the values found when fitting Swift to the pre-necking data of the UTT. Given that FEMU is driven by a gradient-based optimization, a good initial parameter guess is important to robustly identify the sought parameters. The pre-necking data of the UTT is used to determine the initial guess values, see Table 2.

Table 2. Initial guess values of the unknown hardening parameters.

Hardening law	Initial guess values
Swift	$K = 1250{,}62\ MPa\ ;\ \varepsilon_0 = 0{,}02409\ ;\ n = 0{,}1533$
Voce	$C = 1231{,}49\ ;\ m = 0{,}4012\ ;\ B = 3{,}52$
p-model	$p = 5\ ;\ \varepsilon_{max} = 0{,}15$

Finite Element Model Updating (FEMU)
With FEMU, the goal is to minimize the discrepancy between the experimental and numerical data and hereby inversely identifying the sought hardening parameters [7], [14]. To compare both datasets, here the torque-angle curves, a cost function $C(\boldsymbol{p})$ is used:

$$C(\boldsymbol{p}) = \frac{1}{2}\left[\boldsymbol{T}_{exp} - \boldsymbol{T}_{num}(\boldsymbol{p})\right]^T \cdot \left[\boldsymbol{T}_{exp} - \boldsymbol{T}_{num}(\boldsymbol{p})\right] \tag{2}$$

Material Forming - ESAFORM 2023
Materials Research Proceedings 28 (2023) 1167-1174

Materials Research Forum LLC
https://doi.org/10.21741/9781644902479-127

With T_{exp} and $T_{num}(p)$ the column matrices of the torque for specific rotation angles obtained through the experiment and the FE model, respectively. The vector of the unknown parameters p is iteratively tuned to minimize $C(p)$.

To find a local minimum of $C(p)$, Cooreman [15] described two local optimization algorithms, namely Gauss-Newton and Levenberg-Marquardt. Both methods rely on the sensitivity matrix S. This sensitivity matrix captures the influence of a parameter perturbation on the numerical response T_{num}. The construction of the sensitivity matrix via finite differentiation requires an additional simulation per unknown parameter. The parameter update following Levenberg-Marquardt in each iteration reads as:

$$\Delta p = [S^T \cdot S + \alpha \cdot I]^{-1} \cdot \left[S^T \cdot \left[T_{exp} - T_{num} \right] \right]$$

(3)

With α a scalar that is strictly positive. If α is zero, then the Gauss-Newton (GN) algorithm is retrieved. Levenberg-Marquardt (LM) is typically used to mitigate stability problems. The implemented optimization strategy follows the work of Denys [16] in which a combination of GN and LM is proposed. The FEMU starts with GN. While looping through the FEMU code the smallest value of the cost function $C(p_{min})$ is stored in the memory. Upon the first increase of the cost function $C(p_k)$ after a parameter update in iteration k, the optimization switches to the LM algorithm using an initial damping factor α equal to 10^{-7}. In iteration $(k+1)$, the parameters obtained in iteration $(k-1)$ are used to predict the parameters update using LM. If $C(p_{k+1})$ is lower than $C(p_{min})$, the damping factor α will be divided by ten. Otherwise, α will be multiplied by ten to increase the damping behaviour. Convergence is assessed based on the relative change between the current and minimum cost function, $C(p)$ and $C_{min}(p)$, respectively. The FEMU code stops when the change is below 0.5 %.

Results

In a first step, the torsion test was simulated using Swift's hardening law calibrated based on the pre-necking data obtained from UTT. The parameters can be found in Table 2. The results are shown in Fig. 4 with the dotted line indicating the rotation angle where the maximum plastic strain in the simulation is equal to the maximum uniform tensile strain during the UTT. It can be inferred that the discrepancy with the experiment increases when the torsion test starts to probe strains beyond the maximum

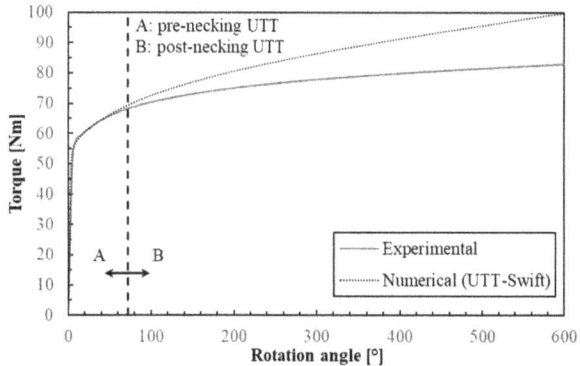

Fig. 4. Comparison between the experimental and the numerical torque-angle curve. With the material behaviour calibrated based on pre-necking data obtained from an UTT.

uniform tensile strain, i.e. beyond a rotation angle of approximately 75 °. This clearly indicates that the extrapolation of the hardening behaviour by Swift's hardening law is not valid, hence the need to identify the large strain flow curve based on experimental data beyond the maximum uniform tensile strain.

Materials Research Forum LLC
https://doi.org/10.21741/9781644902479-127

The FEMU code is fed with the torque-angle curve from 25° until 600°, ensuring that the torsion sample is sufficiently plastically deformed. The rotation angle of 25° is chosen arbitrarily and corresponds to a maximum ε_{eq}^{pl} of 0.04. The upper bound of 600° corresponds to a maximum ε_{eq}^{pl} of 1 without any visible damage. Each 2.5° of rotation, the torque value is probed to accurately capture the material response.

Table 3. The identified hardening parameters.

Hardening law	Hardening parameters	Accuracy Rank
Swift	$K = 1013,51\ MPa$; $\varepsilon_0 = 0,003643$; $n = 0,05792$	3
Voce	$C = 980,5754$; $m = 0,2302$; $B = 8,48$	1
p-model	$p = 13,65$; $\varepsilon_{max} = 0,066$	2

The FEMU started with the initial parameters shown in Table 2. The inversely identified hardening parameters are shown in Table *3*. The accuracy rank shown in Table 3 is based on the value of the cost function upon convergence, also referred to as the cost function residual. The lower the cost function residual, the better the similarity between the experimentally acquired and numerically computed torque-angle curves. Consequently, the cost function residual gives an indication of the accuracy of the identified large strain flow curve. It can be inferred that Voce yields the lowest cost function residual, hence deemed to be the best choice for capturing the overall work hardening behaviour of S700MC during a torsion test. Fig. 5 shows the inversely identified hardening laws along with the extrapolated Swift law (UTT). It can be seen that the inversely identified hardening behaviour exhibits saturation at an equivalent stress of approximately 980MPa. Since Swift's hardening law cannot describe such saturation behaviour, and FEMU merely seeks for lowest cost function residual, a trade-off is found for the inversely identified Swift law. Indeed, the accuracy in the pre-necking region is lost to compensate for the inherent inaccuracy of Swift's hardening law in the post-necking regime. In this regard, the p-model is more flexible than Swift's hardening law. Nevertheless, the p-model cannot reproduce the Voce model in the complete strain range when only ε_{max} and p are considered to be the unknowns. Fig. 6 shows the predicted torque-rotation angle curve using the inversely identified hardening laws along with the

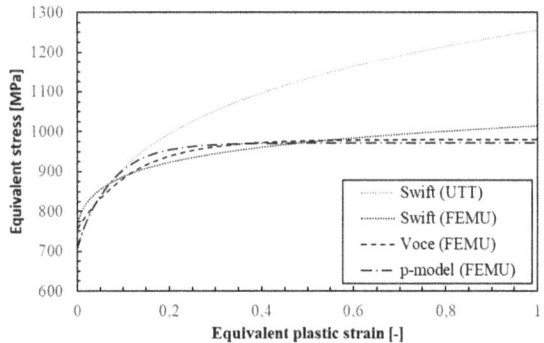

Fig. 5. The inversely identified flow curves compared with the extrapolated Swift law using the UTT.

experiment. It can be inferred that both the Swift law and the p-model predict an overshoot in a particular strain range, while Voce accurately predicts the experiment in the complete strain range.

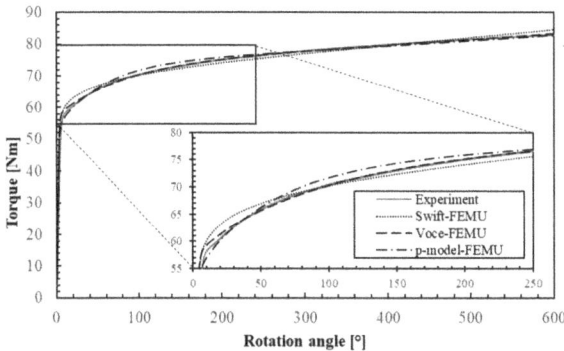

Fig. 6. Torque-rotation angle predicted by the FE model using the inversely identified hardening law. The solid grey line shows the experimental data.

Summary

The torsion test is used to inversely identify the large strain flow curve of a thick S700MC steel sheet. The FEMU approach uses the torque-twist curve to identify the parameters of three phenomenological work hardening laws. The methodology assumes the availability of a standard tensile test in the rolling direction to determine a reasonable initial guess of the sought parameters. It is shown that the hardening behaviour of S700MC, with a nominal thickness of 12mm, during a torsion test up to an equivalent plastic strain of 1 can be accurately described by Voce's hardening law. The following observations will be considered in future work:

- During the torsion test, limited spiral necking in the gauge length could be observed. It must be noted that the adopted FE model did not reproduce this phenomenon. It will be investigated if this is stemming from ignoring plastic anisotropy or an experimental error, e.g. an eccentricity problem.
- The influence of plastic anisotropy on the inverse identification of the large strain flow curve will be investigated.
- A multi-linear hardening law will be implemented to mitigate limitation imposed by the predefined character of phenomenological hardening laws.
- With a direct current potential drop (DCPD) measuring system, the objective is to determine the onset of damage. This would then enables to guarantee that only the undamaged part of the torque-twist curve is fed to the FEMU code.

Acknowledgements

This research was supported by the Research Fund for Coal and Steel under grant agreement No. 888153.

References

[1] S. Coppieters, H. Traphöner, F. Stiebert, T. Balan, T. Kuwabara, A.E. Tekkaya, Large strain flow curve identification for sheet metal, J. Mater. Process. Technol. 308 (2022) 117725. https://doi.org/10.1016/J.JMATPROTEC.2022.117725

[2] H. Zhang, S. Coppieters, C. Jiménez-Peña, D. Debruyne, Inverse identification of the post-necking work hardening behaviour of thick HSS through full-field strain measurements during diffuse necking, Mech. Mater. 129 (2019) 361-374. https://doi.org/10.1016/j.mechmat.2018.12.014

[3] M.V. Erpalov, E.A. Kungurov, Examination of Hardening Curves Definition Methods in Torsion Test, Solid State Phen. 284 (2018) 598. https://doi.org/10.4028/www.scientific.net/SSP.284.598

[4] N. Pardis, R. Ebrahimi, H.S. Kim, Equivalent strain at large shear deformation: Theoretical, numerical and finite element analysis, J. Appl. Res. Technol. 15 (2017) 442-448. https://doi.org/10.1016/j.jart.2017.05.002

[5] S.C. Shrivastava, J.J. Jonas, G. Canova, Equivalent strain in large deformation torsion testing : theoretical and practical considerations, J. Mech. Phys. Solids 30 (1982) 75-90. https://doi.org/10.1016/0022-5096(82)90014-X

[6] A.H. Stang, W. Ramberg, G. Back, Torsion Tests of Tubes, Jan. 1937, Accessed: Nov. 15, 2022. [Online]. Available: https://digital.library.unt.edu/ark:/67531/metadc66259/m1/14/

[7] P. Petrov, D. Shishkin, Y. Kalpin, I. Burlakov, D. Vydumkina, D. Kapitanenko, Determination of the flow curve based on the torsion of conical specimen, Procedia Manuf. 50 (2020) 520-528. https://doi.org/10.1016/J.PROMFG.2020.08.094

[8] A. Gavrus, E. Massoni, J.L. Chenot, An inverse analysis using a finite element model for identification of rheological parameters, J. Mater. Process. Technol. 60 (1996) 447-454. https://doi.org/10.1016/0924-0136(96)02369-2

[9] D. Debruyne et al., Towards Best Practice for Bolted Connections in High Strength Steels, 2019.

[10] S. Coppieters, T. Kuwabara, Identification of Post-Necking Hardening Phenomena in Ductile Sheet Metal, Exp. Mech. 54 (2014) 1355–1371. https://doi.org/10.1007/S11340-014-9900-4/FIGURES/16

[11] H. Shang, C. Zhang, S. Wang, Y. Lou, Large strain flow curve characterization considering strain rate and thermal effect for 5182-O aluminum alloy, Int. J. Mater. Forming 16 (2023) 1-20. https://doi.org/10.1007/S12289-022-01721-4/FIGURES/16

[12] H.W. Swift, Plastic instability under plane stress, J. Mech. Phys. Solids 1 (1952) 1-18. https://doi.org/10.1016/0022-5096(52)90002-1

[13] E. Voce, The relationship between stress and strain for homogeneous deformation, J. Inst. Metals 74 (1948) 537-562.

[14] M. di Donato, S. Bruschi, G. Hirt, M. Franzke, Flow curve determination by torsion tests using inverse modelling, 2016.

[15] S. Cooreman, Identification of the plastic material behaviour through full-field displacement measurements and inverse methods, PhD Thesis, VRIJE Universiteit Brussel, (2008).

[16] K. Denys, Investigation into the plastic material behaviour up to fracture of thick HSS using multi-DIC and FEMU, PhD Thesis, KU Leuven, (2017).

Material Forming - ESAFORM 2023
Materials Research Proceedings 28 (2023) 1175-1182

Materials Research Forum LLC
https://doi.org/10.21741/9781644902479-128

Investigating the suitability of using a single heat transfer coefficient in metal casting simulation: An inverse approach

VASILEIOU Anastasia[1,a] *, VOSNIAKOS George-Christopher[2,b] and PANTELIS Dimitrios I.[3,c]

[1]The University of Manchester, Department of Mechanical, Aerospace & Civil Engineering, Dalton Nuclear Institute, Oxford Rd, Manchester M13 9PL, United Kingdom

[2]National Technical University of Athens, School of Mechanical Engineering, Manufacturing Technology Laboratory, Heroon Politehniou 9, 15773 Athens, Greece

[3]National Technical University of Athens, School of Naval Architecture and Marine Engineering, Shipbuilding Technology Laboratory, Heroon Politehniou 9, 15773 Athens, Greece

[a]anastasia.vasileiou@manchester.ac.uk, [b]vosniak@central.ntua.gr, [c]pantelis@central.ntua.gr

Keywords: Metal Casting, Casting Modulus, Heat Transfer Coefficient, Genetic Algorithm

Abstract. In metal casting simulation the Heat Transfer Coefficient (HTC) is unknown as it depends on melt and mold materials, on the casting modulus at different regions of the casting and on local conditions at the mold-casting gap. In this paper, thermocouple measurements at three regions of a brass investment casting provided reference cooling curves. A genetic algorithm (GA) determined the optimum 3-step time-dependent HTC for the whole of the casting in a simulation program for which cooling curves are as close as possible to the reference curves. The resulting prediction of solidification times is satisfactory but prediction of qualitative characteristics such as start / end of solidification in different regions was not accurate enough.

Introduction

Casting simulation can assist in selection of optimal process parameters, namely melt temperature, mold preheat temperature, melt inlet pressure and velocity, as well as design of the feeding system. The computational power which is available nowadays allows 'virtual casting' involving thermal, mechanical and flow domains, thus replacing costly real-life experiments [1]. Although the definition of the model is dictated by experimental conditions [2], heat transfer through the casting-mold interface is to be guessed; this is described by the "Interfacial Heat Transfer Coefficient (IHTC)" or simply "Heat Transfer Coefficient (HTC)". Heat transfer across the interface is determined by solid-to-solid conduction, conduction through the gas phase and radiation [3]. HTC is affected by the thermo-physical properties of the materials at the interface (cast metal, mold, coating) such as fluidity or solidification range [4]. At least three different representations of the HTC evolution have been proposed, namely: a step function of temperature, a step function of time (which is implicitly associated with solidification phases) [4] and an exponential function of time as a continuous approximation of the latter [5,6].

The well-known Chvorinov's rule [7] depicts a strong relation between solidification time and casting modulus (volume over surface area of the casting) implying that the different regions of the casting exhibit different solidification times [8]. Although it is commonly acknowledged that different HTCs should be assigned to regions of the casting with "significantly different" geometrical characteristics [9], only few publications investigate this issue [1,4,10,11].

In determining HTC, most commonly, temperature transients are measured close to the metal/mold interface using thermocouples despite their embedding close to the interface. This data can then be used combined with either an inverse calculation [12] or with repetitive simulations

Material Forming - ESAFORM 2023   Materials Research Forum LLC
Materials Research Proceedings 28 (2023) 1175-1182   https://doi.org/10.21741/9781644902479-128

[13], until an HTC is identified that results in good agreement with the experiment. These methods can become substantially complex by employing, e.g., 3D instead of 2D models, various "local" HTCs tied to corresponding casting moduli or intricate optimization algorithms. Partitioning the casting shape into different complementary regions differing to a significant enough extent in casting modulus might be challenging, therefore an approach adopting a single HTC is a welcome simplification.

In order to avoid such complexity the present paper aims to answer the following questions: (a) Is it feasible to determine one 'equivalent' HTC and apply this to the whole casting domain during simulation instead of applying several different HTCs for different sub-domains? (b) How efficient is the use of genetic algorithms in the determination of HTCs?

The next section describes the experiment that provided the reference cooling curves. This is followed by two sections presenting the numerical simulation setup that was used in the evaluation function of the genetic algorithm that determines the optimum HTC and the setting up of this algorithm as well. Results and their discussion in the context of alternative genetic algorithm hyperparameters follow. Last, conclusions and further work are briefly summarized.

Experimental

A particular experiment was conducted to provide reference cooling curves to be used in the simulation through which optimum HTC is to be determined.

Lost wax investment casting was performed for a part with a stepwise increase in cross section made of CuZn33 alloy (brass), Fig. 1(a,b). The casting consists of three coaxial consecutive cylinders, namely a large (L), a medium (M) and a small (S) one. Length of the cylinders is 8 mm and their diameter measures 16 mm, 10 mm and 4 mm, respectively. Thus, casting modulus results as 2.30, 1.98 and 0.89, for L, M and S regions respectively.

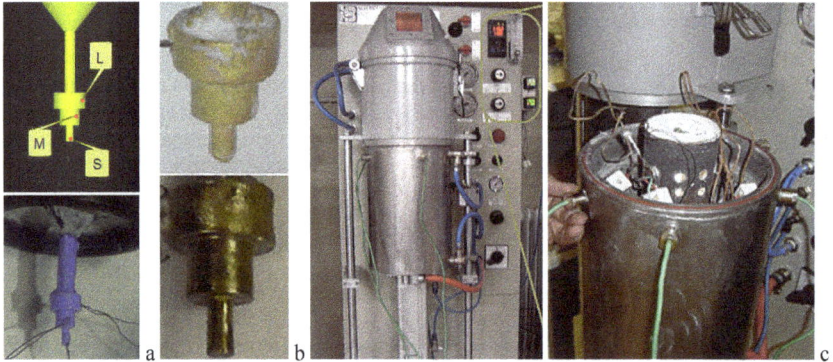

Fig. 1. Experiment setup (a) part/feeder: design (up), wax model with thermocouples (down) (b) casting: rough (up), cleaned (down). (c) casting machine: open with flask (left), closed (right)

The vacuum-pressure casting machine employed (NILAS BROS™) comprises a melting and a casting chamber, see Fig. 1(c). The former contains a graphite crucible with embedded thermocouple openings for temperature monitoring. Melting takes place under inert atmosphere of Ar. The mold is placed in the casting chamber which is sealed so that vacuum can be created and maintained. The machine is equipped with temperature and pressure displays allowing monitoring of the casting conditions in real time and performing corrections, as needed.

Material Forming - ESAFORM 2023
Materials Research Proceedings 28 (2023) 1175-1182

Materials Research Forum LLC
https://doi.org/10.21741/9781644902479-128

The casting parameters were as follows: casting chamber pressure: 0.03 bar (99.7% vacuum), pouring temperature: 990°C and mold preheating temperature: 600°C. One K-type thermocouple for each region was embedded in the mold cavity, as close to the metal-mold interface as possible, Fig. 1(a).

An A/D converter (Personal Daq/55 ™) fed raw temperature measurements to a laptop computer where they were processed by the Savitzky-Golay smoothing filter followed by cubic spline interpolation yielding the experimental temperature evolution curves. As shown in Fig. 2. each curve consists, as expected, of a short period of abrupt cooling (1st stage), a long period of mild cooling (2nd stage – solidification) and a third period of more pronounced cooling (3rd stage – post solidification).

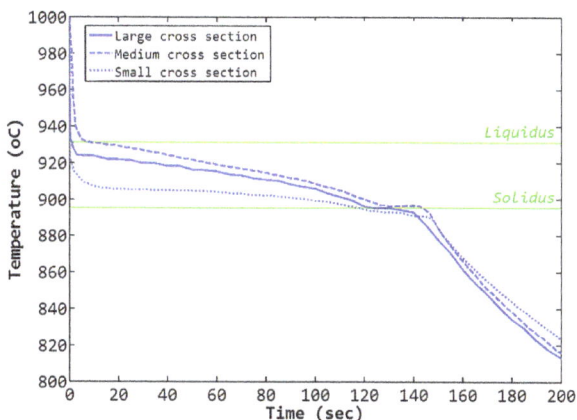

Fig. 2. Experimental temperature evolution curves.

Numerical Simulation Setup

ProCAST 2004.1® was used as the dedicated simulation platform implementing Finite Element Analysis to model coupled heat transfer, fluid flow and stress.

Determination of the initial conditions and the boundary conditions of the problem, i.e. temperature and pressure values during casting (mold preheating temperature, temperature of the outer surface of the mould, vacuum chamber pressure) followed the pertinent experimental settings, see previous Section.

HTC in the part of the sprue section filled by the melt was assumed to be the same as that of the small cylinder part of the casting (S), Fig. 1(a). This is a valid assumption as only a very small height of the sprue is actually filled by melt. The free surface of the melt is assigned radiation losses with an emissivity coefficient equal to 0.3 [14] and convective losses corresponding to air at 600°C with a film coefficient of 20 W/m^{2}K [15]. The pouring cup as such is treated exactly as the outer wall of the flask, i.e. it is assigned a Dirichlet boundary condition.

Melt flow rate was calculated as 1.13 kg/s using Bernoulli equation and considering the dimensions of the vacuum casting machine and of the casting tree.

Cast material properties were defined by interpolating the corresponding data available in literature. For brass CuZn33, liquidus was 931°C and solidus was 886°C; latent heat was 205 KJ/kg. Variation curves of thermal conductivity, density, specific heat, viscosity and fraction solid as a function of temperature are taken from [11].

HTC is modelled as a step function of time correspond ng to three stages as first documented in [16] and is depicted in Fig. 3(a) and represented by the five element vector [h_1 t_1 h_2 t_2 h_3].

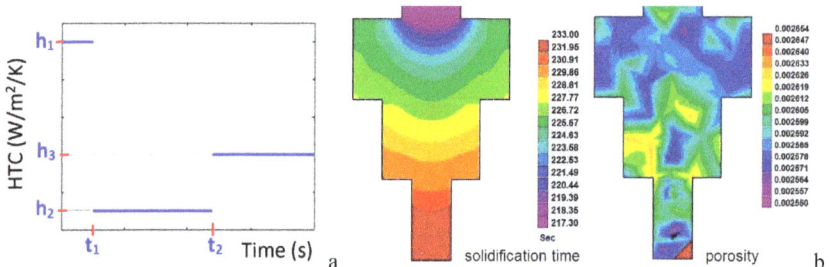

Fig. 3. (a) HTC representation (b) solidification time and porosity indicative simulation results.

Indicative simulation results regarding solidification time and porosity are shown in Fig. 3(b). Note that the right HTC needs to be employed if these results are to be credible, hence the need for inverse optimization approach.

The most valuable simulation result is temperature evolution at the three finite element mesh nodes corresponding to thermocouple locations by analogy to the experimentally obtained counterpart shown in Fig. 2.

Genetic Optimisation Setup

A classic GA was employed in order to determine the optimum HTC values that would cause the predicted temperature evolution curves to be as close as possible to the experimentally obtained ones. It was implemented on Matlab and made use of integer encoding of the chromosome.

The chromosome consists of the five variables that are necessary to define the HTC as a function of time, namely [h1 t1 h2 t2 h3], see Fig. 3(a). For each one of these variables its definition domain was discretised into 150 intervals. Each value is thus mapped to the corresponding interval, i.e. to a number from 1 to 150. The range of values for each variable was determined approximately from literature [11], thus for h_1, which is applicable during filling and generally above solidus, it is 200-30000 W/m^{2}K, whilst for h_2, which is applicable during solidification, it is 10 – 1500 W/m^{2}K; for h_3, which is applicable after the end of solidification, it is 20-3000 W/m^{2}K. The stepwise transitions occur at times t_1 and t_2 roughly corresponding to transition under liquidus and under solidus temperature respectively. Their value ranges are estimated by observing the experimental temperature evolution curves, Fig. 2, as 0.2-30 sec, and 105 – 165 sec respectively.

The fitness of any chromosome is calculated as the RMS metric of the difference in ordinates between 200 points defined on the experimental cooling curve and its simulation counterpart. This is done at the same time for all three curves corresponding to the small medium and large sections, see Fig. 2 and Fig. 1(b). The points are denser in regions where more sudden changes of temperature are expected. The fitness value is stored in a database so that the former does not have to be calculated anew each time the same input vector values are required. Obviously, fitness evaluation presumes that simulation curves of temperature evolution are obtained, which requires running the casting simulation programme with the particular HTC.

Several values were tried for the GA's hyperparameters, namely: for mutation rate: [0.1, 0.4, 0.7] for crossover rate: [0.2, 0.4, 0.5 , 0.8], for maximum iterations: [10, 20, 40, 100] and for population size: [4, 10, 20, 100]. The best result (best coincidence of experimental and simulation cooling curves) was obtained for mutation rate=0.1, population size = 20, crossover rate = 0.8, max iterations=100 and the evolution of the objective function is shown in Fig. 4. Note that if a

Material Forming - ESAFORM 2023   Materials Research Forum LLC
Materials Research Proceedings 28 (2023) 1175-1182   https://doi.org/10.21741/9781644902479-128

much higher number of generations were allowed, e.g. 1000, a further decrease in the value of the objective function might have resulted, since GAs cannot guarantee convergence to the minimum. However, such extravagance is not practical, in view of the high computational cost of each evaluation, as this requires execution of a casting simulation scenario typically lasting several minutes depending on the hardware employed.

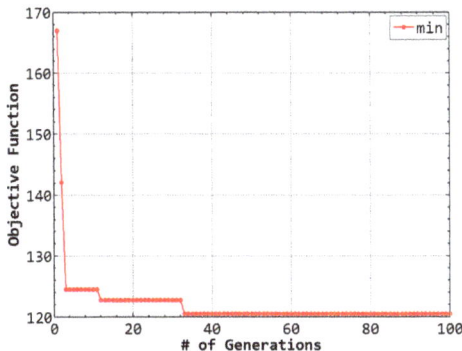

Fig. 4. Genetic algorithm cost evolution for the best chromosome in each generation.

HTC Optimisation Results and Discussion

The optimum HTC value determined by the GA is: [h1 t1 h2 t2 h3] = [102 38 70 3 35] in integer form, whilst in real number terms it is: h1=20400 W/m^2K, h2= 6 W/m^2K, h3=700 W/m^2K, t1=7 sec, t2=117 sec. This solution corresponds to a minimum in objective function equal to 120.4.

Fig. 5 depicts the cooling curves obtained by running the casting simulation for the optimum HTC in comparison to the experimental curves obtained by the corresponding thermocouples.

Referring to Fig. 5 comparison of experimental and simulation curves is generally satisfactory. The slope of the experimental and simulation cooling curves in the 2nd stage (during solidification) and in particular in the 3rd stage (post solidification) are similar.

Regarding the μ section, approximation of the experimental by the simulation curve is fairly good. Regarding the M section, approximation is good for the 2nd stage (solidification) but not so good for the 3rd stage (post solidification). Regarding the m section, approximation is unsatisfactory in the 1st and 2nd stages but fairly good in the 3rd stage. Furthermore, the lag in cooling of the m section with respect to the μ and M sections is not clearly captured by the simulation.

The end of solidification stage as determined by simulation at all three points is not far from the corresponding one that has been experimentally determined. However, the solidification start is not predicted so accurately by simulation.

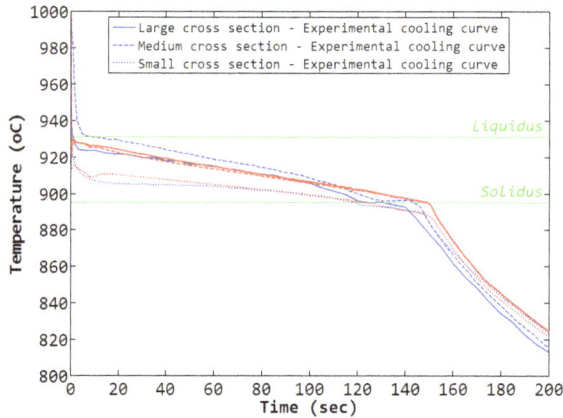

Fig. 5. Experimental and simulation cooling curves comparison.

Table 1 provides quantitative evidence for these observations. Note that the first few seconds after pouring are very important due to their influence on microstructure formation of the casting. Yet, the predicted duration of the 2nd stage (solidification) is in relatively good agreement with the experiment.

For comparison purposes, Table 1 also provides the respective results from a similar exercise considering three separate HTCs that have been reported earlier [11]. Three HTCs are a better alternative to a single HTC but the difference is tolerable especially for the larger sections. The corresponding objective function attains a significantly lower value than that attained in the single HTC case, i.e. 38.50 compared to 120.4. Recalling that this metric denotes the RMS of the ordinate differences between experimental and simulation cooling curves, i.e. a representation of temperature difference in time at the particular points monitored, this improvement is important but not spectacular.

Table 1. Solidification start (S) and end (E) times and duration (D) for large, medium and small (M, m, μ index) according to experiment and simulation.

	Single HTC				3 separate HTCs		
	Experiment	Simulation	Error (sec)	Error (%)	Simulation	Error (sec)	Error (%)
S_M (sec)	0.97	0.13	0.84	86.6%	1.97	-1.10	-103.09%
E_M (sec)	131.8	149.7	-17.9	-13.6%	147.5	-147.64	-11.91%
D_M (sec)	130.83	149.57	-18.74	-14.3%	145.53	-145.67	-11.24%
S_m (sec)	8.5	0.12	8.38	98.6%	2.75	-1.76	67.65%
E_m (sec)	144.1	148.7	-4.6	-3.2%	149.35	-149.38	-3.64%
D_m (sec)	135.6	148.58	-12.98	-9.6%	146.6	-146.70	-8.11%
S_μ (sec)	0.28	0.09	0.19	67.9%	0.24	0.44	14.29%
E_μ (sec)	119.2	124.1	-4.9	-4.1%	119.2	-119.24	0.00%
D_μ (sec)	118.92	124.01	-5.09	-4.3%	118.96	-119.00	-0.03%

Material Forming - ESAFORM 2023
Materials Research Proceedings 28 (2023) 1175-1182

Materials Research Forum LLC
https://doi.org/10.21741/9781644902479-128

Summary

In this work the suitability of simulation modelling vacuum casting of brass with a single HTC that is applicable throughout a casting different casting moduli of different regions was investigated.

An acceptable single HTC can be determined using a single casting experiment for reference, but this of course is case-dependent, i.e. it should be repeated every time any of the process parameters change, e.g. melt temperature, mold temperature and even pouring rate and, of course, if casting or mold geometry or materials change. However, the methodology of determining the HTC is generic and universally applicable

In general, agreement between simulation predictions and experimental measurements concerning temperature evolution at characteristic points of the different regions is fair. In particular, prediction of solidification duration is acceptable. However, qualitative characteristics such as the differences in solidification starting time between different regions are not captured accurately. Yet, predictive performance of the simulation when using a single HTC is tolerably inferior to that achieved when using 3 HTCs as revealed by comparison to pertinent work reported before. However, this is certainly connected to the relatively moderate differences in casting moduli.

In cases of intense differences in casting moduli, most probably different HTCs should be assigned to different regions, but this has to be proved in future extension of this work.

Acknowledgments

This work is funded by the European Commission in the framework HORIZON-WIDERA-2021-ACCESS-03, project 101079398 'New Approach to Innovative Technologies in Manufacturing (NEPTUN)'

References

[1] N. Pagratis, N. Karagiannis, G.-C. Vosniakos, D. Pantelis, P. Benardos, A holistic approach to the exploitation of simulation in solid investment casting, Proc. Inst. Mech. Engs, Part B: J. Eng. Manuf. 221/6 (2007) 967–979. https://doi.org/10.1243/09544054JEM465

[2] D. Furer, S.L. Semiatin (eds.), ASM Handbook Vol. 22B: Metals Process Simulation, ASM Intern., Ohio, 2010.

[3] K. Ho, R.D. Pehlke, Metal-mold interfacial heat transfer, Metall. Trans. B 16 (1985) 585-594. https://doi.org/10.1007/BF02654857

[4] G Palumbo, V. Piglionico, A. Piccininni, P. Guglielmi, D. Sorgente, L. Tricarico, Determination of interfacial heat transfer coefficients in a sand mould casting process using an optimised inverse analysis, Appl. Therm. Eng. 78 (2015) 682-694. https://doi.org/10.1016/j.applthermaleng.2014.11.046

[5] D. O'Mahoney, D.J. Browne, Use of experiment and an inverse method to study interface heat transfer during solidification in the investment casting process, Exp. Therm. Fluid Sci. 22 (2000) 111–122. https://doi.org/10.1016/S0894-1777(00)00014-5

[6] Y. Dong, K. Bu, Y. Dou, D. Zhang, Determination of interfacial heat-transfer coefficient during investment-casting process of single-crystal blades, J. Mater. Process. Technol 211 (2011) 2123-2131. https://doi.org/10.1016/j.jmatprotec.2011.07.012

[7] N. Chvorinov, Theorie der Erstarrung von Gussstucken, Giesserei, 27 (1940) 177-188.

[8] F. Havlicek, T. Elbel, Geometrical modulus of a casting and its influence on solidification process. Arch Foundry Eng. 11 (2011) 170-176.

[9] G. Zhi-peng, S.-M. Xiong, B.-C. Liu, L. Mei, A. John, Determination of the heat transfer coefficient at metal–die interface of high pressure die casting process of AM50 alloy, Int. J. Heat Mass Transfer 51 (2008) 6032-6038. https://doi.org/10.1016/j.ijheatmasstransfer.2008.04.029

[10] Z. Sun, H. Hu, X. Niu, Determination of heat transfer coefficients by extrapolation and numerical inverse methods in squeeze casting of magnesium alloy AM60, J. Mater. Proc. Technol. 211 (2011) 1432-1440. https://doi.org/10.1016/j.jmatprotec.2011.03.014

[11] A. Vasileiou, G.-C. Vosniakos, D. Pantelis, Determination of local heat transfer coefficients in precision castings by genetic optimisation aided by numerical simulation, Proc. Inst. Mech. Eng. Part C: J. Mech. Eng. Sci. 229 (2014) 735-750. https://doi.org/10.1177/0954406214539468

[12] W. Zhang, G. Xie, D. Zhang, Application of an optimization method and experiment in inverse determination of interfacial heat transfer coefficients in the blade casting process, Exp. Therm. Fluid Sci. 34 (2010) 1068–1076.

[13] A. Long, D. Thornhill, C. Armstrong, D. Watson, Determination of the heat transfer coefficient at the metal–die interface for high pressure die cast AlSi9Cu3Fe, Appl. Therm. Eng. 31 (2011) 3996-4006. https://doi.org/10.1016/j.applthermaleng.2011.07.052

[14] ASM, ASM Handbook Volume 02 - Properties and Selection: Nonferrous Alloys and Special-Purpose Materials, ASM International, Ohio, 1990.

[15] L.C. Burmeister, Convective Heat Transfer, 2nd ed, New York, Wiley-Interscience, 1993.

[16] F. Lau, W.B. Lee, S.M. Xiong, B.C. Liu, A study of the interfacial heat transfer between an iron casting and a metallic mould, J. Mater. Process. Technol. 79 (1998) 25-29.

Material Forming - ESAFORM 2023
Materials Research Proceedings 28 (2023) 1183-1192

Materials Research Forum LLC
https://doi.org/10.21741/9781644902479-129

Sensitivity analysis of the of the square cup stamping process using a polynomial chaos expansion

PEREIRA André F. G.[1,2,a]*, MARQUES Armando E.[1,b], OLIVEIRA Marta C.[1,2,c] and PRATES Pedro A.[1,3,4,d]

[1] Centre for Mechanical Engineering, Materials and Processes (CEMMPRE), Department of Mechanical Engineering, University of Coimbra, Portugal

[2] Advanced Production and Intelligent Systems Associated Laboratory (ARISE), 4200-465 Porto, Portugal

[3] Centre for Mechanical Technology and Automation (TEMA), Department of Mechanical Engineering, University of Aveiro, Portugal

[4] Intelligent Systems Associate Laboratory (LASI), 4800-058 Guimarães, Portugal

[a] andre.pereira@uc.pt, [b]armando.marques@uc.pt, [c]marta.oliveira@dem.uc.pt, [d]prates@ua.pt

Keywords: Sobol's Indices, Polynomial Chaos Expansion, Square Cup, Uncertainty

Abstract. The stochastic modelling and quantification of the various sources of uncertainty associated with sheet metal forming processes, usually requires a large computational cost to obtain accurate results. In this work, a polynomial chaos expansion metamodel is used in order to reduce the computational cost of the uncertainty quantification (through Sobol's indices). The metamodel allows to establish mathematical relationships between the square cup forming results and the uncertainty sources associated with the material behaviour and process conditions. Then, sensitivity indices are estimated with the trained metamodel, without resorting to additional numerical simulations. The indices obtained with the metamodel were compared to those obtained with the traditional approach based on a quasi-Monte Carlo method. The metamodel allowed to reduce the computational cost in about 90% when compared to the traditional approach, without compromising the accuracy of the results.

Introduction

Sheet metal forming processes are among the most common and important metal working operations associated with the automotive, aeronautics and metalworking industries [1]. Numerical simulation is a well-established tool for the design and optimization of these processes [2]. However, the traditional use of the finite element method (FEM) is based on a deterministic approach [3], which does not take into account the various sources of uncertainty that are inevitable in a real industrial environment. These sources of uncertainty have a significant effect on the quality of the final product [4,5], leading to an inefficient production and, eventually, to the expensive redesign of the forming process. For all these reasons and due to the increasing availability of big data coupled with the growth of computer performance, the uncertainty analysis of these processes is a current scientific and industrial interest [1,6-8].

In recent years, distinct methods, such as, Monte Carlo Simulation [7-9], design of experiments [3,8] and metamodels [1,10] have been used to model the influence of uncertainty. Sensitivity analyses are used to quantify the influence of each uncertainty source in the variability of the forming results [3,11-13]. Variance-based sensitivity analysis (Sobol's indices [14]) is one of the most common methods to quantify this influence [6]. However, this analysis usually requires a large computational cost in order to obtain accurate sensitivity results [6]. This drawback contributes to the computational inefficiency of the uncertainty analysis, delaying its full and suitable employment in industry.

Material Forming - ESAFORM 2023   Materials Research Forum LLC
Materials Research Proceedings 28 (2023) 1183-1192   https://doi.org/10.21741/9781644902479-129

This work presents a numerical study on the influence of the material and process uncertainty in the results of a square cup forming process. The square cup test was chosen for two main reasons: (i) it is a commonly used benchmark test to represent sheet metal forming processes [1,6,15-18]; (ii) it is a relatively fast process to simulate, which is suitable for performing a large number of numerical simulations. A polynomial chaos expansion (PCE) metamodel is used for reducing the computational cost of the Sobol's indices assessment [19]. The PCE metamodel establishes mathematical relationships between the square cup forming results and the uncertainty sources associated with the elastoplastic material properties of the blank (Hooke's law parameters, hardening law parameters, anisotropy coefficients) and process conditions (blank thickness, friction coefficient and the blank holder force). Then, Sobol's indices are estimated with the trained PCE metamodel, without resorting to additional numerical simulations. The indices obtained with the PCE metamodel were compared to those obtained with the traditional approach based on a quasi-Monte Carlo (q-MC) method.

Stochastic Model

Numerical Model. The square cup forming process was modelled with the same numerical model used in a previous work [6], as shown in Fig. 1. The geometry of the tools was adapted from the NUMISHEET' 93 benchmark [18]. The square blank has an initial thickness t_0, and a side length of 75 mm. The numerical simulation of the forming process consists of three phases: (i) First, the blank holder moves downwards, pressing the blank against the die, until a prearranged black holder force (BHF) is reached; (ii) Then, the punch moves 40 mm downwards, drawing the blank into the die, with a constant BHF; (iii) The final step consists in removing the tools (black holder, punch and die), resulting in the springback of the square cup. The numerical simulations were performed with the software DD3IMP (Deep Drawing 3D Implicit Code) [20]. Only a quarter of the model is simulated due to symmetries in the material, geometry and boundary conditions, and to reduce the computational cost. The blank is discretized with 1800 (8-node hexahedral solid) elements, with 2 elements in thickness and 30x30 elements in the sheet plane. The contact between the blank and the tools is described by the Coulomb's law with a constant friction coefficient, μ_0. In average, the duration of each simulation is approximately 4 minutes and 34 seconds in a computer equipped with an Intel® Core™ i7-8700K Hexa-Core processor (4.7 GHz).

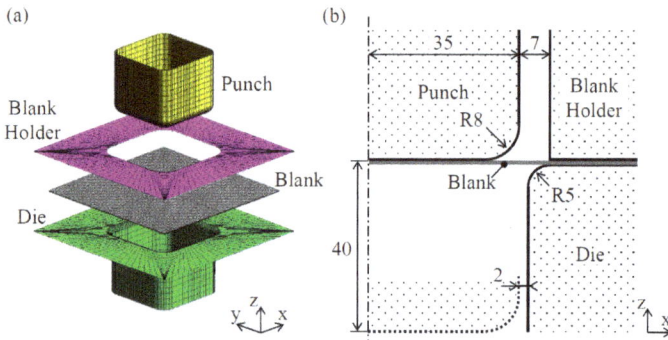

Fig. 1. Square cup forming process: (a) Numerical model [6]; (b) Dimensions of the tools in mm [6].

The mechanical behaviour of the metal sheet is modelled by: (i) the generalized Hooke's law, where E and v are the Young's modulus and the Poisson's ratio, respectively; (ii) the Swift hardening law; (ii) and the Hill'48 yield criterion. The yield criterion is defined by:

$$F(\sigma_{yy} - \sigma_{zz})^2 + G(\sigma_{zz} - \sigma_{xx})^2 + H(\sigma_{xx} - \sigma_{yy})^2 + 2L\tau_{yz}^2 + 2M\tau_{xz}^2 + 2N\tau_{xy}^2 = Y^2, \tag{1}$$

where σ_{xx}, σ_{yy}, σ_{zz}, τ_{xy}, τ_{yz} and τ_{xz} are the components of the Cauchy stress tensor; Y is the yield stress; F, G, H, L, M and N are anisotropy parameters. The parameters follow the condition $G + H = 1$ and $L = M = 1.5$ (von Mises). F, G, H and N are obtained from the anisotropy coefficients r_0, r_{45} and r_{90}, by:

$$F = {r_0}/{r_{90}(r_0 + 1)}; \; G = {1}/{r_0 + 1}; \; H = {r_0}/{r_0 + 1}$$

$$N = {(r_0 + r_{90})(2r_{45} + 1)}/{2r_{90}(r_0 + 1)} \tag{2}$$

The Swift hardening law is given by:

$$Y = C(\varepsilon_0 + \bar{\varepsilon}^p)^n \tag{3}$$

where $\bar{\varepsilon}^p$ is the equivalent plastic strain; n, C and ε_0 are hardening parameters. The initial yield stress is $Y_0 = C(\varepsilon_0)^n$.

Input and Output Parameters. The sensitivity analysis is focused on 11 input parameters, 8 associated with the material behaviour (E, v, n, C, Y_0, r_0, r_{45} and r_{90}) and 3 associated with the blank thickness, t_0, friction coefficient, μ_0, and blank holder force, BHF. The uncertainty of the input parameters is assumed to follow a normal distribution characterized by a mean, μ, and a standard deviation, σ, whose values are given in Table 1 [6].

Table 1. Mean and standard deviation of the normal distribution associated to the uncertainty of each input parameter [6].

	E [GPa]	v	n	C [MPa]	Y_0 [MPa]	r_0	r_{45}	r_{90}	t_0 [mm]	μ_0	BHF [N]
μ	206.00	0.300	0.259	565.32	157.12	1.790	1.510	2.270	0.780	0.1440	2450.0
σ	3.85	0.015	0.018	26.85	7.16	0.051	0.037	0.121	0.013	0.0288	122.5

The uncertainty influence was analysed for 4 output parameters associated with the forming results, namely, the punch force (PF), the equivalent plastic strain ($\bar{\varepsilon}^p$), the thickness change (TC) and the geometry change (GC). The PF and the $\bar{\varepsilon}^p$ values are directly obtained from the numerical simulation, while the TC and the GC are defined by [6]:

$$TC \, [\%] = 100 \times (t_0 - t_f)/t_0 \tag{4}$$

$$GC \, [mm] = \sqrt{\left(\bar{x}_f - x_f\right)^2 + \left(\bar{y}_f - y_f\right)^2 + \left(\bar{z}_f - z_f\right)^2} \tag{5}$$

where t_0 and t_f are the initial and final sheet thickness, respectively, in a given region of the square cup; (x_f, y_f, z_f) and $(\bar{x}_f, \bar{y}_f, \bar{z}_f)$ are, respectively, the final spatial position of a given node for the numerical simulation with and without uncertainty (i.e., using the mean values of Table 1). The GC quantifies the positional difference of a given node between the deterministic and the stochastic simulation. In this work, only the maximum values of the four outputs were analysed.

Sensitivity Analysis

Sobol's Indices. Sobol's Indices are a sensitivity measure of the influence of the input parameters on the output parameters [14]. Two distinct sensitivity indices can be used to quantify this influence, the 1st order indices, S_i, and the total sensitivity indices, S_i^T, which can be defined as follows [14]:

$$S_i = V[E(y|x_i)]\Big/V(y) \tag{6}$$

$$S_i^T = 1 - \left(V[E(y|x_{\sim i})]\Big/V(y)\right) \tag{7}$$

where $V(y)$ is the unconditional variance of the result y; $V[E(y|x_i)]$ is the conditional variance of the expected value of y when all input parameters, but x_i, are fixed; and $V[(y|x_{\sim i})]$ is the conditional variance of the expected value of y when only the input parameter x_i is fixed. The 1st order indices, S_i, quantify the individual influence of each input parameter, x_i, on the result y; while the total sensitivity indices, S_i^T, quantify not only the individual influence of each input parameter, x_i, on the result y, but also the influence of the interactions between the input parameter x_i and the remaining, on the result y.

The indices S_i and S_i^T were already computed and published for the above model of the square cup forming process [6]. These indices were computed with the traditional method proposed in [21], and using the estimators proposed by [22], which allow to significantly improve the stabilization of the indices for a lower number of numerical simulations. A base sample of 3000 simulations was generated with a Sobol's sequence [23], in order to also achieve a faster stabilization. For a base sample of 3000 simulations, a total of 39000 simulations were needed to evaluate the sensitivity indices for the 11 input parameters, accordingly to the traditional procedure [21]. The chosen size of the base sample guarantees the stabilization of the sensitivity indices [6].

Polynomial Chaos Expansion. A Polynomial Chaos Expansion (PCE) metamodel is used to reduce the computational cost of the Sobol's indices evaluation [21]. The PCE metamodel allows to estimate the outputs (i.e., forming results), $y^{PCE}(\mathbf{x})$, as a function of the input parameters (i.e., uncertainty sources), $\mathbf{x}$, by using an orthogonal polynomial basis, Ψ_α. The output value predicted by the PCE metamodel is given by [19]:

$$y^{PCE}(\mathbf{x}) = \sum_{\alpha \in A} \beta_\alpha \Psi_\alpha(\mathbf{x}) \tag{8}$$

where β_α are expansion coefficients and $\mathbf{A}$ is a set of pre-selected multi-index $\alpha = [\alpha_1, \alpha_2, \dots, \alpha_k]$ (k is the number of input parameters). The elements α_i indicates the degree of the polynomial associated with the input parameter x_i. Hermite polynomials are used to build the polynomial basis, Ψ_α, since the input variables follow a gaussian distribution (see Table 1) [19]. The set β of expansion coefficients β_α is determined with the ordinary least squares method [24]:

$$\beta = (\Psi(\mathbf{x})\Psi(\mathbf{x})^T)^{-1}\Psi(\mathbf{x})y^*(\mathbf{x}) \tag{9}$$

where, $y^*(\mathbf{x})$ is a set of q output results obtained with the q training simulations of the numerical model; and $\Psi(\mathbf{x})$ is a $q \times q$ matrix of Hermitian polynomials of degree m. More details about the construction of $\Psi(\mathbf{x})$ can be found in [19]. To avoid a high computational cost, only polynomials up to degree $m \leq 3$ and low order iterations between input variables are considered, following a hyperbolic truncation scheme [25].

Due to the orthogonality property of the polynomial basis, it is possible to directly evaluated the 1st order Sobol's indices by [26]:

$$S_i^{PCE} = \left. \sum_{\alpha \in A^*}(\beta_\alpha^2) \middle/ \sum_{\alpha \in A}(\beta_\alpha^2) \right. \tag{10}$$

where A^* is a subset of A in which the multi-index α is only associated to the input variable x_i (i.e., no other input variable is associated to the multi-index). The total Sobol's indices can be evaluated by [26]:

$$S_i^{T\ PCE} = \left. \sum_{\alpha \in A^T}(\beta_\alpha^2) \middle/ \sum_{\alpha \in A}(\beta_\alpha^2) \right. \tag{11}$$

where A^T is a subset of A in which the multi-index α is associated to the input variable x_i, even if α is simultaneously associated with other input variables. Based on the above equations it is evident that the Sobol's indices are instantaneously calculated after the evaluation of the expansion coefficients β_α, i.e., after the metamodel training.

The metamodel was trained with the same 3000 base simulations, previously used to compute the indices S_i and S_i^T with the traditional approach. Four metamodels were trained each one for a given output PF, $\bar{\varepsilon}^p$, TC and GC. The metamodel was tested for other 1000 simulations by comparing the predicted PCE output, $y^{PCE}(\mathbf{x}^*)$, with the one assessed with the testing simulations $y(\mathbf{x}^*)$. The performance of each metamodel was evaluated with the root-mean-square error, $\sqrt{\epsilon}$, and the coefficient of determination, R^2, given by [26]:

$$\sqrt{\epsilon} = \sqrt{{}^1\!/_{q^*} \sum_{i=1}^{q^*} \left(y(\mathbf{x}_i^*) - y^{PCE}(\mathbf{x}_i^*) \right)^2} \tag{12}$$

$$R^2 = 1 - {}^\epsilon\!/_{V(\mathbf{y}(\mathbf{x}^*))} \tag{13}$$

where q^* is the number of testing simulations, $y(\mathbf{x}_i^*)$ and $y^{PCE}(\mathbf{x}_i^*)$ are the simulation and predicted output for the set of input parameters, $\mathbf{x}_i^*$, of the ith testing simulation. $V(\mathbf{y}(\mathbf{x}^*))$ is the variance of the outputs evaluated for the q^* testing simulations. The root-mean-square error, $\sqrt{\epsilon}$, and the coefficient of determination, R^2, of the metamodels trained for each output, are indicated in Table 2.

The PCE metamodels for the outputs PF, $\bar{\varepsilon}^p$ and TC achieved the best performances, with R^2 values close to 1. On the other hand, the PCE metamodel for the output GC had the poorest performance, with a R^2 value of 0.8834. Fig. 2 compares the simulated outputs of the testing dataset, $y(\mathbf{x}_i^*)$, with those predicted by the PCE, $y^{PCE}(\mathbf{x}_i^*)$. It can be observed that the PCE metamodels were able to accurately predict the simulation outputs, with the exception of GC.

Table 2. Root-mean-square error and coefficient of determination of the metamodels trained for the outputs PF, $\bar{\varepsilon}^p$, TC and GC.

	PF	$\bar{\varepsilon}^p$	TC	GC
$\sqrt{\epsilon}$	0.2207[kN]	0.0036	0.0841%	0.0412[mm]
R^2	0.9956	0.9734	0.9845	0.8834

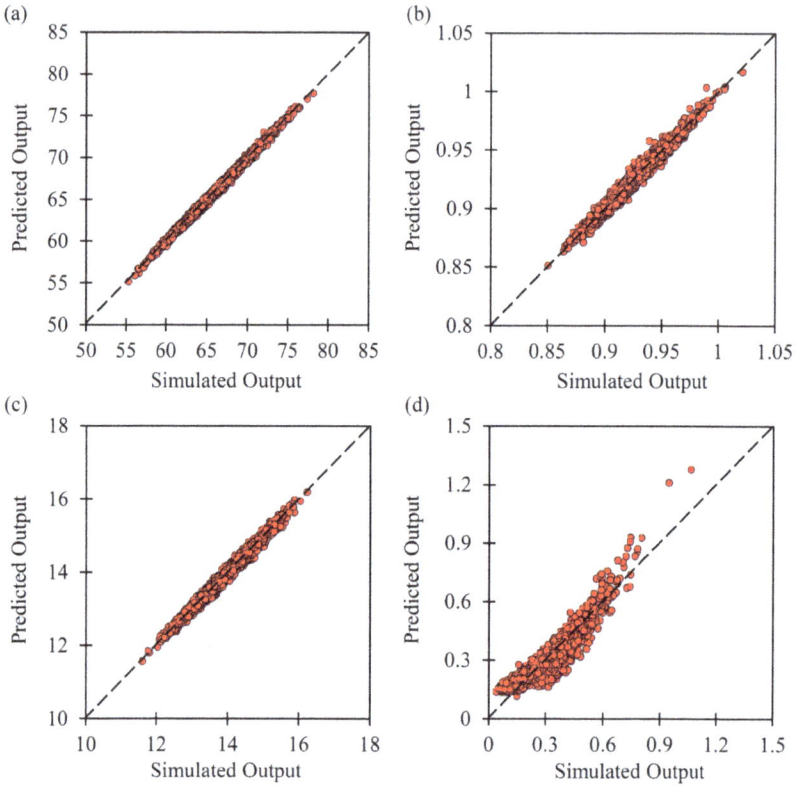

Fig. 2. Predicted (PCE metamodel) and simulated outputs: a) PF [kN]; b) $\bar{\varepsilon}^p$; c) TC [%]; and d) GC [mm]. The dashed line represents the optimal metamodel response, in which predicted outputs are equal to simulated outputs.

Sensitivity Results

In this section, the Sobol's sensitivity indices evaluated with the traditional approach are compared with those evaluated using the PCE metamodel. In this context, Fig. 3 and Fig. 4 show the 1st order and the total Sobol's sensitive indices, respectively, for the outputs PF, $\bar{\varepsilon}^p$, TC and GC.

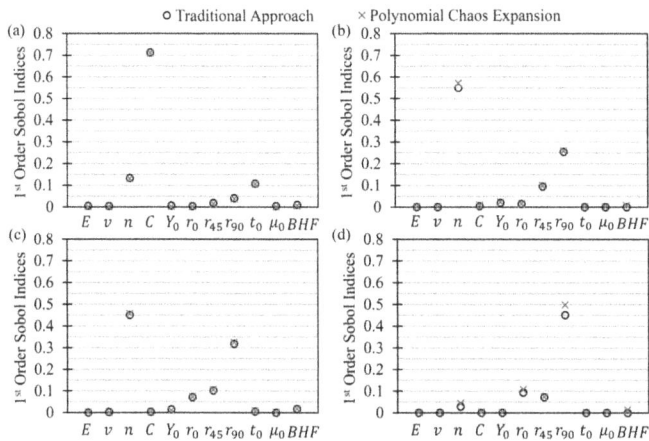

Fig. 3. 1st order Sobol's indices, computed with the traditional approach and with the PCE metamodel, for the outputs: a) PF; b) $\bar{\varepsilon}^p$; c) TC; and d) GC.

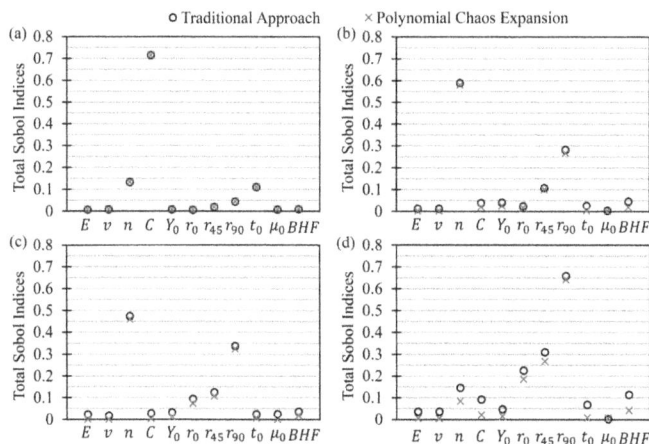

Fig. 4. Total Sobol's indices, computed with the traditional approach and with the PCE metamodel, for the outputs: a) PF; b) $\bar{\varepsilon}^p$; c) TC; and d) GC.

Based on Fig. 3 and Fig. 4, it can be observed that the PCE metamodels are able to accurately predict both sensitivity indices. The only significant difference between the indices computed with the traditional and PCE metamodel is observed for the output GC, where the maximum absolute difference between the sensitivity indices of both methodologies is 0.073. This occurs due to the lower accuracy of the metamodel to predict the GC output, as shown in Fig. 3 (d). Nevertheless, even in this case the accuracy of the sensitivity indices computed by the PCE metamodels is suitable to quantify and rank the influence of the input parameters. It is noteworthy that, that the computation of the Sobol indices with the traditional approach required 39000 simulations, while the computation with the PCE metamodel only required 4000 simulations (3000 for training and 1000 for testing the metamodel).

In summary, the PCE metamodel allowed to evaluate the Sobol's indices with accuracy and computational efficiency, requiring about 90% less numerical simulations when compared to the traditional approach.

Summary

In this work, a polynomial chaos expansion (PCE) metamodel is used to compute sensitivity indices, with the goal of reducing the computational cost associated with the traditional approach. The sensitivity indices were assessed with both methodologies for 4 outputs/results: the punch force (PF), the equivalent plastic strain ($\bar{\varepsilon}^p$), the thickness change (TC) and the geometry change (GC) of the square cup forming process. In this study was assumed uncertainty in 11 input parameters, namely, elasticity parameters, anisotropy coefficients, hardening parameters, blank thickness, friction coefficient and blank holder force.

The PCE metamodel allowed to establish mathematical relationships between the square cup forming results and the sources of uncertainty. The predictive performance of the PCE metamodels was tested, and it was concluded that the metamodels were able to accurately predict the simulation outputs. Then, Sobol's indices were estimated with the trained PCE metamodels and the traditional approach based on a quasi-Monte Carlo (q-MC) method. Both methodologies obtained similar 1st order and total Sobol's indices for the PF, the $\bar{\varepsilon}^p$ and the TC. Small differences in the Sobol's indices were observed for the GC output, but without compromising the sensitivity results.

In summary, the PCE metamodel allowed to reduce the computational cost in 90%, when compared to the traditional approach, without compromising the results accurately. In future works, other metamodel techniques will be tested to further improve the prediction accuracy, particularly, in the case of the geometry change. Furthermore, it is also intended to optimise the number of base simulations required to train and test the metamodel to further reduce the computational cost.

Acknowledgements

This work was sponsored by FEDER funds through the program COMPETE (Programa Operacional Factores de Competitividade), by national funds through FCT (Fundação para a Ciência e a Tecnologia) under the projects UIDB/00285/2020, UIDB/00481/2020, UIDP/00481/2020, CENTRO-01- 0145-FEDER-022083, LA/P/0104/2020 and LA/P/0112/2020. It was also supported by the project RealForm (reference 2022.02370.PTDC), funded by Portuguese Foundation for Science and Technology. A.E. Marques was supported by a grant for scientific research from the Portuguese Foundation for Science and Technology (ref. 2020.08449.BD). All supports are gratefully acknowledged.

References

[1] A.E. Marques, P.A. Prates, A.F.G. Pereira, M.C. Oliveira, J.V. Fernandes, B.M. Ribeiro, Performance Comparison of Parametric and Non-Parametric Regression Models for Uncertainty Analysis of Sheet Metal Forming Processes, Metals 10 (2020) 457. https://doi.org/10.3390/met10040457

[2] A.F.G. Pereira, P.A. Prates, M.C. Oliveira, J.V. Fernandes, Normal Stress Components during Shear Tests of Metal Sheets, Int. J. Mech. Sci. 164 (2019) 105169. https://doi.org/10.1016/j.ijmecsci.2019.105169

[3] P.A. Prates, A.S. Adaixo, M.C. Oliveira, J.V. Fernandes, Numerical Study on the Effect of Mechanical Properties Variability in Sheet Metal Forming Processes, Int. J. Adv. Manuf. Technol. 96 (2018) 561-580. https://doi.org/10.1007/s00170-018-1604-y

[4] W. Hancock, M. Zayko, M. Autio, D. Ponagajba, Analysis of components of variation in automotive stamping processes, Qual. Eng. 10 (1997) 115-124. https://doi.org/10.1080/08982119708919114

[5] K.D. Majeske, P.C. Hammett, Identifying Sources of Variation in Sheet Metal Stamping, Int. J. Flexib. Manuf. Syst. 15 (2003) 5-18. https://doi.org/10.1023/A:1023993806025

[6] A.F.G. Pereira, M.F. Ruivo, M.C. Oliveira, J.V. Fernandes, P.A. Prates, Numerical Study of the Square Cup Stamping Process: A Stochastic Analysis, ESAFORM 2021 (2021). https://doi.org/10.25518/esaform21.2158

[7] M. Arnst, J.-P. Ponthot, R. Boman, Comparison of Stochastic and Interval Methods for Uncertainty Quantification of Metal Forming Processes, Comptes Rendus Mécanique 346 (2018) 634-646. https://doi.org/10.1016/j.crme.2018.06.007

[8] M. Dwivedy, V. Kalluri, The Effect of Process Parameters on Forming Forces in Single Point Incremental Forming, Procedia Manuf. 29 (2019) 120-128. https://doi.org/10.1016/j.promfg.2019.02.116

[9] V.J. Shahi, A. Masoumi, P. Franciosa, D. Ceglarek, Quality-Driven Optimization of Assembly Line Configuration for Multi-Station Assembly Systems with Compliant Non-Ideal Sheet Metal Parts, Procedia CIRP 75 (2018) 45-50. https://doi.org/10.1016/j.procir.2018.02.022

[10] M.A. Dib, N.J. Oliveira, A.E. Marques, M.C. Oliveira, J.V. Fernandes, B.M. Ribeiro, P.A. Prates, Single and Ensemble Classifiers for Defect Prediction in Sheet Metal Forming under Variability, Neural Comput. Appl. 32 (2019) 12335-12349. https://doi.org/10.1007/s00521-019-04651-6

[11] J. Fruth, O. Roustant, Kuhnt, S. Support Indices: Measuring the Effect of Input Variables over Their Supports, Reliab. Eng. Syst. Saf. 187 (2019) 17-27. https://doi.org/10.1016/j.ress.2018.07.026

[12] P. Zhu, L. Zhang, R. Zhou, L. Chen, B. Yu, Q. Xie, A Novel Sensitivity Analysis Method in Structural Performance of Hydraulic Press, Math. Probl. Eng. 2012 (2012) 1-21. https://doi.org/10.1155/2012/647127

[13] S.H. Kim, H. Huh, Design Sensitivity Analysis of Sheet Metal Forming Processes with a Direct Differentiation Method, J. Mater. Process. Technol. 130-131 (2002) 504-510. https://doi.org/10.1016/S0924-0136(02)00797-5

[14] I.M. Sobol', Global Sensitivity Indices for Nonlinear Mathematical Models and Their Monte Carlo Estimates, Math. Comput. Simul. 55 (2001) 271-280. https://doi.org/10.1016/S0378-4754(00)00270-6

[15] T. Hama, M. Takamura, A. Makinouchi, C. Teodosiu, H. Takuda, Effect of Tool-Modeling Accuracy on Square-Cup Deep-Drawing Simulation, Mater. Trans. 49 (2008) 278-283. https://doi.org/10.2320/matertrans.P-MRA2007885

[16] Y.Q. Li, Z.S. Cui, X.Y. Ruan, D.J. Zhang, CAE-Based Six Sigma Robust Optimization for Deep- Drawing Process of Sheet Metal, Int. J. Adv. Manuf. Technol. 30 (2006) 631-637. https://doi.org/10.1007/s00170-005-0121-y

[17] J.M. Gutiérrez Regueras, A.M. Camacho López, Investigations on the Influence of Blank Thickness (t) and Length/Wide Punch Ratio (LD) in Rectangular Deep Drawing of Dual-Phase Steels, Comput. Mater. Sci. 91 (2014) 134-145. https://doi.org/10.1016/j.commatsci.2014.04.024

[18] E. Bayraktar, S. Altintaş, Square Cup Deep Drawing Experiments, Proceedings of the NUMISHEET '93: Proceedings of the 2nd International Conference Numerical Simulation of 3-D Sheet Metal Forming Processes, Isehara, Japan, 1993, p. 441.

[19] B. Sudret, Global Sensitivity Analysis Using Polynomial Chaos Expansions, Reliab. Eng. Syst. Saf. 93 (2008) 964-979. https://doi.org/10.1016/J.RESS.2007.04.002

[20] L.F. Menezes, C. Teodosiu, Three-Dimensional Numerical Simulation of the Deep-Drawing Process Using Solid Finite Elements, J. Mater. Process. Technol. 97 (2000) 100-106. https://doi.org/10.1016/S0924-0136(99)00345-3

[21] A. Saltelli, M. Ratto, T. Andres, F. Campolongo, J. Cariboni, D. Gatelli, M. Saisana, S. Tarantola, Global Sensitivity Analysis, The Primer, 2008.

[22] A. Janon, T. Klein, A. Lagnoux, M. Nodet, C. Prieur, Asymptotic Normality and Efficiency of Two Sobol Index Estimators, ESAIM - Probability and Statistics 18 (2014) 342-364. https://doi.org/10.1051/ps/2013040

[23] I.M. Sobol', On the Distribution of Points in a Cube and the Approximate Evaluation of Integrals, USSR Computat. Math. Math. Phys. 7 (1967) 86-112. https://doi.org/10.1016/0041-5553(67)90144-9

[24] J. Lebon, G. le Quilliec, R.F. Coelho, P. Breitkopf, P. Villon, Variability and Sensitivity Analysis of U-Shaped Deep Drawn Metal Sheet, Proceedings of the 11e colloque national en calcul des structures, Giens, France, May 2013.

[25] G. Blatman, B. Sudret, Adaptive Sparse Polynomial Chaos Expansion Based on Least Angle Regression, J. Comput. Phys. 230 (2011) 2345-2367. https://doi.org/10.1016/j.jcp.2010.12.021

[26] B. Sudret, Risk and Reliability in Geotechnical Engineering, In Risk and Reliability in Geotechnical Engineering, K.-K. Phoon, J. Ching (Eds.), CRC Press: Boca Raton, 2018, ISBN 9781482227222.

Material Forming - ESAFORM 2023   Materials Research Forum LLC
Materials Research Proceedings 28 (2023) 1193-1202   https://doi.org/10.21741/9781644902479-130

Coupling machine learning and synthetic image DIC-based techniques for the calibration of elastoplastic constitutive models

PRATES Pedro A.[1,2,a] *, HENRIQUES Joan D. F.[1,b], PINTO Jose[1,c],
BASTOS Nelson [1,d] and ANDRADE-CAMPOS Antonio[1,2,e]

[1]TEMA – Centre for Mechanical Technology and Automation, Department of Mechanical Engineering, University of Aveiro, 3810-193 Aveiro, Portugal

[2]LASI – Intelligent Systems Associate Laboratory, 4800-058 Guimarães, Portugal

[a]prates@ua.pt, [b]joaodiogofh@ua.pt, [c]josemiguel99@ua.pt, [d]kevinbastos@ua.pt, [e]gilac@ua.pt

Keywords: Constitutive Model Calibration, Elastoplasticity, Machine Learning, DIC

Abstract. Today, most design tasks are based on simulation tools. However, the success of the simulation depends on the accurate calibration of constitutive models. Inverse-based calibration methods, such as the Finite Element Model Updating and the Virtual Fields Method, have been developed for identifying constitutive parameters. These methods are based on mechanical tests that allow heterogeneous strain fields under the "Material Testing 2.0" paradigm in which digital image correlation plays a vital role. Although these methods have been proven effective, constitutive model calibration is still a complex task. A machine learning approach is developed and implemented to calibrate elastoplastic constitutive models for metal sheets, using datasets populated with finite element simulation results of strain field data from mechanical tests. Feature importance analysis is conducted to understand the importance of the different input features and to reduce the computational cost related with model training. Synthetic image DIC-based techniques were coupled with the numerically generated database, enabling the construction of a virtual experiments database that accounts for sources of uncertainty that can influence experimental DIC measurements. A robustness analysis of the methodology is performed for the boundary conditions of the test.

Introduction

The numerical simulation of sheet metal forming processes has become an essential tool to obtain high-performance components for automotive and aerospace industries. The quality of numerical simulations depends on the quality of the constitutive modelling that describes the material behaviour, which is related to the type of constitutive laws and the strategy to identify their parameters. Inverse identification strategies based on Finite Element Model Updating (FEMU) and Virtual Fields Method (VFM) have been proposed (e.g. [1–3]), which make use of full-field measurements of heterogeneous strain field data collected from non-standardized mechanical tests. However, the computational time and a complex implementation process are still major drawbacks of such identification strategies [4–5]. More recently, machine learning based approaches have been explored to identify material parameters (e.g. [6–8]), showing to be a promising alternative to both FEMU and VFM [3]. The authors of the current work have previously proposed a machine learning approach to identify material parameters (isotropic hardening law + orthotropic yield criterion), using a numerically generated database populated with finite element simulation results of cruciform tensile tests and the XG-Boost algorithm [6]. In this work, synthetic image DIC-based techniques are coupled with the numerically generated database, enabling the construction of a virtual experiments database that accounts for sources of uncertainty that can influence experimental DIC measurements. A robustness analysis on the ML approach is performed for the boundary conditions of the finite element model. In particular, the orientation of the material axes was varied with reference to the global axes system of the cruciform test to emulate specimen misalignments that may occur in material testing.

Methodology and Implementation

General Approach.

Fig. 1 schematizes the proposed approach for calibrating elastoplastic constitutive models. First, a database was established using numerically generated features (i.e. strain fields and loads) obtained from FEA simulations of an heterogenous mechanical test. Different combinations of constitutive parameters were considered while maintaining the same sample geometry and boundary conditions; a Python script was created to enhance the whole process and automatize the generation of the database. Then, a feature importance step was performed to evaluate how useful the data is at predicting the constitutive parameters. A virtual experiments database was created from the FEA-generated database, where synthetic images were generated and processed via DIC. The virtual experiments database was then split into training (90%) and testing (10%) sets. The training set was used to establish an ML model for predicting the material parameters from the strain fields and loads. Finally, the predictive performance of the ML model was evaluated using the testing set.

Numerical model.

The mechanical test selected for this study is the biaxial tensile test on a cruciform sample. The sample geometry was designed in a previous work, enabling heterogeneous stress and strain fields and a wide range of stress and strain paths that are commonly observed in sheet metal forming processes [9]. Fig. 2(a) shows the geometry and the dimensions of the cruciform sample in the sheet plane. The numerical simulation model only considers one fourth of the sample (see grey region in Fig. 2(a)) due to symmetries in the boundary conditions, sample geometry and material behaviour. Moreover, plane stress conditions and a constant thickness of the sheet equationual to 1 mm are assumed. Fig. 2(b) shows the boundary conditions and finite element mesh of the numerical model. Symmetry boundary conditions were prescribed on the 0x and 0y axes ($u_x = u_y = 0$ mm), and displacement boundary conditions were applied to the nodes located at the ends of both arms of the sample to promote equal displacements along both 0x and 0y axes ($u_x = u_y = 2$ mm). The numerical model is discretized with a regular mesh made of 405 CPS4R elements (bilinear shape functions and reduced integration). All FEA simulations were performed using ABAQUS CAE software [10]. Each simulation was carried out for twenty equally spaced time-steps, where the force and the strain field (ε_{xx}, ε_{yy}, and ε_{xy}) was obtained for each time-step.

Material Forming - ESAFORM 2023
Materials Research Forum LLC
Materials Research Proceedings 28 (2023) 1193-1202
https://doi.org/10.21741/9781644902479-130

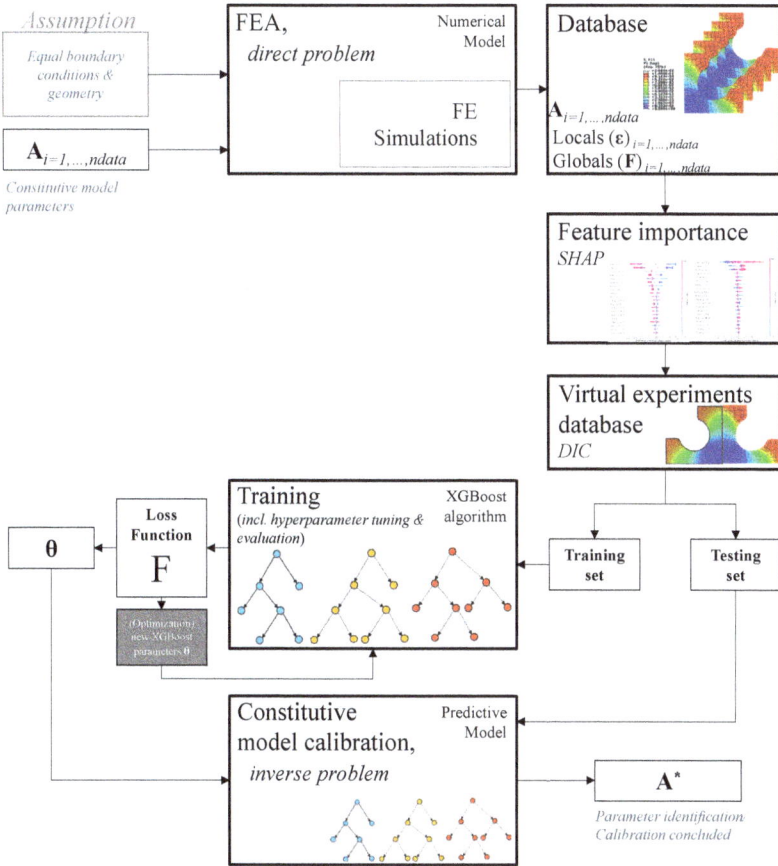

Fig. 1. Flowchart of the proposed identification approach.

The material constitutive model assumes an isotropic elastic behavior, described by the Hooke's law (with Young's modulus $E = 210$ GPa and Poisson's ratio $v = 0.3$), and an orthotropic plastic behavior, described by Hill'48 yield criterion with isotropic hardening described by Swift law under an associated flow rule. Under plane stress conditions, the Hill'48 yield criterion can be written as follows:

$$(F + H)\sigma_{yy}^2 + (G + H)\sigma_{xx}^2 - 2H\sigma_{xx}\sigma_{yy} + 2N\tau_{xy}^2 = Y^2 \tag{1}$$

where F, G, H and N are anisotropy coefficients, σ_{xx}, σ_{yy} and τ_{xy} are the components of the Cauchy stress tensor in the material axes system of the metal sheet, and Y is the yield stress.

The condition $G+H=1$ (i.e. $\sigma_{xx} = Y$) was assumed, which corresponds to the following relationships:

Material Forming - ESAFORM 2023   Materials Research Forum LLC
Materials Research Proceedings 28 (2023) 1193-1202   https://doi.org/10.21741/9781644902479-130

$$F = \frac{r_0}{r_{90}(r_0+1)}; \ G = \frac{1}{r_0+1}; \ H = \frac{r_0}{r_0+1}; \ N = \frac{1}{2}\frac{(r_0+r_{90})(2r_{45}+1)}{r_{90}(r_0+1)} \tag{2}$$

where r_0, r_{45} and r_{90} are the Lankford ratios obtained at $0°$, $45°$ and $90°$ w.r.t. the rolling direction of the sheet, respectively. The Swift law describes the yield stress evolution during plastic deformation as follows:

$$Y = K\left[\left(\frac{Y_0}{K}\right)^{\frac{1}{n}} + \overline{\varepsilon}^{\,\mathrm{p}}\right]^n \tag{3}$$

where $\overline{\varepsilon}^{\,\mathrm{p}}$ is the equivalent plastic strain and Y_0, K and n are material parameters.

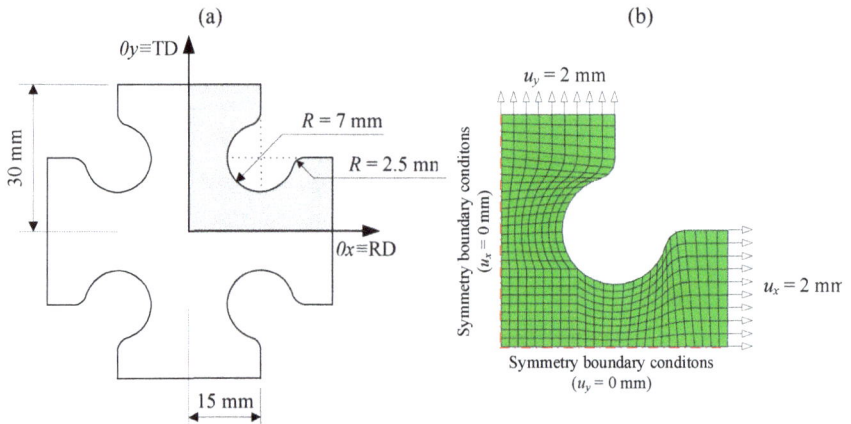

Fig. 2. Biaxial tensile test on a cruciform sample: (a) geometry and dimensions [9]; (b) boundary conditions and finite element mesh.

The database was populated with synthetic data generated by FEA simulations of the cruciform tensile test with different combinations of plasticity material parameters while maintaining the same geometry, boundary conditions and elastic properties. Table 1 shows the range of values of the plasticity material parameters considered as input space for the numerical simulations. Then, 2000 sets of parameters were generated using the Latin Hypercube Sample method, and numerical simulations of the cruciform tensile test were performed for each set. Fig. 3 presents an example of numerical results of the cruciform test that were used to build the database.

Material Forming - ESAFORM 2023
Materials Research Proceedings 28 (2023) 1193-1202

Materials Research Forum LLC
https://doi.org/10.21741/9781644902479-130

Table 1. Input space of plasticity material parameters.

Plasticity material parameters	Range
Y_0 [MPa]	80-300
n	0.1-0.3
K [MPa]	280-700
r_0	1-2.5
r_{45}	1-2.5
r_{90}	1-2.5

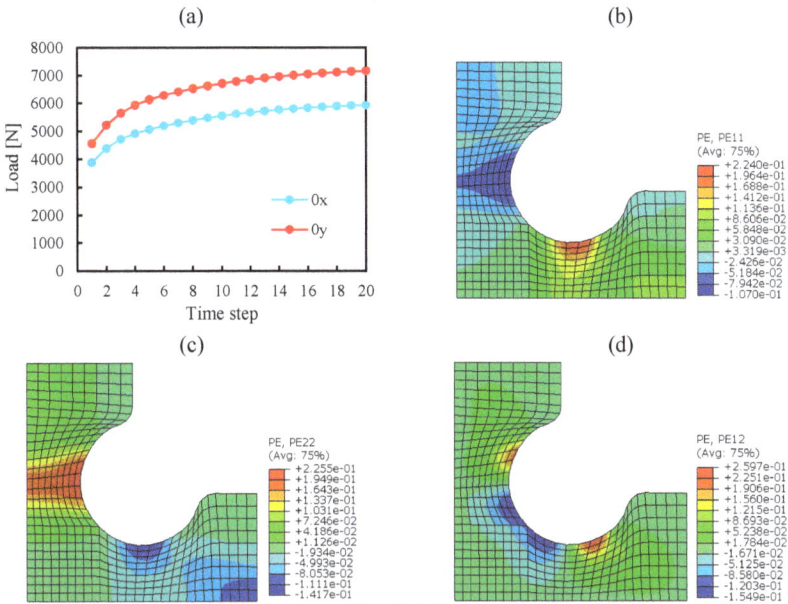

Fig. 3. FEA results of the cruciform test (Y_0=172MPa, n=0.16; K=486MPa; r_0=2.38; r_{45}=1.8; r_{90}=1.06): (a) load vs. displacement along the 0x and 0y axes; strain fields (b) ε_{xx}, (c) ε_{yy}, and (d) ε_{xy}. The strain fields (ε_{xx}, ε_{yy}, and ε_{xy}) were obtained for $u_x = u_y = 2$ mm.

Feature importance.

Many ML systems are essentially considered black boxes, as it is hard to understand and explain how they work after training. Shapley Additive exPlanations (SHAP) are considered state of the art in Machine Learning explainability [11]. SHAP analysis explain how each feature (i.e. loads, strain fields) affects the outputs (i.e. material parameters) of the ML model. As an example, Fig. 4 presents the 20 most important features for predicting the hardening parameters Y_0, n and K. In Fig. 4a, the feature "Force_y_1" (i.e. force 0y at time step 1) is the most relevant feature to predict the initial yield stress, Y_0; moreover, higher values of "Force_y_1" (towards red color) have a positive contribution on the prediction of Y_0 (i.e. higher SHAP values), leading to higher values of Y_0. SHAP analysis confirmed that it is possible to substantially reduce the number of features in

Material Forming - ESAFORM 2023
Materials Research Proceedings 28 (2023) 1193-1202

Materials Research Forum LLC
https://doi.org/10.21741/9781644902479-130

the dataset, by excluding the less important ones and reducing the computational cost without compromising predictive performance.

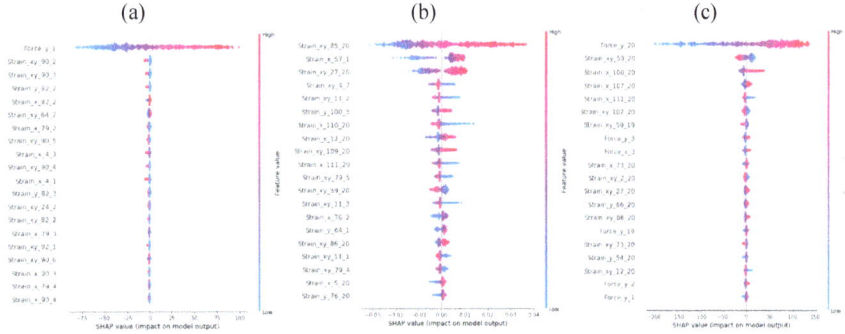

Fig. 4. SHAP analysis of the 20 most important features for predicting the hardening parameters (a) Y_0; (b) n and (c) K.

Virtual experiments.

Virtual experiments enable a proper comparison between simulations and experiments by overcoming issues that arise from the direct numerical-experimental comparison (e.g. alignment, data locations, discretization, filtering effects, experimental uncertainties, etc.) and making the database features one step closer to measurements collected from real experiments. After finishing the feature importance step, the numerical results of the cruciform test were used to generate synthetic images that were then processed with DIC. The DIC setting were selected using the Performance Analysis module from MatchID [12], representing a good compromise between spatial and measurement resolution, as shown in Fig. 5. Each sample was divided into 6455 subsets. In each subset, the location in pixels and the strains ε_{xx}, ε_{yy}, and ε_{xy} were recorded. 2000 samples were generated, 1800 for training and 200 for testing.

XGBoost algorithm.

Extreme Gradient Boosting (XGBoost) is an efficient open-source implementation of the gradient boosted trees algorithm [13]. Gradient boosting is an algorithm in which new models are created from previous models' residuals (weak models) and then combined to make the final prediction (strong models). When adding new models, it uses a gradient descent algorithm to minimize the loss. XGBoost attempts to minimize the regularized objective as follows:

$$obj(\boldsymbol{\theta}) = \sum_i L\left(\hat{y}_i(\boldsymbol{\theta}), y_i\right) + \sum_k \Omega(\boldsymbol{\theta}) \tag{4}$$

where L is the training loss function that measures the deviation between the value $\hat{y}_i$ predicted by the model and the actual value y_i of sample i; Ω is the regularization function (i.e. a penalty term) that measures the complexity of the model, which tends to prevent overfitting; $\boldsymbol{\theta}$ represents the set of parameters to be calibrated during training.

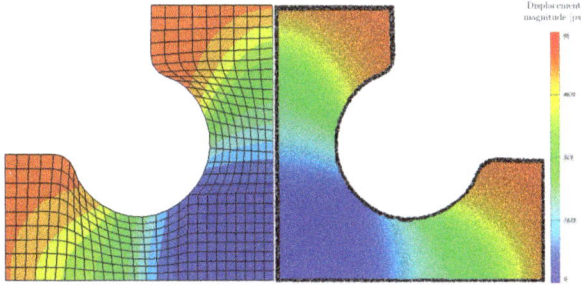

Fig. 5. Comparison of the displacement magnitude in the cruciform test, between FEA (left) and virtual experiments (right).

Performance metrics.

The performance of the ML model was evaluated by comparing the predicted and simulated parameters using the Coefficient of Determination (R^2), given by:

$$R^2 = 1 - \frac{\sum_{i=1}^{j} \left(y_i - y_i^*\right)^2}{\sum_{i=1}^{j} \left(y_i - \bar{y}\right)^2} \tag{5}$$

where y_i and y_i^* are respectively the real and predicted values for the constitutive parameters of simulation i from the test set, $\bar{y}$ is the average of the real values of the constitutive parameters, and j is the total number of simulations of the test set.

Results and Discussion

Performance evaluation.

Fig. 6 compares the Y_0, n and K, r_0, r_{45} and r_{90} values predicted by the ML model with the real values considered in the FEA simulations. In general, the ML model has superior predictive performance, which is confirmed by the values of R^2 presented in the figures.

Robustness Analysis.

The robustness of the ML model was analyzed in relation to the boundary conditions of the cruciform test. In fact, the ML model was trained keeping the boundary conditions unchanged; however, during the performance of experimental tests there may be, to a greater or lesser extent, misalignments in the placement/test of the specimen, which may influence the measurements and consequently the calibration of the constitutive models. In this context, one of the samples from the testing set was chosen (Y_0=172 MPa; n=0.16; K=486 MPa; r_0=2.38; r_{45}=1.8; r_{90}=1.06) and the orientation of the material axes was varied w.r.t. the global axes system 0xy (see Fig. 2) to simulate various cases of sample misalignments that may occur in material testing. Fig. 7 shows the relative error in the prediction of each material parameter, for rotations of the material axes ranging between 0° and 10° w.r.t. the global axes system (the 0° case is taken as reference for calculating the relative error in predictions); this figure also includes an extreme case of 45°. In general, the parameter predictions remain robust regarding possible sample misalignments in-between 0° and 10°; in this interval, the maximum variation of the relative error occurs for parameter r_0 (about 2.65%). This preliminary robustness analysis suggests that sample misalignments do not significantly influence the predictive performance of the ML model, regardless of the overall quality of predictions obtained for each parameter (poor quality for parameters n, r_{45} and Y_0, which requires careful analysis to assess the cause).

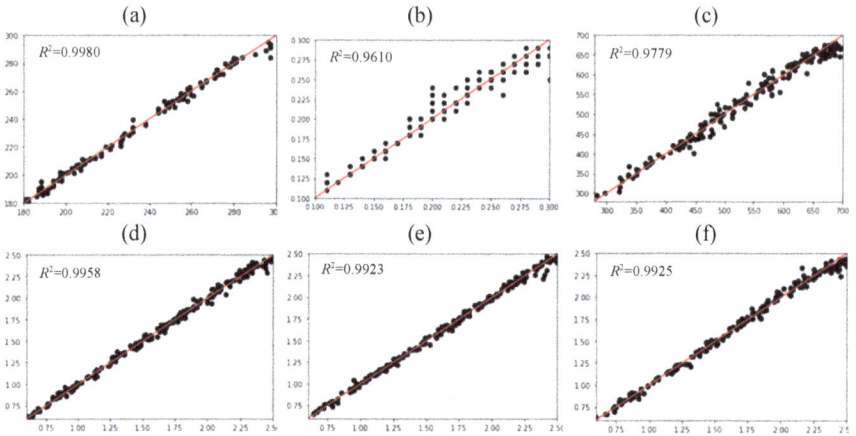

Fig. 6. Predicted (vertical axis) vs. real (horizontal axis) values of the constitutive parameters: (a) Y_0; (b) n; (c) K; (d) r_0; (e) r_{45}; (f) r_{90}. The R^2 values are also presented.

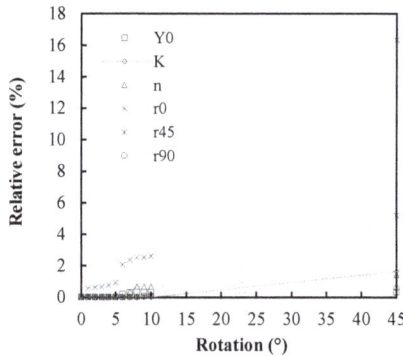

Fig. 7. Relative error in the prediction of the material parameters as a function of the rotation of the material axes w.r.t. the global axes system.

Summary

A machine learning approach was developed for calibrating constitutive models that describe the plastic behaviour of metal sheets. The database was populated with numerically generated features (strain fields and loads) collected from FEA simulations of the biaxial tensile test on a cruciform sample. Feature importance analysis allowed to identify the most relevant features in the database for identifying the constitutive parameters of the Swift hardening law and the Hill'48 yield criterion. A virtual experiments database was established from the FEA-generated database using synthetic images and DIC processing, making the numerically generated features one step closer to measurements collected from real experiments. A robustness analysis on the ML approach was performed for the boundary conditions to simulate possible misalignments that may occur in material testing, namely on the rotation of the material axes w.r.t. the global axes system. It was concluded:

- Feature importance enabled the identification of the most important features of the database for identifying constitutive parameters. ML training was greatly accelerated, without compromising predictive performance;
- Virtual experiments allow for accurate model predictions of the constitutive parameters, bridging the gap between simulation results and experimental measurements;
- Sample misalignments do not significantly influence the performance of the ML model.

The trained ML model enables an almost instantaneous parameter identification, without being significantly influenced by sample misalignments. On the other hand, it is limited to the identification of Hill'48+Swift law parameters, which can be unsuitable for materials whose plastic behaviour is best described using more advanced constitutive models; also, the trained ML model considers uniform displacements imposed at the end of both arms of the sample. Therefore, it is envisaged to extend the robustness analysis to the displacement boundary conditions (i.e. when it is not possible to achieve uniform displacement at both arms of the sample), and to the type of material constitutive model that describes the plastic behaviour of the material.

Acknowledgments

This research has received funding from the Research Fund for Coal and Steel under grant agreement No 888153. This research was also sponsored by FEDER funds through the program COMPETE (Programa Operacional Factores de Competitividade), by national funds through FCT (Fundação para a Ciência e a Tecnologia) under the projects UIDB/00481/2020, UIDP/00481/2020, CENTRO-01-0145-FEDER-022083 and LA/P/0104/2020. J. Henriques was supported by a grant for scientific research from the Portuguese Foundation for Science and Technology (ref. 2021.05692.BD). All supports are gratefully acknowledged.

References

[1] S. Avril, M. Bonnet, A.-S. Bretelle, M. Grédiac, F. Hild, P. Ienny, F. Latourte, D. Lemosse, S. Pagano, E. Pagnacco, F. Pierron, Overview of identification methods of mechanical parameters based on full-field measurements, Exp. Mech. 48 (2008) 381–402. https://doi.org/10.1007/s11340-008-9148-y

[2] P.A. Prates, A.F.G. Pereira, N.A. Sakharova, M.C. Oliveira, J.V. Fernandes, Inverse strategies for identifying the parameters of constitutive laws of metal sheets, Adv. Mater. Sci. Eng. 2016 (2016) 4152963. https://doi.org/10.1155/2016/4152963

[3] A. Andrade-Campos, N. Bastos, M. Conde, M. Gonçalves, J. Henriques, R. Lourenço, J.M.P. Martins, M.G. Oliveira, P. Prates, L. Rumor, On the inverse identification methods for forming plasticity models using full-field measurements, IOP Conf. Ser. Mater. Sci. Eng. 1238 (2022) 012059. https://doi.org/10.1088/1757-899X/1238/1/012059

[4] J. Martins, A. Andrade-Campos, S. Thuillier, Comparison of inverse identification strategies for constitutive mechanical models using full-field measurements, Int. J. Mech. Sci. 145 (2018) 330–345. https://doi.org/10.1016/j.ijmecsci.2018.07.013

[5] N. Souto, A. Andrade-Campos, S. Thuillier, Mechanical design of a heterogeneous test for material parameters identification, Int. J. Mater. Form. 10 (2017) 353-367. https://doi.org/10.1007/s12289-016-1284-9

[6] N. Bastos, P. Prates, A. Andrade-Campos, Material parameter identification of elastoplastic constitutive models using machine learning approaches, Key Eng. Mater. 926 (2022) 2193-2200. https://doi.org/10.4028/p-zr575d

[7] A.E. Marques, A.F.G. Pereira, B.M. Ribeiro, P.A. Prates, On the identification of material constitutive model parameters using machine learning algorithms, Key Eng. Mater. 926 (2022) 2146-2153. https://doi.org/10.4028/p-5hf550

[8] R. Schulte, C. Karca, R. Ostwald, A. Menzel, Machine learning-assisted parameter identification for constitutive models based on concatenated loading path sequences, Eur. J. Mech. A-Solid. 98 (2023) 104854. https://doi.org/10.1016/j.euromechsol.2022.104854

[9] J. Martins, A. Andrade-Campos, S. Thuillier, Calibration of anisotropic plasticity models using a biaxial test and the virtual fields method, Int. J. Solids Struct. 172 (2019) 21-37. https://doi.org/10.1016/j.ijsolstr.2019.05.019

[10] Dassault Systèmes. Abaqus 2017 documentation, 2017.

[11] S. Lundberg, S. Lee, A Unified approach to interpreting model predictions, 2017.

[12] MatchID: Metrology beyond colors. MatchID version 2022.2, 2022.

[13] T. Chen, C. Guestrin, XGBoost: A scalable tree boosting system, in Proceedings of the 22nd ACM SIGKDD International Conference on Knowledge Discovery and Data Mining, 2016, pp. 785-794.

Material Forming - ESAFORM 2023
Materials Research Forum LLC
Materials Research Proceedings 28 (2023) 1203-1210
https://doi.org/10.21741/9781644902479-131

A hybrid VFM-FEMU approach to calibrate 3D anisotropic plasticity models for sheet metal forming

ROSSI Marco[1,a] *, LATTANZI Attilio[1,b] and AMODIO Dario[1,c]

[1]Department of Mechanical Engineering and Mathematical Sciences, Università Politecnica delle Marche, via brecce bianche 12, 60131, Ancona, Italy

[a]m.rossi@staff.univpm.it, [b]a.lattanzi@staff.unvipm.it, [c]a.amodio@staff.univpm.it

Keywords: Inverse Methods, 3D Anisotropic Plasticity, Full-Field Measurements

Abstract. Recently, inverse methods such as the Virtual Fields Method (VFM) or the Finite Element Model Updating (FEMU), coupled with a full-field measurement technique, have been distinguished as efficient strategies for the calibration of complex plasticity models [1]. The use of heterogeneous strain fields, in fact, offers a larger amount of material information compared to the classical standard test, enriching the identification process and, in general, reducing the experimental effort for the calibration [2]. Here, an inverse identification framework is proposed for the calibration of a full-scale anisotropic plasticity model. The inverse identification procedure employs full-field information from two main experiments: a tensile test on double notched specimens for the calibration of the coefficients expressing the planar anisotropy, and an innovative Iosipescu-like test for the through-thickness shear ones. A hybrid approach is used with the VFM employed to identify the planar coefficients and the FEMU for the through thickness ones.

Introduction

Sheet metals are characterized by an anisotropic behavior which plays a crucial role in the prediction of their plastic deformation and failure. Generally, their constitutive response is modeled and calibrated by mainly considering the in-plane material behavior, while the through-thickness one is often neglected based on the plane stress assumption. However, in some applications - for instance the sheet metal forming of complex geometries - the state of stress can deviate from the plane stress assumption and a complete 3D description is necessary [1].

Over the years, several testing protocols have been developed to capture in detail the complex mechanical response under different types of loading conditions and to infer a comprehensive description of the material deformation. Traditionally, the common material testing approach relies on the use of quasi-homogeneous tests, where the relation between stress and strains can be directly obtained from experiments properly designed; on the other side, the application of full-field techniques to material testing has allowed to analyze and simultaneously exploit multiple stress and strain conditions produced through heterogenous tests [2,3].

However, mechanical data from heterogeneous tests cannot be directly used in the calibration process, and are generally coupled with inverse methods. Inverse methods have been already used to identify, for instance, the plastic behavior of metals by resorting to numerical simulations [4,5]: the method is often referred as Finite Element Model Updating (FEMU) since, essentially, performs the identification by iteratively changing the constitutive parameters of a numerical simulation of the test until the difference between the numerical and experimental results, in terms of loading force and strain fields, is minimized. Other examples for the identification of the hardening behavior can be found in [6,7] using an energy balance approach called the Virtual Fields Method (VFM) [8]. The VFM has been applied also to anisotropic plasticity [9-11] and to investigate multiaxial loading conditions such as cruciform specimens [12] and the bulge test [13].

Material Forming - ESAFORM 2023   Materials Research Forum LLC
Materials Research Proceedings 28 (2023) 1203-1210   https://doi.org/10.21741/9781644902479-131

However, experimental procedures to calibrate full 3D anisotropic plasticity models of sheet metals are still not well addressed. The aim of this work is to demonstrate that such calibration can be effectively carried out using inverse methods and simple experiments.

Methods

The aim of this paper is the identification of a full 3D anisotropic model trying to reduce the experimental effort and the number of tests required to make the identification. A similar problem was already tackled in [14] by Denys et al., who introduced a double drilled specimen employed for the full calibration of the Hill48 yield surface, by means of the FEMU approach. However, especially for sheet metals, often it is not possible to perform a hole along the thickness of the specimen, so the previous method can only be applied to thick materials.

A different approach is used here, employing specimens that can be easily machined from a sheet metal blank and tested using a standard uniaxial machine, so that the experimental procedure can be readily implemented in almost each material testing lab. The proposed method also requires the use of a full-field optical measurement technique, e.g. digital image correlation (DIC), to obtain the strain field in a region of interest (ROI) of the specimen, following the Material Testing 2.0 logic [2].

Virtual experiments are used to verify the feasibility of the developed procedure; the three-dimensional anisotropic behavior of the material was reproduced using the Hill48 yield function:

$$f(\boldsymbol{\sigma}) = \sqrt{F(\sigma_y - \sigma_z)^2 + G(\sigma_z - \sigma_x)^2 + H(\sigma_x - \sigma_y)^2 + 2L\tau_{yz}^2 + 2M\tau_{xz}^2 + 2N\tau_{xy}^2} \quad (1)$$

where F,G,H,LM and N are constants that must be identified from experiments, in particular, in this case, we set L=M so that the shear behavior through the thickness in the y-z plane is equal to the one in the x-z plane. The hardening was described by a Swift law:

$$\sigma_{eq} = k \left(\varepsilon_0 + \varepsilon_{eq}\right)^n \qquad (2)$$

where σ_{eq} and ε_{eq} are the equivalent stress and strain, respectively, and k, ε_0 and n are parameters that must be identified. Summarizing, to fully characterize the three-dimensional anisotropic behavior of the material, it is necessary to identify a total of 8 parameters, 5 for the yield function and 3 for the hardening law.

A two-steps identification process was developed using a double notched (DN) specimen to identify the in-plane anisotropic properties and the hardening law, i.e. parameters F, G, H, N, k, ε_0 and n; and a through-thickness shear (TTS) test to identify the shear behavior along the thickness, i.e. parameter L.

FE models of the two tests were developed using ABAQUS and used to generate synthetic data that replicate the load history and the strain maps obtained during a real test. The synthetic data were accordingly used as input for the identification procedure.

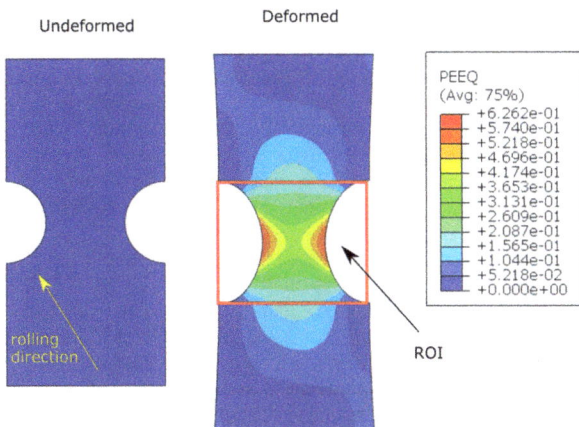

Fig. 1. Double notched specimen used to evaluate the in-plane properties.

Fig. 1 shows the geometry of the DN specimen and the ROI where the full-field strain measurement is performed through DIC. In this case, the FEM data were used to simulate a DIC measurement with 151×51 strain points in the ROI. The rolling direction of the material was inclined with an angle of $22.5°$ with respect to the force direction.

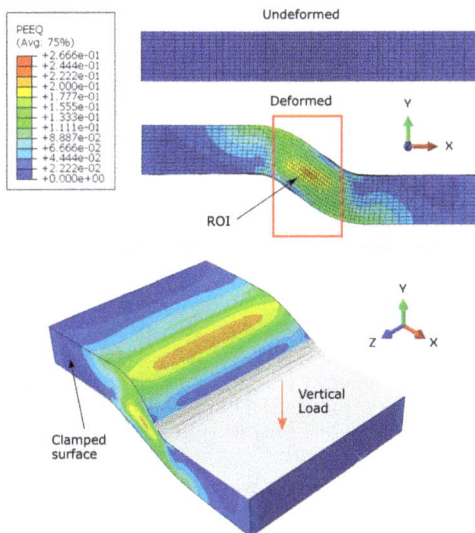

Fig. 2. Through-thickness shear test.

Material Forming - ESAFORM 2023
Materials Research Proceedings 28 (2023) 1203-1210

Materials Research Forum LLC
https://doi.org/10.21741/9781644902479-131

Fig. 2 illustrates the TTS test used to identify the parameters governing the through-thickness shear behavior. In this case, the ROI was placed along the thickness surface with 200 × 200 simulated measurement points, see Fig. 2. For both tests, the thickness of the sheet metal is 2 mm.

The identification of the constitutive parameters was performed using a hybrid VFM-FEMU approach, where the VFM was used with the DN specimen to identify 7 parameters and the FEMU was used with the TTS test to identify 1 parameter. Indeed, the VFM cannot be employed on the TTS test because the problem is three-dimensional and the plane stress or plane strain assumption is not valid, i.e. the strain measured along the thickness surface with DIC is different from the strain inside the material.

The advantage of combining the two methods is that VFM is usually less computationally expensive, so it can be used to quickly identify 7 parameters while FEMU is restricted to the identification of 1 parameter. The hybrid procedure is sequential, first the VFM is applied and then the FEMU using different cost functions, this is possible because the behavior of the DN specimen is not influenced by the parameter L, which is subsequently identified with FEMU. Theoretically, VFM and FEMU could also be put in the same optimization loop, using a common cost function, but such approach will be more complex and less efficient. Test design is essential to develop tests that are uncoupled with respect to different parameters, in order to simplify the identification procedure.

Both methods are implemented in Matlab using the in-built minimization functions to solve the inverse problem. A brief description of the two methods is given below, more details can be found in the references.

VFM Approach

The Virtual Fields Method (VFM) is an inverse method based on the weak form of the equilibrium through the Principle of Virtual Work (PVW) and allows to identify the coefficients of a given constitutive model starting from full-field kinematic and loading data. In the case of quasi-static problems where the body forces are neglected, the VFM is generally expressed for large deformations using the following cost function:

$$\Psi(\xi) = \sum_{i=1}^{N_{vf}} \sum_{j=1}^{N_{step}} \left| \int_V \boldsymbol{T}_j^{1PK} : \delta \boldsymbol{F}_i^* dV - \int_{\partial V} \left(\boldsymbol{T}_j^{1PK} \boldsymbol{n} \right) \cdot \delta \boldsymbol{u}_i dS \right| \tag{3}$$

where $\boldsymbol{T}_j^{1PK}$ indicates the 1st Piola-Kirchhoff stress tensor, $\delta \boldsymbol{u}_i$ is any kinematically admissible virtual displacement field, $\delta \boldsymbol{F}_i^*$ is the corresponding virtual displacement gradient tensor, V is the volume of the inspected solid, ∂V is the boundary surface and $\boldsymbol{n}$ the surface normal. The first integral term represents the Internal Virtual Work, where the stress tensor is calculated from the full-field strain data of the test according to the model constitutive parameters ξ; the second integral indicates the External Virtual Work and accounts for the loading condition on the boundaries. The cost function is evaluated for all the N_{step} timesteps of the test and for all N_{vf} virtual fields introduced. In this study, the constitutive behaviour is described by a non-linear relation, therefore, the identification is achieved through the minimization of the cost function $\Psi(\xi)$ until the equilibrium equation of the PVW is satisfied. The method is widely described in the literature and more details can be found in the book and papers cited in the introduction see [6-13].

The selection of the virtual fields (VFs) directly affects the identification results since they activate and weight the constitutive information contained in each material point. For this reason, the definition of the VFs employed in the cost function minimization is not trivial. Different approaches can be found in literature [15, 16] which, basically, classify the VFs in manually defined VFs and automatically generated VFs. In this work, we adopt the latter approach, where the virtual kinematic fields are produced starting from the sensitivity of the computed stress field to the single material parameter:

$$\delta T_{(i)}^{1TPK}(\xi, t) = T^{1TPK}(\xi - \delta\xi_i, t) - T^{1TPK}(\xi, t) \tag{4}$$

in other words, by perturbing the material parameter ξ_i, it is possible to highlight the material points where ξ_i effectively affects the stress calculation; the sensitivity distribution can be used as $\delta F_i{}^{\cdot}$ and can be integrated to get the δu_I through a piecewise approach, as discussed in detail in [17].

FEMU Approach

The FEMU is based on the minimization of a cost function that represents a weighted difference between the numerical and experimental results. In this particular case, the compared results are the vertical load force applied by the tensile machine (see Fig. 2) and the full-field strain map measured in the ROI. In a real test, the boundary conditions (BC) are a critical factor, in fact possible misalignments or sliding on the clamping zone of the specimen can influence the deformation history. To reduce the bias due to BC, the FEMU is conducted only on a portion of the material within the ROI, see Fig. 3. The boundary conditions are applied using the displacement measured by the full-field measurement at the border. The central zone is not subjected to friction and the vertical load force measured by the machine can always be obtained from the FE model as sum of the nodal reactions along the vertical direction.

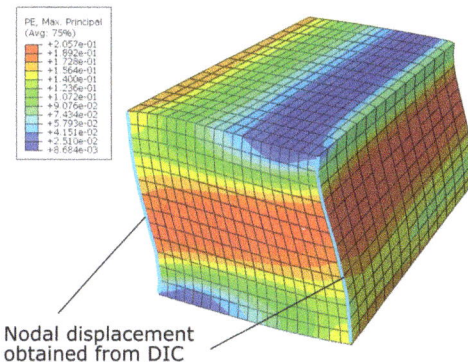

Fig. 3. Portion of the specimen used in the FEMU.

Results and Discussion

The results of the identification are listed in Table 1 in terms of identified parameters and percentage error. The reference parameters are the ones input in the FE model used to create the simulated experiment and can be viewed as the ground truth of the inverse problem. The identification is reasonably good except for parameter ε_0, however it is well known that this parameter has a minor impact in the description of the hardening curve. Moreover, often, the sole parameter identification error is not a proper criterion to evaluate the accuracy of the identification method, instead, it is preferable to verify the results in terms of plastic behavior, i.e. yield surface and hardening curve.

Table 1. Reference and identified parameters.

	VFM							FEMU
	K	ε_0	n	F	G	H	N	L=M
Reference	1000	0.02	0.5	0.3819	0.3125	0.6875	1.389	2.5
Identified	1041	0.0123	0.4846	0.3980	0.3308	0.6692	1.4033	2.66
Error %	-4.1	38.5	3.0	-4.2	-5.8	2.6	-1.03	-6.4

Fig. 4 shows the results in terms of yield surface and hardening curve. Since it is not possible to represent a 6-dimensional anisotropic yield surface, Fig. 4a shows the yield surface in terms of the normal stress components, i.e. $\sigma_x \sigma_y \sigma_z$, and Fig. 4b shows the yield surface for the shear stress components, i.e. in-plane (τ_{xy}) vs out of plane (τ_{xz} or τ_{yz}). Finally, Fig. 4c illustrates the identified hardening curve. It is worth noting that the computation was performed on a workstation and the VFM algorithm took around 20 seconds to solve the inverse problem while FEMU more than 5 hours. Such huge difference is due to the fact that each iteration of FEMU needs to run a complex three-dimensional FE simulation, on the other hand, parameter L can only be identified with FEMU. Moreover, the hybrid approach allows to simplify the FEMU part since only 1 parameter needs to be identified, reducing the computational effort of the minimization algorithm.

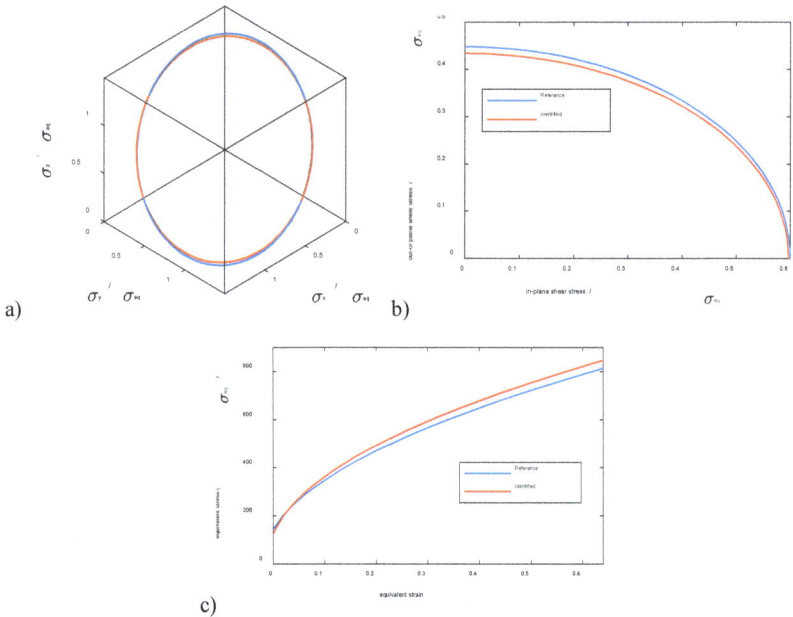

Fig. 4. Comparison of the identification results: a) yield surface for the normal stress components, b) yield surface for the shear components, c) hardening law.

Material Forming - ESAFORM 2023
Materials Research Proceedings 28 (2023) 1203-1210

Materials Research Forum LLC
https://doi.org/10.21741/9781644902479-131

Acknowledgment

This project has received funding from the Research Fund for Coal and Steel under grant agreement No. 888153.

Summary

In this paper an inverse identification procedure to characterize the 3D anisotropic behavior of a metal is proposed and validated through simulated experiments. Two tests are used to identify 8 constitutive parameters combining two well-known inverse methods, i.e. the VFM and FEMU. The identified parameters provides an accurate description of the plastic behavior, with an average error below 5%. In the future, an experimental validation will be conducted and the possibility of identify more complex 3D constitutive model will be evaluated.

References

[1] L. Ma, Z. Wang, The effects of through-thickness shear stress on the formability of sheet metal- A review, J. Manuf. Process. 71 (2021) 269-289. https://doi.org/10.1016/j.jmapro.2021.09.019

[2] F. Pierron, M. Grédiac, Towards Material Testing 2.0. A review of test design for identification of constitutive parameters from full-field measurements, Strain 57 (2021) 12370. https://doi.org/10.1111/str.12370

[3] M. Rossi, A. Lattanzi, L. Morichelli, J.M. Martins, S. Thullier, A. Andrade-Campos, S. Coppieters, Testing methodologies for the calibration of advanced plasticity models for sheet metals: A review, Strain 58 (2022) 12426. https://doi.org/10.1111/str.12426

[4] D. Lecompte, S. Cooreman, S. Coppieters, J. Vantomme, H. Sol, D. Debruyne, Parameter identification for anisotropic plasticity model using digital image correlation, Eur. J. Comput. Mech. 18 (2009) 393-418. https://doi.org/10.3166/ejcm.18.393-418

[5] F. Grytten, B. Holmedal, O.S. Hopperstad, T. Børvik, Evaluation of identification methods for YLD2004-18p, Int. J. Plast. 24 (2008) 2248-2277. https://doi.org/10.1016/j.ijplas.2007.11.005

[6] M. Grédiac, F. Pierron, Applying the Virtual Fields Method to the identification of elasto-plastic constitutive parameters, Int. J. Plast. 22 (2006) 602-627. https://doi.org/10.1016/j.ijplas.2005.04.007

[7] M. Rossi, A. Lattanzi, F. Barlat, A general linear method to evaluate the hardening behaviour of metals at large strain with full-field measurements, Strain 54 (2018) 12265. https://doi.org/10.1111/str.12265

[8] F. Pierron, M. Grediac, The virtual fields method: Extracting constitutive mechanical parameters from full-field deformation measurements, 2012.

[9] M. Rossi, F. Pierron, Identification of plastic constitutive parameters at large deformations from three dimensional displacement fields, Comput. Mech. 49 (2012) 53-71. https://doi.org/10.1007/s00466-011-0627-0

[10] M. Rossi, F. Pierron, M. Štamborská, Application of the virtual fields method to large strain anisotropic plasticity, Int. J. Solids Struct. 97-98 (2016) 322-335. https://doi.org/10.1016/j.ijsolstr.2016.07.015

[11] A. Lattanzi, F. Barlat, F. Pierron, A. Marek, M. Rossi, Inverse identification strategies for the characterization of transformation-based anisotropic plasticity models with the non-linear VFM, Int. J. Mech. Sci. 173 (2020) 105422. https://doi.org/10.1016/j.ijmecsci.2020.105422

[12] J.M.P. Martins, A. Andrade-Campos, S. Thuillier, Calibration of anisotropic plasticity models using a biaxial test and the virtual fields method, Int. J. Solid. Struct. 172-173 (2019) 21-37. https://doi.org/10.1016/j.ijsolstr.2019.05.019

[13] M. Rossi, A. Lattanzi, F. Barlat, J.H. Kim, Inverse identification of large strain plasticity using the hydraulic bulge-test and full-field measurements, Int. J. Solid. Struct. 242 (2022) 111532. https://doi.org/10.1016/j.ijsolstr.2022.111532

Material Forming - ESAFORM 2023   Materials Research Forum LLC
Materials Research Proceedings 28 (2023) 1203-1210   https://doi.org/10.21741/9781644902479-131

[14] K. Denys, S. Coppieters, M. Seefeldt, D. Debruyne, Multi-DIC setup for the identification of a 3D anisotropic yield surface of thick high strength steel using a double perforated specimen, Mech. Mater. 100 (2016) 96-108. http://doi.org/10.1016/j.mechmat.2016.06.011

[15] S. Avril, M. Grédiac, F. Pierron, Sensitivity of the Virtual Fields Method to noisy data, Comp. Mech. 34 (2004) 439-452.

[16] A. Marek, F. Davis, F. Pierron, Sensitivity-based virtual fields for the non-linear Virtual Fields Method, Comp. Mech. 60 (2017) 409-431. https://doi.org/10.1007/s00466-017-1411-6

Machining and cutting

Material Forming - ESAFORM 2023
Materials Research Proceedings 28 (2023) 1213-1222

Materials Research Forum LLC
https://doi.org/10.21741/9781644902479-132

Manufacturing of graded grinding wheels for flute grinding

DENKENA Berend[1,a], BERGMANN Benjamin[1,b] and RAFFALT Daniel[1,c*]

[1]Institute of Production Engineering and Machine Tools (IFW), Leibniz University Hannover, An der Universität 2, 30823 Garbsen, Germany

[a]denkenaifw@uni-hannover.de, [b]Bergmannifw@uni-hannover.de, [c]raffalt@ifw.uni-hannover.de

Keywords: Graded Grinding Wheels, Field-Assisted-Sintering, Flute Grinding, Cemented Carbide

Abstract. In this paper, two different methods for manufacturing of graded grinding wheels for two different metal bonds are presented. One method is based on the use of a mask and manual moulding and the other on a height-adjustable holder for moulding. For this purpose, a brittle and a ductile bronze bond are compared. The graded grinding wheels are fabricated through sintering with Field Assisted Sintering Technology (FAST). An analysis of the grain distribution is used to demonstrate the reproducibility of the manufacturing methodology. For analysis, light microscope images of cross-sections of the abrasive layers are taken. The grain distribution is determined using image processing software and a greyscale method. Finally, the advantages of each method are compared. As a result, both manufacturing methods are evaluated in terms of precision, feasibility and efficiency. From this, a recommendation on the implementation and further development of the methods is derived. This method enables the manufacturing of graded grinding wheels for an effective reduction of wear differences for grinding cemented carbide end mill cutters.

Introduction

Cemented carbide end mill cutters are used in a wide range of applications. These include, for example, the aerospace industry, medical technology or electrical engineering [1]. High demands are therefore placed on the quality of the workpieces that are machined with these end mill cutters. In particular, this requires a high level of precision and surface quality. To ensure this, the end mill cutters used must also meet the highest standards. The quality of the flutes of the end mill cutters has a major influence on their performance [2,3]. However, the deep grinding process of the flutes is characterised by high and varying thermomechanical loads. The high thermomechanical loads result from the high depth of cut associated with the process kinematics and the properties of the cemented carbide [4-6]. Cemented carbide possesses a high degree of hardness and wear resistance [4,5]. In addition, the engagement conditions vary significantly along the width of the grinding tool [6,7]. This is a consequence of the flute geometry. These factors result in a high and at the same time uneven radial wear.

This causes an increasing number of defects in the flute bottom and the cutting edges as well. A defect-free flute, however, is decisive for the operational behaviour of the end mill cutters [2]. To counteract this effect, the dressing intervals must be shortened. This ensures the contour accuracy of the grinding wheel. That is decisive for the necessary manufacturing accuracy. However, the frequent dressing intervals reduce the efficiency of the grinding process. This is due to higher downtimes caused by the more frequent dressing on the one hand and the resulting increased dressing wear on the other. The lifetime of the grinding tools is thus shortened.

Another method of ensuring contour accuracy is to adapt the abrasive layer properties to the local wear conditions. By levelling the radial wear occurring, this allows the dressing intervals to be increased. Economic efficiency will be increased as a result, too. The radial wear can be reduced, for example, by increasing the grain retention forces [8]. Likewise, an adjustment of the

Material Forming - ESAFORM 2023   Materials Research Forum LLC
Materials Research Proceedings 28 (2023) 1213-1222   https://doi.org/10.21741/9781644902479-132

bond properties can be made. By using a more ductile bronze bond, the wear can be reduced compared to a brittle bronze [9]. Both methodologies are not applicable in the present case due to quantification limitations. The reason for this is that in the state of the art there is no sufficient accurate model of the influence of the bond properties as well as the grain retention forces on the resulting wear. Another approach is to adjust the number of abrasive grains. An increasing number of abrasive grains reduces the thermomechanical load per individual grain. As a result, this decreases the radial wear occurring [7,10,11]. Depending on the bond properties, the maximum number of abrasive grains is limited. The grains act as defects in the bond. If the number of abrasive grains is too high, the forces acting on the abrasive layer will cause the bond to fail. Wear then increases in turn. These so-called percolation limits depend on the toughness of the bond and the load occurring during the process [12]. Influencing the number of abrasive grains can therefore be used to affect the radial wear.

First approaches have already shown that the adjustment of the grain concentration in the abrasive layer in two steps can level the resulting radial wear [13]. The authors of the present paper have introduced a model in previous work enabling the radial wear differences to be reduced by up to 50 % compared to conventional grinding wheels on the basis of adapting the number of grains to the occurring loads. The model is based on a simulation-supported analysis of the locally occurring load on the grinding wheel. The resulting abrasive grain numbers for a defined application are shown in Fig. 1 [14].

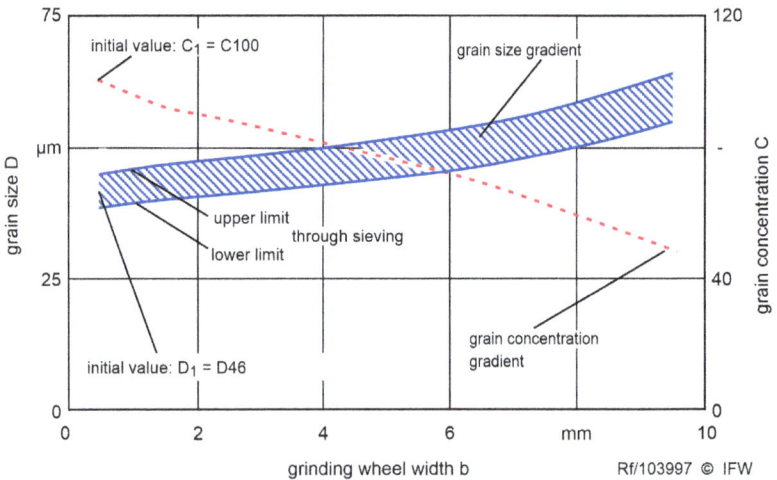

Fig. 1. Comparison of different types of grain number gradients, based on [14].

Two adaptation options are shown. One is via the grain size and the other via the grain concentration. The adjustment via the grain size turned out to be too inaccurate due to the differences in grain sizes after sieving them. This can be seen in the scatter band marked in blue. The grain concentration, on the other hand, can be applied individually in the necessary increments. This is shown in the red line. It has been shown that the implementation of a continuous gradient of the grain concentration is not possible. There are sharp transitions between the concentration ranges [15]. The present work follows on from the preceding research work. So far, only the production of test specimens has been investigated. Therefore, a methodology for the efficient and process-safe production of graded grinding wheels is presented in the following. At

Material Forming - ESAFORM 2023   Materials Research Forum LLC
Materials Research Proceedings 28 (2023) 1213-1222   https://doi.org/10.21741/9781644902479-132

the same time, the moulding method is optimised in order to accurately adjust the heights of the grain concentration ranges. This has led to fluctuations in grain concentration differences in the previously used method based on bulk density.

Materials and Methods

First, grinding layer specimens were sintered. These were used to adjust the moulding accuracy through powder dosing. Specimens are cylindrical with a diameter of 22 mm and a height of 6 mm. The process is comparable to the production of grinding wheels and therefore offers good transferability. The manufacturing of the specimens and the later grinding wheels is shown in Fig. 2.

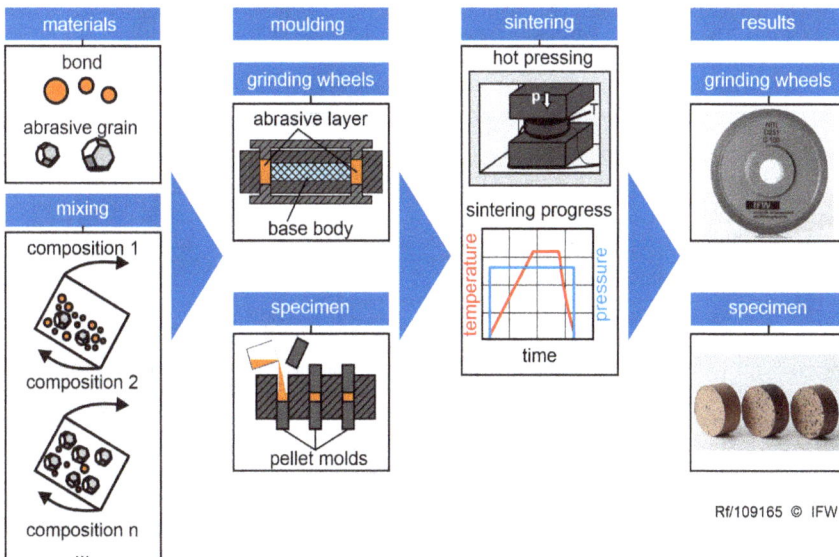

Fig. 2. Grinding wheel and specimen manufacturing.

From this, Fig. 2, also follows the comparability of the two methods. For their production, the components of the abrasive layer must first be mixed and pre-compacted to form a green body.

A Turbula mixer is used for mixing. The mixing time was 30 minutes. The components consist of a bronze bond and diamond grains. Two different pre-alloyed powders were used to evaluate influences on the manufacturing accuracy depending on the components. The composition was 60 wt.% copper and 40 wt.% tin (60/40 bronze), respectively 80 wt.% copper and 20 wt.% tin (80/20 bronze) with an irregular particle shape and an average diameter of 40 μm. Alloys similar to those used are also used in the industrial environment. The sintering temperature used was 420°C for the 60/40 bronze and 620°C for the 80/20 bronze. The holding time was 240 seconds in each case at a pressure of 35 MPa. The diamond grain chosen was an irregular shape grain of size D54 according to the FEPA standard. For each bond type, test specimens with the concentration C0, C25, C50, C75, C100 and C125 were produced. The C specification also complies with the FEPA standard. C100, for example, corresponds to a volume fraction of 25 % of the used grain. Weighting was carried out according to the following equation:

$$m_{bond} = V_{Specimen} \cdot \left(100 - V_{grain}\right) \cdot \rho_{bond} \cdot 0.98 \qquad (1)$$

The target parameter is the mass of the bond m_{bond}. $V_{specimen}$ describes the target volume of the grinding layer specimen, respective the grinding wheel abrasive layer. The volume of the bond is calculated by the difference to the volume of the grains, corresponding to the grain concentration, V_{grain}. The density of the used bond material is given by ρ_{bond}. Based on the available knowledge, porosities smaller than 2 % occur during FAST sintering of metallic bonds. Therefore, the correction factor 0.98 was chosen for the metal content to compensate for its larger volume due to porosity. The test specimens were then measured to ensure the accuracy of the process. The method was verified by sintering graded specimens of composition C100/C0/C50 (gradient 1) and the composition C0/C50/C100 (gradient 2). Each layer was 2 mm thick. Accuracy was evaluated using light microscope images.

Based on the results, the graded grinding wheels were manufactured. Both bonding systems were used for this. The variation of the grain concentrations is shown in Table 1:

Table 1. Composition of the graded grinding wheels for the 60/40 bond.

Layer number and position	Grain concentration	Bronze powder mass	Diamond grain mass	Total mass
1 (0 – 1 mm)	C125	11.974 g	1.949 g	13.923 g
2 (1 – 2 mm)	C121	12.149 g	1.887 g	14.035 g
3 (2 – 3 mm)	C114	12.453 g	1.778 g	14.231 g
4 (3 – 4 mm)	C102	12.976 g	1.590 g	14.566 g
5 (4 – 5 mm)	C83	13.803 g	1.294 g	15.097 g
6 (5 – 10 mm)	C47	76.854 g	3.664 g	80.518 g

The gradients were chosen for future application tests so that they would match the corresponding application. There, they should lead to uniform radial wear when grinding a cemented carbide rod with a diameter of 16.8 mm at an engagement width and depth of 6 mm. The calculation was carried out according to the results from [14]. For comparison, a non-graded reference wheel was manufactured for each bond system. The concentration of the reference wheels was set at C125 over the total grinding wheel with. The composition of the graded grinding wheel with the 80/20 bond corresponds to the same concentration steps. The mass of the bond and diamonds were adjusted concerning this.

Moulding Methods

Two different moulding methods were investigated. For both, the sinter die was placed on a turntable to avoid unsteady rotation. This is to prevent segregation of the abrasive layer components. In the first moulding method, the powder for each abrasive layer was weighed out in i (i = 6; 12; 24) portions. These were filled into the mould along a taped-on mask in the corresponding portioning. The smoothing of the powder layers was done by manual drawing using a metallic plate. The number of steps i is used to evaluate how high the effort must be for reproducible filling. A too high number reduces the economic efficiency. The mould was filled using an aluminium hopper. This counteracted the adhesion of powder.

A height-adjustable holder was used for the second moulding method. The powder for a segment was filled in evenly by turning the sintering mould as required by eye. Subsequently, the powder was smoothed by means of the holder and an appropriate metal plate. In the first step, the height of the holder was adjusted in such a way that a pile of powder was first pushed in front of the plate when the powder was being smoothed out. By raising the height of the plate, an even surface was gradually generated. This was repeated for all layers. Both methods were first carried out using an analogue sintering mould made of acrylic glass. This allowed a visual evaluation of

Material Forming - ESAFORM 2023   Materials Research Forum LLC
Materials Research Proceedings 28 (2023) 1213-1222   https://doi.org/10.21741/9781644902479-132

the moulding process for adhesions on the mould edge or the grinding wheel base body. Concept 2 with the analogue mould can be seen in Fig. 3.

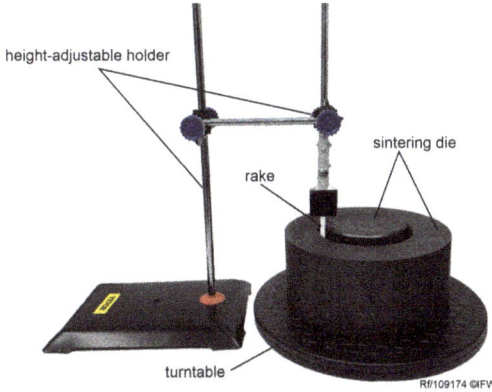

Fig. 3. Manufacturing method using a height-adjustable holder.

All grinding wheels were sintered with the parameters of the test specimens. The evaluation was carried out on the basis of cross-sections of the abrasive layers. Images of these were taken using an optical microscope and a scanning electron microscope. The evaluation of the grain concentration curves was carried out using the evaluation software ImageJ. The method is shown in Fig. 4.

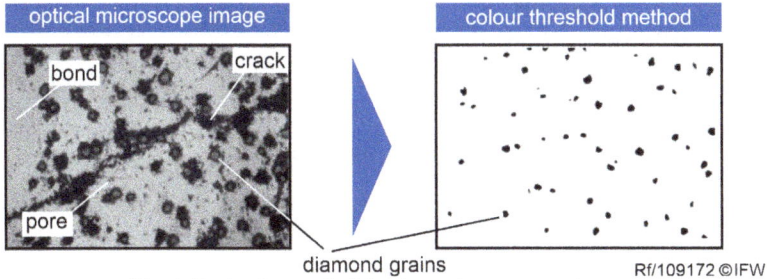

Fig. 4. Evaluation method for the grain concentration.

Analysis of the Specimen Height

As a first step, an assessment of the achieved heights of the abrasive layer specimen was carried out in order to evaluate Eq. 1. This is necessary in order to be able to accurately reproduce the gradients of the grinding wheels. The results of the evaluation are shown in Fig. 5.

Material Forming - ESAFORM 2023   Materials Research Forum LLC
Materials Research Proceedings 28 (2023) 1213-1222   https://doi.org/10.21741/9781644902479-132

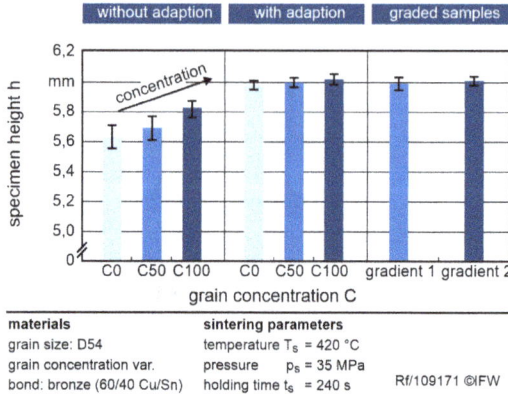

Fig. 5. Evaluation of the sample height adjustment.

The figure shows the inaccuracy in the height of the abrasive layer, as demonstrated in a previous study [15]. Moulding based on bulk density, or a high final porosity, shows that the specimen height decreases correspondingly with an increasing proportion of bond compared to the diamond grain. This is due to the increasing influence of the bond on the specimen height after sintering. The deviations from the target width of the segments can be considered uncritical if the grain concentration differences between them are small. This is the case due to the continuous load progression during deep grinding of the flutes. With larger grain concentration differences, the deviations from the target width can lead to increased radial wear. If, for example, the segment with a lower grain concentration is in the area of a higher load due to the higher shrinkage, the radial wear will increase there. This then counteracts the load design. The deviations should therefore not be greater than a few grain layers. The adjustment of the filling quantities according to Eq. 1, or based on a remaining porosity of 2 %, allows the abrasive layer specimens to be produced at the required target height. This can also be repeated for graded specimens, gradient 1 and gradient 2. Nevertheless, small deviations occur due to inaccuracies caused by the manual manufacturing process and the process chain. Since the deviations correspond to only a few grain layers, it can be assumed that these have only a minor influence on the operational behaviour.

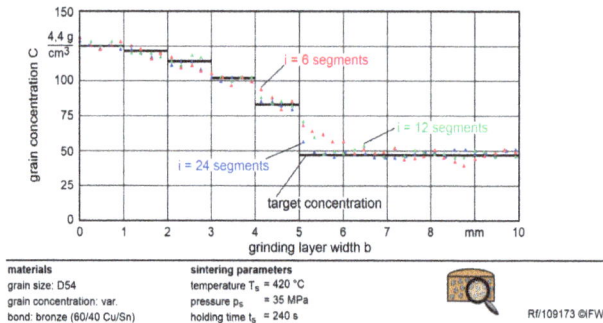

Fig. 6. Concentration gradients of the manufacturing

Fig. 6 shows the process as a microscope image for the evaluation of the single layer heights within the specimen of gradient 1 and 2. From this it is visible that the presented moulding methodology allows an accurate adjustment of the gradients as well. Within the specimen, the deviations from the target concentration are quite small. The presented methodology is therefore suitable for the dimensioning of the moulding-in amounts for graded grinding wheels.

Comparison of the moulding methods

Using the previously presented moulding quantity calculation, the two moulding methods from the previous chapter were compared. Fig. 7 shows the grain concentration curves for the moulding method using the taped mask and manual smoothing for the different divisions of the masks in i segments (i = 6; 12; 24).

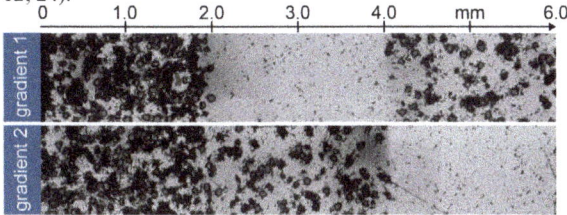

Fig. 7. Evaluation of the gradient course within the samples.

The figure shows that the method with filling based on 6 up to 24 portions of the respective abrasive layer amount enables the production of graded grinding wheels. A smaller quantity of subdivisions (i = 6) results in less accurate production compared to 12 or 24 subdivisions. This can be seen in the more pronounced transitions between the concentration zones. The deviation from the target concentration is particularly noticeable in the transition between C83 and C47. However, the gradient is still within the range of the target gradient. Especially the transition areas between the concentration levels are not to be regarded as disadvantageous. Softer transitions between them are to be aimed for, as the load in the application case is also continuous and not discrete (see Fig. 1). However, a continuous course cannot be implemented in production with current methods. But the inaccuracy, as in the case of production by means of 6 subdivisions, enables an approximation to a continuous course. In addition, the method with fewer subdivisions is advantageous because the moulding time is reduced. If high manufacturing accuracy is required, this can be reached by using 24 subdivisions. The transitions between the concentration zones are strongly pronounced. The fluctuations occurring are in the range of concentration differences >C5. These also appear in non-graded abrasive layers. The outlier in the transition area from C83 to C47 is explained by the smoothing of the layers. Since this was done manually, there are slight variations in their heights. When evaluating the grain concentration in image sections, it is therefore possible that both layers are measured partially in the corresponding section. Nevertheless, the overall small number of outliers also shows that the width of the concentration ranges was fulfilled for all grinding wheels.

Equally, the method based on a height-adjustable holder without individual single weighted amounts was investigated. The results of these investigations are shown in Fig. 8.

Material Forming - ESAFORM 2023   Materials Research Forum LLC
Materials Research Proceedings 28 (2023) 1213-1222   https://doi.org/10.21741/9781644902479-132

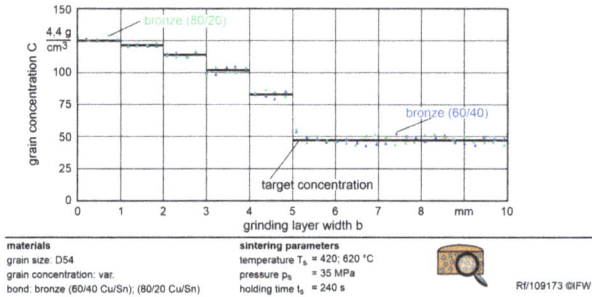

Fig. 8. Concentration gradients of the manufacturing method with the height-adjustable holder.

Two main observations can be drawn from this. First, it shows, as was expected, that the bond has no effect on manufacturing accuracy. Both the (60/40) bronze and the (80/20) bronze can be moulded and sintered into a graded abrasive layer. The differences exist only in local and small variations of the grain concentration. Secondly, the method with the height-adjustable holder shows that the transitions between the grain concentration ranges are even more accurate than the method with the mask. The higher accuracy is due to the fact that the layer heights can be set to the appropriate height without manual influence. A small inaccuracy is again only seen in the transition from C82 to C47. However, this is very slight for both bonds. The grain concentration steps, that can be quantitatively evaluated, are visually indistinguishable. This can be seen in the section of a microscope image of the grinding wheel with the (60/40) bronze. Regardless of the manufacturing method, the more brittle bronze (60/40) has been found to have a tendency to crack. This can be seen in Fig. 9.

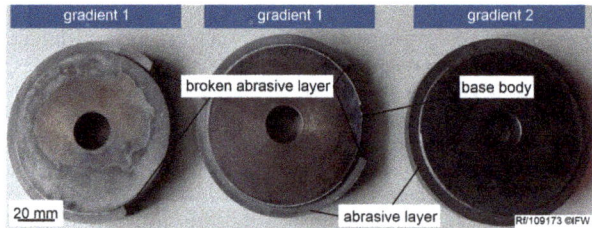

Fig. 9. Graded grinding wheels manufactured on the basis of (60/40) bronze.

The cracks only occurred in the brittle (60/40) bronze. Due to the different diamond concentrations, it can be assumed that the shrinkage of the segments varies with different grain concentrations. Segments with high concentrations are expected to shrink less. This is the case because diamond has a very low coefficient of thermal expansion compared to bronze. As a result, residual stresses in the abrasive layer are to be expected. Unfortunately, these cannot be determined analytically. The reason for this is the geometry of the abrasive layer, the large diamond grains and the fact that some of the residual stresses are already removed after fracture. It was also observed that the non-graded grinding wheels of the (60/40) bronze with C125 also break more frequently than those with C100. The reason for this could be the additionally reduced abrasive layer strength due to the increased number of grains, as discussed in the introduction [12]. As a

Material Forming - ESAFORM 2023   Materials Research Forum LLC
Materials Research Proceedings 28 (2023) 1213-1222   https://doi.org/10.21741/9781644902479-132

solution, an additional gradient was manufactured, starting at C100 (Fig. 9 gradient 2). The structure of the segments is analogous to the gradients from the methods chapter with the steps (C100/C96/C91/C82/C66/C47). As can be seen in Fig. 9, graded grinding wheels with the (60/40) bronze can be manufactured with these gradations. The last segment with C47 was deliberately not chosen lower in order to keep the shrinkage differences small and because concentrations significantly below C50 would lead to increased bond friction when using the grinding wheels.

Summary

In the present work, two methods for moulding and one method for weighing graded grinding wheels were investigated. In the investigation of the weighing-in method, it was demonstrated that the presented Eq. 1 enables an exact implementation of the gradients. This is the case regardless of the bond chosen. With regard to the methods for moulding in, it was shown that the method based on the height-adjustable holder produces the highest precision for moulding in. In addition, this method significantly reduces the time required compared to the method of filling in portions and manual smoothing. The latter method also achieves a high degree of accuracy when using 12 or 24 filling portions per segment. The reduction to 6 portions reduces the time required, but also the interface sharpness. This can be considered favourable with regard to the operational behaviour, as the loads also proceed in non-discrete steps. With regard to the moulding-in method, the use of the height-adjustable holder is therefore nonetheless considered appropriate for the highest reproducibility and productivity. The adjustment of smooth transitions can be achieved, for example, by using a notched metal plate or a rake for smoothing. Future investigations can take this as a starting point.

The layer cracking occurring with brittle bronze could be countered by reducing the grain concentration. Due to the brittleness, the bronze bond (60/40) is more challenging in the manufacturing process than the more ductile bond (80/20). Nevertheless, the more brittle bond results in better operational behaviour due to the improved operational preparation, the self-sharpening during the process and the higher contour accuracy due to the lower ductility. In order to reduce the rejects during the production of graded grinding wheels based on brittle bronzes, two approaches are conceivable for future investigations: First, the use of coated abrasive grains to increase bond strength. Second, the use of specific annealing steps during sintering or quenching to optimise the residual stress states.

In conclusion, it can be stated that the present work represents a functional approach to the production of metal-bonded graded grinding wheels. Furthermore, attempts were presented that will be followed up by subsequent investigations in order to optimise the manufacturing. In addition, the profiling and sharpening of the graded grinding wheels will be the subject of future research.

Acknowledgment

The authors thank the German Research Foundation (DFG) for financially supporting the research project "Operational behavior of sintered metal-bonded diamond grinding wheels with abrasive grain concentration gradient" under grant number "DE 447/228-1".

References

[1] H. Ortner, H. Kolaska, P. Ettmayer, The history of the technological progress of hardmetals, Int. J. Refract. Met. Hard Mater. 44 (2014) 148-159. https://doi.org/10.1016/j.ijrmhm.2013.07.014
[2] B. Denkena, Lasertechnologie für die Generierung und Messung der Mikrogeometrie an Zerspanwerkzeugen, Ergebnisbericht des BMBF Verbundprojektes GEOSPAN, 2005.
[3] K. Dröder, B. Karpuschweski, E. Uhlmann, A comparative analysis of ceramic and cemented carbide end mills, Prod. Eng. 14 (2020) 355-364. https://doi.org/10.1007/s11740-020-00966-9

Material Forming - ESAFORM 2023   Materials Research Forum LLC
Materials Research Proceedings 28 (2023) 1213-1222   https://doi.org/10.21741/9781644902479-132

[4] S. Malkin, C. Guo, Grinding technology: theory and application of machining with abrasives, 2nd Edition, Industrial Press Inc, 2008. ISBN: 9780831132477, 0831132477

[5] J. Mayr, R. Barbist, Untersuchung und Bewertung der Schleifbarkeit von Hartmetall, Der Stahlformenbauer 31 (2014) 74 – 79.

[6] E. Uhlmann, C. Hübert, Tool grinding of end mill cutting tools made from high performance ceramics and cemented carbides, CIRP Annals 60 (2011) 359-362. https://doi.org/10.1016/j.cirp.2011.03.106

[7] T. Heymann, Gezielte Nut- und Schneidkantenpräparation von Vollhartmetall - Zerspanwerkzeugen durch Polierschleifen, Spanende Fertigung / Prozesse – Innovation – Werkstoffe, Vulkan-Verlag Essen 6. Ausgabe, 2012, pp. 104-110.

[8] B. Denkena, A. Krödel, R. Lang, Fabrication and use of Cu-Cr-diamond composites for the application in deep feed grinding of tungsten carbide, Diam. Relat. Mater. 120 (2021). https://doi.org/10.1016/j.diamond.2021.108668

[9] B. Bergmann, P. Dzierzawa, Understanding the properties of bronze-bonded diamond grinding wheels on process behaviour, CIRP Annals 71 (2022) 293-296. https://doi.org/10.1016/j.cirp.2022.04.014

[10] P. Brevern, Untersuchungen zum Tiefschleifen von Hartmetall unter besonderer Berücksichtigung von Schleiföl als Kühlschmierstoff. VDI Fortschrittsberichte Reihe 2, 1996.

[11] T. Friemuth, Schleifen hartstoffverstärkter keramischer Werkzeuge. Dr.-Ing. Dissertation, Universität Hannover, 1999.

[12] F.L. Kempf, A. Bouabid, P. Dzierzawa, T. Grove, B. Denkena, Methods for the analysis of grinding wheel properties, 7. WGP Jahreskongress, 2017, pp. 87-96.

[13] E. Uhlmann, N. Schröer, A. Muthulingam, B. Gülzow, Increasing the productivity and quality of flute grinding processes through the use of layered grinding wheels, Procedia Manuf. 33 (2019) 754-761. https://doi.org/10.1016/j.promfg.2019.04.095

[14] B. Denkena, B. Bergmann, D. Raffalt, Operational behaviour of graded diamond grinding wheels for end mill cutter machining, SN Appl. Sci. 4 (2022). https://doi.org/10.1007/s42452-022-04970-9

[15] B. Denkena, B. Bergmann, D. Raffalt, Manufacturing Of Graded Grinding Layers, World PM2022 - Session 37: Hard metals, cermets and diamond tools - Processing II, 2022.

Material Forming - ESAFORM 2023
Materials Research Proceedings 28 (2023) 1223-1234

Materials Research Forum LLC
https://doi.org/10.21741/9781644902479-133

Modeling of surface hardening in burnishing process

TEIMOURI Reza[1,a] *, GRABOWSKI Marcin[1,b] and SKOCZYPIEC Sebastian[1,c]

[1] Cracow University of Technology, Faculty of Mechanical Engineering, Chair of Production Engineering, al. Jana Pawła II 37, 31-864, Kraków, Poland

[a]reza.teimouri@pk.edu.pl, [b]marcin.grabowski@pk.edu.pl, [c]sebastain.skoczypiec@pk.edu.pl

Keywords: Burnishing, Twining-Induced Hardening, Phase Change, Simulation

Abstract. In the present paper, a multiphysic model was developed to identify the underlying mechanism of surface layer hardening in burnishing of stainless steel 304. The mechanic of burnishing process was firstly modeled to obtain deformation parameters i.e. strain and strain rate by incremental plasticity. Then the strengthening mechanisms were identified association of constitutive equations regarding twinning-induced hardening and phase change. It was found that the twining-induced hardening has greatest contribution in strengthening the surface layer. Moreover, among the process parameters, the burnishing depth has greatest effect on hardness magnitude and corresponding depth.

Introduction

Burnishing is categorized as surface finishing operations which are being used for surface smoothening, and mechanical properties enhancement of the engineering materials [1,2]. Compared to other mechanical surface treatment processes such as shot peening, laser shock peening, or ultrasonic peening, burnishing results in better surface quality and induces compressive residual stress at deeper layers [3]. Moreover, it is easily implemented and does not require complex instrumentations and machines. Thus, it is finding its importance at industrial levels such as aerospace and automotive for the superfinishing of hard materials.

During burnishing, the microstructure evolution is corresponded to different microstructure phenomena that can result in hardening of burnished surface layer. However, in majority of conducted research, the underlying mechanism of surface strengthening has been reported as grain refinement Based on the type of material, hardening mechanisms such as dislocation accumulation, crystal twinning and phase change can be also effective in strengthening of surface layer. Starman et al. [4] reported that the martensitic phase transformation corresponds to the surface hardening of AISI 304 processed by shot-peening and laser shock peening. Laine et al. [5] revealed that the twining-induced hardening beside grain size evolution has contribution in surface hardening of shot-peened Ti64 alloy. Rinaldi et al. [6] identified that the strengthening mechanism of CP-T alloy in hard machining process is mainly attributed to twining induced plasticity and grain refinement.

In order to identify the main strengthening mechanism, complex microscopic examination and mechanical testing is required. However, providing such examination techniques for considerable amount of experiments are too expensive and time consuming. In this context, development of material model based on multiphysics of different metallurgical phenomena can be considered as a fundamental means to afford the lack of capabilities of measurement instrumentation and to identify the underlying mechanisms.

In the present study, a theoretical approach based on multiphysics of contact mechanic and multiscale micro-mechanical material model is developed to identify the underlying strengthening mechanism AISI 304 samples processed by burnishing process. Here, firstly the deformation parameters of process like stress, strain and strain rate are modeled by expanding cavity model. Then the strengthening mechanisms are identified by developing a material model including grain

Material Forming - ESAFORM 2023
Materials Research Proceedings 28 (2023) 1223-1234

Materials Research Forum LLC
https://doi.org/10.21741/9781644902479-133

refinement, twinning-induced hardening and phase change. Confirmation of developed theoretical model has been carried out by series of burnishing experiments including different processing parameters.

Theoretical model

Deformation parameter.

Fig. 1 illustrates the schematic diagram of the rolling of a roller on the flat surface that is the main concept of roller burnishing process. In order to calculate the deformation parameters, theory of expanded cavity model (ECM) proposed by Gao et al. [7] utilized here. Accordingly, the mean contact pressure applied by burnishing roller to the surface of the power law hardening material can be calculated by following equation:

$$p_m = \frac{\sigma_y}{\sqrt{3}}\left[1 + \frac{1}{n}\left(\frac{r_p^{2n}}{a^{2n}} - 1\right)\right] + k\varepsilon_{eq}^n \tag{1}$$

where r_e, n and k are, respectively, the outer radius of cylinder, strain hardening exponent and material constant for power law. Also, r_p and ε_{eq} are the the plastic boundary of deformation and equivalant plastic strain of contact which can be calculated by following formulation [8].

$$\varepsilon_{eq} = \sqrt{6}\frac{\sigma_y}{E}\frac{r_p^2}{r^2} \quad , \quad r_p = \sqrt{\frac{2Ea^3}{\sqrt{3}R\sigma_y}} \tag{2}$$

Once the mean contact pressure is identified, the burnishing force of an each individual roller induces to a surface can be obtained by following formula:

$$F = 2aLp_m , \quad a = \sqrt{2R\delta} \tag{3}$$

where L is roller length, a is the penetration radius and δ is the burnishing depth.

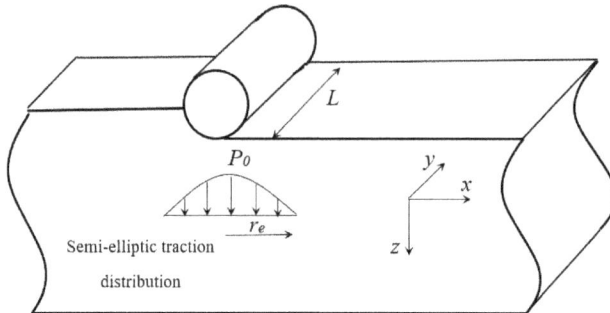

Fig. 1. Schematic illustration of contact of a cylindrical roller to flat surface.

In order to calculate the deformation parameters during motion of the roller over the surface, the theory of incremental plasticity for rolling contact that is suggested by McDowel is utilized [9]. Based on this theory, the loading initiates by elastic stresses and it incrementally increases up to reaching plastic region. Finally, by moving the tool too far from the point of interest, the unloading occurs. The elastic stresses at point of interest $M(x,z)$ can be calculated using integration of Boussineq solution for the normal traction $p(s)$ and tangential traction $q(s)$ in semi-infinite half space over the region of contact [10].

$$
\begin{cases}
\sigma_{xx}(x,z) = -\dfrac{2z}{\pi}\displaystyle\int_{-a}^{a}\dfrac{p(s)(x-s)^2}{[(x-s)^2+z^2]^2}\,ds - \dfrac{2}{\pi}\displaystyle\int_{-a}^{a}\dfrac{q(s)(x-s)^3}{[(x-s)^2+z^2]^2}\,ds \\[2ex]
\sigma_{zz}(x,z) = -\dfrac{2z^3}{\pi}\displaystyle\int_{-a}^{a}\dfrac{p(s)}{[(x-s)^2+z^2]^2}\,ds - \dfrac{2z^2}{\pi}\displaystyle\int_{-a}^{a}\dfrac{q(s)(x-s)}{[(x-s)^2+z^2]^2}\,ds \\[2ex]
\tau_{xz}(x,z) = -\dfrac{2z^2}{\pi}\displaystyle\int_{-a}^{a}\dfrac{p(s)(x-s)}{[(x-s)^2+z^2]^2}\,ds - \dfrac{2z}{\pi}\displaystyle\int_{-a}^{a}\dfrac{q(s)(x-s)^2}{[(x-s)^2+z^2]^2}\,ds
\end{cases}
\tag{4}
$$

More specifically, by assuming semi-elliptic distribution of tangential and normal traction based on boussinesque formulation, following relationship between the axial and tangential tractions and burnishing force are established.

$$
p(s) = \frac{F}{a}\sqrt{1-\left(\frac{2s}{a}-1\right)^2} \ , \quad q(s) = \frac{\mu F}{a}\sqrt{1-\left(\frac{2s}{a}-1\right)^2}
\tag{5}
$$

By identifying the values of elastic stresses, the elastic strain can be obtained using Hook's law taking into account the plain strain condition i.e. $\varepsilon_{yy}=0$.

In order to calculate the value of plastic stress and strains at any individual poit, the yielding criteria by comparing the value of effective stress and Johnson-Cook constitutive model is identified as follows:

$$
\begin{cases}
g = \dfrac{1}{2}(S_{ij}-a_{ij})(S_{ij}-a_{ij}) - Y^2 \\[2ex]
Y = \dfrac{1}{\sqrt{3}}(A+B\varepsilon_{eff}{}^n)\left[1+C\ln\left(\dfrac{\varepsilon^{\bullet}_{eff}}{\varepsilon^{\bullet}_{0}}\right)\right]\left(\dfrac{T-T_0}{T_m-T_0}\right)^m \\[2ex]
S_{ij} = \sigma_{ij} - \dfrac{1}{3}\delta_{ij}\sigma_{kk}
\end{cases}
\tag{6}
$$

where S_{ij} is deviatoric stress, and a_{ij} is back stress tensors. Also, Y is the effective stress based on Johnson-Cook (J-C) yield criterion. A, B and C are the J-C constants, ε_{eff} and $\varepsilon^{\bullet}_{eff}$ are effective strain and effective strain rate, respectively, which are calculated based on incremental plasticity theory that will be described in following part. Furthermore, T is the temperature term which can be neglected in burnishing process [11]. According to Eq. 6, when the g>0, the loading at the point of interest is elastic, and the stress and strain can be calculated using Eq. 4, and Hook's law. When g>0, the plastic loading occurs, and to obtain values of plastic strain and stress, the systems of three equations suggested by McDowel need to be solved.

$$
\begin{cases}
d\varepsilon_{xx} = \dfrac{1}{E}\left[d\sigma_{xx}{}^P - \vartheta(d\sigma_{yy}{}^P + d\sigma_{zz}{}^P)\right] + \dfrac{n_{xx}}{h+c}\left(d\sigma_{xx}{}^P n_{xx} + d\sigma_{yy}{}^P n_{yy} + d\sigma_{zz}{}^P n_{zz} + 2d\tau_{xz}{}^e n_{xz}\right) = \\[1ex]
\quad \Psi\left\{\dfrac{1}{E}\left[d\sigma_{xx}{}^e - \vartheta(d\sigma_{yy}{}^P + d\sigma_{zz}{}^e)\right] + \dfrac{n_{xx}}{h+c}\left(d\sigma_{xx}{}^e n_{xx} + d\sigma_{yy}{}^P n_{yy} + d\sigma_{zz}{}^e n_{zz} + 2d\tau_{xz}{}^e n_{xz}\right)\right\} \\[2ex]
d\varepsilon_{yy} = \dfrac{1}{E}\left[d\sigma_{yy}{}^P - \vartheta(d\sigma_{xx}{}^P + d\sigma_{zz}{}^P)\right] + \dfrac{n_{yy}}{h+c}\left(d\sigma_{xx}{}^P n_{xx} + d\sigma_{yy}{}^P n_{yy} + d\sigma_{zz}{}^P n_{zz} + 2d\tau_{xz}{}^e n_{xz}\right) = 0 \\[2ex]
d\sigma_{yy}{}^P = \dfrac{1}{2}(d\sigma_{xx}{}^P + d\sigma_{zz}{}^P)
\end{cases}
\tag{7}
$$

The other terms of above formula can be calculated

$$n_{ij} = \frac{S_{ij} - a_{ij}}{\sqrt{2}k} , \; da_{ij} = cd\varepsilon_{ij}^{\;p} , \; \Psi = 1 - \exp\left(-\kappa \frac{3h}{2G}\right), \; G = \frac{E}{2(1+\vartheta)} \tag{8}$$

where h and is isotropic and kinematic hardening coefficients.

Our pilot experiemnts on processing of stainless steel 304 showed that the main contributed mechanisms on strengthening the surface layer after burnishing proess are formation of twinning and alpha ferrite. The next step after the calculation of the deformation parameter is to calculate the stress as results of twinning-induced hardening and phase changes.

Twinning-induced hardening.

The stress related to twining induced material strengthening has been presented in following equation [12]:

$$\begin{cases} \sigma_{TW} = M\beta_{TW}Gb\left(\frac{1}{d_{cell}} + \frac{1}{l}\right) \\ \frac{1}{l} = \frac{1}{2t} + \frac{f_{TW}}{1 - f_{TW}} \end{cases} \tag{9}$$

where the σ_{TW} is the term of strengthening stress as result of twinning, M is Taylor factor, β_{TW} is the constant, t is the twin lamellae (that is 18nm) and f_{TW} and is the twining volume fraction that is calculated by following differential equation:

$$df_{TW} = (1 - f_{TW})A_f d\varepsilon \tag{10}$$

where A_f is the constant that equals to 1.916 for twining induced plasticity austenitic stainless steels [13].

Phase change.

Transformation of austenite to ferrite usually occurs in cold working of austenitic stainless steel. The increasing rate of the ferrite during deformation is obtained by [14]:

$$f_m^{\bullet} = (1 - f_m)v_m N_m^{\bullet} \tag{11}$$

where $f_m^{\bullet}$ is the rate of increase of ferrite volume fraction, f_m is the ferrite volume fraction, v_m is the average ferrite volume and $N_m^{\bullet}$ is the increase rate of ferrite that is related to its nucleation probability (P) on shear bands and the total number of ferrite nucleation (N_I).

$$N_m^{\bullet} = PN_I^{\bullet} + N_I P^{\bullet} H(P^{\bullet}) \tag{12}$$

where H is Heaviside step function. Possible nucleation of ferrite is also affected by number of shear bands. Accordingly;

$$N_I^{\bullet} = \frac{rC\left(f_{sb}^{\bullet}\right)^{r-1}}{v_I} , \; f_{sb}^{\bullet} = \alpha(1 - f_{sb})\varepsilon^{\bullet} \tag{13}$$

where r and C are materials constant, $f_{sb}^{\bullet}$ is the increase rate of shear band volume fraction. The ferrite nucleation probability was assumed to follow a Gaussian distribution,

Materials Research Forum LLC
https://doi.org/10.21741/9781644902479-133

$$P = \frac{1}{\sqrt{2\pi}\sigma_g} \int_{-\infty}^{g} \exp\left[-\frac{1}{2}\left(\frac{g'-\bar{g}}{\sigma_g}\right)^2\right] dg' \tag{14}$$

where the terms σ_g and $\bar{g}$ are dimensionless mean and standard deviation, respectively. The thermodynamic driving force g is related to the stress states (Σ) and the temperature (T), and is approximated by:

$$g = g_0 + g_1 T + g_2 \Sigma \tag{15}$$

where g_0, g_1, and g_2 are constants. A fraction function (χ) of the plastic deformation is assumed cause temperature rise (ΔT) as:

$$\Delta T = \int_0^\varepsilon \frac{\chi}{\rho C_p} \sigma(\varepsilon) d\varepsilon \tag{16}$$

where ρ is material density and C_P is the specific heat capacity.

As per the foresaid strain-induced phase transformation kinetics, the increase rate of the ferrite can then be obtained through combination of above-calculated equations:

$$\begin{cases} f_m^\bullet = (1 - f_m)(A\varepsilon^\bullet + B\Sigma^\bullet) \\ A = r\alpha\beta P(1 - f_{sb})(f_{sb})^{r-1} - B_f\frac{g_1\chi}{\rho C_p}\sigma(\varepsilon) \quad B = B_f g_2 \\ B_f = \frac{\beta}{\sqrt{2\pi}\sigma_g}(f_{sb})^r \exp\left[-\frac{1}{2}\left(\frac{g'-\bar{g}}{\sigma_g}\right)^2\right] H(P^\bullet) \\ \alpha = \alpha_0 + \alpha_1 T + \alpha_2 T^2 + \alpha_3 \Sigma \end{cases} \tag{17}$$

By identifying the ferrite volume fraction, the hardness development can be obtained as:

$$\Delta h = \sum_{i=1}^{v} f_i h_i - h_0 \tag{18}$$

where f_i is the fraction of phase i in the element, h_0 is the initial bulk microhardness, h_i is the hardness of phase i, and v represents the number of phases present in the element.

Accordingly, the final hardness of the sample after SSPD can be calculated by:

$$h = h_0 + k_D \sigma_D + k_{TW} \sigma_{TW} + \Delta h \tag{19}$$

Experiments

In the present investigation, surface sever plastic deformation experiments carried out by multi-roller rotary burnishing tool. The tool includes four rollers with diameter of 3mm and length of 10mm as shown in Fig. 2. The tool, as shown in the figure, has been mounted on a CNC milling machine to process the surface of the samples made of AISI 304. During the experiments, the burnishing force is controlled by a force dynamometer KISTLER 9257B as shown in Fig. 2. Samples were prepared in sizes of 70mm length, 30mm width and 10mm thickness. Before conducting the experiments, all the samples were milled with a same machining parameters to assure their flatness. The properties of the AISI 304, including the material's constitutive models, i.e. Johnson-Cook, Isotropic-Kinematic hardening, dislocation density model, twinning-induced plasticity, and ferrite phase transformation constants, have been listed in Tables 1-4, respectively.

Table 1. Johnson-Cook constants of AISI 304 [15].

A (MPa)	B (MPa	C	n
452	694	0.0067	0.311

Table 2. Isotropic-kinematic hardening constants of AISI 304 [15].

h (GPa)	c (GPa)
2.7	0.8

Table 3. Coefficients of austenite-ferrite phase change constitutive equation [14].

α_0	α_1	α_2	α_3	g_0	g_1	g_2	β
-2e-	4.52e-2	11.8	0.8	0.05	-1.2	78.3	4.5

The samples were mechanically polished on a suspension of ethylene glycol and diamond pastes with a gradation of 1 and 6μm. Then, they were electrolytically etched using a voltage of 12V and a exposure time of 30s in a solution of 10g $CrO3$ and 100 ml of distilled water. The microstructure was observed using a scanning electron microscope (SEM) JSM-IT200. The microhardness measurements were carried out using an Innovatest microhardness tester at a load of 0.245 N.

Fig. 2. Experimental setup.

A series of experiments was carried out to confirm the results obtained by the analytical models. Among the process factors, it was found that the burnishing depth has the most prominent effect on burnishing force and hardness. Thus, a total of three experiments under different burnishing

Material Forming - ESAFORM 2023   Materials Research Forum LLC
Materials Research Proceedings 28 (2023) 1223-1234   https://doi.org/10.21741/9781644902479-133

depths, i.e. 0.2mm, 0.25 mm, and 0.3mm were carried out, and the values of force and hardness distribution were taken into account to compare with the analytical model and understand the effect of factors on the hardening mechanism. During the experiments, spindle speed and feed rate were kept constant at 500 RPM and 200 mm/min, respectively. It needs to be pointed out that each experiment has been repeated three times, and the average values of performance measures (i.e., force and hardness) were reported in the paper.

Results and Discussion

In order to confirm the analytical model, the results of burnishing forces which were derived from the developed model was compared with those experimentally measured values. Fig. 3 demonstrates the comparison of measured and predicted values of burnishing axial force for the experimental data presented in Table 6. It is seen from the figure that there are good agreements between the measured and predicted values of burnishing force. According to the quantified results, the prediction errors vary between 5.2% and 8.7%. It can also be seen that the trend of variation of the main force with respect to process factors for those data calculated by the analytical model is well consistent with the experimental value. It is seen that, as a result of higher values of burnishing depth, the burnishing force increases. The higher value of burnishing depth increases the engagement radius and results in greater force.

Influence of burnishing depth on microhardness and influential parameters have been shown in Fig. 4. According to the figure, it is seen that for all three samples, there is close agreement between the measured and predicted values of microhardness. The error values are in the range of 14.3% to 21.5%, which seems acceptable based on the previous works carried out by different researchers [15]. Moreovr, it is clear in the figure that the microhardness hardness significantly increases by increasing the burnishing depth. This trend can be observed in both experimental results and modeled values. Therefore, the developed analytical model is accurate enough and can be utilized to find the influence of process factors on the microhardness.

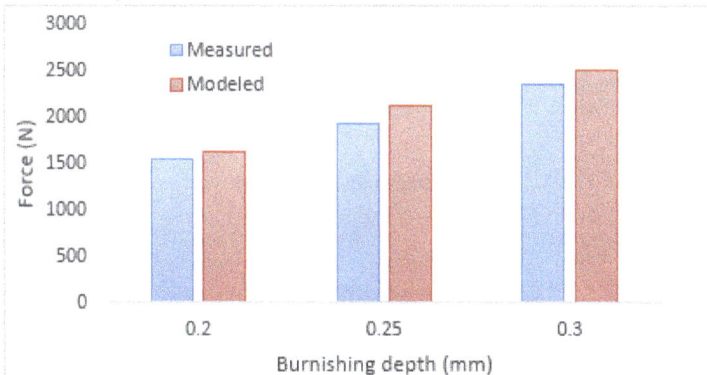

Fig. 3. Comparison between measured and predicted values of burnishing force.

As a result of increasing the burnishing depth, the burnishing axial force increases correspondingly (Fig. 3), which results in further values of plastic strain as shown in Fig. 5a. The greater value of plastic strain corresponds to a higher plastic strain rate under constant values of spindle speed and feed velocity, as shown in Fig. 5b. As result of further values of plastic strain and strain rate, the twinning volume fraction and ferrite volume fraction increase consequently as shown in Fig. 6a and 6b.

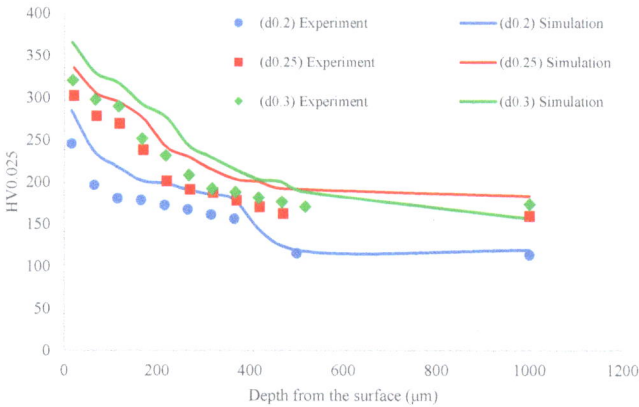

Fig. 4. Comparison between measured and predicted values of microhardness.

In order to check the results derived from the analytical model, scanning electron microscopic images of the surface and subsurface of the samples processed by burnishing at depths of 0.2 mm and 0.3 mm were captured and presented in Fig. 7. According to the figure, it is seen for the sample processed by 0.2mm burnishing depth, amount of twinning and ferrite contents which distributed in subsurface layers (as shown in Fig. 7a) are less than those of sample processed by burnishing depth of 0.3mm (as shown in Fig. 7b). It is seen that for the sample that was processed by further burnishing depth, the twinning, re-twinning, and ferrite content have been distributed up to the entire cross section of the sample that yields hardness development. It is also seen in Fig. 7a that in the deeper surface layer, the twinning induced hardening and austenite-ferrite phase change exist up to a certain depth, which can be considered a validation of the analytic model. This trend was predicted by the analytical model, where the twinning volume fraction is distributed at further depth (Fig. 6a), while the ferrite volume fraction is only available at a limited depth from the surface of the sample (Fig. 6b).

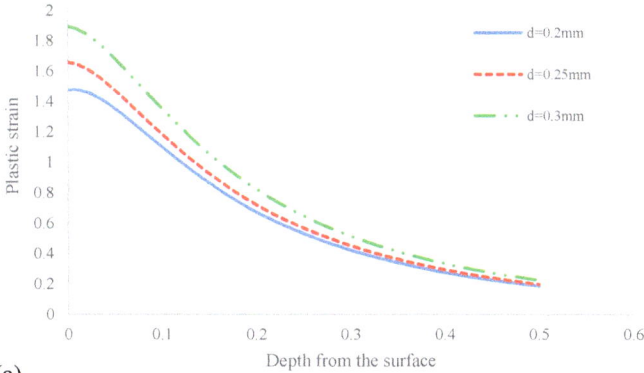

(a)

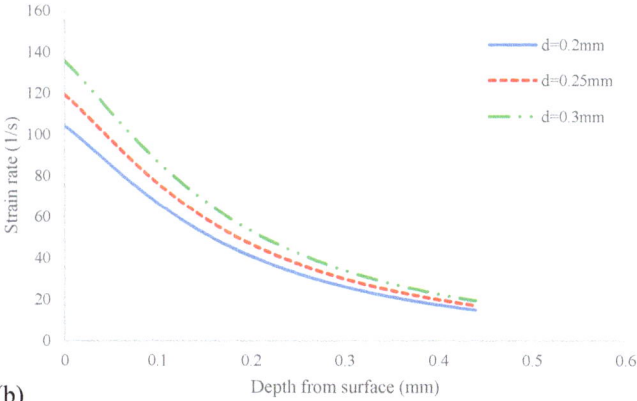

(b)

Fig. 5. Effect of burnishing depth on (a) strain and (b) strain rate.

Summary

The obtained results can be summarized as follows:

- There are close agreements between the measured and predicted values of burnishing force where the prediction error in worst case was around 9.1%. Moreover, microhardness distributions modeled by analytical approach were compatible with experiemental values where the maximum prediction error was 21.3%. Also, the trend of variations of hardness and burnishing force which were modeled by analytical approach were consistent well with experiemtnal findings.

- It was found that thanks to contributions of different hardening mechanisms, the micorhardness of the samples can be improved up to 200%. Among the different hardening mechanisms,

Material Forming - ESAFORM 2023   Materials Research Forum LLC
Materials Research Proceedings 28 (2023) 1223-1234   https://doi.org/10.21741/9781644902479-133

twinning-induced plasticity followed by phase change have the gretest effect on hardness development.

- It was found from the results that the phase change-induced hardening is only available in the layers very close to the surface e.g. 50μm in the best case; however, the twining volume fraction exists in much deeper surface layers.

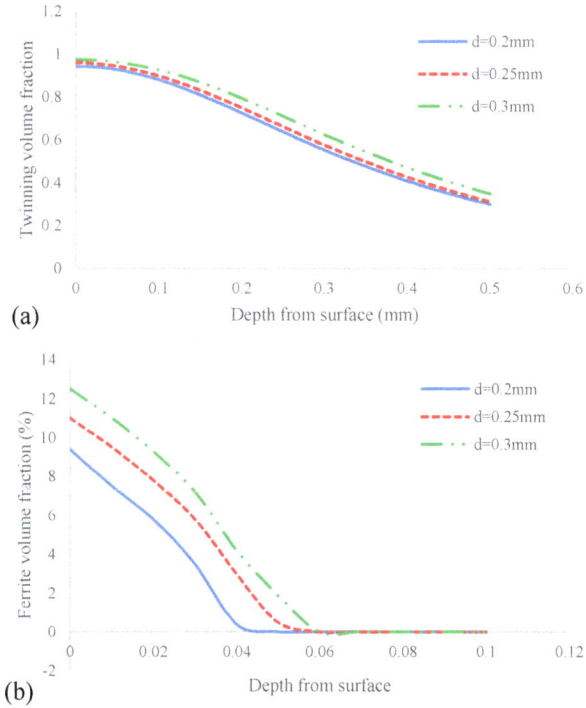

Fig. 6. *Effect of burnishing depth on (a) twinning volume fraction and (b) ferrite volume fraction.*

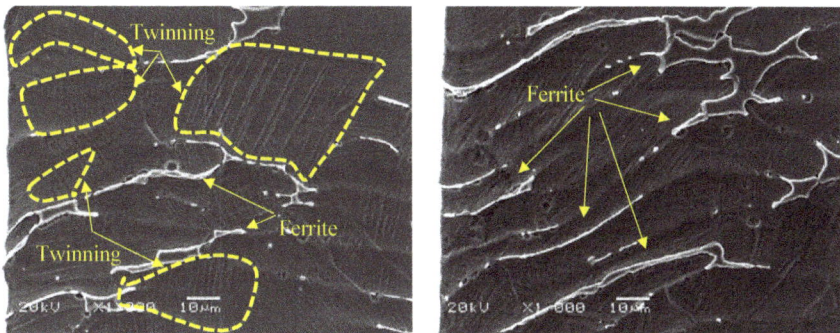

Fig. 7. SEM image of cross section of near surface layers processed by different burnishing depth (a) 0.2mm (b) 0.3 mm (Spindle speed=500 RPM and feed velocity=200mm/min).

References

[1] S. Attabi, A. Himour, L. Laouar, A. Motallebzadeh, Mechanical and wear behaviors of 316L stainless steel after ball burnishing treatment, J. Mat. Res. Tech. 15 (2021) 3255-3267. https://doi.org/10.1016/j.jmrt.2021.09.081

[2] Z.Y. Zhou, G.L. Yu, Q.Y. Zheng, G.Z. Ma, S.B. Ye, C. Ding, Z.Y. Piao, Wear behavior of 7075-aluminum after ultrasonic-assisted surface burnishing, J. Manuf. Process. 51 (2020) 1-9. https://doi.org/10.1016/j.jmapro.2020.01.026

[3] Y. Hua, Z. Liu, B. Wang, X. Hou, Surface modification through combination of finish turning with low plasticity burnishing and its effect on fatigue performance for Inconel 718, Surf. Coat. Technol. 375 (2019) 508-517. https://doi.org/10.1016/j.surfcoat.2019.07.057

[4] B. Starman, H. Hallberg, M. Wallin, M. Ristinmaa, M. Halilovič, Differences in phase transformation in laser peened and shot peened 304 austenitic steel, Int. J. Mech. Sci. 176 (2020) 105535. https://doi.org/10.1016/j.ijmecsci.2020.105535

[5] S.J. Lainé, K.M. Knowles, P.J. Doorbar, R.D. Cutts, D. Rugg, Microstructural characterisation of metallic shot peened and laser shock peened Ti–6Al–4V, Acta Mater. 123 (2017) 350-361. https://doi.org/10.1016/j.actamat.2016.10.044

[6] S. Rinaldi, D. Umbrello, S.N. Melkote, Modelling the effects of twinning and dislocation induced strengthening in orthogonal micro and macro cutting of commercially pure titanium, Int. J. Mech. Sci. 190 (2021) 106045. https://doi.org/10.1016/j.ijmecsci.2020.106045

[7] X.L. Gao, X.N. Jing, G. Subhash, Two new expanding cavity models for indentation deformations of elastic strain-hardening materials, Int. J. Solid Struct. 43 (2006) 2193-2208. https://doi.org/10.1016/j.ijsolstr.2005.03.062

[8] R. Teimouri, M. Grabowski, R. Bogucki, Ł. Ślusarczyk, S. Skoczypiec, Modeling of strengthening mechanisms of surface layers in burnishing process, Mater. Des. 223 (2022) 111114. https://doi.org/10.1016/j.matdes.2022.111114

[9] D.L. McDowell, G.J. Moyar, Effects of non-linear kinematic hardening on plastic deformation and residual stresses in rolling line contact, Wear 144 (1991) 19-37. https://doi.org/10.1016/0043-1648(91)90004-E

[10] K.L. Johnson, K.L. Johnson, Contact mechanics, Cambridge university press, 1987.

[11] Y. Liu, L. Wang, D. Wang, Finite element modeling of ultrasonic surface rolling process, J. Mater. Process. Technol. 211 (2011) 2106-2113. https://doi.org/10.1016/j.jmatprotec.2011.07.009

[12] F. Liu, W.J. Dan, W.G. Zhang, Strain hardening model of twinning induced plasticity steel at different temperatures, Mater. Des. 65 (2015) 737-742. https://doi.org/10.1016/j.matdes.2014.10.008

[13] G. Dini, R. Ueji, A. Najafizadeh, S.M. Monir-Vaghefi, Flow stress analysis of TWIP steel via the XRD measurement of dislocation density, Mater. Sci. Eng. A, 527 (2010) 2759-2763. https://doi.org/10.1016/j.matdes.2014.10.008

[14] W.J. Dan, W.G. Zhang, S.H. Li, Z.Q. Lin, A model for strain-induced martensitic transformation of TRIP steel with strain rate, Comput. Mater. Sci. 40 (2007) 101-107. https://doi.org/10.1016/j.commatsci.2006.11.006

[15] W. Zhang, X. Wang, Y. Hu, S.Wang, Predictive modelling of microstructure changes, micro-hardness and residual stress in machining of 304 austenitic stainless steel, Int. J. Mach. Tool. Manuf. 130 (2018) 36-48. https://doi.org/10.1016/j.ijmachtools.2018.03.008

Material Forming - ESAFORM 2023
Materials Research Proceedings 28 (2023) 1235-1244

Materials Research Forum LLC
https://doi.org/10.21741/9781644902479-134

Orthogonal cutting with additively manufactured grooving inserts made from HS6-5-3-8 high-speed steel

KELLIGER Tobias[1,a] *, MEURER Markus[1,b] and BERGS Thomas.[1,2,c]

[1]Laboratory for Machine Tools and Production Engineering (WZL) of RWTH Aachen University, Campus-Boulevard 30, 52074 Aachen, Germany

[2]Fraunhofer Institute for Production Technology (IPT), Steinbachstr. 17, 52074 Aachen, Germany

[a]t.kelliger@wzl.rwth-aachen.de, [b]M.Meurer@wzl.rwth-aachen.de, [c]t.bergs@wzl.rwth-aachen.de

Keywords: Orthogonal Cutting, Chip Formation, Additive Manufacturing, LPBF, HSS, ASP 2030, HS6-5-3-8

Abstract. Additive manufacturing (AM) of cutting materials such as high-speed steel (HSS) is very challenging. So far, the impact of the layer-by-layer manufacturing technique onto the AM tool performance during machining is widely unknown. In this study, the performance characteristics of AM grooving inserts manufactured from HS6-5-3-8 (ASP 2030) in AM Laser Powder Bed Fusion (LPBF) process were investigated in fundamental cutting experiments. Six different workpiece materials were analyzed and two different parameter sets for the LPBF process investigated. All AM grooving inserts withstood the thermal and mechanical stresses during machining of the investigated materials. Based on these results, AM threading tools manufactured from HS6-5-3-8 will be investigated in a next step, using the geometrical freedom of the AM process for an adapted channel and outlet nozzle design of the internal cutting fluid supply.

Introduction

Additively manufactured (AM) cutting tools enable new design concepts in tool development including benefits such as lightweight construction [1], damping behavior [2] and internal cutting fluid supply [3]. So far, mainly basic bodies of indexable cutting tools were additively manufactured from steel materials, with conventional cutting inserts mounted. AM tools such as drills, grooving toolholders or indexable milling tools can be already found in industrial applications [4-6]. AM processing of cutting materials is rather rare due to the high requirements to geometrical accuracy, defect-free production and the high demands for the material regarding thermal and mechanical resistance in use. Even though some studies dealing with the processing of cutting materials in different AM techniques exist, only very few works investigating the performance behavior of AM processed cutting tools during machining are known. Examples for AM cutting materials can be found in processing of tungsten-carbide in Binder Jetting [7], slurry-based three-dimensional printing (3DP) technology [8] as well as in Laser Powder Bed Fusion (LPBF) [9], processing of ceramics in Lithography-based Ceramic-Manufacturing (LCM) [10] or LPBF processing of HSS [11].

High-speed steel (HSS) still is a very relevant cutting material. Due to its favorable characteristics of high ductility and bending strength, HSS is widely used for drilling and threading tools but also in broaching of difficult-to-cut high-temperature resistant materials, fine blanking tools and high-performance bearings. [12]

LPBF (also known as SLM or DMLS) is the most commonly used AM technique [13]. Due to LPBF machine concepts with preheated base plate, processing of HSS materials became possible during the last years. Thus, even in hard-to-weld materials with high carbon content, such as HSS, crack formation can be prevented. Kempen et al. [14] manufactured AISI M2 HSS with a relative

Material Forming - ESAFORM 2023   Materials Research Forum LLC
Materials Research Proceedings 28 (2023) 1235-1244   https://doi.org/10.21741/9781644902479-134

density of 99.8 % and a hardness of 57 HRC. Zumofen et al. [15] used higher preheating temperatures to further reduce residual stresses in processing of AISI M2 HSS. Saewe et al. [11] processed defect-free AISI M50 HSS with a relative density up to 99.5 % and a hardness of up to 65 HRC for the application in roller bearings.

So far, no studies regarding the performance behavior of LPBF-processed HSS cutting materials in machining experiments are known. In this work, as-built grooving inserts manufactured from HS6-5-3-8 (ASP 2030) were investigated in orthogonal cutting experiments. Main evaluation criteria were cutting force components, tool wear and chip form. Thus, the potential of the AM HSS for an application in geometrical adapted AM threading tools with individualized cutting fluid supply should be assessed.

Tool Characteristics
In a first step, grooving insert test specimens were additively manufactured by LPBF from HS6-5-3-8 on a modified machine Aconity Midi at Fraunhofer Institute for Laser Technology (ILT). The cutting edges had a slight allowance for the downstream grinding process. Two different LPBF parameter settings were applied with focus on high density (Standard, relative density of 99.96 %) and high productivity (Productive, relative density of 99.87 %), see Fig. 1. As the inserts with a total height of 26 mm were built vertically, a slight increase of hardness between +3.2 % to +4.9 % could be detected between edge 1 close to the build plate and edge 2 in distance to the build plate. The hardness of the specimens built with the parameter setting Standard was around 24 % higher than for the parameter setting Productive. This can be explained by the longer exposure time of the laser per area and the resulting differences in heat generation and dissipation. As no heat-treatment was carried out, this difference in mechanical properties had to be considered for the machining experiments.

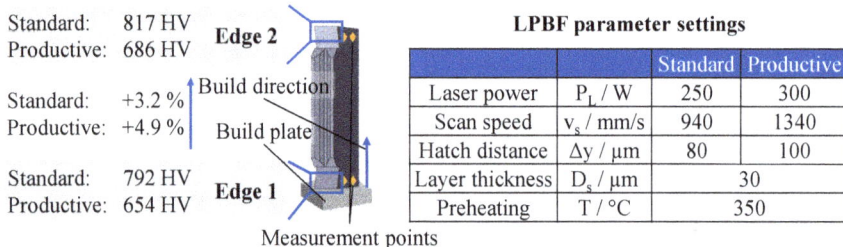

LPBF parameter settings			
		Standard	Productive
Laser power	P_L / W	250	300
Scan speed	v_s / mm/s	940	1340
Hatch distance	Δy / µm	80	100
Layer thickness	D_s / µm	30	
Preheating	T / °C	350	

Standard: 817 HV — Edge 2
Productive: 686 HV

Standard: +3.2 % — Build direction
Productive: +4.9 % — Build plate

Standard: 792 HV — Edge 1
Productive: 654 HV

Measurement points

Fig. 1. Micro hardness measurement in dependency of build height and LPBF process parameters.

Both cutting edges of all inserts were then uniformly grinded by the company Meyer + Dörner Räumwerkzeuge GmbH. The cutting edges were measured on an optical microscope Algona Mikro CAD (cutting edge radius r_β) and a tactile contour measurement device MarSurf PCV (rake angle γ and clearance angle α). The mean values are given in Fig. 2, a. The tools were uncoated.

Material Forming - ESAFORM 2023
Materials Research Proceedings 28 (2023) 1235-1244

Materials Research Forum LLC
https://doi.org/10.21741/9781644902479-134

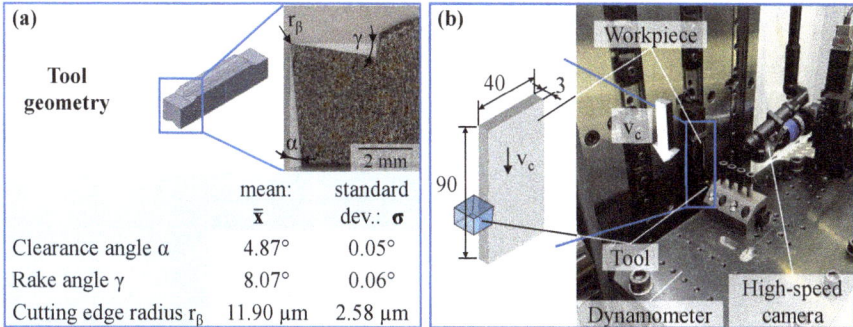

*Fig. 2. Measured tool geometry over all investigated cutting edges (a)
and experimental setup (b).*

Experimental Setup

The experimental setup inside an external vertical broaching machine Forst RASX 2200x800x600 M / CNC is shown in Fig. 2, b. The tool holder was mounted on a dynamometer Kistler Z21289 on the machine table. The workpiece was clamped into the broaching stroke. The chip formation was captured with a high-speed camera Vision Research Phantom v7.3. In order to evaluate the performance behavior of the AM tools, chip formation, chip shape, cutting force and tool wear were investigated. Six different workpiece materials were analyzed with different cutting speed v_c and undeformed chip thickness h each. The cutting parameters are given in Table 1. The applied materials were two steel alloys (X5CrNiMo17-12-2 (1.4401) and 42CrMo4+QT), one copper alloy (CuZn21Si3P (Ecobrass)), one cast iron (GJL 250) as well as two difficult-to-cut materials (Ti6Al4V (β-annealed) and Inconel 718).

Table 1. Cutting parameters for all investigated workpiece materials.

Workpiece material	v_c [m/min]	h [mm]
42CrMo4+QT	30	0.15
X5CrNiMo17-12-2 (1.4401)	30	0.15
CuZn21Si3P	60	0.2
GJL 250	60	0.2
Ti6Al4V	15	0.1
Inconel 718	5	0.05

For the evaluation of the force signal, cutting force F_c and cutting normal force F_{cN} (respective thrust force) were averaged within the range from 30 % to 70 % of the total signal of one cut (see Fig. 3). Each insert was only used for nine consecutive cuts within one workpiece material. Tool wear was analyzed by optical light microscopy and optical 3D measurement Alicona FocusG5.

Material Forming - ESAFORM 2023   Materials Research Forum LLC
Materials Research Proceedings 28 (2023) 1235-1244   https://doi.org/10.21741/9781644902479-134

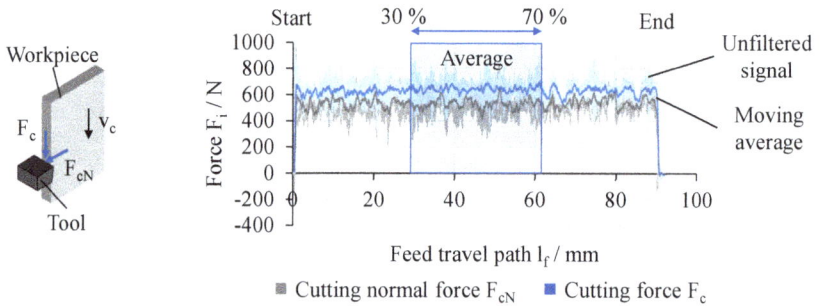

Fig. 3. Evaluation of force signal.

Results and Discussion

In the following section, different comparisons are made in order to evaluate influencing factors onto the tool performance within the scope of the experiments. Not all inserts were tested in each combination of edge number, LPBF parameter setting and workpiece material. A comparison between different materials is not useful as every material was machined with individual cutting parameters. The focus of the analysis was on the comparison between built height and LPBF parameter settings as well as on information regarding the performance behavior of the AM inserts in use with a wide range of different workpiece materials.

In Fig. 4 the influence of the build height (edge 1 and 2) onto the force development is shown for cutting of 42CrMo4 and Inconel 718 with inserts "Standard". A significant change of the force amplitudes between cut no. 1 and no. 2 to 9 was observed for all experiments. This can be traced back to inaccuracies in the positioning of the machine table and thus the cutting edge relative to the workpiece surface for the first cut, leading to a slightly deviating undeformed chip thickness. For all following analyses, the mean value of cut 2 to 9 is taken into account. The growing mean forces F_c and F_{cN} can be explained by a changing cutting edge microgeometry and blunting of the cutting edge. For 42CrMo4, no significant change in the mean forces could be detected, whereas for hard-to-cut Inconel 718 the mean cutting force F_c was around 7 % lower for edge 2. The higher hardness in bigger distances to the build plate (see Fig. 1, edge 2) seems to increase the edges resistance to blunting. For the application in threading tools, which are also build up in vertical direction, the functional part should thus be positioned on the top relative to the build plate. As the force difference for Inconel 718 could be in the deviation range, further tests with a higher number of inserts should be conducted in the future.

Material Forming - ESAFORM 2023   Materials Research Forum LLC
Materials Research Proceedings 28 (2023) 1235-1244   https://doi.org/10.21741/9781644902479-134

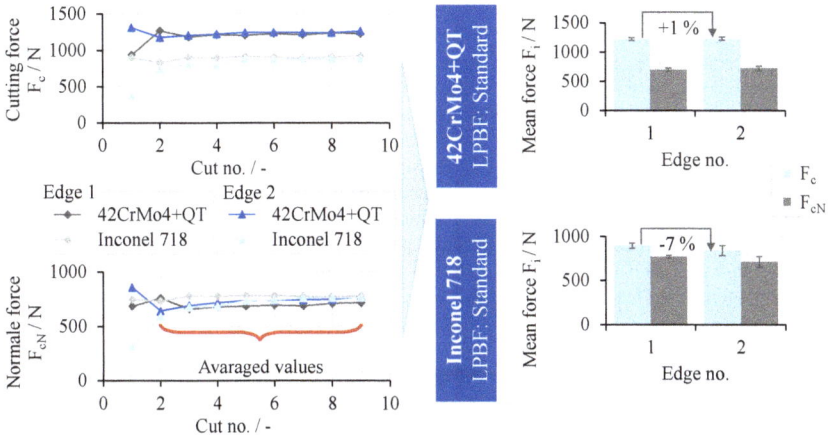

Fig. 4. Influence of build height onto force development (LPBF: Standard).

The LPBF parameter settings Standard and Productive did not influence the development and amplitude of the force signals for cutting of 42CrMo4 (Fig. 5). For cutting of Inconel 718, F_c and F_{cN} did grow along the number of cuts, with F_c being 11 % and F_{cN} 21 % higher for the insert produced with the productive LPBF parameters. The lower hardness of the productive parameters (see Fig. 1) led to a faster wear progression of the cutting edge in the hard-to-cut material even though forces for the standard parameters were comparable (F_{cN}) or lower (F_c).

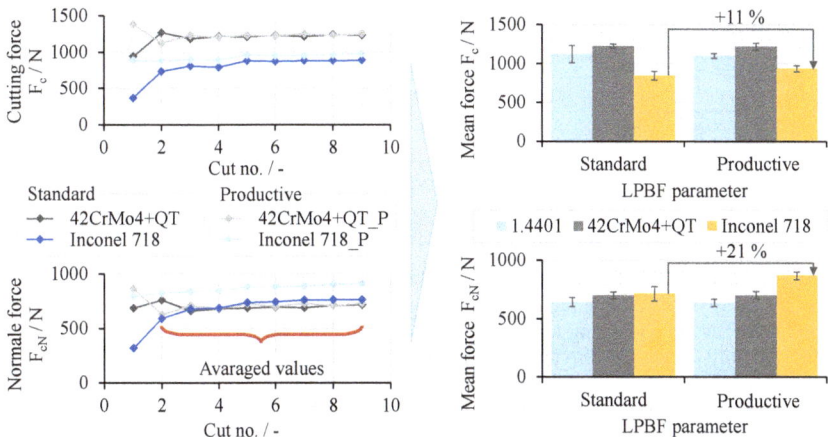

Fig. 5. Influence of LPBF parameter setting onto force development
(1.4401: edge 1, 42CrMo4: edge 1, Inconel 718: edge 2).

Material Forming - ESAFORM 2023 Materials Research Forum LLC
Materials Research Proceedings 28 (2023) 1235-1244 https://doi.org/10.21741/9781644902479-134

A visual influence of the LPBF parameter setting and the build height on tool wear and cutting force progression for machining of 42CrMo4 could not be detected neither in the records of flank and rake face nor in the 3D scans, Fig. 6. A thermal influence, similar in size and shape and visible as an annealed rake face and flank face surface, was detected for all four inserts. The high thermal load can be explained by the high tensile strength of the quenched and tempered workpiece material. Chipping occurred on several inserts due to the high mechanical load and abrasive wear caused by the martensitic microstructure. The rising force amplitudes along the feed travel path l_f, which occurred during every cut, could not be explained by growing tool wear. The mean forces F_c and F_{cN} remained nearly constant along the number of cuts, see Fig. 4. A possible explanation might be the rising contact area between chip and rake face as visible in the high-speed video recording in Fig. 7 (42CrMo4+QT). Due to the rising chip up-curl radius, the contact length is constantly increasing, leading to more friction-induced force in and perpendicular to the direction of cut.

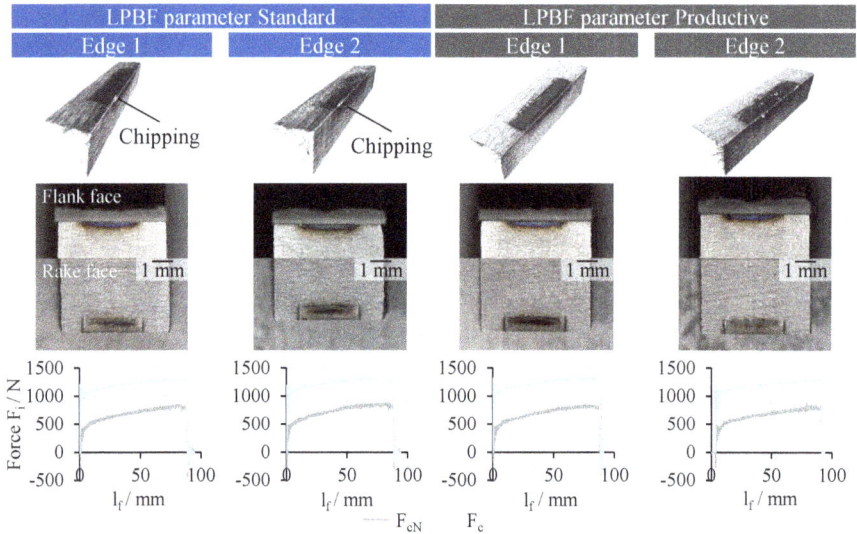

Fig. 6. Tool wear and force signal after nine consecutive cuts in 42CrMo4+QT.

Fig. 7 shows the chip formation captured by the high-speed camera as well as the resulting chip forms. 42CrMo4 exhibited a mainly continuous chip. In the first section of the chip, annealing colors were visible, indicating a high temperature evolution during cutting of the material. For 1.4401, mainly segmented chip formation occurred with material adhesions on the backside of the chip. CuZn21Si3P offered short, discontinuous chips due to the chipbreaking characteristics of the alloy elements. Cast iron GJL 250 with its brittle characteristics of the pearlite lamellae led to very short, dusty chips, whereas for Ti6Al4V a discontinuous lamellar chip formation was observed. Inconel 718 caused a thin, segmented chip with very regular chip up-curl radius.

Material Forming - ESAFORM 2023
Materials Research Proceedings 28 (2023) 1235-1244

Materials Research Forum LLC
https://doi.org/10.21741/9781644902479-134

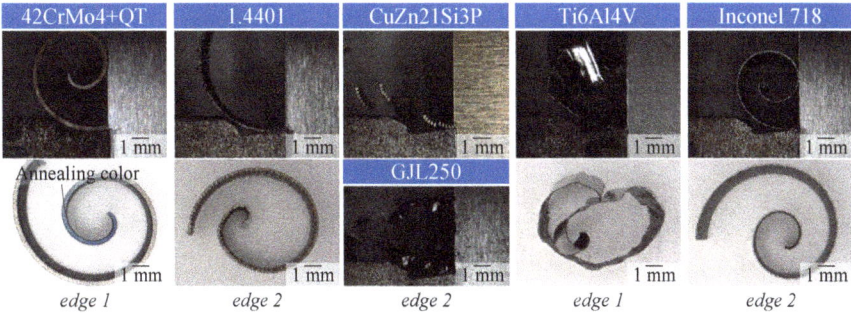

Fig. 7. Chip formation and chip form for all investigated materials (cut no. 3, LPBF: Standard).

As visible in Fig. 8, the forces F_c and F_{cN} did only grow significantly for hard-to-cut materials Inconel 718 and Ti6Al4V along the number of cuts. The cutting inserts seemed to withstand the mechanical load applied during the tool-workpiece-contact for all steel, cast and copper alloys.

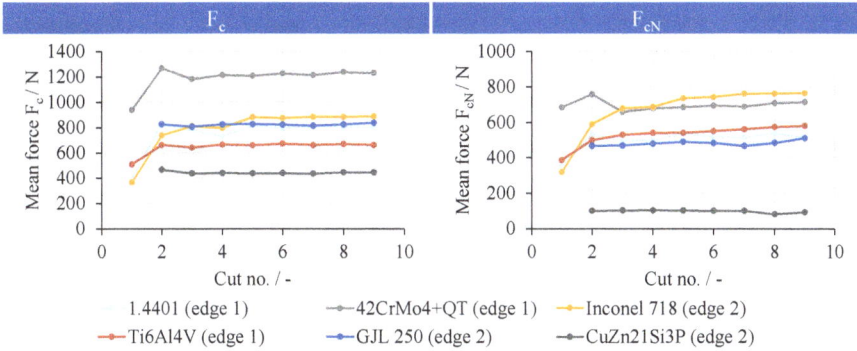

Fig. 8. Force growth along all number of cuts (LPBF: Standard).

The tool wear and force progression for all other materials is presented in Fig. 9. Adhering material could be detected on the rake face for cutting of 1.4401. The thermal load led to discolorations especially on the flank face for cutting of 1.4401 and Ti6Al4V caused by the low thermal conductivity of the materials. A relatively high flank wear was visible for Ti6Al4V and Inconel 718 due to high abrasive wear. The insert used for cutting of CuZn21Si3P appeared nearly unused after nine cuts, whereas strong material adhesions became visible on the flank and rake face for GJL 250. The continuous or discontinuous chip forms correlated with the force signal in terms of fluctuation. Especially for cutting of CuZn21Si3P and GJL 250, a high signal noise appeared in the force measurement, whereas for Inconel 718 nearly no fluctuations were detected.

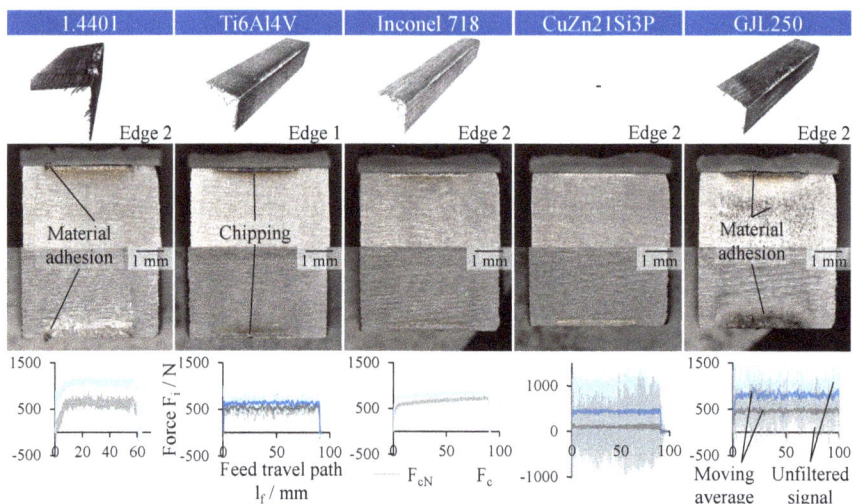

Fig. 9. Tool wear and force signal after nine consecutive cuts for different workpiece materials (LPBF: Standard).

Summary

All as-built, uncoated AM grooving inserts withstood the thermal and mechanical load during machining of the investigated materials within the scope of the experiments. The following findings were made:

- The hardness of the as-built test specimens varied along the distance to the build plate as well as between the two LPBF parameter settings with reduced hardness for small distances and productive LPBF parameters.
- The differences in the material properties of the cutting edge were negligible for machining of the steel alloys 42CrMo4+QT and 1.4401. An influence of the LPBF parameters and build height could be detected only during machining of Inconel 718 with a rising cutting force and cutting normal force over various cuts. The more productive LPBF parameter setting and a small distance to the build plate led to higher forces and faster tool wear development, respective faster tool wear rate of the cutting edge.
- For hard-to-cut materials Inconel 718 and Ti6Al4V, the cutting force rose over various cuts. Chipping phenomena appeared for cutting of Ti6Al4V, whereas high thermal loads resulting in discoloration on the tool flank and rake face, as well as on the chip surface occurred for cutting of 42CrMo4+QT. Adhesion phenomena on the rake face could be detected for alloyed steel 1.4401.

The findings show the capabilities of AM tools made from HS6-5-3-8 even in hard-to-cut materials. As only a limited amount of cuts was performed, tool wear development and thus tool life should be closer analyzed in future investigations, taking into account long-term phenomena. Based on these results, the use of coatings for the AM processed HSS substrate appears reasonable. Thus, thermal influences and material adhesion can be reduced. In future works, the influence of a heat treatment on microstructure and mechanical properties of the AM HSS will be analyzed and compared to conventionally manufactured HSS. In a next step, AM threading tools manufactured from HS6-5-3-8 will be investigated, using the geometrical freedom for an adapted channel and

Material Forming - ESAFORM 2023   Materials Research Forum LLC
Materials Research Proceedings 28 (2023) 1235-1244   https://doi.org/10.21741/9781644902479-134

outlet nozzle design of the internal cutting fluid supply, extending tool life and reducing scrap parts. Another possible application for additively processed HSS can be the manufacturing or repairing of AM broaching tools.

Acknowledgements

The IGF-research project 21581 N (Acronym: "AddBo") of the Forschungsgemeinschaft Werkzeuge und Werkstoffe e.V. (FGW) is funded by the AiF within the program to promote joint industrial research (IGF) by the Federal Ministry for Economic Affairs and Climate Action (BMWK), following a decision of the German Bundestag. The LPBF manufactured parts within the research project were provided by the Fraunhofer Institute for Laser Technology (ILT) in Aachen, Germany with kind support from Tim Lücke. The authors would like to thank the companies Deutsche Edelstahlwerke Specialty Steel GmbH & Co. KG and Meyer + Dörner Räumwerkzeuge GmbH for their support within the project.

References

[1] T. Scherer, Beanspruchungs- und fertigungsgerechte Gestaltung additiv gefertigter Zerspanwerkzeuge, Dissertation, TU Darmstadt, Darmstadt, 2020.

[2] F.A.M. Vogel, S. Berger, E. Özkaya, D. Biermann, Vibration Suppression in Turning TiAl6V4 Using Additively Manufactured Tool Holders with Specially Structured, Particle Filled Hollow Elements, Procedia Manuf. 40 (2019) 32-37. https://doi.org/10.1016/j.promfg.2020.02.007

[3] T. Lakner, High-pressure cutting fluid supply in milling, Dissertation, RWTH Aachen University, Aachen, 2021.

[4] Dr. Mapal, K.G. Kress, Lasersintern erweitert Fertigungsmöglichkeiten von Präzisionswerkzeugen, Diamond Business, 2017.

[5] A.G. Urma Werkzeugfabrik, Serien-Drehwerkzeug mit innenliegenden Kanalstrukturen, Maschinenbau Schweiz No. 11 (2019) 14-15.

[6] T. Kulmala, Geringeeres Gewicht - höhere Leistung, 2019. Available: https://www.sandvik.coromant.com/de-de/mww/pages/t_cm390am.aspx. (accessed 20 July 2020).

[7] M. Padmakumar, Additive Manufacturing of Tungsten Carbide Hardmetal Parts by Selective Laser Melting (SLM), Selective Laser Sintering (SLS) and Binder Jet 3D Printing (BJ3DP) Techniques, Laser. Manuf. Mater. Process. 7 (2020) 338-371. https://doi.org/10.1007/s40516-020-00124-0

[8] B.D. Kernan, E.M. Sachs, M.A. Oliveira, M.J. Cima, Three-dimensional printing of tungsten carbide-10wt% cobalt using a cobalt oxide precursor, Int. J. Refract. Metal. Hard Mater. 25 (2007) 82-94. https://doi.org/10.1016/j.ijrmhm.2006.02.002

[9] T. Schwanekamp, Pulverbettbasiertes Laserstrahlschmelzen von Hartmetallen zur additiven Herstellung von Zerspanwerkzeugen, Dissertation, Ruhr-Universität Bochum, 2021. https://doi.org/10.13154/294-7802

[10] M. Weigold, T. Scherer, E. Schmidt, M. Schwentenwein, T. Prochaska, Additive Fertigung keramischer Schneidstoffe, wt Werkstattstechnik online 110 (2020) 2-6.

[11] J. Saewe, C. Gayer, A. Vogelpoth, J.H. Schleifenbaum, Feasability Investigation for Laser Powder Bed Fusion of High-Speed Steel AISI M50 with Base Preheating System, BHM Berg- und Hüttenmännische Monatshefte 164 (2019) 101-107. https://doi.org/10.1007/s00501-019-0828-y

[12] F. Klocke, Manufacturing Processes 1: Cutting, Springer, Berlin, Heidelberg, 2011.

[13] Wohlers Associates, Wohlers Report 2022: 3D printing and additive manufacturing global state of the industry, Wohlers Associates, Fort Collins (Colorado), 2022.

[14] K. Kempen, B. Vrancken, S. Buls, L. Thijs, J. van Humbeeck, J.-P. Kruth, Selective Laser Melting of Crack-Free High Density M2 High Speed Steel Parts by Baseplate Preheating, J. Manuf. Sci. Eng. 136 (2014). https://doi.org/10.1115/1.4028513

[15] L. Zumofen, C. Beck, A. Kirchheim, H.-J. Dennig, Quality Related Effects of the Preheating Temperature on Laser Melted High Carbon Content Steels, Industrializing Additive Manufacturing - Proceedings of Additive Manufacturing in Products and Applications - AMPA2017 (2018) pp. 210-219. https://doi.org/10.1007/978-3-319-66866-6_21

Material Forming - ESAFORM 2023   Materials Research Forum LLC
Materials Research Proceedings 28 (2023) 1245-1253   https://doi.org/10.21741/9781644902479-135

Machining of PAM green Y-TZP: Influence of build and in-plane directions on cutting forces and surface topography

SPITAELS Laurent[1,a] *, RIVIÈRE-LORPHÈVRE Edouard[1,b]
MARTIC Grégory[2,c], JUSTE Enrique[2,d] and DUCOBU François[1,c]

[1] Machine Design and Production Engineering Lab - Research Institute for Science and Material Engineering - University of Mons, Place du Parc 20, Mons, Belgium

[2] Belgian Ceramic Research Centre, Avenue Gouverneur Cornez 4, Mons, Belgium

[a] laurent.spitaels@umons.ac.be, [b] edouard.rivierelorphevre@umons.ac.be, [c] g.martic@bcrc.be, [d] e.juste@bcrc.be, [e] francois.ducobu@umons.ac.be

Keywords: Green Ceramics, Pellet Additive Manufacturing, Milling, Surface Topography, Hybrid Manufacturing

Abstract. The combination of the pellet additive manufacturing (PAM) process and green ceramic machining within the same hybrid machine is a very promising route to obtain green ceramic parts with complex shapes, smooth surface topography and tight tolerances. However, there is still a lack of data due to the novelty of this manufacturing route. This article studies the possible influence of the build and in-plane directions on the cutting forces and surface topography during the milling of Y-TZP green ceramic parts obtained by the PAM process. The RMS cutting forces, arithmetic and total roughness (Ra and Rt, respectively) were measured. The in-plane direction (aligned with one of the horizontal part edges) did not have a significant influence neither on the cutting forces nor on the surface topography. Conversely, the build direction has a significant effect on the cutting forces recorded. The layers deposited the furthest from the build platform required 57.5% less force to be milled than those in contact with it. The surface topography was not significantly modified across the build direction, all values of Ra were within the 0.8 μm Ra class while all Rt values were < 5 μm.

Introduction

Context. Advanced ceramic materials (such as zirconia) are essential for a large set of sectors thanks to their properties (very high melting point and hardness, chemical inertness, *etc.*) [1,2]. However, the conventional manufacturing routes for ceramics are limited to relatively simple designs and require finishing operations (machining, polishing, grinding or lapping) which can represent up to 80% of the total manufacturing costs [2]. Indeed, these operations are usually performed on the fully sintered part which has the final properties of the material.

On the one hand, green ceramic machining has demonstrated its potential to ease the finishing of the part while reducing the costs and risk to generate macro defects as cracks [3,4]. On the other hand, additive manufacturing (AM) processes open new possibilities to generate near net shape parts with complex design while enabling the production of small or unique part in an economical way [5,6].

Material extrusion process, one of the seven AM processes defined in ISO 52900 standard, is feeding great hopes in a ten year horizon [7]. Indeed, it allows the generation of parts made of metal or ceramics at low cost [8]. Moreover, its variant relying on pellets and screw extrusion, Pellet Additive Manufacturing (PAM) can be fed with feedstock developed for the Ceramic Injection Molding (CIM) or Metal Injection Molding (MIM) industry to shape green parts with complex design. Though, the surfaces generated by the PAM process still suffer from the staircase effect and exhibits high arithmetic roughness (Ra ranging from 9 μm to 40 μm) impacting their fatigue resistance and tribological properties [9-11]. Finishing these parts is then required before foreseeing their usage for demanding applications such as contact (requiring Ra < 1.6 µm, *e.g.*). Nonetheless, even if green ceramic machining can be very attractive thanks to its advantages, it is still limited to simple designs. Lattice, internal surfaces and channels, which can be required in the freeform designs of biomedical implants, are then impossible to finish by machining operations [4,12].

The combination of the green ceramic machining and PAM process inside the same hybrid machine can overcome the disadvantages of both processes [4,13]. Indeed, the subtractive and additive processes can be successively executed so that the milling tool can reach the surfaces to machine when they are still accessible, while the PAM process can ensure an enhanced freedom of design. Hybrid machines are already commercialized, but they are all relying on AM processes which directly produce a fully dense part as Selective Laser Sintering (SLS) or Direct Energy Deposition (DED) [4,13]. As a result, the machining is carried out in fully dense state and leads to higher cutting stresses, wear of the tool and lower material removal rates, making this technology expensive. Developing a machine able to obtain green parts and to machine them at the green state is then an elegant solution to reduce costs, while ensuring new design possibilities for difficult materials such as ceramics. However, except for highly used alloys (Ti6Al4V, *e.g.*), only few data are available for the machining of additively manufactured materials [12].

Goal and motivation of the study. This study aims to determine if the build or in-plane directions within a zirconia green part obtained by the PAM process can have a non-negligible influence on the cutting forces and surface topography generated during finishing operation performed with milling.

Material and Method

Part geometry and printing. The geometry of the part consists of a cube (side of 20 mm) on top of a cylinder (diameter of 15 mm and height of 15 mm). Both are linked using a 3 mm fillet radius. The milling operations are performed on the cube while the cylinder surface is used for the part fixture. Fig. 1 gives the geometry of the part as well as the reference frame (located at one of the part top corners) which was used for the experiments and the build direction selected for the part printing. The X and Y axes are aligned with two horizontal edges of the cube, while the Z axis is aligned with a vertical edge (Z axis is the inverse of the build direction as shown in Fig. 1).

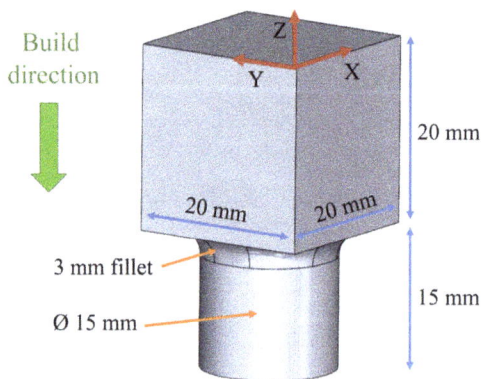

Fig. 1. Part geometry and reference frame.

Material Forming - ESAFORM 2023   Materials Research Forum LLC
Materials Research Proceedings 28 (2023) 1245-1253   https://doi.org/10.21741/9781644902479-135

The part was printed using a Pollen AM Series MC PAM printer by setting the build direction across the part Z axis (see Fig. 1). K2015 pellets from Inmatec were used as feedstock. They are composed of 15wt% polyamide and 85wt% zirconium oxide. The cube is first printed, followed by the cylinder to avoid the need of support structure. The nozzle diameter and layer thickness selected were of 1 mm and 0.35 mm, respectively. These parameters allowed to have a printing time of 25 minutes for the part.

Part machining. A three-jaw chuck was used to clamp the part and the milling operations were conducted with a Stäubli TX200 fitted with a Teknomotor ATC71 electrospindle (maximal power and speed of 7.8 kW and 24000 rpm, respectively). The milling tool used for the operation is supplied by Hoffmann (reference 209425-6, 6 mm diameter, 3 teeth, maximal depth of cut of 19 mm). The cutting conditions were chosen to comply with finishing operations of AM parts. As a result, the axial depth of cut (a_p) was set at 3 mm while the radial depth of cut (a_e) chosen was 0.5 mm. According to a preliminary study, the cutting speed and feed rate were of 339 m/min and 1458 mm/min, respectively.

As depicted in Fig. 2, the part was divided into six zones (each of 3 mm thick) across the build direction (Z axis) and six other zones (each of 3 mm wide) across its in-plane direction (Y axis). Each zone of the in-plane direction corresponds to six different passes. All passes were machined along the part X axis. In total, 216 passes were performed totalizing about 3 minutes of cut within the material.

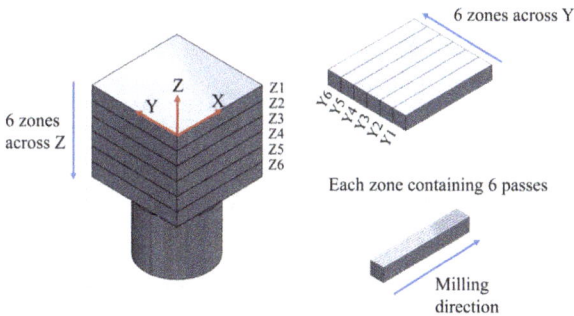

Fig. 2. Zones considered for the milling operations.

Cutting forces and surface topography evaluation. The cutting forces were recorded using a Kistler 9256C2 force sensor coupled with a Kistler charge amplifier 5070A. The DynoWare software executed on a computer as well as a Kistler 5697A2 data acquisition system sampling data at 5 kHz complete the acquisition chain. The reference frame of the cutting sensor is not the same as the frame of the part. Consequently, the total value (Eq. 1) of the X, Y and Z cutting forces components was considered. Fig. 3 gives an example of the total cutting forces signal for three passes of the tests. Finally, the RMS value of the total cutting forces was computed for each pass to have a value representing each pass. These RMS values were then averaged over the six passes contained in each defined zone.

$$F_{tot} = \sqrt{\frac{1}{3} \cdot (F_x^2 + F_y^2 + F_z^2)} \qquad (1)$$

Material Forming - ESAFORM 2023
Materials Research Proceedings 28 (2023) 1245-1253

Materials Research Forum LLC
https://doi.org/10.21741/9781644902479-135

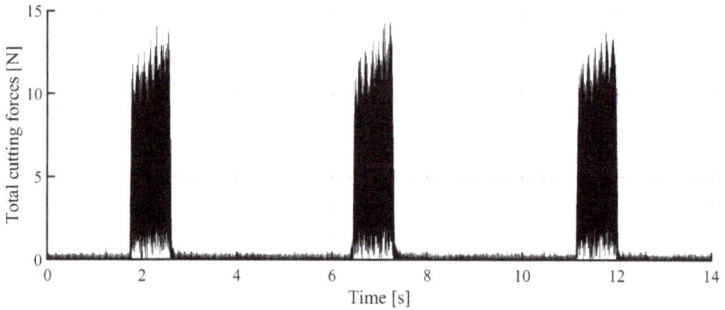

Fig. 3. Example of total cutting forces signal for three passes of the tests.

The surface topography was analyzed qualitatively using a Dino-Lite digital microscope AM7013MZT with DinoCapture software (monitoring of the generation of material pull-out and the existence of porosities inside the parts) and quantitatively with a Diavite DH6 roughness measuring instrument (measurements of Ra and Rt). The evaluation and cut-off lengths (4.8 mm and 0.8 mm, respectively) were selected according to the ISO 4288 standard. The surface topography was evaluated every three passes on the vertical surface generated by side-milling and following the X axis direction.

Results and Discussion
Cutting forces analysis. The cutting forces across the Y and Z zones of the part are depicted in Fig. 4 and 5, respectively. Each graph is given with error bars corresponding to $\pm \sigma$.

As it can be seen in Fig. 4, the RMS total cutting forces remain stable across the different Y zones with an average of 3.97 N. However, the relative standard deviation ranges between about 29.7% and 38.5%, which is very high. Every bar on the graph represents a total of 6 different tests of 6 passes at different Z heights for a given Y zone. Consequently, the build direction (Z axis) has a non-negligible effect on the cutting forces because of the large standard deviations recorded between the measurement of a given Y zone. Conversely, the in-plane direction (Y axis) does not influence the results since the mean value across the different Y zones is nearly constant.

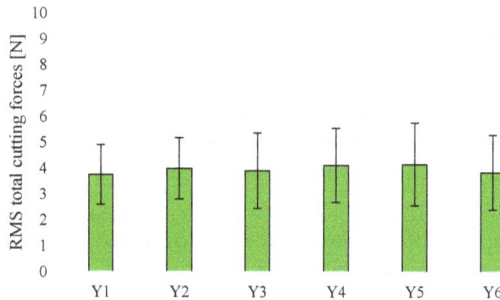

Fig. 4. RMS total cutting forces across the Y zones.

Material Forming - ESAFORM 2023   Materials Research Forum LLC
Materials Research Proceedings 28 (2023) 1245-1253   https://doi.org/10.21741/9781644902479-135

The same conclusions can be retrieved from Fig. 5. Indeed, the graph shows a global decreasing trend across the different Z zones. The maximal value is 6.35 N while the minimal value stands at 2.70 N. The cutting forces in the Z6 zone are 57.5% lower than in the Z1 zone.

This confirms the influence of the build direction (Z axis) on the cutting forces. Moreover, it shows that the zones of the part in contact with the build platform of the printer need more force to be cut than those further away. Indeed, Z1 zone corresponds to the first layers deposited on the build platform while Z6 zone is related to the layers near the cylindric part. This difference of cutting forces may originate from the different thermal history applied to the layers in contact with the build platform in comparison with the higher layers. The deposited layers can then exhibit different properties in terms of mechanical properties (micro-hardness, for example). At the knowledge of the authors, no study in literature mentioned this influence for green ceramics parts obtained neither by the PAM process, nor by other AM processes.

Conversely, the average relative standard deviation is stable through all Z zones with values between 9.5% and 13.6%. This confirms that in-plane direction (Y axis) does not influence much the cutting forces. This decreasing tendency is not asymptotical and, since the Z6 zone is near the cylindric part of the part, this change of section and geometry may have an influence on the required cutting forces because of stresses distribution.

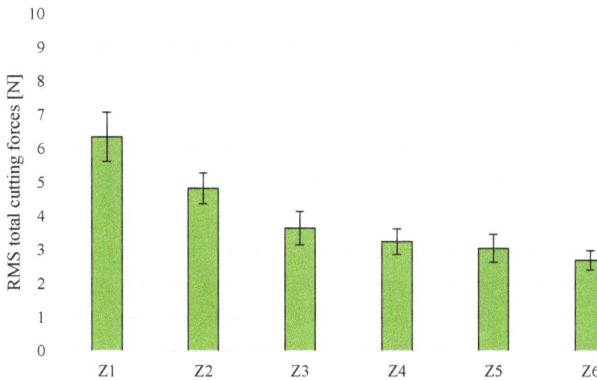

Fig. 5. RMS total cutting across the Z zones.

Surface topography quantitative analysis. Fig. 6 and 7 show the arithmetic roughness evolution across the Y and Z zones of the part. Each graph was given ± σ error bars as well as two red horizontal bars showing the boundaries of the 0.8 μm Ra class, to which the results belong.

As shown in Fig. 6, the Ra results across the Y zones were all within the 0.8 μm Ra class. The values ranged from 0.45 μm to 0.49 μm with a relative standard deviation between 8.5% and 13.4%. Each bar represents the average Ra measured for 6 different tests of 6 passes realized at six different Z positions. Consequently, it shows that neither the in-plane direction (Y axis), nor the build direction (Z axis) significantly influence the arithmetic roughness results.

Material Forming - ESAFORM 2023   Materials Research Forum LLC
Materials Research Proceedings 28 (2023) 1245-1253   https://doi.org/10.21741/9781644902479-135

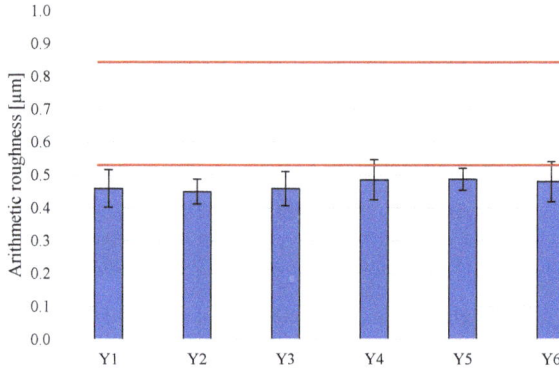

Fig. 6. Arithmetic roughness evolution across the Y zones.

This is also confirmed in Fig. 7 with Ra results across the Z zones between 0.45 µm and 0.49 µm and a relative standard deviation representing from 6.3% to 15.0% of the mean value. This confirms the non-influence of the in-plane and build directions on the arithmetic roughness results. Both graphs also demonstrate the adequacy of cutting parameters to generate a smooth surface topography. Indeed, every pass generates a Ra lower than 1.6 µm with results in the 0.8 µm Ra class. The relatively low standard deviation shows that results are repeatable.

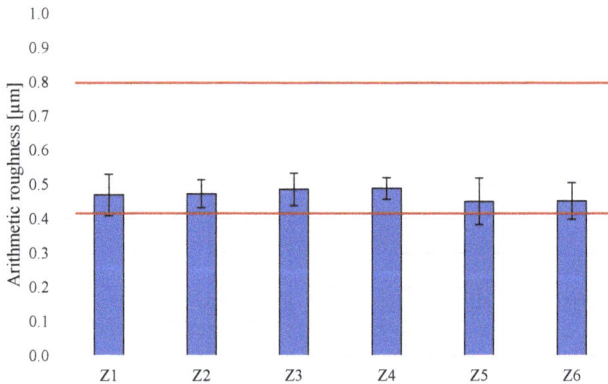

Fig. 7. Arithmetic roughness evolution across the Z zones.

Fig. 8 and 9 give the total roughness measurements obtained across the Y and Z part zones, respectively. Again, each bar of the graph was given a ± σ error bar.

As depicted in Fig. 8, the Rt results across the Y zones were between 2.76 µm and 3.54 µm. The relative standard deviation ranged between 8.6% and 16.6%, except for the Y4 zone where it reached 22.6%. The Y4 zone corresponds to the center of the part and exhibited higher results of Rt compared to the other zones (about 20% higher on average). Some porosities were detected at the center of the part and may influence slightly the Rt measurements.

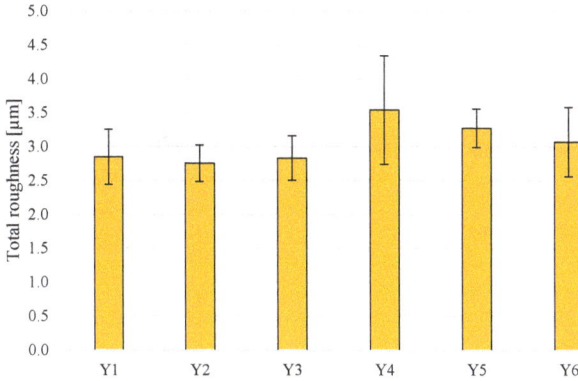

Fig. 8. Total roughness evolution across the Y zones.

By following the build direction (Z axis), the Rt results were between 2.88 μm and 3.30 μm while the relative standard deviation ranged from 7.8% to 19% (see Fig. 9). Even if the mean value varies, all the results are of the same order of magnitude (< 5 μm). Consequently, neither the in-plane nor the build direction (Y axis and Z axis, respectively) influence the total roughness.

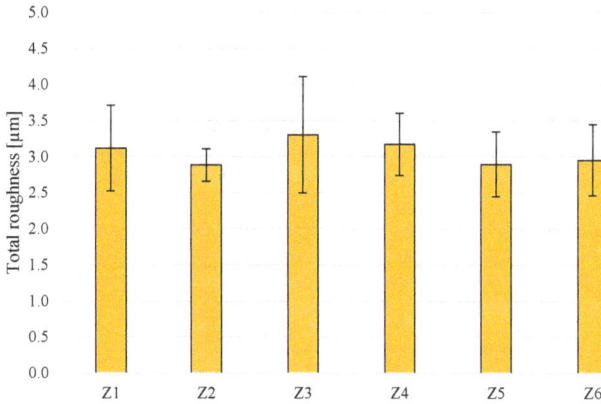

Fig. 9. Total roughness evolution across the Z zones.

Surface topography qualitative analysis. Fig. 10 shows the typical surface topography generated by the milling operation (bottom) compared to the as-built part (top). The bottom part of the picture corresponds to the zones Y1 and Z4. The surface topography obtained after milling is shiny and very smooth.

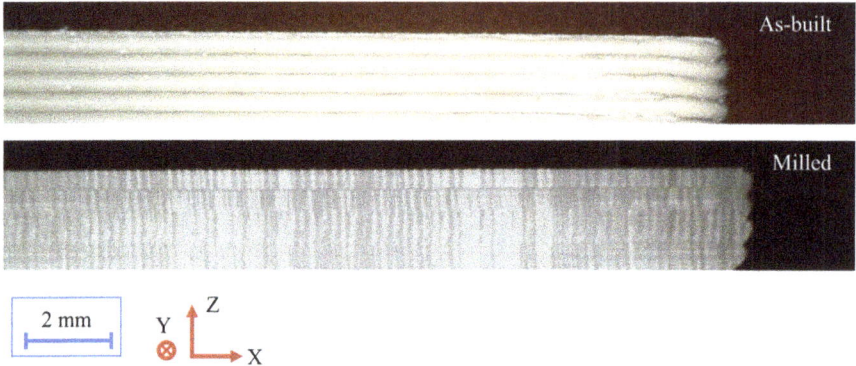

Fig. 10. As-built surface (top) and milled surface of zones Y1 and Z4 (bottom).

Summary

Conclusion. The main findings of this study are:

- The build direction of the part (Z axis) has a significant influence on the cutting forces required to carry out the milling operations. The nearest from the build platform, the higher the cutting forces. When the milling is performed further from the build platform, the cutting forces decrease (up to 57.5% in this study). The experiments do not allow to foresee if this decreasing tendency is asymptotical. The geometry change occurring in Z6 zone (last cubic zone before the transition to the cylindric part) may have an influence on the cutting forces. The in-plane direction (Y axis) does not have a significant influence on the cutting forces.
- From a qualitative point of view, the surface topography generated by the milling operations is smooth and light reflective.
- The arithmetic and total roughness are neither influenced by the build direction (Z axis), nor by the in-plane direction (Y axis) across the part. Indeed, all Ra results belonged to the 0.8 µm Ra class with a maximal relative standard deviation of 15%. So do the Rt results with all values in the same order of magnitude (< 5 µm) with a maximal standard deviation of 22.6%. The center of the part (Y4 zone) exhibited 20% higher Rt results than in the other zones. This may originate from the existence of porosities inside the part.

Perspectives. These are the main perspectives of the work:

- The cutting forces decreasing tendency across the build direction when moving away from the build platform can be further studied. The use of a different part designs with only one geometrical shape (a cube with higher dimensions, *e.g.*) will allow to determine if the cutting forces reach a plateau after a dedicated number of layers deposited. Moreover, micro hardness measurements as well as thermal monitoring can help understanding what physically causes this tendency.
- The influence of the transition between geometrical features (the cube and the cylinder of the part design presented) on the cutting forces can be further investigated by using a different part design.

Acknowledgements

The Wallonian regional government funded this research thanks to a Win²Wal funding instrument (HyProPAM research project, grant number: 2110084). The authors are grateful to the Belgium Ceramic Research Centre (BCRC) for the usage of their Pollen AM Series MC printer and especially to Julien Bossu (UMONS) who printed the part.

References

[1] D. Galusek, K. Ghillányová, Ceramic Oxides, in: R. Riedel, I.-W. Chen (Eds.), Ceramics Science and Technology, Wiley-VCH Verlag GmbH & Co., Weinheim, Germany, 2014, pp. 1-58. https://doi.org/10.1002/9783527631940.ch13

[2] E. Ferraris, J. Vleugels, Y. Guo, D. Bourell, J.P. Kruth, B. Lauwers, Shaping of engineering ceramics by electro, chemical and physical processes, CIRP Ann. 65 (2016) 761-784. https://doi.org/10.1016/j.cirp.2016.06.001

[3] A. Demarbaix, M. Mulliez, E. Rivière-Lorphèvre, L. Spitaels, C. Duterte, N. Preux, F. Petit, F. Ducobu, Green Ceramic Machining: Determination of the Recommended Feed Rate for Y-TZP Milling, J Compos. Sci. 5 (2021) 231. https://doi.org/10.3390/jcs5090231

[4] P. Parenti, S. Cataldo, A. Grigis, M. Covelli, M. Annoni, Implementation of hybrid additive manufacturing based on extrusion of feedstock and milling, Procedia Manuf. 34 (2019) 738-746. https://doi.org/10.1016/j.promfg.2019.06.230

[5] M.K. Thompson, G. Moroni, T. Vaneker, G. Fadel, R.I. Campbell, I. Gibson, A. Bernard, J. Schulz, P. Graf, B. Ahuja, F. Martina, Design for Additive Manufacturing: Trends, opportunities, considerations, and constraints, CIRP Ann. 65 (2016) 737-760. https://doi.org/10.1016/j.cirp.2016.05.004

[6] J. Gonzalez-Gutierrez, S. Cano, S. Schuschnigg, C. Kukla, J. Sapkota, C. Holzer, Additive manufacturing of metallic and ceramic components by the material extrusion of highly-filled polymers: A review and future perspectives, Materials 11 (2018) 840. https://doi.org/10.3390/ma11050840

[7] Smartech Analysis, Ceramics Additive Manufacturing Markets 2017-2028, an opportunity analysis and ten-year market forecast, 2018. https://www.smartechanalysis.com/reports/ceramics-additive-manufacturing-markets-2017-2028/

[8] S.C. Altıparmak, V.A. Yardley, Z. Shi, J. Lin, Extrusion-based additive manufacturing technologies: State of the art and future perspectives, J. Manuf. Processes 83 (2022) 607-636. https://doi.org/10.1016/j.jmapro.2022.09.032

[9] M.A. Krolikowski, M.B. Krawczyk, Does Metal Additive Manufacturing in Industry 4.0 Reinforce the Role of Substractive Machining, in: J. Trojanowska, O. Ciszak, J.M. Machado, I. Pavlenko (Eds.), Advances in Manufacturing II, Springer International Publishing, Cham, 2019, pp. 150-64. https://doi.org/10.1007/978-3-030-18715-6_13

[10] B.N. Turner, S.A. Gold, A review of melt extrusion additive manufacturing processes: II. Materials, dimensional accuracy, and surface roughness, Rapid Prototyp. J. 21 (2015) 250-261. https://doi.org/10.1108/RPJ-02-2013-0017

[11] W. Hung, Post-Processing of Additively Manufactured Metal Parts, in: D.L. Bourell, W. Frazier, H. Kuhn, M. Seifi (Eds.), Additive Manufacturing Processes, ASM International, 2020, pp. 298-315 https://doi.org/10.31399/asm.hb.v24.a0006570

[12] N. Uçak, A. Çiçek, K. Aslantas, Machinability of 3D printed metallic materials fabricated by selective laser melting and electron beam melting: A review, J. Manuf. Process. 80 (2022) 414-457. https://doi.org/10.1016/j.jmapro.2022.06.023

[13] J.M. Flynn, A. Shokrani, S.T. Newman, V. Dhokia, Hybrid additive and subtractive machine tools - Research and industrial developments, Int. J. Mach. Tool. Manuf. 101 (2016) 79-101. https://doi.org/10.1016/j.ijmachtools.2015.11.007

Material Forming - ESAFORM 2023   Materials Research Forum LLC
Materials Research Proceedings 28 (2023) 1255-1264   https://doi.org/10.21741/9781644902479-136

Analysis of Ti-6Al-4V micro-milling resulting surface roughness for osteointegration enhancement

CAPPELLINI Cristian[1,a] *, MALANDRUCCOLO Alessio[1,b], KIEM Sonja [1,c] and ABENI Andrea[2,d]

[1]University of Bergamo,Via Pasubio 7/b, 24044, Dalmine, Bergamo, Italy

[2]University of Brescia, Via Branze 38, 25123, Brescia, Italy

[a]cristian.cappellini@unibg.it, [b]alessio.malandruccolo@unibz.it, [c]sonja.kiem@natec.unibz.it, [d]andrea.abeni@unibs.it

Keywords: Ti-6Al-4V, Osseointegration, Roughness, Micro-Milling, Additive Manufacturing

Abstract. In the last period, the demand of prostheses has massively increased. To guarantee their reliability, properties of durability, biocompatibility, and osseointegration results to be mandatory. Possessing these attributes, Ti-6Al-4V alloy represents the most employed material for implants realization, and because of its microstructure, it can be manufactured by different processing methods, i.e., machining, and additive manufacturing. Considering the necessity of patient-tailored implants, and the capability of additive manufacturing to produce single batch, and complex shapes, at relatively low cost and short time, this latter represents a rewarding process. Compared to biocompatibility, that is mainly function of material chemistry, durability and osteointegration concern mostly surface roughness that affects cells growth at bone-prosthesis interface. After additive manufacturing process and prior to be inserted in the human body, a prosthetic implant is finished by machining operations, hence, the attainment of an appropriate resulting surface roughness is crucial for obtaining a successful implant. Thus, roughness forecasting capability, as a function of the employed finishing process, permits its optimization, avoiding expensive scraps. For this reason, this paper deals with the development of predictive models of surface roughness when micro-milling Ti-6Al-4V alloy specimens.

Introduction

Considering the substitution predictions of actually world-widespread prosthetic implants, such as orthopaedical and dental ones, that are estimated around 3 billion by 2030, they are requested to endure in the human body for a long period [1,2]. For guaranteeing this permanence, implants' materials must own intense resistance to corrosion in body fluids, high strength, wear and fatigue resistance, together with low density and elastic modulus [3]. Amongst metallic biocompatible materials answering to these requirements, Ti-6Al-4V titanium alloy is the most employed for bone and dental replacements [4]. This is related to its characteristics of bio-inertness [5], passivation by titanium dioxide (TiO2) layer formation [6], for avoiding inflammatory reactions, appropriate weight distribution in the human body by high strength-weight ratio [7], paramagnetism, limited stress yielding phenomena [3], osseointegration enhancement due to cementless joints creation [7], and high fatigue strength, mandatory in the normal body-cyclic load conditions [3]. Ti-6Al-4V microstructure is characterized by the coexistence of α-phase with a hexagonal close-packed (hcp) lattice and β-phase having a body-centered cubic (bcc) one, providing good ductility and making it processable by different technologies, such as forming and additive manufacturing (AM) [8]. Considering its capabilities in complex shapes realization and porosity ratios control [9], AM of titanium alloys, in particular Electron Beam Melting (EBM) methodology [10], is largely employed in biomedical field, permitting the production of optimized

Material Forming - ESAFORM 2023
Materials Research Proceedings 28 (2023) 1255-1264

Materials Research Forum LLC
https://doi.org/10.21741/9781644902479-136

and patience-tailored implants at reasonably costs [11]. AM is identified as a net-shape process, but often a milling or micro-milling operation is required for cleaning external surface defects such as lack of fusion and not-melted particles [12,13]. These processes must ensure a suitable surface roughness of the implant for facilitating osseointegration [14]. Contingently from the implant typology, roughness needs are different, and can range from values smaller than a micron up to few microns of the parameters S_a [15]. S_a parameter is usually considered instead of the R_a one since less influenced by scratches and measurement noise [16]. Beyond roughness, there are other functional parameters, such as fatigue, wear, and corrosion resistance, to be taken into account for guaranteeing implant permanence inside human body [17]. Table 1 shows these latter and their trend as a function of the implant type.

Table 1. Functional parameters qualitative values for implant permanence enhancement.

| | Implant type | | |
Parameters	Dental implants	Bone plates and screws	Ball joints
Surface roughness	↑	↔	↓
Mechanical and fatigue resistance	↓	↔	↑
Wear resistance	↓	↑	↑
Pitting corrosion resistance	↑	↓	↓
Fretting resistance	↓	↔	↑
Crevice corrosion resistance	↓	↑	↓
Table legend: ↑ = high values; ↔ = medium values; ↓ = low values			

Pondering all these aspects, the analysis of micro-milling resulting surface roughness is essential to evaluate the capability of this process for promoting osteointegration. This study analyses the achievable roughness in micro-milling of Ti-6Al-4V EBM AM-ed specimens, with different cutting parameters. An Analysis of Variance (ANOVA) of S_a experimental results individuated the most affecting parameters, and mathematical models for S_a calculation were derived. Once validated by further micro-milling tests, these latter can be employed as predictive tools for S_a estimation starting from a set of known process parameters, or to optimize them as a function of the desired S_a values.

Materials and Methods
The analyzed results were attained form an experimental campaign of micro-milling of differently manufactured Ti-6Al-4V specimens. Two groups of cubic samples with a dimension of 10x10x10 mm³ were processed: the first one extracted from 20 mm diameter bars resulting from rolling process, and the second one produced by EBM process, named As-received and EBM respectively. EBM specimens were fabricated with an EBM SYSTEM MODEL A2 machine (ARCAM, Designvägen 2 SE-435 33 Mölnlycke Sweden) in vacuum conditions, from ARCAM Ti-6Al-4V powders. Chemical composition, density, and particle size of these latter are reported in Table 2, where the particle size has been assessed as a function of its distribution for different percentiles (d10, d50, and d90), as indicated by ASTM B214-16 standard. Printing parameters were beam power 1250 W with a focus of 80 μm, printing speed 4530 mm/s, hatch spacing 100 μm, slice thickness 50 μm, and alternating the deposition angle amongst layers 90°. After their realization, the specimens were sonically cleaned in an acetone-isopropanol solution [18].

Cutting operations were performed on a five axis Nano Precision Machining Center KERN Pyramid Nano (Kern Micro Technik, Olympiastr. 2, D-82418 Murnau-Westried Germany) having a Heidenhain iTCN 530 numeric control. Each specimen was previously roughed by face milling it with a three flutes flat bottom mill (nominal diameter of 3 mm) at a depth of cut of 100 μm, a

Material Forming - ESAFORM 2023
Materials Research Proceedings 28 (2023) 1255-1264

Materials Research Forum LLC
https://doi.org/10.21741/9781644902479-136

cutting speed of 100 m/min, and a feed of 7.5 μm/tooth/rev for reaching a planar surface. Then, micro-milling of micro-channels along the whole length in the central part of the samples, by a two-flutes micro-mill, RIME HM79/05, was accomplished. The tool had a diameter of 0.5 mm, an edge radius of 5 μm, a helix angle of 30°, and the material was tungsten carbide (WC) coated with titanium aluminum nitride (TiAlN).

Table 2. Dimensional and chemical properties of Ti-6Al-4V powders.

Percentile of particle size distribution[1]	Particle size [μm]
d10	50
d50	68
d90	98
Powder apparent density [g/cm³]	2.57
Chemical composition [%wt]	
Al	6.42
V	3.88
O	0.13
Fe	0.18
Ti	Balance
[1] ASTM B214-16 Standard Test Method for Sieve Analysis of Metal Powders	

Micro-channels were realized maintaining a constant axial depth of cut a_p equal to 0.03 mm and changing cutting speed V_C and feed f_z in a range of 30-50 m/min and 2-4 μm/tooth/rev respectively, following the central composite design (CCD) experimental plan [19], with an α value of 2 (Fig. 1). The CCD central point with $V_C = 40$ m/min and $f_z = 3.0$ μm/tooth/rev was repeated three times for statistical reliability purposes [20]. The 11 tests shown in Fig. 1 were performed for both the As-received and EBM specimens, giving a total number of 22 experiments.

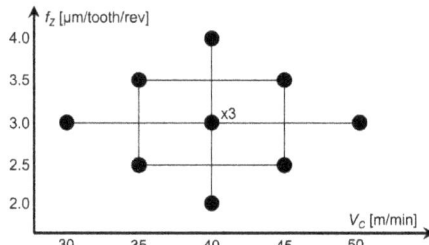

Fig. 1. The employed CCD experimental plan.

A new tool was utilized for each micro-machined channel for avoiding undesired effects of the surface roughness related to tool wear occurrence.

Surface roughness parameters S_a were measured exploiting an optical microscope, namely Mitaka PF60 (Mitaka Kohki.Co.,Ltd., Japan). Consistent with ISO 25178 standard for three-dimensional parametric definition of surface texture, irregularities' heights S_a can be evaluated by Eq. 1:

$$S_a = \frac{1}{LB} \int_0^L \int_0^B |\eta(x,y)| \, dx \, dy \qquad (1)$$

Material Forming - ESAFORM 2023
Materials Research Proceedings 28 (2023) 1255-1264

Materials Research Forum LLC
https://doi.org/10.21741/9781644902479-136

where $\eta(x,y)$ is the surface irregularities deviation from base plane, L is the length and B the width of the explored section. In this work, this latter was positioned at the center of the micro-channel with $L = 0.245$ mm and $B = 2.3$ mm, with a magnification of 400x. Fig. 2 shows a S_a measurement example, while Table 3 the acquired S_a results.

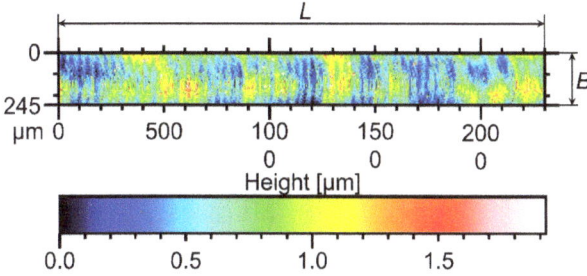

Fig. 2. Example of the acquired S_a measurement (V_C = 40 m/min, f_z = 30 μm/tooth/rev).

Table 3. Summary of the S_a experimental values for all the tests.

V_C [m/min]	f_z [μm/tooth/rev]	As-received S_a [μm]	EBM S_a [μm]
40	4.0	0.36	0.27
45	3.5	0.27	0.20
35	3.5	0.29	0.24
50	3.0	0.27	0.25
40	3.0	0.37	0.26
40	3.0	0.34	0.34
40	3.0	0.35	0.29
30	3.0	0.44	0.25
45	2.5	0.37	0.25
35	2.5	0.32	0.29
40	2.0	0.39	0.23

Results and Discussion

With the intent of evaluating which are the process parameters most influencing S_a, and how it is affected by their variation, an ANOVA of the whole set of experimental values, for both As-received and EBM, was performed. The ANOVA results and the related developed regression models' consistency is directly associated to data normality [21]. Therefore, for checking the normal distribution of S_a data, their probability plots were determined. These are reported in Fig. 3, where the central line denotes the cumulative probability, while the upper and lower curves delimit the 95 % confidence interval (CI) boundaries. For verifying the normality assumption, the analyzed data must remain within the CI curves and close to the probability line. Fig. 3a and 3b, representing the probability plots of As-received and EBM S_a values respectively, underline the data normality.

Analysis of As-received S_a.

Table 4 reports ANOVA results for As-received S_a values. Source column indicates the examined process parameters, where it can be observed that their single effects, interactions, and squared contributions were considered. The other columns represent the number of degrees of freedom (DoF), the adjusted sum of squares (Adj. SS), the adjusted mean of squares (Adj. MS), the F-value, and the p-value of each source. The ANOVA individuated that the S_a value related to the parameter combination $V_C = 35$ m/min and $f_z = 3.5$ µm/tooth/rev had a high standardized residual, indicating it as an outlier. As described in [20], the presence of outliers can be related to several factors, such as the measurer's experience, or the randomization and sequence of the measurements. Anyway, if the value of standardized residual is greater than 4 times the standard deviation, the outlier can be removed without compromising the ANOVA reliability [20]. Since the standardized residual of the outlier had a value of 2.12, 40 times the standard deviation, it was removed in the Response Surface Methodology (RSM) regression analysis.

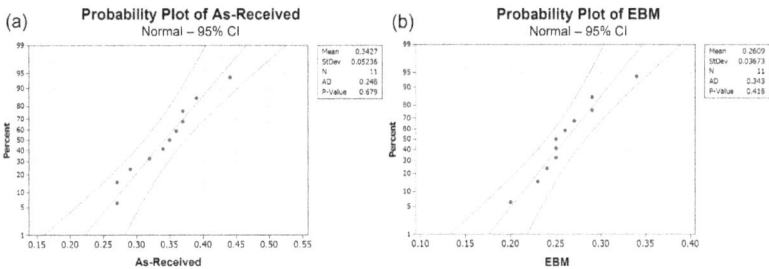

Fig. 3. Probability plots of S_a for (a) As-received, and (b) EBM specimens.

Table 4. ANOVA results for As-received S_a values.

Source	DoF	Adj SS	Adj MS	F-Value	p-Value
f_z	1	0.000057	0.000057	0.16	0.712
V_C	1	0.015858	0.015858	43.93	0.003
$f_z \times f_z$	1	0.000600	0.000600	1.66	0.267
$V_C \times V_C$	1	0.000006	0.000006	0.02	0.900
$f_z \times V_C$	1	0.008901	0.008901	24.66	0.008
Error	5	0.001444	0.000361		
Total	10	0.024360			

The assessment of the source parameter significant influence on response (S_a) is indicated by the p-value. In the common practice, in fact, if p-value is lower than the significance level, the null hypothesis H0, stating that there is no source-response relation, is rejected. Significance level is directly correlated to CI, thus, considering a CI of 95 % for the present analysis, its value is equal to $1 - 95\% = 0.05$. Therefore, when p-value is lower than 0.05 the alternative hypothesis H1 is accepted, concluding that there is a statistically significant source-response relationship [21].

Table 4 results indicate that the variations of V_C and its interaction with f_z have significant effects on S_a for the As-received material. This outcome is confirmed by the main effects plots of Fig. 4a as well, where an increase of V_C improves the surface quality, reducing the S_a value. A similar trend is also detected when considering the f_z variation (Fig. 4b), even if a significant influence is not revealed. The surface and contour plots of Fig. 5a and 5b evidently depict the V_C - f_z interaction effect, while report, once again, that how f_z affects S_a is not clear. At high values of V_C, in fact, an increase of f_z lead to a S_a reduction, but at low V_C values, S_a behavior is the opposite. Due to this, further investigations to clarify the f_z impact are needed.

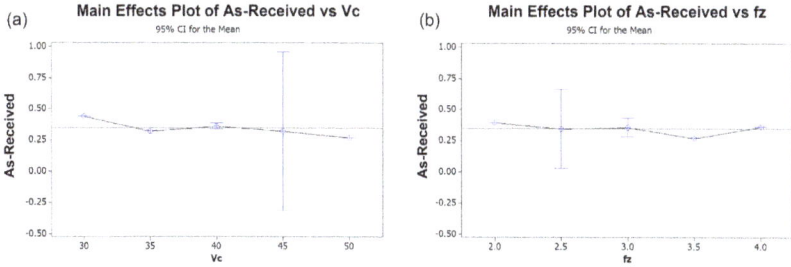

Fig. 4. Main effects plots for As-received S_a.

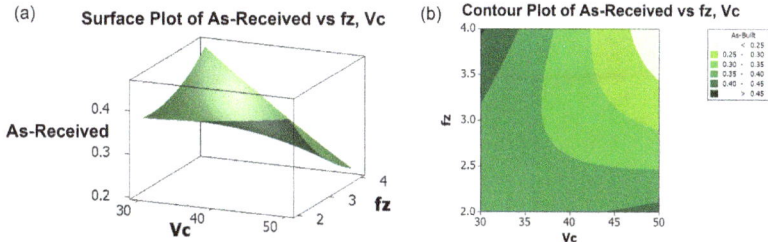

Fig. 5. Surface (a) and contour (b) plots for As-received S_a.

Analysis of EBM S_a.

The ANOVA results of S_a data for EBM micro-milled specimens are reported in Table 5.

Table 5. ANOVA results for EBM S_a values.

Source	DoF	Adj SS	Adj MS	F-Value	p-Value
fz	1	0.001408	0.001408	3.05	0.141
V_C	1	0.005208	0.005208	11.29	0.020
$fz \times fz$	1	0.000647	0.000647	1.40	0.290
$V_C \times V_C$	1	0.000056	0.000056	0.12	0.741
$fz \times V_C$	1	0.000225	0.000225	0.49	0.516
Error	5	0.6598	0.13196		
Total	10	12.7408			

Table 5 highlights that only V_C significantly influences S_a. In particular, when augmenting V_C the surface roughness decreases, and this comportment is represented by the negative slope of the related main effects plot in Fig. 6a. Feed shows an analogous behavior (Fig. 6b), but, as in the case of As-received specimens, a statistical significance of it on S_a is not detectable by ANOVA, therefore its final contribution requires to be further explored. Surface and contour plots in Fig. 7 reveal the deep effect of V_C on S_a for EBM samples, pointing to the negligible influence of fz, mainly at the higher V_C values.

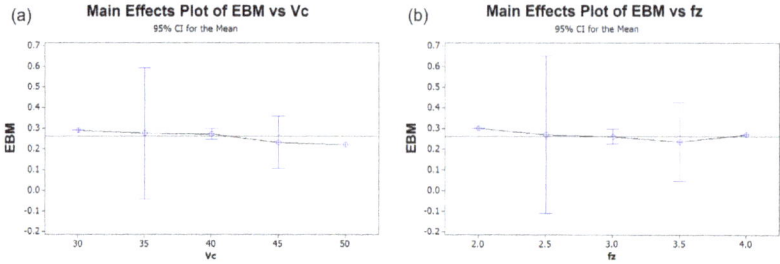

Fig. 6. Main effects plots for EBM S_a.

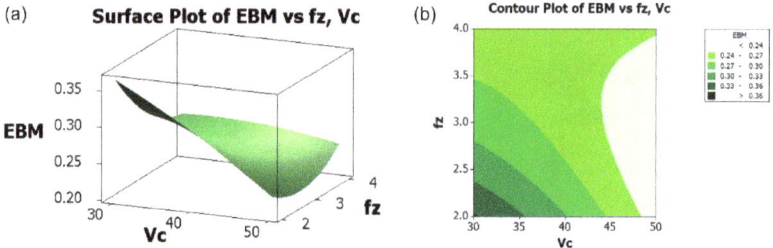

Fig. 7. Surface (a) and contour (b) plots for EBM S_a.

Response Surface Methodology (RSM) regression models.

With the aim of developing mathematical models for predicting S_a as a function of the micro-milling parameters, response surface methodology (RSM) on the whole set of S_a experimental measurements was employed. The developed regression models are reported in Eq. 2 and Eq. 3 for the calculation of As-received (S_{a_AR}) and EBM (S_{a_EBM}) resulting surface roughness respectively.

$$S_{a\_AR} = 1.9 + 0.79f_Z + 0.060V_C + 0.022f_Z^2 + 0.000023V_C^2 - 0.23f_ZV_C \qquad (2)$$

$$S_{a\_EBM} = 0.23 - 0.041f_Z + 0.011V_C + 0.023f_Z^2 - 0.000068V_C^2 - 0.003f_ZV_C \qquad (3)$$

The calculated S_a values for each combination of V_C and f_Z, for As-received and EBM materials, are reported in Table 6. For comparison, Table 6 shows the experimental measurements and the percentage error ($e_\%$) of the calculations as well.

Discussion.

ANOVA results disclose that, freely from the initial furniture status of the material, V_C heavily influences the final part surface quality. A diminution of V_C increases S_a, and this can be correlated to the vibrational effect induced by the non-homogeneous properties of the material, and to the built-up-edges (BUEs) generation at lower V_C [22]. The V_C-f_Z interaction effect can be argued considering that a f_Z increment reduces the micro-tool vibrations permitting to achieve low roughness and higher quality [23]. On the other hand, a reduction of f_Z increases the contribution of ploughing mode cutting mechanism that enlarges the surface plastic deformation lowering the paths' traces and S_a [24]. However, the analysis of further experimental tests is mandatory to clarify f_Z influence.

Table 6. Experimental and estimated S_a values comparison for As-received and EBM specimens.

V_C [m/min]	f_Z [μm/tooth/rev]	Exp S_{a_AR} [μm]	S_{a_AR} [μm]	$e_{\%\_AR}$	Exp S_{a_EBM} [μm]	S_{a_EBM} [μm]	$e_{\%\_EBM}$
40	4.0	0.36	0.369	2.4	0.27	0.258	4.3
45	3.5	0.27	0.257	4.7	0.20	0.222	11.0
35	3.5	0.29	0.452	56.0	0.24	0.278	3.1
50	3.0	0.27	0.274	1.6	0.25	0.209	5.1
40	3.0	0.37	0.351	5.2	0.26	0.257	1.2
40	3.0	0.34	0.351	3.2	0.34	0.257	7.1
40	3.0	0.35	0.351	0.3	0.29	0.257	11.4
30	3.0	0.44	0.432	1.8	0.25	0.292	0.6
45	2.5	0.37	0.378	2.1	0.25	0.259	3.5
35	2.5	0.32	0.341	6.5	0.29	0.285	1.7
40	2.0	0.39	0.378	3.2	0.23	0.302	0.7

The EBM S_a values result to be lower than the ones of As-received specimens. This is ascribable to the higher brittleness, and lower ductility, of EBM samples, related to the localized higher cooling rate of EBM process. Therefore, the superior ductility of As-received material causes an increased material adhesion on the tool cutting edge that negatively affects quality [25].

The calculation errors in Table 6 stand below the 11 %, the only exception is related to a $e_\%$ of 56 % related to the detected outlier, for which the significance of the value must not be taken into consideration. Overall, the low error values achieved highlight the applicability of the proposed models. Further tests need to be performed to finally validate Eq. 2 and 3 and permitting to exploit a reliable tool to forecast the attainable S_a once V_C and f_Z have been selected, or to optimize the cutting parameters as a function of the desired S_a. It should be noted that the assessment of S_a for promoting osseointegration concerns surface chemistry and morphology as well, giving the necessity of a deeper analysis to establish optimal conditions [26]. Additionally, different S_a requirements are crucial for different biomedical applications [27]. As an example, the outcomes of the accomplished micro-milling tests do not warrant to achieve the optimal combination of material, production process, and cutting parameters for a S_a value suitable for dental applications ($S_a \approx 1.5$ μm). Therefore, additional investigations should be made for analyzing the possibility of micro-machining processes employment in the dental field. Anyway, the achieved S_a values are applicable for prosthetic implants and culture cell grows.

Summary

This paper presents the study of micro-milling cutting parameters effects on the resulting surface roughness S_a as a function of different states of furniture of the material, with the aim of promoting osseointegration of prosthetic implants made of Ti-6Al-4V biocompatible alloy. The ANOVA of the data, obtained by means of an extensive experimental campaign, allowed to identify the most affecting cutting parameters, revealing V_C as the most significant. The application of RSM on data lead to the development of regression models capable to estimate S_a with good approximation, underlining their potential to predict and optimize roughness. A general increase of S_a has been observed when both V_C and f_Z decrease, even if, other tests should be performed to clarify f_Z contribution. The optimum roughness value establishment remains, in fact, a difficult task, being it not merely manufacturing process dependent, but implant application, material chemistry, mechanical and corrosion resistance related as well (Fig. 8). Consequently, ulterior studies and in vitro osseointegration tests are needed to identify the optimal topography combining all requirements.

Material Forming - ESAFORM 2023   Materials Research Forum LLC
Materials Research Proceedings 28 (2023) 1255-1264   https://doi.org/10.21741/9781644902479-136

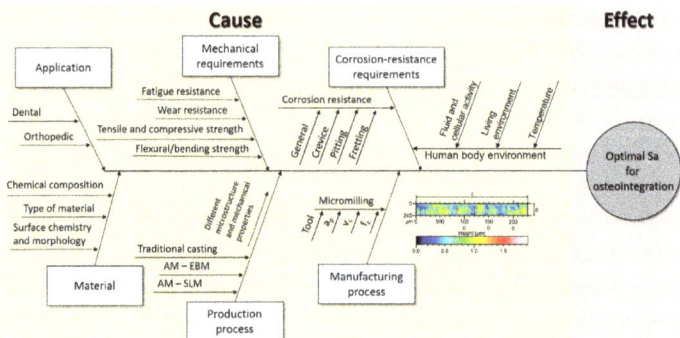

Fig. 8. Summary of influencing parameters on optimal S_a for osseointegration.

References

[1] Y.L. Hao, S.J. Li, R. Yang, Biomedical titanium alloys and their additive manufacturing, Rare Met. 36 (2016) 661-671. http://doi.org/10.1007%2Fs12598-016-0793-5

[2] M. Sarraf, G.E. Rezvani, S. Alipour, S. Ramakrishna, N.L. Sukiman, A state of the art review of the fabrication and characteristics of titanium and its alloys for biomedical applications, Biodes. Manuf. 5 (2022) 371–395. https://doi.org/10.1007/s42242-021-00170-3

[3] R.B. Heimann, Biomaterials – characteristics, history, applications, in: R.B. Heimann (Eds.), Materials for Medical Applications, Walter de Gruyter, Berlin/Boston, 2020, pp. 1-74.

[4] S.A. Aghili, K. Hassani, M. Nikkhoo, A finite element study of fatigue load effects on total hip joint prosthesis, Comput. Methods Biomech. Biomed. Eng. 24 (2021) 1545-1551. https://doi.org/10.1080/10255842.2021.1900133

[5] G.S. Kaliaraj, T. Siva, A. Ramadoss, Surface functionalized bioceramics coated on metallic implants for biomedical and anticorrosion performance-a review, J. Mater. Chem. B 9 (2021) 9433-9460. https://doi.org/10.1039/D1TB01301G

[6] P. Pedeferri, Corrosion in the human body, in: L. Lazzari, M.P. Pedeferri (Eds.), Corrosion Science and Engineering, Springer Cham, Switzerland, 2018, pp. 575-587.

[7] J.W. Nicholson, A.J. Connor, Biological interactions with materials, in: J.W. Nicholson (Eds.), The Chemistry of Medical and Dental Materials, RSC, 2020, pp. 186-226.

[8] S. Gialanella, A. Malandruccolo, Titanium and Titanium Alloys, in: Aerospace Alloys, Springer Cham, Switzerland, 2020, pp. 129-189.

[9] C. Cappellini, Y. Borgianni, L. Maccioni, C. Nezzi, The effect of process parameters on geometric deviations in 3D printing with fused deposition modelling, Int. J. Adv. Manuf. Technol. 122 (2022) 1763-1803. https://doi.org/10.1007/s00170-022-09924-4

[10] J. Tong, C.R. Bowen, J. Persson, A. Plummer, Mechanical properties of titanium-based Ti–6Al–4V alloys manufactured by powder bed additive manufacture, Mater. Sci. Technol. 33 (2017) 138-148. https://doi.org/10.1080/02670836.2016.1172787

[11] W.Y. Yeong, C.K. Chua, A quality management framework for implementing additive manufacturing of medical devices, Virtual Phys. Prototyp. 8 (2013) 193-199. https://doi.org/10.1080/17452759.2013.838053

[12] T. Özel, E. Ceretti, T. Thepsonthi, A. Attanasio, Machining Applications, in: T. Özel, P.J. Bartolo, E. Ceretti, J. De Ciurana Gay, C.A. Rodrigez, J.V. Lopes Da Silva (Eds.), Biomedical devices – Design, prototyping and manufacturing, Wiley, Hoboken, NJ, US, 2017, pp. 99-120.

[13] A. Abeni, C. Cappellini, P.S. Ginestra, A. Attanasio, Analytical modeling of micro-milling operations on biocompatible Ti6Al4V titanium alloy, Procedia CIRP 110 (2022) 8-13. https://doi.org/10.1016/j.procir.2022.06.004

[14] A. Wennerberg, T. Albrektsson, Effects of titanium surface topography on bone integration: A systematic review, Clin. Oral Implants Res. 20 (2009) 172-184. https://doi.org/10.1111/j.1600-0501.2009.01775.x

[15] T. Albrektsson, A. Wennerberg, On osseointegration in relation to implant surfaces, Clin. Implant Dent. Relat. Res. 21 (2019) 4-7. https://doi.org/10.1111/cid.12742

[16] A. Abeni, A. Metelli, C. Cappellini, A. Attanasio, Experimental optimization of process parameters in CuNi18Zn20 micromachining, Micromachines 12 (2021) 1293. https://doi.org/10.3390/mi12111293

[17] L. Lin, H. Wang, M. Ni, Y. Rui, T.Y. Cheng, C.K. Cheng, X. Pan, G. Li, C. Lin, Enhanced osteointegration of medical titanium implant with surface modifications in micro/nanoscale structures, J. Orthop. Translat. 2 (2014) 35-42. https://doi.org/10.1016/j.jot.2013.08.001

[18] P.S. Ginestra, R.M. Ferraro, K. Zohar-Haubert, A. Abeni, S. Giliani, E. Ceretti, Selective laser melting and electron beam melting of Ti6Al4V for orthopedic applications: a comparative study on the applied building direction, Materials 13 2020 5584. https://doi.org/10.3390/ma13235584

[19] Y. Men, J. Liu, W. Chen, X. Wang, L. Liu, J. Ye, P. Jia, Y. Wang, Material parameters identification of 3D printed titanium alloy prosthesis stem based on response surface method, Comput. Methods Biomech. Biomed. Eng. (2022). https://doi.org/10.1080/10255842.2022.2089023

[20] D.C. Montgomery, Design and Analysis of Experiments, 10th ed.; John Wiley & Sons, Hoboken, NJ, US, 2019.

[21] F. Concli, L. Maccioni, L. Fraccaroli, C. Cappellini, Effect of Gear Design Parameters on Stress Histories Induced by Different Tooth Bending Fatigue Tests: A Numerical-Statistical Investigation, Appl. Sci. 12 (2022) 3950. https://doi.org/10.3390/app12083950

[22] G. Kiswanto, A. Mandala, M. Azmi, T.J. Ko, The effects of cutting parameters to the surface roughness in high speed cutting of micro-milling titanium alloy Ti-6Al-4V, Key Eng. Mater. 846 (2020) 133-138. https://doi.org/10.4028/www.scientific.net/KEM.846.133

[23] A. Roushan, U. Srinivas Rao, K. Patra, P. Sahoo, Multi-Characteristics Optimization in Micro-milling of Ti6Al4V Alloy, J. Phys.: Conf. Ser. 1950 (2012) 012046. https://doi.org/10.1088/1742-6596/1950/1/012046

[24] H.K. Rafi, N.V. Karthik, H. Gong, T.L. Starr, B.E. Stucker, Microstructures and mechanical properties of Ti6Al4V parts fabricated by selective laser melting and electron beam melting, J. Mater. Eng. Perform. 22 (2013) 3872-3883. https://doi.org/10.1007/s11665-013-0658-0

[25] S.P.L. Kumar, D. Avinash, Experimental biocompatibility investigations of Ti–6Al–7Nb alloy in micromilling operation in terms of corrosion behavior and surface characteristics study, J. Braz. Soc. Mech. Sci. Eng. 41 (2019) 364. https://doi.org/10.1080/10667857.2021.1903671

[26] A. Kemény, I. Hajdu, D. Károly, D. Pammer, Osseointegration specified grit blasting parameters, Mater. Today: Proc. 5 (2018) 26622-26627.

[27] R. Krishna Alla, K. Ginjupalli, N. Upadhya, M. Shammas, R.R. Krishna, R. Sekhar, Surface roughness of implants: A review, Trends Biomater. Artif. Organs 25 (2011) 112-118.

Material Forming - ESAFORM 2023
Materials Research Proceedings 28 (2023) 1265-1274

Materials Research Forum LLC
https://doi.org/10.21741/9781644902479-137

Comparison of cutting tool wear classification performance with artificial intelligence techniques

COLANTONIO Lorenzo[1,a] *, EQUETER Lucas[1,b], DEHOMBREUX Pierre[1,c] and DUCOBU François[1,d]

[1]Machine Design and Production Engineering Lab, University of Mons, 7000 Mons, Belgium

[a]lorenzo.colantonio@umons.ac.be, [b]lucas.equeter@umons.ac.be, [c]pierre.dehombreux@umons.ac.be, [d]francois.ducobu@umons.ac.be

Keywords: Machining, Turning, Cutting Tool, Artificial Intelligence, Monitoring, Classification

Abstract. Optimal replacement of machining cutting tools is a major challenge in today's manufacturing industry. Due to the degradation of the tool during machining, late replacement of the tool leads to the risk of producing parts that do not meet technical specifications, while early replacement increases machine downtime and tool costs. To replace tools at the right time, it is necessary to monitor their degradation. Therefore, this paper compares the classification performance of different artificial intelligence approaches to classify the condition of cutting tools from cutting signals. Different approaches, namely: Artificial Neural Network (ANN), Support Vector Classifier (SVC), Random Forest (RF) and k-Nearest Neighbour (k-NN) are tested, and their performance is compared. It is highlighted that ANN and RF methods obtain better classification performances (88.8% and 86.4%, respectively) than the rest of the approaches (80%). Nevertheless, all approaches can monitor the degradation of cutting tools in a satisfactory manner (i.e., 80% accuracy). A comparison of training times highlights that training a neural network takes longer than the other approaches. However, with the computational power currently available, this is not an obstacle for their implementation in real applications as this training can still be achieved in a couple of minutes.

Introduction

The condition of a cutting tool is of critical importance to the machined surface and the associated machining tolerances. A worn tool or a tool in an unsatisfactory condition does not allow the creation of machined surfaces of sufficient quality, which in consequence increases the cost of production [1]. Different tool replacement policies exist [2], but often tool replacement maintenance policies attempt to address this problem by replacing the tool well before its end of life which creates waste. This results in higher tool costs and increased machine downtime, further increasing production costs. As tool wear is an extremely complex and non-stationary phenomenon [3], the determination of the tool condition can be complex. Monitoring the degradation of the tool is thus necessary.

There are two types of monitoring: direct and indirect. The direct approach consists of measuring tool wear directly on the tool, but this requires the machining process to be stopped and results in increased machine downtime [4]. The indirect approach consists of measuring signals during the machining process to try to predict the state of the tool [5]. This has the advantage of allowing a continuous machining process with the least intrusive sensor installation possible. A review of the type of signals that can be recovered during machining is available in [6]. Several indirect monitoring methods exist, but lately, artificial intelligence methods are predominant as they can learn from machining data and adjust their prediction on a case-by-case basis. A systematic literature review describing all approaches present in the literature shows that there are mainly two approaches with AI: classification and regression [7]. Classification aims to monitor

Material Forming - ESAFORM 2023   Materials Research Forum LLC
Materials Research Proceedings 28 (2023) 1265-1274   https://doi.org/10.21741/9781644902479-137

the state of the tool via discrete values listing the state of the tool. Regression aims to follow the evolution of the tool by directly monitoring the wear. In some applications, it is not necessary to know exactly how the tool wear is evolving, so classification methods are used because of their ease of understanding.

In the literature, there are only a few comparisons of performance between different classification approaches. These comparisons are often made to highlight the particularity of one model compared to another, but a more general comparison is almost never made. This paper, therefore, proposes to compare the performance of some common artificial intelligence methods, namely: Artificial Neural Network (ANN), Random Forest (RF), Support Vector Classifier (SVC) and k-Nearest Neighbour (k-NN) implemented on the same database. The choice of these approaches is based on their ease of implementation and their ability to perform a classification for this application. All approaches are optimised and tested on identical data that homogeneously represents the different degradation states of the tool. The comparison of the quality of the results and the efficiency of the approach is realised.

Methodology

To compare the performance of the different AI approaches, the database and the experimental conditions are described. Signals from the database correlated with tool wear are identified and used as input for the different AI approaches. The database is then divided into classes and a split between training and test data is made. Finally, all approaches, namely: ANN, RF, SVC and k-NN, are presented and optimised, and their results are highlighted. A comparison of the results is then made.

Experimental Setup and Database

The database comes from experimental tests carried out on a CNC lathe (Weiler E35), which ensures a constant cutting speed throughout the machining process (Fig. 1). It is used to machine (longitudinal turning operation), C45 steel bars at variable cutting speeds with a CNMG120404-MF3 TP40 tool from SECO. This tool is one of the lower-grade tools to limit the amount of material wasted during testing. Table 1 shows the different cutting conditions used during the tests, only variations in cutting speeds are considered. The machine is instrumented with a force sensor (Kistler 9257B) that collect the cutting forces (Fi) and torques (Mi) during the machining operation. The sensor is mounted at the base of the tool and in the machine frame of reference, Fx corresponds to the feed force, Fy is the radial force and Fz is the cutting force. This sensor is mounted for indirect monitoring, i.e., to be as minimally intrusive as possible. Wear and cutting forces are measured every 2.8 minutes (corresponding to one piece). Wear is assessed according to ISO 3685 [8], which defines wear as the value of Vb (Fig. 2) measured directly on the tool using a microscope (Byameyee EU-1000X 3). Vb is measured as the size of the wear in zone B. This B area is located between the corner radius on one side and 1/4 of the wear area (area C where is located the notch wear) on the other side (Fig. 2)

A total of 30 tools are used to create the database, with the degradation of each tool being measured 6.4 times on average during its lifetime. The database therefore consists of 192 data points evenly distributed over the tool degradation. The measured cutting force corresponds to a 20 s signal sampled at 10 kHz and is processed to recover statistical and frequency values. The statistical analysis corresponds to the calculation of the average, the RMS value and the frequency analysis identifies the frequency and the maximum amplitude of the power spectral density.

To identify the relationships between the signals measured during machining and tool degradation, a Spearman correlation analysis is used on all signals. This correlation analysis is adapted to the size and non-normality of the data. The correlation analysis shows that the features most correlated with wear are the following: Mz RMS (correlation indicator: 0.89), Fx RMS (0.87), machining duration (0.84), chip length (total length machined) (0.84) and Fz RMS (0.79).

As cutting forces are strongly correlated with tool wear, they will be used in the following as inputs for the AI methods. It should be noted that the machining time and the chip length have the same correlation score as they are both dependent. The chip length is also dependent on the cutting speed. In this case, since only variations in the machining speed are taken into account, this indicator allows the method to indicate the change in cutting conditions.

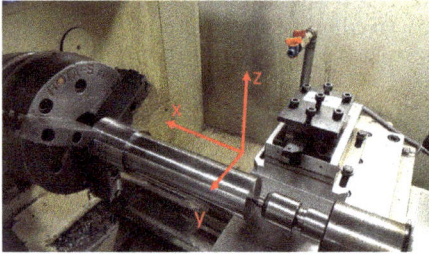

Fig. 1. Experimental turning setup, the tool is mounted into the cutting force and torque sensor.

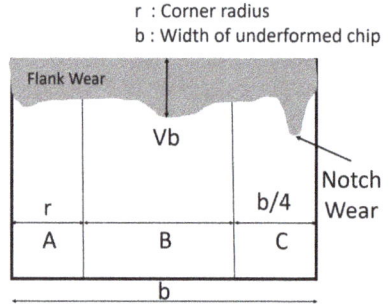

Fig. 2. Flank wear degradation Vb on the flank face.

Table 1. Experimental Cutting Condition.

Tool n°	Feed [mm/rev]	Cutting Speed [m/min]	Depth of cut [mm]
1 to 10	0.2	260	1
11 to 15	0.2	250	1
16	0.2	240	1
17 to 20	0.2	265	1
21 to 30	0.2	Variable: 240 to 260 (for each tool)	1

Features Preparation for Tool Wear Classification

In a classification problem, it is necessary to define classes whose purpose is to define the different possible states of the tool. The degradation of a cutting tool is divided into 3 successive phases (Fig. 3): the first phase corresponds to the beginning of the tool's life, it shows little wear but degrades rapidly. This phase is generally short in relation to the life of the tool. The second phase is the longest and consists of a quasi-linear degradation of the tool, it is a regime phase in which the tool will spend most of the time. Finally, the third phase is the end of the tool's life, which can last more or less time depending on the cutting conditions.

Based on the ISO 3685 standard, it is often accepted that the end-of-life criterion for a cutting tool is when the size of the flank wear reaches 300 μm [8]. This flank wear is called Vb and is presented in Fig. 2. It is therefore proposed that class 1 corresponds to wear between 0 and 150 μm, class 2 corresponds to a wear between 150 and 300 μm and class 3 corresponds to the end of the tool's life with wear exceeding 300 μm. These values (150 and 300 microns) were chosen as they are generally used in the literature as end-of-life criteria.

The database is not uniform across all these classes, indeed, there are significantly more points in class 1 than in class 3 for example. To increase the number of points in each class, data augmentation is performed. This data augmentation consists of linearly interpolating 2 points of

Material Forming - ESAFORM 2023   Materials Research Forum LLC
Materials Research Proceedings 28 (2023) 1265-1274   https://doi.org/10.21741/9781644902479-137

the tool degradation and calculating an intermediate measurement point. This simple type of interpolation is sufficient given the number of measurements points in a complete trajectory; the interpolation error is low. This increases the number of possible points for training the AI but does not change the distribution of points in the different classes (73 % in class 1, 19% in class 2 and 8% in class 3). Nevertheless, with more data points, there will be more points to propose for training AI methods, which generally allows for faster convergence [9].

Artificial intelligence methods need training data to learn the relationships between the data and test data to verify that the model has learned correctly. It was chosen to select 15 points randomly per class to create the test database. For class 3, 15 data points correspond to 40% of the data in this class. This value is quite high (usually 25% of the data is used in testing) but this value does not impact the results presented in the following. By ensuring that each class contains the same number of data points, this ensures that an error in one class has the same overall importance regardless of the class.

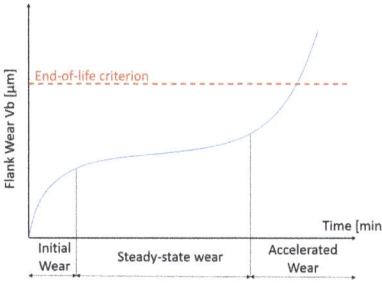

Fig. 3. Typical tool degradation for a given cutting condition.

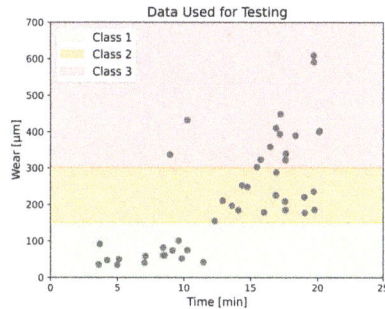

Fig. 4. Testing data randomly selected and their repartition in different classes.

Artificial Neural Network

Neural networks are certainly the most popular intelligence method. Inspired by the way the brain works, this AI can represent highly non-linear relationships between input and output and is extensively represented in the literature [10]. There are multiple hyperparameters that influence the quality of the results, the most common are: the network architecture, the activation function used in each layer, the batch size, and the number of epochs. The most important hyperparameter is the network architecture which will create the relationship between the input and output of the network and is often chosen by trial and error. Table 2 shows the different hyperparameters used to obtain the best classification results. These and all other parameters presented in this paper are optimised by testing the influence of each parameter individually. In this case, for example, different architectures have been tested but the one chosen is the one that gives the best results. The network architecture is presented in Fig. 5.

The results are presented Fig. 6, the overall accuracy is 88.88%. Class 1 is always correctly classified but classes 2 and 3 present an accuracy of 86.7% and 80% respectively.

Material Forming - ESAFORM 2023
Materials Research Proceedings 28 (2023) 1265-1274

Materials Research Forum LLC
https://doi.org/10.21741/9781644902479-137

Table 2. Optimized hyperparameters for ANN.

Hyperparameter	Value
Algorithm	TensorFlow (Python) [11]
Architecture	2 hidden layers each containing 8 neurons
Activation function	1st layer: Tanh, 2nd layer: Relu, output: Softmax
Batch size	5
Epoch	1000

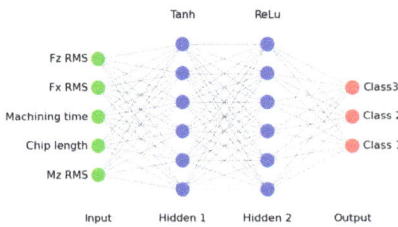

Fig. 5. Best Artificial Neural Network architecture.

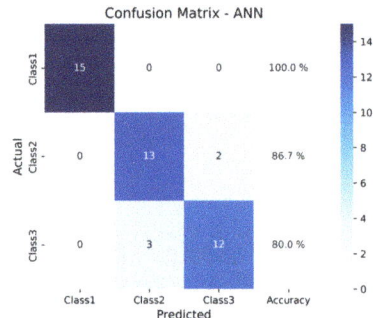

Fig. 6. ANN Confusion matrix – Overall accuracy : 88.8 %.

Random Forest

The random forest classifier is a combination of classical tree classifiers. The Random Forest classifier consists at producing a set of tree classifiers (i.e., Number of estimator) to create an ensemble of classifiers, called forest [12]. To classify a state, each tree of the forest is interrogated, and the most given class is the predicted class (Fig. 7). Combining the results has the advantage of being more accurate than if each tree were taken individually, as the probability of a tree being wrong is higher than the probability of most trees being wrong.

Table 3 shows the different hyperparameters used to create the classifier. The number of estimators corresponds to the number of trees in the forest. The variation of this value can change the accuracy of the approach. The number of 500 is chosen as more estimators do not improve the overall accuracy. The maximum complexity of the trees is controlled by the "maximum depth", the variation of this parameter slightly improves the results, a value too low leads to bad accuracy while a high value does not improve the results. To measure the quality of a split, the Gini impurity criteria is used [13].

The results obtained with the combination of hyperparameters previously identified are presented Fig. 8. The overall accuracy of the approach is 86.6%. Performance is consistent across all classes 2 and 3 with 80% accuracy.

Material Forming - ESAFORM 2023
Materials Research Forum LLC
Materials Research Proceedings 28 (2023) 1265-1274
https://doi.org/10.21741/9781644902479-137

Table 3. Optimized hyperparameters for RF.

Hyperparameter	Value
Algorithm	Sci kit Learn (Python) [13]
Number of estimators	500
Maximum depth	6
Criterion	Gini

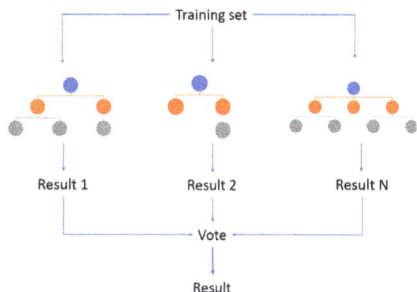

Fig. 7. Random Forest principle.

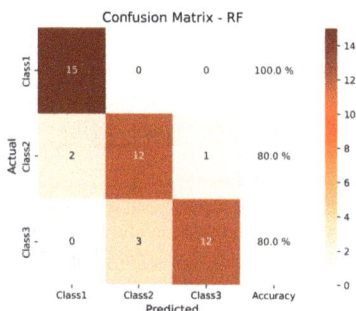

Fig. 8. Random Forest Confusion Matrix – Overall accuracy: 86.8%.

Support Vector Classifier

Support vector machines, and more specifically the support vector classifiers, aim at finding an optimal hyperplane to separate different classes of data [14] (Fig. 9). In this application the kernel function that defines the different classes is the Radial Basis Function (RBF). This approach uses mainly 2 parameters: gamma and C. Gamma controls the curvature of the data separation and so controls the influence of samples selected by the model to be support vectors. The parameter C control the error rate by compromising between the correct classification against the maximization of the decision function. A compromise between these two parameters allows an efficient classification by controlling the shape of the classification area and the influence of outliers. Different combinations of parameters were tested but the combination of parameters that gives the best results is listed Table 4.

Fig. 10 shows the results obtained by the approach described above. The overall accuracy is 80%. Class 1 has no error, class 2 has the highest error rate with an accuracy of 66.7%, and class 3 has a performance of 73.3%.

Table 4. Optimized hyperparameters for SVC.

Hyperparameter	Value
Algorithm	Sci Kit Learn (Python) [13]
C	2
Gamma	1
Kernel	Radial Basis Function

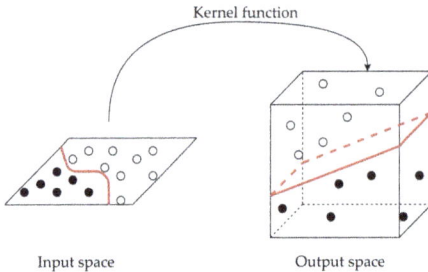

Fig. 9. SVC approach, under Creative
Commons Attribution License [7].

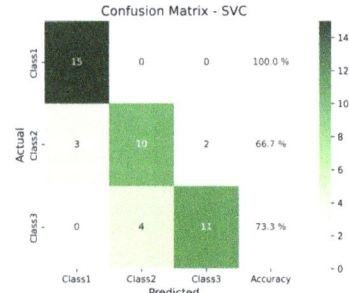

Fig. 10. SVC confusion matrix – Overall
accuracy : 80%.

k-Nearest Neighbour

K-Nearest Neighbour in classification uses the k nearest neighbour in the dataset reference of the input to determine the class of a new input [15]. The weight of each neighbour generally depends on the distance to the new inputs. Fig. 11 shows the example of the classification of a new element with a k value of 5. The 5 nearest data items and their distances are used to determine the class of the new item. Table 5 shows the most important parameters for this approach: the number of neighbour and the weight of the connection. In this approach, the best results were obtained with a number of neighbours of 8 and weight depending on the distance.

Fig. 12 presents the results. The overall accuracy is 80%. Only class 1 is correctly classified, class 2 and 3 have 66.7% and 73.3% accuracy respectively.

Table 5. Optimized hyperparameters for k-NN.

Hyperparameter	Value
Algorithm	Sci Kit Learn (Python) [13]
Number of neighbours	8
Weight	Distance
Algorithm	Auto

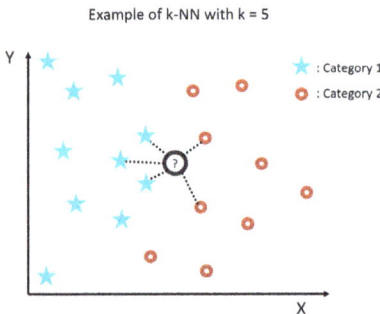

Fig. 11. k-NN approach with k = 5.

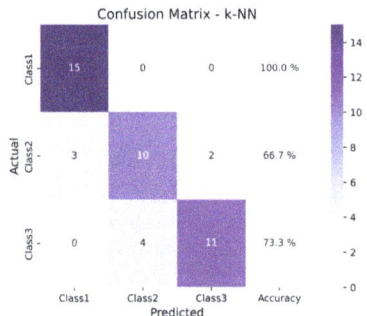

Fig. 12. k-NN confusion matrix – Overall
accuracy: 80%

Material Forming - ESAFORM 2023
Materials Research Proceedings 28 (2023) 1265-1274

Materials Research Forum LLC
https://doi.org/10.21741/9781644902479-137

Discussion

Table 6 compares the results obtained by the different approaches and the one with the highest overall score is the ANN method. It can be noted, however, that all approaches score well, with an average accuracy above 80%. Each method has its advantages and disadvantages, and the simple comparison of results does not allow to highlight them. For example, approaches such as k-Nearest Neighbour does not give good results if the dataset is not complete for all cutting conditions. On the other hand, approaches such as Neural Networks generally allow for a generalisation of the results when faced with never seen before cutting conditions.

Table 6. Comparison of the overall accuracy, training and inference time for different AI approach.

Approach	Overall accuracy	Training Time	Inference Time
ANN	88.8%	84.870s	0.080s
RF	86.6%	0.650s	0.050s
SVC	80%	<0.008s	<0.001s
k-NN	80%	<0.008s	<0.001s

An important point when using these approaches is also the computational time and resources required to obtain the results. All results presented in this paper are obtained on a single core of an Intel I7-9750H @ 2.6 GHz CPU. The different computational times, including training and validation are presented in Table **6**. The training time considers the approach initialisation and computing time to fit the dataset. The inference time refers to the time needed by the approach to predict the classes presented in this paper. It is observed that the ANN approach has the longest training time compared to the other approaches. This is due to its training scheme. These considerations must be considered in relation to the available computational resources. However, it is important to note that the training only must be done once. With current computational resources, this is not a limitation as even ANN only takes less than 2 minutes to converge. It should also be noted that this interpretation is only valid for the amount of data in this database. For larger databases, the inference time of methods other than ANNs can increases considerably. In general, ANN methods have a longer training time but a relatively short inference time, which is not the case for other methods. Since training only needs to be done once, it is preferable to use methods with a constant low inference time such as ANN.

The position of the misclassification is also important. If the state of a tool is misclassified at the transition between two classes, this misinterpretation is less critical than if the tool is misclassified at middle of the class interval. Indeed, the misclassification reflects an error of some μm which does not significantly impact the quality of the machined surface. Fig. 13 shows the different positions of the wear incorrectly classified for each approach presented in this paper. From the figure, it appears that there are some data that are often misclassified in the middle of class 2. On this aspect, ANN has the lowest error than any other approaches. For each approach, the transition from class 2 to 3 leads to misclassification of the wear. As stated previously, this misclassification is not critical as this is only an error of less than 25 μm in the estimation of Vb. SVC is the only approach that misclassified a very worn tool.

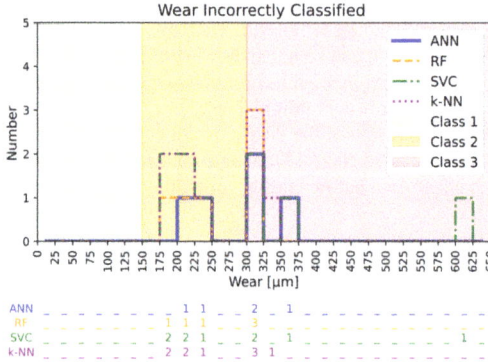

Fig. 13. Misclassified wear class for different techniques.

Summary

Tool condition classification is an effective approach when the tool condition needs to be monitored. artificial intelligence methods are suitable for this purpose, but their performance can vary depending on the approach used. In this paper, a turning database from which cutting signals highly correlated with the wear are used with the following AI classification methods: ANN, RF, SVC, and k-NN. The tool condition is divided into 3 classes, each representing a phase of the tool life. The classes are defined based on the flank wear degradation of the tool (Vb). Each AI approach is optimised to obtain the best achievable results and to be able to compare their relative performances. From the results, it appears that the ANN approach has the best accuracy with an overall accuracy of **88.8%**. The second-best approach that achieves similar results is the RF with an accuracy of **86.6%**. The other approaches have an average accuracy of 80%. Despite having the best accuracy, neural networks have by far the longest learning time (85s) compared to less than 1 s for all the other approaches. This learning time must be considered as with a larger and more complete database, the learning and inference time can greatly grow. Nonetheless, with the actual computing power, these considerations should not represent any obstacle to real applications as the training can be realised off-line. The results presented in this paper are limited to an ideal case to compare the performance of the different approaches presented. In industrial practice, it is necessary to consider the changes in cutting conditions as well as other variables that can influence the process. Nevertheless, the results of this paper allow a comparison of the approaches presented.

Conflicts of interest

The authors declare that they have no conflicts of interest.

References

[1] W. Xu, L. Cao, Optimal tool replacement with product quality deterioration and random tool failure, Int. J. Prod. Res. 53 (2015) 1736-1745. https://doi.org/10.1080/00207543.2014.957878

[2] A. Jeang, Reliable tool replacement policy for quality and cost, European J. Operat. Res. 108, (1998) 334-344. https://doi.org/10.1016/S0377-2217(96)00368-2

[3] F. Klocke, Manufacturing Processes 1, RWTHedition, Springer, Berlin/Heidelberg, Germany, 2011.

[4] D.E.D. Snr, Sensor signals for tool-wear monitoring in metal cutting operations—a review of methods, Int. J. Mach. Tool. Manuf. 40 (2000) 1073-1098. https://doi.org/10.1016/S0890-6955(99)00122-4

[5] M. Kuntoğlu, A. Aslan, D.Y. Pimenov, Ü.A. Usca, E. Salur, M.K. Gupta, T. Mikolajczyk, K. Giasin, W. Kaplonek, S. Sharma, A review of indirect tool condition monitoring systems and decision-making methods in turning: Critical analysis and trends, Sensors 21 (2020) 108. https://doi.org/10.3390/s21010108

[6] A. Siddhpura, R. Paurobally, A review of flank wear prediction methods for tool condition monitoring in a turning process, Int. J. Adv. Manuf. Technol. 65 (2013) 371-393. https://doi.org/10.1007/s00170-012-4177-1

[7] L. Colantonio, L. Equeter, P. Dehombreux, F. Ducobu, A systematic literature review of cutting tool wear monitoring in turning by using artificial intelligence techniques, Machines 9 (2021) 351. https://doi.org/10.3390/machines9120351

[8] ISO 3685—Tool Life Testing with Single-Point Turning Tools. 1993, Available online: https://www.iso.org/fr/standard/9151.html (last accessed 01/12/2022)

[9] T. Zhao, Y. Liu, L. Neves, O. Woodford, M. Jiang, N. Shah, Data augmentation for graph neural networks, in Proceedings of the AAAI Conference on Artificial Intelligence 35 (2021) 11015-11023.

[10] A. Abraham, Artificial Neural Networks, in: Handbook of Measuring System Design, John Wiley & Sons, Ltd.: Chichester, UK, 2005.

[11] Tensorflow, Available online: https://www.tensorflow.org/?hl=en (accessed 01 December 22)

[12] A. Parmar, R. Katariya, V. Patel, A review on random forest: An ensemble classifier, in: International Conference on Intelligent Data Communication Technologies and Internet of Things, Springer, Cham, 2018, pp. 758-763.

[13] Sci-Kit Learn, Available online: https://scikit-learn.org/stable/ (accessed 01 December 22)

[14] I. Steinwart, A. Christmann, Support vector machines, Springer Science & Business Media, 2008.

[15] L. Jiang, Z. Cai, D. Wang, S. Jiang, Survey of improving k-nearest-neighbor for classification, in: Fourth international conference on fuzzy systems and knowledge discovery (FSKD 2007), 1 (2017) 679-683).

Material Forming - ESAFORM 2023
Materials Research Proceedings 28 (2023) 1275-1284

Materials Research Forum LLC
https://doi.org/10.21741/9781644902479-138

Hardware implementation of Monte Carlo and RSM method for the optimization of cutting force during turning of NiTi shape memory alloy

KOWALCZYK Małgorzata[1,a] *

[1]Chair of Production Engineering, Mechanical Faculty, Cracow University of Technology, Avenue Jana Pawła II 37, PL 31-864 Cracow, Poland

[a]malgorzata.kowalczyk@pk.edu.pl

Keywords: NiTi Shape Memory Alloy, Monte Carlo and RSM Methods, Optimization

Abstract. This study was conducted to understand the exact turning of the NiTi shape memory alloy and consisted of four stages: experimental work, function modelling using RSM method, Monte Carlo method optimization and hardware implementation of Monte Carlo method . This article has the following main objectives: to develop a framework for solving machining optimization problems using the Monte Carlo method and hardware implementation of MC method. The solutions presented in this paper are important from the point of view of practical solutions related to the prediction and optimization of the cutting forces components F_c, F_p and F_f during turning of NiTi shape memory alloy.

Introduction

Cutting forces components which have an impact on the tool and on the workpiece causes a certain change in the position of the tool relative to the workpiece. Any displacement of the tool relative to of the workpiece occurring during machining, can adversely affect accuracy dimension of the machined surface. Prediction and optimization of cutting force before real machining can provide significant guidelines to the planning of turning process in particular such difficult to machine of NiTi alloys [1-4].

The shape memory materials are innovative materials with the high application potential. These materials can be used as the smart materials and multifunctional materials due to their memory shape and superelasticity properties [5-7]. But due to their specific properties NiTi alloys are known to be difficult-to-cut materials particularly by using conventional techniques. Their high ductility, high degree of strain hardening, poor thermal conductivity, very low "effective" elastic modulus and unconventional stress–strain behavior are the main properties responsible for their poor machinability [5,6,8,9].

In order to manufacture new products from difficult-to-machine materials, such as shape memory alloys, there is a need to search for more and more effective treatment methods that exceed technological barriers [10-12]. Therefore, an adapted process strategy, prediction and optimization of cutting force is very important when machining NiTi [2,3,13].

The ability to predict and optimization cutting force before machining has attracted great interest from many scientists, being the main goals of many research studies. The prediction and optimization of cutting force is currently determined by using various techniques such as theoretical models [14-15], FE method [4,15-17], the Taguchi procedure [1,2,11,17-20], response surface methodology (RSM) [13, 21-23], the Multi-Objective Ant Lion Optimizer MOALO [21] the multi-response TOPSIS method [3,19], artificial intelligence through the use of the artificial neural networks (ANNs) [15,2-25], genetic algorithms (GAs) [18] and fuzzy logic (FL) [26]. any research works show the use of these methods in the forecasting and also optimization of cutting force [13]. Researchers usually do not use only one modeling approach in their works, but look for

Material Forming - ESAFORM 2023   Materials Research Forum LLC
Materials Research Proceedings 28 (2023) 1275-1284   https://doi.org/10.21741/9781644902479-138

a mutual compilation of the above strategies [18,21,23]. The benefits of using cutting force prediction methods include an increase in the productivity and competitiveness of the production process [1,2,27].

Analyzing the relevant literature regarding the prediction and optimization of machining processes, it can be easily noticed that the current trend is the use of RSM, ANNs, GAs, and the Taguchi procedure for these purposes. The author of this study note that despite the many application possibilities of the MC method [5,10], its application for solving the problems related to the prediction and optimization of machining has not been given much attention in the literature.

Taking into account the above literature review, this paper presents a procedure for model prediction cutting components force (F_f – the feed force, F_p – the thrust force, F_c – the cutting force) in the turning of NiTi alloy (A_f =60°C) with PCD tool. The main objective is to develop a model based on response surface methodology (RSM) to the cutting force in terms of machining parameters such as depth of cut (a_p), cutting speed (v_c) and feed rate (f). The computational model enabled to select optimal machining parameters for minimizing cutting forces values. The generation of mathematical models was necessary for the subsequent optimization with the Monte Carlo method, the purpose of which was to find for which cutting parameters the minimum values of the cutting force components are obtained. The algorithm consisted of performing random draws of input parameters from specified ranges for depth of cut, feed rate and cutting speed in a loop.

The next the hardware implementation of the Monte Carlo method was written in the form of code in the software environment, which was uploaded to the microcontroller.

Material Properties

The NiTi alloy used for the experiment was 57.88Ni-42.12Ti (wt%) obtained from Baoji Hanz Metal Material Co.,Ltd. (China). The diameter of the workpiece was 20 mm. The austenite finish temperature was, A_f =60° C. Table 1 shows the chemical composition (wt. %) of NiTi. The physical, thermal and mechanical properties of the materials of β-TiNi: Tensile Strength, Ultimate, 1364 MPa, Tensile Strength, Yield, 649 MPa, Modulus of Elasticity, 28 GPa, Thermal conductivity, 18 W/m·°C, Hardness, 231 HV, Density, 6500 kg/m³ , Structure (phase), hi-temp B2 [28-29].

Table 1. The spectroscopy (EDS) analysis results.

β-TiNi		
Element	wt.%	at.%
TiK	42.12	47.15
NiK	57.88	52.85
Total	100	100

Cutting tool

The 80° rhombic insert of PCD with single-top corner, brazed tip and a positive rake angle was selected for turning of NiTi. The symbol of insert: CCMT 060202 ID5. The new cutting edge was used for each test sample. Each cutting insert was attached to the designated tool holder SCACR 1616K – 06S. The recommends cutting conditions for insert: depth of cut a_p = 0.08- 3.0 mm, feed f = 0.05 – 0.3 mm/rev;. The literature recommends cutting conditions during precise turning of NiTi alloy; cutting speed v_c = 10-50 m/min; feed rate f <0.2 mm/rev; depth of cut a_p<0.5 mm [6-9].

Experimental design

Turning tests were performed in according Taguchi experiment design at three different cutting parameters (feed f, cutting speeds v_c and depth of cut a_p); Taguchi L9 orthogonal array. For the cutting parameters in turning of NiTi, three factors and three levels, as shown in Table 2 [11]. The parametric design of Taguchi is a very useful technique for elastic design since it provides a simple and systematic empirical effective design at a low cost. It is a method to design an experiment, as it creates the set of arrays with variable factors arranged in such a way that only significant variation is pointed out and the rest insignificant set of variations are neglected thus reducing the number of experiments [2].

Table 2. Cutting parameters.

Parameters	Code	Levels		
		1	2	3
Feed rate, f [mm/rev]	A	0.038	0.058	0.077
Depth of cut, a_p [mm]	B	0.03	0.08	0.13
Cutting speed, v_c [m/min]	C	30	40	50

The following testing equipment was prepared: precise lathe (Masterturn 400), the workpiece (NiTi alloy), CCMT 060202 PCD insert, tool holder SCACR 1616K – 06S, Kistler dynamometer with DynoWere software to visualize measurements [Fig. 1].

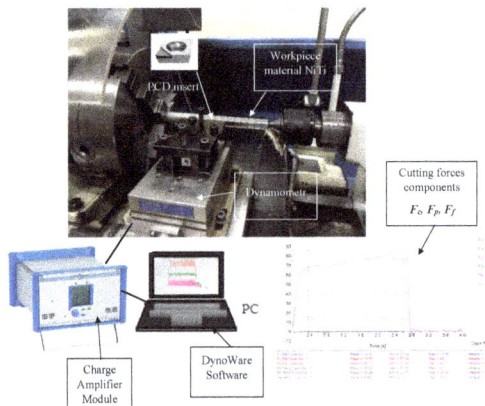

Fig. 1. Experimental set-up for turning of NiTi.

Method

This study was worked out in order to optimize the cutting forces components F_c, F_p and F_f during turning of NiTi shape memory alloy and includes four main stages: experimental work, function modelling using RSM method, Monte Carlo method optimization and hardware implementation of Monte Carlo method. Fig. 2 shows the flow chart which constitutes the overall representation of the methodology developed in the paper. This research has the following main objectives: to develop a framework for solving machining optimization problems using the Monte Carlo method and hardware implementation of MC method.

Experimental work.

The following testing equipment was prepared: precise lathe, work piece (NiTi alloy), tool holder and insert, Kistler dynamometer. Turning tests were carried out in according to Taguchi experiment design (with different feed, cutting speeds and depth of cut). The values of the cutting force components were recorded by the dynamometer and then the cutting force components were in DynoWere software processed (Table 3).

Material Forming - ESAFORM 2023   Materials Research Forum LLC
Materials Research Proceedings 28 (2023) 1275-1284   https://doi.org/10.21741/9781644902479-138

Modelling.

The mathematical relationships between input data and the output parameter-response i.e. cutting force components (F_f, F_p, F_c) Taguchi design was selected. The model for finding parameters of cutting force was appointed using the Response Surface Methodology algorithm.

Monte Carlo Method.

The generation of mathematical models was necessary for the subsequent optimization with the Monte Carlo method, the purpose of which was to find for which cutting parameters the minimum values of the cutting force components are obtained. The algorithm consisted of performing draws of input parameters from specified ranges for depth of cut, feed rate and cutting speed, in a loop.

Hardware implementation.

the Monte Carlo method was written in the form of code in the software environment, which was uploaded to the microcontroller.

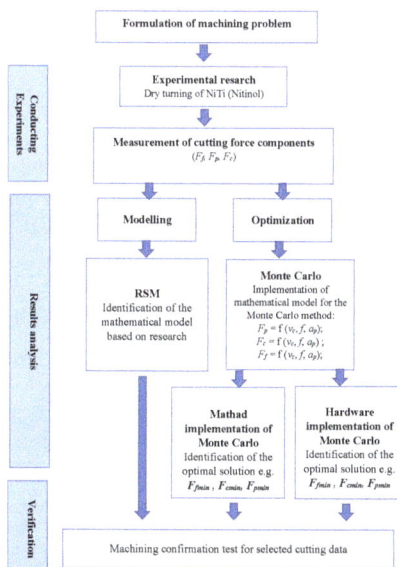

Fig. 2. Flow chart for overall representation of the work methodology.

Results and Discussion

Response surface methodology (RSM).

Response Surface Methodology (RSM) is a series of statistical and mathematical techniques useful for optimizing processes [1,30]. This method is a very effective tool for prediction and modeling the manufacturing problems. It provides more information with a small number of investigations. It is a research strategy to study the limits of input parameters and the emerging experimental statistical model for the measured response, by approximating the existing correlation between the response surface and input process parameters. The limit of the process parameters has to be defined in response surface method, and the first set of experiments was performed to identify machining parameters, which have an impact on the cutting force and to find selected range of cutting parameters [30]. In this study, the cutting speed v_c, depth of cut a_p and the feed f were considered as cutting parameters for monitoring cutting conditions, and the cutting forces components are measured as a response variable (Table 3).

Material Forming - ESAFORM 2023   Materials Research Forum LLC
Materials Research Proceedings 28 (2023) 1275-1284   https://doi.org/10.21741/9781644902479-138

The mathematical model suitable for predicting suitable value is Quadratics model (Eq. 1) where y is a parameter value, describing the investigated physical phenomenon, i.e., the cutting force F_c; b1, b2, …, bi are constant coefficients; and x1, x2, …, xi are factors, which influence the investigated parameter, by considering the full quadratics model as shown in Eq. 2 (F_f – the feed force), Eq. 3 (F_c – the cutting force) and Eq. 4 (F_p – the thrust force).

$$y = bo + b1x1 + b2x2 + b3x3 + b4x4 + b11x12 + b22x22 + b33x32 + b44x42 + b12x1x2 + b13x1x3 + b14x1x4 + b23x3x3 + b24x2x4 + b34x3x4 \tag{1}$$

The mathematical model was established by neglecting the insignificant coefficients of the cuttings forces components (Eq. 2-4):

$$F_f(f, a_p, v_c) = 1.6 + 60.15 \cdot f + 494.26 \cdot a_p - 0.82 \cdot v_c - 5596.04 \cdot f^2 - 3783.73 \cdot a_p^2 \cdot 0.003593 \cdot v_c^2 + 4826.64 \cdot f \cdot a_p + 10.27 \cdot f \cdot v_c \tag{2}$$

$$F_c(f, a_p, v_c) = -381 - 900.2 \cdot f + 2460.6 \cdot a_p + 19.4 \cdot v_c - 14727.8 \cdot f^2 + 7257.6 \cdot a_p^2 - 0.4174 \cdot v_c^2 - 55636.6 \cdot f \cdot a_p + 214.6 \cdot f \cdot v_c \tag{3}$$

$$F_p(f, a_p, v_c) = -24.2 + 1045.5 \cdot f + 1318.1 \cdot a_p - 1.228 \cdot v_c - 17431.4 \cdot f^2 - 4092.9 \cdot a_p^2 \; 0.01768 \cdot v_c^2 - 3756.8 \cdot f \cdot a_p + 39.46 \cdot f \cdot v_c \tag{4}$$

The mathematical model was used to for prediction cutting components force (F_f– the feed force, F_p – the thrust force, F_c – the cutting force) in the turning of NiTi alloy (A_f=60°C) with PCD tool by replacing the values of the machining parameters. The impact of cutting cutting parameters was examined using the developed model.

The information in Table 3 shows the result from the comparison between actual value and forecasting value which found that the forecasting values of mean of the feed force F_f has the maximum error of only 1.67%, for mean of the cutting force F_c – 3,93% and for mean of thrust force F_p – 0.23%.

Table 3. Experimental results.

Test no.	Control factors			Measured value (Mean)			Predicted value RSM			% Error		
	f [mm/rev]	a_p [mm]	v_c [m/min]	F_f [N]	F_c [N]	F_p [N]	F_{fRSM} [N]	F_{cRSM} [N]	F_{pRSM} [N]	F_f	F_c	F_p
1	0.038	0.03	30	3.10	32.70	14.20	3.07	31.42	14,17	0.92	3.90	0.23
2	0.038	0.08	40	14.40	73.70	40.80	14.36	72.01	40,76	0.26	2.29	0.09
3	0.038	0.13	50	7.50	67.50	43.40	7.45	65.41	43,36	0.63	3.09	0.10
4	0.058	0.03	40	2.90	108.50	21.30	2.86	106.81	21,26	1.39	1.56	0.18
5	0.058	0.08	50	21.80	53.30	48.50	21.75	51.20	48.46	0.23	3.93	0.09
6	0.058	0.13	30	19.50	121.30	67.60	19.47	119.98	67.56	0.16	1.09	0.05
7	0.077	0.03	50	3.20	169.00	26.50	3.15	166.89	26.45	1.67	1.25	0.17
8	0.077	0.08	30	20.50	66.30	47.50	20.47	64.98	47.46	0.16	1.99	0.07
9	0.077	0.13	40	26.30	118.80	61.70	26.26	117.06	61.66	0.17	1.46	0.07

Monte Carlo method.

The Monte Carlo method is often used in engineering, finance, statistics and other fields of science. Fig. 3 shows a block diagram of the applied MC method for determining the optimal (minimum) values of the factors as shown in Eq.2 (F_f – the feed force), Eq. 3 (F_c – the cutting force) and Eq. 4 (F_p – the thrust force). The optimization task was solved using Mathcad software and hardware implementation. After optimization using the Monte Carlo method, the optimization results were verified experimentally.

Material Forming - ESAFORM 2023   Materials Research Forum LLC
Materials Research Proceedings 28 (2023) 1275-1284   https://doi.org/10.21741/9781644902479-138

The program implementing the assumptions of the Monte Carlo algorithm was written in the Arduino IDE environment, and then uploaded to the Arduino UNO board with the Atmega 328P microprocessor, which is responsible for performing the calculations necessary to carry out the optimization process. In the first part of the program code 'string' and 'int' variables were initialized. They were used to send numbers (in the form of characters) to the microprocessor, which will define the ranges of input parameters (depth of cut a_p, cutting speed v_c, feed rate f, and number of MC draws). Then 'float' values are declared, which will determine the output values (cutting components force F_f– the feed force, F_p – the thrust force, F_c – the cutting force and the values of optimal cutting parameters (v_c, a_p, f). Auxiliary variables have also been introduced, i.e. m - number of MC draws and time - variable responsible for calculating the time program duration. The 'random' function is responsible for generating pseudo-random numbers from a specific range. The program initialization was included in the void setup() function. The void loop function is defined in the following part( ) in which the main part of the code was placed. It contains the way in which the microcontroller reads character type values entered into its memory. The functions responsible for displaying the ranges of input parameters on the terminal were generated. Next a loop implementing the algorithm was placed in the function void loop() Monte Carlo in the final part. The code contains functions responsible for displaying the results received from the microcontroller on the terminal.

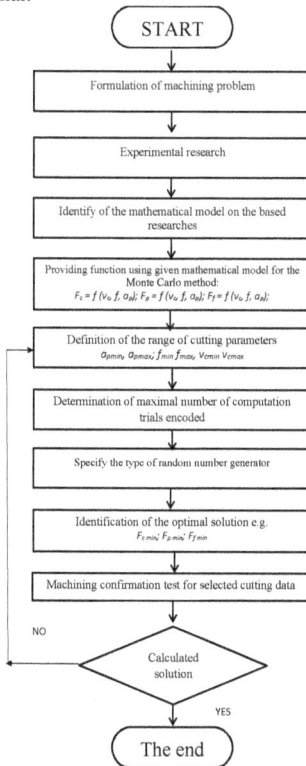

Fig. 3. Monte Carlo method algorithm for solving machining optimization problems [10].
Monte Carlo optimization result.

Table 4 summarizes the results of the calculations of the minimum values (rows 2, 9 and 16) of the functions described by Eq. 2, 3 and 4 and defined by F_f MC min, F_c MC min and F_p MC min for the case of three different numbers of MC draws (the first row): 10^4, 10^5, and 10^6 for Mathcad software and hardware implementation. For the values of the functions F_f MC min, F_c MC min and F_p MC min for all Monte Carlo trials, the corresponding drawing number m (rows 3,10 and 17), the values of parameters f, a_p, and v_c (rows 4–6, 11-13 and 18–20), the measure values of F_f, F_p and F_c (rows 7, 14 and 21) and error of method are determined for Mathcad software and hardware implementation.

Table 4. Calculation results of the minimum value of Eq. 2, Eq. 3 and 4 by using the MC method.

No.	Parameters/Factors	Results MC (Mathcad)			Hardware implementation		
1	MC[no.]	10^4	10^5	10^6	10^4	10^5	10^6
2	$F_f{}^{MC}{}_{min}$	1.028	1.608	1.859	1.037	1.698	1.889
3	m [no.]	4.3	4.787	9.326	4.3	4.787	9.326
4	f [mm/rev]	0.077	0.077	0.077	0.077	0.077	0.077
5	a_p [mm]	0.031	0.03	0.03	0.031	0.03	0.03
6	v_c [m/min]	32.64	30.064	30.042	32.64	30.064	30.042
7	F_f measure	1.81	1.81	1.81	1.81	1.81	1.81
8	%Error	43.20	11.16	2.71	42.71	6.19	4.36
9	$F_c{}^{MC}{}_{min}$	71.697	72.503	82.307	71.89	72.78	82.205
10	m [no.]	3.876	8.194	1.169	3.876	8.194	1.169
11	f [mm/rev]	0.038	0.039	0.038	0.038	0.038	0.038
12	a_p [mm]	0.031	0.034	0.032	0.03	0.03	0.03
13	v_c [m/min]	49.285	49.55	49.967	48.94	48.49	48.87
14	F_c measure	79.25	79.25	79.25	79.25	79.25	79.25
15	%Error	9.53	8.51	3.86	9.29	8.16	3.73
16	$F_p{}^{MC}{}_{min}$	6.229	6.13	7.75	6.34	6.18	7.83
17	m [no.]	3.876	1.291	4.492	3.876	1.291	4.492
18	f [mm/rev]	0.038	0.039	0.038	0.038	0.038	0.038
19	a_p [mm]	0.031	0.03	0.03	0.03	0.03	0.03
20	v_c [m/min]	49.285	49.766	49.792	48.83	48.26	48.95
21	F_p measure	7.71	7.71	7.71	7.71	7.71	7.71
22	%Error	19.21	20.49	0.52	17.77	19.84	1.56

Based on the mathematical model, cutting components force F_f– the feed force, F_p – the thrust force, F_c – the cutting force were optimized using Monte Carlo method. The optimal solution is the following for :
For the feed force F_f *(Mathcad)*: number Monte Carlo trials: 9.326 x10^6, F_f = 1.859 N, f = 0.077 mm/rev, a_p = 0.03 mm, v_c = 30.042 m/min, %error of method: 2.71%.
For the feed force F_f *(hardware implementation)*: number Monte Carlo trials: 9.326 x10^6, F_f = 1.91 N, f = 0.077 mm/rev, a_p = 0.03 mm, v_c = 30.042 m/min, %error of method 4.36% .
For the thrust force F_p *(Mathcad)*: number Monte Carlo trials: 4.492 x10^6, F_p= 7.75N, f = 0.038 mm/rev, a_p = 0.03 mm, v_c = 49.766 m/min, %error of method: 0,52%
For the thrust force F_p *(hardware implementation)*: number Monte Carlo trials: 4.492 x10^6, F_p= 7.82N, f = 0.077 mm/rev, a_p = 0.03 mm, v_c = 48.95 m/min, %error of method: 1.56%
For the cutting force F_c *(Mathcad)*: number Monte Carlo trials: 1.169 x10^6, F_c= 82.307 N, f = 0.038 mm/rev, a_p = 0.032 mm, v_c = 49.967 m/min, %error of method: 3.86%
For the cutting force F_c *(hardware implementation)*: number Monte Carlo trials: 1.169 x10^6, F_c= 81.95 N, f = 0.038 mm/rev, a_p = 0.03 mm, v_c = 48.87 m/min, %error of method: 3.73%

Material Forming - ESAFORM 2023
Materials Research Proceedings 28 (2023) 1275-1284

Materials Research Forum LLC
https://doi.org/10.21741/9781644902479-138

Summary

This paper proposes the Monte Carlo method to predicted and optimized cutting forces based on cutting parameters (cutting speed, feed rate and depth of cut).

The procedure presented in this paper, which is devoted use of Monte Carlo Method for optimization of the cuting forces components (F_c, F_f, F_p) during turning of NiTi alloys, allows for an easy and quick assessment of the obtained results related to the prediction and optimization carried out in the process of conventional machining of NiTi alloys.

The work demonstrates that this method is a quick and effective tool, that enables defining optimal cutting parameters to minimize the cutting force during turning of NiTi alloys.

The important findings are summarized as follows:

1) Prediction accuracy of cutting forces by MC method developed models is efficient both for implementation of MC method in Mathcad software like and hardware implementation of MC method.

2) Comparison of experimental and predicted values of the cutting forces show that a good agreement has been achieved between them (Table 4). The maximum method error was 4.36% for MC draws 10^6. The method is more effective for more MC draws.

3) The Monte Carlo optimization show that the optimal combination of cutting parameters are found to be cutting speed of 50 m/min, feed rate of 0.038 mm/rev, cutting depth of 0.03 mm for the thrust force F_p and the cutting force F_c and cutting speed of 30 m/min, feed rate of 0.077 mm/rev, cutting depth of 0.03 mm for feed force F_f.

In this study, the optimization methodology proposed is a powerful approach and can offer to scientific researchers as well industrial metalworking a helpful optimization procedure for various combinations of the workpiece and the cut material tool. The procedure for the cutting force optimization during the turning proposed in this paper can be used to optimize costs and ensure maximum efficiency during applications of turning for NiTi alloy.

References

[1] M. Kowalczyk, Cutting force prediction in ball-end milling of Ni-Ti alloy, Proc. of SPIE 11176 (2019) 11176650-1-11176650-12. https://doi.org/10.1117/12.2536757

[2] R. Agrawal, N. Kumar, K. Parvez, A. Srivastava, M. Sarfaraz Alam, Optimization of cutting force via variable feed rate in dry turning lathe of AISI 304, Mater. Today Proc. 64 (2022) 1182-1187. https://doi.org/10.1016/j.matpr.2022.03.479

[3] S. Lakshmanan, M. Pradeep Kumar, M. Dhananchezian, Optimization of turning parameter on surface roughness, cutting force and temperature through TOPSIS, Mater. Today Proc. in press (2022). https://doi.org/10.1016/j.matpr.2022.09.209

[4] A. Tzotzis, N. Tapoglou, R. K. Verma, P. Kyratsis, 3D-FEM Approach of AISI-52100 Hard Turning: Modelling of Cutting Forces and Cutting Condition Optimization, Machines 10 (2022) 1-16. https://doi.org/10.3390/machines10020074

[5] M. Kowalczyk, K. Tomczyk, Procedure for Determining the Uncertainties in the Modeling of Surface Roughness in the Turning, Mater. 13 (2020) 1-14. https://doi.org/10.3390/ma13194338

[6] B. Dash, M. Das, M. Das, T.R. Mahapatra, D. Mishra, A concise review on machinability of NiTi shape memory alloys, Mater. Today Proc. 18 (2019) 5141-5150. https://doi.org/10.1016/j.matpr.2019.07.511

[7] E. Kaya, I. Kaya, A review on machining of NiTi shape memory alloys: the process and post process perspective, Int. J. Adv. Manuf. Technol. 100 (2019) 2045-2087. https://doi.org/10.1007/s00170-018-2818-8

[8] H. Shizuka, K. Sakai, H. Yang, K. Sonoda, T. Nagare, Y. Kurebayashi, K. Hayakawa, Difficult cutting property of NiTi alloy and its mechanism, J. Manuf. Mater. Process. 4 (2020) 1-15. https://doi.org/10.3390/jmmp4040124

[9] Y. Zhao, K. Guo, J. Li, J. Sun, Investigation on machinability of NiTi shape memory alloys under different cooling conditions, Int. J. Adv. Manuf. Technol. 116 (2021) 1913-1923. https://doi.org/10.1007/s00170-021-07563-9

[10] M. Kowalczyk, Application of the Monte Carlo method for the optimization of surface roughness during precise turning of NiTi shape memory alloy, Proc. of SPIE 10808 (2018) 58 108084P-1 - 108084P-9. https://doi.org/10.1117/12.2501421

[11] M. Kowalczyk, Application of Taguchi method to optimization of surface roughness during precise turning of NiTi shape memory alloy, Proc. of SPIE 10445 (2017) 104455G-1 104455G-11. https://doi.org/10.1117/12.2281062

[12] Y.Z. Zhao. K. Guo, V. Sivalingam, J.-F. Li, Q.-D. Sun, Z.-J. Zhu, J. Sun, Surface integrity evolution of machined NiTi shape memory alloys after turning process, Adv. Manuf. 9 (2021) 446-456. https://doi.org/10.1007/s40436-020-00330-1

[13] H. Zahia, Y.M. Athmane, B. Lakhdar, M. Tarek, On the application of response surface methodology for predicting and optimizing surface roughness and cutting forces in hard turning by PVD coated insert, Int. J. Ind. Eng. Comput. 6 (2015) 267-284. https://doi.org/10.5267/j.ijiec.2014.10.003

[14] Y. Sun, F. Ren, D. Guo, Z. Jia, Estimation and experimental validation of cutting forces in ball-end milling of sculptured surfaces, Int. J. Mach. Tools Manuf. 49 (2009) 1238-1244. https://doi.org/10.1016/j.ijmachtools.2009.07.015

[15] Y. Song, H. Cao, W. Zheng, D. Qu, L. Liu, C. Yan, Cutting force modeling of machining carbon fiber reinforced polymer (CFRP) composites: A review, Compos. Struct. 299 (2022). https://doi.org/10.1016/j.compstruct.2022.116096

[16] R.A. Mali, M.D. Agrahari, T.V.K. Gupta, FE based simulation and experimental validation of forces in dry turning of aluminium 7075, Mater. Today Proc. 27 (2019) 2319-2323. https://doi.org/10.1016/j.matpr.2019.09.120

[17] C.S. Sumesh, A. Ramesh, Optimization and finite element modeling of orthogonal turning of Ti6Al4V alloys: A comparative study of different optimization techniques, Eng. Solid Mech. 11 (2023) 11-22. https://doi.org/10.5267/j.esm.2022.11.002

[18] S.M. Abdur Rob, A.K. Srivastava, Turning of Carbon Fiber Reinforced Polymer (CFRP) Composites: Process Modeling and Optimization using Taguchi Analysis and Multi-Objective Genetic Algorithm, Manuf. Lett. 33 (2022) 29-40. https://doi.org/10.1016/j.mfglet.2022.07.012

[19] N. A. Sristi, P. B. Zaman, N. R. Dhar, Multi-response optimization of hard turning parameters: a comparison between different hybrid Taguchi-based MCDM methods, Int. J. Interact. Des. Manuf. 16 (2022) 1779-1795. https://doi.org/10.1007/s12008-022-00849-6

[20] N. Li, Y.J. Chen, D.D. Kong, Multi-response optimization of Ti-6Al-4V turning operations using Taguchi-based grey relational analysis coupled with kernel principal component analysis, Adv. Manuf. 7 (2019) 142-154. https://doi.org/10.1007/s40436-019-00251-8

[21] B. Hamadi, M.A. Yallese, L. Boulanouar, M. Nouioua, A. Hammoudi, RSM-based MOALO optimization and cutting inserts evaluation in dry turning of AISI 4140 steel, Struct. Eng. Mech. 84 (2022) 17-33. https://doi.org/10.12989/sem.2022.84.1.017

[22] A. Thakur, V. Guleria, R. Lal, Multi-response optimization in turning of EN-24 steel under MQL, Eng. Res. Express. 4 (2022). https://doi.org/10.1088/2631-8695/ac7a0c

[23] R. Kluz, W. Habrat, M. Bucior, K. Krupa, J. Sęp, Multi-criteria optimization of the turning parameters of Ti-6Al-4V titanium alloy using the Response Surface Methodology, Eksploat. i Niezawodn. 24 (2022) 668-676. https://doi.org/10.17531/ein.2022.4.7

[24] T. Palaniappan, P. Subramaniam, Experimental Investigation and Prediction of Mild Steel Turning Performances Using Hybrid Deep Convolutional Neural Network-Based Manta-Ray Foraging Optimizer, J. Mater. Eng. Perform. 31 (2022) 4848-4863. https://doi.org/10.1007/s11665-021-06552-z

Material Forming - ESAFORM 2023   Materials Research Forum LLC
Materials Research Proceedings 28 (2023) 1275-1284   https://doi.org/10.21741/9781644902479-138

[25] D. Peng, H. Li, Y. Dai, Z. Wang, J. Ou, Prediction of milling force based on spindle current signal by neural networks, Measurement 205 (2022) 112153. https://doi.org/10.1016/j.measurement.2022.112153

[26] A. Alalawin, W.H. AlAlaween, M.A. Shbool, O. Abdallah, L. Al-Qatawneh, An interpretable predictive modelling framework for the turning process by the use of a compensated fuzzy logic system, Prod. Manuf. Res. 10 (2022) 89-107. https://doi.org/10.1080/21693277.2022.2064359

[27] E. Altas, O. Erkan, D. Ozkan, H. Gokkaya, Optimization of Cutting Conditions, Parameters, and Cryogenic Heat Treatment for Surface Roughness in Milling of NiTi Shape Memory Alloy, J. Mater. Eng. Perform. 31 (2022) 7315–7327. https://doi.org/10.1007/s11665-022-06769-6

[28] Y. Kaynak, H.E. Karaca, R.D. Noebe, I.S. Jawahir, Tool-wear analysis in cryogenic machining of NiTi shape memory alloys: A comparison of tool-wear performance with dry and MQL machining, Wear 306 (2013) 51-63. https://doi.org/10.1016/j.wear.2013.05.011

[29] K. Weinert, V. Petzoldt, D. Kötter, Turning and drilling of NiTi shape memory alloys, CIRP Ann. - Manuf. Technol. 53 (2004) 65-68. https://doi.org/10.1016/S0007-8506(07)60646-5

[30] M. Subramanian, M. Sakthivel, K. Sooryaprakash, R. Sudhakaran, Optimization of cutting parameters for cutting force in shoulder milling of Al7075-T6 using response surface methodology and genetic algorithm, Procedia Eng. 64 (2013) 690-700. https://doi.org/10.1016/j.proeng.2013.09.144

Material Forming - ESAFORM 2023   Materials Research Forum LLC
Materials Research Proceedings 28 (2023) 1285-1294   https://doi.org/10.21741/9781644902479-139

Investigation of surface finish and chip morphology in cryogenic machining biomedical grade polyetheretherketone

SAJJA Nikhil Teja[1,a*], BERTOLINI Rachele[1,b], GHIOTTI Andrea[1,c] and BRUSCHI Stefania[1,d]

[1]Department of Industrial Engineering, University of Padova, Via Venezia 1, 35131, Padova, Italy

[a]nikhilteja.sajja@studenti.unipd.it, [b]rachele.bertolini@unipd.it, [c]andrea.ghiotti@unipd.it, [d]stefania.bruschi@unipd.it

Keywords: Cryogenic Machining, PEEK, Surface Finish, Chip Morphology

Abstract. Polyetheretherketone (PEEK) is attracting the attention of the biomedical field, thanks to its high biocompatibility and wear resistance. Nevertheless, the attainment of good quality surfaces when machining PEEK is still challenging. In this framework, the aim of the paper is to investigate the viability of using cryogenic machining to enhance the surface finish of PEEK compared to the outcomes of dry cutting. To do that, turning trails were executed at varying cooling strategy and depth of cut, and the resultant surface finish and chip morphology were evaluated. The obtained results indicated that cryogenic machining carried out at the highest depth of cut greatly enhanced the PEEK machinability.

Introduction

Polyetheretherketone (PEEK) is a thermoplastic semi-crystalline polymer characterized by an elasticmodulus close to that of human bones, high biocompatibility, stable chemical resistance, and good wearresistance [1]. Thanks to these characteristics, PEEK has been used as a biomaterial in the biomedical area for orthopedic implants since 1987 [2] and is still assessed as one of the most promising polymers for biomedical applications. The conventional route to manufacture PEEK parts comprises machining steps carried out without any cutting fluid to avoid oil residues on the machined surface. Davim and Mata

[3] performed dry turning trials on PEEK bars and studied the effect of PCD tool inserts at varying cuttingspeed and feed. They found that the cutting speed increase and feed decrease contributed to reduce the surface roughness. Abdullah et al. [4] dry-turned the biomedical grade PEEK using carbide tools and optimized the machining parameters, namely cutting speed, feed, and depth of cut to minimize the surfaceroughness. However, dry machining of PEEK is still challenging as the increase in temperature at the cutting zone usually prompts a drastic reduction of the surface quality. A reduced surface finish increasesfriction and accelerates polymer wear rate during the lifespan of the implant: an example is given by acetabular cups, which must be characterized by smooth surfaces when coupled with femoral heads to increase the hip implant performances [5].

To this extent, cryogenic machining can be seen as a possible alternative to dry cutting, which can effectively remove the cutting-generated heat, whilst still preserving the cleanliness of the machined surfaces.

Bertolini et al. [6] investigated the machinability of the polyamide 6 under cryogenic cooling conditions and showed smoother and harder surfaces compared to the ones obtained using a conventionalcutting fluid. Aldwell et al. [7] submerged ultra-high molecular weight polyethylene (UHMWPE) in liquid nitrogen for nearly 24 hours before machining: this process chain was found to lead to a stiffer surface compared to dry conditions. Dhokia et al. [8] froze ethylene-vinyl acetate using liquid nitrogen showing the polymer surface hardening, which, in turn, allowed a 1% dimensional error after machining.Putz et al. [9] studied the influence of cryogenic cooling on

Material Forming - ESAFORM 2023   Materials Research Forum LLC
Materials Research Proceedings 28 (2023) 1285-1294   https://doi.org/10.21741/9781644902479-139

cutting forces and surface integrity of the nitrile-butadiene-rubber elastomer. It was found that, by using the cryogenic coolant, the cutting forceswere increased, but the surface deformation was reduced, promoting a better surface quality compared to that obtained under dry cutting conditions.

According to the literature survey, cryogenic machining of PEEK has been scarcely researched, in particular as regards the impact of the depth of cut on the polymer machinability. In this context, the paper investigates the effect of cryogenic cooling on the PEEK surface finish and the onset of surface defects in comparison with the outcomes from dry cutting. Further, the chip morphology was analyzed and correlated to the machined surface characteristics.

Experimental - Material

The polymer material under investigation was the TECAPEEK^{TM} supplied by Ensinger Plastics. Thisis a biomedical grade used specifically to fix spinal traumas, cruciate ligaments, and meniscal tears. Thematerial was purchased in form of a bar of 42 mm diameter. The PEEK mechanical and thermal properties in the as-received state are listed in Table 1.

Table 1. PEEK mechanical and thermal properties in the as-received state [10].

Property	PEEK
Modulus of elasticity (MPa)	4200
Tensile strength at yield (MPa)	116
Hardness ASTM D2240 (Shore D)	85
Glass transition temperature, T_g (°C)	150
Melting temperature, T_m (°C)	341
Density (g/cm^3)	1.31
Thermal conductivity (W/(k·m))	0.27
Specific heat (J/(g·K))	1.1

Machining Trials

The Mori Seiki^{TM} NL 1500 CNC lathe with the experimental apparatus shown in Fig. 1 (a) was usedto carry out the turning tests. A left-hand tool insert, namely the VCEX 11 03 01L-F 1125 purchased from Sandvik Coromant^{TM}, was used, see Table 2 for its nomenclature. The cutting speed (Vc) and feed (f) were selected on the basis of the tool manufacturer's guidelines and kept fixed for all the turning trials.On the contrary, it was chosen to vary the depth of cut (ap) to evaluate its impact on the PEEK machinability.

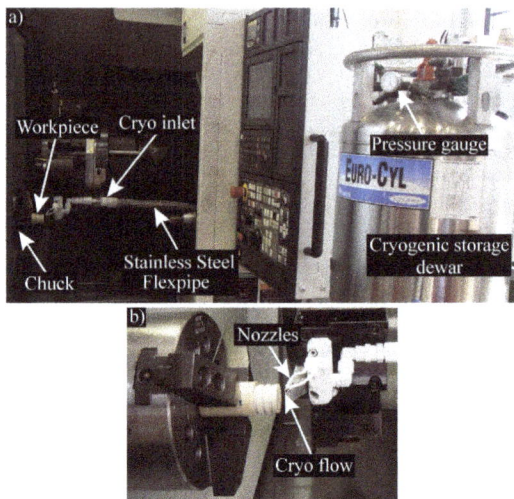

Fig. 1. (a) Machining setup; (b) Closer view of the cutting zone when applying cryogenic cooling.

Table 2. Cutting tool nomenclature and specifications.

ISO code	VCEX 11 03 01L-F 1125
Cutting edge effective length (mm)	10.971
Corner radius (mm)	0.1
Major cutting edge angle (°)	93
Insert thickness (mm)	3.175
Rake angle (°)	5.5
Clearance angle major (°)	7

At first, the external layer of the as-received PEEK bar was removed to eliminate any defect and dimensional error. On this basis, a turning step was performed to achieve a 40 mm diameter by using a cutting speed of 100 m/min, feed of 0.1 mm/rev, and depth of cut of 1 mm.

The turning tests were conducted under both dry and cryogenic cooling conditions with three repetitions for each experimental condition. The liquid nitrogen was stored in a Dewar tank at 15 bars. Two copper nozzles of 0.9 mm diameter were used to spray the liquid nitrogen to the tool flank and rakefaces simultaneously (see Fig. 1 (b)). In each experimental run, the approximate lead time for liquid nitrogen to stabilize was 30 seconds. Table 3 reports the experimental plan of the machining trails.

Material Forming - ESAFORM 2023   Materials Research Forum LLC
Materials Research Proceedings 28 (2023) 1285-1294   https://doi.org/10.21741/9781644902479-139

Table 3. Cutting parameters.

Cutting speed (m/min)	200
Feed (mm/rev)	0.1
Depth of cut (mm)	0.15, 0.25, 0.5
Cooling condition	Dry, Cryogenic
Liquid nitrogen mass flow rate (Kg/s)	0.0058
Length of cut (mm)	5

Surface Quality and Chip Morphology Characterization

The Sensofar PLu Neox^{TM} optical profiler with a 20x magnification Nikon^{TM} confocal objective wasused to analyze the machined surface finish. The scanned surfaces were then assessed using the SensoView^{TM} software.

According to the ISO 25178-2:2012 standard, the surface texture was evaluated in terms of surface roughness, namely the arithmetical mean height (*Sa*), reduced peak height (*Spk*), and reduced valley depth (*Svk*). These surface texture parameters were selected as indicative of the machining quality. According to [11], *Spk* can be related to the wear properties of a surface, while *Svk* to characteristics such as fluid retention properties. The *Spk*/*Svk* ratio was also evaluated to assess the distribution of valleys and peaks on the machined surface.

Scanning electron microscope (SEM) analysis was carried out using a FEI^{TM} QUANTA 450 to evaluate the possible presence of defects on the machined surfaces. To do such analysis, the samples were gold-sputtered with 25 mA for 3 minutes using the Denton Vacuum^{TM} Desk V machine.

After each turning trial, the chips were collected and analyzed using light optical microscope (LOM)and SEM with the aim of evaluating their morphology at varying cutting parameters. In addition, the cutting ratio (*r*) was calculated on the basis of Eq. 1:

$$r = \frac{t_0}{t_c}$$

(1)

where t_o is the uncut chip thickness, equal to the feed, and t_c the chip thickness measured using LOM.

Results and Discussion

Surface texture of the machined samples.

Fig. 2 reports the average values of *Sa* along with their standard deviation bars as a function of the depth of cut and cooling strategy.

In the case of dry cutting, a clear trend cannot be identified since the surface roughness was improvedby varying *ap* from = 0.15 mm to = 0.25 mm, and later it was increased at *ap*=0.5 mm, On the contrary,when using the cryogenic cooling strategy, a strong reduction of the surface roughness was observed at increasing depth of cut. 25% and 36% decreases were found for *ap*=0.25 mm and *ap*=0.5 mm with respectto the lowest depth of cut, respectively. At the highest depth of cut, the section of the chip was increasedas well as its brittleness when deformed at lower temperatures. This made its breakage easier, as it will be later documented.

Regardless of the depth of cut, cryogenic cooling was effective in reducing the surface roughness compared to dry cutting. This is due to the fact that the cryogenic fluid cools down the polymer materialvery rapidly, reducing its elongation at rupture and, therefore, overall ductility.

Material Forming - ESAFORM 2023   Materials Research Forum LLC
Materials Research Proceedings 28 (2023) 1285-1294   https://doi.org/10.21741/9781644902479-139

On the contrary, in case of dry cutting, the much higher ductility induced by the higher cutting temperatures results in uneven surfaces, as the material is too easily deformed.

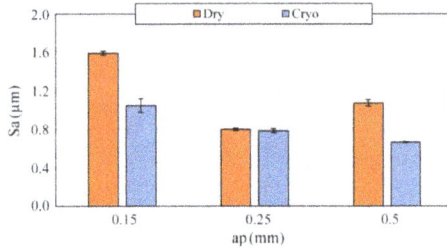

Fig. 2. Sa under dry and cryogenic cooling conditions as a function of the depth of cut.

Further, *Svk* and *Spk* values were analyzed and the results are reported in Fig. 3. For both the coolingstrategies, except at the highest depth of cut that induced both peaks and valleys almost characterized bythe same height, wider valleys and shallow peaks were produced compared to the lowest depth of cut. Both the height of the valleys and depths of the peaks were reduced at increasing depth of cut, regardlessof the cooling strategy. Specifically, the cryogenic strategy assured the heights of the peaks increased by37% and 6% for ap=0.15 mm and ap=0.25 mm compared to ap=0.5 mm, respectively. Similarly, the depths of the valleys increased by 271% and 124% for ap=0.15 mm and ap=0.25 mm compared to ap=0.5mm, respectively.

Cryogenic machining helped in reducing the height of the peaks and especially in increasing the depthof the valleys. The latter was increased by 35%, 12% and 7% at ap=0.15 mm, ap=0.25 mm, and ap=0.5mm compared to the dry cases, respectively.

Having a surface characterized by shallow peaks and deep valleys may help to increase the surface wear resistance as this kind of surface can assure both fluid accessibility and accumulation of the wear debris. In general, deep valleys function as fluid reservoirs to decrease friction and wear by reserving and supplying fluid to the bearing surfaces. Additionally, wear debris can accumulate within deep valleyfeatures. Consequently, abrasive wear produced by third-body wear particles can be reduced [12].

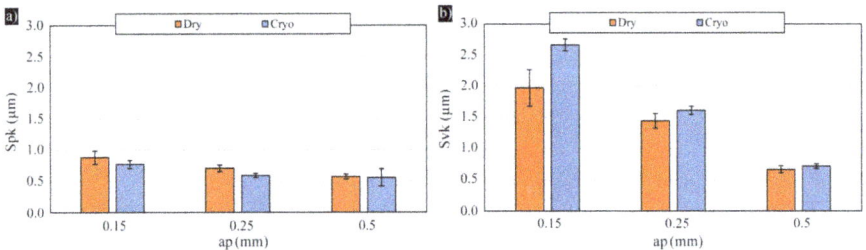

Fig. 3. Spk and Svk under dry and cryogenic cooling conditions as a function of the depth of cut.

The *Spk/Svk* ratio was calculated to understand the peculiar nature of the surface texture. In particularthis ratio refers to the average distributed ratio of the peaks and valleys on the machined surfaces. In all the investigated cases, the lowest depth of cut and the use of cryogenic cooling led to the *Spk/Svk* reduction, meaning that the surface is dominated by valleys, see Fig. 4.

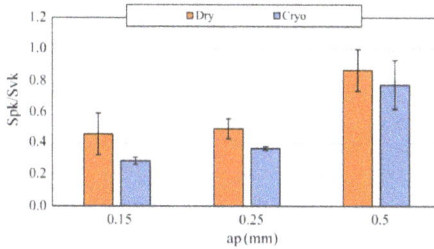

Fig. 4. Spk/Svk under dry and cryogenic cooling conditions as a function of the depth of cut.

The SEM images of the machined surfaces at varying depth of cut are given in Fig. 5. Regardless of the depth of cut, grooves adjacent to feed marks, flakes, and tearing caused by micro-peeling were observed in dry conditions. The highest density of defects was found at the lowest depth of cut, whereas the lowest density of defects was at the intermediate depth of cut.

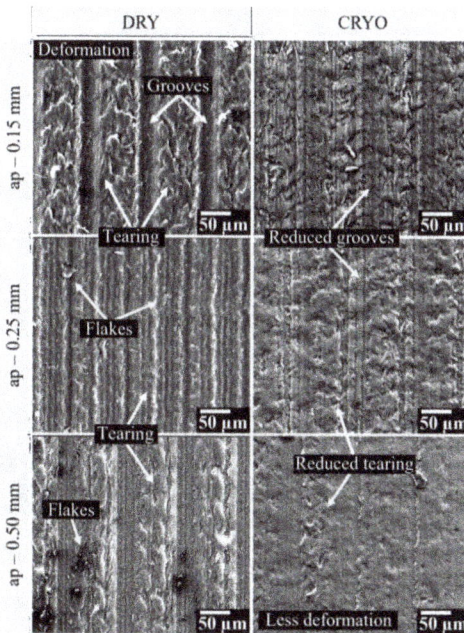

Fig. 5. Dry and cryogenic machined surfaces appearance as a function of the depth of cut.

When applying cryogenic cooling, defects were significantly reduced: the grooves disappeared as well as the tearings. On the other hand, the cryogenic machined surfaces were characterized by bump-shaped features. A nearly defects-free surface was obtained at the highest depth of cut. These findings on the surface appearance well match the surface roughness data shown in Fig. 2.

The attainment of surfaces with a lower amount of defects in the case of cryogenic machining can beascribed to the cutting temperature. In fact, dry machining can drastically increase the polymer temperature in the cutting zone, which induces a significant increase in its ductility, leading to the material tearing instead of a proper cutting [13].

The chip morphology was assessed as well to be correlated with the surface finish. Fig. 6 and Fig. 7 show the chips at different magnifications as a function of the cutting conditions. The chip morphology was determined according to the ISO 3685-1977 (E) standard [14]. In dry machining, the increase in thedepth of cut led to a change of the chip morphology from continuous tubular snarled to continuous ribbonsnarled chips. This can be ascribed to the increase of ductility given by the temperature increases when adopting more severe process parameters. This is confirmed by the higher magnification images reportedin Fig. 7, which show the presence of a higher amount of shear marks that are indicative of attainment ofhigh strain in the chip.

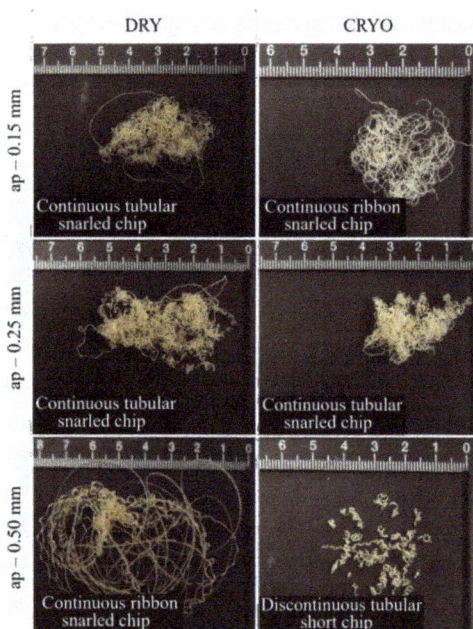

Fig. 6. Morphology of the chips obtained under dry and cryogenic cooling conditions as a function of the depth of cut.

In the case of cryogenic machining, continuous ribbon snarled chips, continuous tubular snarled chips,and discontinuous tubular short chips were formed at ap=0.15 mm, ap=0.15 mm, and ap=0.5 mm, respectively. The breakage of the chip at the highest depth of cut can be ascribed to the synergistic presence of low temperatures and high stresses due to the highest material removal rate. Discontinuous chips can indeed be produced when high compressive stresses are involved or when a brittle material is machined [15].

Actually, Fig. 7 evidences that the shear marks are more emphasized and more frequently observed in the case of the highest depth of cut as a result of the brittleness of the material as a

Material Forming - ESAFORM 2023   Materials Research Forum LLC
Materials Research Proceedings 28 (2023) 1285-1294   https://doi.org/10.21741/9781644902479-139

consequence of the low temperature during cryogenic machining. A fracture can be more likely to nucleate from these shearbands, and then propagate leading to the macroscopic rupture of the chip.

Similar, even if not equal, evidences were found at the lowest and intermediate depths of cut: even though cryogenic machining did not induce a change in the chip morphology, it reduced the bulkiness and lengths of the formed chips with the respect to the dry case.

The presence of discontinuous chips assures the formation of a surface characterized by a lower amount of defects as shown in Fig. 5. If the cutting temperature is kept well below the polymer rubbery region, the polymer is subjected to a lower deformation, which reduces defects like tearings.

It is worth underlining that discontinuous chips are usually desired in turning operations because they are less likely to entangle on the machine surfaces and machine tools [5].

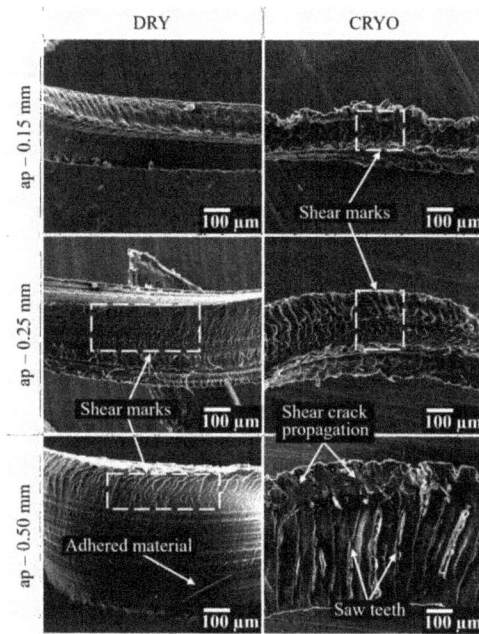

Fig. 7. SEM images of free surfaces of the chips obtained under dry and cryogenic cooling conditions as a function of the depth of cut.

The cutting ratio (r) can be considered a machinability index, being influenced by the nature of the chip-tool interaction, chip contact length, and morphology.

Fig. 8 reports the cutting ratio as a function of the process parameters. Regardless of the depth of cut, the cutting ratio is higher under cryogenic cooling conditions compared to dry ones. To a higher cutting ratio corresponds a higher shear angle and a lower shear plane area that leads to a decrease in the shear force needed to form the chip. This can be attributed to a reduction in the friction angle thanks to the application of cryogenic cooling. This in accordance with [16], where it was stated that the lower the friction coefficient the higher the shear angle.

Only in the case of dry cutting, a slight cutting ratio increase was observed at increasing depth of cut. This confirms the predominance of the temperature effects over the mechanical ones in a dry environment.

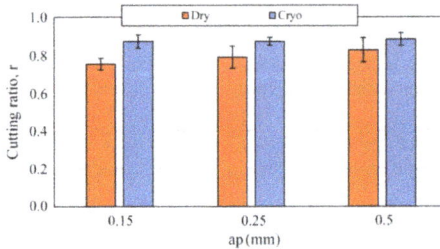

Fig. 8. Cutting ratio under dry and cryogenic cooling conditions as a function of the depth of cut.

Summary

The paper investigated the influence of the depth of cut and coolant strategy on the surface finish andchip morphology in machining biomedical grade PEEK.
The following results can be drawn:

- At increasing depth of cut, smoother surfaces were obtained in cryogenic machining, whereas noclear effect was found in dry machining.
- Cryogenic machining led to a decrease in the height of the peaks, resulting in a surface finish improvement, compared to dry machining. On the other hand, at increasing depth of cut, the height of the peaks was reduced, with consequent improvement of surface finish, regardless of the cooling condition.
- Under cryogenic cooling condition and at the lowest depth of cut, a reduction in the *Spk/Svk* ratiowas observed.
- Cryogenic machining lowered the amount of surface defects compared to dry machining. An increase in the depth of cut under cryogenic cooling condition contributed to improve the machined surface quality.
- The chip breakage was obtained solely when the highest depth of cut and cryogenic cooling condition were applied. For the other depths of cut considered, cryogenic machining reduced the bulkiness and lengths of the formed chips with the respect to the dry case.
- The cutting ratio values under cryogenic cooling condition were higher than those obtained underdry conditions at all the depths of cuts.
- A slight increase in the cutting ratio was observed at increasing depth of cut, solely in the case ofdry cutting.

Acknowledgements

This project has received funding from the European Union's Horizon 2020 research and innovationprogram under the Marie Skłodowska-Curie grant agreement No. 956097.

References

[1] Y. Li, X. Cheng, S. Ling, G. Zheng, H. Liu, He, Study of milling process basics for the biocompatible PEEK material, Mater. Res. Exp. 7 (2020) 015412. https://doi.org/10.1088/2053-1591/AB6A5A

[2] S. Kurtz, PEEK Biomaterials Handbook, 2012. https://doi.org/10.1016/C2010-0-66334-6

Material Forming - ESAFORM 2023
Materials Research Proceedings 28 (2023) 1285-1294

Materials Research Forum LLC
https://doi.org/10.21741/9781644902479-139

[3] J.P. Davim, F. Mata, Chemical vapour deposition (CVD) diamond coated tools performance in machining of PEEK composites, Mater. Des. 29 (2008) 1568-1574. https://doi.org/10.1016/j.matdes.2007.11.002

[4] R. Izamshah, A. Long, M.A. Md Ali, M. Kasim, M.H. Abu Bakar, S. Subramonian, Optimisation of Machining Parameters for Milling Polyetheretherketones (PEEK) Biomaterial, Appl. Mech. Mater. 699 (2014) 198-203. https://doi.org/10.4028/www.scientific.net/AMM.699.198

[5] K. Arora, A.K. Singh, Magnetorheological finishing of UHMWPE acetabular cup surface and its performance analysis, Mater. Manuf. Process. 35 (2020) 1631-1649. https://doi.org/10.1080/10426914.2020.1784928

[6] R. Bertolini, A. Ghiotti, S. Bruschi, Machinability Of Polyamide 6 Under Cryogenic Cooling Conditions, Procedia Manuf. 48 (2020) 419-427. https://doi.org/10.1016/j.promfg.2020.05.064

[7] B. Aldwell, J. O'Mahony, G.E. O'Donnell, The Effect of Workpiece Cooling on the Machining of Biomedical Grade Polymers, Procedia CIRP 33 (2015) 305-310. https://doi.org/10.1016/j.procir.2015.06.058

[8] V.G. Dhokia, S.T. Newman, P. Crabtree, M.P. Ansell, A methodology for the determination of foamed polymer contraction rates as a result of cryogenic CNC machining, Robot. Comput.-Integr. Manuf. 26 (2010) 665-670. https://doi.org/10.1016/j.rcim.2010.08.003

[9] M. Putz, M. Dix, M. Neubert, G. Schmidt, R. Wertheim, Investigation of Turning Elastomers Assisted with Cryogenic Cooling, Procedia CIRP 40 (2016) 631-636. https://doi.org/10.1016/j.procir.2016.01.146

[10] Biocompatible PEEK medical grade - TECAPEEK MT natural, Ensinger, Available: https://www.ensingerplastics.com/en/shapes/products/medical-peek-tecapeek-mt-natural (accessed 06 February 2023)

[11] M. Giordano, F. Villeneuve, L. Mathieu, Product Lifecycle Management: Geometric Variations. 2013. https://doi.org/10.1002/9781118557921

[12] Q. Allen, B. Raeymaekers, Surface Texturing of Prosthetic Hip Implant Bearing Surfaces: A Review, J. Tribol. 143 (2021) 040801. https://doi.org/10.1115/1.4048409

[13] R. Ghosh, J. Knopf, D. Gibson, T. Mebrahtu, G. Currie, Cryogenic Machining of Polymeric Biomaterials: An Intraocular Lens Case Study, in: In Medical Device Materials IV: Proceedings of the Materials &Processes for Medical Devices Conference, 2007, p. 54. https://doi.org/10.1361/cp2007mpmd054

[14] Z. Viharos, S. Markos, C. Szekeres, ANN-based chip-form classification in turning, In Proceedings of the XVII. IMEKO World Congress-Metrology in the 3rd Millennium. (2003) 1469-1473.

[15] M. Kaiser, F. Fazlullah, S. Ahmed, A comparative study of characterization of machined surfaces of some commercial polymeric materials under varying machining parameters, Journal of Mechanical Engineering, Automation and Control Systems. 1 (2020) 75-88. https://doi.org/10.21595/jmeacs.2020.21643

[16] H. Autenrieth, M. Weber, J. Kotschenreuther, V. Schulze, D. Löhe, P. Gumbsch, J. Fleischer, Influence of friction and process parameters on the specific cutting force and surface characteristics in micro cutting, Proceedings of the 10th CIRP International Workshop on Modeling of Machining Operations 6 (2007) 539-548. https://doi.org/10.1080/10910340802518728

Material Forming - ESAFORM 2023
Materials Research Proceedings 28 (2023) 1295-1302

Materials Research Forum LLC
https://doi.org/10.21741/9781644902479-140

Cutting force in peripheral milling of cold work tool steel

TAMURA Shoichi [1,a] * and MATSUMURA Takashi [2,b]

[1]Ashikaga University, 268-1 Omae-cho, Ashikaga-shi, Tochigi, 326-8558, Japan

[2]Tokyo Denki University, 5 Senjyu Asahi-cho, Adachi-ku, Tokyo, 120-8551, Japan

[a]tamura.shoichi@g.ashikaga.ac.jp, [b]tmatsumu@cck.dendai.ac.jp

Keywords: Cutting Force, Residual Stress, Surface Finish, Heat Treatment

Abstract. Tool steels have been commonly applied to die and mold parts because of the high strength and the high abrasive wear resistance obtained by the heat treatments. In order to achieve high machining rate, the heat-treated tool steels have recently finished with end mills coated by hard thin layers. The cutting force in milling of the tool steel should be controlled to improve the fatigue lives of die and mold, which are associated with not only the surface qualities but also the microstructure in the subsurface. This paper discusses the cutting process in milling of a cold work tool steel in terms of the cutting force and the residual stress in finished subsurface.

Introduction

Tool steels have been applied to die and mold manufacturing due to their excellent mechanical strengths, and wear resistances. Milling of tool steel has been usually performed before quenching of workpiece, then the hardened product has been finished in polishing so far. Because hard milling without polishing is required to improve productivity and product quality, advanced cutting tool and coating materials, such as gradient cemented carbide with different grain sizes, has been developed [1].

Durakbasa et al. [2] performed milling of AISI H13 hot work tool steel to optimize end cutting parameters by Taguchi method. The authors summarized that an end mill coated with AlTiN showed better performance regarding the surface finish and the tool wear resistance in comparison to TiAlCN and ZrN coatings. Aramcharoen et al. investigated the effects of coating materials in milling of hardened tool steel H13 with a 0.5 mm diameter micro end mill [3]. They concluded TiN coated tool was effective to control the tool wear, the chipping, the surface finish, and the burr formation.

Furthermore, the die and molding manufacturing has recently required for control of the affected layer and the residual stress in subsurface in terms of lives of die and mold. Denkena et al. conducted hard millings for AISI H13 [4]. They reported that the undercut geometry of the flank face reduced the tensile residual stress and the tool wear. Caruso et al. investigated the residual stress in subsurface in dry orthogonal cutting of AISI 52100 steel [5]. The large compressive residual stress was obtained in cutting of a harder workpiece with a chamfered tool. Although a large number of experiments have been conducted to investigate the cutting characteristics of tool steels so far, the analytical works have not yet been done so much to discuss the cutting process of hardened tool steels in terms of the properties of surface and subsurface.

The paper studies the cutting process in peripheral milling of a hardened cold tool steel comparing with a carbon steel. The cutting tests are conducted to measure the cutting force and the residual stresses and observe the surface finish. The cutting processes, then, are discussed in an analytical force model. The residual stress in subsurface was associated with the shear plane cutting model.

Material Forming - ESAFORM 2023   Materials Research Forum LLC
Materials Research Proceedings 28 (2023) 1295-1302   https://doi.org/10.21741/9781644902479-140

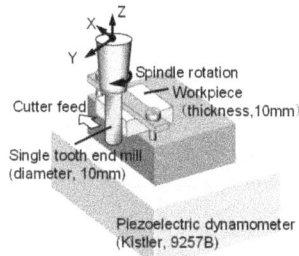

Fig. 1 Cutting test in peripheral milling.

Table 1 Cutting conditions.

Axial depth of cut [mm]	10
Radial depth of cut [mm]	0.1
Cutting speed [m/min]	25
Feed rate [mm/tooth]	0.05
Cutting direction	Down-cut
Lubrication	Dry

Cutting Test

The cutting tests in peripheral milling were conducted on a 3-axis machining center (FANUC Robodrill α-T14iF), as shown in Fig. 1. A 10 mm thick plate of tool steel (JIS, SKD11) was employed as a workpiece for the cutting test. The workpiece was quenched and tempered at 1020°C and 220°C in a vacuum furnace, and controlled the hardness of 54 HRC before the cutting test. In addition, a plate of carbon steel (JIS, S50C) with a controlled hardness of 14 HRC was also used as a reference. The workpiece was clamped on a piezoelectric dynamometer (Kistler, 9257B) mounted on the machine table to measure the cutting force. A single tooth square end mill coated with AlCN thin layer, which was ground off one tooth from the original two teeth end mill, was employed to exclude the influence of the cutter runout. The tool diameter and the helix angle were 10 mm and 30°, respectively. The end mill was clamped to the spindle of the machine tool with a collet chuck. The reference surface was machined to control the radial depth of cut before the cutting test. The end mill was fed along +X direction in down-cut peripheral milling at an axial depth of cut of 10 mm, and a radial depth of cut of 0.1 mm, as shown in Table 1. In order to reduce the vibration occurring in the initial cutting state with the sharp edge, the flank wear land was control so as to contact the workpiece in a width of approximately 20 μm by preliminary cutting, where the cutting speed, the feed rate, and the feed travel were 25 m/min, 0.1 mm/tooth, and 30 mm, respectively. The low cutting speed for milling of steel was set to reduce tool vibration, even though the cutting speed was possibly associated with the surface finish. Then, the residual stresses of the machined surfaces were measured in the feed direction (X-axis) and the axial direction (Z-axis) to compare with the residual stress measured before machining, where the machined surfaces were not processed for the measurements. The residual stresses were measured with an X-ray residual stress analyzer (Pulstec, μ-X360s) based on the cos α method [6]. The measurements for each workpiece were conducted at three points on the center line of workpiece thickness.

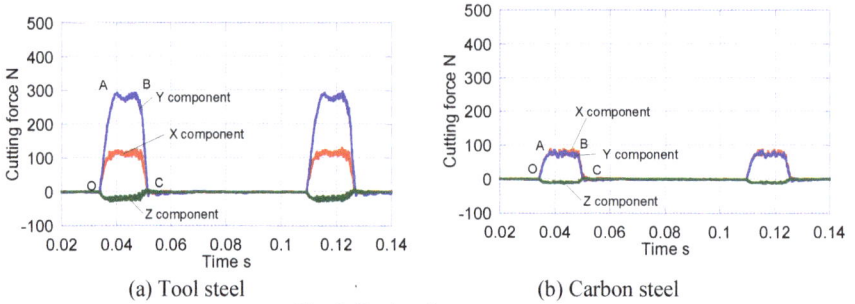

(a) Tool steel   (b) Carbon steel

Fig. 2 Cutting forces.

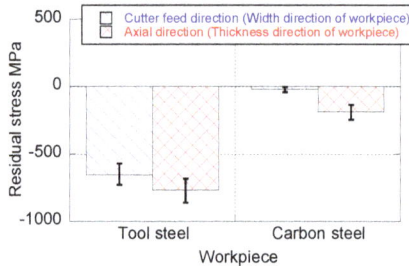

Fig. 3 Residual stresses in subsrface.

Cutting Force in Milling of Cold Work Tool Steel

Figure 2 compares the measured cutting forces in milling of tool steel and carbon steel at a cutting speed of 25 m/min and a feed rate of 0.05 mm/tooth. The cutting force changes periodically in 0.75 s of the tool rotation cycle. Since the cutting area of the helical end mill changes with the cutter rotation, the cutting force changes as:

1) the lowest point on the edge penetrates the bottom of the workpiece plate at O;
2) cutting force increases with the cutting area in O−A;
3) the cutting area moves upward and the cutting force becomes constant without changing the cutting area in A−B;
4) the cutting force decreases in B−C with the cutting area after the cutting edge exits from the top of the workpiece plate.

The large cutting force in milling of tool steel is confirmed because of the higher hardness, where the X and Y components of tool steel are 1.5 and 3.9 times larger than those of carbon steel. The Y component becomes significantly large in comparison to the X component in milling of tool steel. It suggests that the indentation effect of the edge radius is greater than that in milling of carbon steel, as described later.

Surface Finish and Residual Stress

Fig. 3 shows the residual stresses of tool steel and carbon steel, where the positive value is estimated as tensile. The compressive residual stress of tool steel is much larger than that of carbon steel. The surface finishes and the height distributions were observed with a laser microscope (Keyence, VK9700), as shown in Fig. 4. The relatively better surface was obtained in milling of tool steel, as shown in Fig. 4(a), even though chatter marks are slightly left on the finished surface. Regarding the height distributions of the surface finishes, as shown in Fig. 4(b) and (d), the height differences are 9 and 20 μm tool steel and carbon steel, respectively. The built-up edge may overcut the workpiece in milling of carbon steel.

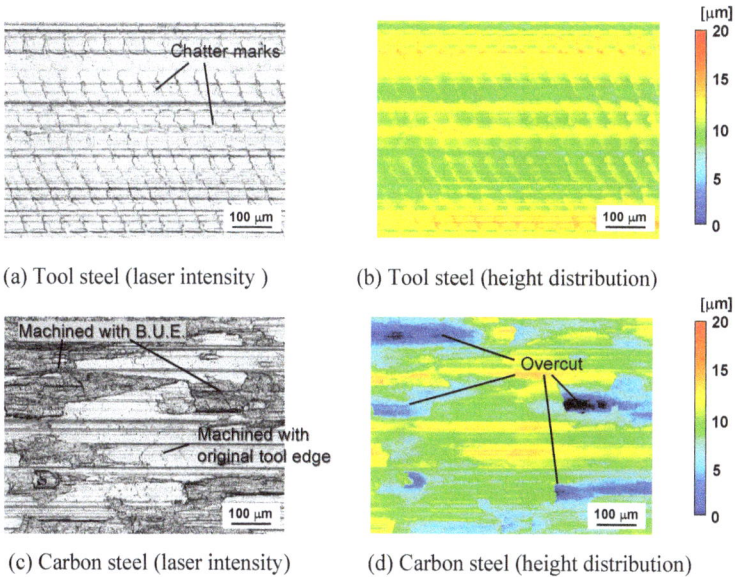

(a) Tool steel (laser intensity)   (b) Tool steel (height distribution)

(c) Carbon steel (laser intensity)   (d) Carbon steel (height distribution)

Fig. 4 Surface finishes.

Analytical Force Model in Milling

A force model is applied to characterize the cutting process in peripheral milling of tool steel. The force model in milling was presented in Reference [7]. The force model is briefly described here. A three-dimensional chip flow in milling is interpreted as a piling up of the orthogonal cuttings in the planes containing the cutting velocities V and the chip flow velocities V_c, as shown in Fig. 5. Although plastic deformation occurs in the chip formation, the interaction between each orthogonal cutting plane is ignored on the assumption that a rigid chip flows at an angular velocity. The orthogonal cutting models are given by the following equations:

$$\left.\begin{array}{l} \phi = \exp(A_{00}V + A_{01}t_1 + A_{02}\alpha + A_{03}) \\ \tau_s = \exp(A_{10}V + A_{11}t_1 + A_{12}\alpha + A_{13}) \\ \beta = \exp(A_{20}V + A_{21}t_1 + A_{22}\alpha + A_{23}) \end{array}\right\} \quad (1)$$

where ϕ, τ_s and β are the shear angle, the shear stress on the shear plane, and the friction angle. V, t_1 and α are the rake angle, the cutting velocity, and the uncut chip thickness in the orthogonal cutting. A_{ij} ($i = 0,1,2$; $j=0,1,2,3$) are parameters acquired in the orthogonal cutting tests. The data can also be identified or refined with referring to the actual cutting force in milling by inverse analysis [8]. The cutting force should be regarded as the sum of the indentation force component due to the ploughing and the chip generation force component due to shearing in the shear zone and friction on the rake face. For the sake of simplicity, this study employs the orthogonal cutting data expressed as Eq. (1) on the assumption that the chip generation force component is the major effect here, where the indentation force component is implicitly associated with the parameters for the uncut chip thickness. Regarding friction on the rake face, the friction angle associated with the friction coefficient is controlled by the third equation in Eq. (1). Because the orthogonal cutting data are acquired in the actual cutting tests for the combination of the tool and the workpiece materials, friction in the interface between the tool face and the chip is characterized in the actual cutting. The shear angle and the shear stress on the shear plane depend on the material behavior. The thermal effect on the material behavior may be mainly controlled by the first term related to cutting velocity dependency, which is associated with the cutting temperature. Because the cutting energy is consumed into the shear energy in the shear plane and the friction energy on the rake face, the cutting energy is estimated in the chip flow model consisting of the orthogonal cuttings. Because the cutting energy depends on the chip flow direction, the chip flow angle is determined to minimize the cutting energy. Therefore, the cutting force is predicted in the determined chip flow model.

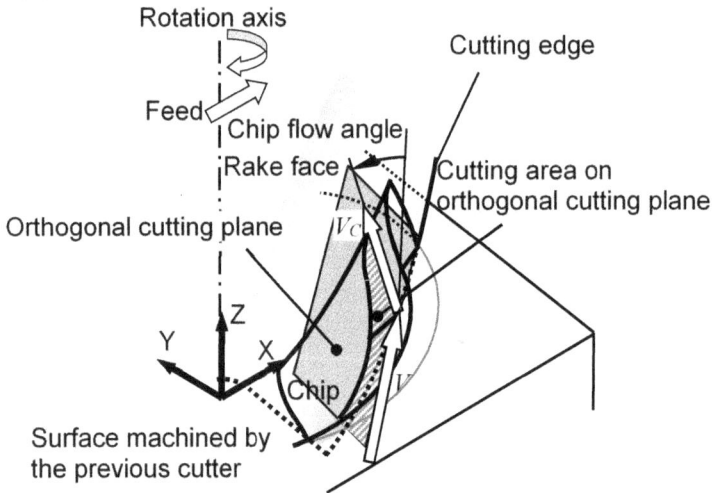

Fig. 5 Chip flow model in analysis.

Material Forming - ESAFORM 2023
Materials Research Proceedings 28 (2023) 1295-1302

Materials Research Forum LLC
https://doi.org/10.21741/9781644902479-140

Cutting Model in Cold Work Tool Steel

Based on the results of the cutting tests, the orthogonal cutting data were acquired as:
Tool steel:

$$
\left.\begin{array}{l}
\phi = \exp(0.189V + 34380t_1 - 6.781\alpha - 0.429) \\
\tau_s = \exp(0.046V - 2172t_1 + 0.002\alpha + 21.25) \\
\beta = \exp(-0.095V - 27280t_1 + 1.795\alpha + 0.1696)
\end{array}\right\} \tag{2}
$$

Carbon steel:

$$
\left.\begin{array}{l}
\phi = \exp(0.254V - 11450t_1 + 0.487\alpha - 3.037) \\
\tau_s = \exp(-0.001V - 2121t_1 + 0.009\alpha + 20.10) \\
\beta = \exp(-0.042V - 5414t_1 - 1.357\alpha - 0.266)
\end{array}\right\} \tag{3}
$$

Figure 6 compares the simulated and the measured cutting forces at the same cutting conditions in Fig. 2. The force model and the orthogonal cutting data are verified in the agreement of the

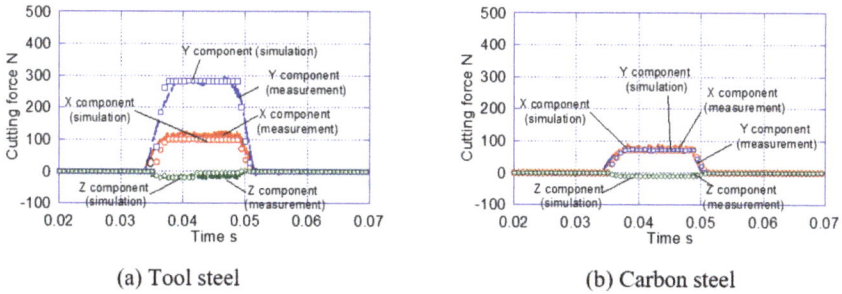

(a) Tool steel   (b) Carbon steel

Fig. 6 Cutting force simulations.

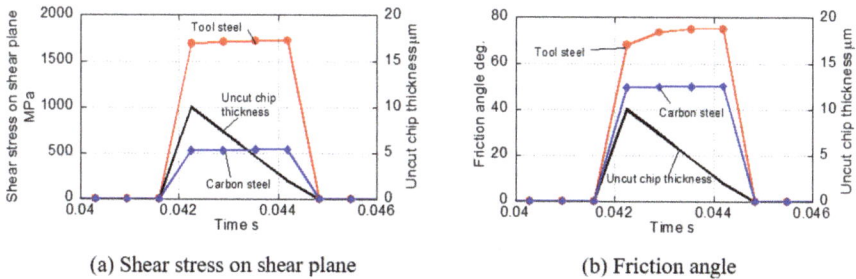

(a) Shear stress on shear plane   (b) Friction angle

Fig. 7 Changes in cutting models at a center of axial depth of cut.

simulated cutting force with the measured force. Figure 7 shows the change in shear stress on the shear plane and the friction angle at the center of the plate thickness of the workpiece, where the uncut chip thickness is also shown as a reference. The uncut chip thickness decreases with the cutter rotation in down-cut process. The shear stress on the shear plane in milling of tool steel reaches about 3.3 times larger than that of carbon steel. The shear stresses of both materials are

Material Forming - ESAFORM 2023   Materials Research Forum LLC
Materials Research Proceedings 28 (2023) 1295-1302   https://doi.org/10.21741/9781644902479-140

nearly constant for the change in the uncut chip thickness. Then, the friction angle of tool steel becomes about 1.4 times larger than that of carbon steel at the edge engagement and increases with decreasing the uncut chip thickness. Meanwhile, the friction angle of carbon steel does not increase significantly. Since the coefficients in the second terms of uncut chip thickness in friction angle in equations (2) and (3) involve the indentation effect of the edge radius, the result shows that the indentation effect of the cutting edge in milling of tool steel is relatively much larger than that of carbon steel. Burnishing on the interface between the finished surface and the cutting edge promotes the compressive residual stress in subsurface [9]. Therefore, in milling of tool steel, a large compressive residual stress is induced by burnishing with a large indentation force on the edge radius. On the other hand, the original tool edge does not finish the machined surface and does not promote the burnishing effect on the machined surface in milling of carbon steel, because the built-up edge generates the overcut surface, as shown in Fig. 4(d). The compressive residual stress in subsurface of carbon steel is not expected with less burnishing.

Summary

The paper has been studied the peripheral milling of hardened cold work tool steel to characterize the cutting process. The results are summarized as:
1) In milling of tool steel, the cutting force components in the feed and radial directions of the tool are 1.5 and 3.9 times greater than those of carbon steel.
2) According to the analytical cutting simulation, the shear stress on the shear plane and the friction angle of tool steel becomes much larger than that of carbon steel. Because of the large negative terms of uncut chip thickness in friction angle, the indentation effect of the cutting edge in milling of tool steel is larger than that of carbon steel
3) The large indentation force on the edge radius enhances burnishing to induce a large compressive residual stress in milling of tool steel. In contrast, the original tool edge does not finish the machined surface and does not promote the effect of burnishing on the machined surface in milling of carbon steel, because the built-up edge generates the overcut surface. As consequence, the effect of burnishing is small in milling of carbon steel.

References

[1] V.F.C. Sousa, F.J.G. Silva, Recent advances on coated milling tool technology-a comprehensive review, Coatings, 10 (2020), 3, https://doi: 10.3390/coatings10030235

[2] M.N. Durakbasa, A. Akdogan, A.S. Vanli, and A.G. Bulutsuz, Optimization of end milling parameters and determination of the effects of edge profile for high surface quality of AISI H13 steel by using precise and fast measurements, Measurement 68 (2015) 92-99, https://doi: 10.1016/j.measurement.2015.02.042

[3] A. Aramcharoen, P.T. Mativenga, S. Yang, K.E. Cooke, and D.G. Teer, Evaluation and selection of hard coatings for micro milling of hardened tool steel, Int J Mach Tools Manuf. 48 (2008), 14, 1578–1584, http://doi: 10.1016/j.ijmachtools.2008.05.011

[4] B. Denkena, J. Köhler, and B. Bergmann, Development of cutting edge geometries for hard milling operations, CIRP J Manuf Sci Technol, 8 (2015), 43-52, https://doi: 10.1016/j.cirpj.2014.10.002

[5] S. Caruso, D. Umbrello, J.C. Outeiro, L. Filice, and F. Micari, An experimental investigation of residual stresses in hard machining of AISI 52100 steel, Procedia Engineering, 19 (2011), 67-72. https://doi: 10.1016/j.proeng.2011.11.081

[6] M. Matsuda, K. Okita, T. Nakagawa, and T. Sasaki, Application of X-ray stress measurement for residual stress analysis by inherent strain method - Comparison of cosα and sin2Ψ method-, Mechanical Engineering Journal, 4 (2017), 5, 17-00022-17–00022, https://doi: 10.1299/mej.17-00022

Material Forming - ESAFORM 2023   Materials Research Forum LLC
Materials Research Proceedings 28 (2023) 1295-1302   https://doi.org/10.21741/9781644902479-140

[7] T. Matsumura and E. Usui, Predictive cutting force model in complex-shaped end milling based on minimum cutting energy, Int J Mach Tools Manuf, 50 (2010), 5, 458-466, https://doi: 10.1016/j.ijmachtools.2010.01.008

[8] T. Matsumura, T. Shirakashi, and E. Usui, Adaptive cutting force prediction in milling processes, International Journal of Automation Technology, 4(2010), 3, 221-228, https://doi: doi.org/10.20965/ijat.2010.p0221

[9] H. Mizutani and M. Wakabayashi, Influence of cutting edge shape on residual stresses of cut surface (Effects of rake angle and contact width of flank face), Transactions of the Japan society of mechanical engineers. C, 72 (2006), 715, 247-251

Material Forming - ESAFORM 2023
Materials Research Proceedings 28 (2023) 1303-1312

Materials Research Forum LLC
https://doi.org/10.21741/9781644902479-141

Evaluation of the surface integrity characteristics of internal threads machined on lead-free brass alloy

ZOGHIPOUR Nima[1,2,a] *, TORAMAN Yaren[1,b], KARA Koray[1,c], BAS Kaan Can[1,2,d] and KAYNAK Yusuf[2,e]

[1]Torun Metal Alloy Ind. & Ltd. Inc., GOSB, İhsan Dede St. No: 116 41480 Gebze,Kocaeli, Turkey

[2]Department of Mechanical Engineering, Marmara University, 34722 Goztepe Campus-Istanbul, Turkey

[a]nima.zoghipour@gmail.com, [b]yaren.toraman@torunmetal.com, [c]koray.kara@torunmetal.com, [d]kaan.bas@torunmetal.com, [e]yusuf.kaynak@marmara.edu.tr

Keywords: Tapping, Threading, Surface Integrity, Lead-Free Brass Alloy

Abstract. Tapping is a typical machining method being utilized in manufacturing internal threads. However, it contains few difficulties to be dealt with since during the process, many cutting edges involve synchronized axial and rotational movements. This work presents the results of an experimental study on the material type considering the influences of the manufacturing process (extrusion and forging) and tap type on the surface integrity characteristics of lead-free brass alloys. Experimental data on the surface quality of the threads, subsurface characteristics including microhardness is presented and analyzed. This study demonstrates that the manufacturing methods have a robust effect on microstructural aspects of low-lead brass alloy and resulting in hardening of the surface and subsurface.

Nomenclature			
Ext.	Extruded	F+Ann.	Forged+Annealed

Introduction

In line with advancing technology, the necessity for brass alloys in industry is rising daily. A variety of these alloys are employed in the production of the parts that are primarily used in the food, water, and pumping industries. These parts are manufactured by extrusion or hot forging processes and go through machining processes such as threading in order to be assembled. Addition of some elements into the chemical composition results in enhanced machinability, one of which is Pb [1]. However, due to its use in drinking water, the element's hazardous properties have resulted in the adoption of rigorous regulations and the widespread use of Pb [2]. As a result, recent years have seen a surge in hopes for the creation of lead-free brass alloys due to the rapid advancement of technology. However, the issue that modern industry must face to stay competitive is the manufacture of consumer goods, which is built on a triangle consisting of low costs, low production times, and excellent quality. A new problem that needs to be resolved in this situation is the poor machinability. Industrial methods for producing profiled parts are constantly evolving and competing for market share based on both product quality and price. The utilization of efficient processes is required to achieve these goals [3]. Using both cut and form taps, internal threads can be produced. Cut tapping is a type of machining; chip removal produces the thread. When form tapping is used, the work material is the sole thing that moves to create the thread.

Material Forming - ESAFORM 2023   Materials Research Forum LLC
Materials Research Proceedings 28 (2023) 1303-1312   https://doi.org/10.21741/9781644902479-141

According to Fromentin et al. [4], the main benefit of form tapping is that no chips are produced. The authors claim that by doing this, time-consuming cleaning of the threaded area can be avoided because chips can cause interference when the components are assembled. In another study, the plastic strain and the high strain stress produced by the form tapping were explored by Fromentin et al. [5]. The authors came to the conclusion that form tapping is a good alternative to the conventional tapping procedure. However, it might be argued that the production of forming threads only has a significant potential for materials with high ductility. This is due to the external form tapping's superior surface and stronger threads compared to conventional tapping for thread profiles [6]. The size of the split crest at the top of the thread following the form tapping operation will depend on the diameter of the hole. Both high ductile alloys, such as non-ferrous metals [7–10], and hardened steels, as demonstrated by [7, 11], can be formed using this method. On AISI 304 austenitic stainless steel, tapping operations were performed by Uzun et al. [12] utilizing taps of various sizes and cutting parameters. The depth of cut for each cutting feed and the cutting torques generated during threading were used to calculate the cutting performances of the taps. In the tapping forming process of the 7075-T6 aluminum alloy, Oliveira et al. [13] examined the effects of the tapping speed, tap coating, and the tapered region of the tap. They used a sophisticated statistical analysis to determine the impact of the key factors on the thrust force, torque, and connection of these factors with microhardness and thread profile quality. Burr development during the form tapping process was investigated by Filho et al. [14]. With uncoated and coated taps, internal threads in workpieces made of the 7075-aluminum alloy were completed, and the burr formation at the entrance and exit was examined. The impact of new vegetable oil formulations on the precision of AISI 316L parts and the integrity of their surfaces during reaming and tapping were examined by Belluco et al. [15]. They studied surface roughness and surface metallurgy as part of their investigation of surface integrity. Profilometry was used to evaluate roughness, while optical metallography and microhardness testing were used to investigate the subsurface. The surface characteristics of the threads produced by form tapping were investigated by Fromentin et al. [4]. The two main factors affecting the process are the work material's tensile strength and the lubricant's influence, and a relationship with tapping torque was suggested. In their study, Coelho et al. [16] concentrated on the usage of a minimum quantity of lubricant and emulsion during the machine and form tapping of 7075-T651 aluminum alloy while using both tapping techniques at three forming/cutting speeds. Numerous studies on the methods of tapping various materials have been conducted [17-20].

A comparison of the various types of tools on brass alloys has not been done, according to the research that is currently available in the literature. This paper presents the findings of an experimental investigation on the material type taking into account the influences of the manufacturing method (extrusion and forging) and tap type on the surface integrity properties of lead-free brass alloys. Experimental data on cutting forces, torques, thread surface quality, geometrical accuracy, and subsurface parameters including microhardness are presented and examined.

Methodology
Materials.
Extruded CuZn40Pb2 (CW617N), CuZn38As (CW511L), and CuZn21Si3P (CW724R), each with a diameter of 60 mm and a length of 30 mm, were the three different types of brass alloys employed as test materials in this study. In a Hydromec 550 eccentric press machine with a press load of 165 tons, a portion of these materials was hot forged at $730\pm10°C$ at a rate of 11 strokes per minute. The forged specimens underwent a 150 min. annealing process at 550°C. Table 1 and 2 list the chemical composition and mechanical properties of the test materials.

Material Forming - ESAFORM 2023   Materials Research Forum LLC
Materials Research Proceedings 28 (2023) 1303-1312   https://doi.org/10.21741/9781644902479-141

Table 1. Chemical composition of the studied brass alloys [21].

Composition		Cu	Zn	Pb	Sn	Fe	Ni	Al	As	Mn	P	Si
CW617N	%Min.	57.0	Rem.	1.6	-	-	-	-	-	-		-
	%Max.	59.0	Rem.	2.5	0.3	0.3	0.3	0.05	-	-	-	-
CW511L	%Min.	61.5	Rem.	-	-	-	-	-	0.02	-	-	-
	%Max.	63.5	Rem.	0.2	0.1	0.1	0.3	0.05	0.15	-	-	-
CW724R	%Min.	75.0	Rem.	-	-	-	-	-	-	-	0.02	2.7
	%Max.	77.0	Rem.	0.09	0.3	0.3	0.2	0.05	-	0.05	0.10	3.5

Table 2. Mechanical properties of the studied brass alloys [21].

Material	E(GPa)	ρ (g/cm^3)	Machinability
CW617N	96	8.43	%95
CW511L	100	8.25	%80
CW724R	100	8.41	%40

Experimental setup.

The internal threading operation using Gühring's C type M10x1.5 form and cut taps on a Fanuc α-D21LiB5 CNC milling center was the main focus of the machining experiments. Fig. 1 depicts the experimental configuration. Through all holes, the experiments were the primary focus and coolant fluid was used. Throughout the testing, the spindle rotational speed was maintained at constant 800 rpm. The cutting forces were measured for 15 seconds at a sampling frequency of 100 Hz using a Kisteler dynamometer type 9129AA. Using a Keyence digital optical microscope, a Mitutoyo Contracer CV-2100M4, and other tools, the integrity of the machined surfaces was assessed. According to ASTM E 384 standards, measurements were taken using an MHVD 1000 IS microhardness tester at various depths beneath the machined surfaces. The dimensional accuracy of the threads was controlled by M10 tap gauge as well as Keyence digital optical microscope according to the illustrated Metric ISO thread dimensions as illustrated in Fig. 2.

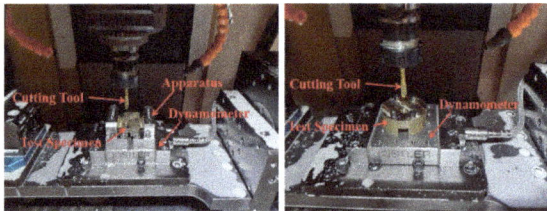

Fig. 1. The experimental setup; a) Ext., b) F+Ann. specimen.

Material Forming - ESAFORM 2023   Materials Research Forum LLC
Materials Research Proceedings 28 (2023) 1303-1312   https://doi.org/10.21741/9781644902479-141

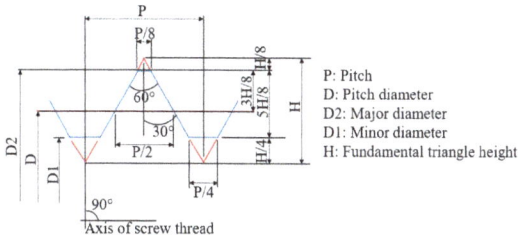

Fig. 2. Metric ISO thread control dimensions.

Results and Discussions

Microstructural Analysis.

Fig. 3 illustrates the microstructure of the Ext. and F+Ann. CW617N, CW511L, and CW724R brass alloys. α and β phases are observed in CW617N and CW511L, while α, γ and κ are seen in CW724R. The β phase content for CW617N has increased from%46 to %51 as a result of multi-directional hot forging and annealing. For CW511L, this rate was gauged between %5 and %17. This augmentation and elongation can be attributed to the phase's superplastic deformation characteristic during the forging process. In CW724R, κ and γ phases have increased from %16 to %19. The looming up of κ phase and the smaller percentages of γ phases compared to κ phases make it seem possible to ignore them and conclude that forging. Table 3 illustrates the measured average grain sizes of two test specimens. The lowest grain size was measured for F+Ann. CW724R as equal to 6.44 µm. The highest grain size was observed for Ext. CW511L as 8.13 µm. It is seen that the average grain size has been reduced by hot forging followed by annealing due to the movement of the dislocations, boundaries, breakages in grain structure and replacement with finer grains.

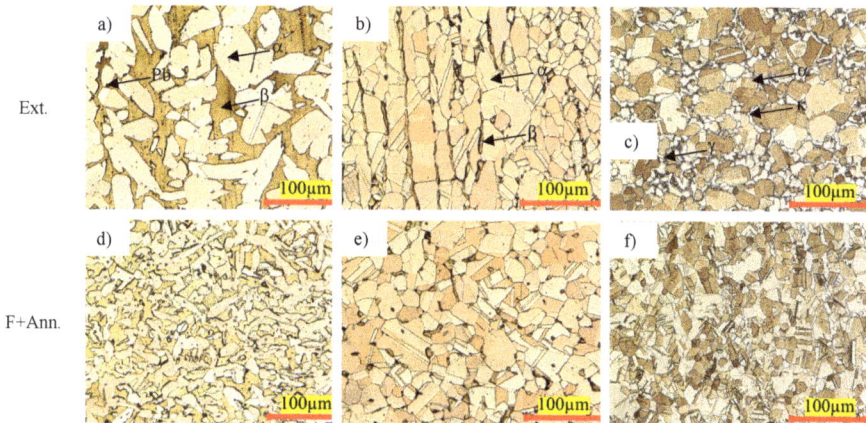

*Fig. 3. Microstructure of the test speciemens; Ext.; a) CW617N, b) CW511L, c) CW724R,
F+Ann.; d) F+Ann. CW617N, e) CW511L, f) CW724R.*

Material Forming - ESAFORM 2023   Materials Research Forum LLC
Materials Research Proceedings 28 (2023) 1303-1312   https://doi.org/10.21741/9781644902479-141

Table 3. Measured average grain sizes and microhardness of the test specimens.

Ave. Grain size (μm)	Ext.	F+Ann.	Ave. Microhardness (HV)	Ext.	F+Ann.
CW617N	7.87	7.14		91.2	112.35
CW511L	8.31	7.74		85.5	102.80
CW724R	6.91	6.44		119	139..35

Cutting Forces and Torques Analysis.

The thrust cutting forces and torques of the tapping operation of the CW724R is illustrated in Fig. 4. Focusing on Fig. 5 b, when the form tap was inserted into the hole, a transition regime was seen for 0.2 to 0.4 seconds. The procedure was halted and reversed in the second of 1.4, leading to the taps returning and a large decrease in the force values. Similar to full immersion of the tapping, the return likewise exhibits a proportional constant zone with a cylindrical component and a slope. The torque increases asymptotically in region I of the torque graph, which represents the point of contact between the surfaces of the tap and the workpiece. Region I's tapered design of the tap produces the thread profiles by causing material deformation. Contrarily, area II experiences a steadying of the torque. The maximum torque is measured at the end of region II, where it shows a slight variance that may have resulted from friction variation. It should be highlighted that area III experiences an abrupt decrease in effort as a result of the process of tapping coming to an end. Region IV has a reversed tap spindle speed and entirely withdrawn tap returns from the hole. Additionally, in area IV, it should be noticed that the spindle's reversal counteracted the force applied in the forward direction before the tap emerged from the hole. On the other hand, when the operation was carried out deploying cut tap, the graph is formed of two regions: the engagement of the conical profile and then the hole diameter. The F+Ann. CW511L was machined utilizing cut and form forces, and the greatest thrust forces were measured as 183.7 N and 872.6 N, respectively. The lowest values, however, for Ext. CW617N were recorded at 110.3 and 660.5 N, respectively, utilizing cut and form taps. The same pattern was seen for the torque values of 711.98 N.m. and 29.83 N.m. in F+Ann. CW511L. Additionally, utilizing cut and form taps, the lowest values were determined to be 6.13 N.m and 14.64 Ext. CW617N, respectively. It is clear that the plastic movement of the material without chip removal in form taps led to larger frictional forces and coefficients, which in turn resulted in thrust forces and torque values that are almost four times those of cut taps. The F+Ann. specimens had greater measured thrust cutting forces and torques than the Ext. specimens. This phenomenon is primarily caused by an increase in the β-phase in the microstructure of CW617N and CW511L acquired by applied forging and annealing processes as well as κ-phase in the CW724R. It is obvious that the presence of Pb results in lower cutting forces and torques by retaining more lubricant on its surface, assisting in the decrease of wear or friction during contact. The measured thrust force and torque values were lower in CW724R than CW511L, despite the decreased Pb content. This is explained by the melting points of CW724R and CW511L, which are 600 and 900°C, respectively. Because of all this, softening during machining may also account for the reduced cutting forces of CW724R compared to CW511L. The silicon-rich κ -phase in the material's microstructure is another factor that contributes to the achieved result. In silicon-free brasses, the κ -phase is harder than the α- and β-phase and also has a BCC lattice structure, supporting the findings of [7].

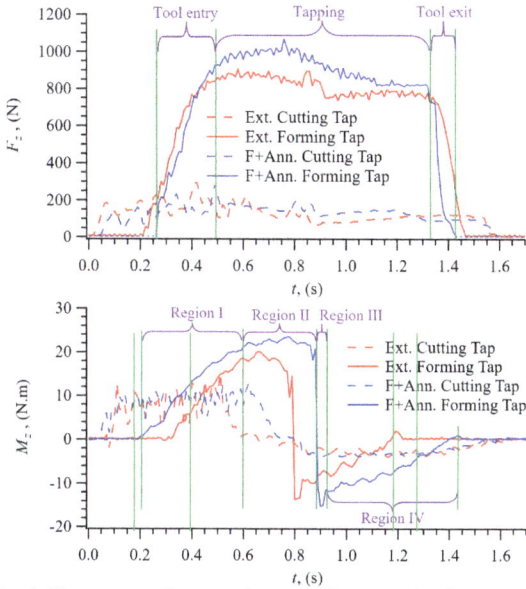

Fig. 4. The measured cutting forces and torques for CW724R.

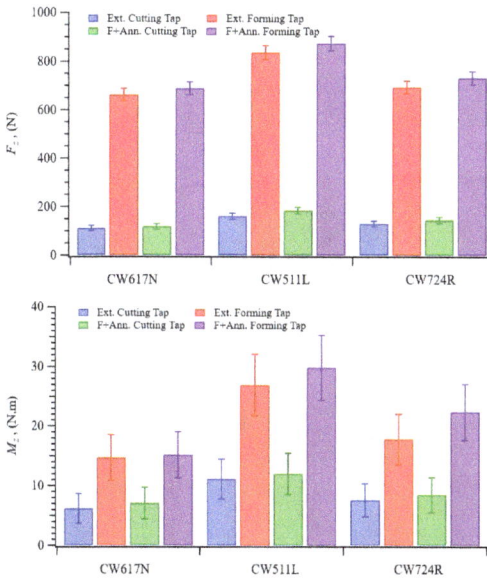

Fig. 5. The measured cutting forces and toques during the experiments.

Material Forming - ESAFORM 2023
Materials Research Proceedings 28 (2023) 1303-1312

Materials Research Forum LLC
https://doi.org/10.21741/9781644902479-141

Surface Quality Analysis and Geometrical Characteristics.

In the form tapping operation, on the top of the threads a split crest was observed due to no metallographic transformation. As demonstrated in Fig. 6b, the highest zone of deformation takes place in Z2 thread's root which during formation is in direct contact with the lobes. The work material flows from Z2 to Z3 on the flank of the thread which is a highly deformed zone and experiences significant displacements. In Z4 no significant deformation is observed. In the cutting tapping operation as shown in Fig. 6a, the thread form was achieved by material removal. Therefore, there were no formed crests on the top of the threads, and the triangle shape was clearly seen. The thread pitch, thread angles, flank straightness and minor, major diameters were inspected by optical Mitutoyo Contracer as well as tap gauges. The geometrical characteristics of all the threads in terms of dimensional tolerances were confirmed by the carried-out inspections and it was determined that the only different zone is Z4 of the threads generated between the tools. The length of Z4 was higher in formed threads than cut ones. It was seen that although in the dimensional tolerance range, the Z4 area in F+Ann. forming threads were much smoother as compared to Ext. specimens which might be attributed to the change in the materials' microstructure in terms of deformation, dislocation, phase content and microhardness. However, no remarkable differences in shape of the threads generated with the cutting tap were observed in all the experiments. Furthermore, some impurities and chip marks were observed on the surfaces Ext. CW511L and CW724R. On the other hand, the quality of the thread surfaces of the CW617N was excellent. The highest impurities belong to CW511L which is also as verification to the force measurements of this paper.

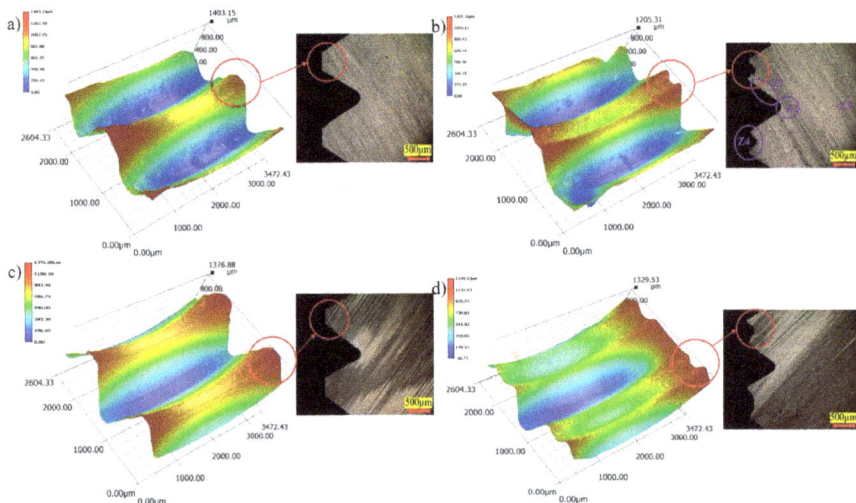

Fig. 6. The formed threads; Ext. CW511L a) cutting, b) forming, F+Ann. CW511L c) cutting, d) forming.

Material Forming - ESAFORM 2023

Materials Research Forum LLC

Materials Research Proceedings 28 (2023) 1303-1312

https://doi.org/10.21741/9781644902479-141

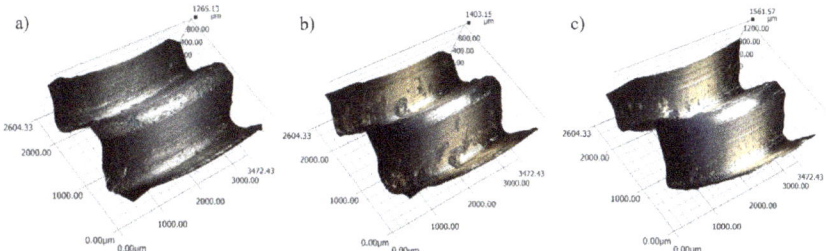

Fig. 7. The cut threads on Ext.; a) CW617N, b) CW511L, c) CW724R.

Surface and Subsurface Microhardness Analysis.

Fig. 8 shows the measured machined surface and sub-surface microhardness of the threads in at 0, 15, 30, 50, 75, 150, 300 and 600 μm distances from the surface. All measurements have been carried out from the most deformed zone, the root of the thread and the average value of two measurements is reported. The properties of the base material could be assessed 600 μm beneath the surface due to the generated strain hardening. The highest microhardness was measured at the machined surface of F+Ann. CW511L with %234.6 increase with respect to the base material using cutting tap. Tapping with cutting tool resulted in higher variations in the surface and sub-surface microhardness. The obtained results show that the tapping tool type plays a significant role in the sub-surface strain hardening. This phenomenon is generated by the deformation of the material structure by the friction and forces between tap and work specimen.

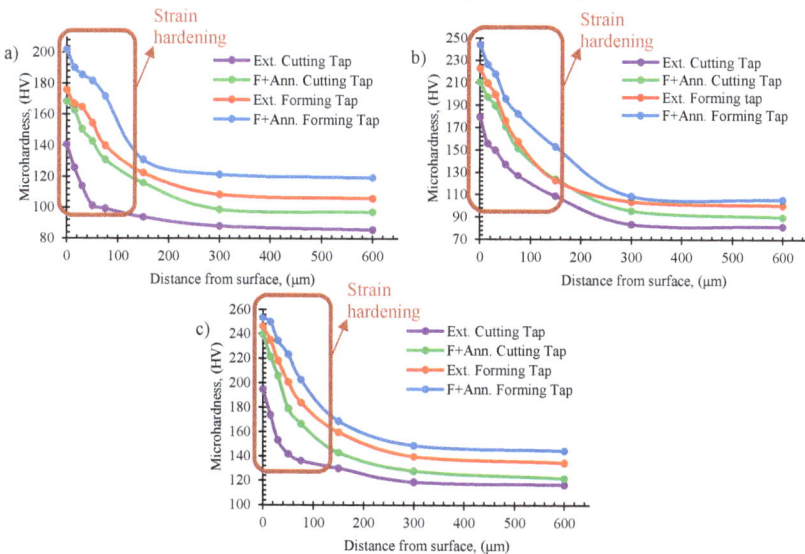

Fig. 8. The measured machined surface and sub-surface microhardness; a) CW617N, b) CW511L, c) CW724R.

Material Forming - ESAFORM 2023
Materials Research Proceedings 28 (2023) 1303-1312

Materials Research Forum LLC
https://doi.org/10.21741/9781644902479-141

Summary

An investigation was carried out to compare cutting and forming tap tool efficiency with respect to thread accuracy, surface quality, generated cutting forces, torques and microhardness in internal threading operation on different types of lead-free brass alloys. Tool type, as well as materials manufacturing method were found to have a significant effect on surface integrity and process responses. In form tapping compared to cutting, higher cutting forces and torques were recorded. It was observed that the phase contents and grain refinement were obtained by multi-directional hot forging operation in the test specimens' microstructure.

The cutting tap tool determined higher strain hardening, but lower thickness of the plastically deformed sub-surface layer and enhanced thread quality in terms of form and dimension.

Acknowledgements

The authors thank TUBITAK (The Scientific and Technological Research Council of Turkey) for partially supporting this work under project number 118C069.

References

[1] N. Zoghipour, E. Tascioglu, G. Atay, Y. Kaynak, Machining-induced surface integrity of holes drilled in lead-free brass alloy, Procedia CIRP 87 (2020) 148-152. https://doi.org/10.1016/j.procir.2020.02.102

[2] Information on: https://rohs.exemptions.oeko.info/fileadmin/user_upload/RoHS_Pack_9/Exemption_6_c_/

[3] G.M. Sicoe, D. Iacomi, M. Iordache, E. Nitu, Research on numerical modelling and simulation of the form tapping process to achieve profiles, Int. J. Modern Manuf. Technol. 5 (2013).

[4] G. Fromentin, G. Poulachon, A. Moisan, Thread forming tapping of alloyed steel, ICME Proceedings, Naples, Italy, 2002, pp. 115-118.

[5] G. Fromentin, G. Poulachon, A. Moisan, B. Julien, J. Giessler, Precision and surface integrity of threads obtained by form tapping, CIRP Annals 54 (2005) 519-522. https://doi.org/10.1016/S0007-8506(07)60159-0

[6] V. Ivanov, V. Kirov, Rolling of internal threads: part 1, J. Mater. Process. Technol. 72 (1996) 214–220.

[7] J.S. Agapiou, Evaluation of the effect of high-speed machining on tapping, J. Manuf. Sci. Eng. Technol. ASME 116 (1994) 457- 462.

[8] V. Ivanov, Rolling of internal threads: Part 2, J. Mater. Process. Technol. 72 (1996) 221-225.

[9] S. Chowdhary, S.G. Kapoor, R.E. DeVor, Modeling forces including elastic recovery for internal thread forming, J. Manuf. Sci. Eng. ASME 125 (2003) 681-688. https://doi.org/10.1115/1.1619178

[10] R. Chandra, S.C. Das, Forming taps and their influence on production, J. India Eng. 1975.

[11] G. Fromentin, G. Poulachon, A. Moisan, Metallurgical aspects in cold forming tapping, NCMR Proceedings, Leeds, UK, 2002, pp. 373-377.

[12] G. Uzun, I. Korkut, The effects of cutting conditions on the cutting torque and tool life in the tapping process for AISI 304 stainless steel, Mater. Technol. 50 (2016) 275–280. http://doi.org/10.17222/mit.2015.044

[13] J.A. de Oliveira, S.L.M. R. Filho, L.C. Brandão, Investigation of the influence of coating and the tapered entry in the internal forming tapping process, Int. J. Adv. Manuf. Technol. 101 (2018) 1051-1063.

[14] S.L.M. R. Filho, J.A. de Oliveira, É.M. Arruda, L.C. Brandão, Analysis of burr formation in form tapping in 7075 aluminum alloy, Int. J. Adv. Manuf. Technol. 84 (2015). http://doi.org/10.1007/s00170-015-7768-9

Material Forming - ESAFORM 2023
Materials Research Proceedings 28 (2023) 1303-1312

Materials Research Forum LLC
https://doi.org/10.21741/9781644902479-141

[15] W. Belluco, L. De Chiffre, Surface integrity and part accuracy in reaming and tapping stainless steel with new vegetable based cutting oils, Tribol. Int. 35 (2002) 865-870. http://doi.org/10.1016/S0301-679X(02)00093-2

[16] C.C.F. Coelho, R.B.D. Pereira, C.H. Lauro, L.C. Brandao, Performance evaluation of tapping processes using a 7075 aluminium alloy with different cooling systems and threading heads, Proc IMechE Part C: J. Mech. Eng. Sci. 203-210 (2019). http://doi.org/10.1177/0954406219867730

[17] H.J. Patel, B,P, Patel, S,M, Patel, A review on thread tapping operation and para-metric studym Int, J, Eng, Res, Appl. 2 (2012) 109–113.

[18] EMUGE-FRANKEN, Threading Technology, InnoForm Cold-forming Taps - Chipless production of internal threads, Germany, 2009, 28.

[19] ISO 68-1:1998 – a (1998a) ISO general purpose screw threads—basic profile—Part 1: metric screw threads, 1ft ed, pp. 3.

[20] ISO 68-2:1998 – b (1998b) ISO general purpose screw threads—basic profile—Part 2: metric screw threads, 1ft ed., pp. 4.

[21] Information on http:\\www.sarbak.com\document\pdf

Material Forming - ESAFORM 2023
Materials Research Proceedings 28 (2023) 1313-1322

Materials Research Forum LLC
https://doi.org/10.21741/9781644902479-142

Analytical method for determining cutting forces during orthogonal turning of C45 steel

ŚLUSARCZYK Łukasz[1,a] and FRANCZYK Emilia[1,b] *

[1]Cracow University of Technology, Faculty of Mechanical Engineering, Chair of Production Engineering, Al. Jana Pawła II 37, 31-864, Cracow, Poland

[a]lukasz.slusarczyk@pk.edu.pl, [b]emilia.franczyk@pk.edu.pl

Keywords: Cutting Forces, Orthogonal Turning, Johnson-Cook Model

Abstract. The authors present a method for determining cutting forces during orthogonal turning of C45 steel. The method utilizes Oxley's chip formation model as well as Johnson-Cook (J-C) constitutive equation and is based on the assumption that the tool is perfectly sharp and the chip formation process is continuous. It is also assumed that the heat exchange between the workpiece, the tool and the chip is carried out by conduction with negligibly small loses caused by convection and radiation and that the thickness of the chip contacting the rake face is constant. The adoption of the above assumptions, together with the knowledge of cutting parameters (including the tool rake angle) as well as of material constants of J-C equation, allows to estimate the thermal-mechanical state of the cutting process and to determine feed and tangential components of the cutting force. Average values of feed and tangent components of the cutting force are calculated using an algorithm implemented in the Matlab environment. The method is based on iterative determination of the minimum difference between stress values in the secondary shear zone. Considered tangential and normal stress values are expressed by formulas based on Oxley's cutting mechanics and the J-C model. The cutting force components obtained in the described method have been compared with the results obtained during experimental studies and with the results obtained in computer simulations using the FEM numerical calculation method.

Introduction

Understanding the phenomena that occur in the cutting zone during machining as well as examination of their impact on process efficiency and quality of machined parts is practically impossible without the use of appropriate models. The issue of modeling of the machining process usually relates to a mechanistic model of the cutting zone, a model of heat generation and distribution or a model of cutting edge wear. Analytical and numerical calculation models are used for forecasting the course of machining process. When describing the behaviour of the machined material at decohesion during the orthogonal turning, it is assumed that there is a slip in the Primary Shear Zone (PSZ), while in the Secondary Shear Zone (SSZ) the friction of the chip against the cutting tool face is taken into account. A good example of using analytical models of the machining process are temperature models determining the partitions of heat penetrating the chip and the workpiece. The examples that can be found in the literature include the Loewen and Shaw's model [1], Trigger and Chao's model [2], the Boothroyd's model [3] and the Oxley's model based on it [4]. The mentioned models allow estimating average temperature the in the slip zone and average contact temperature at the chip-cutting edge interface.

Numerical calculation models for simulating the machining process are a separate group used for process forecasting. Paper [5] describes the use of the ABAQUS application and the FEM method for determination of the impact of friction between the workpiece, chip and the tool on the distribution of forces, strain and temperature in the cutting zone during the turning of AISI 4340 steel. In [6], Özel used the DEFORM application for modelling the friction on the chip-cutting edge interface. The author showed the impact of the friction model used on the chip shape. Another

Material Forming - ESAFORM 2023
Materials Research Proceedings 28 (2023) 1313-1322

Materials Research Forum LLC
https://doi.org/10.21741/9781644902479-142

example of using numerical calculation methods for machining process simulation is [7] where authors used the AdvantEdge application to investigate the impact of protective coatings applied on the cutting tools on heat partition in the cutting zone. Computer simulations are now considered as one of the most important aspects of research on machining processes. They allow for a safe and non-invasive view of the process, also in places inaccessible to conventional monitoring methods. Thus, they make it possible to check how individual changes introduced in the cutting zone or the process itself affect the tested output characteristics. Further, through parameter studies, computer simulations allow to analyze various solutions without the need to conduct costly research.

The aforementioned analytical methods can also be used in order to determine the cutting forces during orthogonal turning. In [8,9], the authors presented methods of determination of the cutting forces based on the known machining parameters and cutting tools geometries. The other input into the method is the workpiece mechanical properties and the constants of J-C constitutive equation. The described method for prediction of cutting forces during orthogonal turning is a less expensive and faster alternative for computer applications based on numerical calculation methods.

Methodology

The described approach is an alternative for numerical calculation methods that use complex calculation algorithms. The presented method is based on the chip forming model proposed by Oxley [9,10]. The principal part of the method is the algorithm which allows determination of the tangential feed components and the of the cutting force during orthogonal turning. In this paper, the method has been used to determine the cutting force components in a wider range of input parameters. The chip forming model during orthogonal turning is presented Fig. 1.

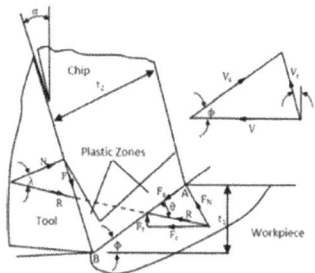

Fig. 1. Chip formation model for orthogonal turning [11].

Where α is the rake angle, ϕ is the shear angle, λ is the average friction angle at tool-chip interface, θ is the angle between resultant cutting force R and primary shear zone AB. Parameters t_1 and t_2 are the depth of cut and the chip thickness respectively. V, V_s, and V_c are the cutting velocity, shear velocity, and chip velocity, respectively. Parameter w is the cutting width that is not shown. The cutting forces are determined by recurrently changing the values: ϕ, C_0 and δ. C_0 is the ratio of shear plane length to the thickness of the PSZ -$l/\Delta s_2$ and δ is the ratio of the thickness of SSZ to chip thickness - $\Delta s_1/t_2$. The described parameters are presented in Fig. 2.

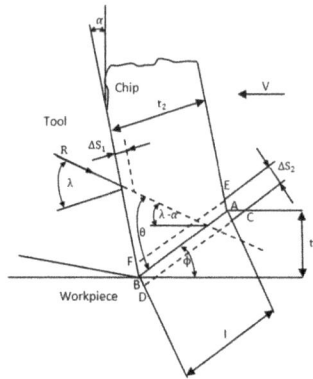

Fig. 2. Parallel-sided shear zone model [11].

Presented method includes calculation of stress values in the PSZ, on the sear plane AB and in the SSZ. Values of the shear flow stress (k_{AB}), strain (ε_{AB}) and strain rate ($\dot{\varepsilon}_{AB}$) on the shear plane in the PSZ are determined from the following equations:

$$k_{AB} = \frac{\sigma_{AB}}{\sqrt{3}} = \frac{1}{\sqrt{3}}(A + B\varepsilon^n{}_{AB})\left(1 + C\ln\frac{\dot{\varepsilon}_{AB}}{\dot{\varepsilon}_0}\right)\left(1 - \left(\frac{T_{AB}-T_r}{T_m-T_r}\right)^m\right) \tag{1}$$

$$\varepsilon_{AB} = \frac{\gamma_{AB}}{\sqrt{3}} = \frac{\cos\alpha}{2\sqrt{3}\sin\phi\cos(\phi-\alpha)} \tag{2}$$

$$\dot{\varepsilon}_{AB} = \frac{\dot{\gamma}_{AB}}{\sqrt{3}} = C_0\frac{V_s}{\sqrt{3}l_{AB}} \tag{3}$$

The length of the shear zone (l_{AB}), the tool-chip interface length (h), the shear velocity (V_s), the chip velocity (V_c) and the chip thickness (t_2) are determined as follows:

$$l_{AB} = \frac{t_1}{\sin\phi} \tag{4}$$

$$h = \frac{t_1\sin\phi}{\cos\lambda\sin\phi}\left(1 + \frac{C_0 n_{eq}}{3\left(1+2\left(\frac{\pi}{4}-\phi\right)C_0 n_{eq}\right)}\right) \tag{5}$$

$$V_s = \frac{V\cos\phi}{\cos(\phi-\alpha)} \tag{6}$$

$$V_c = \frac{V\sin(\phi)}{\cos(\varphi-\alpha)} \tag{7}$$

$$t_2 = \frac{t_1\cos(\phi-\alpha)}{\sin\phi} \tag{8}$$

Secondly, in the SSZ, the shear flow stress in the chip with von Mises criterion are determined on the basis of the equation:

$$k_{chip} = \frac{1}{\sqrt{3}}(A + B\varepsilon_{int}^n)\left(1 + Cln\frac{\varepsilon_{int}}{\varepsilon_0}\right)\left(1 - \left(\frac{T_{int}-T_0}{T_m-T_0}\right)^m\right) \tag{9}$$

while strain (ε_{chip}) and strain rate ($\dot{\varepsilon}_{chip}$) are calculated as stated below:

$$\varepsilon_{chip} = \frac{\gamma_{int}}{\sqrt{3}} = 2\varepsilon_{AB} + \frac{h}{2\sqrt{3}\delta t_2} \tag{10}$$

$$\dot{\varepsilon}_{chip} = \frac{\dot{\gamma}_{int}}{\sqrt{3}} = \frac{V_c}{2\sqrt{3}\delta t_2} \tag{11}$$

Basing on [12], a portion of heat conducted to the workpiece from the shear zone (β) is expressed as:

$$\beta = 0.5 - 0.35log_{10}(R_T tan\phi) \qquad for\ 0.04 \leq R_T tan\phi \leq 10$$

$$\beta = 0.3 - 0.15 = log_{10}(R_T tan\phi) \qquad for\ R_T tan\phi \geq 10 \tag{12}$$

where R_T is a non - dimensional thermal number.

$$R_T = \frac{\rho C_p V t_1}{K} \tag{13}$$

The average temperature in PSZ along AB is given by [12]:

$$T_{AB} = T_0 + \eta\Delta T_{sz} \tag{14}$$

$$\Delta T_{sz} = \frac{(1-\beta)F_s V_s}{\rho V t_1 w C_p} \tag{15}$$

where: T_0 - initial work temperature (25°C in present analysis), η - heat partition factor (90% in present analysis), ΔT_{sz} - the temperature rise in the PSZ.

The temperature at the SSZ (T_{int}) is calculated using a heat partition Eq. 16.

$$T_{int} = T_0\Delta T_{sz} + \psi\Delta T_M \tag{16}$$

Where ψ - is a factor to allow for the fact that the T_{int} is average value ($\psi = 0.9$ in present analysis), ΔT_M is the maximum temperature rise in the chip occurring at the interface. By assuming a rectangular heat source at the interface Boothroyd developed the following equation to find ΔT_M:

$$log_{10}\left(\frac{\Delta T_M}{\Delta T_C}\right) = 0.06 - 0.195\delta\sqrt{\frac{R_T t_2}{t_1}} + 0.5log_{10}\left(\frac{R_T t_2}{h}\right) \tag{17}$$

$$\Delta T_C = \frac{FV_c}{\rho V t_1 w C_p} \tag{18}$$

where ΔT_C is the average temperature rise in the chip.
The angles and forces in Fig. 1 are expressed by formulas:

$$\theta = arctan\left(1 + 2\left(\frac{\pi}{4} - \phi\right) - C_0 n_{eq}\right) \tag{19}$$

$$\lambda = \theta - \phi + \alpha \tag{20}$$

$$n_{eq} \approx \frac{nB\varepsilon_{AB}^n}{(A + B\varepsilon_{AB}^n)} \tag{21}$$

$$F = Rsin\,\lambda \tag{22}$$

$$N = Rcos\,\lambda \tag{23}$$

$$F_s = Rcos\big((\phi + \lambda - \alpha)\big) = k_{AB}l_{AB}w \tag{24}$$

$$N_s = Rsin(\phi + \lambda - \alpha) \tag{25}$$

$$F_c = Rcos(\lambda - \alpha) \tag{26}$$

$$F_t = Rsin(\lambda - \alpha) \tag{27}$$

where: F and N are the shear force and normal force at the tool-chip interface, respectively, F_s and N_s are the shear force and normal force at the PSZ, respectively, F_c is the cutting force and F_t is the thrust force and n_{eq} is the strain hardening constant. The normal stress (σ_N) and the shear stress (τ_{int}) and at the tool-chip interface are calculated from the Eq. 28 and 29:

$$\sigma_N = \frac{N}{hw} \tag{28}$$

$$\tau_{int} = \frac{F}{hw} \tag{29}$$

The normal stress $\acute{\sigma}_N$ is determined from J-C model and stress boundary conditions on the primary shear zone - Eq. 30,

$$\acute{\sigma}_N = k_{AB}\left(1 + \frac{\pi}{2} - 2\alpha - 2C_0 n_{eq}\right) \tag{30}$$

Materials Research Forum LLC
https://doi.org/10.21741/9781644902479-142

Values of the shear angle φ and the strain-rate constant C_0 are determined when the differences between τ_{int} and k_{chip} as well as σ_N and σ_0, respectively, are minimal. Also the strain-rate constant δ is determined when the calculated F_c is minimal. With determined shear angle φ, and strain rate constants C_0 and δ, the machining forces are calculated. A detailed scheme of the algorithm for determining the tangential (F_c) and the feed (F_f) components of the cuttning force has been presented in the article [12]. The results obtained by means of the presented method were compared with the results of experimental studies as well as with the results obtained from computer simulations based on FEM (Finite Element Method).

Laboratory Test
Laboratory tests included performing a series of orthogonal turning tests. The workpiece was a AISI1045 steel pipe with outer diameter of D=50 mm and wall thickness of d=3mm. The test stand consisted of a TUJ 50 conventional lathe and a measuring track enabling the measurement of cutting forces, which consisted the following elements:
- Kistler (Winterthur, Switzerland) 9257B piezoelectric dynamometer together with a Kistler 9403 tool holder (sensitivity: F_f: -7.70 pC/N; F_p: -7.82 pC/N; F_c: -3.71 pC/ N; threshold: <0.01 N) attached to the lathe support,
- Kistler multichannel charge amplifier type 5019B with a set of filters (scale: F_f, F_p: 50 N/V, F_c: 200 N/V),
- PC with DynoWare Software (Kistler) for archiving and analysis of cutting force components.

The measuring stand with the parameters given above made it possible to measure the components of the total cutting force with a frequency of 1000 Hz and with the following inaccuracies: F_f, F_p: ±0.25 N, F_c: ±1N. The schematic diagram of the test stand is shown in Figure 3. The figure also shows the distribution of cutting force components occurring in the process of orthogonal turning.

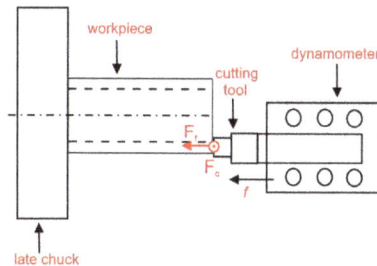

Fig. 3. The schematic diagram of the test stand.

The chemical composition of the AISI 1045 steel in accordance with the PN-EN ISO 683-1:2018-09 standard is presented in Table 1.

Table 1. The percentage chemical composition of AISI 1045 steel.

Symbol	C	Si	Mn	Cr	Ni	Mo	Cu	S	P
	0.42-0.50	0.10-0.40	0.50-0.80	≤0.30	≤0.30	≤0.10	≤0.30	≤0.40	≤0.40

The parameters variability range was chosen on the basis of catalogue data and is presented in Table 2.

Table 2. Values of cutting parameters.

Cutting parameters	Parameter values		
f [mm/rev]	0.1		
V [m/min]	100	200	300
Rake angle α [deg]	-12	-6	0

The tools used in the tests, shown in Fig. 4, were made of P10 carbide without a protective coating. They had a flat rake surface and the length of the cutting edge was 5 mm.

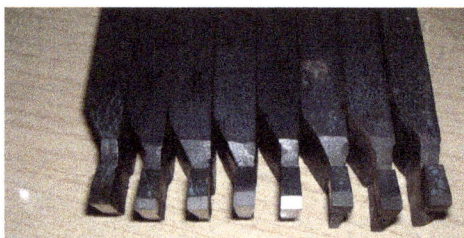

Fig. 4. The tools used in the tests.

Simulation Researches

Computer simulation of the cutting process was carried out using FEM. The simulation was carried out from the moment of first contact of the tool and the workpiece and was continued until a steady state was obtained, that is, until stable stress and temperature values were observed. Both the workpiece and the cutting edge were discretized. Stress of the tool material was calculated using a software package based on a numerical model of the chip formation process, the so-called two-dimensional Lagrange model. In the initial stage of building the material model of the workpiece, three sets of material constants were used. Tables 3 and 4 below present constans of J-C flow stress model and thermo physical properties of AISI 1045 steel, respectively.

Table 3. Constans of J-C flow stress model for AISI 1045 steel.[10,13]

Set number	A(MPa)	B(MPa)	C	n	m	T_0[°C]	T_m[°C]
1	553.1	600.8	0.0134	0.234	1	25	1460
2	731.63	518.7	0.00571	0.94054	0.3241	25	1460
3	546.83	609.35	0.01376	0.94053	0.2127	25	1460

Table 4. Thermo physical properties of AISI 1045 steel.[10]

Specific heat, S (J/kgK)	Thermal conductivity, λ_c (W/mK)	Density ρ (kg/m³)
420+0.504T	52.61-0.0281T	7850

After verifying the results and comparing them with empirical data, set no. 1 was selected for further research. This set of constants was also used in the analytical method for determining the cutting forces described in this article.

Material Forming - ESAFORM 2023
Materials Research Proceedings 28 (2023) 1313-1322

Materials Research Forum LLC
https://doi.org/10.21741/9781644902479-142

Results

Fig. 5 includes mean values of cutting forces determined by means of:

- experiment: F_{c_EXP}, is the tangential component, F_{f_EXP} is the feed component,
- simulation (FEM): F_{c_FEM}, F_{f_FEM},
- analytical method: F_{c_CALC}, F_{f_CALC}.

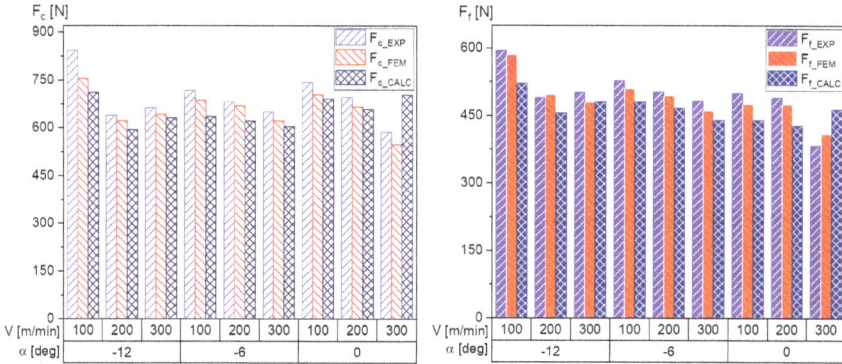

Fig. 5. Mean values of cutting forces determined by various means.

Values of the relative error for the results obtained by means of the FEM and analytical methods as well as for the results obtained from the laboratory test are presented in Figure 6 and were determined according to the formulas:

$$\Delta F_{c\_FEM} = abs(F_{c\_EXP} - F_{c\_FEM})/F_{c\_EXP} * 100\%$$

$$\Delta F_{f\_FEM} = abs(F_{f\_EXP} - F_{f\_FEM})/F_{f\_EXP} * 100\%$$

$$\Delta F_{c\_CALC} = abs(F_{c\_EXP} - F_{c\_CALC})/F_{c\_EXP} * 100\%$$

$$\Delta F_{f\_CALC} = abs(F_{f\_EXP} - F_{f\_CALC})/F_{f\_EXP} * 100\%$$

(31)

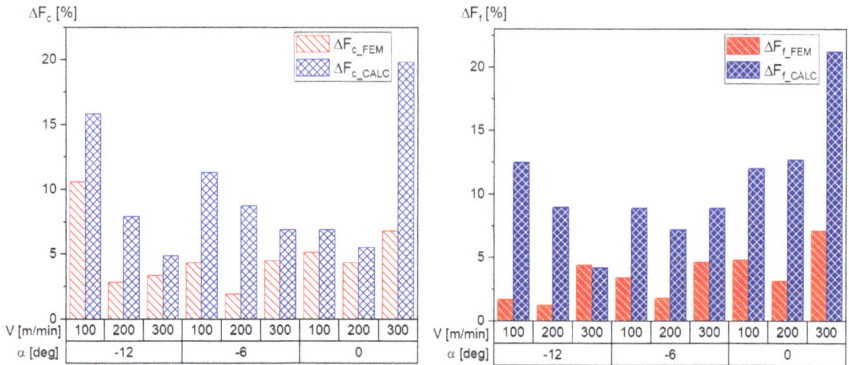

Fig. 6. Relative error for obtained results.

Summary

The presented analytical method for determination of cutting forces during orthogonal turning is characterized by a low calculational complexity and can be an alternative for very complex numerical programs. The input parameters include cutting parameters, tool geometry and the constants of J-C constitutive equation. The analysis of forces obtained in experimental studies allows the following conclusions to be drawn:

- The cutting forces obtained using the FEM simulation have a good conformity with the forces obtained experimentally. The maximum relative error between the results is 10.56% for the F_c component and 7.1% for the F_f component;
- The cutting force values used in the analytical method are closest to the experimental results for the rake angle of $\alpha=-6°$. In this case, the maximum relative error at V=100 m/min is 11.3% for F_c and 8.9% for F_f;
- In the remaining trials, where rake angles were -12° and 0°, respectively, the relative error between the results obtained experimentally and with the analytical method was greater;
- The maximum relative error for angle α =-12 and V =300 m/min was ΔF_{c_CALC}=19.8%, ΔF_{f_CALC}=21.2%.

The proposed analytical method is sensitive to the values of the constants of J-C constitutive equation.

References

[1] Goyal, S. Kumar, D. Shailendra, R. Suresh, R. Sharma, Studying Methods of Estimating Heat Generation at Three Different Zones in Metal Cutting, A Review of Analytical models, Int. J. Eng. Trend. Technol. 8 (2014) 532-545. https://doi.org/10.14445/22315381/IJETT-V8P296

[2] G. Hao, L. Zhanqiang, The heat partition into cutting tool at tool-chip contact interface during cutting process: a review, Int. J. Adv. Manuf. Technol. 108 (2020) 393-411. https://doi.org/10.1007/s00170-020-05404-9

[3] J. Davim, C. Maranhão, P. Faria, A. Abrao, J. Campos Rubio, L. Silva, Precision radial turning of AISI D2 steel, Int. J. Adv. Manuf. Technol. 42 (2009) 842-849. https://doi.org/10.1007/s00170-008-1644-9

[4] Yun, L. Huaizhong, W. Jun, Further Development of Oxley's Predictive Force Model for Orthogonal Cutting, Machin. Sci. Technol. 19 (2015) 86-111. https://doi.org/10.1080/10910344.2014.991026

[5] P.J. Arrazola, T. Özel, Investigations on the effects of friction modeling in finite element simulation of machining, Int. J. Mech. Sci. 52 (2010) 31-42. https://doi.org/10.1016/j.ijmecsci.2009.10.001

[6] T. Özel, The influence of friction models on finite element simulations of machining, Int. J. Mach. Tool. Manuf. 46 (2006) 518-530. https://doi.org/10.1016/j.ijmachtools.2005.07.001

[7] W. Grzesik, M. Bartoszuk, P. Nieslony, Finite element modelling of temperature distribution in the cutting zone in turning processes with differently coated tools, J. Mater. Process. Technol. 164-165 (2005) 1204-1211. https://doi.org/10.1016/j.jmatprotec.2005.02.136

[8] J. Ning, S. Y.Liang, Inverse identification of Johnson-Cook material constants based on modified chip formation model and iterative gradient search using temperature and force measurements, Int. J. Adv. Manuf. Technol. 102 (2019) 2865-2876. https://doi.org/10.1007/s00170-019-03286-0

[9] L. Xiong, J. Wang, Y. Gan, B. Li, N. Fang, Improvement of algorithm and prediction precision of an extended Oxley's theoretical model, Int. J. Adv. Manuf. Technol. 77 (2015) 1-13. https://doi.org/10.1007/s00170-014-6361-y

[10] M. Aydın, Cutting temperature analysis considering the improved Oxley's predictive machining theory, J. Braz. Soc. Mech. Sci. Eng. 38 (2016) 2435-2448. https://doi.org/10.1007/s40430-016-0514-x

[11] J.A. Williams, Mechanics of machining: an analytical approach to assessing machinability: By P.L.B. Oxley, published by Ellis Horwood, Chichester, 1989.

[12] D.I. Lalwani, N.K. Mehta, P.K. Jain, Extension of Oxley's predictive machining theory for Johnson and Cook flow stress model, J. Mater. Process. Technol. 209 (2009) 5305-5312. https://doi.org/10.1016/j.jmatprotec.2009.03.020

[13] J.Ning, S.Y. Liang, Article Predictive Modeling of Machining Temperatures with Force-Temperature Correlation Using Cutting Mechanics and Constitutive Relation, Materials 12 (2019) 284. https://doi.org/10.3390/ma12020284

Material Forming - ESAFORM 2023
Materials Research Proceedings 28 (2023) 1323-1330

Materials Research Forum LLC
https://doi.org/10.21741/9781644902479-143

Novel methodology for burr extension estimation on machined SLM surfaces

GINESTRA Paola[1,a]*, QUARTO Mariangela[2,b], ABENI Andrea[1,c],
ATTANASIO Aldo[1,d], CERETTI Elisabetta[1,e], D'URSO Gianluca [2,f]
and GIARDINI Claudio[2,g]

[1]Department of Management, Information and Production Engineering, University of Bergamo,
via Pasubio 7/b, 24044 Dalmine (BG)

[2]Department of Mechanical and Industrial Engineering, University of Brescia, via Branze 38,
25123 Brescia (BS)

[a]paola.ginestra@unibs.it, [b]mariangela.quarto@unibg.it, [c]andrea.abeni@unibs.it,
[d]aldo.attanasio@unibs.it, [e]elisabetta.ceretti@unibs.it, [f]gianluca.d-urso@unibg.it,
[g]claudio.giardini@unibg.it

Keywords: Additive Manufacturing, Metals, Micromachining, Surface Characterization

Abstract. Additive manufacturing techniques can cover the needs of these components thanks to their high level of customization of the products and the ability to realize complex shapes without high effort. Despite the benefits derived from AM processes, these techniques are characterized by a low-quality of surface finish, one of the most important requirements for medical devices. Considering this aspect, it is important to develop a solution able to improve the surface finish in order to enjoy the low lead times and the high level of customization typical of these processes. A characteristic element of micro-machining is the presence of burrs on the machined surface which can affect the surface texture of micro-features. For this reason, a novel technique for the evaluation of burr was defined. The burrs were evaluated by means of an autofocus variation digital microscope Keyence VHX-7100, on 17-4 PH steel samples produced by laser bed fusion. A 3D reconstruction of the channels and holes surfaces was performed and through the analysis of the peaks distribution, a threshold was defined to discriminate between the original surface of the part and the burrs. In this way, it was possible to estimate and approximate the value of the extension of the burrs in terms of volume and area, where the area is referred to the projected area. Specifically, in micro-machining, not only the extension of the burrs is significant, but also the variability is particularly high, preventing a priori consideration of the defect.

Introduction

Laser bed fusion technologies are widely used to produce complex metal parts not obtainable with traditional and conventional machining processes. Selective Laser Melting (SLM) is a popular and suitable technique for the production of metal components from powder melting by the passage of a laser source in a controlled atmosphere. This technology has been highlighted for the acceptable structural integrity (i.e. density) of the produced parts and the wide range of materials available for this process [1]. On the contrary, such technology, as all the powder bed processes, suffers from an inadequate final surface finishing that is often characterized by high variability affecting the dimensional accuracy of the specimens.

The poor surface quality derived by powder bed fusion is influencing the technical properties of the parts and compromising their tolerances, leading to a limited range of applicability, extremely related to the post-processing requirements [2]. Other defects, such uncontrolled internal porosities, are deeply under study, to exploit the possible applications of these products. The high surface roughness reached by the final parts produced with SLM is easily explained by the

Material Forming - ESAFORM 2023
Materials Research Proceedings 28 (2023) 1323-1330

Materials Research Forum LLC
https://doi.org/10.21741/9781644902479-143

presence of partially melted powder and layer thickness requirements typical of the building process.

The surface finishing and mechanical properties of the metal 3D printed parts are highly related and therefore, an adequate post-processing technique needs to be identified and applied. Among the wide range of traditional processes used for high precision surface finishing, considering the final applications of 3D components that can require low roughness values (i.e. biomedical applications where the surface roughness can influence bacterial adhesion), micro milling is the most convenient manufacturing process when considering the volume and cost ratio [3,4].

Micro-size features of considerable complexity can be obtained by the mechanical removal of material using micro tools on a variety of substrates and materials. Micro milling is a high efficiency process but requires an extended selection of tools and part miniaturization as well as an optimal process optimization before being applied to complex parts especially 3D printed ones. Before considering the usage of this technology on a SLM sample, the micro milling forces and chip thickness formation have to be analyzed and optimized to avoid any material removal behavior that could decrease the precision of the technique [5]. Specifically, the cutting energy and forces are highly influenced by the undeformed chip thickness causing side effects. When the uncut chip thickness is too low, the cutting process is not anymore guided by an elastic deformation (shearing) but by a ploughing regime, which is an elasto-plastic deformation that causes an incorrect chip formation.

The undeformed chip thickness has been identified as one of the causes of the burrs formation during micro milling operations, affecting the quality of the product and the specified dimensional accuracy [6]. Burrs removal is often cost and time consuming since it is extremely difficult on small size components and rarely performed by conventional processes [7]. Therefore, the controls on burrs formation has assumed a lot of importance for a lot of applications, especially in the biomedical field that requires burrs free components.

For this reason, numerous studies focus on the timely and direct characterization of the burrs to provide prompt feedback on the process. Most of the optical systems used for burrs estimation are not able to provide accurate information on geometry or surface roughness and thus specific additional systems are often required to access this information (profilometers, micro-probes or SEM) [8]. Usually, studies on obtaining burrs data by SEM images are conducted manually, meaning that time and practice are required [9].

In this work, a prior evaluation of the machinability and material removal behavior on 17-4PH steel SLM samples has been performed, and a fast and user-friendly solution for measuring burr formation at the micro-scale requiring only one image as input, is proposed. The use of the data acquired through the optical profilometer allows a geometrical characterization of the machined sample with high accuracy, including the burrs geometrical characteristics in terms of morphology and volume evaluation.

Materials and Methods
Micromilling assessment

17-4 PH samples were produced by the laser powder bed fusion machine ProX 100 (© 3D Systems, Rock Hill, South Carolina, United States) as $25 \times 25 \times 5$ mm^3 square samples, with four holes on the corners for the micro milling machine positioning.

The process parameters were: 50W of laser power, 30 microns of layer thickness and 300 m/s of scan speed. The samples were micro-machined by using the five axes Nano Precision Machining Center Kern Pyramid Nano (© Kern Microtechnik GmbH, 82438 Eschenlohe, Germany). The machining tests consist in channel fabrication with a single tool pass from the outer to the center of the sample at a constant depth (a_p) of 200 μm (Fig.1).

Material Forming - ESAFORM 2023   Materials Research Forum LLC
Materials Research Proceedings 28 (2023) 1323-1330   https://doi.org/10.21741/9781644902479-143

Fig. 1. Reference picture of the machining setup.

Tests can be grouped in two different categories: (i) micro machining at constant cutting speed and different feed per tooth (f_z) values as reported elsewhere [10]; (ii) micro machining with a constant feed rate and three different cutting speeds in order to evaluate the dynamic effects on the burrs developing.

In the first group of tests, a cutting speed (v_c) of 40 m/min was kept constant for each cut. Twenty micro channels were machined by using twenty feed per tooth values ranging between 10 µm and 0.5 µm. The microchannels were produced using a coated two flutes micro mill with a nominal diameter of 0.8 mm.

These tests were designed to identify the MUCT as a function of the feed per tooth. A Labview code was implemented to calculate the cutting force by the cutting load components and the signal was filtered in order to identify the cutting force maximum peak on the flutes (z=2) for each rotation.

The tool run-out has been taken into account when normalizing the cutting force: with two flutes (A and B), the maximum chip thickness for one flute (h_{Amax}) will be different than the thickness for the other flute (h_{Bmax}). This asymmetric condition causes two cutting force peaks (F_{c_maxA}; F_{c_maxB}) which should be normalized by considering the effective thickness (for further details please refer to [10]). The second group of channels were fabricated with a constant depth of cut and by changing the cutting speed on three different levels. It is well known that the cutting speed has a major effect on the size of the burrs in micromachining [11,12].

The tests were finalized to evaluate the burrs formation on the inner and outer side of the channel. Two different dimensional scales were investigated by using a coated two flutes micro mill with a nominal diameter of 0.8 mm and a coated two flutes micro mill with a nominal diameter of 0.2 mm. Two sizes were tested in order to verify the procedure to estimate the burr dimensions at different dimensional scales. With the 0.2 mm micro mill a feed per tooth of 1 µm/tooth and cutting speed of 18, 22 and 26 m/min were tested, while with the 0.8 mm micro mill was chosen a feed per tooth of 3 µm/tooth and cutting speed of 30, 40 and 50 m/min. Each test was repeated three times to statistically validate the results.

Burrs estimation methodology

The new developed method assumes that it is possible to extract more information from surface topography. Specifically, the purpose is to define a method able to quantify burrs extensions exploiting the capabilities of an optical profilometer.

Material Forming - ESAFORM 2023
Materials Research Proceedings 28 (2023) 1323-1330

Materials Research Forum LLC
https://doi.org/10.21741/9781644902479-143

The use of this instrument enables a complete characterization of the machined micro-channels, since the data acquired on the bottom part of the channels can be used for the surface texture evaluation (roughness and amplitude parameters), profiles on different sections can be used for the geometrical characterization of the channels, the data acquired on the top part can be used for the burrs evaluation.

In this paper the focus is on this last point. Since the burrs have complex and irregular geometry, their surface results are difficult to be measured with direct measurement techniques. The proposed method for burrs evaluation leverages the 3D topography enabling the evaluation of burrs volume and projected area. The process flow-chart describing the proposed methodology is reported in Fig. 2.

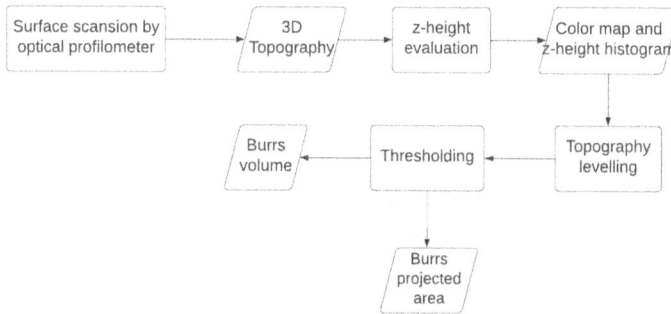

Fig. 2. Flow-chart of proposed method.

The optical profilometer (Sensofar S-neox© Copyright 2021 by Sensofar, Barcelona, Spain) allows the acquisition of a points cloud containing spatial distribution of the scanned points along the 3 main directions (x,y,z). An example of the 3D topography of a micro-machined channel is reported in Fig. 3 where it is possible to distinguish two main regions: the bottom of the channel and the top plane which indicate the rough surface.

The top region is used for the 3D alignment procedure. A roto-translation matrix is determined by applying a least-square algorithm between the top region points and a plane. A linear least-square fitting is applied, using a polynomial plane equation. The coefficients of the equation are then used for the matrix creation, referring to the x-y plane. The obtained matrix is then applied to the entire point cloud along the three main directions x,y,z. The alignment of the top plane is necessary for the thresholding procedure. This step is applied on the height map in order to exclude all the points below the top plane.

The thresholding criterion is based on the points height distribution around the fitted plane, specifically, the limits of the threshold was set in correspondence of the starting point of the positive queue of the z-height distribution which allows to identified only the points characterized by a z-height higher than the as-built surface.

Material Forming - ESAFORM 2023   Materials Research Forum LLC
Materials Research Proceedings 28 (2023) 1323-1330   https://doi.org/10.21741/9781644902479-143

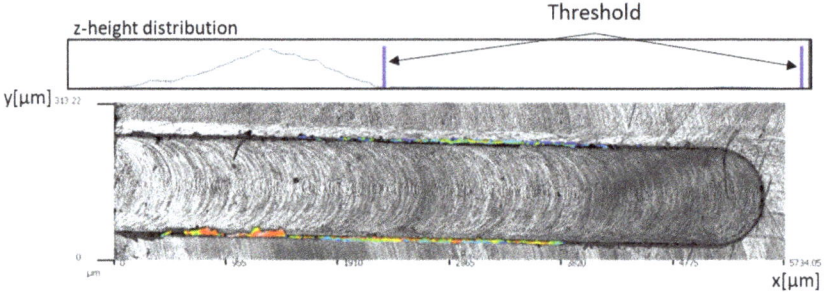

Fig. 3. Portion of burrs identification on the 3D reconstructed machined samples.

Results and Discussion

Identification of the MUCT

As shown in Fig. 4 with a feed per tooth equal or lower than 2 microns, the difference between the cutting force peaks of the two flutes increases. This behavior can be related to a ploughing condition involving flute B that causes an increment of the undeformed depth of cut for flute A. The normalization performed on the resulting cutting force values allowed us to investigate the cutting regime of the material by making the tool run out effects negligible.

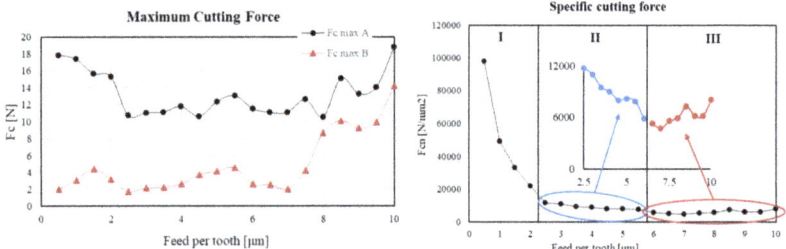

Fig. 4. Cutting force peaks as a function of the feed per tooth (left) and Specific cutting force maximum values for different feed rates (right).

Furthermore, from Fig. 4 it is evident that the specific cutting force changes during the process due to different deformation regimes. Below a f_z of 2.5 μm/tooth, the normalized cutting force increases significantly. For this reason, the MUCT for this analysis was set at 2.5 μm, which falls within the range of the 20-40% of the cutting-edge radius of the tool (0.8 mm). In fact, the region I reported in Fig. 4 is indicating a ploughing deformation region.

The region II instead, is a transition region between the deformation mechanism where the ploughing is progressively substituted by the shearing regime of region III, characterized by the independence of the cutting force in relation to the feed per tooth values.

Influence of cutting speed on burr formation

An overview of the machined channels is reported in Fig. 5, specifically Fig. 5(a) reports the 0.2 mm micro-channels, whilst the Fig. 5(b) shows the 0.8 mm micro-channels.

Material Forming - ESAFORM 2023   Materials Research Forum LLC
Materials Research Proceedings 28 (2023) 1323-1330   https://doi.org/10.21741/9781644902479-143

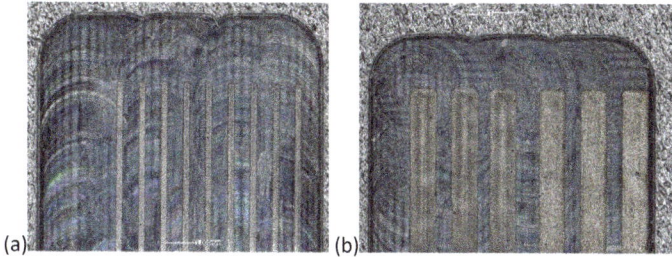

Fig. 5. Reference image of the burrs on the machined samples with 0.2 mm (left) and 0.8 mm (right) tools.

As reported in Fig. 6 the extension of the burrs for this micro milling operation is significant, and the variability between the measured values is particularly high, preventing a priori consideration of the defect.

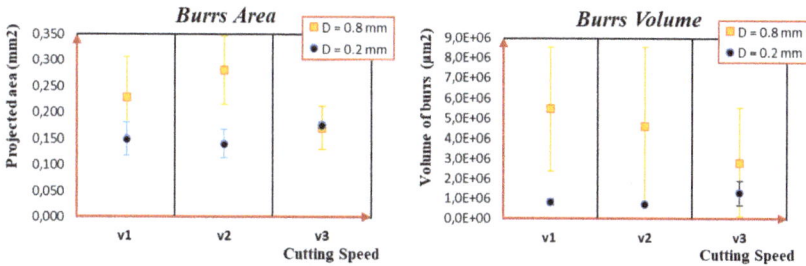

Fig. 6. Burrs Area (left) and volume (right) reported as a function of the cutting speed values for the two used tools.

As visible in Fig. 6, the burrs area on 0.8 mm channels is higher than the 0.2 mm channels, except with the higher cutting speed when it is comparable. There is not a significant dependence of burrs area on cutting speed. The variability of data is calculated as the standard deviation of the measurements, and it resulted lower with 0.2 mm channels. About the volume, the gap between the burrs on 0.2- and 0.8-mm channels increases. The data show a great variability with the 0.8 mm channels and a decrease of the average volume as cutting speed increases. Higher cutting speeds promote higher deformation rates, with a consequent work-hardening of the material. The resulting decrease of the workpiece ductility facilitates the chip evacuation with a consequent reduction of burrs. With 0.2 mm channels, the volume of burrs is almost constant, and it does not depend on cutting speed because on the micro scale, other phenomena are prominent, such as the development of undesired cutting regimes.

The ratio between the burrs area and cross section of the channels (*ratio A*), and the ratio between the burrs volume and the channel volume (*ratio V*), were calculated in order to compare the dimensions of the burrs on channels with different diameters.

$$Ratio\ A = \frac{Burrs\ Area}{D * a_p} \tag{1}$$

$$Ratio\ V = \frac{Burrs\ Volume}{D*a_p*L_C} \qquad (2)$$

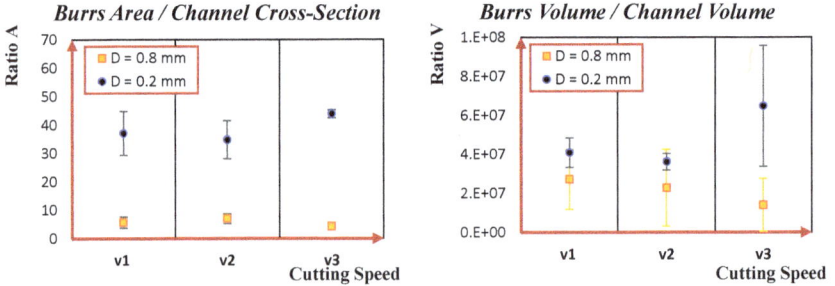

Fig. 7. Ratio A (left) and Ratio V (right) reported as a function of the cutting speed values for the two used tools.

As visible in Fig. 7, the coefficients A and V are higher for the 0.2 mm channels if compared with 0.8 mm channels. It means that the relative dimension of the burrs is higher with the channels machined with 0.2 mm micro-mill. The miniaturization of the machined features determines a worsening of the surface integrity due to the size effect. It is related to the promotion of negative rake angles due to the decrease of the ratio between the chip thickness and the cutting-edge radius. The compression of the material, known as ploughing, avoids the correct chip formation with a consequent increasement of the materials collected on both sides of the channels. The size effect has more impact on the ratio A then the ration V. It means that the morphology of burrs strictly depends on the size of the machining process. With the 0.2 mm diameter micro-mill, the burrs are more developed on the xy plane (see Fig. 3), while they are less developed on the z-direction. On the other hand, as the size increases, the burr are more developed in the z-direction. Each micro-mill has the same helix angle, consequently this trend can be related with the compression of the material in ploughing regime.

Summary

This paper deals with the micro milling of a 3D printed stainless steel material for the post-processing of poorly finished samples. Moreover, this paper introduces an easy-to-handle and time efficient procedure for the burrs identification and characterization after micro milling. From the reported results on the micro milling process, it was possible to notice the occurrence of two deformation regimes as a function of the feed per tooth values. Specifically, at low fz, ploughing dominates the material removal behavior while at higher feed rates, the shearing deformation regime becomes predominant. As a result, the cutting forces during shearing are ten times lower than the forces calculated during ploughing.

Moreover, during shearing the cutting force is not related to the feed per tooth values. The measured transition corresponds to a MUCT of 2.5 microns which is 30-35% of the actual tool edge radius, according to literature [13]. The optimal feed rates have been highlighted to reduce the cutting forces and reduce the tool damage, which causes severe tool wear. We moved beyond the state of the art by introducing a quick and easy-to-handle method for burrs estimation on micro-machined samples. Specifically, a threshold has been established and the software automatically selected the volume and the projected areas based on the z-height of the samples.

Moreover, our method avoids the binary transformation of the values and the consequent projection of the area, which leads to an approximation of the area extension in relation to the

Material Forming - ESAFORM 2023
Materials Research Proceedings 28 (2023) 1323-1330

Materials Research Forum LLC
https://doi.org/10.21741/9781644902479-143

projection. Therefore, we based our work on the estimation of the burrs 3D extension and not only the burrs 2D dimension.

References

[1] C. Ransenigo, M. Tocci, F. Palo, P. Ginestra, E. Ceretti, M. Gelfi, A. Pola, Evolution of Melt Pool and Porosity During Laser Powder Bed Fusion of Ti6Al4V Alloy: Numerical Modelling and Experimental Validation, Laser. Manuf. Mater. Process. 9 (2022) 481-502. https://doi.org/10.1007/s40516-022-00185-3

[2] P. Ginestra, L. Riva, E. Ceretti, D. Lobo, S. Mountcastle, V. Villapun, S. Cox, L. Grover, M. Attallah, O. Addison, D. Shepherd, M. Webber, Surface finish of additively manufactured metals: Biofilm formation and cellular attachment. ESAFORM 2021 - 24th International Conference on Material Forming (2021) art. no. 2089. https://doi.org/10.25518/esaform21.2089

[3] M. Carminati, M. Quarto, G. D'Urso, C. Giardini, C. Borriello, A Comprehensive Analysis of AISI 316L Samples Printed via FDM: Structural and Mechanical Characterization, Key Eng. Mater. 25th International Conference on Material Forming, ESAFORM (2022) 926:46-55. https://doi.org/10.4028/p-szzd04

[4] G. D'Urso, C. Giardini, G. Maccarini, M. Quarto, C. Ravasio, Analysis of the surface quality of steel and ceramic materials machined by micro-EDM. European Society for Precision Engineering and Nanotechnology, Conference Proceedings - 18th International Conference and Exhibition, EUSPEN, 2018.

[5] A. Abeni, P.S. Ginestra, A. Attanasio, Comparison Between Micro Machining of Additively Manufactured and Conventionally Formed Samples of Ti6Al4V Alloy, Lecture Notes in Mech. Eng. (2022) 91 - 106. https://doi.org/10.1007/978-3-030-82627-7_6

[6] P. Cardoso, J.P. Davim, A brief review on micromachining of materials, Rev. Adv. Mater. Sci. 30 (2012) 98-102.

[7] B.Z. Balázs, N. Geier, M. Takács, J.P. Davim, A review on micro-milling: recent advances and future trends, Int. J. Adv. Manuf. Technol. 112 (2021) 655–684. https://doi.org/10.1007/s00170-020-06445-w

[8] F. Medeossi, M. Sorgato, S. Bruschi, E. Savio, Novel method for burrs quantitative evaluation in micro-milling, Precis. Eng. 54 (2018) 379-387. https://doi.org/10.1016/j.precisioneng.2018.07.007

[9] F. Akkoyun, A. Ercetin, K. Aslantas, D.Y. Pimenov, K. Giasin, A. Lakshmikanthan, M. Aamir, Measurement of Micro Burr and Slot Widths through Image Processing: Comparison of Manual and Automated Measurements in Micro-Milling, Sensors. 21 (2021) 4432. https://doi.org/10.3390/s21134432

[10] A. Abeni, P.S. Ginestra, A. Attanasio, Micro-milling of Selective Laser Melted Stainless Steel, Lect. Note. Mech. Eng. (2021) 1 - 12. https://doi.org/10.1007/978-3-030-57729-2_1

[11] A.K. Yadav, M. Kumar, V. Bajpai, N.K. Singh, R.K. Singh, FE modeling of burr size in high-speed micro-milling of Ti6Al4V, Precis. Eng. 49 (2017) 287-292. https://doi.org/10.1016/j.precisioneng.2017.02.017

[12] Z. Kou, Y. Wan, Y. Cai, X. Liang, Z. Liu, Burr controlling in micro milling with supporting material method, Procedia Manuf. 1 (2015) 501-511. https://doi.org/10.1016/j.promfg.2015.09.015

[13] M. Malekian, M.G. Mostofa, S.S. Park, M.B.G. Jun, Modeling of minimum uncut chip thickness in micro machining of aluminum, J. Mater. Process. Technol. 212 (2012) 553-559. https://doi.org/10.1016/j.jmatprotec.2011.05.022

Material Forming - ESAFORM 2023
Materials Research Proceedings 28 (2023) 1331-1340

Materials Research Forum LLC
https://doi.org/10.21741/9781644902479-144

On the impact of tool material and lubrication in ball end milling of ceramic foams

ROTELLA Giovanna[1,a], SANGUEDOLCE Michela[2,b*],
SAFFIOTI Maria Rosaria[2,c], TESTA Flaviano[3,d], UMBRELLO Domenico[2,e]
and FILICE Luigino[2,f]

[1]Department of Management, Finance and Technology, University LUM Giuseppe Degennaro, Casamassima-Bari BA 70100, Italy

[2]Department of Mechanical, Energy and Management Engineering, University of Calabria, CS 87036, Italy

[3]Department of Computer Engineering, Modeling, Electronics and Systems Engineering, University of Calabria, CS 87036, Italy

[a]giovanna.rotella@unical.it, [b]michela.sanguedolce@unical.it, [c]mariarosaria.saffioti@unical.it, [d]f.testa@unical.it, [e]domenico.umbrello@unical.it, [f]luigino.filice@unical.it

Keywords: Milling, Ceramic Foams, Tool Wear, Surface Quality

Abstract. High porosity materials, such as ceramic foams, can be used for several applications spanning from thermal insulators, biomedical implants, molten metal filters, and others. The general practice is to adjust the shape of those ceramic foams in a pre-sintering stage to obtain complex shapes. This work aims to investigate the workability of ceramic foams in a post-sintering condition as an alternative way to overcome premature product failure during production and inhomogeneous shrinkage during sintering. An experimental campaign of ball end milling of alumina-based ceramic foams in a sintered state was carried out herein. Two different tool materials (aluminum oxide-based, diamond-coated) have been tested using two levels of spindle speed under minimum quantity lubrication (MQL) and flood lubrication regimes. The most important findings are: (i) the influence of lubricant is more pronounced analyzing the tool wear, but it has a smaller effect on surface characteristics of the workpiece, (ii) higher spindle speed improves workpiece surface quality (ii) diamond coated tools are the best available choice in terms of both tool wear and surface quality.

Introduction

Porous ceramics are classified as advanced materials which, in general, are capable to supply additional functionalities compared to conventionally used ones; these materials include a wide range of structures based on several morphologies and compositions. High porosity materials can be used for different applications embracing thermal insulators [1], biomedical implants [2], molten metal and other fluids filters [3-7], gases porous burners [6], monolithic catalyst supports [7-9] because of their elevated surface-volume ratio, coatings to enhance lubrication during mechanical processing [10]; some of these applications may require a high level of refractoriness, creep and corrosion endurance, as properties belonging to ceramic materials.

To manufacture ceramic foams several processes are available, some of them involve other materials like polymers as precursors, to provide the lattice shape after sintering [11]. Usually, to change the structure of the obtained ceramic foams, it is possible to modify the precursor shape or work the material ahead of sintering [12]. Nowadays, many studies are being carried out about new sustainable production processes and complex shape development [13], also involving extrusion [14] and additive manufacturing [15], to avoid premature product failure during the production process and inhomogeneous shrinkage in the sintering phase [16,17]. However,

Material Forming - ESAFORM 2023   Materials Research Forum LLC
Materials Research Proceedings 28 (2023) 1331-1340   https://doi.org/10.21741/9781644902479-144

additive manufacturing still shows limitations (e.g. thermal shrinkage, surface quality, resolution), thus it is increasing the use of additive processes in combination with subtractive ones [18]. Processing is often supported by sustainable techniques, such as cryogenic and minimum quantity of lubricant (MQL), allowing to perform machining using a lower amount of cutting fluids to reduce waste and the spread of hazardous substances [19]. Concerning conventional and non-conventional machining methods of bulk ceramic materials [18] many studies were carried out while there is still poor knowledge about the machinability of porous ceramic materials. These latter are considered difficult to machine, since they lead to accelerated tool wear [19], together with poor final surface quality. In fact, they are characterized by a high brittleness combined with the complex lattice shape and porosity, that impact on workability.

Ceramic foam components are mainly used as molten metal filters due to their specific properties in terms of creep and thermal-shock resistance, high functional porosity, low molten metal wetting and specific heat capacity, erosion and corrosion resistance, chemical affinity with inclusions to be removed. Thus, the microstructure and the associated mechanical and thermal properties have a critical influence on the suitability of materials for these types of filters [20]. Many studies demonstrated that the performance of porous ceramics depends on pore size and shape, matrix grain size, intergrain bonding, and pore volume fraction [21-23]. Further analysis has shown that geometric parameters and roughness affect the performance of the ceramic foam in the molten metal filtering process [24], as a result of different inclusion filtering mechanisms such as direct particle impact to the filter, adhesion, and entrapment into recirculation areas [25].

It is very necessary to find innovative, cheap, and easy to perform production methods to guarantee high performance, reliable, long lifetime ceramic filters. At the same time, it is important to optimize resources minimizing the maintenance costs and downtime arising from the failure of a filter element ensuring compliance with the environmental constraints.

This paper aims to assess the machinability of alumina based ceramic foams for molten metal filter applications. In fact, it is worth noting that the effective machining of ceramic foams has not been yet comprehensively assessed. Most commonly used machining techniques for bulk ceramics involve conventional methods (e.g. abrasive wheel cutting, diamond or wire saw cutting), non-conventional machining (e.g. wire electrical discharge machining WEDM, laser beam machining LBM, abrasive water jet machining AWJM) and hybrid machining (e.g. hybrid laser waterjet machining, electrochemical discharge assisted wire machining) and the most commonly used tools are diamond-based ones, for their durability [26]. However, studies based on these techniques [20] demonstrated that the machinability of bulk ceramics still represents an issue for both tool wear and surface integrity of the obtained components.

Thus, this paper presents the evaluation of the machinability of alumina based ceramic foams by spherical end milling process using different lubrication methods and tools materials.

Materials and Methods

The samples under investigation are cylindrical alumina-based ceramic foams, with a diameter of 30 mm, the height of 30 mm, and pore density of 30 pores per inch (ppi). The foams, produced via the replica process [16,27], consist of a network of randomly-oriented dodecahedral-shaped cells interconnected through struts.

Two different tool materials (Fig. 1) were chosen for the experimental campaign on spherical end milling process: vitrified hard bond pink aluminum oxide and electroplated diamonds on a metal substrate with an average grit size of around 120 μm. The reason of these choices lies within to the common use of diamond-coated tools to process ceramics, as explained in the review of the state of the art, and the opportunity to differentiate material removal mechanisms by comparison with abrasive tools. Aluminum oxide tools were provided by Meusburger Georg GmbH & Co KG while diamond coated ones were provided by SICUTOOL S.p.A.

Material Forming - ESAFORM 2023
Materials Research Proceedings 28 (2023) 1331-1340

Materials Research Forum LLC
https://doi.org/10.21741/9781644902479-144

A full factorial experimental campaign, as a complement to a preliminary campaign performed under the same conditions [28] was carried out at fixed axial depth of cut, feed rate, and number of passes and varying tool material, lubrication conditions (flood and minimum quantity of lubricant - MQL) and spindle speed. The parameters of the experimental campaign were recommended by the tool manufacturer since a comprehensive literature on the machining of porous ceramics does not exist.

The setup for milling tests is shown in Fig. 2 while the factors considered for the experimental campaign and their levels are reported in Table 1.

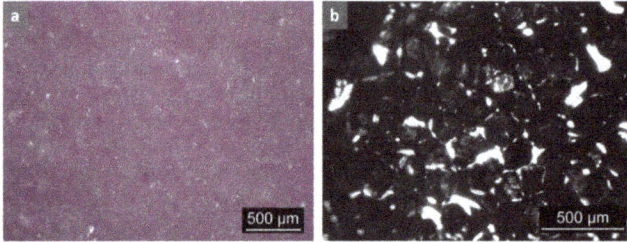

Fig. 1. Micrographs of as received pink aluminum oxide (a) and diamond coated (b) tools.

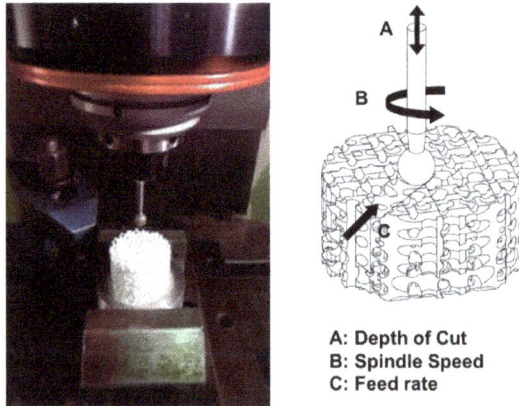

A: Depth of Cut
B: Spindle Speed
C: Feed rate

Fig. 2. Schematic of the experimental setup and spherical end milling process.

Table 1. Design of experiments for the full factorial experimental campaign.

Factor	Level 1	Level 2
Tool material	Aluminum oxide	Diamond coated
Lubrication conditions	MQL	Flood
Spindle speed [RPM]	10000	40000
Number of Passes	2	
Depth of Cut [mm]	0.5	
Feed Rate [mm/min]	50	

After processing, samples and tools were inspected through a Scanning Electron Microscope to assess the surface characteristics before and after milling, and the tool wear mechanisms.

Results and Discussion

Tool wear is a direct expression of tool life, and it is strictly related to machined surface and subsurface quality. Fig. 3 shows the surfaces of aluminum oxide tools, and the damage is immediately recognizable. For the test at 40000 RPM and using MQL, the tool surface appears flattened (Fig. 3d), the same phenomenon appears to be attenuated for the remaining tests. Also, adhesive wear takes place during the process since a certain amount of adhered foam material was found on the tool surface.

Tool surface flattening and materials mixing can be explained by tool and sample material affinity in terms of composition and mechanical properties. Fig. 4 shows the aluminum oxide worn tools at varying spindle speed, highlighting how different parameters can affect the tool life. Lower spindle speed (Fig. 4a) shows a more marked wearing effect in terms of build-up edge and plastic deformation while rising the speed (Fig. 4b) leads to a lower content of workpiece material adhering to the tool, even though brittle cracks are detectable. Also, adhesion is reduced in the case of flood lubrication which results, by nature, in a more efficient lubrication effect.

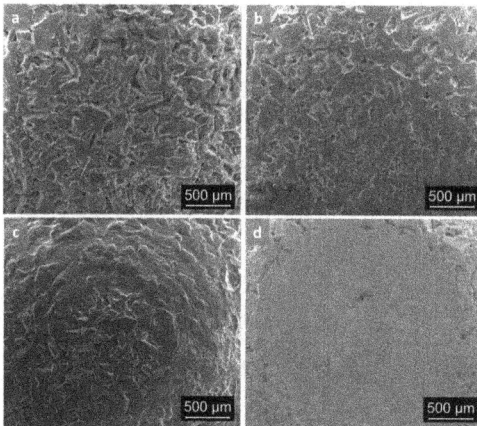

Fig. 3. Aluminum oxide tools after milling at (a) spindle speed 10000 rpm, flood lubrication [28], (b) spindle speed 40000 rpm, flood lubrication, (c) spindle speed 10000 rpm, MQL, (d) spindle speed 40000 rpm, MQL.

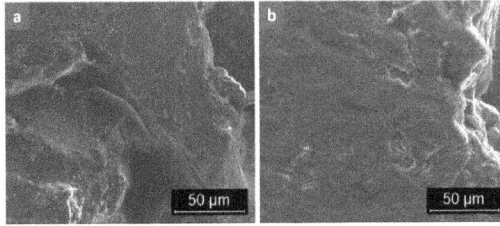

Fig. 4. Details of aluminum oxide tools after milling at different spindle speeds
(a) 10000 rpm, (b) 40000 rpm.

As regards diamond coated tools, the damage is not easily recognizable at low magnification, as shown in Fig. 5. A low amount of residual adhered foam material can be seen (Fig. 5d), greater in the case of tests involving MQL. Furthermore, Fig. 6a shows a void left by a diamond being pulled out while Fig. 6b the adhesion phenomenon [28].

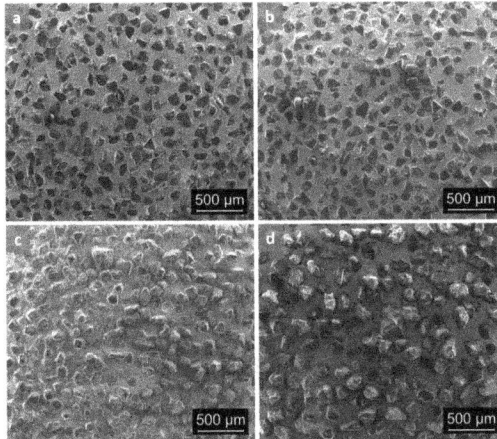

Fig. 5. Diamond coated tools after milling at (a) spindle speed 10000 rpm, flood lubrication
[28], (b) spindle speed 40000 rpm, flood lubrication, (c) spindle speed 10000 rpm, MQL,
(d) spindle speed 40000 rpm, MQL.

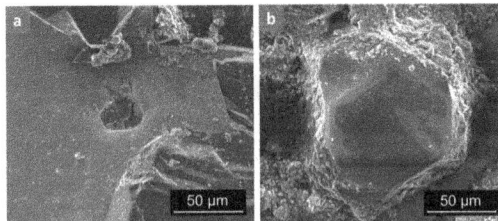

Fig. 6. Details of diamond coated tools after milling (a) diamond pull out, (b) adhesion
phenomenon [28].

Material Forming - ESAFORM 2023   Materials Research Forum LLC
Materials Research Proceedings 28 (2023) 1331-1340   https://doi.org/10.21741/9781644902479-144

The process described, inevitably brings both brittle and ductile fractures in workpieces due to the multitude of scratches and their relative interactions when abrasive grains cut into ceramic specimens. The machined surface results ideally deformed similarly to that left by an indentation process, but other mechanisms also take place, like adhesion between tool/workpiece surfaces, as reported above. Furthermore, excessive local forces can generate different defects like chips, fissures, and cracks. The above issues are worsened by the geometric characteristics of the investigated workpieces.

When aluminum oxide tools are used, the adhesion effect evidently affects the fracture mechanism, as shown in Fig. 7. In fact, the chemical affinity between the foam and the tool brings to machined surfaces characterized by a higher content of brittle fractures. The analysis here reported also confirms adhesive wear on the tool.

The effect of lubrication methods can also be evaluated by analyzing Fig. 7 and 8, where no appreciable difference in the final quality of the ceramic surface can be detected. This finding suggests the possibility of effectively machining the samples by avoiding the massive use of lubricants.

On the other hand, the spindle speed influences the machined surface quality as shown by Fig. 7. However, such difference cannot be appreciated at higher magnifications, as Fig. 8 highlights.

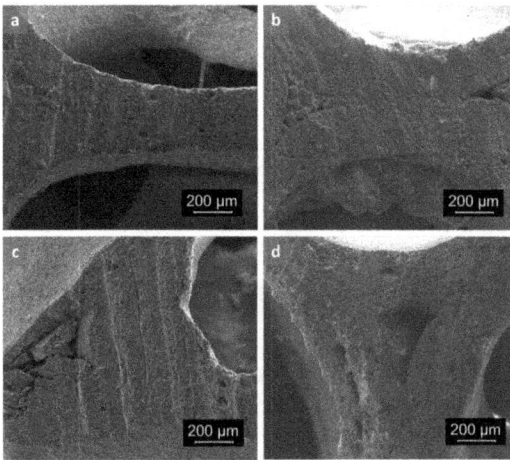

Fig. 7. Ceramic foam surface after milling with aluminum oxide tools at (a) spindle speed 10000 rpm, flood lubrication [28], (b) spindle speed 40000 rpm, flood lubrication, (c) spindle speed 10000 rpm, MQL, (d) spindle speed 40000 rpm, MQL.

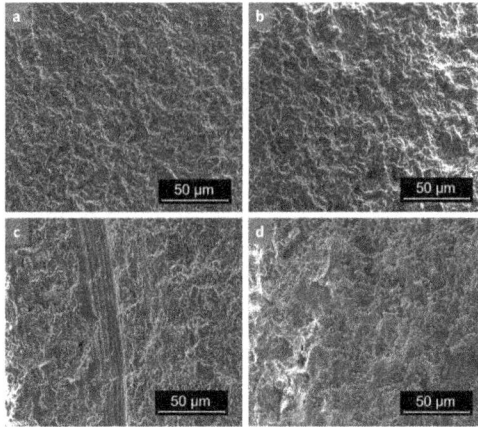

Fig. 8. Details of ceramic foam surface after milling with aluminum oxide tools at (a) spindle speed 10000 rpm, flood lubrication [28], (b) spindle speed 40000 rpm, flood lubrication, (c) spindle speed 10000 rpm, MQL, (d) spindle speed 40000 rpm, MQL.

Diamond coated tools result in a more efficient cutting process since less adhesion effect is present. Fig. 9 and 10 show both ductile and brittle fracture areas, which are typical of the process. The diamond abrasive grains exposed on the milling tool are of random distribution, with a consequent difference in the exposure highness of each diamond. The overall machining is achieved by the combined actions of what can be thought of as several micro-cutting edges with locally different depths of cut, causing areas of both ductile and brittle fracture when machining ceramics. Furthermore, it is possible to state that the effect of lubricant is minimum and that the spindle speed has the same effects as those reported for the aluminum oxide tool.

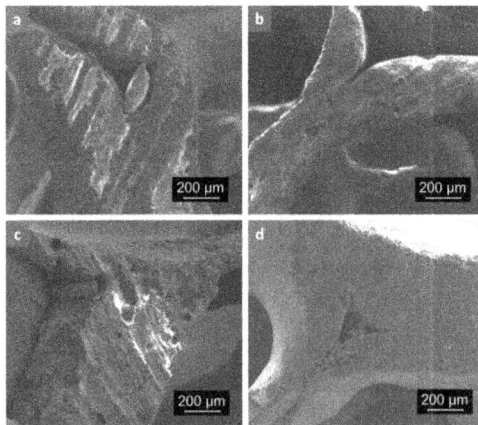

Fig. 9. Ceramic foam surface after milling with diamond coated tools at (a) spindle speed 10000 rpm, flood lubrication [28], (b) spindle speed 40000 rpm, flood lubrication, (c) spindle speed 10000 rpm, MQL, (d) spindle speed 40000 rpm, MQL.

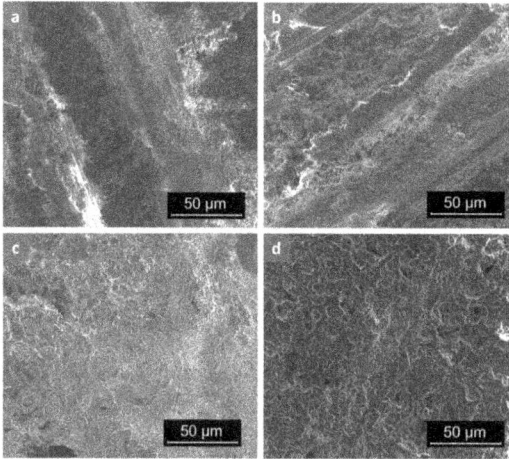

Fig. 10. Details of ceramic foam surface after milling with diamond coated tools at (a) spindle speed 10000 rpm, flood lubrication [28], (b) spindle speed 40000 rpm, flood lubrication, (c) spindle speed 10000 rpm, MQL, (d) spindle speed 40000 rpm, MQL.

Summary

This paper presents an analysis of the workability of alumina based ceramic foam. Such materials combine the issues of machining conventional ceramics with those related to the machinability of geometrically complex parts. The experimental campaign involved the spherical end milling of the workpieces at varying cutting tools, lubricant conditions, and spindle speed.

The results highlighted the tendency of such a process to fast wearing the tools and the combination of ductile and brittle deformation modes on the machined surfaces. The influence of lubricant is more pronounced analyzing the tool wear, where the flood lubrication ensures a longer tool life. Concerning the surface characteristics, lubrication has a smaller effect. On the other hand, the spindle speed plays instead a key role in the deformation mechanism showing a better surface quality when higher speeds are employed.

The results obtained allow the authors to state that ceramic foams can be machined by spherical end milling to obtain a variety of complex shapes and that diamond coated tools are, up to now, the best available choice. In future works, a more comprehensive study of the involved phenomena will be investigated, involving the transition from ductile to brittle material removal, related to locally different depths of cut.

It should be noted that the overall process needs to be optimized to define a workability window able to minimize the tool wear and maximize the surface quality.

References

[1] M. Fukushima, Y.I. Yoshizawa, Fabrication of highly porous silica thermal insulators prepared by gelation-freezing route, J. Am. Ceram. Soc. 97 (2014) 713-717. https://doi.org/10.1111/jace.12723

[2] K. Sharma, M.A. Mujawar, A. Kaushik, State-of-Art Functional Biomaterials for Tissue Engineering, Front. Mater. 6 (2019) 1-10. https://doi.org/10.3389/fmats.2019.00172

[3] J.R. Brown, Foseco Non-Ferrous Foundryman's Handbook, Butterworth-Heinemann, 2000.

[4] J.R. Brown, Foseco Ferrous Foundryman's Handbook, Butterworth-Heinemann, 2000.

[5] A. Salomon, M. Dopita, M. Emmel, S. Dudczig, C.G. Aneziris, D. Rafaja, Reaction mechanism between the carbon bonded magnesia coatings deposited on carbon bonded alumina and a steel melt, J. Eur. Ceram. Soc. 35 (2015) 795-802. https://doi.org/10.1016/j.jeurceramsoc.2014.09.033

[6] P. Wang, K. Zhang, R. Liu, Influence of air supply pressure on atomization characteristics and dust-suppression efficiency of internal-mixing air-assisted atomizing nozzle, Powder Technol. 355 (2019) 393-407. https://doi.org/10.1016/j.powtec.2019.07.040

[7] T. Granato, F. Le Piane, F. Testa, A. Katović, R. Aiello, MCM-41 deposition on alumina ceramic foams, Mater. Lett. 64 (2010) 1622-1625. https://doi.org/10.1016/j.matlet.2010.03.063

[8] T. Granato, F. Testa, R. Olivo, Catalytic activity of HKUST-1 coated on ceramic foam, Microporous Mesoporous Mater. 153 (2012) 236-246. https://doi.org/10.1016/j.micromeso.2011.12.055

[9] F. Jun, Z. ZengFeng, C. Wei, M. Hong, L. JianXing, Computational fluid dynamics simulations of the flow field characteristics in a novel exhaust purification muffler of diesel engine, J. Low Freq. Noise Vib. Act. Control. 37 (2018) 816-833. https://doi.org/10.1177/1461348418790488

[10] G. Ambrogio, F. Gagliardi, Temperature variation during high speed incremental forming on different lightweight alloys, Int. J. Adv. Manuf. Technol. 76 (2015) 1819-1825. https://doi.org/10.1007/s00170-014-6398-y

[11] A.R. Studart, U.T. Gonzenbach, E. Tervoort, L.J. Gauckler, Processing routes to macroporous ceramics: A review, J. Am. Ceram. Soc. 89 (2006) 1771-1789. https://doi.org/10.1111/j.1551-2916.2006.01044.x

[12] M.V. Twigg, J.T. Richardson, Fundamentals and applications of structured ceramic foam catalysts, Ind. Eng. Chem. Res. 46 (2007) 4166-4177. https://doi.org/10.1021/ie061122o

[13] G.V. Franks, C. Tallon, A.R. Studart, M.L. Sesso, S. Leo, Colloidal processing: enabling complex shaped ceramics with unique multiscale structures, J. Am. Ceram. Soc. 100 (2017) 458-490. https://doi.org/10.1111/jace.14705

[14] F. Gagliardi, C. Ciancio, G. Ambrogio, Optimization of porthole die extrusion by Grey-Taguchi relational analysis, Int. J. Adv. Manuf. Technol. 94 (2018) 719-728. https://doi.org/10.1007/s00170-017-0917-6

[15] G. Liu, Y. Zhao, G. Wu, J. Lu, Origami and 4D printing of elastomer-derived ceramic structures, Sci. Adv. 4 (2018) 1-11. https://doi.org/10.1126/sciadv.aat0641

[16] P. Colombo, H.P. Degischer, Highly porous metals and ceramics, Mater. Sci. Technol. 26 (2010) 1145-1158. https://doi.org/10.1179/026708310X12756557336157

[17] B. Su, S. Dhara, L. Wang, Green ceramic machining: A top-down approach for the rapid fabrication of complex-shaped ceramics, J. Eur. Ceram. Soc. 28 (2008) 2109-2115. https://doi.org/10.1016/j.jeurceramsoc.2008.02.023

[18] R. Rakshit, A.K. Das, A review on cutting of industrial ceramic materials, Precis. Eng. 59 (2019) 90-109. https://doi.org/10.1016/j.precisioneng.2019.05.009

[19] G. Rotella, M. Sanguedolce, M.R. Saffioti, L. Filice, F. Testa, Strategies for Shaping of Different Ceramic Foams, Procedia Manuf. 47 (2020) 493-497. https://doi.org/10.1016/j.promfg.2020.04.345

[20] D.A. Hirschfeld, T.K. Li, D.M. Liu, Processing of porous oxide ceramics, Key Eng. Mater. 115 (1996) 65-80. https://doi.org/10.4028/www.scientific.net/kem.115.65

[21] Porosity of Ceramics by Roy W. Rice (z-lib.org).pdf, (n.d.).

[22] H. Deng, B. Latella, T. Liu, Y. Ke, L. Zhang, Processing and Property Assessment of Porous Alumina and Mullite-Alumina Ceramics, Mater. Sci. Forum. 437-438 (2003) 423-426. https://doi.org/10.4028/www.scientific.net/msf.437-438.423

[23] J.H. She, T. Ohji, Porous mullite ceramics with high strength, J. Mater. Sci. Lett. 21 (2002) 1833-1834. https://doi.org/10.1023/A:1021576104859

[24] E. Werzner, M. Abendroth, C. Demuth, C. Settgast, D. Trimis, H. Krause, S. Ray, Influence of Foam Morphology on Effective Properties Related to Metal Melt Filtration, Adv. Eng. Mater. 19 (2017) 1-10. https://doi.org/10.1002/adem.201700240

[25] O. Dávila-Maldonado, A. Adams, L. Oliveira, B. Alquist, R.D. Morales, Simulation of fluid and inclusions dynamics during filtration operations of ductile iron melts using foam filters, Metall. Mater. Trans. B Process Metall. Mater. Process. Sci. 39 (2008) 818-839. https://doi.org/10.1007/s11663-008-9190-2

[26] R. Rakshit, A.K. Das, A review on cutting of industrial ceramic materials, Precis. Eng. 59 (2019) 90-109. https://doi.org/10.1016/j.precisioneng.2019.05.009

[27] P. Colombo, Conventional and novel processing methods for cellular ceramics, Philos. Trans. R. Soc. A Math. Phys. Eng. Sci. 364 (2006) 109-124. https://doi.org/10.1098/rsta.2005.1683

[28] G. Rotella, M. Rosaria, M. Sanguedolce, F. Testa, L. Filice, F. Micari, Milling of Alumina-Based Ceramic Foams: Tool Material Effects, Meccanica.

Material Forming - ESAFORM 2023
Materials Research Proceedings 28 (2023) 1341-1346

Materials Research Forum LLC
https://doi.org/10.21741/9781644902479-145

Superfinishing processes applied on the biomedical implants surface to improve their performance

SAFFIOTI Maria R.[1,a] *, ROTELLA Giovanna[2,b] and UMBRELLO Domenico[1,c]

[1]Dept. of Mechanical, Energy and Management Engineering, Univ. of Calabria, CS 87036, Italy

[2]Dept. of Management, Finance and Technology, Univ. LUM, 70010 Casamassima BA, Italy

[a]mariarosaria.saffioti@unical.it, [b]giovanna.rotella@unical.it, [c]domenico.umbrello@unical.it

Keywords: Burnishing, Biomechanics, Surface Integrity

Abstract. The demand of biomedical implants characterized by more and more advanced performances is expected to grow as well as their service life is expected to increase. Therefore, the necessity to research new material properties and production techniques in producing biomedical implants, characterized by higher durability, is strongly required. Regarding permanent prosthesis, the surface conditions play a key role on the prostheses overall mechanical performance. This work aims, through the application of burnishing process, to combine innovative machining finishing operations coupled with environmentally friendly lubricooling techniques to improve the surface integrity of biomedical devices in order to increase their quality, durability and reliability.

Introduction

Nowadays, the demand of products characterized by advanced performances that can provide great societal and environmental benefits is continuously increasing, especially in the biomedical field [1]. Surface integrity modification is one of the available strategies that can be used to enhance the in-service life of a component. As a matter of fact, Severe Plastic Deformation (SPD) processes have been extensively reported for modifying surface properties by altering the surface metallurgical characteristics (e.g. microstructure, hardness, precipitates amount and characteristics, roughness, residual stresses) [2]. Not only mechanical processes are used to change surface characteristics but also thermal processes as laser are employed to modify the surface properties, for example in adhesive joints of non-ferrous materials as shown in this study [3].

Burnishing is a SPD process since it is capable to greatly affect surface integrity due to the high level of strain induced in the workpiece material.

Several studies on different metals of industrial interest were carried out to understand the relationship between the burnishing parameters and post-processed surface integrity. As example in [4] roller burnishing tests were performed on the *Ti6Al4V* titanium alloy, enabling to find a set of processing parameters and lubrication conditions capable to significantly improve the surface quality of the final component. Furthermore, according to [4], the burnishing process aided to improve the surface finish and hardness, also generating compressive residual stresses that contributed to extend the fatigue life [5].

Burnishing is applied to titanium alloys, which represent the most used material for producing orthopedic prostheses like hip and knee implants. Generally, these prostheses, which stay permanently in the human body, need to be revised as a consequence of wear, corrosion and oxidation, which all initiate at the implant surface and may lead to leakage of ions dangerous for the human body [6]. In this case, burnishing is applied in order to increase the wear resistance of titanium alloys. An experimental campaign of burnishing process on cylindrical bars of *Ti6Al4V* was carried out. In particular, the tests were performed under MQL lubrication conditions and by varying process parameters, such as burnishing force, at fixed burnishing speed, feed rate and tool

Material Forming - ESAFORM 2023
Materials Research Proceedings 28 (2023) 1341-1346

Materials Research Forum LLC
https://doi.org/10.21741/9781644902479-145

radius. The choice of using MQL lubricant condition was taken in order to achieve sustainable machining [7] and because the provider of the burnishing tool recommended to not work in dry conditions but to use lubricant as in this case vegetable oil emulsion. Afterwards, the surface integrity was assessed by means of microstructural analysis, roughness and micro-hardness measurements below the machined surface with the aim to analyze the mechanical property improvements after burnishing.

Results showed that the advanced manufacturing technology was able to thoroughly modify surface integrity of the investigated biomaterials. Furthermore, burnishing process enhances surface integrity in correspondence of the subsurface region of *Ti6Al4V*, in terms of grain refinement, surface roughness reduction, and hardness increase, in accordance with the fundamental characteristics required in the biomedical field to achieve the highest product performance and increasingly durable implants.

Experimental Procedure

The titanium alloy involved into the ongoing experimental campaign was the *Ti6Al4V*, grade 5. The as-received material was in form of a bar, with microstructure consisting of α equiaxed grains and intergranular β with 5 µm mean grain size. The bar roughness before processing was equal to 1.2 µm while after the turning process reached a value of 0.8 µm and up to a minimum value of 0.4 µm post burnishing.

Burnishing is a chipless SPD process, which allows increasing the product performance modifying the surface quality by smoothing the roughness, refining the grain structure and increasing the hardness [8. 9].

A deeper analysis concerning the influence of the burnishing process parameters on the *Ti6Al4V* surface integrity was carried out. The burnishing tests were performed using cylindrical bars of *Ti6Al4V* with 30 mm starting diameter, under MQL lubrication conditions. Roller burnishing tests were carried out on a high-speed CNC turning center equipped in particular with Quick Turn Nexus 200-II CNC machine. The lubricant, consisting of a vegetable oil, was delivered at the tool-workpiece interface through an external nozzle.

Table 1. Experimental plan for burnishing.

Burnishing	Value
Cutting speed v (m/min)	150
Feed f (mm/rev)	0.05
Force F (N)	1000 – 1500 – 2000 – 2500
Tool radius R (mm)	2.5
Number of passes	2

The burnishing tool was the SKUV20 (Yamato) with the roller made of hardened steel. Before performing the burnishing process, the bar was previously turned with standard semi finishing parameters, namely 30 m/min of cutting speed, feed of 0.15 mm/rev and 0.1 mm of depth of cut, while the burnishing parameters are reported in Table 1.

In order to control the effective achievement of the fixed burnishing force and monitor the temperature evolution, a three components piezoelectric dynamometer and an infrared thermo-camera were used during the tests. Fig. 1 shows the burnishing process set-up.

The burnished samples were cut in the transverse direction. The cross section was firstly mounted using a thermosetting black epoxy hot mounting resin, and then mechanically polished. These operations were carried out to obtain a mirror-like surface suitable for the metallographic analysis. Before performing the metallographic analysis with the optical microscope (Leica), the samples were etched using the Kroll's reagent able to oxidize the samples in order to highlight and measure the grain edges, then using an automatic procedure provided by ImageJ software, the grain size was measured. Furthermore, a portable surface profilometer (Pocket Surf®) was used to measure the mean surface roughness (Ra) while the Vickers micro-hardness (HV0.1) of the surface and subsurface layer was measured by means of an instrumented micro indenter.

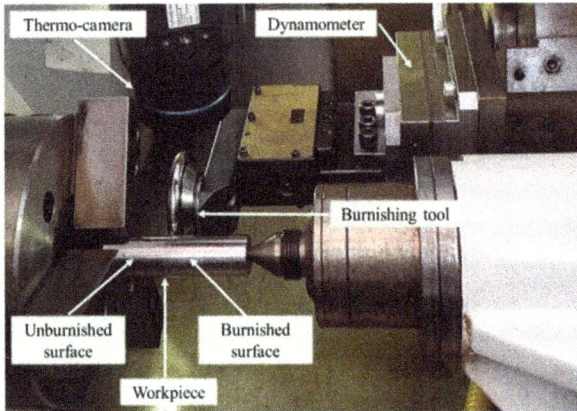

Fig. 1. Burnishing process set-up.

Results

In this preliminary analysis, the overall results were classified analyzing the influence of the burnishing force (B1000, B1500, B2000, B2500), which, according to literature [10], it is considered one of the parameters that mostly affects the surface integrity. Concerning the microstructure, Fig. 2 shows the changes between the different process conditions, in particular the As-Received (AR), As-Turned (AT), and As-Burnished (AB) samples, with 150 m/min burnishing speed, 0.05 mm/rev feed, 2500 N burnishing force, and 2.5 mm tool radius. Focusing on the AB sample, the grain distortion is visible along the working direction. Comparing the AB and AT samples, the AB one appears more deformed due to the larger amount of plastic strain induced by the burnishing process. Fig. 3 allows a better evaluation of the microstructural modifications, outlining the grain size evolution, showing that the surface grain size reduced during machining and even more during burnishing at the highest burnishing force. Although, during burnishing process significant rise in temperature was observed (temperature varied from 20°C to a maximum of 30°C), and therefore dynamic recrystallization did not occur, nevertheless the plastic deformation caused a noticeable grain refinement. This grain refinement is often the outcome of improvements in wear and corrosion resistance, generating compressive residual stresses enabling fatigue life enhancement. In terms of surface roughness, as Fig. 4 shows, an increase in the burnishing force led to a gradually decrease of the surface roughness up to 50% compare to AR sample. Burnishing process was able to give high brightness at the worked surface, also thanks to the MQL lubrication condition, which according to [4] is the optimal solution to obtain a lower roughness.

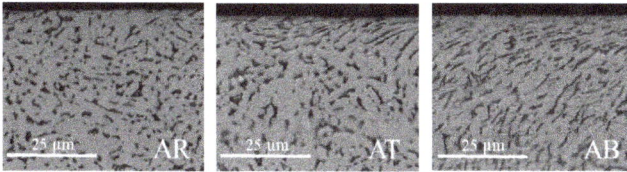

Fig. 2. Microstructure of the AR, AT and AB samples.

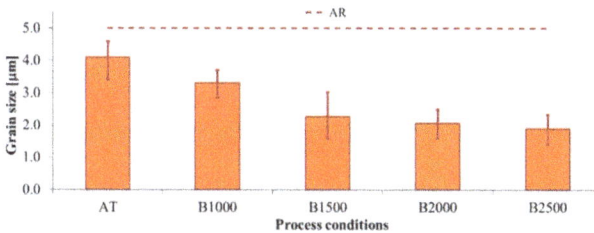

Fig. 3. Grain size evolution at varying burnishing force.

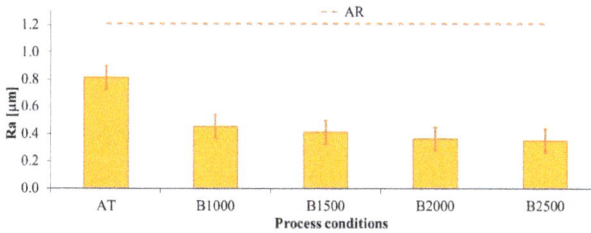

Fig. 4. Roughness at varying burnishing force.

Fig. 5 presents the hardness trend below the machined surface at varying burnishing forces. The hardness of the AR sample was measured equal to 390 HV. Thus, a slight change in the surface hardness resulted after burnishing compared to turning. Furthermore, the deformed layer was deeper after burnishing, which proves an increased work hardening, up to 400 µm from the machined surface, which increased at increasing burnishing force.

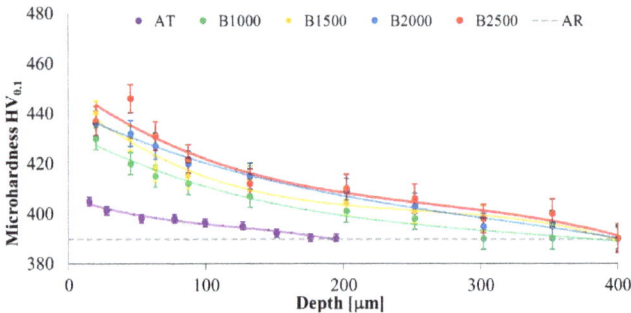

Fig. 5. Microhardness below the machined surface at varying burnishing force.

Summary

Burnishing process is shown to be viable SPD routes for achieving surface integrity enhancement in correspondence of the subsurface region of the biomedical grade *Ti6Al4V* alloy.

The burnishing process of *Ti6Al4V* highlighted enhancements in terms of grain refinement, surface roughness reduction, and hardness increase, which are fundamental characteristics required especially in the biomedical field to achieve the highest products performance.

Future studies will be devoted to investigate the influence of the different processing technologies on in-service life performances of the investigated biomaterials after being processed through burnishing. Specifically, further analysis concerning the influence of other burnishing process parameters as for instance the influence of the feed or tool radius on the *Ti6Al4V* wear resistance will be performed.

Acknowledgements

This research was carried out in the framework of the project PRIN 201742RB8R_002 "Bionic" funded by the Italian Ministry of University and Research.

References

[1] S. Affatato, Recent developments and future trends in total hip arthroplasty (THA), in: Perspectives in Total Hip Arthroplasty, Elsevier, 2014, pp. 76-95.

[2] J.S. Ansari, G.S. Matharu, H. Pandit, Metal-on-metal hips: current status, Orthop. Trauma 32 (2018) 54-60. https://doi.org/10.1016/j.mporth.2017.11.010

[3] M. Alfano, G. Ambrogio, F. Crea, L. Filice, F. Furgiuele, Influence of Laser Surface Modification on Bonding Strength of Al/Mg Adhesive Joints, J. Adhes. Sci. Technol. 25 (2011) 1261-1276. https://doi.org/10.1163/016942410X533381

[4] G. Rotella, L. Filice, Surface modifications induced by roller burnishing of Ti6Al4V under different Cooling/Lubrication conditions, Lect. Note. Mech. Eng. (2021) 141-151.

[5] S. Thamizhmnaii, B. Bin Omar, S. Saparudin, S. Hasan, Surface roughness investigation and hardness by burnishing on titanium alloy, J. Achiev. Mater. Manuf. Eng. 28 (2008) 139-142.

[6] K.L. Urish, N.J. Giori, J.E. Lemons, W.M. Mihalko, N. Hallab, Trunnion Corrosion in Total Hip Arthroplasty—Basic Concepts, Orthop. Clin. North Am. 50 (2019) 281-288. https://doi.org/10.1016/j.ocl.2019.02.001

[7] R.S. Revuru, N.R. Posinasetti, V.R. Vsn, M. Amrita, Application of cutting fluids in machining of titanium alloys—a review, Int. J. Adv. Manuf. Technol. 91 (2017) 2477-2498. https://doi.org/10.1007/s00170-016-9883-7

[8] V. Schulze, F. Bleicher, P. Groche, Y.B. Guo, Y.S. Pyun, Surface modification by machine hammer peening and burnishing, CIRP Ann. - Manuf. Technol. 65 (2016) 809-832. http://doi.org/10.1016%2Fj.cirp.2016.05.005

[9] I.S. Jawahir, Y. Kaynak, T. Lu, The impact of novel material processing methods on component quality, life and performance, Procedia CIRP 22 (2014) 33-44. https://doi.org/10.1016/j.procir.2014.09.001

[10] J. Caudill, J. Schoop, I.S. Jawahir, Correlation of surface integrity with processing parameters and advanced interface cooling/lubrication in burnishing of Ti-6Al-4V alloy, Adv. Mater. Process. Technol. 5 (2019) 53-66. https://doi.org/10.1080/2374068X.2018.1511215

Material Forming - ESAFORM 2023
Materials Research Proceedings 28 (2023) 1347-1356

Materials Research Forum LLC
https://doi.org/10.21741/9781644902479-146

Evolution of contact lengths during the turning of treated Ti64β

WAGNER Vincent[1,a] * and DESSEIN Gilles[1,b]

[1]Laboratoire Génie de Production, Ecole Nationale d'Ingénieurs de Tarbes, Université de Toulouse, Tarbes Cedex, France

[a]vincent.wagner@enit.fr, [b]gilles.dessein@enit.fr

Keywords: Contact Length, Ti64β, Tool Wear

Abstract. One of the main ways to increase productivity during the machining of titanium alloy parts is to control the wear mechanisms and consequently the life of the cutting tools. The state-of-the-art shows wear phenomena generated mainly by diffusion. The latter is due to the intrinsic physical properties of titanium generating high temperatures on the cutting and relief faces. Concerning the secondary cutting zone, the wear phenomena result from thermomechanical actions induced by the contact between the secondary zone of the chip and the cutting face of the tool. The analysis of the contact lengths, divided into two parts (sliding contact and sticking contact), is then a strong indicator to be privileged in order to allow the understanding of the present mechanisms. This analysis is even more important when machining processed Ti64 β in which chip formation is no longer periodic but a function of the angle between the primary shear band and the orientation of the individual colonies comprising the titanium. The purpose of this article is the analysis of the contact lengths during the machining of Ti64 treated β . After having presented the experimental device allowing visualising the evolution of the contact length between the chip and the cutting face, a statistical analysis based on the collected images makes it possible to put forward the differences of sliding behaviour on the cutting face. It appears that the distribution of the collected values allows explaining and extrapolating the wear appearing on the cutting face.

Introduction

When machining titanium alloys, two types of tool wear occur. Because of their high mechanical properties and low thermal properties, the wear encountered is flank wear and edge collapse. Tool wear has been studied for many years [1] and [2]. To avoid or limit tool wear, one of the solutions is then to define an optimal tool geometry and reduce the cutting conditions. The aim is to reduce the phenomena of diffusion. These phenomena appear mainly on the rake face and result from the high thermomechanical stresses applied to the cutting face over a distance called the contact length. Some studies concern the analysis of the parameter but it is always considered as constant and function of the cutting conditions, the cutting tool geometry and the workpiece behaviour. For example, Lee and Shaffer [3] and Abdulaze and al. [4] some pioneers in this filed developed an analytical tool-chip contact length models for standard steel. With this model, they link the geometrical aspect such as rake angle, uncut chip thickness or shear angle. Zhang et al. introduced the effect of cutting speed on the model of these lengths and in application for low-carbon steel [5]. Oxley provide a model based on coefficients traducing the strain hardening of the material [6]. Moufki et al attempt to take into account the effect of temperature of the friction law between the chip and the tool, where most of the studies use a Coulomb contact [7]. Moreover, it is commonly accepted that the contact length decreases with increasing cutting speed and rake angle and decreases with the feed rate [8]. A major part of these models are dedicated to a sharp cutting edge. However, when machining with chamfered tools, the cutting process is relatively different. Indeed, the chamfered edge will often lead to the formation of a dead metal zone under itself. This dead metal zone will replace the missing nose of the tool and form a small edge radius, changing the

Material Forming - ESAFORM 2023   Materials Research Forum LLC
Materials Research Proceedings 28 (2023) 1347-1356   https://doi.org/10.21741/9781644902479-146

shear angle. With this specific shape will appear a stagnation point around which the material will flow. Recently, Barelli and al. [9] showed there is a particular evolution of contact lengths during machining. They show from a statistical analysis of the contact lengths over a short period of time that the contact evolves and generates more or less significant deposits. However, this analysis, although accurate, is limited to a single alloy.

Wagner and al. [10] shows that chip formation during the machining of Ti64 β does not follow the cutting process introduced by [11]. It then appears that chip formation is not constant over time but is a function of the orientation of the structure encountered during the creation of the shear band. The morphology of the harvested chips is then very variable.

The objective of this study is to provide an analysis of the contact lengths during the machining of Ti64 β in order to identify if the particular process of chip formation, which is not constant, impacts the contact lengths. After having presented the alloy and the chip formation process, this study is based on an experimental device allowing measuring at each moment the contact lengths during machining. The images collected and analysed allow correlating the contact lengths with the cutting conditions and to compare their variability with the more common structure.

Studied Material

The objective of this part is to present the machined material as well as the experimental device. The alloy used is a Ti64 composed of titanium, 6% aluminium and 4% vanadium. The recrystallisation process is defined by different steps (Fig. 1). The difference between this structure and the more common structure results in the addition of a solution phase in the βdomain between the forging and tempering phases. Indeed, the microstructure of Ti64 is a function of the different stages of the recrystallisation process. The heat treatment is divided in four steps. The first one corresponds to the homogenisation of the structure. The second step is the forging carried out only in the domain α +β . This is followed by solution heat treatment starting in the β range and followed by a long cooling phase. Ultimately, a tempering is performed at a temperature below β transus.

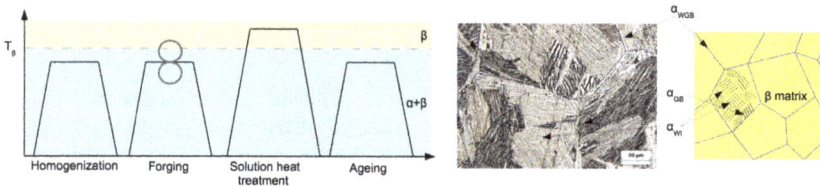

Fig. 1. Heat treatment (left) and structure of Ti64β (two figures on the right).

In contrast to the classical Ti64 structure (with nodular grain [11]), the structure of the alloy studied is fully lamellar (Fig. 2). The α phase is transformed into an acicular phase α ' in the matrix β. It is also possible to observe the α_{BG} phase near the grain boundaries. It appears as the continuous line around the old grains β. Concerning the α_{WBG} phase, it germinates from the grain boundary to the centre and appears as lamellae. Finally, the α_{WI} phase germinates inside the old grain β. The whole phases are immersed in the matrix β.

Material Forming - ESAFORM 2023   Materials Research Forum LLC
Materials Research Proceedings 28 (2023) 1347-1356   https://doi.org/10.21741/9781644902479-146

Experimental Device

For these tests, Genymab 900 turning lathe was used. The inserts used were CCMX 120408 with TiN PVD coating, the rake angle was 20° and inserts had a chamfer edge preparation of 0.02 mm. The tests were carried out on tubes in order to be in simple, generalisable cutting conditions. The chip flow will not be unidirectional insofar as a slight angle of inclination of the edge appears. These tests were carried out without lubrication in order not to disturb the image acquisition. The tubes used for these tests have a thickness of 3 mm. Force measurements during machining were performed using a Kistler 9129A dynamometer, a 5019A amplifier and Dynoware software.

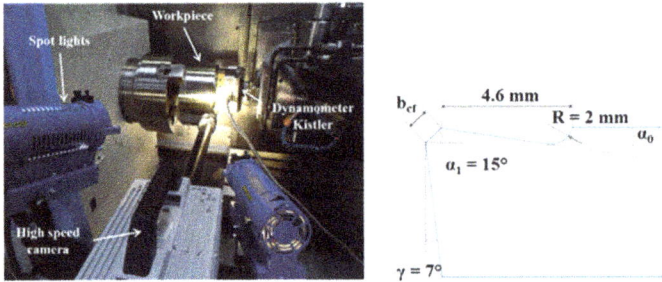

Fig. 2. Experimental setup.

The observations of the chip flow during the machining process were carried out with the help of the PhotronFastcam SA1 high-speed camera whose acquisition characteristics of 5000 images/s^{-1} is chosen. The resolution used is 1024x1024 pixels2 allowing an acquisition time of about 1s. For these tests, two cold light projectors were used to limit their influence on the cutting temperature. Definitely, rings were added between the cell and the x100 zoom lens. The resolution obtained is 5µ m. The experimental setup is shown in Fig. 2.

Chip Formation

For this alloy, the study shows that chip formation does not follow the classical process developed in many works [11-1]. Indeed, the chip formation is not only due to shears generated between the tool tip and the root of the chip. Several phenomena mainly due to the structure composed of lamellar colonies $\alpha + \beta$ oriented whose orientation varies appear (Fig. 3). The purpose of this section is to highlight these phenomena in order to explain, subsequently, the evolution of the contact lengths.

Material Forming - ESAFORM 2023   Materials Research Forum LLC
Materials Research Proceedings 28 (2023) 1347-1356   https://doi.org/10.21741/9781644902479-146

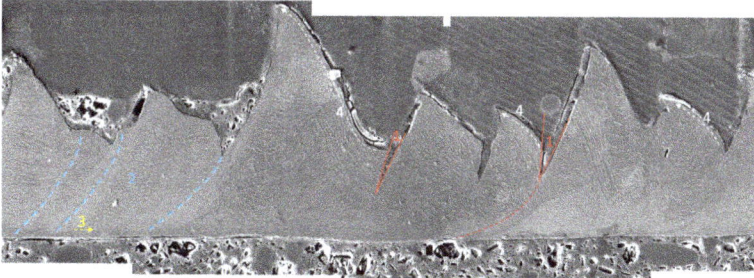

Fig. 3. Example of a micrograph of a Ti64 β chip.

The study of the chip formation allowed to put forward 4 configurations (Table 1).

Configuration 1

Whatever the cutting conditions, cracks can appear near the primary shear bands. There are two possible causes. The first one is similar to the one observed on more classical structures. The crack is then explained by intense shears. The second cause is specific to this alloy. The crack is then explained by a strong accumulation of plastic deformation at the interface of two strongly disoriented colonies (Colony 3 compared to Colony 4 and Colony 1 compared to Colony 2) (Table 1). In this case, the stress accumulation is explained by the difficulty of providing dislocations between two differently oriented colonies.

Configuration 2

The shear bands will appear in a privileged way in colonies whose orientation is transverse to the shear propagation. If the interfaces α /β are not a brake to the propagation of dislocations, a stress gradient exists between the different lamellaeβ going towards an increase of the stress near the interface. Consequently, the strain required to shear the interfaces is greater than that required to shear the laminae α .

Configuration 3

It appears that colonies relatively collinear to the main shear direction can be sheared transversely. The latter is initiated by the movement between the various colonies surrounding it.

Configuration 4

In this configuration, it appears that a continuous shearing of the colony takes place near the free edges of the chip. The chip thus follows the movement imposed between the end and the interface of the closest colony.

This initial study thus shows the randomness of chip formation. A study from [10] concerning the influence of cutting conditions on chip formation then shows that adiabatic shear and cracking become regular when the feed rate and therefore the shear forces are high enough to generate them despite the orientation of the colonies.

The objective of the next part is then to observe the effect of this particular chip formation on the chip flow length.

Table 1. Summary of the different configurations found.

Cutting Length Measurements

The images are then processed using Matlab © software following an algorithm detailed in Fig. 4. The first step consists in locating the cutting edge by defining in X and from the left of the image, the last black pixel (grey level 0). This same operation is performed in Y. From the coordinates of the cutting edge, the algorithm recreates the cutting edge. For this and for each column, the program defines the last pixel with gray levels equal to 255 (white). Then, the chip being one of the elements reflecting the lightest, it is easy to determine a threshold value of gray level delimiting the lower surface of the chip in the background. This value is specific to each serie of images, the algorithm artificially defined that all gray levels below this threshold value are equal to 0. From the points belonging to the edge, the algorithm searches column by column, in the direction of the positive Y, for the last pixel with a gray level of 0. These pixels then belong to the lower surface of the chip. The set of coordinates of the edge and the cutting face being known, it is then possible by difference to calculate the contact lengths between the chip and the cutting face.

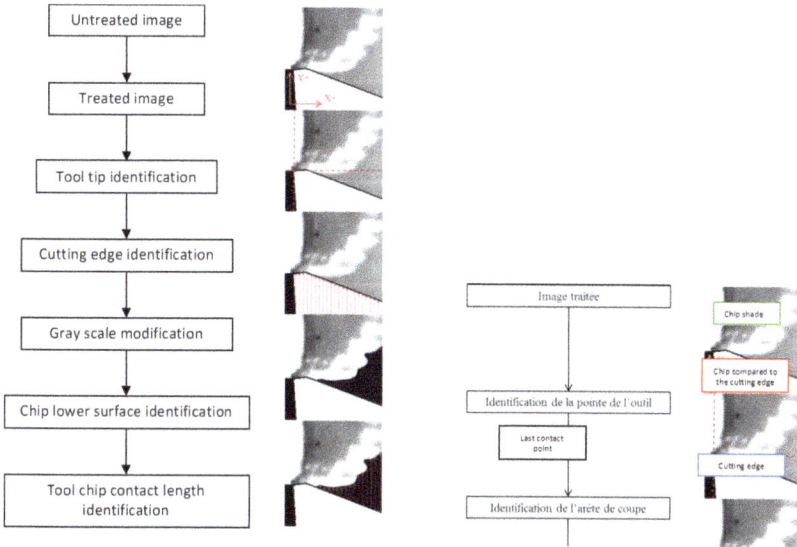

Fig. 4. Image processing.

Given the large number of values obtained, a statistical analysis was carried out. The results obtained are presented in the form of a moustache box (Fig. 5). This choice makes it possible to highlight the distribution of the data and the preferred values taken by the contact lengths. On the representations exposed in this article will be exposed the 5th percentile (the low end of the segment), the 1 st quartile (the low part of the box), the median (the line in the middle of the box), the 3rd quartile (the high part of the box) and finally the 95th percentile (the end of the segment). Each result sums up measurements made on 1000 images of tool-chip contact lengths within cutting process.

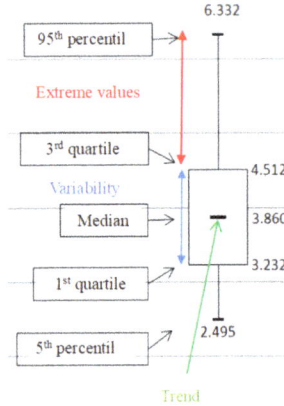

Fig. 5. Moustache box.

When machining Ti64 with a classical structure, the general trend observed is a quasi-linear increase in contact lengths with the feed rate. The increase of the cutting speed has the effect of slightly reducing the contact lengths but in lesser proportions compared to the feed rate.

For this microstructure, the trends are slightly different (Fig. 6). The classical relationship between contact length and feed rate does not appear. For all three cutting speeds tested, increasing the feed rate results in an increase in contact length. From these tests, it is not possible to define the conditions giving the greatest contact length. Concerning their extent, the most variable conditions appear for Vc=30 m/min. Indeed, whatever the quantile or the percentile observed, the greatest differences appear for Vc=30 m/min.

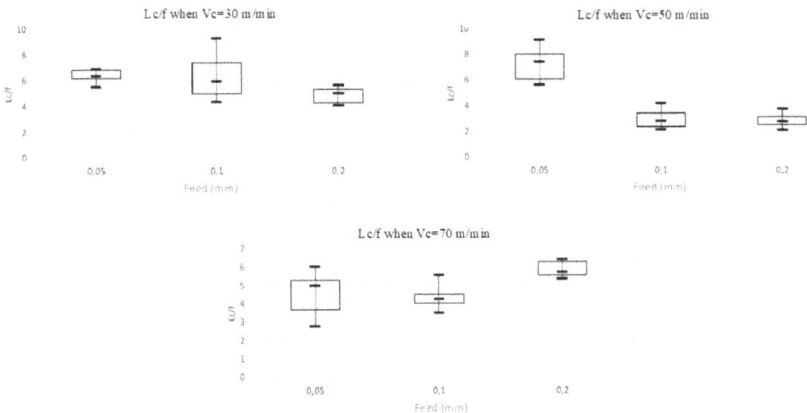

Fig. 6. Evolution of contact lengths for Vc=30 m/min, Vc=50 m/min and Vc=70 m/min.

Material Forming - ESAFORM 2023   Materials Research Forum LLC
Materials Research Proceedings 28 (2023) 1347-1356   https://doi.org/10.21741/9781644902479-146

This trend appears to follow the cutting process and chip formation developed earlier. The variability also seems to correlate with the effect of cutting conditions on chip formation.

Contrary to the analysis made on the more classical structure, it appears that whatever the analysis time, the radius of curvature is not impacted. Table 2 shows the chip formation observed at different times (at the beginning of the analysis, in the first quarter, in the middle, in the third quarter and at the end). It appears that the chip never touches the chip breaker, never segments and because of the edge inclination never meets the machined face (in this example out of the field). Moreover, the tests showed that over the recording period, the chip weight did not influence the contact lengths. Indeed, as the chip does not segment, the chip weight could impact the radius of curvature and consequently the contact lengths. The change in contact lengths can then be explained by the formation of chips.

Table 2. Evolution of chip formation over the duration of the statistical analysis.

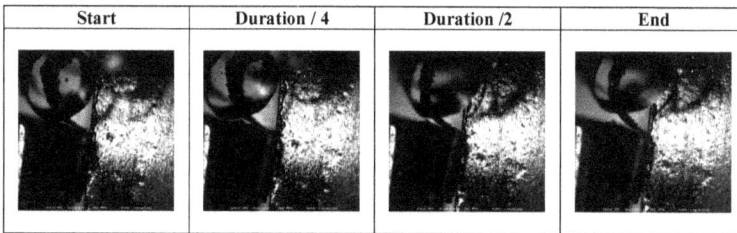

Start	Duration / 4	Duration /2	End

The study of the effect of cutting conditions on chip formation shows a strong correlation between cutting speed, feed rate and chip geometry. In contrast to the more classical structures where the chips remain, whatever the cutting conditions, more or less scalloped chips, the shape of the chips is much more evolutive. Table 3 shows the effect of cutting conditions on chip formation. At low cutting speeds, the chips are continuous. As the cutting conditions increase, it appears that the chips become scalloped with increasingly constant distances between scallops. However, unlike other alloys, the scalloping becomes more regular as the cutting conditions increase. This relationship is explained by the level of stress that appears at high cutting conditions. As said before, the creation of shear bands is initiated especially when the colonies are oriented perpendicular to the band formation. This initiation is the result of high thermomechanical stress accumulation at the boundary of highly disoriented colonies. Higher stress accumulation then facilitates the creation of primary shear bands. An increase in cutting conditions results in higher stresses and temperatures which then facilitate the appearance of shear bands. Wagner also shows that for certain cutting conditions, shear bands are created regardless of colony orientation [1].

In the study of contact lengths, it would appear that the lengths and especially the variability are a function of chip formation.

For Vc=30 m/min and when fz=0.05 mm, the variability is the lowest. For these conditions, the thermomechanical stresses are so low that the chip is continuous, whatever the structure. In this case, the variability is the lowest. When the feed rate increases, the stresses also increase and the chip formation becomes a function of the orientation of the microstructure. Its morphology becomes variable as well as the contact length. When Vc=50 m/min, there is a significant variability regardless of the feed rate. This disparity is explained by transient phenomena. For the lowest feed rate, the stresses are too low to generate a shear band each time. Conversely, when the feed rate increases, the stresses are greater but above all the probability of the primary bands crossing colonies increases and the chips, like the contact lengths, remain stable.

Table 3. Chip morphology as a function of cutting conditions.

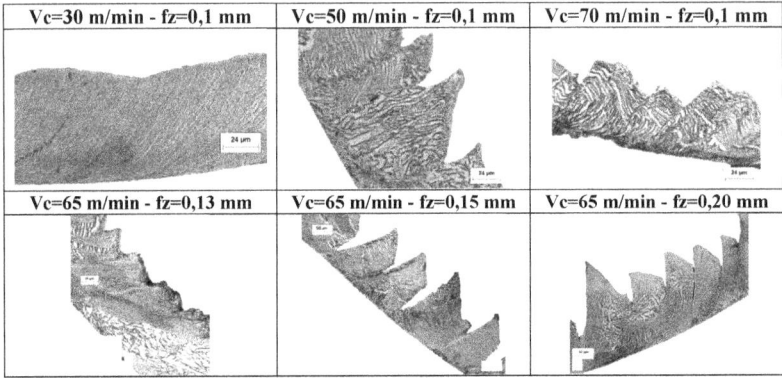

Vc=30 m/min - fz=0,1 mm	Vc=50 m/min - fz=0,1 mm	Vc=70 m/min - fz=0,1 mm
Vc=65 m/min - fz=0,13 mm	Vc=65 m/min - fz=0,15 mm	Vc=65 m/min - fz=0,20 mm

Summary

Through an innovative experimental device, the objective of this study is to observe the evolution of contact lengths during the machining of Ti64β . After having detailed the microstructure and the particularities of the chip formation, the study develops the method of statistical analysis of the collected images. It appears that this evolution is mainly a function of the chip formation process. The study also shows a strong variability of the contact lengths explained by various points like, for example, the weak cutting conditions obtained during certain tests. In order to complete this analysis and to clearly define the relationship between contact length and tool wear, the next step is to chemically analyse the inserts and to propose a diffusion model complementary to those already existing based on a large experimental campaign.

References

[1] V. Wagner, E. Duc, Study of Ti-1023 milling with toroidal tool, Int. J. Adv. Manuf. Technol. 75 (2014) 1473-1491. https://doi.org/10.1007/s00170-014-6217-5

[2] V. Wagner, M. Baili, G. Dessein, The relationship between the cutting speed, tool wear, and chip formation during Ti-5553 dry cutting, Int. J. Adv. Manuf. Technol. 76 (2015) 893-912. https://doi.org/10.1007/s00170-014-6326-1

[3] E.H. Lee, W. Shaffer; The theory of plasticity applied to the problem of machining, J. Appl. Mech. T ASME 18 (1951) 405-413.

[4] N.G. Abuladze. Character and the length of tool-chip contact (in Russian), in: Proceedings of the machinability of heat-resistant and titanium alloys, Kuibyshev, 1962, pp. 68-78.

[5] H.T. Zhang, P.D. Liu, R.S. Hu, A three zone model and solution of shear angle in orthogonal machining, Wear 143 (1991) 29-43. https://doi.org/10.1016/0043-1648(91)90083-7

[6] P.L.B. Oxley, Mechanics of machining: an analytical approach to assessing machinability, Chichester: Ellis Horwood Limited, 1989.

[7] A. Moufki, A. Molinari, D. Dudzinski, Modelling of orthogonal cutting with a temperature dependent friction law, J. Mech. Phys. Solid. 46 (1998) 2103-2138. https://doi.org/10.1016/S0022-5096(98)00032-5

[8] N.N. Zorev, Metal cutting mechanics, Oxford: Pergamon Press, 1966.

[9] F. Barelli, V. Wagner, R. Laheurte, G. Dessein, P. Darnis, O. Cahuc, M. Mousseigne, Orthogonal cutting of TA6V alloys with chamfered tools: Analysis of tool-chip contact lengths,

Material Forming - ESAFORM 2023
Materials Research Forum LLC
Materials Research Proceedings 28 (2023) 1347-1356
https://doi.org/10.21741/9781644902479-146

Proceedings of the Institution of Mechanical Engineers, Part B: J. Eng. Fabric. 231 (2017) 2384-2395.

[10] V. Wagner, F. Barelli, R. Laheurte, G. Dessein, P. Darnis, O. Cahuc, M. Mousseigne, Thermal and Microstructure Study of the Chip Formation During Turning of Ti64 β Lamellar Titanium Structure, J. Manuf. Sci. Eng. 140 (2018) 031010. http://doi.org/10.1115/1.4038597

[11] R. Komanduri, B.F Von Turkovicz, New Observations on the Mechanism of Chip Formation When Machining Titanium Alloys, Wear 69 (1981) 179-188. https://doi.org/10.1016/0043-1648(81)90242-8

Material Forming - ESAFORM 2023
Materials Research Proceedings 28 (2023) 1357-1366

Materials Research Forum LLC
https://doi.org/10.21741/9781644902479-147

Investigation of machining performance of lead-free brass materials forged in different conditions after cooling with liquid nitrogen

ATAY Gokhan[1,a] *, ZOGHIPOUR Nima[1,2,b] and KAYNAK Yusuf[2,c]

[1]Torun Metal Alloy Ind. & Ltd. Inc., GOSB, İhsan Dede St. Gebze,Kocaeli, Turkey

[2]Department of Mechanical Engineering, Marmara University, Istanbul, Turkey

[a]gokhan.atay@torunmetal.com, [b]nima.zoghipour@gmail.com, [c]yusuf.kaynak@marmara.edu.tr

Keywords: Lead-Free Brass, Microstructure, Hot Forging, Machining, Cryogenic Cooling

Abstract. The forging process, which is one of the hot forming methods, is used to produce plumbing systems, valves, batteries, fittings, condensers, pipes, etc. from brass alloys in duplex structures. Compared to leaded brasses, machining becomes very difficult due to reasons such as long and continuous chips, burrs, built-up edge in the tool, and unacceptable surface quality problems. In this study, studies have been carried out to improve the machining process that minimizes the mentioned machining problems by controlling the microstructural of lead-free brass materials. Therefore, by examining the effect of the copper content of the forged material, forging temperature, the cryogenic cooling method input parameters after forging on the phase ratio of the material, investigations were carried out on obtaining a high-performance machining process close to the machinability performance of leaded brass materials with good chip breakability. The change in the distribution of beta phases in the material after forging was investigated and machining tests were performed as a hole drilling operation. Cutting forces, surface roughness, and dimensional accuracy analyzes were evaluated within this purpose. The relationship between the beta phase ratio in the material and the machinability has been revealed and it has been observed that the CW511L alloy produced in three different ratios in the standard range and although they are the same material, the Cu% ratio causes significant changes in terms of machinability.

Introduction

Forging process is one of the widely used plastic deformation methods in terms of shaping the material. In this method, in which components from copper alloys are widely produced, the deformability of the billet material increases by heating it in the furnace and it is easily shaped. Zinc element in body-centered cubic crystal lattice structure, which is one of the elements that increase the forging ability. Brass is generally known as copper alloys containing 5% to 40% zinc in concentrations. While brass is manufactured, strength, corrosion resistance, machinability etc. Different alloying elements can be added to change the properties [1]. The most important element in this context is lead, which is associated with excellent chip breaking, low tool wear, and high cutting parameters. However, lead poses a danger to humans and nature and is included in the 2nd class carcinogen group, which cannot be eliminated by human body [2,3]. Pipes, fittings, etc., are produced from brass alloys, especially in contact with drinking water. The lead element ratio in the alloy in the components should not be more than 0.20%. CW511L material standard alloy ratios also support this [4]. In this study, the material that was shaped by hot forging and on which the machining tests were carried out is CW511L from lead-free brass materials. In the brass alloy, the semi-fluid lead layer reduces friction during cutting, acts as a lubricant, and encourages the formation of small discontinuous chips, improving machinability [5,6], cutting forces generated during machining, and thus cutting tool life are improved. The fact that lead is such an important element for machinability and its elimination causes various problems in terms of machinability

Material Forming - ESAFORM 2023   Materials Research Forum LLC
Materials Research Proceedings 28 (2023) 1357-1366   https://doi.org/10.21741/9781644902479-147

in lead-free brass alloys and results in a decrease in productivity and disruption of processes with long chips and reduced tool life that occur during the processing of these materials [6].

Pantazopoulos et al. [7] investigated the microstructure analysis, mechanical behavior and fracture mechanisms of traditional leaded brass materials CuZn39Pb3 (CW614N) and CuZn36Pb2As (CW602N). It has been revealed in their studies that CW602N material with high copper content contains lower β phase compared to the other, has lower hardness, and decreases in toughness due to phase structure. Nobel et al. [6] examined and compared the machinability of lead-free alloys CW510L, CW511L, CW724R, CW508L and leaded alloys CW614N. After the machining of these materials by turning using carbide cutting tools were used in different coatings, temperature and force results investigations were carried out in terms of chip breakability, chip form and tool life. Toulfatzis et al. [8] evaluated the microstructures and β phase ratios of extruded cylindrical lead-free brass materials after heat treatment and looked at the impact of these outputs on the material's machinability in their study. In a different investigation [9], they used heat treatment to lead-free brass alloys to explore the chip breakability, the change in β phase following this heat treatment, and the machining test that followed. The fracture energy and fatigue life of lead-free brass alloy (CW510L and CW511L) materials as determined by the material's notch impact test were examined by Pantazopoulos et al. [10]. The drilling process of hot forged lead-free brass alloys was investigated experimentally by Zoghipour et al. [11] using a form cutting tool. The cutting forces, dimensional accuracy, and surface quality of the holes were tested while taking into account the tools' varied geometries, feed rates, and rotating speeds on hot-forged lead-free brass alloys with different copper contents. Then, to predict and improve the machining process, they applied genetic algorithm-based optimization techniques and modeling of artificial neural networks.

It is anticipated that changes in the material's copper ratio in the Cu-Zn equilibrium diagram will have a significant impact on the material's forgeability and recrystallization temperature, as well as on its microstructural distribution, α and β phase distribution, and consequently its machinability performance. After being hot forged into the desired shape, the brass alloy material is cooled either on its own or in fan conveyor systems, depending on the desired microstructural and mechanical qualities. There are several studies in the literature that demonstrate the use of the cryogenic cooling method in various operations for the machining of different materials [12–14], that it improves machinability performance [15], and that it has no negative effects on human health or the environment [16]. It is crucial to conduct systematic studies to enhance the machinability performance of lead-free brass materials because there are not any available studies examining the impact of forging parameters and cooling process on β phases in the material and the change in machinability performance as a result. In the study, a major breakthrough was made in terms of machinability since the β phases of the cryogenic cooling approach were restrained in the structure without dissolving with rapid cooling. Therefore, it has been attempted to investigate the effects of the forged material's copper content, the sample temperature prior to forging, the cryogenic cooling method input parameters following forging, and the material's phase ratio on achieving a high-performance machining process that is comparable to the machinability performance of materials containing lead. The microstructural characteristics of the material are significantly affected by forging process parameters including the material's percentage Cu alloy, the post-forging cooling technique, the forging temperatures, etc. These consequences have been shown along with this study. Several process characteristics, such as machinability, have a substantial impact on the β%, which are crucial in terms of machinability.

Experimental Setup

Hot forged extruded bars with dimensions of Ø50x66 mm and three different Cu% rates were put through machining testing. The shaped workpieces formed by the hot forging technique are shown in Fig. 1a. Before being employed in the machining experiments, Fig. 1b depicts the burr cutting

Material Forming - ESAFORM 2023
Materials Research Proceedings 28 (2023) 1357-1366

Materials Research Forum LLC
https://doi.org/10.21741/9781644902479-147

procedure and the final shape shown in Fig. 1c. Based on the results of the spectral analysis, the chemical composition of the CW511L alloy material generated at three different Cu% ratios is shown in Table 1.

Fig. 1. a) Forged CW511L sample with burrs, b) burr cutting process c) final shape.

Table 1. Chemical composition of the CW511L material with different %Cu.

Material	Cu	Zn	Pb	As	Al	Sn	Ni	Bi	Si	Sb
%61.7 Cu	61.7	Rem.	0.181	0.106	0.003	0.00	0.000	0.003	0.011	0.004
%62.9 Cu	62.9	Rem.	0.174	0.117	0.003	0.00	0.001	0.003	0.002	0.003
%63.4 Cu	63.4	Rem.	0.164	0.106	0.003	0.00	0.002	0.003	0.006	0.002
EN12165 Standart [4]	61.5 63.5	Rem.	- 0.2	0.02 0.15	- 0.05	- 0.1	- 0.3	-	-	-

In a Hydromec-550 tons automated press, forging tests were performed. Forging temperatures of 750, 780 and 810°C were selected during the forging process. All test samples were subjected to a 1 minute cryogenic cooling process in a chamber made of 304L stainless steel, which has a low heat transfer coefficient and is resistant to cryogenic temperatures, after forging. This has been performed in order to apply the coolant (liquid nitrogen) to the component uniformly during cryogenic cooling. During this time, the test samples at forging temperatures between 750 and 810°C were cooled to an average temperature of 200°C. The machining tests were conducted on Fanuc-Robodrill α-D21MiB5 CNC machine as shown in Fig. 3. Boron coolant flood was used during the machining tests. Ø12 spherical carbide cutting tools were particularly made and utilized in the tests as part of the experimental investigation. A thermal camera was used to monitor this process for each sample. Table 2 provides the experiments specific conditions and parameters. Following the machining tests, the CMM was used to measure the dimensional accuracy in the hole diameters.

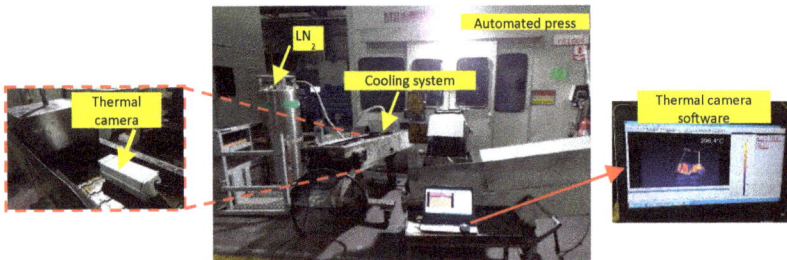

Fig. 2. Forging press and experimental setup.

Material Forming - ESAFORM 2023
Materials Research Proceedings 28 (2023) 1357-1366

Materials Research Forum LLC
https://doi.org/10.21741/9781644902479-147

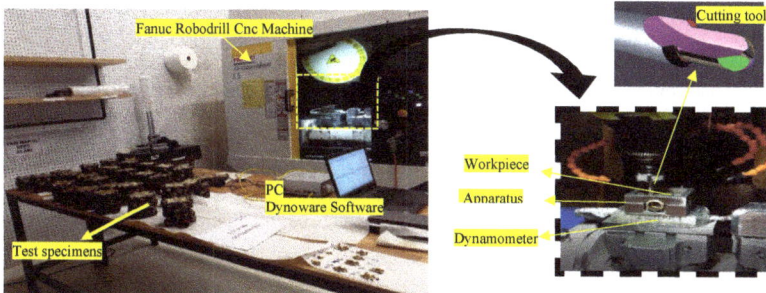

Fig. 3. The used experimental setup for the machining tests.

Table 2. The utilized parameters in hot forging and machining experiments.

Cu (%)	Forging Temperature (°C)	Cooling Duration (min.)	Rotational Speed (rpm)	Feed rate (mm/min.)
61,7	750			
62,9	780	1 (LN$_2$)	3000	300
63,4	810			

Results and Discussions

β Phase Analysis.

The β% phase ratio and its distribution are the most important factors affecting machinability. In this section, β phases are more clearly expressed by the red color in phase analysis microstructural distributions while the yellowish colors represent α phase. In the case of the material with the lowest Cu content of 61.7%, its intermetallic β% phase is higher compared to the other materials. In the Cu-Zn equilibrium diagram, the percentage of the β phase is controlled by the percentage of Zn, subsequently the Cu% content, and the thermomechanical process. The β phase percentage increased with Zn content, as expected, in the case where the zinc-rich and associated Cu% was the lowest. The β phase (Cu-Zn), an ordered type of intermetallic phase with a body-centered cubic crystal structure, (Cu) is a face-centered cubic crystal which expresses the solid solution, is structurally harder and less ductile than the α-phase. Zinc has the hexagonal close-packed (HCP) lattice structure [17]. The distribution and sizes of these phases have a significant effect on machinability [18]. The presence of the beta phase reduces the ductility of the alloy and thus the segmentation of the chips formed after machining. The morphology exhibited by the β phases and the increase in the percentage distribution in the structure cause a slight increase in hardness and low ductility, as a result, machinability, and chip breakability during machining increase with the increasing beta phase [8, 19]. The measured results from the microstructural analysis of β phase distributions in Fig. 4 demonstrate the effect of the %Cu and forging temperature on the distribution of β% phases in the material. The β phases in the structure showed a decreasing trend in terms of distribution and percentage with the increase of Cu%. This situation can be explained by the decreasing zinc distribution with the increase of Cu% in the structure. β phase is rich in zinc elements. Especially at high forging temperatures, the 61.7% Cu material rises to the point where the percent β phases are at their maximum during the process. By rapid cryogenic cooling, these β phases are being restrained in the brass material's internal structure [20]. In other words, the

Material Forming - ESAFORM 2023
Materials Research Proceedings 28 (2023) 1357-1366

Materials Research Forum LLC
https://doi.org/10.21741/9781644902479-147

necessary time duration and temperature are not given for the β phases in the structure, such as slow cooling, to dissolve and turn into the α phase. In the case of forging at 750 °C forging temperature, approximately 70% less β phase distribution is obtained from the 61.7% Cu sample, whereas in 63.4% Cu specimen this value is around 63%. While there was no significant change in the β% phase ratios of the forging temperature in the 61.7% Cu sample, there was a decrease of approximately 22% from 750 °C to 810°C in the 62.9% Cu specimen, and an increase of approximately 25% was observed in the 63.4% Cu workpieces. There were no significant changes in the β phase distribution of the forging temperature in the samples with a low %Cu, but no clear distributions in the β% phase in the material with an increase in the Cu% could be obtained. It can be said that this is caused by the decrease in the percentage of zinc in the structure with the increase of %Cu.

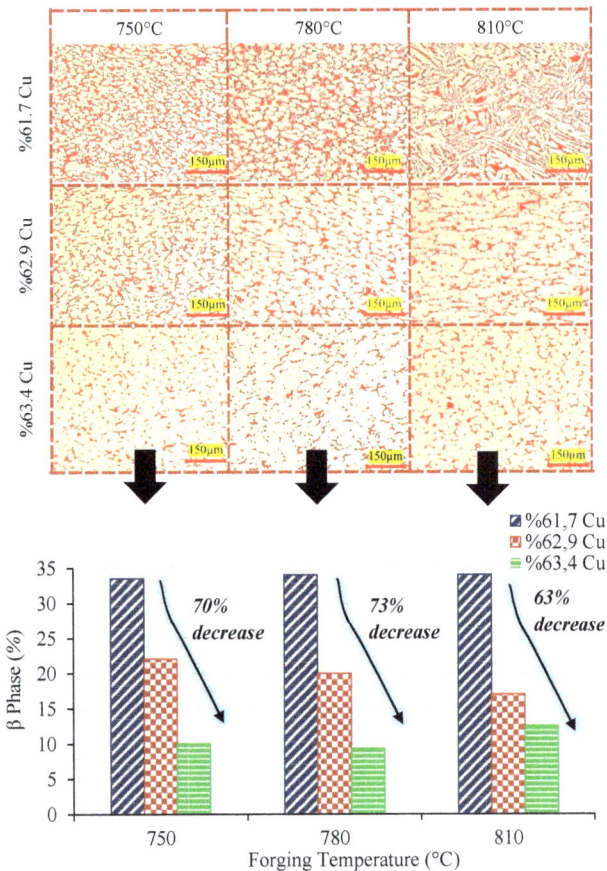

Fig. 4. Effect of Copper content and forging temperature on the beta phases of material.

Material Forming - ESAFORM 2023   Materials Research Forum LLC
Materials Research Proceedings 28 (2023) 1357-1366   https://doi.org/10.21741/9781644902479-147

Cutting Forces.

One of the copper alloys, CW511L, is a low lead alloy that is difficult to machine than other brass alloys [15]. This may therefore result in excessive cutting forces and other machining operations [2]. The feed force is the primary cutting force affecting the cutting tool. Under normal conditions, major variations in cutting forces and machinability qualities are primarily influenced by machining cutting parameters such as speed, feed, and cutting tool geometry. Alloy microstructure and mechanical characteristics affect an alloy's machinability. While the machinability of the metal depends on the metal's microstructural changes, the power consumption depends on the metal's mechanical characteristics [20]. In this instance, β phases and hence hardness is also impacted. The size of the interphase boundaries, which are influenced by the β-phase content and the size of the α-phase crystals, can be increased to enhance machining. In general, soft materials use less energy than hard metals, and tool wear is reduced. Finer grain sizes are advantageous for mechanical qualities, and since they are stronger and more ductile than coarser grains and possess higher machinability. Additionally, imperfections in the material's structure may have an impact on chip breakability and therefore machinability [18].

The average of three repetition cutting forces generated during drilling operation are presented in Fig. 5 according to the Cu% and the forging temperature. A close examination of the graph reveals a considerable impact of the Cu% ratio. It has been noted that the alterations made result in considerable variations in terms of machinability, especially given that these three Cu% ratios are CW511L despite being the same material. The dispersion of the β phases increased with the reduction of Cu%, and as a result, greater β phase distribution and finer α crystals tend to produce longer interphase boundaries, and microstructural distributions significantly influence on the machinability [18]. With an increase in the Cu%, it is seen that the cutting forces increase and the machinability decreases. An increase in cutting forces with significant friction in the contact area may occur in alloys with a ductile phase and low lead content in high FCC lattice structures [6]. This is confirmed by the fact that cutting forces rise with higher phases due to an increase in the Cu%. Once again, in different research [21], machinability tests on various copper alloys were conducted, and it was found that the alloys that we may refer to as zinc-poor or copper-rich yielded the greatest results in terms of cutting forces. At 750°C of forging temperature, 885 N force was attained at a rate of 61.7% copper, a rise of about 23% at a rate of 62.9% copper, an average of 1092 N, and an increase of roughly 36% at a rate of 63.4% copper, resulting in a force value of 1207 N. As can be observed, when the Cu ratio is lowest, the dominating β phases, and smaller grain sizes of the α phases facilitate cutting. In comparison to coarse grains, grains with a finer structure in terms of grain sizes enhance the structure and improve workability [20]. At 780°C forging temperature, 864 N force was measured in 61.7% Cu material, while an increase of roughly 27% at 62.9% Cu, an average of 1099 N, and 63.4% Cu, an increase of nearly 42%, and a force value of 1224 N was recorded. According to the obtained results, forging temperature has no impact on the cutting forces needed for machinability. 869 N force was obtained at 810°C forging temperature at a rate of 61.7% Cu, whereas 1183 N average force was obtained with an increase of almost 36% at a rate of 62.9% Cu, and this ratio was the highest percent increase compared to lower forging temperatures. There was a rise of about 43% and a force value of almost 1240 N was generated at the rate of 63.4% Cu.

Material Forming - ESAFORM 2023
Materials Research Proceedings 28 (2023) 1357-1366

Materials Research Forum LLC
https://doi.org/10.21741/9781644902479-147

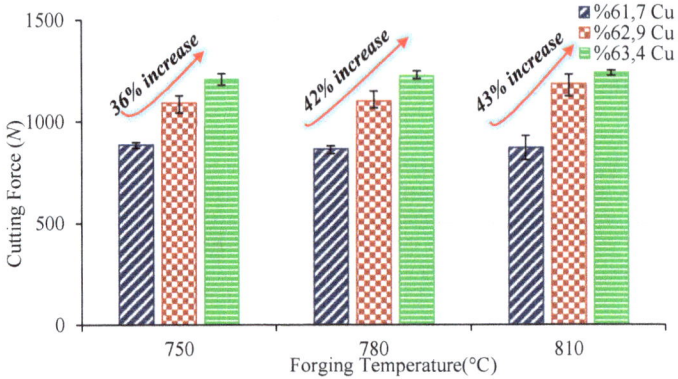

Fig. 5. The average measured cutting forces.

Surface Roughness.

The surface quality of the machined components is a desired characteristic and maintaining high surface quality is always one of the most crucial duties after machining. Surface roughness is known to impact a part's heat conductivity, friction, corrosion, and fatigue life [22]. In specific situations, machining factors such as feed and cutting tool tip radius have an impact on surface roughness [23]. However, since consistent speed, feed, and cutting tools were employed in this investigation, additional knowledge was gained about how the forging parameters and therefore the change in the material effect the machining and hence the change in the desired surface quality. Cutting tools that have been improperly ground have a propensity to scratch soft materials, like brass, leading to a poor surface finish. Fig. 6 shows the variations in surface roughness (Ra) values of the three different Cu% samples formed at 750-780 and 810°C forging temperatures as a result of changes in parameters following the cryogenic cooling process. The illustrated results represent the average of five measurements. When the graph is inspected, the roughness values decrease as Cu% increases. This means that improving the ductility of the outer surface with an increase in Cu% will result in a higher surface quality. While hard materials produce more clastic chips, it may be claimed that ductile materials provide superior surfaces with somewhat more continuous chip formation. Considering the roughness values at 750°C forging temperature, the highest roughness value was found in the material with a Cu content of 61.7%. The measured value in this condition is typically 0.52 μm. With an average roughness value of 0.43 μm measured at 62.9% Cu, it produces a value that is around 17% less than that of 61.7% Cu, while a 17–18% change was seen at 63.4% Cu ratio. At 780°C forging temperature, the highest roughness value was found in the material with a Cu content of 61.7%. In this condition, the measured value is around 0.49 μm on average. With an average roughness value of 0.37 μm measured at 62.9% Cu, a value was approximately 24% less than that of 61.7% Cu material, while a decrease of approximately 18% was observed in 63.4% Cu material. It can be interpreted that the rate of change in the roughness values of Cu% ratio decreases to minimum levels under the condition of the highest forging temperature in the experiments.

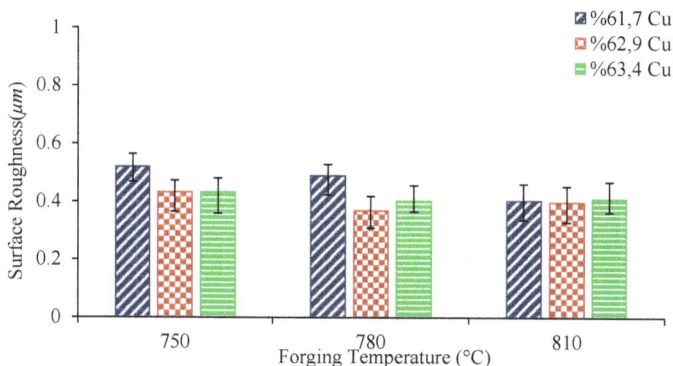

Fig. 6. The average values of the surface roughness (Ra) of the generated holes.

Dimensional Accuracy.

The dimensions and geometric tolerances in the component geometry are extremely important, much as the surface quality attributes of the machined item. After the manufacturing of parts, significant efforts are undertaken, particularly in industry, to obtain products with the appropriate dimensions and quality. Softer materials are more prone to bend when subjected to cutting tool pressure, decreasing dimensional accuracy. High-strength alloys may require carbide tools, grinding, and in certain conditions heat treatment for optimal manufacturing, whereas soft coppers and brasses may be machined rather readily [20]. The diametrical variations of the Ø12 holes that were produced after drilling in materials with various Cu contents of test specimens that were forged at different forging temperatures and treated to cryogenic cooling are shown in Fig. 7. Examining the graph reveals that the Cu% has some influence because, under all conditions, the hole sizes fall within the specified tolerance range. Additionally, it can be observed that holes in samples with a Cu ratio of 63.4% take values that are closer to the nominal size. The ductility of the material allows for a superior surface topography in the scenario when the Cu% ratio is at its highest, and the surface roughness values shown in the preceding section further corroborate this claim. However, experiments with the cutting tool in sequential conditions are required before it can be said that this is long-term. Because issues like lengthy chips, continuous chip status, and surface roughness that can result from chip removal from ductile materials can damage cutting tools and result in additional quality issues. It is typical for these graphical findings to show little fluctuations since the correct cutting tool dimensions are the most crucial aspect of dimensional variations. However, even if there is a change at low levels, the impact of this will be assessed as the impact of material and forging parameters will be explored within the scope of the research.

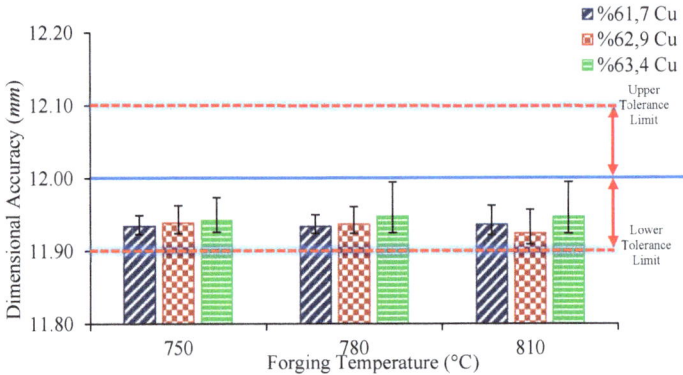

Fig. 7. The variation of the dimensional accuracy values of the generated holes.

Summary

In this research, the machining performance of CW511L material—one of the copper alloy-based lead-free brass materials—was explored by comparing its microstructural characteristics after forging. With the hot forging of the test sample at the correct temperature values, the β% phases in the structure increase to extremely high levels, and in this instance, the Cu% ratio has a major impact. This is a crucial step for machining because it ensures that most of these β phases remain in the structure without transformation (starting with the regular structure) by the action of cryogenic cooling. Additionally, the cutting forces clearly indicate the influence of the forging temperature, the Cu% ratio, and indirectly the distribution of β phases, as well as the influence of the input parameters on the outcomes for roughness and dimensional accuracy. The study is notable because, despite the fact that all of the samples used for the testing were CW511L, there was a substantial variation in the machinability of three distinct CW511L materials that were chosen from the standard range and in the machining tests.

References

[1] R. Francis, The corrosion of copper and its alloys: a practical guide for engineers, NACE International, 2010.

[2] X. Chen, A. Hu, M. Li, D. Mao, Study on the properties of Sn–9Zn–xCr lead-free solder, J. Alloy. Compd. 460 (2008) 478-484. https://doi.org/10.1016/j.jallcom.2007.05.087

[3] H. Atsumi, H. Imai, S. Li, K. Kondoh, Y. Kousaka, A. Kojima, High-strength, lead-free machinable α–β duplex phase brass Cu–40Zn–Cr–Fe–Sn–Bi alloys, Mater. Sci. Eng. A 529 (2011) 275-281. http://doi.org/10.1016/j.msea.2011.09.029

[4] TSE, TS EN 12165, in Copper and copper alloys - Wrought and unwrought forging stock, 2016.

[5] A.L. Fontaine, V. Keast, Compositional distributions in classical and lead-free brasses, Mater. Charact. 57 (2006) 424-429.

[6] C. Nobel, F. Klocke, D. Lung, S. Wolf, Machinability enhancement of lead-free brass alloys, Procedia CIRP 14 (2014) 95-100. https://doi.org/10.1016/j.procir.2014.03.018

[7] G.A. Pantazopoulos, A.I. Toulfatzis, Fracture modes and mechanical characteristics of machinable brass rods, Metallogr. Microstruct. Anal. 1 (2012) 106-114. https://doi.org/10.1007/s13632-012-0019-7

Material Forming - ESAFORM 2023
Materials Research Proceedings 28 (2023) 1357-1366

Materials Research Forum LLC
https://doi.org/10.21741/9781644902479-147

[8] A.I. Toulfatzis, G. Pantazopoulos, A. Paipetis, Microstructure and properties of lead-free brasses using post-processing heat treatment cycles, Mater. Sci. Technol. 32 (2016) 1771-1781. https://doi.org/10.1080/02670836.2016.1221493

[9] A.I. Toulfatzis, G. Pantazopoulos, C. David, D.S. Sargis, A. Paipetis, Final heat treatment as a possible solution for the improvement of machinability of pb-free brass alloys, Metals 8 (2018) 575. https://doi.org/10.3390/met8080575

[10] A.I. Toulfatzis, G. Pantazopoulos, A. Paipetis, Fracture mechanics properties and failure mechanisms of environmental-friendly brass alloys under impact, cyclic and monotonic loading conditions, Eng. Fail. Anal. 90 (2018) 497-517. https://doi.org/10.1016/j.engfailanal.2018.04.001

[11] N. Zoghipour, G. Atay, Y. Kaynak, Modeling and optimization of drilling operation of lead-free brass alloys considering various cutting tool geometries and copper content, Procedia CIRP 102 (2021) 246-251. https://doi.org/10.1016/j.procir.2021.09.042

[12] A. Gharibi, Y. Kaynak, The influence of depth of cut on cryogenic machining performance of hardened steel, J. Faculty Eng. Archit. Gazi University 34 (2019) 582-596.

[13] Y. Kaynak, H. Karaca, I. Jawahir, Cutting speed dependent microstructure and transformation behavior of NiTi alloy in dry and cryogenic machining, J. Mater. Eng. Perform. 24 (2015) 452-460. https://doi.org/10.1007/s11665-014-1247-6

[14] Y. Kaynak, H. Karaca, R.D. Noebe, I. Jawahir, Analysis of tool-wear and cutting force components in dry, preheated, and cryogenic machining of NiTi shape memory alloys, Procedia CIRP 8 (2013) 498-503. https://doi.org/10.1016/j.procir.2013.06.140

[15] Y. Kaynak, H. Karaca, R.D. Noebe, I. Jawahir, Tool-wear analysis in cryogenic machining of NiTi shape memory alloys: A comparison of tool-wear performance with dry and MQL machining, Wear 306 (2013) 51-63. https://doi.org/10.1016/j.wear.2013.05.011

[16] O. Pereira, A. Rodríguez, A.I. Fernández-Abia, J. Barreiro, L.N. López de Lacalle, Cryogenic and minimum quantity lubrication for an eco-efficiency turning of AISI 304, J. Clean. Prod. 139 (2016) 440-449. https://doi.org/10.1016/j.jclepro.2016.08.030

[17] L. Collini, Copper Alloys: Early Applications and Current Performance-Enhancing Processes, 2012, BoD–Books on Demand.

[18] A.I. Toulfatzis, G.J. Besseris, G. Pantazopoulos, C. Stergiou, Characterization and comparative machinability investigation of extruded and drawn copper alloys using non-parametric multi-response optimization and orthogonal arrays, Int. J. Adv. Manuf. Technol. 57 (2011) 811-826. https://doi.org/10.1007/s00170-011-3319-1

[19] G. Pantazopoulos, A. Vazdirvanidis, Characterization of the microstructural aspects of machinable-phase brass, Microscopy and Analysis (2008) 13.

[20] J.R. Davis, Copper and copper alloys, ASM international, 2001.

[21] C. Nobel, U. Hofmann, F. Klocke, D. Veselovac, Experimental investigation of chip formation, flow, and breakage in free orthogonal cutting of copper-zinc alloys, Int. J. Adv. Manuf. Technol. 84 (2016) 1127-1140. https://doi.org/10.1007/s00170-015-7749-z

[22] M. Kurt, Y. Kaynak, E. Bagci, H. Demirer, M. Kurt, Dimensional analyses and surface quality of the laser cutting process for engineering plastics, Int. J. Adv. Manuf. Technol. 41 (2009) 259-267. https://doi.org/10.1007/s00170-008-1468-7

[23] S. Vajpayee, Analytical study of surface roughness in turning, Wear 70 (1981) 165-175. https://doi.org/10.1016/0043-1648(81)90151-4

Keyword Index
(includes keywords from Part 1-3)

e

i

About the Editors

Lukasz Madej is a full professor and the head of the Industrial Digitalization and Multiscale Modelling Division at the Faculty of Metal Engineering and Industrial Computer Science at AGH University. His research focuses on the development of modern numerical solutions for the industry and, in particular full-field multiscale modelling models for materials science. He is a member of various scientific organizations, e.g., Polish Association of Computational Mechanics and Polish Materials Science Society. In 2022 he was elected as a Fellow of the CIRP - The International Academy for Production Engineering. Since 2020 he is also a member of the Board of Directors of the ESAFORM. At the same time, he is editor-in-chief of Computer Methods in Materials Science journal and a member of editorial boards in various academic journals, e.g., Steel Research International, International Journal of Material Forming, Journal of Materials Processing Technology or Production Engineering - Research and Development. He is the author and co-author of more than 250 published works, including 5 books. Besides scientific activities, he is also active in the Polish Forging Association, where he holds the position of vice president and in the EUROFORGE as an Executive Board member. More information can be found at the following website: http://home.agh.edu.pl/~lmadej

Mateusz Sitko is an assistant professor at the Faculty of Metal Engineering and Industrial Computer Science at AGH University. His research focuses on the development of multiscale models for microstructure evolution dedicated to high-performance computing environments and the incorporation of virtual and augmented reality tools into industrial applications. He is a member of various scientific organizations, e.g., Polish Association of Computational Mechanics, Polish Forging Association or ESAFORM. At the same time, he is the assistant editor of Computer Methods in Materials Science journal. He is the author and co-author of one book chapter and 35 journal articles. More information is available at the following website: http://home.agh.edu.pl/~msitko/en/

Konrad Perzynski is an assistant professor at the Faculty of Metal Engineering and Industrial Computer Science at AGH University. His research focuses on the numerical modelling of a wide range of industrial processes, including the development of full-field multiscale models for materials science. He is also involved in the modelling and analysis of thin film nanostructures. He is a member of several scientific organizations, e.g., Polish Association of Computational Mechanics or ESAFORM. He is also a core group member of the ECCOMAS Young Investigators committee. At the same time, he is the assistant editor of Computer Methods in Materials Science journal. He is the author and co-author of more than 70 journal articles. More information can be found at the following website: https://home.agh.edu.pl/~kperzyns/

www.ingramcontent.com/pod-product-compliance
Lightning Source LLC
Chambersburg PA
CBHW071312210326
41597CB00015B/1209